METHODS IN MOLECULAR BIOLOGY

Injury Models of the Central Nervous System

Methods and Protocols

Edited by

Firas Kobeissy

Banyan Biomarkers, Inc., Alachua, FL, USA

C. Edward Dixon

Safar Center for Resuscitation Research, University of Pittsburgh, Pittsburgh, PA, USA

Ronald L. Hayes

Banyan Biomarkers, Inc., Alachua, FL, USA

Stefania Mondello

Banyan Biomarkers, Inc., Alachua, FL, USA

Humana Press

Editors
Firas Kobeissy
Banyan Biomarkers, Inc.
Alachua, FL, USA

Ronald L. Hayes
Banyan Biomarkers, Inc.
Alachua, FL, USA

C. Edward Dixon
Safar Center for Resuscitation Research
University of Pittsburgh
Pittsburgh, PA, USA

Stefania Mondello
Banyan Biomarkers, Inc.
Alachua, FL, USA

ISSN 1064-3745 ISSN 1940-6029 (electronic)
Methods in Molecular Biology
ISBN 978-1-4939-3814-8 ISBN 978-1-4939-3816-2 (eBook)
DOI 10.1007/978-1-4939-3816-2

Library of Congress Control Number: 2016943079

Printed on acid-free paper

This Humana Press imprint is published by Springer Nature
The registered company is Springer Science+Business Media LLC New York

A mentor empowers a person to see possible future, and believe it can be obtained. To Prof. Emanuele Scribano.

Stefania Mondello

To my life-long mentor, Dr. Kevin K. Wang & his wife Miss Alice Wang, I dedicate this humble work. I will be forever grateful to Kevin for all his kind-heartedness and for sharing his wise knowledge

Firas Kobeissy

Foreword

It had been a universally acknowledged truth that animal models provide the means for understanding pathogenesis and for guiding treatment of human diseases and conditions. That is, until the recently recognized failure of apparently successful treatment of animal models of traumatic brain injury (TBI) to translate into effective treatment of human TBI in over 30 clinical trials. One might rightly ask: what, at this time, motivated the publication of this detailed compendium of varying animal models in several species? At the least, abundantly detailed information in this volume permits accurate replication of experimental data by investigators working in separate laboratories. At best, the attention to detail in scaling particular animal injuries to the human responses to similar injuries, in magnitude, time lines, and outcomes, is made apparent as a critical need in current TBI research. The answer to the question, based on history of scientific approaches and rationale, would appear obvious, but yet, preclinical treatment approaches in animal TBI, as pointed out by the introductory chapter in this book, continue to fail when applied in human trials.

Animal models for understanding and treating human disease began in 1894 using Koch's four postulates to prove that infectious agents cause particular diseases. Animal models have been essential for scientific progress. Yet even with infectious disease models, the relationships between cause and effect are neither obvious nor simple. For example Koch's first postulate, 'abundance of microorganisms or infectious agents' required modification in that carrier or asymptomatic states were found to exist. The peculiar susceptibility of the guinea pig to *mycobacterium tuberculosis* was fortuitous. As few as five microorganisms introduced by an intranasal route are capable of initiating pulmonary lesions closely mimicking those occurring in man. Yet rabbits resist infection with *m tuberculosis*. The extensive use of rodent models, mainly mice, for immunologic study and treatment *of pneumococcal, staphylococcal* and *streptococcal* infections contributed immensely to successful development of antibiotic and immunologically based treatments.

As an investigator supported by the American Heart Association and the National Institutes of Health, I have had the privilege of using animal models to study shock and atherosclerosis. These studies proved remarkably effective in delineating pathogenesis and providing insights into treatment. In the first instance, the use of crystalloid combinations with red cells for fluid replacement in hemorrhage; for atherosclerosis, canine and rhesus subhuman primate models provided critical understanding of the central role of lipids in promoting arterial plaques similar to those occurring in human disease. Importantly, lipid levels were found to relate to progression and regression of atherosclerotic plaques in arteries. Lately, animal models to measure the key role of the inflammatory responses within the arterial wall have been most informative. The lessons learned from these experimental models have been successfully translated into clinical practice with ongoing interventions based on level 1 clinical outcomes.

What then is needed for TBI 'preclinical modeling'? We require more attention to details and repeatability along with the relevance of particular models to particular aspects of human injury. This volume provides that kind of detail, whether it be the mass effects of rotational injury using pigs with differing ratios of gray to white matter or the widely variant effects of a multitude of shock tube configurations to mimic primary blast injury favored by disparate investigators. The details of shock wave interactions in open field explosions, near, intermediate and remote are essential for understanding effects of such exposures to blasts occurring in combat. How to standardize explosive blast static, dynamic, total and reflective pressures along with loading effects (depending on angle of incidence and geometry) requires delineation. While two chapters reference modeling actual explosions, using actual blast exposures in scalable animal models remains an unfulfilled need. Rather than overemphasizing difficulties, the practicalities of how to model future blast research require more attention. Modeling of injury modes: impact, rotational and non-impact blast injury needs essential attention to detail to achieve and document consistent results. Once TBI models are standardized and widely replicable, then and possibly only then, can neuroprotective treatments be assayed with confidence using focused human trials.

This book highlights modeling TBI using larger animals. Difficulties and barriers to subhuman primate modeling are also mentioned. In this author's opinion, subhuman primate models, currently used for study of prevention and treatment of HIV and Ebola viral infections, must be considered for study of TBI. Such studies should be done with due respect to all animal species used given the context of human risks of the devastating effects of TBI, particularly combat TBI. The classic studies of subhuman primates which yielded critical quantitative information in terms of measuring rotational acceleration are now described in a porcine model. First steps toward progress have been provided: a taxonomy and detailed description of various animal models of TBI. Hopefully, future clinical trials of TBI will be based upon validated replicable models of the heterogeneous effects of traumatic brain injury as related to differing injury mechanisms and modalities.

Ralph G. DePalma
VA Office of Research and Development
Washington, DC, USA

Uniformed University of the Health Sciences
Bethesda, MD, USA

Preface

"The use of animals in medical research and safety testing is a vital part of the quest to improve human health. It always has been and probably always will be, despite the alternatives available. Indeed, in this era of genomics and proteomics, more rather than fewer animals will be needed. Without animal testing, there will be no new drugs for new or hard-to-treat diseases."

Source: The Lancet. Vol 364 September 4, 2004

Traumatic brain injury (TBI) is a major and growing public health concern and a leading cause of mortality and morbidity worldwide, both in civilian life and on the battlefield worldwide. It is now considered as a complex spectrum of diseases associated with structural damage and functional deficits that are consequences of both primary and secondary injury mechanisms.

Most notably, over the past 50 years, TBI field has matured tremendously, and our understanding of the cellular and molecular biology along with the pathophysiological processes underlying acute and chronic TBI (and its recovery) has evolved. Throughout, this time research involving animal models of TBI has grown from a few individual laboratories to a global effort. Experimental TBI models have provided a driving force and undeniable contribution for much of this research allowing us to test hypotheses, to investigate basic mechanisms as well as to explore and determine the efficacy of novel potential treatments and therapeutic strategies. As a consequence, we have witnessed an explosion of "potentially useful" models replicating the various aspects of human TBI. However, it is now evident that to be effective in our quest for answers and cures, and to prevent failure in clinical translation of preclinical studies, methodological rigor, extensive standardization of outcome assessment, functional tests, pre-analytical and analytical aspects, and approach to treatment, as well as reproducibility are all essential.

The current volume represents an attempt to provide a list of the state-of-the-art currently established animal models of TBI thoroughly describing practical procedures, experimental methods, and protocols. While specific TBI models are discussed in different chapters, we view this as a strength because it exposes the reader to different viewpoints regarding model implementation and highlights the need for greater standardization in terms of reporting methods and findings. One noteworthy section of the book is dedicated to the *Combined Neurotrauma Models*—models combining a standardized TBI with systemic insults and multimodality monitoring—that can more adequately mimic the heterogeneity of clinical TBI.

As the experimental TBI research has grown, the diversity of outcomes continues to expand. The part on *Outcome Measures in Brain Injury Models* provides researchers with comprehensive methods of detecting behavioral, histopathological, cerebrovascular, imaging, and biochemical outcomes that aim at advancing the search for injury mechanisms and clinical translation.

This volume also aims to identify the issues and challenges related to TBI preclinical studies and offers guidance and solutions from renowned experts in this field that can certainly be of use for young investigators. We hope it can also be helpful to clinicians and laboratory researchers interested in the field and in fresh ways of addressing both experimental and clinical questions.

In this volume, we have assembled 40 chapters which have been authored by renowned experts in the areas of CNS neurotrauma and who are among the pioneers in the development of several of the currently used CNS injury models. The book opens with an introductory overview by Professor John Povlishock discussing the history and the evolution of experimental brain injury models. This volume is divided into six parts. Part II (General Consideration in Using Animal Laboratory in CNS Injury Research) describes general experimental concepts and approaches used in assessing CNS injuries, with emphasis on clinical and experimental models currently available and a summary of the development of these models along with a handy overview comparison of their different/respective characteristics. In the following part (Classical TBI Models and Their Link with Pathophysiological Features of CNS Injury—Models), the authors describe the utility of different injury models and their general pathological characteristics. This part elaborates on the different modified and novel experimental models ranging from pediatric rodent models of TBI to more recent models of diffuse axonal injuries. In Part IV (Special Topics in CNS Trauma: Comorbid Conditions in CNS Injury), we discuss the experimental combination of comorbid conditions in CNS injury models including hypoxia, hypoxemia, and epilepsy. In Part V (Outcome Measures in Brain Injury Models), assessment and outcome measures involved in CNS injury are systematically compiled to include a comprehensive summary of the behavioral and cognitive methods utilized. In addition, this part includes dedicated chapters discussing the utility of novel techniques and approaches that can be applied both in neurotherapeutics (stem cell therapy, small molecule inhibitors) and in evaluating indices in CNS injury models including proteomics, pressure myography, and mitochondrial bioenergetics. The volume closes with an overview chapter by Professor Frank Tortella who contributed a very nice piece discussing challenges that face preclinical models of TBI and translating them to clinical practice.

Finally, this book would not have been possible without the assistance of a large group of TBI experts, each of whom has particular expertise related to a given model. The contributing authors have done a superb job in presenting experimentally useful information and applied technical approaches of TBI-relevant models. We have also charged a set of investigators with the task of providing an overview of various aspects, current knowledge, advantages, and limitations of animal TBI models. Special gratitude goes to Professor Ralph Depalma who agreed to write the Foreword to this volume; his comments and suggestions have been always an excellent guide for researchers in the field of neurotrauma. We hope that the reader will find that these descriptions provide useful help and support as well as foster further innovation, development, and optimization of experimental TBI research.

Alachua, FL, USA — *Firas Kobeissy*
Pittsburgh, PA, USA — *C. Edward Dixon*
Alachua, FL, USA — *Ronald L. Hayes*
Alachua, FL, USA — *Stefania Mondello*

Acknowledgement

First, we would like to express our sincere appreciation to everyone who contributed in the making of this book. The motivation and support among the authors was impressive and I hope to see it continuously. Without the authors' expertise, contribution, and leadership qualities, finalizing this book would not have been feasible.

This volume of Molecular Methods of Molecular Biology discussing experimental brain injury models contains valuable information carefully chosen to widen the researchers' horizon about neurotrauma and constitutes a highly needed reference for the progress of this discipline. It also contains the relevant experimental design approaches that have been meticulously adapted and ready for application in laboratory settings.

I wish to thank Hibah Ghandour, Edwyn Assaf, Alaa Kassir, and Cherine Fawaz for their editorial support and meticulous corrections in chapter editing. Their help has added a special value to this book.

Contents

Contributors

HUSSEIN ABOU ABBASS • *Faculty of Medicine, Beirut Arab University, Beirut, Lebanon; Department of Biochemistry and Molecular Genetics, Faculty of Medicine, American University of Beirut, Beirut, Lebanon*

HASSAN AL-ALI • *Miami Project to Cure Paralysis, University of Miami Miller School of Medicine, Miami, FL, USA*

SAWSAN AL LAFI • *Department of Biology, American University of Beirut, Beirut, Lebanon*

ALI ALAWIEH • *Department of Microbiology and Immunology, and Neuroscience Institute, Medical University of South Carolina, Charleston, SC, USA; Neuroscience Institute, Medical University of South Carolina, Charleston, SC, USA*

HADI ALIGHOLI • *Department of Neuroscience, School of Advances Medical Sciences and Technologies, Shiraz University of Medical Sciences, Shiraz, Iran; Shefa Neuroscience Research Center, Khatam-al-Anbia Hospital, Tehran, Iran*

MOHAMMAD A. ANWAR • *Department of Biological & Environmental Sciences, College of Arts and Sciences, Qatar University, Doha, Qatar*

PATRICK AOUAD • *Department of Biology, Faculty of Arts and Sciences, American University of Beirut, Beirut, Lebanon*

HIBAH O. AWWAD • *Department of Pharmaceutical Sciences, College of Pharmacy, Oklahoma City, OK, USA;*
Oklahoma Center for Neuroscience, College of Medicine, University of Oklahoma Health Sciences Center, Oklahoma City, OK, USA

HASSAN AZARI • *Neural Stem Cell and Regenerative Neuroscience Laboratory, Department of Anatomical Sciences, Shiraz School of Medicine, Shiraz University of Medical Sciences, Shiraz, Iran; Shiraz Stem Cell Institute, Shiraz University of Medical Sciences, Shiraz, Iran*

ZEINA AZRAK • *Department of Pharmacology and Toxicology, American University of Beirut, Beirut, Lebanon*

HISHAM BAHMAD • *Department of Anatomy, Cell Biology, and Physiological Sciences, Faculty of Medicine, American University of Beirut, Beirut, Lebanon; Faculty of Medicine, Beirut Arab University, Beirut, Lebanon*

ZACHARY S. BAILEY • *School of Biomedical Engineering and Sciences, Virginia Tech, Blacksburg, VA, USA*

JUSTIN BERKNER • *Division of Emergency Medicine, Boston Children's Hospital, Boston, MA, USA*

JOHN L. BIXBY • *Miami Project to Cure Paralysis, University of Miami Miller School of Medicine, Miami, FL, USA; Center for Computational Science, University of Miami Miller School of Medicine, Miami, FL, USA; Molecular & Cellular Pharmacology, University of Miami Miller School of Medicine, Miami, FL, USA*

PRODIP K. BOSE • *Brain Rehabilitation Research Center, North Florida/South Georgia Veterans Health Center, Gainesville, FL, USA; Department of Physiological Sciences, College of Veterinary Medicine, University of Florida, Gainesville, FL, USA; Department of Neurology, University of Florida, Gainesville, FL, USA*

Rose-Mary Boustany • *Department of Biochemistry and Molecular Genetics, American University of Beirut, Beirut, Lebanon; Neurogenetics Program and Division of Pediatric Neurology, Departments of Pediatrics and Adolescent Medicine, American University of Beirut Medical Center Special Kids Clinic, Beirut, Lebanon*

Angela M. Boutté • *Brain Trauma Neuroprotection and Neurorestoration Branch, Center for Military Psychiatry and Neuroscience, Walter Reed Army Institute of Research, Silver Spring, MD, USA*

Kevin D. Browne • *Center for Brain Injury & Repair, Department of Neurosurgery, Perelman School of Medicine, University of Pennsylvania, Philadelphia, PA, USA; Department of Neurology, Perelman School of Medicine, Philadelphia Veterans Affairs Medical Center, Philadelphia, PA, USA*

Daniele Bruschetta • *Department of Biomedical Sciences and Morphological and Functional Images, University of Messina, Messina, Italy*

Ross M. Bullock • *Department of Neurosurgery, University of Miami Miller School of Medicine, Miami, FL, USA*

Jaclyn Carlson • *Department of Medicine, Royal Melbourne Hospital, The University of Melbourne, Parkville, VIC, Australia*

Paul R. Carney • *Departments of Pediatrics, Neurology, Neuroscience, and Biomedical Engineering, University of Florida College of Medicine, Gainesville, FL, USA*

Krista Caudle • *Brain Trauma Neuroprotection and Neurorestoration Branch, Center for Military Psychiatry and Neuroscience, Walter Reed Army Institute of Research, Silver Spring, MD, USA*

Robert S.B. Clark • *Department of Critical Care Medicine, Children's Hospital of Pittsburgh of UPMC, University of Pittsburgh School of Medicine, Pittsburgh, PA, USA; Department of Pediatrics, Children's Hospital of Pittsburgh of UPMC, University of Pittsburgh School of Medicine, Pittsburgh, PA, USA; Safar Center for Resuscitation Research, University of Pittsburgh School of Medicine, Pittsburgh, PA, USA; The Clinical and Translational Science Institute, University of Pittsburgh School of Medicine, Pittsburgh, PA, USA*

D. Kacy Cullen • *Center for Brain Injury & Repair, Department of Neurosurgery, Perelman School of Medicine, University of Pennsylvania, Philadelphia, PA, USA; Philadelphia Veterans Affairs Medical Center, Philadelphia, PA, USA*

Salvatore Cuzzocrea • *Department of Biological and Environmental Sciences, University of Messina, Messina, Italy; Manchester Biomedical Research Center, University of Manchester, Manchester, UK*

Hala Darwish • *Hariri School of Nursing, Abou Haidar Neuroscience Institute, American University of Beirut, Beirut, Lebanon*

Jitendra R. Dave • *Brain Trauma Neuroprotection and Neurorestoration Branch, Center for Military Psychiatry and Neuroscience, Walter Reed Army Institute of Research, Silver Spring, MD, USA*

Johan Davidsson • *Vehicle Safety Division, Department of Applied Mechanics, Chalmers University of Technology, Göteborg, Sweden*

Ying Deng-Bryant • *Brain Trauma Neuroprotection and Neurorestoration Branch, Center for Military Psychiatry and Neuroscience, Walter Reed Institute of Research, Silver Spring, MD, USA*

Yuchuan Ding • *Department of Neurosurgery, Wayne State University School of Medicine, Detroit, MI, USA*

C. Edward Dixon • *Safar Center for Resuscitation Research, University of Pittsburgh, Pittsburgh, PA, USA; Department of Neurological Surgery, University of Pittsburgh, Pittsburgh, PA, USA; V.A. Pittsburgh Healthcare System, Pittsburgh, PA, USA*

Carter P. Dodge • *Department of Pediatric Anesthesia, Dartmouth Hitchcock, Lebanon, NH, USA; Dartmouth Hitchcock Medical Center, Lebanon, NH, USA*

Sylvain Doré • *Department of Anesthesiology, University of Florida College of Medicine, Gainesville, FL, USA; Center for Translational Research in Neurodegenerative Disease, University of Florida College of Medicine, Gainesville, FL, USA; Departments of Neuroscience, Neurology, and Psychiatry, University of Florida College of Medicine, Gainesville, FL, USA*

John E. Duda • *Department of Neurology, Perelman School of Medicine, Philadelphia Veterans Affairs Medical Center, Philadelphia, PA, USA*

Ann-Christine Duhaime • *Department of Neurosurgery, Harvard Medical School, Boston, MA, USA; Massachusetts General Hospital, Boston, MA, USA*

Timothy Q. Duong • *Research Imaging Institute, University of Texas Health Science Center, San Antonio, TX, USA; Department of Ophthalmology, University of Texas Health Science Center, San Antonio, TX, USA; Department of Radiology, University of Texas Health Science Center, San Antonio, TX, USA; Department of Veterans Affairs, South Texas Veterans Health Care System, San Antonio, TX, USA*

Ali H. Eid • *Department of Biological & Environmental Sciences, College of Arts and Sciences, Qatar University, Doha, Qatar; Department of Pharmacology and Toxicology, Faculty of Medicine, American University of Beirut, Beirut, Lebanon*

Sally El Sitt • *Department of Biochemistry and Molecular Genetics, American University of Beirut, Beirut, Lebanon*

Lama Fawaz • *Department of Experimental Pathology, Microbiology, and Immunology, American University of Beirut, Beirut, Lebanon*

Shyam Gajavelli • *Department of Neurosurgery, University of Miami Miller School of Medicine, Miami, FL, USA*

Jinhuan Gao • *Institute of Hypoxia Medicine, Xuanwu Hospital, Capital Medical University, Beijing, China*

Hussein Gazalah • *Department of Biochemistry and Molecular Genetics, American University of Beirut, Beirut, Lebanon*

Hiba Ghandour • *Faculty of Medicine, American University of Beirut Medical Center, Beirut, Lebanon*

Noël Ghanem • *Department of Biology, American University of Beirut, Beirut, Lebanon*

Crystal M. Ghantous • *Department of Anatomy, Cell Biology, and Physiology, American University of Beirut, Beirut, Lebanon*

Alexander V. Glushakov • *Department of Anesthesiology, University of Florida College of Medicine, Gainesville, FL, USA; Center for Translational Research in Neurodegenerative Disease, University of Florida College of Medicine, Gainesville, FL, USA*

Olena Y. Glushakova • *Banyan Biomarkers, Inc., Alachua, FL, USA*

Ali Gorji • *Epilepsy Research Center, Westfälische Wilhelms-Universität Münster, Münster, Germany; Department of Neurosurgery and Neurology, Westfälische Wilhelms-Universität Münster, Münster, Germany*

Shonnette F. Grant • *Brain Trauma Neuroprotection and Neurorestoration Branch, Center for Military Psychiatry and Neuroscience, Walter Reed Army Institute of Research, Silver Spring, MD, USA*

DANIEL R. GRIFFITHS • *BARROW Neurological Institute at Phoenix Children's Hospital, Phoenix, AZ, USA; Department of Child Health, University of Arizona College of Medicine, Phoenix, AZ, USA*
JAMES P. HARRIS • *Center for Brain Injury & Repair, Department of Neurosurgery, Perelman School of Medicine, University of Pennsylvania, Philadelphia, PA, USA; Department of Neurology, Perelman School of Medicine, Philadelphia Veterans Affairs Medical Center, Philadelphia, PA, USA*
GHOLAMREZA HASSANZADEH • *Department of Neuroscience, School of Advances Technologies in Medicine, Tehran University of Medical Sciences, Tehran, Iran; Department of Anatomy, School of Medicine, Tehran University of Medical Sciences, Tehran, Iran*
RONALD L. HAYES • *Banyan Biomarkers, Inc., Alachua, FL, USA*
SARAH C. HELLEWELL • *Canadian Military and Veterans' Clinical Rehabilitation Research Program, Faculty of Rehabilitation Medicine, University of Alberta, Edmonton, AB, Canada*
JIAMEI HOU • *Department of Physiological Sciences, College of Veterinary Medicine, University of Florida, Gainesville, FL, USA*
W. BRAD HUBBARD • *School of Biomedical Engineering and Sciences, Virginia Tech, Blacksburg, VA, USA*
HANA A. ITANI • *Division of Clinical Pharmacology, Department of Medicine, Vanderbilt University Medical Center, Nashville, TN, USA*
ZAYNAB JABER • *Department of Biochemistry, Graduate School and University Center of CUNY, New York, NY, USA; Department of Biochemistry and Molecular Genetics, Faculty of Medicine, American University of Beirut, Beirut, Lebanon*
XUNMING JI • *Institute of Hypoxia Medicine, Xuanwu Hospital, Capital Medical University, Beijing, China*
BRIAN T. KALISH • *Department of Newborn Medicine, Boston Children's Hospital, Boston, MA, USA*
KRISTEN KEELEY • *School of Medicine, Creighton University, Omaha, NE, USA*
FIRAS KOBEISSY • *Banyan Biomarkers, Inc., Alachua, FL, USA; Department of Biochemistry and Molecular Genetics, Faculty of Medicine, American University of Beirut, Beirut, Lebanon; Department of Psychiatry Center for Neuroproteomics and Biomarkers Research, University of Florida, Gainesville, FL, USA*
PATRICK M. KOCHANEK • *Department of Critical Care Medicine, Children's Hospital of Pittsburgh of UPMC, University of Pittsburgh School of Medicine, Pittsburgh, PA, USA; Department Pediatrics, Children's Hospital of Pittsburgh of UPMC, University of Pittsburgh School of Medicine, Pittsburgh, PA, USA; Safar Center for Resuscitation Research, University of Pittsburgh School of Medicine, Pittsburgh, PA, USA; Department Anesthesiology, Children's Hospital of Pittsburgh of UPMC, University of Pittsburgh School of Medicine, Pittsburgh, PA, USA*
OLGA N. KOKIKO-COCHRAN • *Department of Neurosciences, Cleveland Clinic, Cleveland, OK, USA*
STEPHANIE W. LEE • *Department of Neurosurgery, University of Miami Miller School of Medicine, Miami, FL, USA*
VANCE P. LEMMON • *Miami Project to Cure Paralysis, University of Miami Miller School of Medicine, Miami, FL, USA; Center for Computational Science, University of Miami Miller School of Medicine, Miami, FL, USA; Department of Neurological Surgery, University of Miami Miller School of Medicine, Miami, FL, USA*

Alessandra Dal Cengio Leonardi • *John D Dingell VAMC, Detroit, MI, USA*
Lai Yee Leung • *Brain Trauma Neuroprotection and Neurorestoration Branch, Center for Military Psychiatry and Neuroscience, Walter Reed Institute of Research, Silver Spring, MD, USA*
Wei Li • *Research Imaging Institute, University of Texas Health Science Center, San Antonio, TX, USA; Department of Ophthalmology, University of Texas Health Science Center, San Antonio, TX, USA*
Jonathan Lifshitz • *BARROW Neurological Institute at Phoenix Children's Hospital, Phoenix, AZ, USA; Department of Child Health, University of Arizona College of Medicine, Phoenix, AZ, USA; Phoenix Veteran Affairs Healthcare System, Phoenix, AZ, USA*
Bruce G. Lyeth • *Department of Neurological Surgery, University of California at Davis, Davis, CA, USA*
Rebekah Mannix • *Division of Emergency Medicine, Boston Children's Hospital, Boston, MA, USA; Harvard Medical School, Boston, MA, USA*
Sarah Mantash • *Department of Biochemistry and Molecular Genetics, American University of Beirut, Beirut, Lebanon; ER045, Laboratory of Stem Cells, DSST, PRASE, Lebanese University, Beirut, Lebanon*
Susan S. Margulies • *Center for Brain Injury & Repair, Department of Neurosurgery, Perelman School of Medicine, University of Pennsylvania, Philadelphia, PA, USA; Department of Bioengineering, School of Engineering and Applied Science, University of Pennsylvania, Philadelphia, PA, USA*
Niklas Marklund • *Division of Neurosurgery, Department of Neuroscience, Uppsala University, Uppsala, Sweden*
David F. Meaney • *Center for Brain Injury & Repair, Department of Neurosurgery, Perelman School of Medicine, University of Pennsylvania, Philadelphia, PA, USA*
Mohamad Al Medawar • *Division of Vascular Endothelium and Microcirculation, Department of Medicine III, University Hospital Carl Gustav Carus, TU Dresden, Dresden, Germany*
Stefania Mondello • *Banyan Biomarkers, Inc., Alachua, FL, USA; Department of Neurosciences, University of Messina, Messina, Italy*
M. Cristina Morganti-Kossmann • *Barrow Neurological Institute at Phoenix Children's Hospital, Phoenix, AZ, USA; Department of Epidemiology and Preventive Medicine, Monash University and Australian New Zealand Intensive Care Research Centre, Melbourne, VIC, Australia; Neuroscience Program, Department of Psychology, Arizona State University, Tempe, AZ, USA*
Jennifer C. Munoz Pareja • *Department of Pediatrics, University of Florida College of Medicine, Gainesville, FL, USA; Shands Children's Hospital, University of Florida, Gainesville, FL, USA*
Aarti Narang • *Department of Microbiology and Immunology, Medical University of South Carolina, Charleston, SC, USA; Neuroscience Institute, Medical University of South Carolina, Charleston, SC, USA*
Linda J. Noble-Haeusslein • *Department of Neurological Surgery, University of California School of Medicine, San Francisco, CA, USA; Department of Physical Therapy and Rehabilitation Science, University of California School of Medicine, San Francisco, CA, USA*
Nicole Osier • *Safar Center for Resuscitation Research, University of Pittsburgh, Pittsburgh, PA, USA; School of Nursing, University of Pittsburgh, Pittsburgh, PA, USA*

Jignesh D. Pandya • *Brain Trauma Neuroprotection and Neurorestoration Branch, Center for Military Psychiatry and Neuroscience, Walter Reed Army Institute of Research, Silver Spring, MD, USA*
Helen W. Phipps • *San Antonio, TX, USA*
John Povlishock • *Department of Anatomy and Neurobiology, Virginia Commonwealth University, Richmond, VA, USA*
Jianhua Qiu • *Division of Emergency, Boston Children's Hospital, Boston, MA, USA; Harvard Medical School, Boston, MA, USA*
Angelo Quartarone • *Department of Neurosciences, Policlinico Universitario, University of Messina, Messina, Italy; Department of Physiology and Pharmacology, City University of New York Medical School, New York, NY, USA; IRCCS Centro Neurolesi "Bonino Pulejo", Messina, Italy*
Farah Abdel Rahman • *Department of Anatomy, Cell Biology, and Physiology, American University of Beirut, Beirut, Lebanon*
Thangavelu Soundara Rajan • *Department of Neurosciences, Policlinico Universitario, University of Messina, Messina, Italy; IRCCS Centro Neurolesi "Bonino Pulejo", Messina, Italy*
Naify Ramadan • *Department of Biochemistry and Molecular Genetics, American University of Beirut, Beirut, Lebanon*
Ryan Readnower • *Brain Trauma Neuroprotection and Neurorestoration Branch, Center for Military Psychiatry and Neuroscience, Walter Reed Institute of Research, Silver Spring, MD, USA*
Changhong Ren • *Institute of Hypoxia Medicine, Xuanwu Hospital, Capital Medical University, Beijing, China*
Mårten Risling • *Experimental Traumatology, Department of Neuroscience, Karolinska Institutet, Stockholm, Sweden*
David Ritzel • *Dyn-FX Consulting Ltd, Amherstburg, Blacksburg, ON, Canada*
Elham Rostami • *Department of Neuroscience, Section of Neurosurgery, Uppsala University, Uppsala, Sweden; Department of Neuroscience, Karolinska Institutet, Stockholm, Sweden; Laboratory of Neurogenetics,NIH, NIAAA, Bethesda, MD, USA*
Rachel K. Rowe • *BARROW Neurological Institute at Phoenix Children's Hospital, Phoenix, AZ, USA; Department of Child Health, University of Arizona College of Medicine, Phoenix, AZ, USA; Phoenix Veteran Affairs Healthcare System, Phoenix, AZ, USA*
Bridgette D. Semple • *Department of Medicine, Royal Melbourne Hospital, The University of Melbourne, Parkville, VIC, Australia*
Deborah Shear • *Brain Trauma Neuroprotection and Neurorestoration Branch, Center for Military Psychiatry and Neuroscience, Walter Reed Institute of Research, Silver Spring, MD, USA*
Qiang Shen • *Research Imaging Institute, University of Texas Health Science Center, San Antonio, TX, USA; Department of Ophthalmology, University of Texas Health Science Center, San Antonio, TX, USA; Department of Radiology, University of Texas Health Science Center, San Antonio, TX, USA*
Dennis W. Simon • *Department of Critical Care Medicine, Children's Hospital of Pittsburgh of UPMC, University of Pittsburgh School of Medicine, Pittsburgh, PA, USA; Department of Pediatrics, Children's Hospital of Pittsburgh of UPMC, University of*

Pittsburgh School of Medicine, Pittsburgh, PA, USA; Safar Center for Resuscitation Research, University of Pittsburgh School of Medicine, Pittsburgh, PA, USA

DOUGLAS H. SMITH • *Center for Brain Injury & Repair, Department of Neurosurgery, Perelman School of Medicine, University of Pennsylvania, Philadelphia, PA, USA*

JIHANE SOUEID • *Neurogenetics Program and Division of Pediatric Neurology, Departments of Pediatrics and Adolescent Medicine, American University of Beirut Medical Center Special Kids Clinic, Beirut, Lebanon*

MARKUS S. SPURLOCK • *Department of Neurosurgery, University of Miami Miller School of Medicine, Miami, FL, USA*

PATRICK G. SULLIVAN • *Spinal Cord and Brain Injury Research Center, University of Kentucky, Lexington, KY, USA*

CHRISTOPHER SY • *Department of Neurosurgery, Wayne State University School of Medicine, Detroit, MI, USA*

ERIC P. THELIN • *Department of Clinical Neuroscience, Section for Neurosurgery, Karolinska Institutet, Stockholm, Sweden*

FLOYD J. THOMPSON • *Brain Rehabilitation Research Center, North Florida/South Georgia Veterans Health Center, Gainesville, FL, USA; Department of Physiological Sciences, College of Veterinary Medicine, University of Florida, Gainesville, FL, USA; Department of Neuroscience, McKnight Brain Institute, Gainesville, FL, USA*

STEPHEN TOMLINSON • *Department of Microbiology and Immunology, Medical University of South Carolina, Charleston, SC, USA; Neuroscience Institute, Medical University of South Carolina, Charleston, SC, USA; Ralph H. Johnson Veteran Affairs Medical Center, Charleston, SC, USA*

FRANK TORTELLA • *Brain Trauma Neuroprotection and Neurorestoration Branch, Center for Military Psychiatry and Neuroscience, Walter Reed Institute of Research, Silver Spring, MD, USA*

VINCENT M. VAGNI • *Safar Center for Resuscitation Research, University of Pittsburgh School of Medicine, Pittsburgh, PA, USA*

KEN C. VAN • *Department of Neurological Surgery, University of California at Davis, Davis, CA, USA*

PAMELA J. VANDEVORD • *School of Biomedical Engineering and Sciences, Virginia Tech, Blacksburg, VA, USA*

WENXUE WANG • *Department of Microbiology and Immunology, Medical University of South Carolina, Charleston, SC, USA; Neuroscience Institute, Medical University of South Carolina, Charleston, SC, USA*

LORA TALLY WATTS • *Research Imaging Institute, University of Texas Health Science Center, San Antonio, TX, USA; Department of Cellular and Structure Biology, University of Texas Health Science Center, San Antonio, TX, USA*

MICHAEL J. WHALEN • *Department of Pediatric Critical Care Medicine, Massachusetts General Hospital, Boston, MA, USA*

MARK D. WHITING • *Neuroscience Program, University of Illinois at Urbana Champaign, Urbana, IL, USA*

JOHN A. WOLF • *Center for Brain Injury & Repair, Department of Neurosurgery, Perelman School of Medicine, University of Pennsylvania, Philadelphia, PA, USA; Department of NeurologyPerelman School of Medicine, Philadelphia Veterans Affairs Medical Center, Philadelphia, PA, USA*

SHOJI YOKOBORI • *Department of Emergency and Critical Care Medicine, Nippon Medical School, Tokyo, Japan*
ASAD ZEIDAN • *Department of Anatomy, Cell Biology, and Physiology, American University of Beirut, Beirut, Lebanon*
KAZEM ZIBARA • *ER045, Laboratory of Stem Cells, DSST, PRASE, Lebanese University, Beirut, Lebanon; Faculty of Sciences-I, Lebanese University, Beirut, Lebanon*
JENNA M. ZIEBELL • *Barrow Neurological Institute at Phoenix Children's Hospital, Phoenix, AZ, USA; Department of Child Health, University of Arizona College of Medicine—Phoenix, Phoenix, AZ, USA*

Part I

Introduction

Chapter 1

The History and Evolution of Experimental Traumatic Brain Injury Models

John Povlishock

Abstract

This narrative provides a brief history of experimental animal model development for the study of traumatic brain injury. It draws upon a relatively rich history of early animal modeling that employed higher order animals to assess concussive brain injury while exploring the importance of head movement versus stabilization in evaluating the animal's response to injury. These themes are extended to the development of angular/rotational acceleration/deceleration models that also exploited brain movement to generate both the morbidity and pathology typically associated with human traumatic brain injury. Despite the significance of these early model systems, their limitations and overall practicality are discussed. Consideration is given to more contemporary rodent animal models that replicate individual/specific features of human injury, while via various transgenic technologies permitting the evaluation of injury-mediated pathways. The narrative closes on a reconsideration of higher order, porcine animal models of injury and their implication for preclinical/translational research.

Key words Animal models, Traumatic brain injury, Acceleration/deceleration injuries, Diffuse axonal injury, Pathophysiology, Outcome

1 Introduction

In considering the various injury models of CNS traumatic brain injury (TBI), it is instructive to recall that our endeavors in this area are not new and span a modern history of over 70 years, from the seminal work of Denny Brown in subhuman primates to the more contemporary studies described in the current volume. Important in the consideration of the development and use of various animal models of TBI is the underlying premise that the chosen model replicates the mechanisms at work in real-life head injuries, with the injuries produced being both neurologically and pathologically comparable to those observed in man. While these goals seem straightforward and certainly appropriate for the study and treatment of TBI, the achievement of these goals has remained elusive.

Firas Kobeissy et al. (eds.), *Injury Models of the Central Nervous System: Methods and Protocols*, Methods in Molecular Biology, vol. 1462, DOI 10.1007/978-1-4939-3816-2_1, © Springer Science+Business Media New York 2016

Historically, the first forays into animal modeling of TBI focused on primarily subhuman primates, with the belief that the gyrencephalic nature of the animals and their more human-like physiological responses would better replicate the behavioral, functional, and structural consequences associated with human injury. In early studies by Denny Brown and Russell, a controlled head impact was used, delivered through a calibrated pendulum drop to strike a freely movable head [1, 2]. In these elegant studies, they demonstrated that the acceleration of the head in its freely movable state was necessary to achieve concussion. This fact was further confirmed by the same investigators who demonstrated that when the head was fixed in place, it proved more difficult to consistently evoke a concussive response. Following upon these important studies, multiple investigators joined the field, incorporating the use of multiple species, including rats and cats, which emphasized the importance of a freely movable head in the injury process, alluding to the importance of acceleration and deceleration injury in influencing the ensuing pathophysiology. While providing extensive information to the field, the major limiting factor of these studies was the fact that their utility was limited by the variability of thc cnsuing pathophysiological response due to the inability to precisely control the impact conditions and the animal's head movement in response to injury. These studies were then followed by the use of impact injuries to fixed animal heads which, although more reproducible, introduced their own subset of interpretive problems [3–5]. Despite their collective limitations, however, these studies provided the first insight into the pathophysiological response of the brain to injury, creating a structural framework upon which subsequent studies could build.

Ommaya and colleagues made major contributions to the field in the early 1970s, studying subhuman primates fixed on a moving sled which could be rapidly accelerated and then, abruptly stopped to generate impulsive rather than contact loading [6–10]. In these studies, the animal's head was free to move and there was no impact or contact phenomena that could be related to any ensuing structural change. This fact led the authors to speculate that the acceleration and deceleration associated with the model employed, coupled with the rotational forces generated, caused distortion in the brain and concussion without contact or the generation of focal loading. These studies introduced the important concept that brain movement and acceleration/deceleration are major contributors to the pathophysiology of TBI, deemphasizing the previously held beliefs that local contact phenomena with contusional and/or local cellular change were the major components of injury and key regulators of concussion.

These important studies by Ommaya and colleagues were further advanced by the equally important contributions of Gennarelli et al. who embraced these overarching concepts to develop a model of angular acceleration, using a high-G force acceleration device affixed to the head of subhuman primates [11–15]. Through rapid angular/rotational acceleration/deceleration injuries, Gennarelli and colleagues consistently evoked physiological responses that were paralleled by pathological changes observed throughout the subcortical white matter. Specifically, using silver salts, they identified axonal damage within the corpus callosum, subcortical white matter, and brainstem. Their findings bore remarkable similarity to pathological findings in humans whose clinical course was not complicated by contusion or mass lesion formation and yet revealed significant morbidity leading to the belief that they had sustained diffuse axonal injury. Gennarelli and colleagues thereby provided the field with an animal model that replicated important features of human brain injury which could then be further exploited in determining pathogenesis and subsequent anterior and retrograde change, as well as potential therapeutic screening. Unfortunately, despite the importance of this model, subsequent progress was limited, influenced by multiple factors including, but not limited to, the technical sophistication and cost of the model system used, emerging complaints by anti-vivisectionists, and the evolving use of molecular biological approaches that defaulted to the use of a wide range of transgenic mice to assess the subcellular molecular responses to injury.

Framed by the current trends in cell and molecular biology as well as the presence of anti-vivisectionist groups, more recent animal modeling in the context of TBI has focused almost exclusively on the use of rodents, together with the use of more select in vitro approaches. Constrained by the realities of inducing TBI in rodents and the difficulties in evoking significant acceleration/deceleration therein, the animal models detailed in this volume have focused on replicating specific features of TBI rather than evaluating the full spectrum of the disease. Specifically, many of the models utilized, including focal fluid percussion injury and cortical impact models, have been used to evoke contusional change, allowing the assessment of the local progression of injury as well as the evaluation of pericontusional damage and its related tissue modification. Although in large part, these rodent models do not generate diffuse components, some relatively modest diffuse change can be identified and followed through various forms of pathophysiological assessment. In addition to these models of injury which employ fluid percussion and/or cortical impact to the exposed brain, closed head injury models have also been employed, primarily utilizing weight drop injury or modifications of the cortical impact model, now using rubber-tipped impactors to strike the intact skull [16, 17]. These model systems have nicely complemented some of the more focal injury models described above and have allowed for the development of repetitive injury studies wherein the damaging structural and functional consequences of repetitive brain injury can be assessed. Importantly, despite their limitations, these rodent models and particularly those involving mice, have opened up the flood gates of opportunity for the use of various knock-in and knockout approaches to better understand at the molecular and subcellular level the acute and chronic sequelae of injury. These studies conducted in mice are providing unprecedented insight into multiple mechanisms of injury and repair including, but not limited to, the importance of stem cells, the role of chronic inflammation, the potential for persisting neurodegenerative change, and the occurrence of altered metabolic and vascular responses that could not be addressed via other approaches. Importantly, although current model development has taken a reductionist position relying almost exclusively on rodents, several recent studies have revisited the use of higher order animals, now focusing on the use of purpose bred pigs to replicate important features of human injury, particularly those focusing on contusion development and diffuse axonal injury. In this regard, important studies launched by Smith and colleagues [18], Margulies et al. [19, 20], and Povlishock and colleagues [21] have incorporated the humane use of purpose bred pigs to better understand the pathogenesis of specific features of the human condition which then may be targeted for therapeutic interventions previously screened in rodent models of injury.

2 Conclusion

Although traumatic brain injury remains the most complex disease known to man in the most complex of the body's organs—the brain, the history of animal model development offers the hope that its continued study will provide new insight into this devastating condition. While one animal model cannot replicate the full spectrum of this disease nor faithfully replicate all the features of human TBI, it can provide important information on specific features of injury relevant to man. There is hope that the continued integration of the structural and functional observations and therapeutic screenings obtained in rodents with the use of higher order animal models, will provide enhanced translational benefit that could ultimately favorably impact upon the care and management of traumatically brain-injured humans.

References

1. Denny-Brown D (1961) Brain trauma and concussion. Arch Neurol 5:1–3
2. Denny-Brown D (1945) Cerebral concussion. Physiol Rev 25:296–325
3. Chason JL, Fernando OU, Hodgson VR et al (1966) Experimental brain concussion: morphologic findings and a new cytologic hypothesis. J Trauma 6(6):767–779
4. Takahashi H, Manaka S, Sano K (1981) Changes in extracellular potassium concentration in cortex and brain stem during the acute phase of experimental closed head injury. J Neurosurg 55(5):708–717
5. Young JW, Fisher RG, Price GT, Chandler RF (1973) Experimental trauma of occipital impacts. Anonymous proceedings of the international conference on the biokinetics of impacts. Lyon, pp 271–283
6. Hirsch AE, Omaya AK (1970) Protection from brain injury: the relative significance of

translational and rotational motions of the head after impact. 14th Stapp Car Crash Conf Proc, SAE. New York, pp 299–328
7. Ommaya AK, Hirsch AE (1971) Tolerances for cerebral concussion from head impact and whiplash in primates. J Biomech 4(1):13–21
8. Ommaya AK, Corrao P, Letcher FS (1973) Head injury in the chimpanzee. 1. Biodynamics of traumatic unconsciousness. J Neurosurg 39(2):152–166
9. Ommaya AK, Corrao PG (1970) Pathologic biomechanics of central nervous system injury in head injury in head impact and whiplash trauma. Accident pathology. Proc of the international conference on accident pathology. Government Printing Office, Washington, DC. p 160
10. Ommaya A, Hirsch A, Martinez J (1966) The role of whiplash in cerebral concussion. 10th Stapp Car Crash Conference Proc, SAE. New York. pp 197–203
11. Gennarelli T, Thibault LE (1982) Acceleration damage to the brain. Proc of Advisory Group for Aerospace Res and Dev, AGARD-CP-322 NASO, NATO
12. Gennarelli T, Thibault LE (1982) Experimental production of prolonged traumatic coma in the primate. In: Villiani R (ed) Advances in neurotraumatology. Excerpta Medica, Amsterdam, pp 31–33
13. Gennarelli TA, Thibault LE, Adams JH et al (1982) Diffuse axonal injury and traumatic coma in the primate. Ann Neurol 12(6):564–574
14. Gennarelli TA, Adams JH, Graham DI (1981) Acceleration induced head injury in the monkey. I. The model, its mechanical and physiological correlates. Acta Neuropathol Suppl 7:23–25
15. Adams J, Graham D, Gennarelli T (1982) Neuropathology of acceleration induced head injury in the subhuman primate. In: Gildenberg P, Grossman RG (eds) Head injury: basic and clinical aspects. Raven Press, New York, pp 141–150
16. Marmarou A, Foda MA, van den Brink W et al (1994) A new model of diffuse brain injury in rats. Part I: pathophysiology and biomechanics. J Neurosurg 80(2):291–300
17. Shitaka Y, Tran HT, Bennett RE et al (2011) Repetitive closed-skull traumatic brain injury in mice causes persistent multifocal axonal injury and microglial reactivity. J Neuropathol Exp Neurol 70(7):551–567
18. Browne KD, Chen XH, Meaney DF et al (2011) Mild traumatic brain injury and diffuse axonal injury in swine. J Neurotrauma 28(9):1747–1755
19. Kilbaugh TJ, Karlsson M, Duhaime AC et al (2015) Mitochondrial response in a toddler-aged swine model following diffuse non-impact traumatic brain injury. Mitochondrion 26:19–25, S1567-7249(15)30033-7 [pii]
20. Jaber SM, Sullivan S, Margulies SS (2015) Noninvasive metrics for identification of brain injury deficits in piglets. Dev Neuropsychol 40(1):34–39
21. Lafrenaye AD, Todani M, Walker SA et al (2015) Microglia processes associate with diffusely injured axons following mild traumatic brain injury in the micro pig. J Neuroinflammation 12:186

Part II

General Consideration in Using Animal Laboratory in CNS Injury Research

Chapter 2

Clinical Traumatic Brain Injury in the Preclinical Setting

Justin Berkner, Rebekah Mannix, and Jianhua Qiu

Abstract

Traumatic brain injury (TBI) is the leading cause of death and disability for people under 45 years of age. Clinical TBI is often the result of disparate forces resulting in heterogeneous injuries. Preclinical modeling of TBI is a vital tool for studying the complex cascade of metabolic, cellular, and molecular post-TBI events collectively termed secondary injury. Preclinical models also provide an important platform for studying therapeutic interventions. However, modeling TBI in the preclinical setting is challenging, and most models replicate only certain aspects of clinical TBI. This chapter details the most widely used models of preclinical TBI, including the controlled cortical impact, fluid percussion, blast, and closed head models. Each of these models replicates particular critical aspects of clinical TBI. Prior to selecting a preclinical TBI model, it is important to address what aspect of human TBI is being sought to evaluate.

Key words Traumatic brain injury, Preclinical models, Controlled cortical impact, Closed head injury, Fluid percussion, Blast injury

1 Introduction

Traumatic brain injury (TBI) is the leading cause of death and disability for people under 45 years of age [1]. Worldwide, in excess of 10 million deaths or hospitalizations are attributable to TBI each year, and more than 57 million people currently are living with the sequelae of TBI [1]. While recent estimates suggest that the disease burden of TBI continues to increase, few targeted therapeutic interventions have proven effective for this common injury. Most therapeutic interventions in TBI are tested in the preclinical setting, using animal models to simulate the pathophysiology of human TBI. Understanding the issues and challenges related to utilizing animal models to study human TBI pathophysiology is a vital first step in translating these models to the clinical setting. This chapter aims to provide a broad overview of some of the most commonly employed animal models of TBI, to identify practical as well as translational issues in both the execution and application of these models.

Firas Kobeissy et al. (eds.), *Injury Models of the Central Nervous System: Methods and Protocols*, Methods in Molecular Biology, vol. 1462, DOI 10.1007/978-1-4939-3816-2_2, © Springer Science+Business Media New York 2016

1.1 Pathophysiology of TBI

TBI is defined as damage to the brain resulting from an external mechanical force, often leading to temporary or permanent impairment of cognitive, physical, and psychosocial functions. The pathophysiology of TBI can be divided into two distinct processes: primary injury and secondary injury. The primary injury is the result of the immediate mechanical force, which can include diverse mechanisms such as blast wave, crush, impact, penetration, or rapid acceleration or deceleration. These diverse mechanisms can manifest in a wide array of primary injuries including contusion, hemorrhage, and axonal shearing. It is important to note that clinical TBI is often the result of a heterogeneous mixture of mechanical forces leading to mixed primary injuries. Interventions targeting primary injury are essentially preventative, and only effective if they preclude or mitigate primary injury. Examples of interventions targeting primary injury would include measures such as mandatory bike helmet laws or improved airbag technology.

In contrast, interventions targeting secondary injury may have a prolonged window in which to act. After the immediate primary injury is sustained, secondary injury develops over minutes to months to possibly years. Secondary injury reflects a complex cascade of metabolic, cellular, and molecular events including glutamate excitotoxicity, disordered cellular calcium homeostasis, mitochondrial dysfunction, inflammation, apoptosis/necroptosis, diffuse axonal injury (DAI), increased free radical generation and lipid peroxidation. In the extreme, secondary injury can lead to cell death, diffuse inflammation, and brain atrophy.

Despite the long window in which mitigation of secondary injury could occur, promising therapeutic interventions in the preclinical setting have failed to translate to clinical trials, with more than 30 failed TBI clinical trials based on successful preclinical studies [2]. While many reasons for failed TBI trials have been cited, including poor central nervous system drug penetration, delayed treatment initiation, heterogeneity across treatment sites, and insensitive outcomes measures [3], one important factor is the heterogeneity of clinical TBI compared to preclinical TBI. Indeed, clinical TBI often reflects a mixture of primary injuries as well as host-specific differences in the secondary injury response. In contrast, preclinical TBI studies rarely account for confounders including location, nature and severity of the primary injury, preexisting medical conditions, genetic background, age, gender, and illicit and prescribed drug use (to name a few). Thus, a promising therapeutic intervention in the preclinical setting may fail in the face of all the clinical confounders for which it was not tested.

Another important caveat to utilizing animal models of TBI is the lack of common terminology around injury severity. In human TBI trials, the Glasgow coma scale (GCS) is the primary means for assessing initial injury severity and the Glasgow outcome scale (GOS), or its extended version (GOSe), is the primary method for assessing outcomes. Whether or not these scales provide the optimal assessment tools is a source of controversy, but irrespective of this controversy, they do provide a common language to describe injury severity and outcomes. No such common language exists in preclinical trials. With no widely adopted common scoring system for injury severity in animal models of TBI, histological changes and functional tests provide the most reliable estimates of severity of TBI. However, subtle changes in injury mechanisms or devices as well as differences in techniques between labs, can sometimes make the comparison of various animal TBI models precarious at best.

Despite these limitations, animal models of TBI are extremely important tools to study the biomechanical, cellular and molecular events after TBI, many of which cannot be feasibly addressed in the clinical setting. The purpose of this chapter is to review some of the most commonly used methods of preclinical TBI both in terms of clinical relevance and potential technical pitfalls. Understanding these models are important prior to developing new models that better recapitulate the spectrum of human TBI.

2 General Experimental Model Approach

Prior to selecting a preclinical TBI model, it is important to address what aspect of human TBI is being sought to evaluate. For example, an investigator studying sports-related TBI would be wise to avoid a model with high-associated mortality, intraparenchymal hemorrhages, and skull fractures. New model development can take months to years, and translating established models to a new laboratory setting may also require significant effort and time. Understanding sources of variability in the model and establishing baseline functional and histological outcomes are particularly important prior to testing therapeutic interventions. In an attempt to help standardize preclinical TBI studies, the National Institutes of Health (NIH) has recommended the use of common data elements for preclinical TBI models (Table 1).

2.1 Controlled Cortical Impact Injury Model

2.1.1 Introduction

The controlled cortical impact (CCI) model has been utilized mostly to model moderate or severe human TBI. The classic CCI device employs a pneumatic or electromagnetic mechanism to drive a rigid impactor onto the exposed, intact dura, resulting in acute subdural hematoma, axonal injury, blood–brain barrier (BBB) dysfunction, and cortical tissue loss [4, 5]. CCI has been applied to species including mice, rats, ferrets, swine, and monkeys. One of the key advantages of CCI is that injury severity can be carefully calibrated by altering mechanical factors such as time, velocity, and depth of impact. In contrast to weight drop models, CCI lacks a rebound injury and produces a more focal injury compared to FPI. Another advantage of CCI is its use in developing novel therapeutic treatments for brain injury [6–11].

Table 1

***Common Data Elements* (*CDEs*) for preclinical TBI research from the NIH**

Animal characteristics	Animal history	Assessments and outcomes
Species	Pre-injury subject housing	Outcome timing
Birthdate	Pre-injury conditions	Assessment date and time
Age	Pre-injury surgical procedures	Acute neurological assessment Apnea indicator
Age group	Injury group	Apnea duration
Sex	Injury date and time	Righting response time
Animal vendor	Anesthetic type	Toe pinch response
Strain/genetic modifications	Anesthetic route	Acute physiological assessments Brain imaging type
Weight measurement	Anesthesia duration	Chronic physiologic assessments
		Memory/retention tests
		Learning/acquisition tests
		Sensory/motor tests
Injury model characteristics	Analgesia type	Anxiety tests
External cause modeled	Injury severity	Social interaction tests
Injury model	Number of injury exposures	Body weight change
Device manufacturer	Interval between injuries	Histopathology
Device manufacturer other text Animal stabilization method	Post-injury surgical procedures	
Impact location side	Post-injury conditions	
Impact location cortical region Impact location coordinates	Post-injury subject housing Treatment group	
	Treatment onset	
	Drug treatment route	
	Treatment or therapy type	
	Treatment control	
	Treatment dose	
	Survival time	
	Euthanasia date and time	
	Euthanasia type	

2.1.2 Methods

In rodent CCI models the device is hooked up to an air tank, injury controller, and mounted on a crossbar where the impactor can be adjusted to a range of desired angles of injury. Animals are induced using isoflurane or a comparable anesthetic before the head is secured in a stereotactic frame. A midline incision is made on the top of the head and the periosteum removed to prevent interference with the drill. Depending on desired area of injury, a unilateral [12] or bilateral [13] craniotomy is performed, most commonly, between the bregma and lambda. In both models the diameter of the craniotomy ranges from 4 to 6 mm. Injured mice can receive a mild–severe injury depending on the depth of brain deformation caused by the piston (0.2–1.2 mm) while sham animals receive only a craniotomy. The impact velocity can be adjusted from 0.5 to 10 m/s and impact duration ranges from 25 to 250 ms.

Once the skull flap is removed, using the injury controller the piston is set in the down position and the exposed brain is raised so that it is just touching the piston. The piston is then set back up, the apparatus is unlocked and the dial adjusted to the desired depth. After relocking, the piston is fired onto the brain using the controller. After completion of the injury, the incision is sutured, antibiotic cream is applied, and the animal is placed in a separate cage to wake up (*see* **Notes 1–4**). Common data elements for CCI studies are found in Table 2.

2.1.3 Clinical Relevance

Animals subjected to CCI have a wide range of functional deficits, which are highly related to both the depth of deformation and the velocity of the impact [12, 14]. Functional deficits after CCI include but are not limited to sensorimotor deficits (wire grip, rotarod, wire grip) memory deficits (Morris Water Maze (MWM) performance) and increased depression, anxiety, and impulsivity (forced swim,

Table 2
Controlled cortical injury (CCI) relevant data elements

Invasive surgery
Craniotomy size
Impactor angle
Impactor angle measurement
Impactor tip/projectile shape
Impactor tip rigidity
Impactor depth setting
Impactor dwell time
Impactor velocity
Surface material

elevated-plus maze, acoustic startle) [14–18]. Some functional deficits can be persistent up to 1 year post-CCI injury. The most marked histopathological change after CCI is the development of a cavitary lesion. However, CCI can result in a spectrum of anatomic injury including diffuse axonal injury (DAI) [5, 19]. Beta-amyloid precursor protein (APP) positive cells are increased in both the contused cortex and ipsilateral hippocampus as early as 6 h. The highest immunoreactivity levels occur on days 1–3 after injury and the positive cells can be seen for weeks [20]. Neuronal degeneration is also observed in early and late stages after injury [21, 22]. Due of the contusive nature of this injury, CCI causes inflammation and a dramatic increase in cytokines and chemokines [15, 20, 23]. Microglia activation is also accumulated in the lesion [15, 24].

2.1.4 Caveats

Most clinical cases of contusive injuries have a concussive injury as well. However, the CCI model lacks a concussive impact because the animals head is fixed in a stereotactic frame. To overcome this the experimental animals can be subjected to a concussion immediately following the contusive injury [25]. The spectrum of functional deficits after CCI is dependent on the localization of injury. Injuries located in the posterior cortex result in transient motor deficits as well as impaired memory and sensory function [15].

2.2 Fluid Percussion Injury

2.2.1 Introduction

Although it was originally created for use in larger animals, Fluid Percussion Injury (FPI) has become of the most commonly used methods of creating a non-penetrating head injury in rodents [26]. In FPI models, TBI is the result of a fluid pressure pulse transmitted to the intact dura through a craniotomy. The FPI device consists of a cylindrical Plexiglas reservoir of sterile saline with one end attached to a transducer which, through a tube, connects to the cap cemented to the skull. The injury is generated by a pendulum striking the piston, which creates a pressure pulse that travels through the transducer and onto the plastic cap causing a deformation of the intact dura. The fluid pulse produces brief displacement and deformation of brain tissue, and the severity of injury depends on the strength of the pressure pulse. In its most severe forms, FPI can result in intracranial hemorrhage, edema and progressive grey matter damage. FPI models can be categorized by the location of the craniotomy, into midline (centered on the sagittal suture), parasagittal (<3.5 mm lateral to midline), and lateral models (>3.5 mm lateral to midline) [26–29]. Midline and parasagittal FPIs cause bilateral cortical damage, while lateral FPI (LFPI) inflicts primarily unilateral cortical damage, rarely involving the contralateral cortices and brainstem.

2.2.2 Methods

Animals are anesthetized using 3% halothane via a vaporizer and a vacuum trap or a comparable anesthetic, placed in a prone position on a warming pad and secured in a stereotactic frame. The head is shaved, eye lubricant is applied, and the area of incision is prepped using three alternating treatments of Betadine and ethanol. A 1.5 cm sagittal incision is made at the midline between the ears and extended toward the nose. The periosteum is removed with a cotton swab, then, using a 5 mm drill bit, a craniotomy is performed either centrally over the midline (FPI) or laterally (LFPI) between the bregma and lambda. A cannula is set in the cranial opening so that it is touching the intact dura and fixed it to the skull using glue. The cannula is filled with saline and any bubbles are removed from the apparatus.

To determine the resting position, set the pendulum at 90° perpendicular to the ground and make sure it is touching the piston. Double check and remove any bubbles from all tubing and connections and ensure that the tubing from FPI device is filled with water. Using forceps to hold the cannula and tubing, attach them together by twisting in opposing directions. This prevents the application of excessive torque, which would cause the head cannula to detach from the rat's skull; Thereby causing leakage and an indeterminable injury. Set the pendulum to the desired height, clear the path from the pendulum to the piston and release the pendulum, allowing it to strike the piston. Severity of the injury depends on the intensity of the pressure pulse, which is controlled by the height of the pendulum drop. There are three levels of injury, mild, moderate, and severe. Mild injury occurs following a percussion of 0.1–1 atm (10.13–101.29 kPa), moderate after 1.5–2.0 atm, and severe after 2.5–3 atm. After the piston strikes, check for any leaking of saline and make sure to catch the pendulum to prevent any inadvertent injury. Remove the cannula, fill the burr hole with bone wax, suture the incision and place the animal on a heating pad to wake up (*see* **Notes 5** and **6**). Common data elements for FPI studies are found in Table 3.

Table 3

Fluid percussion injury (FP) relevant data elements

Craniotomy size
Connector angle
Connector tube
Connector tube length
Connector tube material
Port distal diameter
Cement
Transducer manufacturer
Cap characteristics
Peak pressure pulse
Pressure wave duration

2.2.3 Clinical Relevance

LFPI, the most commonly utilized FPI procedure in rats, produces a combination of focal cortical contusion and diffuse subcortical neuronal injury (including injury in the hippocampus and thalamus). As in CCI, the contused cortex beneath the injury site becomes a cavitary lesion surrounded by reactive gliosis. Over days to months, progressive degenerative changes may be found in ipsilateral hippocampus, thalamus, medial septum, striatum and amygdala. Widespread β-APP expression, comparable to human DAI, may be seen in more severe FPI models [30–32]. Functional deficits in motor and memory outcomes are similar to those seen after CCI, and can persist for up to 1 year after injury.

2.2.4 Caveats

FPI models are associated with high mortality, probably due to brainstem-mediated apnea. Post-injury seizures are common. The FPI model can be difficult to calibrate, since the height of pendulum is the only adjustable mechanical parameter. Reproducibility has been a challenge of FPI models until recently, when a microprocessor-controlled, pneumatically driven instrument has been developed. Though LFPI is the most commonly employed FPI model, there has been recent interest in midline FPI models, which are believed to better represent sports-related and military TBI.

2.3 Penetrating/ Direct Impact TBI

2.3.1 Introduction

A wide variety of experimental penetrating TBI models have been designed to generate cerebral deformation including microinjection [33], cryolesion [34], mechanical suction [35], vacuum pulse [36, 37], and inflation of a balloon [38, 39] and others [40]. However, because of their clinical relevance, the most commonly used are penetrating ballistic like blast injury (PBBI) and fragment penetration models. In PBBI models, the insult is caused by transmission of a high energy projectile that produces a cavitary lesion much larger than the projectile. There may be marked intracerebral hemorrhage, but less diffuse axonal injury than other models. Outcomes are directly related to the anatomical path and energy transfer of the projectile.

2.3.2 Methods

In Carey et al.'s [41–43] model anesthetized animals are placed in a stereotactic frame and the sloping outer wall of the right frontal sinus is removed; thereby allowing the missile like object to penetrate the intact and vertically disposed sinus wall. A 2-mm, 31-mg steel sphere fired from 80 cm at 220 or 280 m/s penetrates the right frontal bone and traverses the right cerebral hemisphere from anterior to posterior. The energy from the projectile can range from 0.9 to 1.4 J. Finnie's [44] model utilizes a 0.22 caliber firearm to inflict a head wound to restrained sheep. The bullet is fired at the temporal region of the skull, causing a right to left transverse injury to the temporal lobes. Recently several new PBBI models have been developed in rodents [45, 46] and fragment blast models are becoming more common as well [47]. Recently a novel, non fatal, low velocity model for PBBI has been established in rats [48] and is similar to CCI in that a cylindrical carbon fiber rod is accelerated and enters the brain creating a cavitary lesion. However, the pin (2 mm in diameter) is attached to a secondary projectile that is accelerated by a pellet fired from a modified rifle. The secondary projectile and pin, guided by a tube, penetrates the brain at a speed of 90 m/s. The base of the projectile is surrounded by a compressible ring (ferrule) that controls the depth of the penetration. Although the depth is usually set at 5 mm, depth, speed and shape of the pin can be altered to obtain different pathological outcomes. This model has also been modified for use in mice [49].

2.3.3 Clinical Relevance

This model mimics gun-related brain injuries caused by a high-velocity penetrating object. The injury can be induced by multiple mechanisms including the high pressures in front of an object; longitudinal shock wave and pressure waves from kinetic energy transfer [42]. Injury phenotypes are dependent on the injured part of brain and the most common consequences after injury are intracerebral hemorrhages, edema, elevated intracranial pressure, decreased cerebral blood flow, significant neuroinflammation, and notable lesion [38, 50]. Additionally, both apoptosis and necrosis are observed in the lesion [50, 51]. Motor function and cognitive impairments are observed in this model [51–53], including deficits in motor (balance beam and rotarod tasks) and memory (MWM test) that correlate with the degree of injury severity as well as clinical moderate-severe TBI [54, 55].

2.3.4 Caveats

The original model was established by Carey et al. using mongrel cats, however, it is difficult to run behavioral tasks on larger animals, thus rodent models have become increasingly popular [48, 49, 56, 57]. However, rodent PBBI models are relatively new and less standardized than other TBI models. Even though they are underdeveloped, rodent PBBI models provide opportunities to use genetic modification to investigate the mechanisms of injury and therapeutic targets. Due to the high velocity of objects entering the brain, heat damage of the tissue is a potential factor, though this may be also relevant to clinical penetrating TBI.

2.4 Weight Drop Models

2.4.1 Introduction

In weight-drop models, a free falling, guided weight is allowed to impact the skull (with or without a craniotomy). One of the major advantages of weight drop models is their relative inexpensiveness. Injury severity in these models can be altered by adjusting the mass of the weight and the height from which the weight falls. In addition, several modifications of the weight-drop model have been made, which markedly alter both functional and histopathological outcomes.

Models such as Feeney's weight-drop, are delivered to the exposed dura which results in a cortical contusion [58] and progresses to a cavitary lesion similar to those found in CCI. Shohami's group later introduced a modified rodent weight drop model, delivering closed head injury (CHI) to the intact but unprotected skull [59–63]. Of *Note*, in both the Feeney and Shohami methods, there is minimal rotational acceleration of the fixed head. In attempt to add rotational acceleration, a prominent feature in the primary injury of human TBI, Marmarou's impact acceleration model was developed to recapitulate this important mechanism. In the Marmarou method, a sectioned brass weight is allowed to fall freely from a designated height through a Plexiglas tube, onto the exposed skull of rats placed on a foam bed. Marmarou's model, which often results in a hight mortality rate in the absence of mechanical ventilation, is characterized by diffuse axonal injury (DAI) and neuronal injury, as well as functional deficits in beam walking and memory [64, 65]. Subsequent modifications of the Marmarou model have been designed to reproduce the frontal impact characteristic of motor vehicle and sports-related injuries [66, 67].

In general, these models do not result in mortality, prolonged apnea, skull fractures or cortical contusions, and are thought to be more representative of the vast majority of human TBI, which is classified as mild. Common data elements for weight drop studies are found in Table 4.

2.4.2 Methodsz

In a recent model of mild TBI [66], animals are induced for 45 s with 4% isoflurane or alternative anesthetic. Immediately following, a timer is started to initiate the loss of consciousness (LOC) and the animal is placed supine on a Kimwipes. The animal's tail and each end of the Kimwipes are secured by hand and the animal's head is placed directly beneath a hollow tube and positioned

Table 4

Weight drop injury (WD) relevant data elements

Invasive surgery	Weight drop height	Impactor retraction
Surface material	Weight drop guidance	WD-specific pre-injury surgical procedures
Craniotomy size	Weight drop characteristics Impactor velocity	WD-specific post-injury surgical procedures
Impactor/projectile mass	Contact surface type	
Impactor/projectile material	Contact surface area	
Impactor tip/projectile shape	Impactor dwell time	
Impactor tip rigidity		

with the tube opening directly posterior to the eyes. A cylindrical metal rod (54 g) is dropped dorsally on the animal's head between the coronal and lambdoid sutures. Upon impact the animals head readily penetrates the Kimwipes, causing the head and body to fall in a circular trajectory while the tail remains secured. The severity of the injury can be varied by altering the weight of the object and the height that it is dropped from. Mild injuries are 54 g dropped from 28″, while severe ones are 54 g at 60″. The mouse is then placed on its side and allowed to wake up and the timer is stopped. The (LOC) is recorded as the latency in seconds to spontaneous ambulation. The control group consists of sham animals given anesthesia but no weight drop and all animals are allowed to recover in room air in their cages.

Other commonly used models vary in the level of surgical invasiveness and are aimed to reproduce different clinical aspects of brain injury. The Feeney et al. model involves dropping a weight onto the intact dura through a craniotomy [68], which produces hemorrhages and cell death. In the Marmarou model, which mimics DAI and diffuse TBI, the animals head rests on a foam pad and a guided brass weight impacts a stainless steel disk mounted on the skull [69]. Shohami's model drops the weight onto one side of an unprotected skull while resting on a hard surface [60] and has been used to investigate BBB disruption. Weight drop models are primarily conducted on mice, rats, and swine (*see* **Notes 7** and **8**).

2.4.3 Clinical Relevance

Each variation of the weight drop model has distinct clinical features that should be evaluated prior to commencing. Model selection can be based on which clinical features of TBI are desired, for example loss of consciousness may be sought after in modeling TBI in military populations, since 4.9 % of US soldiers returning from Iraq had loss of consciousness [70]. However, the majority of patients with sports-related concussions do not experience loss of consciousness and a model minimizing this outcome might be desirable in modeling sport related TBI [71]. Many of the current weight drop models feature a rapid acceleration and deceleration of head movement [72]. The drastic change in speed and direction of the head is the main cause of concussions and axonal injury [73–75] in closed head traumas. The concussive injury does not induce immediate cell death; however, it does result in axonal injury and DAI at the extreme end of the spectrum. Tau pathology, an area of much clinical investigation in the field of TBI, may or may not be present. Reactive gliosis may also be a prominent feature of these models. The closed head model has been shown to result in impaired neurological and cognitive outcomes including motor, learning, memory, and anxiety [72, 76, 77].

2.4.4 Caveats

While weight-drop models are generally easy and inexpensive to employ, they can result in relatively high variability in injury severity. Another concern is the possibility of rebound injury. However, in general, weight-drop procedures are capable of producing graded axonal injury that is highly relevant to the vast majority of human TBI.

2.5 Non Impact Blast

2.5.1 Introduction

Blast models of TBI have been developed to better understand the effects of primary blast waves on the CNS. Some estimates suggest that nearly one in five veterans of the conflicts in Afghanistan and Iraq were exposed to TBI, much of which was blast-associated TBI. Initial efforts at simulating blast associated TBI were challenged by resultant systemic injuries, but the use of thoracic/abdominal protective strategies have minimized these types of associated injuries, allowing investigators to study the effects of blast associated TBI in isolation. Most commonly, blast models delivery primary injury using a compression-driven shock tube to simulate blast effects.

2.5.2 Methods

The most commonly used blast models have been established in rodents [78–82] and swine [83, 84] and utilize explosives [80, 83, 84] or compressed air [82, 85] to create a shock wave. In these models, anesthetized animals are individually fixed in a cylindrical metal tube in such a way that prevents any movement of the body. However, in some cases, the head and neck are able to freely to flex, extend, and rotate during the injury [86]. Additionally Kevlar vests, which encase the thorax and part of the abdomen, have been used to test acute mortality in rats [79]. The blast is generated by a controlled detonation or by the release of compressed air. Blast parameters can be regulated to generate mild to severe injury. Intensity of the blast and subsequent injury is controlled by altering the amount of explosives or pressure of the compressed air; or by changing the animal's distance from the blast point. When using compressed air more complex waveforms can be created by adding additional chambers of compressed air. Although reproducible, the blast tube and other models lack the clinical relevance and unpredictable conditions of real life event. In a recent open field study [87], 12 anesthetized animals are loosely fixed in individual compartments using a plastic net and placed at 4 or 7 m from the detonation point. The net allows for their body to be exposed to the blast without being injured by shrapnel or debris. The platforms are then elevated to a height of 1 m to prevent any interference from blast wave reflection off the ground. Pressure meters are placed at each distance to detect the intensity of the blast. After the detonation of 500 g of TNT, the mice are allowed to wake up and the pressure sensors data are recorded (*see* **Notes 9** and **10**).

2.5.3 Clinical Relevance

Non-impact models of blast injury are characterized by cerebral edema, hyperemia and delayed vasospasm, the degree of which corresponds to injury severity. The most prominent histopathological finding in blast models is DAI, which corresponds to changes in diffuse tractography imaging seen in blast exposed military veterans [88]. Associated with DAI, blast-exposed mice also demonstrate phosphorylated tauopathy, myelinated axonopathy, chronic neuroinflammation and neurodegeneration, consistent with clinical reports of chronic traumatic encephalopathy in military populations. Functional deficits reported after blast models of TBI include deficits in social recognition, spatial memory, and motor coordination [89].

2.5.4 Caveats

Clinically, blast injuries consist of primary, secondary, tertiary, and quaternary mechanisms, of which the blast preclinical models really only recapitulates primary mechanism. In addition, animal placement in relation to the shock tube (that is, inside, outside or near the exit of the tube) can significantly modify injury type and severity. Blast models, while highly clinically relevant, also suffer from the most variability within and between models, making comparisons between studies very difficult. Utilization of common data elements (Table 5) is particularly important in this setting. Additionally shock tube models are incredibly complex and require the use of heavy machinery.

Table 5

Blast-induced neurotrauma (BIN) relevant data elements

Blast-induced delivery device Pressure wave type	Distance between animal and tube	Reflective surfaces
Detonation type	Animal orientation to blast wave	Primary blast effects
Detonation material quantity	Overpressure peak	
Driver gas	Overpressure rise time	Secondary blast effect type
Pressure wave medium	Overpressure wave duration	Secondary blast effect specifications
Distance from detonation	Impulse	Tertiary blast effects
Blast tube or column area	Reflective wave overpressure	Tertiary blast effect specifications
Blast tube length	Blast wind pressure	Quaternary blast effects
Shock tube driven section length	Pressure sensor orientation	Systemic injury
Membrane thickness	Pressure sensor type	Extracranial injuries
Membrane burst method	Pressure sensor sampling frequency	BIN-specific pre-injury surgical procedures
Membrane burst pressure	Incident pressure time history	BIN-specific post-injury surgical procedures
Tube end configuration	Body exposure	
Placement relative to shock tube	Protective shielding location	
	Protective shielding type	

3 Notes

1. Ensuring a snug fit in stereotactic frame is essential before starting as it makes drilling difficult and keeping the head still ensures consistent injuries.
2. Make sure head and stereotactic frame are out of the way when setting the impactor.
3. If the impactor is set at 90 degrees perpendicular to the ground the piston will not be flush with the brain because of the curvature of the brain. Make sure to make the contact between the piston and brain the same between animals.
4. Do not replace skull flap from craniotomy as there will be edema which will cause an increase in ICP.
5. The cannula should fit the burr hole snugly and there should be nothing obscuring visualization of the dura through the center of the cannula. Optionally, a few drops of cresyl violet dye can be added to the saline to facilitate observation of any leaks around the burr hole site during the injury.
6. Bubbles can be removed by disconnecting the distal end of the tubing from the rat and forcing fluid through the reservoir and tubing until all air bubbles have been expelled. The tubing is then reconnected to the rat.
7. When taking mouse and Kimwipes off of the table, hold one end of Kimwipes and tail and place mouse vertical in the air then grab the other end and flatten out. This will stretch the mouse's body out and make placing the head under the tube easier. Hold the Kimwipes tight but not too tight as to rip it.
8. In order to get a more direct hit on top of the head and not over the cerebellum/spinal cord, tilting the fore body upward towards the pipe.
9. When using compressed air model, it is critical that the mylar membrane is mounted correctly. If the sheet gets folded at all, the membrane doubles in thickness and creates inconsistent blasts.
10. Production of mylar sheets is sometimes inconsistent. Therefore, test each new box before running an experiment by generating and measuring the blast.

References

1. Langlois JA, Rutland-Brown W, Wald MM (2006) The epidemiology and impact of traumatic brain injury: a brief overview. J Head Trauma Rehabil 21:375–378
2. Loane DJ, Faden AI (2010) Neuroprotection for traumatic brain injury: translational challenges and emerging therapeutic strategies. Trends Pharmacol Sci 31:596–604
3. Narayan RK, Michel ME, Ansell B, Baethmann A, Biegon A, Bracken MB, Bullock MR, Choi SC, Clifton GL, Contant CF, Coplin WM, Dietrich WD, Ghajar J, Grady SM, Grossman RG, Hall ED, Heetderks W, Hovda DA, Jallo J, Katz RL, Knoller N, Kochanek PM, Maas AI, Majde J, Marion DW, Marmarou A, Marshall LF, McIntosh TK, Miller E, Mohberg N, Muizelaar

JP, Pitts LH, Quinn P, Riesenfeld G, Robertson CS, Strauss KI, Teasdale G, Temkin N, Tuma R, Wade C, Walker MD, Weinrich M, Whyte J, Wilberger J, Young AB, Yurkewicz L (2002) Clinical trials in head injury. J Neurotrauma 19:503–557
4. Dixon CE, Clifton GL, Lighthall JW, Yaghmai AA, Hayes RL (1991) A controlled cortical impact model of traumatic brain injury in the rat. J Neurosci Methods 39:253–262
5. Lighthall JW (1988) Controlled cortical impact: a new experimental brain injury model. J Neurotrauma 5:1–15
6. Kawamata T, Katayama Y, Maeda T, Mori T, Aoyama N, Kikuchi T, Uwahodo Y (1997) Antioxidant, OPC-14117, attenuates edema formation and behavioral deficits following cortical contusion in rats. Acta Neurochir Suppl 70:191–193
7. Verweij BH, Muizelaar JP, Vinas FC, Peterson PL, Xiong Y, Lee CP (1997) Mitochondrial dysfunction after experimental and human brain injury and its possible reversal with a selective N-type calcium channel antagonist (SNX-111). Neurol Res 19:334–339
8. Faden AI, Fox GB, Fan L, Araldi GL, Qiao L, Wang S, Kozikowski AP (1999) Novel TRH analog improves motor and cognitive recovery after traumatic brain injury in rodents. Am J Physiol 277:R1196–R1204
9. Kroppenstedt SN, Stroop R, Kern M, Thomale UW, Schneider GH, Unterberg AW (1999) Lubeluzole following traumatic brain injury in the rat. J Neurotrauma 16:629–637
10. Dempsey RJ, Baskaya MK, Dogan A (2000) Attenuation of brain edema, blood-brain barrier breakdown, and injury volume by ifenprodil, a polyamine-site N-methyl-D-aspartate receptor antagonist, after experimental traumatic brain injury in rats. Neurosurgery 47:399–404, discussion 404–406
11. Sullivan PG, Thompson M, Scheff SW (2000) Continuous infusion of cyclosporin A postinjury significantly ameliorates cortical damage following traumatic brain injury. Exp Neurol 161:631–637
12. Washington PM, Forcelli PA, Wilkins T, Zapple DN, Parsadanian M, Burns MP (2012) The effect of injury severity on behavior: a phenotypic study of cognitive and emotional deficits after mild, moderate, and severe controlled cortical impact injury in mice. J Neurotrauma 29:2283–2296
13. Geddes RI, Sribnick EA, Sayeed I, Stein DG (2014) Progesterone treatment shows benefit in a pediatric model of moderate to severe bilateral brain injury. PLoS One 9:e87252
14. Fox GB, Fan L, Levasseur RA, Faden AI (1998) Sustained sensory/motor and cognitive deficits with neuronal apoptosis following controlled cortical impact brain injury in the mouse. J Neurotrauma 15:599–614
15. Bermpohl D, You Z, Lo EH, Kim HH, Whalen MJ (2007) TNF alpha and Fas mediate tissue damage and functional outcome after traumatic brain injury in mice. J Cereb Blood Flow Metab 27:1806–1818
16. Mahmood A, Wu H, Qu C, Xiong Y, Chopp M (2013) Effects of treating traumatic brain injury with collagen scaffolds and human bone marrow stromal cells on sprouting of corticospinal tract axons into the denervated side of the spinal cord. J Neurosurg 118:381–389
17. Wagner AK, Postal BA, Darrah SD, Chen X, Khan AS (2007) Deficits in novelty exploration after controlled cortical impact. J Neurotrauma 24:1308–1320
18. Ajao DO, Pop V, Kamper JE, Adami A, Rudobeck E, Huang L, Vlkolinsky R, Hartman RE, Ashwal S, Obenaus A, Badaut J (2012) Traumatic brain injury in young rats leads to progressive behavioral deficits coincident with altered tissue properties in adulthood. J Neurotrauma 29:2060–2074
19. Tran HT, LaFerla FM, Holtzman DM, Brody DL (2011) Controlled cortical impact traumatic brain injury in 3xTg-AD mice causes acute intra-axonal amyloid-beta accumulation and independently accelerates the development of tau abnormalities. J Neurosci 31:9513–9525
20. Ciallella JR, Ikonomovic MD, Paljug WR, Wilbur YI, Dixon CE, Kochanek PM, Marion DW, DeKosky ST (2002) Changes in expression of amyloid precursor protein and interleukin-1beta after experimental traumatic brain injury in rats. J Neurotrauma 19:1555–1567
21. Colicos MA, Dixon CE, Dash PK (1996) Delayed, selective neuronal death following experimental cortical impact injury in rats: possible role in memory deficits. Brain Res 739:111–119
22. Pohl D, Bittigau P, Ishimaru MJ, Stadthaus D, Hubner C, Olney JW, Turski L, Ikonomidou C (1999) N-Methyl-D-aspartate antagonists and apoptotic cell death triggered by head trauma in developing rat brain. Proc Natl Acad Sci U S A 96:2508–2513
23. Dohi K, Kraemer BC, Erickson MA, McMillan PJ, Kovac A, Flachbartova Z, Hansen KM, Shah GN, Sheibani N, Salameh T, Banks WA (2014) Molecular hydrogen in drinking water protects against neurodegenerative changes induced by traumatic brain injury. PLoS One 9:e108034

24. Kabadi SV, Stoica BA, Byrnes KR, Hanscom M, Loane DJ, Faden AI (2012) Selective CDK inhibitor limits neuroinflammation and progressive neurodegeneration after brain trauma. J Cereb Blood Flow Metab 32:137–149
25. Dapul HR, Park J, Zhang J, Lee C, DanEshmand A, Lok J, Ayata C, Gray T, Scalzo A, Qiu J, Lo EH, Whalen MJ (2013) Concussive injury before or after controlled cortical impact exacerbates histopathology and functional outcome in a mixed traumatic brain injury model in mice. J Neurotrauma 30:382–391
26. McIntosh TK, Vink R, Noble L, Yamakami I, Fernyak S, Soares H, Faden AL (1989) Traumatic brain injury in the rat: characterization of a lateral fluid-percussion model. Neuroscience 28:233–244
27. Vink R, Mullins PG, Temple MD, Bao W, Faden AI (2001) Small shifts in craniotomy position in the lateral fluid percussion injury model are associated with differential lesion development. J Neurotrauma 18:839–847
28. Sanders MJ, Dietrich WD, Green EJ (1999) Cognitive function following traumatic brain injury: effects of injury severity and recovery period in a parasagittal fluid-percussive injury model. J Neurotrauma 16:915–925
29. Floyd CL, Golden KM, Black RT, Hamm RJ, Lyeth BG (2002) Craniectomy position affects morris water maze performance and hippocampal cell loss after parasagittal fluid percussion. J Neurotrauma 19:303–316
30. Wang E, Gao J, Yang Q, Parsley MO, Dunn TJ, Zhang L, DeWitt DS, Denner L, Prough DS, Wu P (2012) Molecular mechanisms underlying effects of neural stem cells against traumatic axonal injury. J Neurotrauma 29:295–312
31. Ekmark-Lewen S, Flygt J, Kiwanuka O, Meyerson BJ, Lewen A, Hillered L, Marklund N (2013) Traumatic axonal injury in the mouse is accompanied by a dynamic inflammatory response, astroglial reactivity and complex behavioral changes. J Neuroinflammation 10:44
32. Hoshino S, Kobayashi S, Furukawa T, Asakura T, Teramoto A (2003) Multiple immunostaining methods to detect traumatic axonal injury in the rat fluid-percussion brain injury model. Neurol Med Chir (Tokyo) 43:165–173, discussion 174
33. Fitch MT, Doller C, Combs CK, Landreth GE, Silver J (1999) Cellular and molecular mechanisms of glial scarring and progressive cavitation: in vivo and in vitro analysis of inflammation-induced secondary injury after CNS trauma. J Neurosci 19:8182–8198
34. Sun D, Tani M, Newman TA, Krivacic K, Phillips M, Chernosky A, Gill P, Wei T, Griswold KJ, Ransohoff RM, Weller RO (2000) Role of chemokines, neuronal projections, and the blood-brain barrier in the enhancement of cerebral EAE following focal brain damage. J Neuropathol Exp Neurol 59:1031–1043
35. Mathew P, Bullock R, Graham DI, Maxwell WL, Teasdale GM, McCulloch J (1996) A new experimental model of contusion in the rat. Histopathological analysis and temporal patterns of cerebral blood flow disturbances. J Neurosurg 85:860–870
36. Shreiber DI, Bain AC, Ross DT, Smith DH, Gennarelli TA, McIntosh TK, Meaney DF (1999) Experimental investigation of cerebral contusion: histopathological and immunohistochemical evaluation of dynamic cortical deformation. J Neuropathol Exp Neurol 58:153–164
37. Shreiber DI, Smith DH, Meaney DF (1999) Immediate in vivo response of the cortex and the blood-brain barrier following dynamic cortical deformation in the rat. Neurosci Lett 259:5–8
38. Williams AJ, Hartings JA, Lu XC, Rolli ML, Dave JR, Tortella FC (2005) Characterization of a new rat model of penetrating ballistic brain injury. J Neurotrauma 22:313–331
39. Williams AJ, Wei HH, Dave JR, Tortella FC (2007) Acute and delayed neuroinflammatory response following experimental penetrating ballistic brain injury in the rat. J Neuroinflammation 4:17
40. Ghirnikar RS, Lee YL, He TR, Eng LF (1996) Chemokine expression in rat stab wound brain injury. J Neurosci Res 46:727–733
41. Carey ME (1995) Experimental missile wounding of the brain. Neurosurg Clin N Am 6:629–642
42. Carey ME, Sarna GS, Farrell JB, Happel LT (1989) Experimental missile wound to the brain. J Neurosurg 71:754–764
43. Carey ME, Sarna GS, Farrell JB (1990) Brain edema after an experimental missile wound. Adv Neurol 52:301–305
44. Finnie JW (1993) Brain damage caused by a captive bolt pistol. J Comp Pathol 109:253–258
45. Shear DA, Lu XC, Bombard MC, Pedersen R, Chen Z, Davis A, Tortella FC (2010) Longitudinal characterization of motor and cognitive deficits in a model of penetrating ballistic-like brain injury. J Neurotrauma 27:1911–1923
46. Wei G, Lu XC, Yang X, Tortella FC (2010) Intracranial pressure following penetrating ballistic-like brain injury in rats. J Neurotrauma 27:1635–1641
47. Risling M, Skold M, Larsson I, Angeria M, Davidsson J (2004) Leakage of S-100 protein

after high velocity penetration injury to the brain. Restor Neurol Neurosci 23:141, 7th international neurotrauma symposium. Medimond (ed) Adelaide, Australia

48. Plantman S, Ng KC, Lu J, Davidsson J, Risling M (2012) Characterization of a novel rat model of penetrating traumatic brain injury. J Neurotrauma 29:1219–1232
49. Cernak I, Wing ID, Davidsson J, Plantman S (2014) A novel mouse model of penetrating brain injury. Front Neurol 5:209
50. Williams AJ, Hartings JA, Lu XC, Rolli ML, Tortella FC (2006) Penetrating ballistic-like brain injury in the rat: differential time courses of hemorrhage, cell death, inflammation, and remote degeneration. J Neurotrauma 23:1828–1846
51. Tompkins P, Tesiram Y, Lerner M, Gonzalez LP, Lightfoot S, Rabb CH, Brackett DJ (2013) Brain injury: neuro-inflammation, cognitive deficit, and magnetic resonance imaging in a model of blast induced traumatic brain injury. J Neurotrauma 30:1888–1897
52. Lu XC, Shear DA, Graham PB, Bridson G, Uttamsingh V, Chen Z, Leung LY, Tortella FC (2015) Dual therapeutic effects of C-10068, a dextromethorphan derivative, against post-traumatic nonconvulsive seizures and neuroinflammation in a rat model of penetrating ballistic-like brain injury. J Neurotrauma 32:1621–1632
53. Shear DA, Lu XC, Pedersen R, Wei G, Chen Z, Davis A, Yao C, Dave J, Tortella FC (2011) Severity profile of penetrating ballistic-like brain injury on neurofunctional outcome, blood-brain barrier permeability, and brain edema formation. J Neurotrauma 28:2185–2195
54. Driscoll DM, Dal Monte O, Solomon J, Krueger F, Grafman J (2012) Empathic deficits in combat veterans with traumatic brain injury: a voxel-based lesion-symptom mapping study. Cogn Behav Neurol 25:160–166
55. Salazar AM, Grafman J, Schlesselman S, Vance SC, Mohr JP, Carpenter M, Pevsner P, Ludlow C, Weingartner H (1986) Penetrating war injuries of the basal forebrain: neurology and cognition. Neurology 36:459–465
56. Elias PZ, Spector M (2012) Characterization of a bilateral penetrating brain injury in rats and evaluation of a collagen biomaterial for potential treatment. J Neurotrauma 29:2086–2102
57. Gajavelli S, Kentaro S, Diaz J, Yokobori S, Spurlock M, Diaz D, Jackson C, Wick A, Zhao W, Leung LY, Shear D, Tortella F, Bullock MR (2015) Glucose and oxygen metabolism after penetrating ballistic-like brain injury. J Cereb Blood Flow Metab 35:773–780
58. Dail WG, Feeney DM, Murray HM, Linn RT, Boyeson MG (1981) Responses to cortical injury: II. Widespread depression of the activity of an enzyme in cortex remote from a focal injury. Brain Res 211:79–89
59. Shapira Y, Shohami E, Sidi A, Soffer D, Freeman S, Cotev S (1988) Experimental closed head injury in rats: mechanical, pathophysiologic, and neurologic properties. Crit Care Med 16:258–265
60. Shohami E, Shapira Y, Cotev S (1988) Experimental closed head injury in rats: prostaglandin production in a noninjured zone. Neurosurgery 22:859–863
61. Chen Y, Constantini S, Trembovler V, Weinstock M, Shohami E (1996) An experimental model of closed head injury in mice: pathophysiology, histopathology, and cognitive deficits. J Neurotrauma 13:557–568
62. Stahel PF, Shohami E, Younis FM, Kariya K, Otto VI, Lenzlinger PM, Grosjean MB, Eugster HP, Trentz O, Kossmann T, Morganti-Kossmann MC (2000) Experimental closed head injury: analysis of neurological outcome, blood-brain barrier dysfunction, intracranial neutrophil infiltration, and neuronal cell death in mice deficient in genes for pro-inflammatory cytokines. J Cereb Blood Flow Metab 20:369–380
63. Flierl MA, Stahel PF, Beauchamp KM, Morgan SJ, Smith WR, Shohami E (2009) Mouse closed head injury model induced by a weight-drop device. Nat Protoc 4:1328–1337
64. Heath DL, Vink R (1995) Impact acceleration-induced severe diffuse axonal injury in rats: characterization of phosphate metabolism and neurologic outcome. J Neurotrauma 12:1027–1034
65. Schmidt RH, Scholten KJ, Maughan PH (2000) Cognitive impairment and synaptosomal choline uptake in rats following impact acceleration injury. J Neurotrauma 17:1129–1139
66. Khuman J, Meehan WP III, Zhu X, Qiu J, Hoffmann U, Zhang J, Giovannone E, Lo EH, Whalen MJ (2010) Tumor necrosis factor alpha and Fas receptor contribute to cognitive deficits independent of cell death after concussive traumatic brain injury in mice. J Cereb Blood Flow Metab 31:778–789
67. Kilbourne M, Kuehn R, Tosun C, Caridi J, Keledjian K, Bochicchio G, Scalea T, Gerzanich V, Simard JM (2009) Novel model of frontal impact closed head injury in the rat. J Neurotrauma 26:2233–2243
68. Feeney DM, Boyeson MG, Linn RT, Murray HM, Dail WG (1981) Responses to cortical

injury: I. Methodology and local effects of contusions in the rat. Brain Res 211:67–77

69. Marmarou A, Foda MA, van den Brink W, Campbell J, Kita H, Demetriadou K (1994) A new model of diffuse brain injury in rats. Part I: pathophysiology and biomechanics. J Neurosurg 80:291–300
70. Hoge CW, McGurk D, Thomas JL, Cox AL, Engel CC, Castro CA (2008) Mild traumatic brain injury in U.S. Soldiers returning from Iraq. N Engl J Med 358:453–463
71. Silverberg ND, Luoto TM, Ohman J, Iverson GL (2014) Assessment of mild traumatic brain injury with the King-Devick Test in an emergency department sample. Brain Inj 28:1590–1593
72. Meehan WP III, Zhang J, Mannix R, Whalen MJ (2012) Increasing recovery time between injuries improves cognitive outcome after repetitive mild concussive brain injuries in mice. Neurosurgery 71:885–891
73. White BC, Krause GS (1993) Brain injury and repair mechanisms: the potential for pharmacologic therapy in closed-head trauma. Ann Emerg Med 22:970–979
74. McAllister TW (1992) Neuropsychiatric sequelae of head injuries. Psychiatr Clin North Am 15:395–413
75. Krave U, Al-Olama M, Hansson HA (2011) Rotational acceleration closed head flexion trauma generates more extensive diffuse brain injury than extension trauma. J Neurotrauma 28:57–70
76. Mannix R, Meehan WP, Mandeville J, Grant PE, Gray T, Berglass J, Zhang J, Bryant J, Rezaie S, Chung JY, Peters NV, Lee C, Tien LW, Kaplan DL, Feany M, Whalen M (2013) Clinical correlates in an experimental model of repetitive mild brain injury. Ann Neurol 74:65–75
77. Mannix R, Berglass J, Berkner J, Moleus P, Qiu J, Andrews N, Gunner G, Berglass L, Jantzie LL, Robinson S, Meehan WP III (2014) Chronic gliosis and behavioral deficits in mice following repetitive mild traumatic brain injury. J Neurosurg 121:1342–1350
78. Cernak I, Savic J, Malicevic Z, Zunic G, Radosevic P, Ivanovic I, Davidovic L (1996) Involvement of the central nervous system in the general response to pulmonary blast injury. J Trauma 40:S100–S104
79. Long JB, Bentley TL, Wessner KA, Cerone C, Sweeney S, Bauman RA (2009) Blast overpressure in rats: recreating a battlefield injury in the laboratory. J Neurotrauma 26:827–840
80. Clemedson CJ (1956) Shock wave transmission to the central nervous system. Acta Physiol Scand 37:204–214
81. Cheng J, Gu J, Ma Y, Yang T, Kuang Y, Li B, Kang J (2010) Development of a rat model for studying blast-induced traumatic brain injury. J Neurol Sci 294:23–28
82. Risling M, Davidsson J (2012) Experimental animal models for studies on the mechanisms of blast-induced neurotrauma. Front Neurol 3:30
83. Bauman RA, Ling G, Tong L, Januszkiewicz A, Agoston D, Delanerolle N, Kim Y, Ritzel D, Bell R, Ecklund J, Armonda R, Bandak F, Parks S (2009) An introductory characterization of a combat-casualty-care relevant swine model of closed head injury resulting from exposure to explosive blast. J Neurotrauma 26:841–860
84. de Lanerolle NC, Bandak F, Kang D, Li AY, Du F, Swauger P, Parks S, Ling G, Kim JH (2011) Characteristics of an explosive blast-induced brain injury in an experimental model. J Neuropathol Exp Neurol 70:1046–1057
85. Wang Y, Wei Y, Oguntayo S, Wilkins W, Arun P, Valiyaveettil M, Song J, Long JB, Nambiar MP (2011) Tightly coupled repetitive blast-induced traumatic brain injury: development and characterization in mice. J Neurotrauma 28:2171–2183
86. McKee AC, Stern RA, Nowinski CJ, Stein TD, Alvarez VE, Daneshvar DH, Lee HS, Wojtowicz SM, Hall G, Baugh CM, Riley DO, Kubilus CA, Cormier KA, Jacobs MA, Martin BR, Abraham CR, Ikezu T, Reichard RR, Wolozin BL, Budson AE, Goldstein LE, Kowall NW, Cantu RC (2013) The spectrum of disease in chronic traumatic encephalopathy. Brain 136:43–64
87. Rubovitch V, Ten-Bosch M, Zohar O, Harrison CR, Tempel-Brami C, Stein E, Hoffer BJ, Balaban CD, Schreiber S, Chiu WT, Pick CG (2011) A mouse model of blast-induced mild traumatic brain injury. Exp Neurol 232:280–289
88. Mac Donald CL, Johnson AM, Cooper D, Nelson EC, Werner NJ, Shimony JS, Snyder AZ, Raichle ME, Witherow JR, Fang R, Flaherty SF, Brody DL (2011) Detection of blast-related traumatic brain injury in U.S. military personnel. N Engl J Med 364:2091–2100
89. Koliatsos VE, Cernak I, Xu L, Song Y, Savonenko A, Crain BJ, Eberhart CG, Frangakis CE, Melnikova T, Kim H, Lee D (2011) A mouse model of blast injury to brain: initial pathological, neuropathological, and behavioral characterization. J Neuropathol Exp Neurol 70:399–416

Chapter 3

Rodent Models of Traumatic Brain Injury: Methods and Challenges

Niklas Marklund

Abstract

Traumatic brain injury (TBI) has been named the most complex disease in the most complex organ of the body. It is the most common cause of death and disability in the Western world in people <40 years old and survivors commonly suffer from persisting cognitive deficits, impaired motor function, depression and personality changes. TBI may vary in severity from uniformly fatal to mild injuries with rapidly resolving symptoms and without doubt, it is a markedly heterogeneous disease. Its different subtypes differs in their pathophysiology, treatment options and long-term consequences and to date, there are no pharmacological treatments with proven clinical benefit available to TBI patients. To enable development of novel treatment options for TBI, clinically relevant animal models are needed. Due to their availability and low costs, numerous rodent models have been developed which have substantially contributed to our current understanding of the pathophysiology of TBI. The most common animal models used in laboratories worldwide are likely the controlled cortical impact (CCI) model, the central and lateral fluid percussion injury (FPI) models, and weight drop/impact acceleration (I/A) models. Each of these models has inherent advantages and disadvantages; these need to be thoroughly considered when selecting the rodent TBI model according to the hypothesis and design of the study. Since TBI is not *one* disease, refined animal models must take into account the clinical features and complexity of human TBI. To enhance the possibility of establishing preclinical efficacy of a novel treatment, the preclinical use of several different experimental models is encouraged as well as varying the species, gender, and age of the animal. In this chapter, the methods, limitations, and challenges of the CCI and FPI models of TBI used in rodents are described.

Key words Traumatic brain injury, Controlled cortical impact (CCI), Fluid percussion injury (FPI), Rats, Mice, Neurodegeneration, Outcome

1 Introduction

Traumatic brain injury (TBI) has been recognized as a silent epidemic and is considered by the World Health Organization to be a health threat at the order of HIV and malaria. Recent epidemiological data suggest that the number of TBI victims worldwide is increasing [1–3] and survivors of TBI commonly suffer from persistent functional deficits, including personality changes, motor impairment and cognitive problems, leading to an impaired capacity for work and a reduced

Firas Kobeissy et al. (eds.), *Injury Models of the Central Nervous System: Methods and Protocols*, Methods in Molecular Biology, vol. 1462, DOI 10.1007/978-1-4939-3816-2_3, © Springer Science+Business Media New York 2016

quality of life. Current treatment options include optimized prehospital stabilization, rapid decompressive surgery when needed, neurocritical care and frequently prolonged neurorehabilitation. Despite extensive preclinical and clinical studies for several decades, there is still a lack of pharmacological alternatives [4]. The enormous heterogeneity of TBI has been noted in several recent overviews [5] which is further exacerbated by individual injury factors, age and gender issues and perhaps also the genetics and/or epigenetics of the individual patient. The clinical reality of TBI makes the discovery of a single treatment or pharmacological tool effective in each of the TBI subtypes highly unlikely. Thus, the preclinical TBI research aiming to develop novel pharmacological compounds or rehabilitative strategies faces a challenge when aiming to mimic the clinical setting and it is crucial that clinically relevant TBI models are used. The question is then—how many animal models of TBI are needed to cover the clinical aspects of TBI?

To date, the search of "traumatic brain injury" AND "animal model" yields 159 hits on PubMed. Several models for repetitive, blast, penetrating, focal and diffuse TBI combined with various secondary insults such as hypotension and hypoxia have been developed [6–11]. Although TBI models developed for higher-order species such as the miniature swine and the primate models are available, rodent models will likely continue to be used in the vast majority of TBI research due to their accessibility and low costs. It may be argued that the task of making rodent models more clinically relevant is not to develop even more animal models—instead it is the refinement of the existing ones [12, 13]. Thus, the selected study treatment may be tested in a range of other relevant parameters such as strain, species, sex, and age of the animal. The use of a rather limited number of animal models also carries the advantage of enabling comparisons among different laboratories. As such, the controlled cortical impact (CCI) and the fluid percussion injury (FPI) models have been used in both rats and mice and all have the advantage of being widely used and mimicking certain aspects of human TBI.

The fluid percussion injury, initially established in the rabbit [14], was in 1987 adopted for use in the rat by Dixon et al., [15], the midline FPI model, and modified in 1989 by McIntosh et al., [16] who established the lateral FPI model. Despite presumed difficulties, both the lateral [17] and recently, the central (midline) fluid percussion model [18, 19], could successfully be adopted to mice. The basic principle behind this TBI model is that a brief pressure pulse is transmitted to the exposed dura, allowing for a degree of brain displacement resulting in cortical, hippocampal, and brain stem injury as well as widespread axonal injury in brain regions such as the corpus callosum, the fimbriae and the brain stem. As described in the following paragraphs below, the brain injury and behavioral deficit created will vary depending on the location of the craniectomy and the delivery and force of the pressure pulse. Two common types of FPI models are in common use—the lateral FPI model [16, 20, 21] where the pressure pulse is delivered over one hemisphere and the central/midline CPI model [15] where the pulse is delivered over the midline, resulting in different brain injuries and behavioral deficits.

The controlled cortical impact (CCI) model of TBI, first used in the ferret [22], was then adopted for use in the rat [23] and then to the mouse [24]. This model uses a pneumatically or electromagnetically [25] driven piston striking (most commonly) the exposed dura allowing for brain penetration of a predetermined depth at a predetermined velocity. This model predominately produces a focal TBI although diffuse neurodegeneration may be observed throughout the rodent brain extending also into the contralateral hemisphere [26]. Although not identical with CCI, the weight-drop model originally developed by Feeney et al., [27], used by many labs including ours [28], reproducibly resulted in a focal contusion extending into the ipsilateral hippocampus when a weight was dropped from a predetermined height onto the exposed dura with the head of the rodent fixed. Arguably, this model may be less precise and adjustable than the CCI, despite the commonly used term "controlled cortical contusion model", and it is less commonly used to date.

Due to the intrinsic differences between rodents and man, a TBI model in the rat or the mouse may never replicate all aspects of the immense complexity of human TBI, particularly with regards to the complex cognitive and behavioral disturbances observed in TBI patients. In order to mimic the clinical situation to the extent possible, a thorough understanding of the behavioral consequences of the TBI models is crucial as well [29]. One example of such a strategy is the evaluation of the long-term clinical problem of post-traumatic epilepsy and both the CCI and lateral FPI models were shown to cause delayed seizures post-injury [30]. The commonly used CCI and FPI TBI models, in the view of this author, may be the most clinically relevant ones available to date, and their histological and behavioral consequences have resemblance to certain aspects of human TBI and their techniques the as well as their advantages and disadvantages are outlined in the following paragraphs.

2 Materials

2.1 Animals

We use Sprague-Dawley male rats at a weight of 350–400 g (9–12 weeks of age, Taconic M&B laboratory, Ry, Denmark) and C57BL/6J mice at a weight of 25–28 g (age 9–12 weeks old; supplied by Taconic, Denmark). The animals should be housed in a colony maintained by trained staff including regular surveillance for common viruses known to potentially infect rodent facilities. All scientists involved in animal research must undergo authorized training in animal welfare, handling and ethics and each study must be approved by the animal ethics board and follow the local and national rules and regulations of animal experiments. Any measure to reduce stress and suffering to the animal is taken (*see* **Notes 1–5**). The animals are housed at 24 °C in cages with three to five mice or two rats per cage with free access to food and water in a 12-h light/dark cycle, and kept in the colony for a minimum of 1 week prior to any surgical procedures.

2.2 Surgical Setup

1. A surgical area, preferably sheltered from other lab spaces, designed and devoted to animal surgery is used. The experimenter involved in the surgery should use a clean lab coat, a surgical face mask, foot covering and gloves. A set of clean, sharpened surgical tools should be available for smooth and controlled surgery.
2. The setup for volatile anesthesia requires specialized equipment including vaporizers. Leakage of volatile gases readily occurs and poses an occupational hazard to the investigator and scavenging systems must be in place and be regularly tested by the responsible authorities and all staff must be trained accordingly. Such scavenging systems for reducing exposure to the researcher is paramount and may include carbon filters, fume hoods or, preferably, a vacuum setup connected to an exhaust system to minimize the risk of inadvertent exposure of volatile anesthetics (*see* **Notes 6** and **7**).
3. A heating pad, coupled to a rectal probe (CMA150, CMA, Stockholm, Sweden), is needed to maintain the core temperature of the animal at 37 °C and is positioned under the animal during surgery. The body of the animal is covered by towels to reduce hypothermia.
4. A stereotaxic head holder (Kopf Instruments, Tujunga, California, USA) including blunt ear pins to avoid ear drum penetration (both rats and mice) and a snout clamp is used for head fixation.
5. A surgical microscope and a light source is needed.
6. An electric razor for minimal hair trimming is used. The area is sterilized by using sterile swabs with 70% alcohol and bupivacaine at 5 mg/mL is used for local anesthesia in the scalp (Marcain, AstraZeneca, Sweden).
7. Polyethylene tubing for arterial catheter and/or venous lines if needed (rat; PE 50, Becton Dickinson, Franklin Lakes, NJ, USA) and for intubation (rat) if needed (PE 205, Becton Dickinson).
8. Artificial tear lubricants (Viscotears, Novartis, Inc., Basel, Switzerland) is carefully placed on the cornea of the animal to keep them from drying during surgery.

9. A pencil is needed to carefully mark the exact location of the craniotomy.
10. Trephines for mice and rats, preferably attached to a dental drill, or a 1.0 mm diamond drill head attached to the dental drill is needed. Q-tips and bone wax may be used to control minor bleedings by gentle pressure.

2.3 Fluid Percussion Brain Injury (FPI)

The basic idea of FPI is that a short pressure pulse is delivered to the brain via the intact dura mater, resulting in a rapid rise in intracranial pressure [31] and a degree of brain displacement, which leads to brain injury by cell death, opening of the blood-brain barrier, axonal injury and a neuroinflammatory response [20]. The pressure pulse is created by a pendulum striking a piston at the proximal end of a fluid-filled cylinder and saline is being rapidly injected into the closed cranial cavity of the animal (Fig. 1). The severity of injury is regulated by varying the height from which the pendulum is released, which results in variations of the pressure pulse transmitted to the rodent.

1. The FPI device consists of a Plexiglas cylinder (Fig. 1) filled with physiologic saline and at its distal end, there is a transducer measuring the extracranial pressure wave and then a male Luer-Lok fitting enabling it to be locked on the trauma cap attached to the rodent's skull for delivering of the pressure pulse. The FPI device needs regular maintenance, including cleaning of the cylinder and replacing the saline as well as lubricating the plunger and its O-rings using for example petroleum jelly or

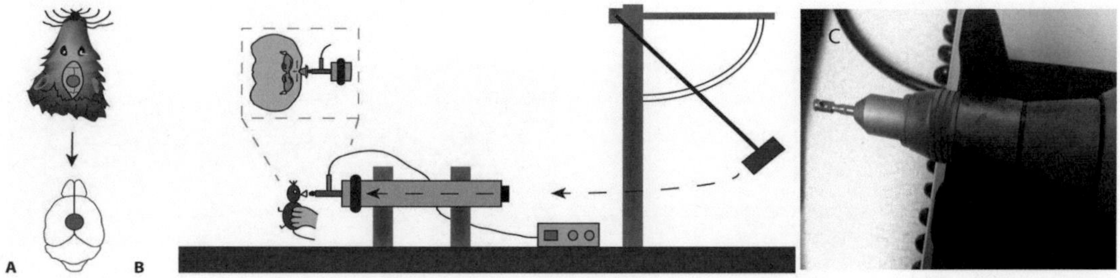

Fig. 1 Cartoon of the basic setup for fluid percussion brain injury. (**a**) The craniectomy for central (midline) FPI is centered over the sagittal suture midway between bregma and lambda. The *lower* image shows the corresponding impact area of the rodent brain. (**b**) When the pendulum (*right*) is released from a predetermined height, it will strike the end of a Plexiglas cylinder filled with isotonic saline. The impact creates a pressure wave of ca. 20 ms which is transmitted to the brain of the rodent using the techniques described in coming sections. The rodent is held by the researcher's hand and gently attached to the distal end of the cylinder via the Luer-Lok setup cemented onto the skull of the rodent, over the craniectomy. *Note* the transducer located at the distal end of the cylinder, reading the pressure changes which is displayed on a computer screen. (**c**) To improve the drilling for the craniectomy, we use a trephine attached to a dental drill at 5000 rpm. For best results of surgery, including control of small bleedings and securing hat the dura is intact, a surgical microscope is preferred

similar (Renolit, Mannheim, Germany). In modern FPI systems, there is less need for calibrating the pressure output from the transducer and amplifier. The pressure is typically measured in atmospheres of pressure per square inch (PSI) where one atmosphere = 14.7 PSI.

2. A hub of a needle needs to be mounted over the craniectomy (Fig. 2) and this injury hub is created from a 23G needle (Becton Dickinson, USA) with a razor blade or a pencil sharpener. Its lower end needs to be angled to be closely fitted into the craniectomy.

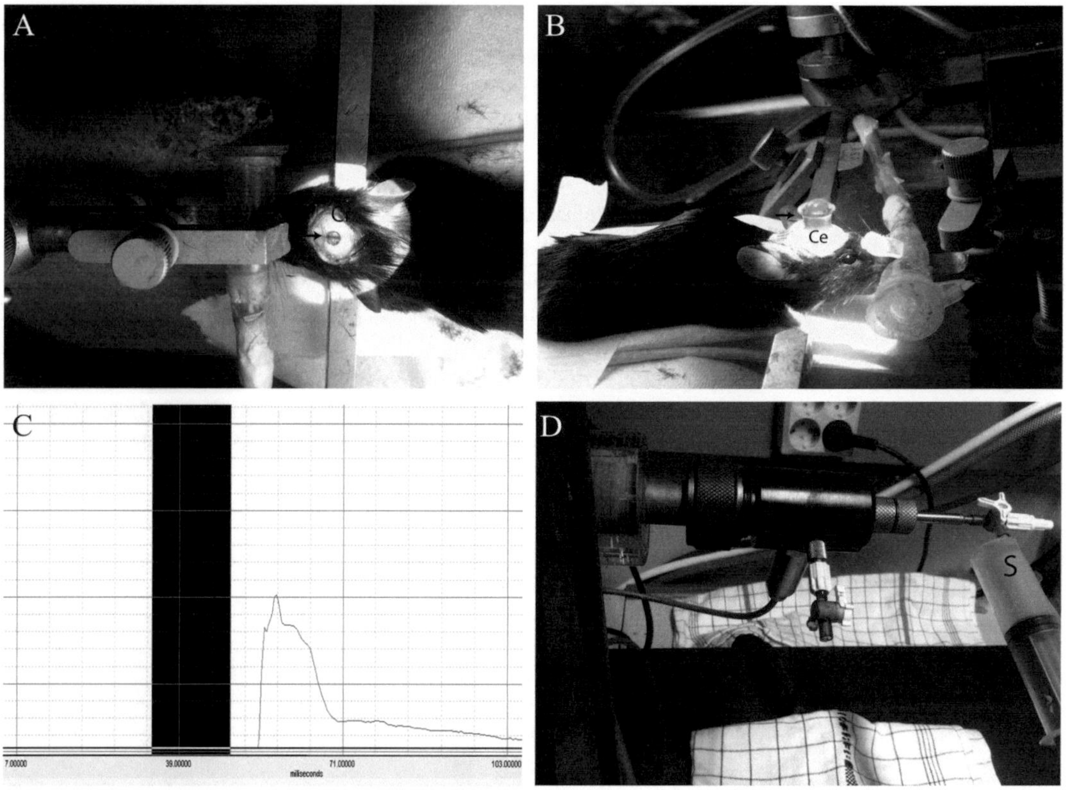

Fig. 2 In vivo, "real-life" central fluid percussion injury in the mouse. (**a**) The surgical draping has been removed to better display the setup and the surgical area. *Note* the craniectomy (Cr) and the *arrow* pointing to the sagittal sinus after removal of the bone. The head is firmly held in place by blunt ear pins and by a snout clamp. (**b**) Lateral view. The mouse is placed on a heating pad to maintain core temperature, which is monitored. *Note* the homemade nose cone delivering volatile anesthetics. The hub of a needle is placed over the craniectomy, held in place by cement (Ce). The hub is then filled with saline (*arrow*) and checked that it remains in place securing the tight closure of this now fluid-filled space. (**c**) Typical reading of the pressure wave from the transducer. In our experience, the peak pressure wave in mice may show some irregularities despite taking all the measures described in the text. In rats, this peak is typically smoother without the "edges" shown in the mouse. (**d**) Close-up of the transducer. It is crucial that this part is devoid of air and this area need to be flushed with saline (S) prior to each injury and the pressure pulse controlled

3. Dental cement (Dentalon Plus, Heraeus Kulzer GmbH, Hanau, Germany, Fig. 2) and two flat-ended anchoring screws (rat) is used to anchor the hub over the craniectomy as well as a strong adhesive (Loctite Precision Super-glue, Loctite, Westlake, Ohio, USA).

2.4 Controlled Cortical Impact

Following a craniotomy centered over one hemisphere, the idea of the CCI model is that a piston will deliver a force to the exposed and intact dura, allowing rapid deformation of the underlying brain and resulting in necrotic and apoptotic cell death. Although this model is clearly a focal TBI model, it also has diffuse injury components [26].

1. We use a pneumatically controlled CCI-device (VCU Biomedical Engineering Facility, Richmond, Virginia, USA) with a flat tip (Fig. 3).
2. To produce the craniotomy, a prefabricated plastic plate may be used to aid in drilling, although it is not used in our lab.

3 Methods

3.1 Animals and Surgery

1. Rodents can be of any age, weight and strain. To reduce the stress reactions associated with the experiments, we recommend pre-injury handling of the animals for five min twice daily for 2 days before taking part in any behavioral experiments (*see* **Note 4**).

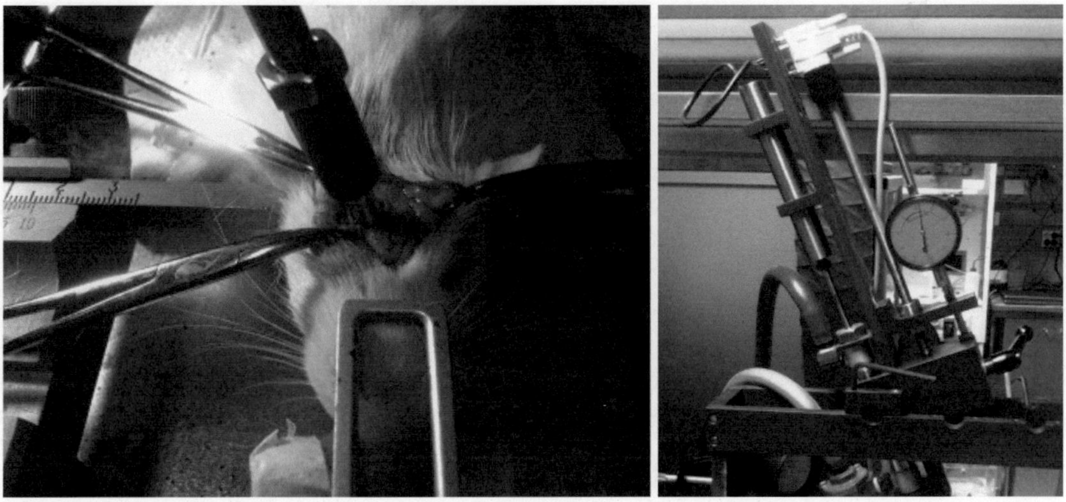

Fig. 3 CCI procedure. Following cleaning of the scalp, a midline incision is made. Following craniotomy made with a dental drill, a 5 × 6 mm (rats) or 4 × 4 mm (mice) craniotomy centered over the right parietal cortex is made. The piston (*left*) is gently attached to the exposed dura. To the *right*, the CCI apparatus including the micrometer gauge is shown

2. The weight and welfare of the animals must be monitored pre- and post-injury, at least daily for the first three post-injury days, then a minimum of twice per week. Animals losing more than ca 15 % body weight following the injury should be excluded from the study (*see* **Note 5**).
3. Carefully chose the anesthetics and remember that they all influence the results of the injury in one way or the other. Common choices are volatile anesthetics such as isoflurane or halothane with sodium pentobarbital or xylazine/ketamine used as alternatives (*see* **Note 6** and 7). A combination of these anesthetic agents may be used, for instance by using sodium pentobarbital to prepare for the injury. Then, after a 24 h recovery, injury may be induced using for example isoflurane or halothane anesthesia [19].
4. We prefer volatile anesthesia where the animal is induced in a chamber with 4 % isoflurane in air and then moved to the stereotaxic frame where a mixture of isoflurane (1.2–1.4 %) and N_2O/O_2 (70/30 %) is delivered through a nose cone or, in long experiments such as when using intracerebral microdialysis, following intubation using PE 205 tubing (Becton Dickinson) attached to a rodent ventilator (Ugo Basile, Comerio, Italy).
5. In rats, when for example electrophysiological experiments are conducted, xylazine in combination with ketamine is an alternative anesthetic option. We use a 1:6 mixture of xylazine (Rompun Vet, 20 mg/mL; Bayer Animal Health, Lyngby, Denmark) and ketamine (Ketaminol Vet, 50 mg/mL, Intervet Int., Boxmeer, the Netherlands), the mixture administered at a volume of 0.14 mL/100 g body weight intramuscularly into the right thigh. When needed, an i.m. supplemental dose of 0.2 mL of ketamine is used [32], *see* **Note 6**.
6. Intraperitoneal administration of 60 mg/kg sodium pentobarbital anesthesia is common particularly in rats where the duration from a single injection is approximately 60–90 min. The depth of anesthesia may be assessed by for example the toe-pinch reflexes prior to the incision and surgical procedures.
7. The animal is then positioned in a stereotaxic system with the head fixed with blunt ear pins and a bite plate/snout clamp firmly fixing the head. In accordance with the limited shaving used in clinical neurosurgery, we only trim the hair of the rodent immediately prior to incision. The scalp is anesthetized at the incision site using bupivacaine (Marcain, AstraZeneca, Sweden) applied subcutaneously.
8. Physiological parameters such as blood gases (pO_2, pCO_2), blood pH or arterial blood pressure can be measured invasively using for example tail or femoral artery cannulations and heparinized blood samples repeatedly analyzed using appropriate

equipment during the procedure. The evaluations of these physiological parameters should particularly be considered when prolonged anesthesia is used, e.g., for experiments using intracerebral microdialysis [28], and if a pharmacological compound is evaluated.

9. For all TBI models described in this chapter, a midline scalp incision is made from between the eyes to the neck using a scalpel. After the scalp is reflected, the fascia is scraped from the skull to the crista temporalis with a sharp forceps.
10. Uninjured, sham animals undergo all of the procedures with the exception of being subjected to the actual fluid pulse or piston impact. For most TBI research, is recommended that naïve animals are also included [33].
11. Although clinical experience suggests that pain from a scalp incision as used in these TBI models is relatively short-lasting, postoperative analgesia may be required and also frequently mandated by the animal ethics committee. Although many analgesic compounds may influence the pathophysiology of TBI, opioids such as buprenorphine are commonly used (*see* **Note 8**).

3.2 Lateral and Midline (Central) Fluid Percussion Injury

1. After the head is fixed in the stereotaxic frame and the scalp has been reflected, a craniotomy is made which is 4.8 mm (rats) or 3.0 mm (mice) for both FPI models. In central (cFPI) models, extreme caution must be taken not to injury the sagittal sinus vessel (Figs. 1 and 2). Carefully elevate the bone flap, which may be slightly adherent towards the midline and gently separate the dura from the bone using a microspatula. Inspect the dura, it must be intact and if it is not, the animal should be excluded from further analysis. Carefully stop any bleeding from the bone edges, typically sufficient with gentle pressure from a Q-tip although a small amount of bone wax may occasionally be needed (*see* **Note 9**).
2. The craniotomy is for lateral FPI placed over the left parietal cortex, the medial edge 1 mm (mice) to 2 mm (rats) from the midline. For central FPI, it is placed at the midline over the superior sagittal sinus, midway between bregma and lambda (Figs. 1 and 2).
3. Anterior to the bregma and immediately posterior to the lambda, two injury screws with flat distal ends are placed (in rats) lateral to the midline to anchor the cement (see below). The skull bone in mice is too thin to allow for the placement of injury screws.
4. The injury hub, "the trauma cap", is carefully placed over the craniectomy by holding it with small forceps and secured using Super-glue placed on the bony edges of the craniectomy. Particularly in mice, the process of avoiding glue on the dura is simplified by placing glue around the lower edges of the cap prior to placing it over the craniectomy. It is of paramount importance that glue is not placed on the dura and/or that is obstructs the opening of the trauma cap—this will markedly attenuate the pressure pulse delivered to the brain of the animal.

5. Using a small plastic cup, a semifluid dental acrylic (Dentalon Plus) solution is mixed. Use a 1 cc syringe to withdraw the solution and to place it over the craniectomy site covering and securing the injury screws and trauma cap. It is imperative that the cement gets solid prior to induction of the injury.
6. An important step is filling of the trauma cap with 37 °C saline and this saline needs to stay in the cap (Fig. 2). If the fluid levels sinks, this indicated incomplete closure between the cap and the bony edges.
7. Prime the FPI device by releasing the pendulum and verifying a smooth curve on the reading from the transducer prior to inducing the trauma (Fig. 2). If the curve is jagged, injection of saline back and forth with a syringe attached to the transducer (Fig. 2) may clear the air from the system and improve the pressure reading.
8. Reduce the isoflurane and when the animals starts breathing more shallowly, the animal is ready for trauma since injuring the animal at a surgical plane of anesthesia may cause excessive mortality. The animal is placed towards the male Luer-Lok end of the FPI device, ensuring that there is a continuous saline pillar by injecting saline over the injury cap when placing the animal towards the Luer-Lok. The animal is held firmly without compromising breathing, the head supported against the FPI device.
9. When the pendulum is released and the brain injury produced, there is commonly brief seizure-like motor activity/tonic posturing and an apnea. The pressure of the pulse must be recorded. We use a cutoff limit of 60s apnea for both mice and rats, if it is longer we exclude the animal from the study since we cannot rule out additional hypoxia which could be a confounding variable. Since the post-injury apnea is an important indicator of injury severity, absence of apnea may imply an insufficient injury. Immediately after the injury, the animal should be placed on its back and the duration of seizures and apnea monitored and recorded (*see* **Note 10**).
10. Only animals with the expected/planned range of apnea should be included and we consider this parameter to be an important assessment of the injury. There is no universally accepted method of determining injury severity when using the FPI method. Usually, the injury response including the apnea as well as mortality ranging from 10 to 20% ("moderate FPI") to ca. 40% ("severe FPI") is considered and there should be evidence of behavioral disturbance post-injury (*see* **Note 11**). In addition, depending on whether lateral or midline FPI is analyzed, there should be histological evidence of an injury—e.g., a cortical contusion, ipsilateral hippocampal injury and ipsilateral white matter injury in the lateral FPI model and widespread axonal injury in important white matter tracts including the brain stem in midline

FPI (*see* **Note 12**). After the injury and resumption of breathing, the animals are taken back to the heating pad and isoflurane anesthesia is reinstated. The trauma cap, injury screws and cement is removed by firmly holding down the head and nose of the animal while gently pulling the block of cement upwards. The dura must then be carefully checked for the presence of breach, and if found the animal should be excluded, and the same holds true if there is glue obstructing the craniectomy.

3.3 Controlled Cortical Impact

The CCI device (Fig. 3) uses a pneumatically [23] or, less commonly, an electromagnetically [25] controlled impact which may be with high precision directed to the exposed dura at almost any chosen angle. The material of the end of the rod may be either rigid or a soft (e.g., silicone) and its shape can be either flat or rounded. Depending on the diameter of the rod, the depth of the impact, the velocity with which the rod strikes and its dwell time, i.e., the time the rod remains at the target depth, the tissue injury will change. The basic idea with CCI is to produce a focal TBI with tissue destruction and cell death ipsilateral to the injury, particularly in the injured cortex but also in the hippocampus and the underlying white matter tracts. A motor impairment and learning and memory deficits are also produced in this model in addition to a large tissue loss, depending on the various parameters described above.

1. The anesthesia, placement of the animal and fixation of the head is similar to what is described in the FPI section. Using the diamond drill, a 4.0 mm diameter (mice) and a 5 × 6 mm craniotomy (rat) is created. It is crucial that the dura is kept intact and should be inspected for tears prior to induction of the injury. For mice, the craniotomy is performed over the left parietal cortex centered at 2.5 mm posterior to bregma and 2.5 mm lateral to the midline, anterior margin 1 mm posterior to bregma with the lateral margin at the crista temporalis. We use a flat 2.5 mm tip impounder striking the dura at 2.8–5.0 m/s allowing a 0.5–1 mm compression of the brain in mice [34]. In rats, we use a flat 4 mm diameter piston delivering the impact at 3 m/s allowing a brain deformation of 2.0 mm [35], *see* **Notes 13–15**.
2. After the injury, a dural breach is commonly seen at least with the more marked compression depths. After the injury, the rather large bone flap should be reattached using tissue adhesive (e.g., Histoacryl®, B.Braun Surgical, S.A., Rubi, Spain).
3. Similarly for all the models mentioned above after TBI or sham injury, the animal is, following closure of the wound by interrupted and resorbable sutures, removed from the stereotaxic frame and placed in a wake-up cage with an overhead heating lamp.
4. The animal remains in this cage until ambulatory after which it is allowed to return to its home cage.

3.4 Concluding Remarks

The animal models discussed in the present chapter are common worldwide and when used correctly, will continue to increase our knowledge of clinical TBI. A vast majority of experimental TBI research is performed on adolescent/young male animals and TBI scientist should more actively expand their research into other age groups, both genders and when pharmacological treatments are evaluated, also attempt the use of prolonged and more clinically relevant therapeutic windows. This author is not a strong believer in attempting the development of yet additional TBI models unless a specific TBI mechanisms not covered by the models described here (e.g., penetrating TBI, blast) is targeted. The CCI model as well as the lateral and central/midline FPI models can all be used in both rats and mice and they likely cover a large proportion of the heterogeneous TBI population (diffuse axonal injury, focal injury, "mixed" TBI) observed clinically. Thus, continuous refinement of the techniques and setup described here will likely continue to be important aspects of future preclinical TBI research.

4 Notes

Despite all efforts put into these TBI models, they do not produce human injuries due to many factors including different gray-white matter ratios in rodents and humans, the lissencephalic design of the rodent brain and the size differences resulting in considerably different biomechanical properties. Importantly, no rodent model adequately produce the long-lasting coma observed in severe human TBI. Regardless of these difficulties, preclinical TBI research is crucial in the development of novel treatment options. For that reason, there are numerous factors which should be considered when designing rodent TBI research to enhance the possibility of successful clinical translation.

(A) *Animals and preoperative considerations*

Any effort should be made to reduce variability in research and avoiding bias. All experiments should be planned to minimize pain and suffering of the animals and if possible, reduce the number of animals without compromising with scientific and statistical quality.

1. Prepare thoroughly prior to the start of the investigations including a careful review of existing literature. Are all ethical permissions granted and is the experiment described therein? What injury model best reflects the hypothesis of the study and at what severity? What age and gender should be used? What outcome measure should be used?
2. The investigator performing the surgeries and/or the outcome evaluation must be blinded to the genotype of

each animal, and describe it in the scientific publications. There are several guidelines providing tools to avoid bias in preclinical research, including the Collaborative Approach to Meta-Analysis and Review of Animal Data from Experimental Studies (CAMARADES) and ARRIVE guidelines [36].

3. Since rodents are social animals, single housing should not be used and each cage should be equipped with sufficient stimuli supporting the normal behavior of a rodent.
4. Handling of the rodents prior to any experiment is recommended to reduce stress and anxiety, particularly when behavioral outcome analysis is used [37].
5. The included animals should preferably have a narrow weight and age range. Any TBI-induced weight loss should be monitored and documented and if it exceeds more than ~15 % of its original weight, or what is mandated by the animal ethics committee, the animal should be excluded from the study and humanely euthanized. Typically, FPI results in a higher degree of weight loss than the CCI models and the more severe TBI, the more profound the weight loss.
6. All anesthetic agents have profound influence on cerebral and systemic physiology and will influence the outcome aimed to study. In general, sodium pentobarbital anesthesia is not ideal due to its influence on brain energy metabolism, its variable duration—particularly when supplemental doses are used—and its poor analgesic effects. In addition, female animals require less pentobarbital doses to achieve the same plane of anesthesia as male animals. Since it is used clinically for refractory intracranial hypertension in humans, it has obvious effects on cerebral physiology post-injury. It is the view of this author that volatile anesthetic agents are to be recommended, not because they are without influence on brain physiology, although since they can be easily titrated and cause rapid induction and resolution of the anesthesia [38, 39].
7. Anesthetics and the surgery/injury itself will alter systemic parameters and influence breathing and circulation. Physiological monitoring is strongly recommended, either noninvasively [40] using pulse oximetry and/or a blood pressure cuff monitor, or invasively for arterial blood pressure and arterial blood gases using a catheter in either the tail or femoral artery [28].
8. Post-operative pain analgesia is frequently mandated where opioids such as buprenorphine are common choices. If used at all, their potential influence on

outcome following experimental TBI and the pathobiology of TBI [41, 42] should be considered. In general, nonsteroidal anti-inflammatory drugs (NSAID) are to be avoided, due to their potential for vasoconstriction, their possibility of increased hemorrhage and since they are not indicated for use in acute clinical TBI.

(B) *Fluid percussion injury*

The possibility of FPI to create widespread axonal injury, a clinically relevant brain displacement, its resulting cognitive deficits and the histological similarities to the human TBI setting makes both the lateral and central/midline FPI models useful and clinically relevant.

9. Many of the technical aspects of creating FPI are mentioned in the Subheading 3. One crucial point is creating the craniotomy and its margin must be very consistent between animals since even small changes in craniotomy position will markedly alter outcome of lateral FPI [43, 44]. The size of the craniectomy must also be consistent since any change alters outcome [45, 46]. When the craniotomy is made using a dental drill, there will be heat generated by the drilling and the site should be regularly flushed with sterile saline at room temperature. During the drilling, the bone plate should regularly be checked and when it feels loose, it may be gently lifted by a pair of micro tweezers. Again, the dura cannot show evidence of a dural tear—if it does, the animal must be excluded from further study.
10. At the moment of impact, there is, at least for an injury at a moderate–severe level, a consistent apnea which we use as a physiological measure of the impact and a "receipt" that the injury was correctly delivered. Thus, we recommend that it be monitored and measured. Unless systemic monitoring ensures that hypoxia does not occur, we suggest a cutoff level of 60s. If the apnea is of longer duration, it may introduce an additional insult to the injured brain and increase the variability of the injury as well as increasing mortality. At impact of experimental TBI, there may be acute electroencephalographic evidence of seizures [47]. Whether the acute motor twitching, which may be clonic in nature, observed at impact are in reality seizures remains to be established although we argue that their duration should be monitored.
11. One important issue of the FPI technique is how to determine injury severity. Righting reflex time, level of atmospheres and—perhaps most commonly—the mortality have all been commonly used [21, 48]. At present there is no consensus on how to assess injury severity. For instance, a 10–20% mortality is commonly used as a "moderate" injury and a pressure in atmospheres of 0.9–2.1 a mild–moderate injury [48]. We aim

for 3.0 ± 0.1 atm for the lateral FPI model and 2.4 ± 0.2 for the central/midline FPI in rats and 1.4 ± 0.1 atm for midline FPI in mice [18] where the aim is a 30–45 s apnea, ca 10 % mortality, a clear cognitive deficit in for example the Morris Water Maze and a consistent and evident histological injury. Thus, it may be argued that one sole parameter may not fully indicate injury severity and comparisons among laboratories using the atm reading also appears difficult. Thus, the FPI should be titrated to the suggested outcome measure of the particular study instead of using one single parameter. Also remember that a "moderate" FPI is not equal to a "moderate" clinical TBI using the Glasgow Coma Scale Score.

12. Histological parameters are important when assessing the quality of the FPI impact and there should be evidence of brain stem and white matter injury in the midline FPI and a cortical contusion, hippocampal injury, and white matter injury evident in for example the corpus callosum and the ipsilateral external capsule following lateral FPI [19, 20, 49].

(C) *Controlled cortical impact*

The CCI is easy to learn and master, it is fast and reproducible, is suitable for both rats and mice and is useful for the study of for example the progression of cortical contusion although it also produces white matter injury and neurodegeneration distant to the impact. Still, its role is mainly to mimic the "focal" TBI observed in the clinical setting.

13. The crucial part of creating the CCI is the craniotomy which can be performed similar to what is described for the FPI. Using a dental drill at the carefully outlined area (some authors use a prefabricated disc glued onto the exposed skull to aid in the drilling), a standardized craniotomy can be produced. Importantly, the commonly used size for the bone flap is large, clinically approaching the size used for a decompressive craniectomy allowing for brain swelling and bulging through the craniotomy area. For that reason and to enhance clinical translation, the bone flap should as soon as possible following the impact be glued back using tissue adhesive (e.g., Histoacryl (see above), or Dermabond, Ethicon, Somerville, NJ; [50, 51]).

14. Similar to the FPI, attempts have been made to characterize the CCI into different severity grades. Similar to our use of the Feeney weight-drop model, a compression depth of 1.5 mm was suggested to be a mild injury, 2.0 mm a moderate, and 2.5 mm a severe injury keeping all other injury parameters (speed, piston diameter, dwell time) the same [52]. However, due to the highly variable use of different piston diameters and materials, piston speed, dwell time and brain compression

depths when the CCI is used, comparisons among laboratories are difficult and a standardized grade of "mild", "moderate," and "severe" has not been established.

15. CCI should not be used at a very high severity level where a large part of the hemisphere is injured, which in the clinical setting is not an injury compatible with survival. Thus, if your CCI produces a massive destruction of one hemisphere, consider lowering not only the depth of brain compression but also the speed of impact.

Acknowledgments

The author would like to thank Dr. Fredrik Clausen, Malin Olsen (for designing the FPI cartoon), and Dr. Johanna Flygt for their invaluable help with the animal models and for comment on the manuscript; Dr. Tracy K. McIntosh and Dr. John Povlishock for teaching the author the animal models described herein. This work was supported by funds from the Swedish Research council, European Union and the Uppsala University Hospital.

References

1. Tagliaferri F, Compagnone C, Korsic M, Servadei F, Kraus J (2006) A systematic review of brain injury epidemiology in Europe. Acta Neurochir (Wien) 148:255–268
2. Corrigan JD, Selassie AW, Orman JA (2010) The epidemiology of traumatic brain injury. J Head Trauma Rehabil 25:72–80
3. Maas AI, Stocchetti N, Bullock R (2008) Moderate and severe traumatic brain injury in adults. Lancet Neurol 7:728–741
4. Diaz-Arrastia R, Kochanek PM, Bergold P, Kenney K, Marx CE, Grimes CJ, Loh LT, Adam LT, Oskvig D, Curley KC, Salzer W (2014) Pharmacotherapy of traumatic brain injury: state of the science and the road forward: report of the Department of Defense Neurotrauma Pharmacology Workgroup. J Neurotrauma 31:135–158
5. Saatman KE, Duhaime AC, Bullock R, Maas AI, Valadka A, Manley GT, Workshop Scientific T, Advisory Panel M (2008) Classification of traumatic brain injury for targeted therapies. J Neurotrauma 25:719–738
6. Brody DL, Benetatos J, Bennett RE, Klemenhagen KC, Mac Donald CL (2015) The pathophysiology of repetitive concussive traumatic brain injury in experimental models; new developments and open questions. Mol Cell Neurosci 66:91–98
7. Xu L, Nguyen JV, Lehar M, Menon A, Rha E, Arena J, Ryu J, Marsh-Armstrong N, Marmarou CR, Koliatsos VE (2016) Repetitive mild traumatic brain injury with impact acceleration in the mouse: multifocal axonopathy, neuroinflammation, and neurodegeneration in the visual system. Exp Neurol 275(Pt 3):436–449
8. Schutz C, Stover JF, Thompson HJ, Hoover RC, Morales DM, Schouten JW, McMillan A, Soltesz K, Motta M, Spangler Z, Neugebauer E, McIntosh TK (2006) Acute, transient hemorrhagic hypotension does not aggravate structural damage or neurologic motor deficits but delays the long-term cognitive recovery following mild to moderate traumatic brain injury. Crit Care Med 34:492–501
9. Feng JF, Zhao X, Gurkoff GG, Van KC, Shahlaie K, Lyeth BG (2012) Post-traumatic hypoxia exacerbates neuronal cell death in the hippocampus. J Neurotrauma 29:1167–1179
10. Risling M, Davidsson J (2012) Experimental animal models for studies on the mechanisms of blast-induced neurotrauma. Front Neurol 3:30
11. Plantman S, Ng KC, Lu J, Davidsson J, Risling M (2012) Characterization of a novel rat model of penetrating traumatic brain injury. J Neurotrauma 29:1219–1232
12. Marklund N, Hillered L (2011) Animal modelling of traumatic brain injury in preclinical drug development: where do we go from here? Br J Pharmacol 164:1207–1229

13. Morales DM, Marklund N, Lebold D, Thompson HJ, Pitkanen A, Maxwell WL, Longhi L, Laurer H, Maegele M, Neugebauer E, Graham DI, Stocchetti N, McIntosh TK (2005) Experimental models of traumatic brain injury: do we really need to build a better mousetrap? Neuroscience 136:971–989
14. Lindgren S, Rinder L (1966) Experimental studies in head injury. II. Pressure propagation in "percussion concussion". Biophysik 3:174–180
15. Dixon CE, Lyeth BG, Povlishock JT, Findling RL, Hamm RJ, Marmarou A, Young HF, Hayes RL (1987) A fluid percussion model of experimental brain injury in the rat. J Neurosurg 67:110–119
16. McIntosh TK, Vink R, Noble L, Yamakami I, Fernyak S, Soares H, Faden AL (1989) Traumatic brain injury in the rat: characterization of a lateral fluid-percussion model. Neuroscience 28:233–244
17. Carbonell WS, Maris DO, McCall T, Grady MS (1998) Adaptation of the fluid percussion injury model to the mouse. J Neurotrauma 15:217–229
18. Ekmark-Lewen S, Flygt J, Kiwanuka O, Meyerson BJ, Lewen A, Hillered L, Marklund N (2013) Traumatic axonal injury in the mouse is accompanied by a dynamic inflammatory response, astroglial reactivity and complex behavioral changes. J Neuroinflammation 10:44
19. Greer JE, McGinn MJ, Povlishock JT (2011) Diffuse traumatic axonal injury in the mouse induces atrophy, c-Jun activation, and axonal outgrowth in the axotomized neuronal population. J Neurosci 31:5089–5105
20. Thompson HJ, Lifshitz J, Marklund N, Grady MS, Graham DI, Hovda DA, McIntosh TK (2005) Lateral fluid percussion brain injury: a 15-year review and evaluation. J Neurotrauma 22:42–75
21. Kabadi SV, Hilton GD, Stoica BA, Zapple DN, Faden AI (2010) Fluid-percussion-induced traumatic brain injury model in rats. Nat Protoc 5:1552–1563
22. Lighthall JW (1988) Controlled cortical impact: a new experimental brain injury model. J Neurotrauma 5:1–15
23. Dixon CE, Clifton GL, Lighthall JW, Yaghmai AA, Hayes RL (1991) A controlled cortical impact model of traumatic brain injury in the rat. J Neurosci Methods 39:253–262
24. Smith DH, Soares HD, Pierce JS, Perlman KG, Saatman KE, Meaney DF, Dixon CE, McIntosh TK (1995) A model of parasagittal controlled cortical impact in the mouse: cognitive and histopathologic effects. J Neurotrauma 12:169–178
25. Brody DL, Mac Donald C, Kessens CC, Yuede C, Parsadanian M, Spinner M, Kim E, Schwetye KE, Holtzman DM, Bayly PV (2007) Electromagnetic controlled cortical impact device for precise, graded experimental traumatic brain injury. J Neurotrauma 24:657–673
26. Hall ED, Bryant YD, Cho W, Sullivan PG (2008) Evolution of post-traumatic neurodegeneration after controlled cortical impact traumatic brain injury in mice and rats as assessed by the de Olmos silver and fluorojade staining methods. J Neurotrauma 25:235–247
27. Feeney DM, Boyeson MG, Linn RT, Murray HM, Dail WG (1981) Responses to cortical injury: I. Methodology and local effects of contusions in the rat. Brain Res 211:67–77
28. Marklund N, Lewander T, Clausen F, Hillered L (2001) Effects of the nitrone radical scavengers PBN and S-PBN on in vivo trapping of reactive oxygen species after traumatic brain injury in rats. J Cereb Blood Flow Metab 21:1259–1267
29. Bondi CO, Semple BD, Noble-Haeusslein LJ, Osier ND, Carlson SW, Dixon CE, Giza CC, Kline AE (2014) Found in translation: Understanding the biology and behavior of experimental traumatic brain injury. Neurosci Biobehav Rev 58:123–146
30. Bolkvadze T, Pitkanen A (2012) Development of post-traumatic epilepsy after controlled cortical impact and lateral fluid-percussion-induced brain injury in the mouse. J Neurotrauma 29:789–812
31. Clausen F, Hillered L (2005) Intracranial pressure changes during fluid percussion, controlled cortical impact and weight drop injury in rats. Acta Neurochir 147:775–780
32. Axelson HW, Winkler T, Flygt J, Djupsjo A, Hanell A, Marklund N (2013) Plasticity of the contralateral motor cortex following focal traumatic brain injury in the rat. Restor Neurol Neurosci 31:73–85
33. Cole JT, Yarnell A, Kean WS, Gold E, Lewis B, Ren M, McMullen DC, Jacobowitz DM, Pollard HB, O'Neill JT, Grunberg NE, Dalgard CL, Frank JA, Watson WD (2011) Craniotomy: true sham for traumatic brain injury, or a sham of a sham? J Neurotrauma 28:359–369
34. Clausen F, Hanell A, Israelsson C, Hedin J, Ebendal T, Mir AK, Gram H, Marklund N (2011) Neutralization of interleukin-1beta reduces cerebral edema and tissue loss and improves late cognitive outcome following traumatic brain injury in mice. Eur J Neurosci 34:110–123
35. Clausen F, Lindh T, Salimi S, Erlandsson A (2014) Combination of growth factor treatment and scaffold deposition following traumatic brain injury has only a temporary effect on regeneration. Brain Res 1588:37–46
36. Kilkenny C, Browne WJ, Cuthill IC, Emerson M, Altman DG (2010) Improving bioscience

research reporting: the ARRIVE guidelines for reporting animal research. J Pharmacol Pharmacother 1:94–99
37. Hanell A, Marklund N (2014) Structured evaluation of rodent behavioral tests used in drug discovery research. Front Behav Neurosci 8:252
38. Todd MM, Weeks J (1996) Comparative effects of propofol, pentobarbital, and isoflurane on cerebral blood flow and blood volume. J Neurosurg Anesthesiol 8:296–303
39. Statler KD, Alexander H, Vagni V, Dixon CE, Clark RS, Jenkins L, Kochanek PM (2006) Comparison of seven anesthetic agents on outcome after experimental traumatic brain injury in adult, male rats. J Neurotrauma 23:97–108
40. Thal SC, Plesnila N (2007) Non-invasive intraoperative monitoring of blood pressure and arterial pCO2 during surgical anesthesia in mice. J Neurosci Methods 159:261–267
41. Gentile NT, McIntosh TK (1993) Antagonists of excitatory amino acids and endogenous opioid peptides in the treatment of experimental central nervous system injury. Ann Emerg Med 22:1028–1034
42. Armstead WM (1997) Role of opioids in the physiologic and pathophysiologic control of the cerebral circulation. Proc Soc Exp Biol Med 214:210–221
43. Vink R, Mullins PG, Temple MD, Bao W, Faden AI (2001) Small shifts in craniotomy position in the lateral fluid percussion injury model are associated with differential lesion development. J Neurotrauma 18:839–847
44. Floyd CL, Golden KM, Black RT, Hamm RJ, Lyeth BG (2002) Craniectomy position affects morris water maze performance and hippocampal cell loss after parasagittal fluid percussion. J Neurotrauma 19:303–316
45. Sato M, Chang E, Igarashi T, Noble LJ (2001) Neuronal injury and loss after traumatic brain injury: time course and regional variability. Brain Res 917:45–54
46. D'Ambrosio R, Fairbanks JP, Fender JS, Born DE, Doyle DL, Miller JW (2004) Post-traumatic epilepsy following fluid percussion injury in the rat. Brain 127:304–314
47. Nilsson P, Ronne-Engstrom E, Flink R, Ungerstedt U, Carlson H, Hillered L (1994) Epileptic seizure activity in the acute phase following cortical impact trauma in rat. Brain Res 637:227–232
48. Alder J, Fujioka W, Lifshitz J, Crockett DP, Thakker-Varia S (2011) Lateral fluid percussion: model of traumatic brain injury in mice. J Vis Exp (54):pii: 3063
49. Flygt J, Djupsjo A, Lenne F, Marklund N (2013) Myelin loss and oligodendrocyte pathology in white matter tracts following traumatic brain injury in the rat. Eur J Neurosci 38:2153–2165
50. Hanell A, Clausen F, Djupsjo A, Vallstedt A, Patra K, Israelsson C, Larhammar M, Bjork M, Paixao S, Kullander K, Marklund N (2012) Functional and histological outcome after focal traumatic brain injury is not improved in conditional EphA4 knockout mice. J Neurotrauma 29:2660–2671
51. Zweckberger K, Stoffel M, Baethmann A, Plesnila N (2003) Effect of decompression craniotomy on increase of contusion volume and functional outcome after controlled cortical impact in mice. J Neurotrauma 20:1307–1314
52. Washington PM, Forcelli PA, Wilkins T, Zapple DN, Parsadanian M, Burns MP (2012) The effect of injury severity on behavior: a phenotypic study of cognitive and emotional deficits after mild, moderate, and severe controlled cortical impact injury in mice. J Neurotrauma 29:2283–2296

Chapter 4

Traumatic Brain Injury Models in Animals

Elham Rostami

Abstract

Traumatic brain injury (TBI) is the leading cause of death in young adults in industrialized nations and in the developing world the WHO considers TBI a silent epidemic caused by an increasing number of traffic accidents. Despite the major improvement of TBI outcome in the acute setting in the past 20 years, the assessment, therapeutic interventions, and prevention of long-term complications remain a challenge. In order to get a deeper insight into the pathology of TBI and advancement of medical understanding and clinical progress experimental animal models are an essential requirement. This chapter provides an overview of most commonly used experimental animal TBI models and the pathobiological findings based on current data. In addition, limitations and advantages of each TBI model are mentioned. This will hopefully give an insight into the possibilities of each model and be of value in choosing one when designing a study.

Key words Traumatic brain injury models, Experimental animal models, Fluid percussion, Controlled cortical impact, Blast brain injury

1 Introduction

Acute traumatic brain injury is characterized by a *primary* and a *secondary injury*. The primary traumatic brain injury is the result of mechanical forces that put a strain on the brain parenchyma at the time of impact. These forces may be of various forms, e.g., penetrating injuries, rotational acceleration, compression, and distension from acceleration or deceleration. This can lead to injuries to vessels, axons, neurons, and glia in a focal or diffuse pattern. The vascular injuries may result in intracerebral, subdural, extradural, or subarachnoid hemorrhage. The damage to the parenchyma may lead to contusions or lacerations. Diffuse injuries may strike the vessels leading to multiple small hemorrhages throughout the brain or affect the white matter with no apparent focal injuries. The primary cause of diffuse injuries is head rotation and the most common microscopic pathology found is axonal injuries [1].

The primary injury inducing TBI is also of different magnitude, usually characterized as mild, moderate, and severe. Currently, in

Firas Kobeissy et al. (eds.), *Injury Models of the Central Nervous System: Methods and Protocols*, Methods in Molecular Biology, vol. 1462, DOI 10.1007/978-1-4939-3816-2_4, © Springer Science+Business Media New York 2016

the clinical management of TBI the Glasgow Coma Scale (GCS), a clinical scale that assesses the level of consciousness after TBI is used to divide the patients into the broad categories of mild, moderate and severe injury [2]. The 15-point GCS is classified as mild GCS 13–15, moderate GCS 9–13, and severe GCS 3–8. The GCS has proved to be extremely useful as a tool that assists neurosurgeons or trauma physicians in the early triaging of TBI patients and it also has a high inter-observer reliability [3]. It does however not provide any specific information about the pathophysiological mechanisms responsible for the neurological deficits [4]. A GCS ≤8 can cover the significant heterogeneity of pathological findings such as epidural hematoma, contusion, diffuse axonal injury, or subarachnoid hemorrhage. It has been shown that outcomes among patients with the same admission GCS were significantly different and were influenced by the mechanism of injury [5].

The classification of TBI based on GCS for trial inclusion and targeted therapies is important but mechanistic classification has great utility in modeling injuries and developing preventive measures. TBI models discussed in this chapter are classified based on the physical mechanisms that produce TBI. Experimental TBI models in rodents that are currently in use are summarized in Table 1.

2 TBI Models

The seminal work of Holbourn made a first classification of a "localized injury due to scull distortion" and "injury due to rotation" [6]. The physical mechanisms have been further developed and can be classified according to "impact loading" which usually results in focal injuries while "inertial loading" generally causes diffuse injuries. The most commonly used TBI models can be classified as described below.

2.1 Focal "Impact Loading"

1. Weight drop model (*Fenney* [7], *Shohami* [8]).
2. Fluid percussion injury model (*Hayes* [9, 10], *McIntosh* [11]).
3. Controlled cortical impact model (*Lighthall* [12], *Dixon* [13], *Hayes*).
4. Missile and ballistic injury models (*Carey* [14], *Williams* [15]).
5. Penetrating TBI model (*Davidsson* and *Risling* [16]).

2.1.1 Diffuse "Inertial Loading"

Impact

1. Inertial acceleration model (*Ono* [17]).
2. Diffuse injury model (*Cernak* [18, 19]).
3. Impact acceleration model (*Marmarou* [20]).
4. Central fluid percussion injury model (*Dixon* [10]).

Table 1
This table represents currently used TBI models in rodents and different characteristics based on reported results

	Focal	Diffuse	Axonal injury[a]	Contusion/hemorrhages[b]	BBB disruption	Neurobehavioral	Gradable	Impact	Non-impact	Reproducibility[c]
Weight drop model [7]	x			x	x	x	x	x		+
Weight drop model [8]	x			x	x	x	x	x		+
Fluid percussion injury model [9–11]	x	x	x	x	x	x	x	x		++[d]
Controlled cortical impact model [12, 13]	x			x	x	x	x	x		+++
Penetrating ballistic injury models [14, 15]	x			x	x	x	x		x	+++[e]
Penetrating TBI model [16]	x			x	x	x	x	x		+++[e]

(continued)

Table 1
(continued)

	Focal	Diffuse	Axonal injury[a]	Contusion/hemorrhages[b]	BBB disruption	Neurobehavioral	Gradable	Impact	Non-impact	Reproducibility[c]
Diffuse injury model [18]		x	x	x	x	x		x		+++[e]
Impact acceleration model [20]		x	x	x	x	x	x	x		+
Central fluid percussion injury model [10]		x	x	x	x	x	x	x		+++
Rotational TBI model [23]		x	x			x	x		x	+++[e]

[a]Most TBI injuries induce some degree of axonal injury; however this column indicates the predominant injury

[b]Most of the TBI models generate contusions/hemorrhages such as subarachnoid hemorrhage. This column provides information regarding the lowest level of injury possible and rotational TBI model is the only one reported no contusions/hemorrhages

[c]Reproducibility accordingly to available literature, (+) low, (++) moderate, (+++) high

[d]There is high variability in injury parameters between different laboratories

[e]Used in fewer than two laboratories

2.1.2 Non-impact

1. Inertial acceleration models (*Genneralli* [21], *Gutierrez* [22]).
2. Rotational TBI model (*Davidsson* and *Risling* [23, 24]).
3. Blast TBI models.

2.2 Weight Drop Model

The weight drop model is considered the original TBI model [7, 25]. The focal impact is produced by a free falling weight guided in a tube that is made to hit the exposed skull. The impact of the weight on the skull produces a contusion type injury. The severity of the injury can be adjusted by the height and the mass of the weight dropped and can be combined with or without craniotomy [8]. The weight drop model is a fast and easy model, hence its popularity. However, there are limitations such as unintentional skull fractures, risk of a second rebound injury, and inaccuracy with regard to the impact site.

2.3 Fluid Percussion Injury Model

A model of closed head injury with fluid pressure was developed already in the 1960s by Lindgren et al. in order to produce an "experimental brain concussion" [26]. A further development of the model, the fluid percussion injury (FPI) is one of the most frequently used focal injury models [9–11, 27]. In this model a craniotomy is made either *centrally* in the midline on the sagittal suture between the bregma and lambda or *laterally* on the parietal cortex midway between bregma and lambda. A cylindrical reservoir filled with saline is attached to a cap cemented on the place of craniotomy on the animal's skull. A strike of a pendulum at the other end of the cylindrical reservoir generates a pressure pulse that is delivered to the intact dura and causes deformation of the underlying brain. Different levels of injury severity can be produced by adjusting the height of the pendulum, which defines the force of the fluid pressure pulse transmitted through the saline reservoir. The injury caused by this model replicates clinical contusion without skull fracture mixed with diffuse injury characteristics [28]. The central FPI is used to induce concussive and diffuse injuries [10]. It generates diffuse axonal injuries, increased blood-brain barrier permeability, inflammatory response, and neurobehavioral deficits that are distinct from the lateral FPI [29]. The placement of the craniotomy has shown to be important in producing a localized ipsilateral injury or an additional contralateral injury and also affects the reproducibility and reliability of this model. A limitation of the fluid percussion model is to generate a reliable continuum of injury severity since it cannot reproduce prolonged unconsciousness. Furthermore there is a disproportional brainstem involvement and injury severity and neurogenic pulmonary edema, adding to increased morbidity.

2.4 Controlled Cortical Impact Model

The controlled cortical impact model (CCI) was developed in order to generate a controlled variable impact causing equivalent deformation in brain parenchyma [12]. The brain injury could be quantified and the biomechanical forces could be analyzed. The model can be used both in smaller animals such as mouse and rats [10] as well as larger ones such as pigs [30]. A compressed air-driven metallic piston produces a controlled impact causing deformation of the brain parenchyma with an intact dura [12, 13]. The injury was graded from mild to severe by using a range of impact velocity of 2–4 m/s and a penetration depth of 2–4 mm. The lower range injury did not result in any changes of vital signs nor any contusions or obvious change in cerebral tissue. Increases in the level of deformation resulted in immediate acute subdural hematoma formation with associated cortical contusion. Deformation level greater than 3.5 mm was lethal in many cases regardless of the level of impact velocity. In order to induce concussive and mild TBI the impact can strike intact skull.

Features seen in clinical TBI such as subdural hematoma, increased ICP, contusion, and axonal injury can be induced in moderate and severe CCI. These findings were absent in the mild level of injury with CCI reported by Lighthall [12]. It should be noted that different groups report different parameters used to generate mild TBI. There can be variability in the injury parameters [31, 32] or physiological parameters such as apnea and hypotension used to define level of injury [33]. This should be considered when interpreting reported data or designing a study for mild TBI using CCI. The availability of data with different injury parameters as well as neurobehavioral outcome in this model provides good ground for designing future studies.

The CCI is superior to the FPI model due to a better control over mechanical factors such as time, velocity of impact and depth of resulting deformation of the brain. An additional strength of this model of TBI is the lack of risk of a rebound injury that can be seen in gravity-driven devices. An elaborate discussion of the CCI model is presented in Chapters 11 and 19.

2.5 Penetrating TBI Models

Several models have been developed to mimic penetrating brain injuries that can be produced by missiles, gunshots or sharp objects in general [14, 34–36]. All of these models are in non-rodent animals and not in use today. With the exception of a ballistic brain injury model using a balloon inflation technique [15], no high-speed penetrating rodent injury model has been available. The main obstacle has been the association of high mortality rate in high-speed penetrating TBI.

In a newly developed rat model a probe is driven into the brain parenchyma at approximately 90 m/s after being hit by a pellet accelerated from a specially designed air rifle. The speed and the depth of penetration is adjustable and highly reproducible [16]. This model is discussed in detail in Chapter 25. The biomechanics of this TBI model enables survival of the animals following a high-speed penetration of brain tissue. Neurodegeneration was detected in the injured cortex 24 h after injury and declined rapidly. The injured area showed a progressive expansion that had developed to a large cavity by day 14. The injured area showed also BBB defect and signs of extracellular perivascular edema. The injured animals displayed sustained deficits in reference memory and transient attention and showed balance and motor disturbances.

2.6 Diffuse Injury Models

Diffuse brain injuries usually arise when the skull is accelerated and the brain mass, due to its inertia, lags behind or continues its motion relative to the skull [6].

Brain tissue is more likely to be injured due to rotational acceleration rather than linear because the brain is relatively incompressible while the shear modulus for the brain tissue is relatively low. The main pathological finding in rotational acceleration injury is diffuse axonal injuries [37–39]. Experiments on primates have demonstrated that the incidence and degree of diffuse axonal injury is strongly correlated with the direction of the head acceleration: coronal plane angular acceleration was the direction causing the longest lasting coma, while sagittal plane angular accelerations and oblique accelerations produced coma for a shorter period [39]. This model is discussed in more detail in Chapters 13 and 14.

The first models producing acceleration injury mass impacts were performed on the unconstrained head of primates [40, 41]. The anesthetized animal is positioned prone on the injury device, the head is tightly fixed, and inertial loading is generated through a biphasic centroidal rotation for 110° within 20 ms. These models reproduced the acceleration-deceleration force seen in human head injuries. Additional models generating acceleration brain injuries, in which different impactors stroke the head of primates, were developed by Ono [17]. This model caused concussion by a frontal or occipital impact over a narrow contact area without the using of a head restraint. Hemorrhages were seen dependent on the severity of the concussion [42]. In contrast to the findings of Gennarelli et al. they did not find any correlation between concussion and angular acceleration.

Other models generated impact on the temporal region of the unrestrained skull of sheep [43]. Although the unrestrained head models may replicate some of the characteristics of human TBI, they lack injury reproducibility. There is no control over the biomechanical forces related to impact and head dynamic response. Additional acceleration models were developed to understand injuries following the movement of the head alone and these models expose the head to acceleration injuries without any impact [21, 22, 44–46]. The majority of these models were used on larger animals such as primates, pigs and rabbits. This nonhuman primate model uses a pneumatic shock tester to generate a nonimpact, controlled, single rotation, which displaces the head 60° within 10–20 ms [21, 44]. In the model using swine, the head is secured to a pneumatic actuator through a snout clamp. The pneumatic actuator produces linear motion that is further converted to angular motion through a linkage assembly directly mounted to the device. Based on position of the head the motion can produce coronal or axial plane rotation [45]. The cost and size of the animals in addition to limitations in behavioral outcome measures make their use difficult. Thus rodent models have been found more convenient.

In order to replicate human concussive and diffuse brain injury in rodents, without any focal damages, haemorrhages, skull fractures and bleeding, several animal models have been developed [18–20, 47]. These models all have in common that they are constrained impact acceleration models that can produce graded brain injury.

The one most frequently used is the Marmarou's weight drop model; it is inexpensive, easy to perform and can produce graded DAI. Despite these advantages there have been concerns regarding a second hit induced by the weight dropped on the skull. Also the movement of the weight during the fall in the Plexiglas can produce a lateralized impact with uneven distribution; this has been discussed in Chapter 12. Cernak et al. developed a constrained impact acceleration model to improve the control and reproducibility of the impact. Although this model succeeded in this matter the impact cannot be graded [18]. Maruichi et al. reported on a model based on the methodology used in the Cernak model mentioned above but made advancements to grade the impact. However, subarachnoid and intraventricular hemorrhages in addition to haemorrhages in corpus callosum were frequently observed [19].

2.7 Rotational TBI Model

Despite the number of models mentioned above, none of them is able to produce a graded DAI without large quantities of contusion or hemorrhage injuries in rodents. Therefore a new model in which the heads of the rats are exposed to sagittal plane rotational accelerations resulting in graded levels of DAI have been developed [24]. In addition, this model is used for studying injury threshold. This model is discussed in more detail in Chapter 17. The range of rotational acceleration studied is 0.3–2.1 Mrad/s^2; β-APP positive axons are seen in all animals exposed to head rotational trauma of 1.0 Mrad/s^2 or above. These β-APP positive axons were detected as early as 2 h post-injury and were found in corpus callosum and the border of white and grey matter. No contusion or hemorrhages were detected. In animals with high acceleration trauma β-APP positive axons were detected in the brain stem. There were also signs of axonal swelling and bulbs in the brain stem of these animals detected by FD silver staining. No signs of BBB changes could be detected. The behavioral outcome of this model was modest and transient indicating a mild TBI [23].

2.8 Blast Injury Models

When an explosive is detonated it generates a high-pressure wave that travels outwards. Recent studies have shown that despite the lack of a direct head injury, a blast trauma can cause significant brain damage [48–51]. In real life the blast injuries are classified according to the forces causing the injury. There are four main categories: primary, secondary, and tertiary, with various additional injuries forming an additional (quaternary) group. Blast injuries are discussed in more detail in Chapters 7 and 8.

Today there are both large-animal models of blast injury [48, 52–54] as well as small-animal models [49, 55–57] where chemical explosives are used as the source of the blast wave. Most of these models produce the primary blast injury type. The models can broadly be classified as open-field exposure, blast tube explosive and shock tube with compressed air or gas [58].

2.8.1 Open-Field Exposure

The first studies of open-field exposure to blast were carried out in the 1960s where both large and small animals were subjected to blasts with simple wave forms [59, 60]. Although the open-field experiments with large animals provide a more realistic and similar setting to that of the real scenario, they require large amounts of explosives. Furthermore, it is difficult to control the physiology of the experimental animals and reproduce the exact same experimental conditions. Today there are newly developed modified open-field models for primates with promising results [54].

2.8.2 Blast Tubes

The blast tubes were developed by Clemendson at the Swedish FOA (Swedish Defense Research Establishment) in the 1950s [61, 62]. Initially they used larger animals such as pigs. The tube was later developed for rodents [63]. In the current rodent models of the blast tube, the anesthetized rats are fixed in special net holders to avoid movement of the body [64]. This model is discussed in more detail in Chapters 7 and 8. The detonation charge is placed at the other end of the tube, 1 m from the rat. The rats can also wear protection or be placed in a holder to avoid pulmonary injuries. The blast waves produced have short duration and a simple form. The blast tube generates mainly primary blast injuries; however the gas and smoke emission might generate a tertiary blast injury. A blast tube for larger animal such as swine is also currently in use [52].

2.8.3 Shock Tubes

The simplest form of shock tubes consists of two chambers, separated by a membrane called the diaphragm. One of the chambers is filled with compressed air or gas and the other chamber contains the animal. The diaphragm is ruptured and the compressed air or gas simulates a propagating blast wave. There are currently several research facilities using shock tubes [55, 56, 63, 65]. The one used by Cernak et al. is complex, with a flexible, multi-chamber shock tube capable of reproducing complex shock waves.

3 Summary

Experimental animal TBI models are essential for understanding the underlying pathology and developing new diagnostics and therapeutics. Despite the existence of various TBI models none of them can reproduce the entire spectrum of TBI seen in humans. However, each model has a well-defined purpose and may mimic a specific relevant clinical feature such as neuropathological hallmark or neurobehavioral deficits.

It is important to identify the purpose of the TBI model used when designing a study. The use of distinct types of TBI models will reveal information about the heterogeneous underlying pathology and can help in identifying signatures that are characteristic or unique for the given type of TBI.

References

1. Adams JH (1992) Head injury. In: Adams JH, Duchen L (eds) Greenfield's neuropathology, 5th edn. Edward Arnold, London, pp 106–152
2. Teasdale G, Jennett B (1974) Assessment of coma and impaired consciousness. A practical scale. Lancet 2:81–84
3. Narayan RK, Michel ME, Ansell B, Baethmann A, Biegon A, Bracken MB, Bullock MR, Choi SC, Clifton GL, Contant CF, Coplin WM, Dietrich WD, Ghajar J, Grady SM, Grossman RG, Hall ED, Heetderks W, Hovda DA, Jallo J, Katz RL, Knoller N, Kochanek PM, Maas AI, Majde J, Marion DW, Marmarou A, Marshall LF, McIntosh TK, Miller E, Mohberg N, Muizelaar JP, Pitts LH, Quinn P, Riesenfeld G, Robertson CS, Strauss KI, Teasdale G, Temkin N, Tuma R, Wade C, Walker MD, Weinrich M, Whyte J, Wilberger J, Young AB, Yurkewicz L (2002) Clinical trials in head injury. J Neurotrauma 19:503–557
4. Saatman KE, Duhaime AC, Bullock R, Maas AI, Valadka A, Manley GT (2008) Classification of traumatic brain injury for targeted therapies. J Neurotrauma 25:719–738
5. Demetriades D, Kuncir E, Murray J, Velmahos GC, Rhee P, Chan L (2004) Mortality prediction of head Abbreviated Injury Score and Glasgow Coma Scale: analysis of 7,764 head injuries. J Am Coll Surg 199:216–222
6. Holbourn H (1943) Mechanics of head injuries. Lancet 2:438–441
7. Feeney DM, Boyeson MG, Linn RT, Murray HM, Dail WG (1981) Responses to cortical injury: I. Methodology and local effects of contusions in the rat. Brain Res 211:67–77
8. Flierl MA, Stahel PF, Beauchamp KM, Morgan SJ, Smith WR, Shohami E (2009) Mouse closed head injury model induced by a weight-drop device. Nat Protoc 4:1328–1337
9. Hayes RL, Stalhammar D, Povlishock JT, Allen AM, Galinat BJ, Becker DP, Stonnington HH (1987) A new model of concussive brain injury in the cat produced by extradural fluid volume loading: II. Physiological and neuropathological observations. Brain Inj 1:93–112
10. Dixon CE, Lyeth BG, Povlishock JT, Findling RL, Hamm RJ, Marmarou A, Young HF, Hayes RL (1987) A fluid percussion model of experimental brain injury in the rat. J Neurosurg 67:110–119
11. McIntosh TK, Noble L, Andrews B, Faden AI (1987) Traumatic brain injury in the rat: characterization of a midline fluid-percussion model. Cent Nerv Syst Trauma 4:119–134
12. Lighthall JW (1988) Controlled cortical impact: a new experimental brain injury model. J Neurotrauma 5:1–15
13. Dixon CE, Clifton GL, Lighthall JW, Yaghmai AA, Hayes RL (1991) A controlled cortical impact model of traumatic brain injury in the rat. J Neurosci Methods 39:253–262
14. Carey ME, Sarna GS, Farrell JB, Happel LT (1989) Experimental missile wound to the brain. J Neurosurg 71:754–764
15. Williams AJ, Hartings JA, Lu XC, Rolli ML, Dave JR, Tortella FC (2005) Characterization of a new rat model of penetrating ballistic brain injury. J Neurotrauma 22:313–331
16. Plantman S, Ng KC, Lu J, Davidsson J, Risling M (2012) Characterization of a novel rat model of penetrating traumatic brain injury. J Neurotrauma 29:1219–1232
17. Ono K, Kikuchi A, Nakamura M, Kobayashi H, Nakamura N (1980) Human head toler-

ance to sagittal impact reliable estimation deduced from experimental head injury using subhuman primates and human cadaver skulls. Proc. of the 24th STAPP Car Crash Conference

18. Cernak I, Vink R, Zapple DN, Cruz MI, Ahmed F, Chang T, Fricke ST, Faden AI (2004) The pathobiology of moderate diffuse traumatic brain injury as identified using a new experimental model of injury in rats. Neurobiol Dis 17:29–43
19. Maruichi K, Kuroda S, Chiba Y, Hokari M, Shichinohe H, Hida K, Iwasaki Y (2009) Graded model of diffuse axonal injury for studying head injury-induced cognitive dysfunction in rats. Neuropathology 29:132–139
20. Marmarou A, Foda MA, van den Brink W, Campbell J, Kita H, Demetriadou K (1994) A new model of diffuse brain injury in rats. Part I: Pathophysiology and biomechanics. J Neurosurg 80:291–300
21. Gennarelli TA, Adams JH, Graham DI (1981) Acceleration induced head injury in the monkey. I. The model, its mechanical and physiological correlates. Acta Neuropathol Suppl 7:23–25
22. Gutierrez E, Huang Y, Haglid K, Bao F, Hansson HA, Hamberger A, Viano D (2001) A new model for diffuse brain injury by rotational acceleration: I model, gross appearance, and astrocytosis. J Neurotrauma 18:247–257
23. Rostami E, Davidsson J, Ng KC, Lu J, Gyorgy A, Walker J, Wingo D, Plantman S, Bellander BM, Agoston DV, Risling M (2012) A model for mild traumatic brain injury that induces limited transient memory impairment and increased levels of axon related serum biomarkers. Front Neurol 3:115
24. Davidsson J, Risling M (2011) A new model to produce sagittal plane rotational induced diffuse axonal injuries. Front Neurol 2:41
25. Dail WG, Feeney DM, Murray HM, Linn RT, Boyeson MG (1981) Responses to cortical injury: II. Widespread depression of the activity of an enzyme in cortex remote from a focal injury. Brain Res 211:79–89
26. Lindgren S, Rinder L (1969) Production and distribution of intracranial and intraspinal pressure changes at sudden extradural fluid volume input in rabbits. Acta Physiol Scand 76:340–351
27. Sullivan HG, Martinez J, Becker DP, Miller JD, Griffith R, Wist AO (1976) Fluid-percussion model of mechanical brain injury in the cat. J Neurosurg 45:521–534
28. Thompson HJ, Lifshitz J, Marklund N, Grady MS, Graham DI, Hovda DA, McIntosh TK (2005) Lateral fluid percussion brain injury: a 15-year review and evaluation. J Neurotrauma 22:42–75
29. Ekmark-Lewen S, Flygt J, Kiwanuka O, Meyerson BJ, Lewen A, Hillered L, Marklund N (2013) Traumatic axonal injury in the mouse is accompanied by a dynamic inflammatory response, astroglial reactivity and complex behavioral changes. J Neuroinflammation 10:44
30. Duhaime AC, Margulies SS, Durham SR, O'Rourke MM, Golden JA, Marwaha S, Raghupathi R (2000) Maturation-dependent response of the piglet brain to scaled cortical impact. J Neurosurg 93:455–462
31. Budinich CS, Chen H, Lowe D, Rosenberger JG, Bernstock JD, McCabe JT (2012) Mouse brain PSA-NCAM levels are altered by graded-controlled cortical impact injury. Neural Plast 2012:378307
32. Washington PM, Forcelli PA, Wilkins T, Zapple DN, Parsadanian M, Burns MP (2012) The effect of injury severity on behavior: a phenotypic study of cognitive and emotional deficits after mild, moderate, and severe controlled cortical impact injury in mice. J Neurotrauma 29:2283–2296
33. Igarashi T, Potts MB, Noble-Haeusslein LJ (2007) Injury severity determines Purkinje cell loss and microglial activation in the cerebellum after cortical contusion injury. Exp Neurol 203:258–268
34. Tan Y, Zhou S, Liu Y, Li Z (1998) A gross and microscopic study of cerebral injuries accompanying maxillofacial high-velocity projectile wounding in dogs. J Oral Maxillofac Surg 56:345–348
35. Crockard HA, Brown FD, Johns LM, Mullan S (1977) An experimental cerebral missile injury model in primates. J Neurosurg 46:776–783
36. Finnie JW (1993) Pathology of experimental traumatic craniocerebral missile injury. J Comp Pathol 108:93–101
37. Adams JH, Doyle D, Ford I, Gennarelli TA, Graham DI, McLellan DR (1989) Diffuse axonal injury in head injury: definition, diagnosis and grading. Histopathology 15:49–59
38. Adams JH, Doyle D, Ford I, Graham DI, McGee M, McLellan DR (1989) Brain damage in fatal non-missile head injury in relation to age and type of injury. Scott Med J 34:399–401
39. Gennarelli TA, Thibault LE, Adams JH, Graham DI, Thompson CJ, Marcincin RP (1982) Diffuse axonal injury and traumatic coma in the primate. Ann Neurol 12:564–574
40. Gurdjian ES, Lissner HR, Webster JE, Latimer FR, Haddad BF (1954) Studies on experimental concussion: relation of physiologic effect to time duration of intracranial pressure increase at impact. Neurology 4:674–681

41. Ommaya AK, Grubb RL Jr, Naumann RA (1971) Coup and contre-coup injury: observations on the mechanics of visible brain injuries in the rhesus monkey. J Neurosurg 35:503–516
42. Kanda R, Nakamura N, Sekino H, Sakai H, Masuzawa H, Mii K, Aoyagi N, Aruga T, Kono H, Sugimori T, Sugiura M, Mori N, Kikuchi A, Ono K, Kobayashi H (1981) Experimental head injury in monkeys—concussion and its tolerance level. Neurol Med Chir 21:645–656
43. Lewis SB, Finnie JW, Blumbergs PC, Scott G, Manavis J, Brown C, Reilly PL, Jones NR, McLean AJ (1996) A head impact model of early axonal injury in the sheep. J Neurotrauma 13:505–514
44. Ross DT, Meaney DF, Sabol MK, Smith DH, Gennarelli TA (1994) Distribution of forebrain diffuse axonal injury following inertial closed head injury in miniature swine. Exp Neurol 126:291–299
45. Smith DH, Chen XH, Xu BN, McIntosh TK, Gennarelli TA, Meaney DF (1997) Characterization of diffuse axonal pathology and selective hippocampal damage following inertial brain trauma in the pig. J Neuropathol Exp Neurol 56:822–834
46. Xiao-Sheng H, Sheng-Yu Y, Xiang Z, Zhou F, Jian-ning Z (2000) Diffuse axonal injury due to lateral head rotation in a rat model. J Neurosurg 93:626–633
47. Goldman H, Hodgson V, Morehead M, Hazlett J, Murphy S (1991) Cerebrovascular changes in a rat model of moderate closed-head injury. J Neurotrauma 8:129–144
48. Saljo A, Arrhen F, Bolouri H, Mayorga M, Hamberger A (2008) Neuropathology and pressure in the pig brain resulting from low-impulse noise exposure. J Neurotrauma 25:1397–1406
49. Cernak I, Wang Z, Jiang J, Bian X, Savic J (2001) Ultrastructural and functional characteristics of blast injury-induced neurotrauma. J Trauma 50:695–706
50. Cernak I, Savic J, Malicevic Z, Zunic G, Radosevic P, Ivanovic I, Davidovic L (1996) Involvement of the central nervous system in the general response to pulmonary blast injury. J Trauma 40:S100–S104
51. Kato K, Fujimura M, Nakagawa A, Saito A, Ohki T, Takayama K, Tominaga T (2007) Pressure-dependent effect of shock waves on rat brain: induction of neuronal apoptosis mediated by a caspase-dependent pathway. J Neurosurg 106:667–676
52. Bauman RA, Ling G, Tong L, Januszkiewicz A, Agoston D, Delanerolle N, Kim Y, Ritzel D, Bell R, Ecklund J, Armonda R, Bandak F, Parks S (2009) An introductory characterization of a combat-casualty-care relevant swine model of closed head injury resulting from exposure to explosive blast. J Neurotrauma 26:841–860
53. Garner JP, Watts S, Parry C, Bird J, Kirkman E (2009) Development of a large animal model for investigating resuscitation after blast and hemorrhage. World J Surg 33:2194–2202
54. Lu J, Ng KC, Ling G, Wu J, Poon DJ, Kan EM, Tan MH, Wu YJ, Li P, Moochhala S, Yap E, Lee LK, Teo M, Yeh IB, Sergio DM, Chua F, Kumar SD, Ling EA (2012) Effect of blast exposure on the brain structure and cognition in Macaca fascicularis. J Neurotrauma 29:1434–1454
55. Long JB, Bentley TL, Wessner KA, Cerone C, Sweeney S, Bauman RA (2009) Blast overpressure in rats: recreating a battlefield injury in the laboratory. J Neurotrauma 26:827–840
56. Chavko M, Koller WA, Prusaczyk WK, McCarron RM (2007) Measurement of blast wave by a miniature fiber optic pressure transducer in the rat brain. J Neurosci Methods 159:277–281
57. Elder GA, Cristian A (2009) Blast-related mild traumatic brain injury: mechanisms of injury and impact on clinical care. Mt Sinai J Med N Y 76:111–118
58. Risling M, Davidsson J (2012) Experimental animal models for studies on the mechanisms of blast-induced neurotrauma. Front Neurol 3:30
59. White CS, Bowen IG, Richmond DR (1965) Biological tolerance to air blast and related biomedical criteria. CEX-65.4. CEX [reports]; civil effects exercise. U.S. Atomic Energy Commission. pp 1–239
60. Richmond DR, Damon EG, Bowen IG, Fletcher ER, White CS (1967) Air-blast studies with eight species of mammals. Techn Progr Rep DASA 1854. Fission product inhalation project [technical progress report]. Lovelace Foundation for Medical Education and Research. pp 1–44
61. Clemedson CJ, Criborn CO (1955) A detonation chamber for physiological blast research. J Aviat Med 26:373–381
62. Clemedson CJ, Hartelius H, Holmberg G (1957) The effect of high explosive blast on the cerebral vascular permeability. Acta Pathol Microbiol Scand 40:89–95
63. Saljo A, Bao F, Haglid KG, Hansson HA (2000) Blast exposure causes redistribution of phosphorylated neurofilament subunits in neurons of the adult rat brain. J Neurotrauma 17:719–726

64. Risling M, Plantman S, Angeria M, Rostami E, Bellander BM, Kirkegaard M, Arborelius U, Davidsson J (2011) Mechanisms of blast induced brain injuries, experimental studies in rats. Neuroimage 54(Suppl 1):S89–S97

65. Cernak I, Merkle AC, Koliatsos VE, Bilik JM, Luong QT, Mahota TM, Xu L, Slack N, Windle D, Ahmed FA (2011) The pathobiology of blast injuries and blast-induced neurotrauma as identified using a new experimental model of injury in mice. Neurobiol Dis 41:538–551

Chapter 5

Systematic Review of Traumatic Brain Injury Animal Models

Helen W. Phipps

Abstract

The goals of this chapter are to provide an introduction into the variety of animal models available for studying traumatic brain injury (TBI) and to provide a concise systematic review of the general materials and methods involved in each model. Materials and methods were obtained from a literature search of relevant peer-reviewed articles. Strengths and weaknesses of each animal choice were presented to include relative cost, anatomical and physiological features, and mechanism of injury desired. Further, a variety of homologous, isomorphic/induced, and predictive animal models were defined, described, and compared with respect to their relative ease of use, characteristics, range, adjustability (e.g., amplitude, duration, mass/size, velocity, and pressure), and rough order of magnitude cost. Just as the primary mechanism of action of TBI is limitless, so are the animal models available to study TBI. With such a wide variety of available animals, types of injury models, along with the research needs, there exists no single "gold standard" model of TBI rendering cross-comparison of data extremely difficult. Therefore, this chapter reflects a representative sampling of the TBI animal models available and is not an exhaustive comparison of every possible model and associated parameters. Throughout this chapter, special considerations for animal choice and TBI animal model classification are discussed. Criteria central to choosing appropriate animal models of TBI include ethics, funding, complexity (ease of use, safety, and controlled access requirements), type of model, model characteristics, and range of control (scope).

Key words Traumatic brain injury TBI, Animal models, Homologous, Predictive, Static dynamic, Isomorphic/induced, Direct indirect, Penetrating, Non-penetrating

1 Introduction

The use and application of in vivo animal models to human traumatic brain injury (TBI) research is widespread and very diverse. With numerous models available, researchers must consider their aims when choosing appropriate animal models. In general, animal models are categorized as homologous, isomorphic/induced [1], or predictive [2]. Homologous animal models mimic all aspects of a condition from causes to symptoms and treatment enabling biophysical, cellular/molecular, and behavioral assessments. Isomorphic or induced animal models share the same cellular/molecular mechanisms,

Firas Kobeissy et al. (eds.), *Injury Models of the Central Nervous System: Methods and Protocols*, Methods in Molecular Biology, vol. 1462, DOI 10.1007/978-1-4939-3816-2_5, © Springer Science+Business Media New York 2016

symptoms and treatments but the cause is inflicted with laboratory equipment versus the actual cause in humans thus limiting biophysical assessments. These are the most common animal models used in TBI research. Predictive animal models only replicate the effect of a treatment or a single aspect of a multifaceted condition. This method is commonly used for screening and validating localized aspects of a condition when researchers do not know the underlying cause(s).

Animal models of TBI are further classified according to the mechanism and characteristics of injury desired. Moshang [3] compared diffuse and mechanical TBI while Cernak [4] further divided mechanical TBI models into static and dynamic TBI. Static TBI is defined as a localized and crushing injury via physical manipulation of tissue such as pinching a nerve with forceps [5] and allows for direct correlation of specific nerve damage to a particular neuropsychological outcome. While the amplitude and duration are predetermined and considered constant in static crushing injuries, acceleration and velocity are not considered as contributing factors to static injuries [4]. Alternatively, a dynamic TBI model may inflict contra coup damage to the brain due to transfer of forces, a ricochet/rebound effect of the brain within the skull, and/or stretching of tissue, including nerves and blood vessels as well as decreased blood flow due to inflammation [6]. Such dynamic TBI models are adjustable in amplitude, duration, velocity, and acceleration that may impact the entire body versus only a single nerve or tissue. Dynamic TBI models can be considered as either direct, or indirect. All models utilize anesthesia to minimize animal suffering during the injury and hence it is not considered as an advantage or limitation to any of the models.

With such a wide variety of available animals, types of models, mechanisms, and characteristics of injury, no single "gold standard" model of TBI exists. Therefore, comparing data across studies is very difficult. Criteria central to choosing appropriate animal models of TBI include: anatomy, physiology, cost, complexity (ease of use, safety, and controlled access requirements), type of model, model characteristics, and range of control (scope). This chapter aims to elaborate on these central criteria.

2 Materials

2.1 Animal Selection and Preparation

Anatomical and physiological features are critical to consider with respect to the aims of the research. The most common candidates for TBI animal models are rodents and porcine animals. Rodents and porcine animals are commonly used in TBI research; however, the larger animals are more expensive than smaller animals. Less common candidates for TBI animal models are nonhuman primates, canine, feline, chinchilla, and ferret. Not only is the initial cost of the animal important, but also ongoing animal care and complex assays which are necessary to inflict, support, and sustain the research.

Anatomically, the mechanism of injury should provide accurate and precise results scalable to the organism chosen. Consider hitting a small rodent with the same force as a football player—survival would be in jeopardy. The elasticity, porosity and thickness of the skull are significant issues to consider animal models seeking to replicate child abuse and/or TBI among the elderly. Additionally, a strong framework of physics is necessary to adequately quantify and validate a model's amplitude, duration, velocity, and acceleration for consistency. These parameters should be correlated with the severity and symptomology of the injury inflicted—the more severe the impact, the more severe the symptoms. The *Lab Animal Buyer's Guide* [7] provides detailed and current information on hundreds of laboratory animal suppliers and associated products such as biohazard equipment, euthanasia products, restraint equipment, and monitoring supplies.

Physiological features such as proteomics need to be considered if the intent of the research is to identify protein effects. If a particular organism does not have homologous biochemistry to humans then an isomorphic or induced model is not appropriate as primary and secondary injury cascades might differ [8]. Alternatively, if a biomechanical study aims to investigate the biophysics of the injury, then the molecular makeup of the organism is not necessarily a significant factor to consider as compared to the specific mechanical insult. For example, a military helmet manufacturer may be interested in the rebound and pressure changes within the skull as a result of blast overpressure, but may not be interested in the secondary molecular cascades that result from such an insult.

2.1.1 Rodents

Rats and mice are used extensively in TBI testing. The specific species used will depend on the aims of the research. For instance, a species should be chosen that does not float if the intent is to conduct water navigation studies. The two primary advantages of using rodents are cost and versatility (e.g., the ability to use in a wide variety of models to include ability to position the animal at the mouth of a shock tube such that they can be enveloped by blast, as a human would be in a free-field exposed to blast. The primary disadvantage is that the rat and mouse brain differs substantially in size and structure from a human brain. Without scaling laws to account for these differences, testing these species may be challenged in yielding relevant load thresholds to an adult human. However, if scaling laws between a rodent and human can be accurately identified, then the significant volume of historical data on rats and mice would be of greater value to the TBI research community. Furthermore, many knockout mice exist and abundant behavioral assessments are available for use with rodents which make them a more appealing choice for many scientists.

Anatomically, rodents are lissencephalic and have very large olfactory bulbs but small cerebral cortex that is arranged differently than the human cortex; therefore, the location of TBI must be considered and controlled. However, only small organisms (less than or equal to approximately 280 g) are able to fit in most high resolution magnetic resonance imaging (MRI) devices. Physiologically, rodents show striking similarities to humans lending to their common use in biomedical research. However, care must be taken to maintain vital signs as close to normal as possible. Average animal vitals are shown in Table 1.

Alternatives to rats and mice still within the rodent species include guinea pigs. Guinea pigs have more gyra than a rat brain. Both rodents and guinea pigs are relatively inexpensive to test and manage, thus enabling evaluation of a large number of subjects in a relatively short time. The relatively inexpensive cost of rodents and guinea pigs also enable the determination of optimum methods prior to application on more expensive animals.

2.1.2 Porcine (and Ovine) Animals

Porcine specimens have been commonly used for TBI testing over the last decade [9–12]. The porcine model provides a larger animal to facilitate scaling to the human brain based upon the gyrencephalic macrostructure and overall size of the brain. This allows thresholds of injury in porcine brain to be more readily translatable to human injury thresholds than rodent. Pigs can be acquired with a similar body mass to humans and although the frontal and

Table 1
Average animal vital signs

Animal	Heart rate range (beats per minute)	Body temperature (°C)
Rodents: rat	250–400	37
Rodent: mice	450–750	37
Rodent: guinea pig	200–300	37–39
Porcine (and ovine)	70–120 (70–80)	38–40
Nonhuman primate (rhesus monkey)	160–330	37–39
Canine	70–120	38–39
Feline	120–140	37–39
Chinchilla	100–150	37–38
Ferret	180–250	37–39
Reference: The Merck Veterinary Manual		

midbrain skull are generally thicker in pigs than in humans [13] porcine and human parietal skull thicknesses are similar. In a CT study of 180-lb female pigs, the average thickness was 9.7 ± 1.7 mm while for humans, the average thickness is 7.2 ± 1.7 mm for females and 6.6 ± 1.3 mm for males [13]. Increased anatomical complexity of the pig brain allows for more sophisticated assessments of regional differences in injury. The distribution of many neurotransmitter uptake sites in pigs corresponds very closely to that in the brain of humans [14]. Overall, pigs are very similar to humans in general physiology (similar homeostatic mechanisms, similar shock response). The primary disadvantage to porcine research is the expense of managing large, hoofed species for animal testing.

2.1.3 Nonhuman Primates

Non-human primates are the most biofidelic surrogates available. The regional and vascular anatomies as well as cervical range are more similar to humans than lower order gyrencephalic animals. Furthermore, non-human primates can accommodate high order neurobehavioral testing, allowing for improved clinical correlation between behavioral deficit and observed tissue or physiological injury. However, despite the advantages of the nonhuman primates, research with these models is not common because it is typically much more expensive. Often, these models will not be approved by institutional animal care and use committees secondary to considerations related to pain and suffering of this high-order animal.

2.1.4 Canines, Felines, Chinchillas, and Ferrets

Canine and feline subjects have some structural similarities, and though they are large enough to perform detailed brain imaging, they present skull thickness and mass scaling challenges. Physiologically, their brains are organized differently than those of humans, thus further complicating translation [15]. However, as companion animals, the use of canine and feline species is accompanied by more significant animal usage concerns [16]. Chinchillas and ferrets present many of the same advantages and issues as do rats, but there is a paucity of historical data on their use in TBI experimentation. Ferrets have similar brain anatomy in that they are gyrencephalic [17], thus enabling determination of scaling factors from brain geometry and structure.

2.2 Animal Preparation

All animals should be anesthetized with a pre-approved agent prior to surgery and maintained under anesthesia throughout the procedure. In addition to anesthetics, each static and direct dynamic animal model has commonly used materials. The materials listed below are typically necessary for every model, regardless of the nature of the model or the type of animal used. Materials that are unique to each model are listed within subsequent sections.

1. Laboratory scale proportional to animal.
2. Electric hair shaver.

3. Autoclave.
4. 10 % Ethanol and/or dry sanitizer.
5. Povidone iodine.
6. 70 % Isopropyl alcohol.
7. Scalpel.
8. Hemostats.
9. Forceps.
10. Retractors.
11. Sterile saline.
12. Sterile cotton-tipped applicators.
13. Sterile gauze.
14. Trephine or some other microdrill and drill-bit.
15. Suture monofilament and suture proportional to animal.

2.3 Static TBI Materials

The first main category of TBI is static. The key differentiators of static from dynamic TBI are the lack of movement and the confinement of the primary mechanism of injury to a specific, defined region of the brain. Compression is the primary static model used in animal research.

2.3.1 Compression Materials

1. Clock to time duration of compression.
2. Iridectomy scissors.
3. Marker (e.g., powdered carbon).

2.4 Direct Dynamic TBI Materials

The second main category of TBI is direct dynamic (here forward referred to as direct). Direct TBI models may or may not impact and/or penetrate the skull and/or cortex of the animal. Direct impact TBI is the typical injury, the layman assumes as the sole mechanism of TBI. Although there are many impact-driven mechanisms of TBI, other models do exist. Impact TBI models mimic mechanisms ranging from penetrating head injury due to automobile accidents and military relevant mechanisms (ballistics) to sports and elderly related concussions. These direct-impact models may or may not penetrate the skull and may or may not involve constraining the animal while enduring the injury.

2.4.1 Penetrating Constrained (Percussion) Direct Brain Deformation

Penetrating constrained (percussion) direct brain deformation is a type of dynamic penetrating injury whereby the organism is constrained while the injury is inflicted. There are many specific models that fall within this category; however, most do not generate long-term TBI signs and symptoms. Instead, many of these models offer localized anatomical evaluations.

2.4.1.1 Fluid Percussion Injury (FPI) Materials

1. Stereotaxic frame.
2. FPI device consisting of a Plexiglas cylindrical reservoir restricted at the distal end by a rubber-covered Plexiglas piston and with the proximal end having a built-in transducer housing as well as a central injury screw or luer-loc needle hub fixed to the animal's skull.
3. Pressure transducer.
4. Oscilloscope.

2.4.1.2 Controlled Cortical Impact (CCI) Materials

1. Stereotaxic frame.
2. CCI device.
3. Dental cement, or bone wax (optional).

2.4.1.3 Vacuum/ Suction (Suction Ablation) Materials

1. Stereotaxic frame.
2. Fine glass Pasteur pipet.
3. Measurable, controlled vacuum.
4. Dental cement, or bone wax (optional).

2.4.1.4 Zymosan Materials

1. Stereotaxic frame with injector and syringe.
2. Hand-held drill.
3. Zymosan.

2.4.1.5 Cryolesion Materials

1. Stereotaxic frame.
2. Metal cylinder (typically copper) proportional to animal size.
3. Liquid nitrogen.
4. Clock to time duration of cryolesion.

2.4.1.6 Stab Materials

1. Stereotaxic frame.
2. Guide cannula.
3. 22 G Needle.
4. Dental cement, or bone wax.
5. Razor blade.
6. Clock to time duration of stab.

2.4.1.7 Epidural Mass Lesion (EML) Materials

1. Stereotaxic frame.
2. Latex surgical balloon with mitered valve.
3. *Penetrating ballistic brain injury* (PBBI) probe or customized needle.
4. Latex balloon with mitred valve.
5. Variable pressure waveform generator.
6. Dental cement, or bone wax.
7. Clock to time duration of inflation and deflation.

2.4.1.8 HVM Materials

1. Stereotaxic frame (if constrained).
2. Ballistics (proportional to animal).
3. Firearm.
4. High-speed video.

2.4.1.9 Penetrating Ballistic Brain Injury (PBBI) Materials

1. Stereotaxic frame.
2. Latex surgical balloon with mitered valve.
3. PBBI probe or customized needle.
4. Latex balloon with mitred valve.
5. Variable pressure waveform generator.
6. Dental cement, or bone wax.
7. Clock to time duration of inflation and deflation.

2.4.2 Direct Dynamic Non-penetrating TBI

Similar to direct dynamic penetrating brain deformation TBI models, direct dynamic *non*-penetrating TBI can be either constrained or not. However, non-penetrating TBI models alleviate the concern of additional damage by craniotomy.

Non-impact models that initiate movement of the brain within the skull without displacing the head laterally are also categorized as direct dynamic *non*-penetrating TBI. Safety concerns are significantly less pronounced in non-impact models. Rapid rotation is such a model which causes the characteristic diffuse axonal injury not replicable in non-rotational models. The nonhuman primate model [18, 19] delivers a single rotation of variable degree and acceleration. Porcine animal models are used in a similar device controlled pneumatically via a snout clamp [20]. Although this model accurately replicates the diffuse axonal injury most characteristic of human TBI, the use of large animals is more expensive and requires more technical equipment. A similar model for small animals is not currently available.

2.4.2.1 Concussion Materials

1. CCI device.
2. Plate or cone to affix to skull.
3. Cushion customized to allow predetermined movements during and after impact.

2.4.2.2 Unconstrained Impact Acceleration Materials

1. Plate or cone to affix to skull.
2. Cushion customized to allow predetermined movements during and after impact.
3. Weight drop device and weight.

2.4.2.3 Weight Drop Materials

1. Plate or cone to affix to skull.
2. Dental cement if investigating a penetrating injury.
3. Weight drop device and weight.

2.5 Indirect Dynamic TBI

The second main category of dynamic TBI is indirect. Blast TBI is the newest addition to the animal models of TBI and is the main type of indirect TBI. Instead of direct impact, blast TBI uses compressed air, live explosives, or an LISW to reproduce overpressure injuries. Original blast TBI models exposed entire organisms to over- and under-pressure. The most commonly used species for evaluating blast exposure are rodents and pigs. Scientists quickly identified multiple traumas to other organs besides the brain—mainly lung damage. With new body armor available to protect the body, the model was modified to only blast the head. Several companies have developed compressed air driven shock tubes that deliver an adjustable pressure over variable durations, distances, and angles to rodents. Many aspects of shock tubes are adjustable. For example, a pressure-sensitive membrane ruptures at specific pressures and releases a rapid overpressure. The type and thickness of the membrane, tube-length, and initial pressure determine the peak pressure. The diameter of the exit can be altered to change the duration of the overpressure. Typical peak overpressure ranges from 154 to 340 kPa [Standard atmospheric pressure is 101 kPa (1 atm)] but can be increased to over 10 kPa. To further complicate the blast TBI model, the organism can be located on or off axis (directly in line, or offset from the blast center) and at varying distances from the blast tube exit. Multiple expensive pressure sensors with associated software are necessary to monitor the resulting pressure. As the membrane ruptures, a loud noise is generated reaching (120–160 Db depending on the dimensions of the shock tube) similar to a gunshot sound. Typically, military personnel are limited to a maximum peak of 140 dB [21]. Therefore, shams must include exposure to anesthesia and sound only with no blast to account for potential effects due to high Db exposure and researchers must wear hearing protection. Alternatively, no incisions are made as in the direct models of TBI thus eliminating unrelated damage and allowing for biomechanical evaluations in addition to cellular, molecular, and behavioral assessments.

The use of live explosives complicates every aspect of experimentation from execution to evaluation by introducing changes in thermal load and gas byproducts of the explosion, not to mention potential fragmentation. An additional challenge faced with live explosives is reducing the scale of blast exposure for small organisms. Existing software, such as Conventional Weapons software and Bowen Curves to determine survivability at various standoff distances, are geared toward calculating acceptable impulse (mechanical energy), peak overpressure, and charge size for humans. Safety precautions include clearance for handling explosives, adequate standoff from the explosive site, and ensuring the area is clear prior to returning to the explosion site, to name a few.

2.5.1 Pneumatic Blast Materials

1. Compressed air or other gas.
2. Previously characterized shock tube and associated accessories to include hearing protection.
3. Pressure transducers.
4. Video camera.

2.5.2 Live Blast Materials

1. Explosive charge.
2. Previously characterized blast tube and associated accessories to include hearing protection.
3. Pressure transducers.
4. Video camera.

2.5.3 Laser-Induced Shock Wave (LISW) Blast Materials

1. Laser source (neodymium: yttrium-aluminum-garnet [Nd:YAG] laser at 1064 μm wavelength for underwater microexplosion LISW [22], or Q-switched ruby laser at 694 μm [23] or Q-switched Nd:YAG laser at 532 nm [24] wavelength for elastic targeting LISW).
2. Previously characterized truncated ellipsoidal reflector (proportional to the size craniotomy) for underwater microexplosion.
3. 1–400 μg Silver azide for underwater microexplosion LISW [22] or 1.0 mm polyethylene terephthalate (PET) and 0.5 mm black rubber for elastic targeting LISW [23, 24].
4. 0.6 mm diameter quartz optical fiber (diameter dependent on animal chosen and craniotomy size) for underwater microexplosion LISW [22].
5. Physiological saline (volume, concentration, and temperature dependent on animal chosen) for underwater microexplosion [22] or ultrasound gel for elastic targeting LISW [23, 24].
6. Polyvinylidene fluoride (PVDF) needle hydrophone (laser fluence sensor to monitor pressure available from Precision Acoustics Ltd, Onda Corp, or Müller Instruments).

3 Methods

3.1 Systematic Review Procedures

To collect data for this systematic review, literature searches were conducted for animal choice and unique methodologies utilized in relevant peer-reviewed TBI research articles published in or after 1980 (*see* Fig. 1). PubMed, ISI Web of Knowledge, and Google Scholar were searched for the name of each model. A representative sampling of 50 peer-reviewed articles with comprehensive methods in animals were systemmatically reviewed for models where hundreds of articles were available. Twenty-five compression, 50 CCI, 50 FPI, 12 suction/ablation, 25 zymosan, 29 cryolesion, 16 stab, 8 EML, 3 HVM, 33 PBBI, 50 concussion, 8 impact acceleration, 5 weight drop, 50

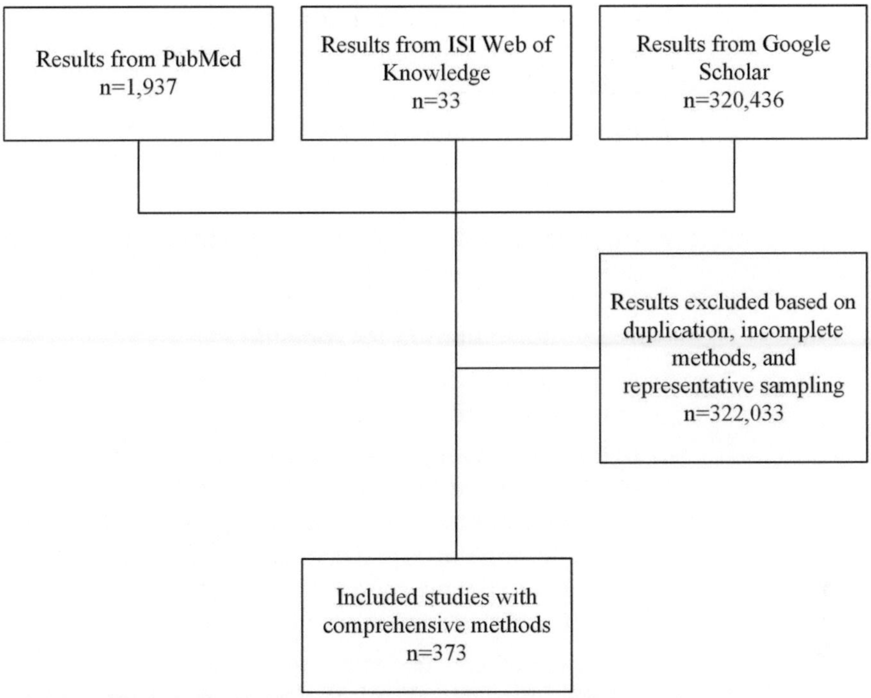

Fig. 1 Systematic review flow chart. The search strategy for this systematic review of animal models of traumatic brain injury entailed screening three major literature databases, excluding duplicates and publications with incomplete methods, and sampling representative publications when the search resulted in hundreds of publications with like models. The search strategy resulted in 373 publications included in the qualitative methods review

pneumatic blast, and 6 live blast brain injury articles were reviewed. The general animal choice and methodologies utilized for each respective model were combined to compare and contrast the approaches used in various animal models of TBI. Considerations for choosing each animal model were identified and factored into the systematic review. Using the parameters of each approach, each animal model was classified with respect to a commonly accepted hierarchy of TBI [4]. To determine the relative ease of use of each model, the volume and complexity of the materials and methods used for each general approach were compared to include safety precautions. The results of the systematic review were summarized, analyzed, and integrated into a summary table (*see* Table 2).

3.2 Animal Adaptation/Preparation

For all protocols detailed in this chapter, it is important to note the following:

1. All animal procedures should be performed with the approval of the local Institution's Animal Care and Ethics Committee and in accordance with the National Institutes of Health guidelines, with measures taken to minimize pain and discomfort. Animals must be handled according to the approved protocol at the research institution as well as the funding agency, if applicable.

2. Similar to behavioral studies, after the arrival of the rats at the animal facility, the rats should acclimate to the new environment for 1 week.
3. Procedures should be executed at the same time each day based on physiological rhythms unless the dependent variable is the time of day.
4. Animals should be housed under temperature-controlled conditions on an animal-appropriate light and dark cycle and diet (e.g., 12-h reverse light and dark cycle, while fed mouse chow and water ad libitum for rodents).
5. All procedures should be accomplished while wearing a mask, gown, and sterile gloves.
6. Before conducting any procedures, the animals to be tested should be moved into the vicinity of the testing area to allow them to adapt to the new environment for 20–30 min in their cages before starting the procedure. *Note*: Only the animal undergoing the procedure should be in the procedure room at any given time to avoid undue stress.
7. The user should become familiar with the procedure and set up the protocols ahead of time to avoid mistakes during the procedure.
8. All surfaces of the surgical tools and devices must be autoclaved (if possible) and cleaned with 10% ethanol solution or sterilized with an instant dry sterilizer before any procedure.
9. All animals should be anesthetized with a pre-approved agent prior to surgery and maintained under anesthesia throughout the procedure.
10. Weigh the animal immediately prior to administering anesthesia to ensure appropriate dosing.
11. To maintain the most sterile conditions possible, the skin at the site of surgery should be shaved. *Note*: This does not apply to indirect blast models where no surgery is involved.
12. Clean the skin using sterile cotton tipped applicators and povidone iodine.
13. Consider application of ophthalmic ointment to the eyes using sterile cotton tipped applicators to avoid dryness if the procedure is expected to last for longer durations.
14. Place the animal on a sterile stainless steel plate with a preheated homeothermic blanket system underneath.
15. Secure all animal limbs taking care to position the hind limbs symmetrically.
16. Cover the surgical field with a sterile drape.

Table 2
Animal models of TBI criteria

Model name	Relative cost[a]	Ease of use[b]	Type of model[c]	Characteristics[d]			Approximate range/adjustability				
							Amplitude	Duration	Mass/size	Pressure	Velocity
Compression	Low	5	Pr	D	P	C	N/A	Adjustable			
FPI	Medium	4	In	D	P	C	Adjustable				
CCI	Medium	4	In	D	P	C	Fixed	Adjustable			
Vacuum/suction	Low	3	Pr	D	P	C	N/A	Adjustable			N/A
Zymosan	Low	3	Pr	D	P	C	N/A		Adjustable	N/A	
Cryolesion	Low	3	Pr	D	P	C	N/A	Adjustable		N/A	N/A
EML	Low	4	Pr	D	P	C	N/A	Adjustable			N/A
Stab	Low	3	H	D	P	C	N/A	Adjustable			N/A
HVM	Varies	5	H	D	P	C, NC	Adjustable				
PBBI	High	5	In	D	P	C, NC	Adjustable				
Concussion	Medium	2	In	I	NP	C, NC	Fixed	Adjustable			
Pneumatic blast	High	2	In	I	NP	C, NC	Adjustable				
Live blast	Varies	5	H	I	NP	C, NC	Adjustable				
LISW blast	High	3	In	I	P	C	Adjustable				

Relative cost was approximated to account for the non-animal equipment that delivers the injury. Ease of use considers the complexity of using the model itself, general preparation (e.g., craniotomy adds complication), safety considerations, and controlled access restrictions (e.g., explosives are highly controlled). Rotational and shaken baby syndrome models were not included due to limited available information. The order of the models is based upon the order of appearance within the chapter

[a]Relative cost scale and references: Relative costs are qualitative and are based on the estimated comparative price to acquire the non-animal materials necessary to conduct the procedure from a variety of sources (AllegroMedical; Amscien Instruments; Dragonfly, Inc.; DuPont; eSutures; Harvard Apparatus; Kopf; Leica Biosystems; Mitre Corp.; Popper and Sons, Inc.; Precision Systems and Instrumentation, LLC; Sigma-Aldrich®; Sportaid). A variable relative cost reflects the range from low through high depending on the chosen characteristics of the model

[b]Ease of use scale of 1–5: 1 being the least invasive, 5 being the most invasive and technically consuming

[c]Type of model nomenclature: Homologous (H), induced (In), or predictive (Pr)

[d]Characteristic nomenclature: Direct (D) or indirect (I), penetrating (P) or not (NP), constrained (C) and/or not (NC)

17. Once the procedure is completed, the animal should be returned to animal-appropriate, temperature controlled recovery housing until they wake from anesthesia, and not directly to their original housing. Note the time the animal was under anesthesia and any deviations from normal physiological parameters (*see* **Notes 1–5**).
18. Return animal to the appropriate housing facility. Note the time until recovery (*see* **Notes 6** and 7).
19. CAUTION: All sharps and tools used for animal studies should be handled as biosafety level-2 waste, collected in special containers and disposed of only after being autoclaved or inactivated with bleach or other disinfectant while wearing protective gloves and a lab coat.

3.3 Compression

3.3.1 Compression Procedures

1. Place the animal in a stereotaxic frame secured by ear bars and an incisor bar.
2. Carefully expose the skull and perform a craniotomy using a trephine drill or some other minidrill at predetermined stereotaxic coordinates. The length of the incision and diameter of the craniotomy are dependent on the animal chosen, the complexity of accessing the tissue to crush, and the experience of the surgeon.
3. Using forceps, iridectomy scissors, and retractors, gently expose the tissue to be crushed. For a negative control, equivalent tissue should be exposed, but not crushed.
4. Place the nerve on the bottom jaw of a hemostatic forceps proportional to the tissue to be crushed (e.g., super-fine if a single, specific nerve). Note the alignment of the tissue and forceps before crushing to ensure consistency between trials and to avoid displacing the tissue beyond the forceps as a result of the crush.
5. Crush the tissue for the predetermined amount of time by securing the forceps to a predetermined setting (e.g., once for 20 s at three clicks). Do not shift the forceps when the tissue is being crushed to avoid unintended consequences.
6. If crushing the tissue multiple times, mark the specific area with a sterile indicator for reference. Powdered carbon is one way to mark the crush site [25] between crush injuries, but care must be taken to avoid marking beyond the injury site. Bauder [25] recommends opening sterile forceps in powdered carbon, gently close but do not click the forceps while still in the carbon, and wipe the carbon on the outside of the forceps off using sterile gauze before marking the area.
7. After crushing the tissue, carefully return the retracted tissue and suture the skin back in place.

3.3.2 Compression Precautions

Safety concerns include the use of a scalpel for the scalp resection, use of a drill for the craniotomy, use of iridectomy scissors to expose the tissue to be crushed, and sutures for closing the scalp after the injury.

3.4 Fluid Percussion Injury

FPI is a highly characterized and widely accepted model for use in numerous organisms (refer to Table 2 for a comparison of all models with corresponding criteria) that falls within the penetrating constrained (percussion) direct brain deformation category. FPI is inflicted as a metal pendulum is released and strikes a piston which drives injection of a small volume of sterile saline into the cranial cavity of the animal through the plastic luer-loc needle hub. In other words, FPI entails a craniotomy and direct saline application to intact dura via the use of a plastic cap and a Plexiglass cylinder. The hammer (pendulum) is controlled electrically and the voltage correlates to the pressure delivered. The location of the craniotomy can be changed; however, with potential consequences. Specifically, a lateral craniotomy (at least 3.5 mm lateral to the sagittal suture) leads to contralateral cortical damage with impact [26]—both focal (such as hemorrhage) and diffuse similar to most human TBI. However, noting the specific location of the craniotomy allows for consistency. Lateral fluid percussion (LFP) allows for a direct unilateral injury with very little to no subsequent contralateral rebound and/or brainstem injury. Central fluid percussion (CFP) impacts both cortices and subsequently shifts the brainstem which results in increased mortality thus limiting the dose–response [8, 27]. Each injury model induces similar signs and symptoms as in humans which allows for study of pathology, physiology, and pharmacology. However, the use of fluid is less mechanically controllable because of its diffusivity. If the fluid evaporates, gas bubbles accumulate, and/or the plastic cap degrades then the fluid may escape, the gas may change the delivered pressure, or the plastic cap may expand differently, thus creating inconsistencies between injuries. To increase the longevity of the model, the various o-rings must be lubricated and inspected.

One notable case of the use of FPI is in modeling epilepsy after severe TBI similar to the case of Henry Gustav Molaison (HM). After receiving a TBI from a supposed bicycle accident, HM suffered ongoing seizures—post-traumatic epilepsy (PTE). To alleviate the seizures, HM's medial temporal lobes were extracted. Severe FPI (greater than 3 atm fluid pressure) initiates such seizures in hippocampal structures [28, 29]. Unfortunately, the onset of the seizures is often 7–12 months post-FPI.

3.4.1 FPI Procedures

1. Place the animal in a stereotaxic frame secured by ear bars and an incisor bar.
2. Carefully expose the skull and perform a craniotomy using a trephine drill or some other minidrill at predetermined stereotaxic coordinates.

3. Carefully secure a sterile luer-loc needle hub over the exposed dura with either skull screws or cyanoacrylate glue.
4. Once secure, fill the luer-loc needle hub with saline and cap until ready to perform the FPI.
5. Prior to preparing the animal for the FPI, ensure the injury settings are appropriate (e.g., pulse duration).
6. At a predetermined time after the preparation of the animal luer-loc needle hub, re-anesthetize the animal and then fill the luer-loc needle hub with saline to ensure consistent pulse and avoid air bubbles. Connect the hub to the FPI device.
7. Deliver the impact while measuring the amplitude of the resulting pressure pulse using a pressure transducer and wave form with an oscilloscope.
8. Remove the animal from the FPI device and cap the hub.

3.4.2 FPI Precautions

Safety concerns include the use of a scalpel for the scalp resection, use of a drill for the craniotomy, use of a syringe to remove gases in fluid cylinder, use of electricity to power the hammer, and sutures for closing the scalp after the injury.

3.5 Controlled Cortical Impact (CCI)

Controlled cortical impact, or rigid percussion model, also falls within the penetrating constrained (percussion) direct brain deformation category. CCI is the attempt to solve the limitations of the FPI model while maintaining consistency with clinical signs and symptoms. Both models utilize a stereotaxic device to constrain the animal (rat or mouse in most cases) and ensure a repeatable alignment with the impact device. Instead of fluid, the CCI is delivered to the intact dura via an adjustable pneumatic- or weight-drop-driven piston (impactor) of variable diameter, thus providing a more localized injury. The impactor is mounted to an adjustable ruled crossbar that can be rotated to regulate it at different angles. In addition to the adjustable diameter and depth, the velocity and dwell time of the impactor are adjustable. Depth and velocity of impact correlate to pathological severity of the injury. Evidence to support a correlation between dwell time and severity of injury is not apparent. Although this model alleviates the controllability limitation of the FPI impact, the CCI model still requires a craniotomy which causes additional damage to the surrounding cortex area.

3.5.1 Controlled Cortical Impact Procedures

1. Prior to preparing the animal for the CCI, ensure that the injury settings are appropriate (e.g., depth, velocity, and duration).
2. Place the animal in a stereotaxic frame secured by ear bars and an incisor bar.
3. Carefully expose the skull and perform a craniotomy using a trephine drill or some other minidrill at predetermined stereotaxic coordinates. Note: It is helpful to align the impactor with the craniotomy site prior to removal of the bone.

4. Deliver the CCI to the intact dura with a predetermined penetration depth, velocity, and duration.
5. Consider replacing the bone from the craniotomy and sealing with either dental cement, or bone wax.
6. Close the scalp incision with sutures following the injury.

3.5.2 CCI Precautions

Safety concerns include the use of a scalpel for the scalp resection, use of a drill for the craniotomy, use of high velocity blunt objects, and sutures for closing the scalp after the injury.

3.6 Vacuum/Suction

Vacuum pulse and/or mechanical suction are another type of direct dynamic penetrating brain deformation which creates a focal cortical injury, without neither hemorrhage nor diffuse injury, by inserting a suction force to the intact dura via a plastic plunger of variable diameter. The main benefit is the reduction of neurochemical cascades due to reduced hemorrhaging. Disadvantages include the lack of an increase in intracranial pressure (ICP) which is characteristic of most TBI and the vacuum may need application for a longer duration than most TBI would sustain [30]. Such a model may be applicable to brain injury due to asphyxia and/or brain deformation without blood vessel damage. For instance, many infants are reported to sustain brain injury during birth. The use of medical forceps and vacuum extraction increase the occurrence of infant brain injury, both focal and diffuse.

3.6.1 Vacuum/Suction Procedures

1. Prior to preparing the animal for vacuum/suction injury, ensure the vacuum is set appropriately (e.g., 200–250 mm of mercury).
2. Place the animal in a stereotaxic frame secured by ear bars and incisor bar.
3. Carefully expose the skull and perform a craniotomy using a trephine drill or some other minidrill at predetermined stereotaxic coordinates.
4. Deflect the dura and remove the cortex underlying the craniotomy site with gentle suction through a fine glass Pasteur pipet. Suction should be continued until a predetermined depth.
5. Consider replacing the bone from the craniotomy and sealing with dental cement, or bone wax.
6. Close the scalp incision with sutures following the injury.

3.6.2 Vacuum/Suction Precautions

Safety concerns include the use of a scalpel for scalp resection, use of a drill for the craniotomy, and sutures for closing the scalp after the injury.

3.7 Zymosan

Other direct dynamic penetrating brain deformation models attempt to bypass the primary injury mechanism in order to model aspects of the secondary injury such as glial scarring and inflammation. Fitch [31] modeled such scarring with a zymosan (macrophage activator) solution delivered to the corpus callosum. Such a model creates cavitation without biophysical components typical of TBI. Not many studies appear to use this technique. One potential application of this model is the in vivo research regarding the glial scarring process and enabling axons to regrow beyond the barrier created by the scar.

3.7.1 Zymosan Procedures

1. Prior to preparing the animal for zymosan injection, ensure the zymosan solution is prepared and loaded in the needle for injection to avoid delays in the procedure.
2. Place the animal in a stereotaxic frame secured by ear bars and incisor bar.
3. Carefully expose the skull and perform a craniotomy at predetermined stereotaxic coordinates by thinning the skull over the area to be injected. The area of thinned skull should be approximately 1 mm by 1 mm per injection site using a hand-held drill. Stop when the bone is very thin and the blood vessels in the dura become clearly visible. Note: Do not drill through the bone to avoid unintended injury to the surface of the brain parenchyma. Keep both the skull and exposed dura moist with saline.
4. Carefully perforate the thinned area of skull with a small needle (e.g., 27 G).
5. Carefully flip up and remove the thinned bone with fine-tipped forceps.
6. Align the craniotomy site with the tip of the stereotaxic injector.
7. Lower the injector tip until it touches the exposed dura. If necessary, cut a small incision in the dura with a 27-G needle held at a flat angle to avoid unanticipated injury.
8. After penetrating the dura, slowly lower the injector tip to the predetermined z coordinate.
9. Slowly and consistently apply pressure to the injector syringe to inject the predetermined volume of zymosan solution.
10. To avoid backflow of the zymosan solution, wait for 2–3 min before withdrawing the injector.
11. Clean the injection site with moist cotton swabs.
12. Consider replacing the bone from the craniotomy and sealing with dental cement, or bone wax.
13. Close the scalp incision with sutures following the injury.

3.7.2 Zymosan Precautions

Safety concerns include use of a scalpel for scalp resection, use of a drill for the craniotomy, use of a syringe to deliver the solution, and sutures for closing the scalp after the injury.

3.8 Cryolesion

Focal cortical cryolesion is a direct dynamic model of TBI which causes a focal penetrating brain deformation using extreme cold [31]. To avoid the spread of injected liquid nitrogen in the freezing process, a pre-cooled probe chilled in liquid nitrogen is applied to the brain. This model is usually used when the primary mechanism of the brain injury is not the primary aim of the study. Instead cryolesion is used to induce local histopathological damage consistent with TBI. This particular model of TBI is effective in very small organisms (infants) where the other TBI models available may not otherwise allow for survival.

3.8.1 Cryolesion Procedures

1. Place the animal in a stereotaxic frame secured by ear bars and incisor bar.
2. Carefully expose the skull and perform a craniotomy using a trephine drill or some other minidrill at predetermined stereotaxic coordinates.
3. Precool a metal cylinder in liquid nitrogen.
4. Stereotactically apply the precooled metal cylinder to the predetermined region of the brain for a predetermined amount of time. Care should be taken to avoid unanticipated injury due to too much pressure applied to the precooled cylinder. If the duration of cryolesion is long, the cylinder should be filled with liquid nitrogen to ensure it remains cold for the entire duration of the injury.
5. Consider replacing the bone from the craniotomy and sealing with dental cement, or bone wax.
6. Close the scalp incision with sutures following the injury.

3.8.2 Cryolesion Precautions

Safety concerns include use of a scalpel for scalp resection, use of a drill for the craniotomy, use of liquid nitrogen may cause burns, and sutures for closing the scalp after the injury.

3.9 Stab Wounds

Stab wounds are easily modeled in many organisms, most popularly in the rodent [32]. The cortical stab injury (CSI) model is homologous to stab TBI in humans, but is seldom used due to the relatively low incidence of stab TBI [The main cause of TBI (non-military) is automobile accident]. Some studies may opt to skip the craniotomy, thus reproducing the effect of stabbing through the skull. The main use of this model appears to be in emergency medicine where careful removal of the blade to reduce hemorrhage is considered paramount for acute TBI management.

3.9.1 Stab Procedures

1. Place the animal in a stereotaxic frame secured by ear bars and incisor bar.
2. Carefully expose the skull and perform a craniotomy using a trephine drill or some other minidrill at predetermined stereotaxic coordinates.
3. Implant and cement a guide cannula at predetermined stereotaxic coordinates using a 22 G needle.

4. At a predetermined time after implantation of the guide cannula, re-anesthetize the animal and thin the skull bone located at the appropriate distance from the guide cannula with a hand-held drill and stereotactically insert a sterilized razor blade to a predetermined depth. Remove the blade after a predetermined amount of time.
5. Remove the guide cannula.
6. Consider replacing the bone from the craniotomy and sealing with dental cement, or bone wax.
7. Close the scalp incision with sutures following the injury.

3.9.2 Stab Precautions

Safety concerns include the use of a scalpel for scalp resection, use of a drill for the craniotomy, use of high-velocity sharp objects, and sutures for closing the scalp after the injury.

3.10 Epidural Mass Lesions (EML)

Epidural mass lesions (EML) are produced with variable inflation and reperfusion of a latex balloon typically inserted epidurally [33]. Such a model mimics the physical effects of hemorrhage and inflammation without the primary mechanism of injury or secondary neurochemical cascades associated with inflammation (e.g., cytokines). A specific application of this model is in bicycle crashes that result in middle meningeal artery transection. A famous case for potential EML application is in the death of Liam Neeson's wife, Natasha Richardson, who likely suffered from "walk and die" syndrome [walk and die" syndrome is typically due to delayed epidural bleeding] due to a skiing accident in March of 2009.

3.10.1 EML Procedures

1. Place the animal in a stereotaxic frame secured by ear bars and incisor bar.
2. Carefully expose the skull and perform a craniotomy using a trephine drill or some other minidrill at predetermined stereotaxic coordinates.
3. Stereotactically inject an uninflated latex balloon with a mitered valve using a microcatheter assembly. Do NOT penetrate the dura.
4. Carefully expand the latex balloon to a predetermined pressure in a predetermined amount of time, using air or water supplied via a variable pressure waveform generator.
5. Carefully deflate the latex balloon at a predetermined rate. Typically, the rate for inflation of the balloon is slower and the duration of maintaining the inflation longer when investigating EML than it is for PBBI.
6. Immediately retract the probe with the latex balloon.
7. Consider replacing the bone from the craniotomy and sealing with either dental cement or bone wax.
8. Close the scalp incision with sutures following the injury.

3.10.2 EML Precautions

Safety concerns include the use of a scalpel for scalp resection, use of a drill for the craniotomy and sutures for closing the scalp after the injury.

3.11 High-Velocity Missile (HVM)

High-velocity missile (HVM) models are also referred to as craniocerebral missile injuries (CMI) and use actual ballistics fired into the skull at varying distances, while the animal is either constrained or unconstrained. This model often utilizes larger organisms such as cats [34, 35] and sheep [36] to mimic the anatomical alterations due to large military weapons. The HVM model can vary in caliber of the firearm used, distance (range) of fire, and direction of impact (entrance and exit). Anterior to posterior paths cause unilateral damage whereas transverse impacts cause bilateral damage. Unfortunately, the size of the animals must be large due to the large size of the ballistics and resulting cavitation. Resulting hemorrhage may limit subsequent molecular monitoring with increased mortality.

3.11.1 HVM Procedures

1. Place the animal in a stereotaxic frame secured by ear bars and incisor bar (if constrained).
2. Stereotactically discharge a firearm at predetermined stereotaxic coordinates.
3. Consider replacing any bone displaced by the ballistic and sealing with dental cement, or bone wax.
4. Close the scalp incision with sutures following the injury.

3.11.2 HVM Precautions

Safety concerns include the use of a controlled weapon and sutures for closing the scalp after the injury. It is necessary to perform any experiments involving ballistics in a controlled range approved for animal use.

3.12 Penetrating Ballistic-Like Brain Injury (PBBI)

Penetrating ballistic-like brain injury (PBBI) [3] is the attempt to solve the limitations of the HVM models by using inflatable stereotaxic implant probes versus actual ballistics. The probe is smaller, but proportional to the ballistics used in HVM models to allow for use in rodents. The inflatable bladder is adjustable in time and percent. For example, 12 % inflation creates a larger cavity than 5 % inflation. Some researchers opt to leave the skull intact thus replicating the effect of the ballistics penetrating the skull on the injury site. A recent application of both the HVM and PBBI models is in understanding the case of Gabrielle Giffords, a Democratic congresswoman who was shot in the head by Jared Laughner. Medical examiners and investigators can apply HVM and PBBI research to aid in diagnosing the specific path of the ballistics, type of ballistics, and resulting pathology.

3.13 PBBI Procedures

1. Place the animal in a stereotaxic frame secured by ear bars and incisor bar.
2. Carefully expose the skull and perform a craniotomy using a trephine drill or some other minidrill at predetermined stereotaxic coordinates.

3. Stereotactically inject an uninflated latex balloon with a mitered valve using a microcatheter assembly.
4. Carefully expand the latex balloon to a predetermined pressure in a predetermined amount of time using air or water supplied via a variable pressure waveform generator.
5. Carefully deflate the latex balloon at a predetermined rate. Typically, the rate for inflation of the balloon is faster and the duration of maintaining the inflation shorter when investigating PBBI than it is for EML.
6. Immediately retract the probe with latex balloon.
7. Consider replacing the bone from the craniotomy and sealing with dental cement, or bone wax.
8. Close the scalp incision with sutures following the injury.

3.14 PBBI Precautions

Safety concerns include the use of a scalpel for scalp resection, use of a drill for the craniotomy, use of high-velocity blunt or sharp objects, and sutures for closing the scalp after the injury.

3.15 Concussion

Controlled concussion is similar to CCI without a craniotomy. Since the impact is to the skull, the injury is not as well localized and often results in subsequent brainstem damage. Many models fail to correlate impact with severity [6, 37].

3.15.1 Concussion Procedures

1. Prior to preparing the animal for the impact, ensure the settings are appropriate (e.g., dwell time, duration, and velocity).
2. Leave the animal unconstrained with its head on the customized cushion.
3. Carefully expose the skull and secure the plate or cone to the predetermined location on the skull.
4. Deliver the impact(s) to the intact skull via the plate or cone.
5. Close the scalp incision with sutures following the injury.

3.15.2 Concussion Precautions

Safety concerns include the use of a scalpel for scalp resection, use of high-velocity blunt objects, and sutures for closing the scalp after the injury.

3.16 Unconstrained Impact Acceleration

Unconstrained primate and ovine impact acceleration models use a piston impactor of variable weight to inflict head motion similar to that of constrained controlled concussion. The rapid acceleration and deceleration of the head allow for biophysical evaluation of the injury, but are less controllable since the head is free to move rotationally and laterally. This head movement creates the characteristic diffuse axonal injury but is difficult to reproduce unless the animal is at least partially restrained. When partially restrained (pseudo-constrained), at the neck, thus only allowing head movement, the animals head can only move in one dimension. Application of this model lies primarily in transportation-related acceleration and deceleration injuries.

3.16.1 Unconstrained Impact Acceleration Procedures

1. Prior to preparing the animal for the impact, ensure that the settings are appropriate (e.g., weight and height of drop).
2. Leave the animal unconstrained with its head on the customized cushion. Typically, unconstrained impact acceleration delivers an impact more laterally, whereas the weight drop impact is typically more along the midline.
3. Carefully expose the skull and secure the plate or cone to the predetermined location on the skull.
4. Deliver the weight drop to the intact skull, or dura if penetrating, with a predetermined weight from a predetermined height.
5. Consider replacing the bone from the craniotomy and sealing with dental cement, or bone wax if investigating a penetrating injury.
6. Close the scalp incision with sutures following the injury.

3.16.2 Unconstrained Impact Acceleration Precautions

Safety concerns include the use of a scalpel for scalp resection and sutures for closing the scalp after the injury.

3.17 Weight Drop

Marmarou's weight drop model [38] is an example of a pseudo-constrained acceleration model. In this case, a variable weight is dropped from a variable height through a Plexiglas tube to impact with the skull of an animal placed on a foam pad. The tube must be slightly larger than the weight which allows for uncontrolled lateral movement of the weight prior to impact. Some argue that additional variability is due to the weight rebounding after the initial drop thus creating a double injury/impact [4]. However, this rebounding effect can be countered with a barrier that restricts the second impact—or bounce. This model is comparable to the controlled concussion in application, but is less controllable.

Cernak's model [39] attempts to solve the limitations of the Marmarou model [38, 40] by using a pneumatic driven steel impactor that is laser guided to a steel disc cemented to the animal's skull. The pneumatic driven impactor is similar to the CCI and is controllable in velocity, dwell time and depth. The force is monitored via software, ensuring accurate correlation of injury severity and delivered impact while still replicating clinical signs and symptoms.

3.17.1 Weight Drop Procedures

1. Prior to preparing the animal for the weight drop, ensure that the settings are appropriate (e.g., weight and height of drop).
2. Place the animal in a stereotaxic frame secured by ear bars and incisor bar, or leave unconstrained.
3. Carefully expose the skull and secure the plate or cone to the predetermined location on the skull. If investigating a penetrating

injury, perform a craniotomy using a trephine drill or some other minidrill at predetermined stereotaxic coordinates. Note: It is helpful to align the impactor with the craniotomy site prior to removal of the bone.

4. Deliver the weight drop to the intact skull, or dura if penetrating, with a predetermined weight from a predetermined height.
5. Consider replacing the bone from the craniotomy and sealing with dental cement, or bone wax if investigating a penetrating injury.
6. Close the scalp incision with sutures following the injury.

3.17.2 Weight Drop Precautions

Safety concerns include the use of a scalpel for scalp resection, use of high-velocity blunt objects, and sutures for closing the scalp after the injury. In addition to the safety concerns of the Marmarou's model, safety concerns of Cernak's model include use of a laser which may cause eye damage.

3.18 Pneumatic Blast

3.18.1 Pneumatic Blast Procedures

1. Prepare the shock tube to the predetermined shockwave and overpressure parameters (*see* **Note 8**).
2. Place the animal within or at a predetermined distance from the shock tube opening. Consider placing the animal at 20–40° lateral to the shock tube axis to avoid unanticipated effects due to the exhaust gasses, or blast wind.
3. Consider constraining the head laterally and inferiorly to prevent head rotational acceleration-induced injury.
4. Consider placing a metal cylinder around the body to limit shockwave overpressure exposure of the torso.
5. Deliver the shockwave.

3.18.2 Pneumatic Blast Precautions

Safety concerns include the use of high pressure air and cut metal membranes which result in sharps. Besides the blast overpressure, the pneumatic blast causes a loud noise capable of disrupting hearing in unprotected ears.

3.19 Live Blast

3.19.1 Live Blast Procedures

1. Prepare the blast tube either with the predetermined explosive charge parameters, or in an explosive range (*see* **Note 8**).
2. Place the animal within or at a predetermined distance from the blast tube opening, or explosive charge. Consider placing the animal at 20–40° lateral to the blast tube axis to avoid unanticipated effects due to the exhaust gasses, or blast wind.
3. Consider constraining the head and neck to prevent head rotational acceleration-induced injury.
4. Consider placing a metal cylinder around the body in order to limit shockwave overpressure exposure of the torso.
5. Detonate the explosive charge.

3.19.2 Live Blast Precautions

1. Safety concerns include use of live explosives. Besides the blast overpressure, the live blast causes a loud noise capable of disrupting hearing in unprotected ears.

3.20 LISW Blast

3.20.1 LISW Blast Procedures

1. Prepare the laser according to the manufacturer's instructions and using the predetermined parameters (wavelength, pulse width) based on the animal and specific aims of your study.
2. Place the animal in a stereotaxic frame secured by ear bars and incisor bar unless the LISW is being inflicting to the ear in which case the animal will need to be secured manually or with a custom restraint.
3. Carefully expose the skull and perform a craniotomy using a trephine drill or some other minidrill at predetermined stereotaxic coordinates (if necessary based on the location of the LISW).
4. For underwater microexplosion LISW [22], fill the truncated ellipsoidal reflector with physiological saline and gently secure against the brain tissue within the craniotomy with the scalp. Care should be taken to avoid unanticipated injury due to too much pressure applied to the truncated ellipsoidal reflector. Deliver the predetermined LISW (single, or multiple pulses).
5. For elastic targeted LISW [23, 24], gently place the black rubber against the brain tissue within the craniotomy. On top of the black rubber, smear a thin layer of ultrasound gel. On top of the ultrasound gel, place the PET. Care should be taken to avoid unanticipated injury due to too much pressure applied to the elastic laser target. Deliver the predetermined LISW (single, or multiple pulses).
6. Gently remove the underwater microexplosion apparatus, or the elastic target. Consider replacing the bone from the craniotomy and sealing with dental cement, or bone wax.
7. Close the scalp incision with sutures following the injury.

3.20.2 LISW Blast Precautions

Safety concerns include use of a scalpel for scalp resection, use of laser which may cause eye damage, and sutures for closing the scalp after the injury.

3.21 Conclusion

Ultimately, no single "gold standard" model of TBI exists and there is overwhelming overlap among the existing models. With the abundance of animal models of TBI and the incomplete characterization of each model, data comparison is complicated between models. Therefore, the use of multiple models and animals as well as common data elements is encouraged, to expedite characterization and to validate results. Although there are a number of limitations to each animal and each model, the use of multiple models and multiple animals in combination provides an opportunity to explore mass and skull thickness among other scaling effects. Each stakeholder is seeking answers to different questions that require different approaches. No one model suffices to fully characterize every type of human TBI possible. Moreover, ethics are driving computerized models to replace animal models. Thus, one day we may rely on a virtual human to conduct our research rather than animal models.

4 Notes

1. The lighting in the room is an important factor to consider while conducting all procedures to avoid shadowing and allow for illumination of the subject.
2. Sterile techniques before, during, and after each procedure are critical to avoiding secondary effects of infection which can interfere with the interpretation of results.
3. Use of pulse oximetry is essential to maintaining a controlled experiment involving animal models. A specific range of physiological parameters needs to be set and adhered to based upon the animal chosen. Should the specific physiological range be exceeded, the animal should be omitted from the study and the rationale should be formally documented.
4. It is advisable to use the same surgeon throughout a study to maintain consistency. Should a study involve administration of a treatment protocol while surgeries are still occurring, a second researcher should be involved to allow the surgeon to continue uninterrupted. The second researcher should also remain the same, for consistency in treatment administration.
5. Consideration to physiological rhythms must be factored into planning procedures to ensure that observed effects are due to the procedures and not due to naturally occurring biological rhythms.
6. Animals should not be returned to the same cage with other animals if treating with strategies that can either transfer between animals or be tampered with by other animals.
7. Animals should be returned to recovery housing before being returned to original housing especially if sharing a cage with other animals to avoid startling any cage mates. Consideration should be given to isolating animals for the entire testing period to avoid acute changes to stress and anxiety-like behavior.
8. Aerodynamics and blast tubes characteristics must be fully delineated prior to application to any animal studies. Video-recording is highly recommended.

Disclaimer

The material presented in this chapter does not necessarily represent the views or opinions of Booz Allen Hamilton.

References

1. Ds S (1997) Biological Aspects of Disease. CRC Press, Netherlands
2. Pinel JPJ (2011) Biopsychology, 8th edn. Allyn & Bacon, Boston
3. Moshang E, L. G. (2002) A model of penetrating traumatic brain injury using air inflation technique. (Army, U. S. ed., U.S. Army Medical Research and Materiel Command, Ft Detrick, MD.
4. Cernak I (2005) Animal models of head trauma. NeuroRx 2:410–422
5. David S, Aguayo AJ (1985) Axonal regeneration after crush injury of rat central nervous system fibres innervating peripheral nerve grafts. Journal of neurocytology 14:1–12
6. Salzman SK, Faden AI (1994) The Neurobiology of central nervous system trauma. Oxford University Press, New York
7. Lab Animal's Buyer Guide. LabAnimal.
8. Povlishock JT, Hayes RL, Michel ME, McIntosh TK (1994) Workshop on animal models of traumatic brain injury. J Neurotrauma 11:723–732
9. Kilbaugh TJ, Lvova M, Karlsson M, Zhang Z, Leipzig J, Wallace DC, Margulies SS (2015) Peripheral Blood Mitochondrial DNA as a Biomarker of Cerebral Mitochondrial Dysfunction following Traumatic Brain Injury in a Porcine Model. PLoS One 10, e0130927
10. Sillesen M, Rasmussen LS, Jin G, Jepsen CH, Imam A, Hwabejire JO, Halaweish I, DeMoya M, Velmahos G, Johansson PI, Alam HB (2014) Assessment of coagulopathy, endothelial injury, and inflammation after traumatic brain injury and hemorrhage in a porcine model. J Trauma Acute Care Surg 76:12–19, discussion 19-20
11. Hwabejire JO, Jin G, Imam AM, Duggan M, Sillesen M, Deperalta D, Jepsen CH, Lu J, Li Y, deMoya MA, Alam HB (2013) Pharmacologic modulation of cerebral metabolic derangement and excitotoxicity in a porcine model of traumatic brain injury and hemorrhagic shock. Surgery 154:234–243
12. Glass TF, Fabian MJ, Schweitzer JB, Weinberg JA, Proctor KG (2001) The impact of hypercarbia on the evolution of brain injury in a porcine model of traumatic brain injury and systemic hemorrhage. J Neurotrauma 18:57–71
13. Ross AH, Jantz RL, McCormick WF (1998) Cranial thickness in American females and males. Journal of forensic sciences 43:267–272
14. Smith DF, Jensen PN, Gee AD, Hansen SB, Danielsen E, Andersen F, Saiz PA, Gjedde A (1997) PET neuroimaging with [11C]venlafaxine: serotonin uptake inhibition, biodistribution and binding in living pig brain. European neuropsychopharmacology : the journal of the European College of Neuropsychopharmacology 7:195–200
15. Roberts T, McGreevy P, Valenzuela M (2010) Human induced rotation and reorganization of the brain of domestic dogs. PLoS One 5, e11946
16. Russow LM, Theran P (2003) Ethical issues concerning animal research outside the laboratory. ILAR J 44:187–190
17. McSherry GM (1984) Mapping of cortical histogenesis in the ferret. J Embryol Exp Morphol 81:239–252
18. Gennarelli TA, Adams JH, Graham DI (1981) Acceleration induced head injury in the monkey. I. The model, its mechanical and physiological correlates. Acta Neuropathol Suppl 7:23–25
19. Adams JH, Graham DI, Gennarelli TA (1981) Acceleration induced head injury in the monkey. II. Neuropathology. Acta Neuropathol Suppl 7:26–28
20. Smith DH, Chen XH, Xu BN, McIntosh TK, Gennarelli TA, Meaney DF (1997) Characterization of diffuse axonal pathology and selective hippocampal damage following inertial brain trauma in the pig. J Neuropathol Exp Neurol 56:822–834
21. USAM Command (1997) Noise limits for military materiels. USAM Command, Redstone Arsenal, AL
22. Nakagawa A, Fujimura M, Kato K, Okuyama H, Hashimoto T, Takayama K, Tominaga T (2008) Shock wave-induced brain injury in rat: novel traumatic brain injury animal model. Acta Neurochir Suppl 102:421–424
23. Kurioka T, Matsunobu T, Niwa K, Tamura A, Kawauchi S, Satoh Y, Sato S, Shiotani A (2014) Characteristics of laser-induced shock wave injury to the inner ear of rats. J Biomed Opt 19:125001

24. Sato S, Kawauchi S, Okuda W, Nishidate I, Nawashiro H, Tsumatori G (2014) Real-time optical diagnosis of the rat brain exposed to a laser-induced shock wave: observation of spreading depolarization, vasoconstriction and hypoxemia-oligemia. PLoS One 9, e82891
25. Bauder AR, Ferguson TA (2012) Reproducible mouse sciatic nerve crush and subsequent assessment of regeneration by whole mount muscle analysis. J Vis Exp 60:pii 3606
26. Vink R, Mullins PG, Temple MD, Bao W, Faden AI (2001) Small shifts in craniotomy position in the lateral fluid percussion injury model are associated with differential lesion development. J Neurotrauma 18:839–847
27. Gennarelli TA (1994) Animate models of human head injury. J Neurotrauma 11:357–368
28. D'Ambrosio R, Fender JS, Fairbanks JP, Simon EA, Born DE, Doyle DL, Miller JW (2005) Progression from frontal-parietal to mesial-temporal epilepsy after fluid percussion injury in the rat. Brain 128:174–188
29. Kharatishvili I, Nissinen JP, McIntosh TK, Pitkanen A (2006) A model of posttraumatic epilepsy induced by lateral fluid-percussion brain injury in rats. Neuroscience 140:685–697
30. Mathew P, Bullock R, Graham DI, Maxwell WL, Teasdale GM, McCulloch J (1996) A new experimental model of contusion in the rat. Histopathological analysis and temporal patterns of cerebral blood flow disturbances. J Neurosurg 85:860–870
31. Fitch MT, Doller C, Combs CK, Landreth GE, Silver J (1999) Cellular and molecular mechanisms of glial scarring and progressive cavitation: in vivo and in vitro analysis of inflammation-induced secondary injury after CNS trauma. J Neurosci 19:8182–8198
32. Ghirnikar RS, Lee YL, He TR, Eng LF (1996) Chemokine expression in rat stab wound brain injury. J Neurosci Res 46:727–733
33. Burger R, Bendszus M, Vince GH, Roosen K, Marmarou A (2002) A new reproducible model of an epidural mass lesion in rodents. Part I: characterization by neurophysiological monitoring, magnetic resonance imaging, and histopathological analysis. J Neurosurg 97: 1410–1418
34. Carey ME (1995) Experimental missile wounding of the brain. Neurosurg Clin N Am 6:629–642
35. Carey ME, Sarna GS, Farrell JB (1990) Brain edema following an experimental missile wound to the brain. J Neurotrauma 7: 13–20
36. Finnie JW (1993) Pathology of experimental traumatic craniocerebral missile injury. J Comp Pathol 108:93–101
37. Lighthall JW, Dixon CE, Anderson TE (1989) Experimental models of brain injury. J Neurotrauma 6:83–97
38. Marmarou A, Foda MA, van den Brink W, Campbell J, Kita H, Demetriadou K (1994) A new model of diffuse brain injury in rats. Part I: pathophysiology and biomechanics. J Neurosurg 80:291–300
39. Cernak I, Vink R, Zapple DN, Cruz MI, Ahmed F, Chang T, Fricke ST, Faden AI (2004) The pathobiology of moderate diffuse traumatic brain injury as identified using a new experimental model of injury in rats. Neurobiol Dis 17:29–43
40. Foda MA, Marmarou A (1994) A new model of diffuse brain injury in rats. Part II: morphological characterization. J Neurosurg 80: 301–313

Chapter 6

Methods of Drug Delivery in Neurotrauma

Ying Deng-Bryant, Ryan Readnower, Lai Yee Leung, Frank Tortella, and Deborah Shear

Abstract

The central nervous system (CNS) is protected by blood–brain barrier (BBB) and blood-cerebrospinal-fluid (CSF) barrier that limit toxic agents and most molecules from penetrating the brain and spinal cord. However, these barriers also prevent most pharmaceuticals from entering into the CNS. Drug delivery to the CNS following neurotrauma is complicated. Although studies have shown BBB permeability increases in various TBI models, it remains as the key mitigating factor for delivering drugs into the CNS. The commonly used methods for drug delivery in preclinical neurotrauma studies include intraperitoneal, subcutaneous, intravenous, and intracerebroventricular delivery. It should be noted that for a drug to be successfully translated into the clinic, it needs to be administered preclinically as it would be anticipated to be administered to patients. And this likely leads to better dose selection of the drug, as well as recognition of any possible side effects, prior to transition into a clinical trial. Additionally, novel approach that is noninvasive and yet circumvents BBB, such as drug delivery through nerve pathways innervating the nasal passages, needs to be investigated in animal models, as it may provide a viable drug delivery method for patients who sustain mild CNS injury or require chronic treatments. Therefore, the focus of this chapter is to present rationales and methods for delivering drugs by IV infusion via the jugular vein, and intranasally in preclinical studies.

Key words Neurotrauma, Traumatic brain injury, Drug delivery, Intravenous, Intranasal

1 Introduction

Clinically, oral delivery is often preferred as it provides controlled and sustained release of drug actions with prolonged dosing intervals, which minimizes patient intervention. After the drug is ingested and disintegrated in the gastrointestinal (GI) track, it is absorbed prior to entering the bloodstream. However, the biochemical properties of the drug, and its permeability, stability, and metabolism in the biological system, can largely affect its absorption, thus leading to decreased bioavailability. In contrast, the most commonly used route of administration for preclinical drug studies for traumatic brain injury (TBI) has been intraperitoneal (IP) or subcutaneous (SubQ) delivery. These routes have been favored because of their technical ease. Pharmacokinetics (PK) of drugs delivered IP is

Firas Kobeissy et al. (eds.), *Injury Models of the Central Nervous System: Methods and Protocols*, Methods in Molecular Biology, vol. 1462, DOI 10.1007/978-1-4939-3816-2_6,

similar to those delivered orally because the drug is mainly absorbed into mesenteric vessels which empty into the hepatic portal system. The drug is then subjected to what is known as the first-pass effect which results in drug metabolism before the drug enters the systemic circulation that leads to reduced plasma drug concentrations. The SubQ route allows for drugs to be absorbed slowly which can achieve prolonged steady-state plasma concentrations of a drug and bypass the first-pass effect resulting in higher bioavailability. It should be noted that although these systemic routes are easier to implement, they carry the caveat of BBB permeability issues.

Alternatively, intracerebroventricular (ICV) or intrathecal (IT) delivery introduces drugs directly into the CSF, bypassing the BBB [1]. Drugs can then be exchanged freely between the CSF and the extracellular fluid of the brain parenchyma, which could result in high drug concentrations in the CNS that cannot be easily achieved by systemic therapy. These routes of administration can also circumvent systemic adverse effects in cases that drugs are highly toxic. However, drug diffusion through the brain parenchyma is very slow, and delivery through these routes risks inadequate CNS exposure due to the rapid turnover of the CSF. More critically, a major drawback of these methods is that they require invasive neurosurgical procedures, which is not practical in the clinical setting. Relevant to this, preclinical studies showing drug efficacy based on data through ICV/IT delivery have not been successful in gaining approval for clinical trials.

In the TBI patients, the most commonly used route of administration of a drug would be intravenous (IV) delivery. A well-practiced IV approach in modern medicine that also plays an important role in the clinics is the IV delivery through jugular catheterization. Using the indwelling jugular catheters, medications, fluid, blood products, and chemical agents (e.g., radiological agents) can be easily administered. Jugular catheterization can also provide critical IV access for trauma patients when the extremity veins are not attainable. In addition to that, the indwelling jugular catheter is a viable route for repeated serial blood sampling both in clinical and animal studies. As indicated in preclinical drug studies using small animals, indwelling jugular catheters have been shown to provide blood samples that are superior to other IV sampling methods [2–5]. However, how do we translate research from the laboratory using IP, SubQ, or ICV/IT route of administration into patients using an IV delivery route presents a conundrum in the current TBI research. While there is no straightforward answer, the most logical approach would be to select drugs which showed the most promise in the IP, SubQ, and ICV/IT studies, and then repeat these studies using the IV route of administration.

A point to emphasize here is the need for intensive PK profiling when switching routes of administration in preclinical studies. IV dosing will result in greater bioavailability, higher maximum concentration (Cmax), and shorter time to maximum concentration

(Tmax) than IP or SubQ. Another consideration is that often TBI patients may have altered drug metabolism compared to normal "healthy" patients which highlights the importance to perform the PK studies in injured and sham animals. For example, Hatton and colleagues reported that TBI patients receiving cyclosporine A displayed an increase in whole blood clearance rates and beta half-life compared to transplant and healthy patients [6]. As such, a viable repetitive blood sampling technique is important for successful preclinical drug studies for the treatment of TBI. Additionally, when dealing with serial bleeding within a relatively narrow time window, the welfare of the animals is to be considered as the first priority. One advantage of the use of indwelling jugular catheters for repeated blood collection is that it causes minimal pain and distress to the animal, and the collection can be done while the subject is awake and freely moving. Furthermore, blood sampling via jugular cannulation can be performed at precise time intervals which are important for PK and biomarker analysis. A well-performed jugular vein catheterization usually remains accessible for up to 2 weeks, during which time period a large volume of blood can be collected, and the lost blood volume can be easily replaced by administering normal saline using the same catheter.

An important observation following neurotrauma is that there is a breakdown of the BBB that would allow for a transient window for non-BBB-permeable drugs to gain entry into the CNS [7–10]. However, the precise timing of this BBB permeability window is not completely understood and regions more distal from the site of injury (penumbral tissue) are less likely to have overt BBB breakdown. Although IV dosing will increase bioavailability, it does not ensure that the drug will cross the BBB, which highlights the need for a delivery route that is noninvasive, yet capable of bypassing the BBB. One of such delivery method is intranasal (IN) delivery. This route of administration delivers drugs directly into the upper nasal passage, requiring no surgical procedures. Drugs then enter the brain and spinal cord mainly through the nerves connecting the nasal passages to the CNS. Therefore, the direct IN delivery of therapeutics to the CNS represents a viable, noninvasive method for bypassing the BBB. Since its first development by William H. Frey as a delivery method directly targeting the CNS in 1989, a growing body of evidence has shown therapeutic efficacy of IN delivery approach across injury and CNS disease platforms [11–17]. This novel delivery method supports the transport of peptides, proteins, and even neural stem cells directly to the brain parenchyma [18–21]. Most notably, growth factors with molecular weight over 20 kDa have been shown to enter the CNS directly via IN delivery [22]. Using the IN delivery method in animal models, researchers have successfully reduced stroke damage, reversed Alzheimer's neurodegeneration, reduced anxiety, and improved memory [19, 23–26].

The key component of the IN delivery route, the olfactory neuronal pathway, traverses the cribiform plate and is surrounded by perineurial spaces containing CSF that drains into the local lymphatic system [27]. Once administered intranasally, the therapeutics reaches the CNS mainly through extracellular bulk flow transport within the channels formed by olfactory sensory neurons, or intracellular axonal transport along nerve connections, involving mechanisms such as passive diffusion, receptor uptake, or endocytosis [28]. In addition to the transport through nerve pathways, other factors, to include nasal vasculature, CSF, and lymphatic system, also contribute to IN drug delivery. Collectively, a drug delivered intranasally can enter the CNS through more than one pathway, and the exact mechanism of how a drug is transported into the brain or spinal cord is relevant to the biochemical property of the drug, and how it is formulated for IN delivery. Therefore, the careful design of administration techniques and drug formulation can significantly impact the therapeutic outcome. Additional consideration for IN drug delivery is the nasal cavity, such as the possibility of mucosal irritation and potential variability caused by nasal pathology. With regard to that, IN drug delivery method would not be feasible in TBI patients who sustained substantial mandibular-maxillary fracture concurrent with neurotrauma. However, the IN approach could be used for chronic therapy after the acute recovery phase is complete, attributing to its technical ease and non-requirement for any surgical procedures. In that regard, IN approach could represent a viable drug delivery method for treating concussions, which emphasizes the need for testing therapeutics using IN delivery in preclinical studies.

2 Materials

2.1 Animals

1. Male adult Sprague Dawley rats (280–320 g; Charles River Labs, Raleigh, VA, USA) were used in these experiments.
2. Animals were housed individually under a 12-h light/dark cycle in a facility accredited by the Association for Assessment and Accreditation of Laboratory Animal Care International.

2.2 Anesthesia

2.2.1 IV Procedure

1. Induction: 4.0 % Isoflurane.
2. Maintenance: 1.5–2 % Isoflurane delivered in breathing air/oxygen mixture (fraction of inspired oxygen $F_iO_2 = 0.26$).

2.2.2 IN Procedure

1. Animals will be induced and maintained with isoflurane delivered in breathing air/oxygen mixture (fraction of inspired oxygen $F_iO_2 = 0.30$).

2.3 Materials for Operating Procedure

2.3.1 IV Procedure

1. Homeothermic heating system (Harvard Apparatus, MA, USA).
2. Stereotaxic frame with dual manipulators (David Kopf Instruments, Model 902).
3. Isoflurane vaporizers (Matrix™ VIP 3000) and isoflurane (Fluriso™ 502017).
4. Rodent clippers (Oster Golden A5).
5. Cotton-tipped applicator and gauze (Covidien Curity, Catalog no. 8884540500), and alcohol pads (VWR Scientific, Catalog no. 82003-856).
6. Cauterizer (Gemini Cautery Kit, Catalog no. 726067).
7. Heparin saline (20 USP units/ml—sterile) (heparin: Sagent Pharmaceuticals, NDC#25021-400-30; 0.85 % saline: RICCA Chemical Company, Catalog no. 7200-1).
8. Sterile water for injection, USP 100 ml (VWR Scientific, Catalog no. 68099-186).
9. Local anesthesia (1 % lidocaine) (MWI/VetOne, Catalog no. 510213).
10. Antibiotic ointment, 1 oz. tube (Fisher Scientific, Catalog no. 19-082-795).
11. Catheters for cannulation (PE50) (VWR Scientific, Catalog no. 63019-047).
12. Silastic tubing 0.020×0.037 (Fisher Scientific, Catalog no. 11-189-15A).
13. Chloroform for the assembly of IV cannula (Sigma Aldrich, Catalog no. 650471).
14. FS-2 silk suture (Fisher Scientific, NC9201704).
15. Sterile bone wax (Ethicon, W31G).
16. Syringes (1 ml) (Beckson, Dickson and Company, Catalog no. 309628).
17. Sterile surgical instruments (fine-tip forceps, scissors, micro-scissors, wound clip applier, microclips, hemostats, blunt-tip needle, etc.) (Biomedical Research Instruments, Inc.).

2.3.2 IN Procedure

1. Homeothermic heating system (Harvard Apparatus, MA, USA).
2. Nose breather (Harvard Apparatus, Catalog no. 723026).
3. Plexiglas rat body holder (Harvard Apparatus, Catalog no. 500108).
4. Parafilm (Fisher Scientific, Catalog no. 13-374-10).
5. Cotton-tipped applicator, and gauze (Covidien Curity, Catalog no. 8884540500).
6. Timer (Fisher Scientific, Catalog no. 06-662-44).
7. Pipette and tips (Fisher Scientific).

3 Methods

3.1 IV Preoperative Procedure

1. Sterilize all surgical materials via autoclave or vaporization with hydrogen peroxide.
2. Assembly of IV cannula: Pre-soak 3.1 cm long segments of 0.020 × 0.037 silastic tubing in chloroform in a glass vial for approximately 5 min under the fume hood. Prepare 13.0 cm long segments of PE50 tubing with perpendicular cuts to avoid bevels. Use the fine-tip forceps to pick up a segment of silastic, tap it on a piece of gauze to remove access chloroform, and insert one end into the PE50 tubing until they overlap by 1.0 cm. Let rest untouched until the silastic leader dries. Re-measure the silastic leader and trim it down to 2.9 cm long for a 300 g rat (*see* **Notes 1** and **2**; Fig. 1a and Fig. 2).

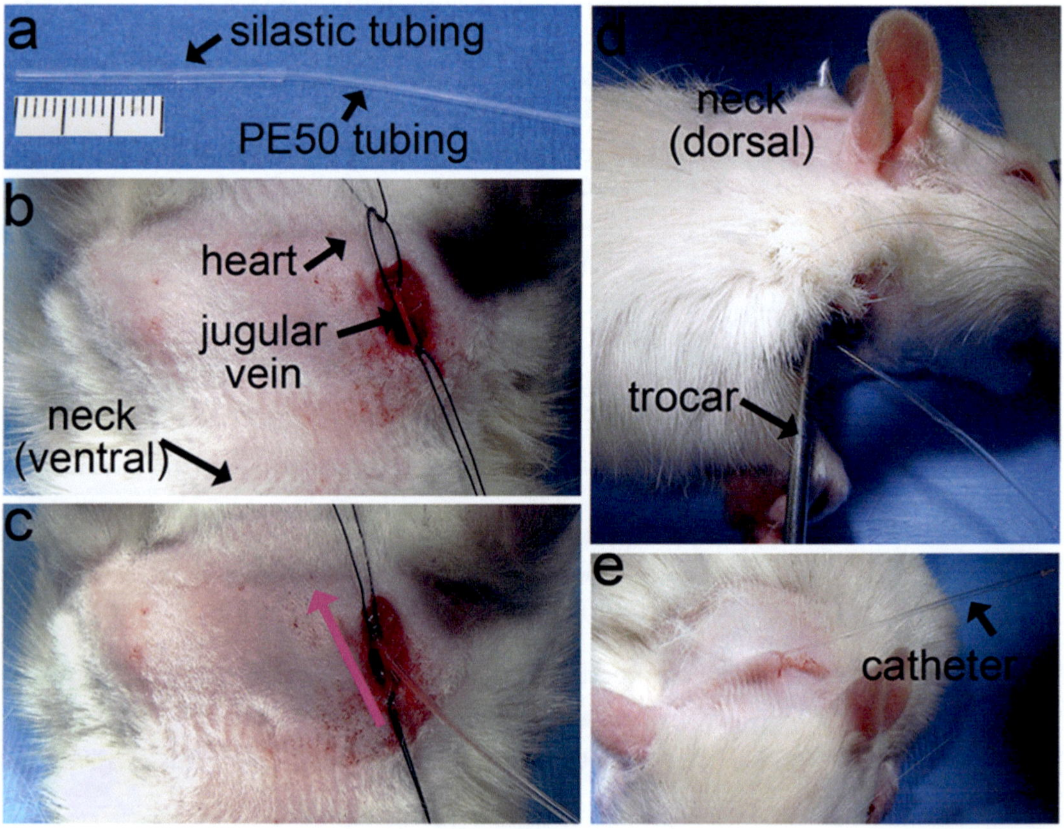

Fig. 1 The IV cannulation via jugular vein. The IV cannula is constructed using a silastic tubing and a PE50 tubing (**a**). Tie two silk sutures around the exposed external jugular vein, one loose and one tight, leaving approximately 1 cm space between the two ligatures (**b**). Insert the silastic leader of the IV cannula into the jugular vein until it passes approximately 1 mm beyond the loose suture knot (**c**). Use a trocar to guide the distal end of the IV catheter subcutaneously around the right side of the body, and to exit at the dorsal neck region (**d**, **e**)

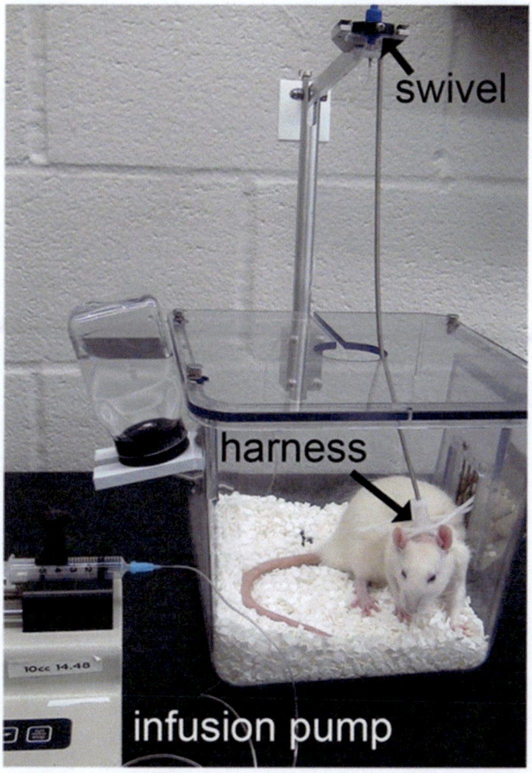

Fig. 2 While conducting continuous IV infusion, one end of the IV catheter will be connected to the infusion pump, and the other end will be connected to a swivel via a tether, and secured onto the rat through a harness, allowing free movement during dosing

3. Fill up 1 ml syringe with heparin saline, and attach the syringe to the PE50 leader of the IV cannula.
4. Turn on heating blanket for maintaining body temperature.
5. Place the rat in an isoflurane chamber connected to the vaporizer for anesthetization.
6. Shave hair over the dorsal head skin, as well as the right inguinal region.
7. Prior to incision, clean the incision sites with swabs of alcohol pads.

3.2 IV Jugular Vein Cannulation

All surgery must be performed with aseptic procedures, including sterile gloves, masks, and aseptic techniques. Research personnel must be qualified and trained in all procedures (*see* **Notes 3** and **4**).

1. Place the rat on a heating pad in a dorsal recumbent position with the head towards the operator. Insert the animal head into a nose cone system connected to the isoflurane vaporizer. Adjust the isoflurane to 2 %.

2. Verify that the animal is completely anesthetized by pinching the foot pad.
3. Locate the right external jugular pulse rostral to the right clavicle, and make a 1 cm cutaneous longitudinal incision at the inguinal region. Remove subcutaneous fat and connective tissues.
4. Locate the external jugular bundle (vein, artery, nerve).
5. Isolate the jugular vein from the bundle using a fine-tip forceps.
6. Using the forceps carefully remove superficial layer of connective tissue surrounding the exposed external jugular between the junction with the cephalic vein and the rostral boundary of the incision site.
7. Place two approximately 9 cm length silk suture underneath the external jugular. Gently and firmly tie one with double knots to occlude the jugular vein anteriorly, and loosely tie the other around the jugular vein near the heart, leaving approximately 1 cm space between the two ligatures (Fig. 1b).
8. Use fine-tip forceps to grasp the external jugular vein, draw it slightly toward yourself, and put a small amount of tension on the vein to temporally halt any blood flow while using microscissors to make a small incision on the vein between the two ligatures (preferably towards the distal end).
9. Insert the silastic leader of the IV cannula which is connected to a syringe filled with heparin saline for venous cannulation. Pass the silastic leader approximately 1 mm beyond the loose suture knot and stop (Fig. 1c).
10. Tie the loose suture knot with gentle tension around the vein to fasten the catheter, and withdraw the cannula until the edge of the silastic leader anchors against the suture tie.
11. Draw back on the syringe attached to the PE50 leader of the IV cannula until a small amount of blood draws up into the cannula, and then push syringe back until no blood is visible in the cannula.
12. Detach the syringe from the IV cannula and quickly seal the end with bone wax before any leak.
13. Guide the distal end of the catheter subcutaneously via a trocar, around the right side of the body, and to exit at the dorsal neck region (Fig. 1d, e).
14. Close the incision wound with sterile wound clips (*see* **Note 3**).

3.3 IN Preoperative Procedure

1. Roll a few pieces of 2 × 2 gauze into a bundle to use as a pillow for positioning the rat head.
2. Wrap parafilm around the cotton-tipped applicator for temporary nostril occlusion.
3. Turn on the electrical heating pads to achieve 37 °C.

3.4 IN Procedure

1. Place the rat in the anesthesia induction chamber with ~2–4 % isoflurane in a carrier gas of 70 % air with 30 % oxygen flowing at a rate of approximately 2–4 l/min until it achieves normal breathing rate.
2. Remove the rat from the induction chamber, and verify that the animal is completely anesthetized by pinching the foot pad.
3. Place the rat on a Plexiglas rat body holder in a dorsal recumbent position with the tail towards the operator. Snuggly fit the nose with the nose breather to ensure constant anesthetic vapor (Fig. 3a).
4. Place a rolled 2 × 2 in. gauze pillow under the neck to maintain a horizontal placement of the head, critical for accurately targeting the cribriform plate (Fig. 3b).
5. Remove the rat away from the nose breather ONLY for a brief period when conducting IN delivery. Rats remain on isoflurane vapor throughout the dosing session until completion.
6. Draw therapeutics using pipette with appropriate pipette tips. The recommended volume per drop is 3–10 μl.
7. Place the pipette tip opposite the targeted nostril without touching.
8. Position the pipette at a 45° angle to the targeted nostril, gently press the pipette plunger to dispense a drop of therapeutics, and allow it to be naturally inhaled (sniffed up) by the rat (Fig. 4a).
9. While administering therapeutics to the targeted nostril, use the cotton-tipped applicator wrapped in parafilm to occlude the other nostril (Fig. 4b).

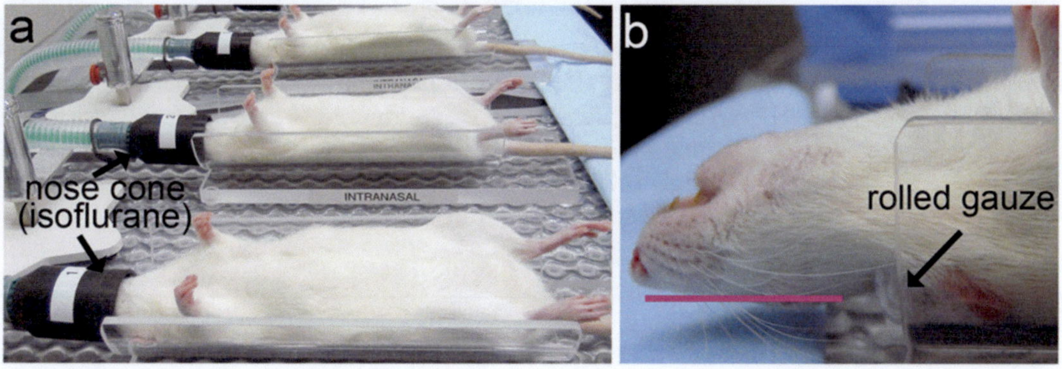

Fig. 3 Rats are placed on a Plexiglas body holder in a dorsal recumbent position with the tail towards the operator during IN procedure. The nose is snuggly fit and aligned with the nose breather to ensure constant anesthetic vapor (**a**). Rats will be positioned on their back with a rolled 2 × 2 in. section of gauze under the neck to maintain a horizontal placement of the head (**b**)

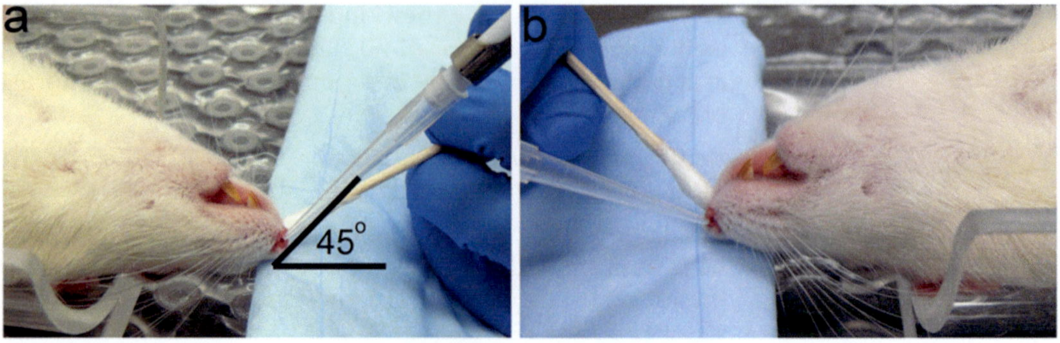

Fig. 4 During IN delivery, the therapeutic agent will be drawn up using a pipette and administered at a 45° angle to each nostril, alternating every 2 min until the full dose is delivered (**a**). A cotton swab wrapped in parafilm will be used to occlude one nostril while a drop of a therapeutic agent be placed onto the opposite nostril, and naturally inhaled (sniffed up) by the rat (**b**)

10. Alternate between the two nostrils with 2-min interval using a timer until the full dose is delivered.
11. The recommended dosing volume for a full session is 30–60 μl.

4 Notes

1. The assembly of IV cannula should be conducted under a fume hood due to chloroform vapors.
2. The length of the silastic leader of the IV cannula can vary due to the size of the rats: for rats weighing approximately 300 g, 2.8–2.9 cm is appropriate; the leader can be shorter for smaller rats (2.6–2.7 cm for approximately 250 g) or left longer for larger rats (3.0–3.1 cm for approximately 350 g or above).
3. Once the IV cannula is placed, it should be either used or flushed daily with 0.5 ml normal saline to avoid blockage of the cannula.
4. All procedures involving animal use must be reviewed and approved by the Institutional Animal Care and Use Committee (IACUC). Research is conducted in compliance with the Animal Welfare Act, Guide for the Care and Use of Laboratory Animals (National Research Council; 8th Edition, 2011), and other federal statutes and regulations.

Disclaimers

The views of the authors do not purport or reflect the position of the Department of the Army or the Department of Defense (para 4-3, AR 360-5). The authors declare that there are no conflicts of interest in this protocol. This research is funded by Combat Casualty Care Research Program.

References

1. Cook AM et al (2009) Intracerebroventricular administration of drugs. Pharmacotherapy 29(7):832–845
2. Thrivikraman KV, Huot RL, Plotsky PM (2002) Jugular vein catheterization for repeated blood sampling in the unrestrained conscious rat. Brain Res Brain Res Protoc 10(2):84–94
3. Florczyk AP, Schurig JE (1981) A technique for chronic jugular catheterization in the ferret. Pharmacol Biochem Behav 14(2):255–257
4. Bakar SK, Niazi S (1983) Simple reliable method for chronic cannulation of the jugular vein for pharmacokinetic studies in rats. J Pharm Sci 72(9):1027–1029
5. Bardelmeijer HA et al (2003) Cannulation of the jugular vein in mice: a method for serial withdrawal of blood samples. Lab Anim 37(3):181–187
6. Cook AM et al (2009) Cyclosporine A for neuroprotection: establishing dosing guidelines for safe and effective use. Expert Opin Drug Saf 8(4):411–419
7. Shen DD, Artru AA, Adkison KK (2004) Principles and applicability of CSF sampling for the assessment of CNS drug delivery and pharmacodynamics. Adv Drug Deliv Rev 56(12):1825–1857
8. de Lange EC, Danhof M (2002) Considerations in the use of cerebrospinal fluid pharmacokinetics to predict brain target concentrations in the clinical setting: implications of the barriers between blood and brain. Clin Pharmacokinet 41(10):691–703
9. Baldwin SA et al (1996) Blood-brain barrier breach following cortical contusion in the rat. J Neurosurg 85(3):476–481
10. Dhillon HS et al (1999) Severity of experimental brain injury on lactate and free fatty acid accumulation and Evans blue extravasation in the rat cortex and hippocampus. J Neurotrauma 16(6):455–469
11. Cunningham TL et al (2014) Correlations between blood-brain barrier disruption and neuroinflammation in an experimental model of penetrating ballistic-like brain injury. J Neurotrauma 31:505
12. Shear DA et al (2011) Severity profile of penetrating ballistic-like brain injury on neurofunctional outcome, blood-brain barrier permeability, and brain edema formation. J Neurotrauma 28(10):2185–2195
13. Benedict C et al (2004) Intranasal insulin improves memory in humans. Psychoneur oendocrinology 29(10):1326–1334
14. Parker KJ et al (2005) Intranasal oxytocin administration attenuates the ACTH stress response in monkeys. Psychoneuroendocri nology 30(9):924–929
15. De Rosa R et al (2005) Intranasal administration of nerve growth factor (NGF) rescues recognition memory deficits in AD11 anti-NGF transgenic mice. Proc Natl Acad Sci U S A 102(10):3811–3816
16. Gozes I, Divinski I (2007) NAP, a neuroprotective drug candidate in clinical trials, stimulates microtubule assembly in the living cell. Curr Alzheimer Res 4(5):507–509
17. Fliedner S, Schulz C, Lehnert H (2006) Brain uptake of intranasally applied radioiodinated leptin in Wistar rats. Endocrinology 147(5):2088–2094
18. Hashizume R et al (2008) New therapeutic approach for brain tumors: intranasal delivery of telomerase inhibitor GRN163. Neuro Oncol 10(2):112–120
19. Hanson LR, Frey WH 2nd (2007) Strategies for intranasal delivery of therapeutics for the prevention and treatment of neuroAIDS. J Neuroimmune Pharmacol 2(1):81–86
20. Hanson LR, Frey WH 2nd (2008) Intranasal delivery bypasses the blood-brain barrier to target therapeutic agents to the central nervous system and treat neurodegenerative disease. BMC Neurosci 9(Suppl 3):S5
21. Danielyan L et al (2009) Intranasal delivery of cells to the brain. Eur J Cell Biol 88(6):315–324
22. Hanson LR et al (2009) Intranasal deferoxamine provides increased brain exposure and significant protection in rat ischemic stroke. J Pharmacol Exp Ther 330(3):679–686
23. Fine JM et al (2012) Intranasal deferoxamine improves performance in radial arm water maze, stabilizes HIF-1alpha, and phosphorylates GSK3beta in P301L tau transgenic mice. Exp Brain Res 219(3):381–390
24. Chen XQ et al (1998) Delivery of nerve growth factor to the brain via the olfactory pathway. J Alzheimers Dis 1(1):35–44
25. Liu XF et al (2001) Non-invasive intranasal insulin-like growth factor-I reduces infarct volume and improves neurologic function in rats following middle cerebral artery occlusion. Neurosci Lett 308(2):91–94
26. Gomez D et al (2012) Intranasal treatment of neurodegenerative diseases and stroke. Front Biosci (Schol Ed) 4:74–89

27. Hanson LR et al (2012) Intranasal delivery of growth differentiation factor 5 to the central nervous system. Drug Deliv 19(3):149–154

28. Wolf DA et al (2012) Lysosomal enzyme can bypass the blood-brain barrier and reach the CNS following intranasal administration. Mol Genet Metab 106(1):131–134

Chapter 7

Bridging the Gap of Standardized Animals Models for Blast Neurotrauma: Methodology for Appropriate Experimental Testing

Pamela J. VandeVord, Alessandra Dal Cengio Leonardi, and David Ritzel

Abstract

Recent military combat has heightened awareness to the complexity of blast-related traumatic brain injuries (bTBI). Experiments using animal, cadaver, or biofidelic physical models remain the primary measures to investigate injury biomechanics as well as validate computational simulations, medical diagnostics and therapies, or protection technologies. However, blast injury research has seen a range of irregular and inconsistent experimental methods for simulating blast insults generating results which may be misleading, cannot be cross-correlated between laboratories, or referenced to any standard for exposure. Both the US Army Medical Research and Materiel Command and the National Institutes of Health have noted that there is a lack of standardized preclinical models of TBI. It is recommended that the blast injury research community converge on a consistent set of experimental procedures and reporting of blast test conditions. This chapter describes the blast conditions which can be recreated within a laboratory setting and methodology for testing in vivo models within the appropriate environment.

Key words Blast, Neurotrauma, Static pressure, Dynamic pressure, Advance blast simulators, Standard methods, Animal models

1 Introduction

Research efforts in blast neurotrauma have expanded significantly in the past decade. Blast-related traumatic brain injury (TBI) has been identified as one of the signature wounds which resulted from recent military conflicts. The Department of Defense provided TBI statistics which showed the increasing number of diagnoses since 2000. In 2011, there were 30,000 reported cases of TBI in America's armed forces, which is nearly three times the amount of reported cases in 2005. Various factors have caused the increasing number of TBI cases in veterans. First, the transition of warfare from traditional tactics, such as bullets and blunt force, to improvised explosive devices (IEDs) has greatly increased the odds of TBI affecting service men and women. About 72% of combat

Firas Kobeissy et al. (eds.), *Injury Models of the Central Nervous System: Methods and Protocols,* Methods in Molecular Biology, vol. 1462, DOI 10.1007/978-1-4939-3816-2_7, © Springer Science+Business Media New York 2016

injuries involve explosives in modern warfare [1]. Next, modern body armor and combat medicine improvements protect military personnel from injuries that would have been fatal in the past. Therefore the soldiers who died from other external injuries most likely also sustained TBI that went undiagnosed due to death. Lastly, the increased awareness of TBI has consequently increased the number of identified cases. Medical professionals are now more aware of TBI symptoms, and Congress has required military medical centers to screen all returning service members for TBI.

Blast TBI (bTBI) is a known possible consequence for both military personnel and civilians who are exposed to blast. Veterans returning from Operations Enduring and Iraqi Freedom, and New Dawn (OEF/OIF/OND), are suffering from closed head injuries due to blast overpressure (BOP) exposure, with prevalence ranging from 15 to 20% depending on the diagnostic criteria and patient population. Blast TBI is the second most cause of injuries from BOP, next to amputations. Studies of personnel returned from the field have shown that injuries from explosive munitions make up increasingly large percentages of casualties, increasing from 50% in the 1990s to more than 73% today [2]. Furthermore, nearly 4 out of 5 veterans suffering from TBI were classified as mild TBI (mTBI), making this injury the hallmark of recent conflicts. A study by Murray et al. reported that 88% of military personnel treated at a medical unit in Iraq were injured by IEDs, with at least 47% of these injuries involving the head, highlighting the importance of continuing the research efforts in this area [3]. To further complicate the injury, combat personnel can be exposed to multiple low-level blasts which could lead to the long-term sequelae [4]. The number of these individuals is increasing and poses a major long-term challenge to society. Thus, the incidence of TBI and psychological health issues in the current veteran population is at the forefront when planning for the long-term healthcare of our wounded soldiers [5].

Major concerns of blast-induced mTBI, from the clinical standpoint, include no outward signs of injury, delayed onset of symptoms, and overlapping symptoms with post-traumatic stress disorder (PTSD) and impact-related TBI. Hoge et al. reported that of 2525 soldiers, 4.9% reported an injury with loss of consciousness [6]. However, in the same cohort 10.3% reported injury with altered mental status, but no loss of consciousness, indicating that milder forms of TBI are more prevalent. Additionally, it was stated that the incidence of mTBI was associated with exposure to high combat intensity situations, one or more explosions, and hospitalization during deployment. Clinical reports have indicated the development of cognitive associated disorders following blast exposure. Majority of these disorders are associated with anxiety, attention deficits, memory issues, and problem-solving skills [6–11]. Most individuals who sustain mTBI return to baseline cognitive function within weeks. However, delayed onset of these symptoms is one of the major issues associated with the diagnosis of those affected by TBI [12–14]. Symptoms such as sleeplessness, irritability, and distress are seen at an acute stage clinically [15]. Levin et al. screened US veterans returning from Iraq and Afghanistan and similarly reported significant increases in PTSD, distress, depression, and loss of verbal memory [16]. Most of the long-term issues associated are memory deficits, anxiety, mood changes, and substance abuse [17].

Thus, short- and long-term preclinical studies are vital to give an oversight in assessing the extent of behavioral changes and their associated long-term pathological sequel. The use of standardized animal models for blast neurotrauma is required in order to advance studies in this field. The lack of standardized preclinical models of TBI was highlighted as the number one research gap in the 2012 Neurotrauma Pharmacology Workgroup established by the US Army Medical Research and Materiel Command (USAMRMC) [18]. While animal research is a cost-effective, well-controlled, and quick method to initially measure drug safety and efficiency, if the testing methodology is not established and validated for proper simulation of the blast environment, the results are futile. Furthermore, variability in experimental approaches among blast researchers makes comparison of results across laboratories and models difficult, if not impossible. The goals of this chapter are to describe the blast conditions which can be recreated within a laboratory setting and methodology for testing in vivo models within the appropriate environment.

1.1 Blast Physics Synopsis

When designing blast experiments, it is vital to understand the fundamental properties of blast waves before attempting to recreate these in a laboratory simulation. Several references are available which provide a good introduction to the basic physics of blast and blast loading of objects [19–25]. Failure to use the well-established knowledge of blast physics to design and validate laboratory simulations will likely yield deceptive results and misinformation regarding blast injury risk and injury outcomes. Misleading data could ultimately cause harm such as leading to inappropriate diagnoses, treatments, and countermeasures. Since only recently there has been acceptance that the brain is damaged when exposed to a blast, there is very limited knowledge on the neurological deficits (behavioral and histological) that occur subsequent to the exposure. Preclinical testing is needed to expand the understanding of blast TBI. This requires laboratory models that are well controlled, reproducible, and cost efficient. Most importantly, the experimental model needs to create a real-world environment within the laboratory. Thus, simulating a free-field blast wave should be undertaken with care and validated before animal research is conducted.

1.1.1 Classic Free-Field Blast

Figure 1 depicts the early stages of blast wave propagation. When a detonation occurs, there is a rapid expansion of the combustion product gases seen as the "fireball." The leading boundary of the expanding fireball is defined as the *contact surface* which can be considered to act like a "spherical piston" driving the shock front into the surrounding atmosphere. During this early stage of the blast wave generation, there is rapid momentum exchange from the expanding products to ambient air. The zone within the extent of the fireball expansion is defined as the near-field regime, where complex and rapidly evolving flow conditions produce highly transient wave dynamics within the fireball. Although individuals can be exposed to near-field blast and survive, considerations for this overpressure regime above 10 atm (~150 psi) are beyond the scope of this chapter. The mid-field regime refers to the zone beyond the fireball expansion yet still having strong nonuniform wave dynamics typically at overpressures from 10 to 1 atm. Beyond 1 atm, the blast wave will generally have equilibrated to a quasi-1D flow and the classical self-similar shape often characterized as the "Friedlander" waveform. The classical blast wave consists of a leading edge shock front (typically of the order of a micron thick), a positive pressure phase with continued expansion to below ambient levels, and a return to ambient conditions. It is important to note that this characteristic wave shape applies to all of the gas dynamic conditions in the mid- to far-fields; that is, flow velocity also has a negative phase in which the flow reverses and is drawn back to ground zero.

A simplified representation of the static overpressure time-waveform developed during a free-field blast wave is depicted in Fig. 2 and known as a Friedlander waveform. Blast waves are described by several characteristics, with peak pressure, positive phase, impulse and duration, and decay constant being the most commonly used parameters as used in the Friedlander equation.

The peak pressure (Ps in the diagram) is the level at the shock front; the positive phase is the time from the shock front to the

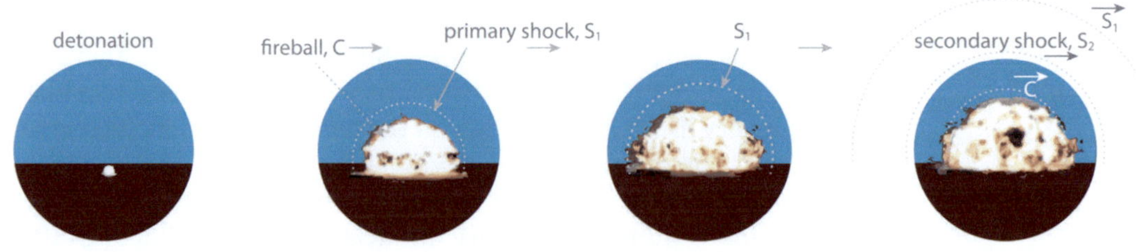

Fig. 1 Schematic of blast wave detonation of an idealized hemispherical high explosive charge showing the early growth of the fireball and air shock wave propagation. The edge of the fireball of expanding detonation products is labeled 'C' denoting a gas dynamic contact surface; the primary and secondary shocks are denoted as S1 and S2, respectively. The mid- to far-field propagation would develop after the depicted stages and generally applies to overpressures below 1 MPa

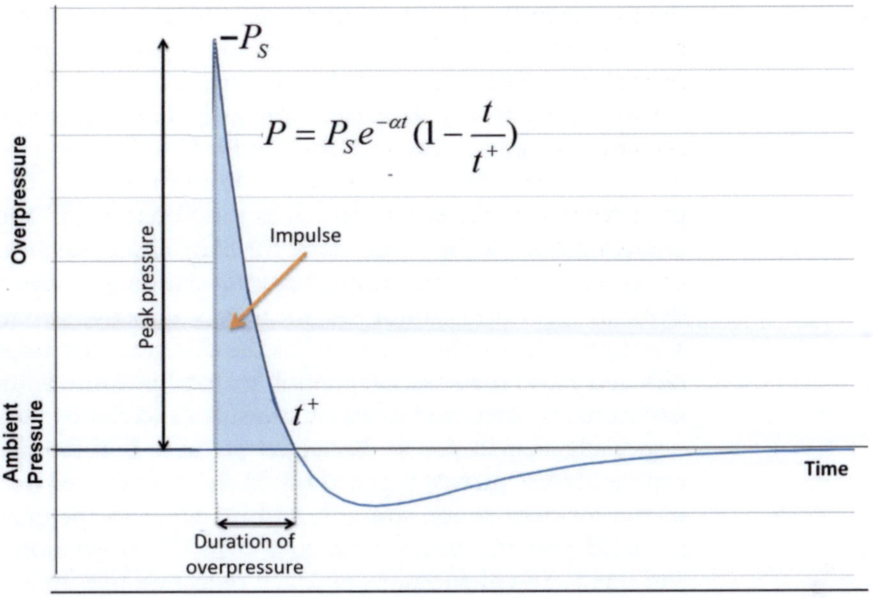

Fig. 2 Classical free-field blast wave, referred to as the "Friedlander waveform," is a simplified representation of the variation through time of static pressure for a classical free-field blast wave as would be measured at a fixed location passed by the blast. It does not account for reflection against the ground or other naturally occurring anomalies. In general, the classical free-field blast wave is characterized by a single high-pressure pulse followed by a rapid exponential decrease of the overpressure. The pulse is terminating with a period of negative overpressure before returning to ambient conditions

crossing below ambient pressure; impulse is measured as the integral of pressure and time during the positive phase. The rise time of the shock from ambient to peak level is exceedingly short, on the order of nanoseconds. A deficiency of the Friedlander waveform is lack of proper resolution of the negative phase including a secondary shock as will be discussed later.

1.1.2 Blast Wave Propagation and Pressures

The violent expansion of combustion gases from the detonation abruptly pushes the surrounding air, causing it to be compressed as well as imparting flow velocity and increased temperature. This disturbance takes the form of a propagating wave much as a stone dropped in a pool will cause a surface wave. Since the fireball expansion from an explosion is so abrupt, the compression wave in air will have a shock front. However, it is important to consider that all the gas dynamic properties including flow velocity, density, and temperature, as well as static pressure, change nearly instantaneously through the shock front and subsequent decay of a blast wave as would be experienced at an exposed location.

The various terms for "pressure" need to be clarified, as this causes confusion when attempting to compare conditions from different experimental methods. Static pressure (also called side-on pressure) refers to the pressure due to the thermodynamic state of the gas, as in $P=\rho RT$ for an ideal gas, where ρ is the density, R the gas constant, and T the temperature. Static pressure can be considered the crushing force of the blast wave and can be measured perpendicular to the flow direction such that it does not impede the flow. Dynamic pressure is the specific kinetic energy of the flow, ½ $\rho v2$, where ρ is the local flow density and v its velocity. Dynamic pressure is sometimes referred to as the "blast wind" and for certain conditions is the major cause for blast displacement of objects as opposed to the static pressure causing crushing action. Although dynamic pressure cannot be measured directly, within certain assumptions it can be calculated as the difference between *stagnation* and *static* pressure. Stagnation or total pressure is the combination of the static and dynamic pressures and can be measured by a specially shaped probe. Reflected pressure is the peak pressure experienced at surface exposed to the shock front and is a function of the incident shock strength and its angle to the surface. The reflected pressure is always higher than the stagnation pressure, and due to the phenomena of Mach reflection maximum reflected pressures can develop at angles less than normal incidence. The imparted loading on a structure is determined by the structure's geometry and the combination of the reflected, stagnation, dynamic, and static pressures, which vary over time during the reflection/diffraction phase. There is a splitting of energy between the static and dynamic pressures in blast waves which change with distance from the source. For example, in the far-field when the blast weakens to its acoustic limit, there is negligible dynamic pressure in the wave; near the fireball dynamic pressure is about 50% higher than the static overpressure. When a blast wave reflects and diffracts over an object, highly spatially and temporally variant pressures develop over the surfaces due to the combination of these incident flow conditions. This transient surface loading imparts stress waves through the materials and ultimately causes damage to the target and global motions. Figure 3 depicts pressures created when a structure is exposed to shock flow.

In terms of physiological importance, the imparted loading is determined by the static (Ps) and dynamic (Pd) pressures of the blast wave and the structure's geometry during the reflection/diffraction phase. This loading will vary dramatically around the structure and through time. Therefore the combined static and dynamic pressure conditions through time are critical. Simplistically, the single best parameter to define the loading conditions imparted by a blast wave is the peak reflected pressure Prefl = 2Ps + 2.4Pd [26].

Shock wave interaction with objects is usually extremely complex involving highly transient, high-rate loading typically in the timescale of milliseconds. The fact that the shock front involves an effective step-increase in pressure is often cause for damage or injury that would not occur had that same pressure been applied rapidly by any other means. As presented above, when a shock

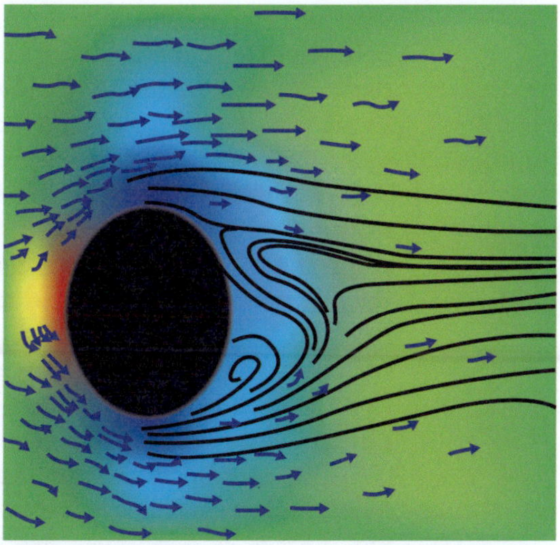

Fig. 3 In classical aerodynamic flow over an obstacle, the dynamic and static pressures vary over the surfaces causing lift and drag

wave interacts with an object, any surface normal to the propagating shock wave will experience localized reflected pressures which will often exceed severalfold the incident pressure. Also, as the shock wave traverses the object, the wave diffracts around and possibly refracts into the material according to its material properties and geometry, causing abrupt nonuniform loading and stresses including zones of underpressure. Depending on the size and shape of the object, the shock wave may cause large shifting pressure differentials on the object's surfaces.

1.2 Laboratory Simulations of Blast

Shock waves cover a broad category of gas dynamic disturbances in compressible gases which can include standing and quasi-steady waves. In fact, most studies of shock waves are in connection with supersonic aerodynamics, other quasi-steady flows, or to assess the molecular kinetics through the shock front itself. Laboratory apparatus devised in the 1950s to study the effects of such gas dynamics include the conventional shock tube and blow-down supersonic wind tunnel. However, a blast wave is a particular type of shock wave distinguished by being a propagating or traveling wave generated by an explosive event. A "classical" blast wave in air, as might be generated from the idealized detonation of TNT, has a characteristic decay profile with time as previously described in Fig. 2. Although a conventional compressed-gas-driven shock tube can generate a traveling shock wave with decaying profile, this can only be achieved with careful setup and only partially simulates blast-wave exposure conditions [27–29]. Different methods can be undertaken to create simulated blast waves within a shock tube, for example detonation of small charges, the ignition of fuel/air mixtures, or rupture of a membrane in a compressed gas driver.

The majority of laboratories conducting shock tube tests use the compressed gas method as it has many advantages such as ease and repeatability of trials. For this reason, this method for creating shock waves within a gas-driven shock tube will be discussed.

1.2.1 Conventional Shock Tubes (ST)

A conventional constant-area shock tube consists of two pipe sections separated by a frangible membrane. The high-pressure region is called the "driver" section and the low-pressure region is called the "driven" section. The driver is filled with gas which is pressurized until the frangible membrane ruptures by overpressurization or means of a puncturing device. Figure 4 shows the consequent development of the shock flow conditions after rupture of the diaphragm and the transition of gas phases during testing.

The expansion of the high-pressure gas generates a shock front, and the boundary between driver and driven gas is the contact surface as previously described in the case of high-explosive gases. However, the conditions of the expanding compressed gas from an ST driver are quite different than those of detonation products, and due to the geometry of the ST the wave dynamics do not replicate those of explosive blast. At a considerable distance from the diaphragm the shock wave profile will evolve such that it can approximate that of the positive phase of explosive blast [27]. When the shock wave reaches the open end of the tube it is reflected as a rarefaction (expansion) wave that propagates back upstream. The pressure reduction due to this rarefaction is associated with an increase in outflow velocity. The open-end rarefaction will be sufficiently strong that the flow becomes overexpanded and a recompression shock develops at its tail traveling upstream. Specimens subjected to the end rarefaction will be exposed to anomalous conditions of exaggerated outflow and a recompression shock striking from the rear (*see* **Note 1**). Therefore, for conventional ST with no end-wave elimination, there is a narrowly defined section within the tube that can be used to approximate conditions of the positive phase of explosive blast.

Therefore, a conventional ST does not intrinsically generate a shock wave replicating that of a free-field explosion. To approximate loading conditions associated with a free-field blast wave, a test specimen has to be placed at a specific location in the tube where the various waves interact to create a profile similar to that of the positive phase of explosive blast. The design of any experiment must account for the anomalies of ST wave-dynamics contact surface, rarefaction from the driver, rarefaction from the open end of the tube, and its trailing recompression shock [30]. In actual high-explosive blast, the initial overexpansion of the detonation products causes a negative phase. At the same time it may also expose the specimen to a secondary shock of lower magnitude and in the same direction as the first shock, away from the blast source.

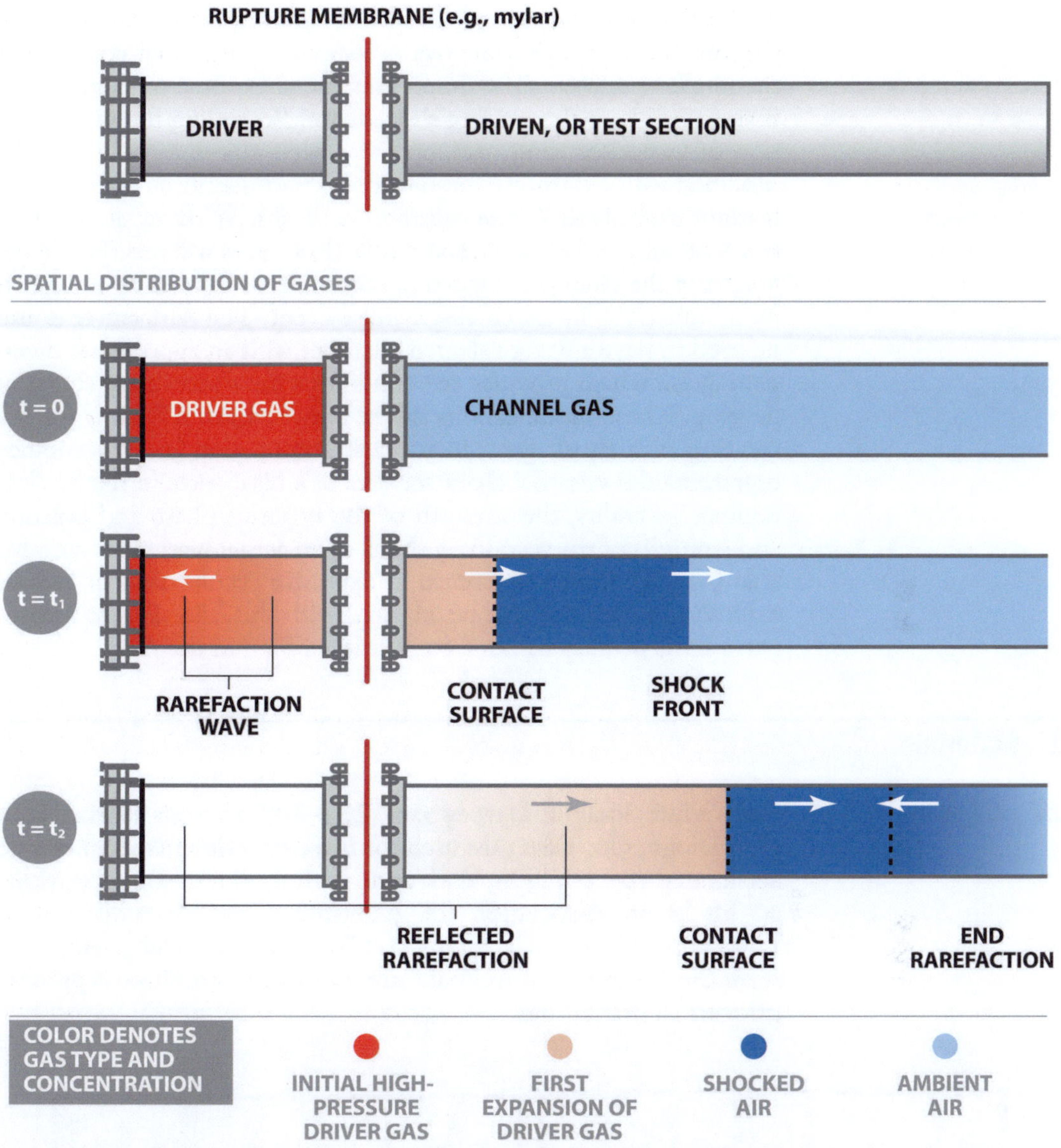

Fig. 4 Schematic depiction of the development of the gas dynamic flow conditions within a constant-area shock tube. The spatial distribution of gases are represented by color. At time 0, the driver is filled with high pressure gas which is separated from the driven section filled with ambient air by a frangible membrane. Once the membrane ruptures (t_1), the high-pressure gas expands into the driven section creating the shock front. In addition, the high-pressure gas reflects off the rear wall of the ST and creates a rarefaction wave

1.2.2 Advanced Blast Simulators

In order to eliminate anomalous artifacts of conventional ST, the advanced blast simulator (ABS) has been recently designed. The ABS is a specially modified ST designed to intrinsically replicate all the key features of blast wave flow conditions, including the negative phase and secondary shock. One specific difference between ST and ABS is the addition of an end-wave eliminator (EWE). The EWE is required to mitigate the immediate expansion of the flow at the end of test section upon exiting. The EWE consists of a dump tank that can contain the expanding gases from entering the lab space and at the same time creates some overpressure reflection that counteracts the rarefaction wave. The dump tank includes some manner of baffling to break up the venting shock front. Regarding the neutralization of waves traveling back up the device, the dump tank can be modified with a diffusing mesh insert to best distribute the resultant loading and divert forces radially (vs. axial). Without an EWE, a series of unintended shock and rarefaction waves will reverberate the length of the chamber imparting accumulated injuries to the specimen, unlike the intended single-pulse event. The ABS driver is also tailored to have a wedge-shaped chamber with an appropriate divergent angle which provides the conditions for a secondary shock to develop. These modifications of the conventional ST allow for the creation of a shock wave in the test section that more accurately represents the essential characteristics of a blast wave in the far-field regime. In reality, the strength of the negative phase and position and strength of the secondary shock of an actual blast wave are variable (~25 %) due to difference in explosive combustion of various explosives which can also be adjusted with the ABS. Figure 5 compares static profiles within a conventional ST and the ABS.

2 Materials

2.1 Animals

Male adult Sprague Dawley rats (250–300 g) from Harlan Labs (San Diego, CA, USA) are used for these experiments. Animals are acclimated to 12-h light/dark cycle with food and water provided ad lib in an Association for Assessment and Accreditation of Laboratory Animal Care (AAALAC)-approved facility. Approval from the Institutional Animal Care and Use Committee is granted prior to all procedures.

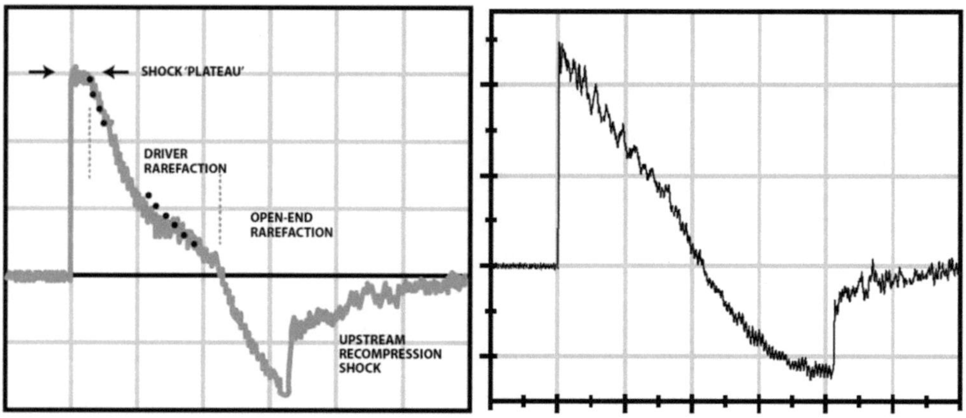

Fig. 5 Comparing static profiles within a conventional ST and the ABS. Pressure profiles at the same monitoring station before (*left*) and after (*right*) the installation of an EWE at the open end of the tube. Several features entirely atypical of blast waves are present in the *ST profile* and disappear in the *ABS* plot

2.2 Anesthesia

The animals are anesthetized with a continuous flow of 3% isoflurane and 60% oxygen for 5 min.

2.3 Materials for Testing

Compressed gas (helium preferred).

- ABS (Driver section, test section, end-wave eliminator).
- Membrane (Acetate, Mylar, Vinyl, Metal; Grafix Plastics, Cleveland, OH, USA).
- Pressure sensors (PCB Piezotronics Inc., Depew, NY, USA).
- Signal conditioner (PCB Piezotronics Inc., Depew, NY, USA).
- Dash 8HF data acquisition system (Astro-Med, Inc, West Warwick, RI, USA).
- Animal sling holder (Custom made from mesh).
- High-speed video camera (Phantom Miro eX2, Vision Research, Wayne, NJ, USA).
- Physiological monitoring system (Nonin PulseSense VET Pulse Oximeter, Henry Schein Inc. Melville, NY, USA).

3 Methods

3.1 Standardizing In Vivo Blast TBI Models

Our established rodent model of primary blast-induced neurotrauma has provided evidence of behavioral and cognitive deficits as well as neuropathology following blast exposures [31–33]. Our translational model uses an ABS located at the Center for Injury Biomechanics at Virginia Tech (Fig. 6).

The ABS maintains the two separate chambers: the driver and the driven section separated by a frangible membrane (for more details on suitable membrane options *see* **Note 2**). An EWE is incorporated at the end of the ABS, preventing reflecting artifacts. The system is activated by helium to generate a shock front and blast overpressure. Animals are anesthetized with an induction of isoflurane

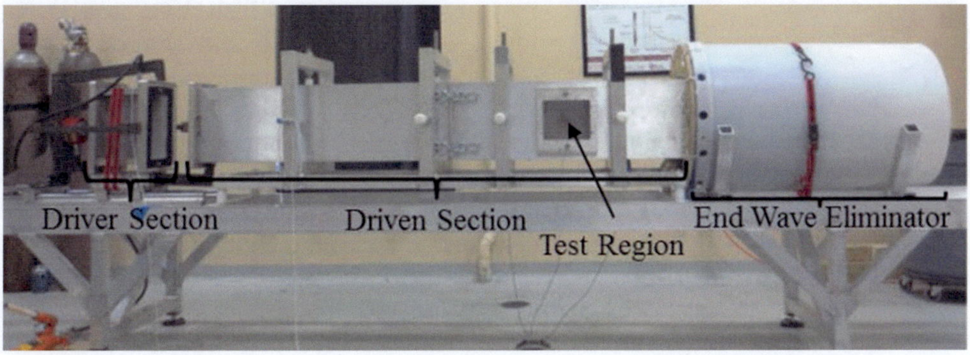

Fig. 6 Advanced blast simulators (ABS) have been specifically designed to intrinsically replicate all the key features of blast wave flow conditions, including the negative phase and secondary shock

and subsequently positioned within mesh sling positioned inside the ABS test section. A blast wave is produced with compressed helium and calibrated acetate sheets (Grafix Plastics, Cleveland, OH). Helium is chosen over compressed air as it will cause a shock wave profile that will result in an initial increased amplitude, a steeper decay rate, reduction of the negative pressure phase, and a delay in the upstream shock [30]. Three static pressure measurements are collected along the test section at 250 kHz using a Dash 8HF data acquisition system (Astro-Med, Inc, West Warwick, RI) and peak overpressure is calculated by determining wave speed (m/s) at the specimen position (for more details on pressure measurement optimization *see* **Note 3**). Wave profiles are verified to maintain consistent exposure pressures between subjects. A mesh sling was used to hold the animal during the exposure that allows for minimal hindrance of the wave through the tube. If an animal is not harnessed properly, there is a chance for tertiary injuries occurring from blunt impact with a rigid surface. However, isolating the primary injury is not always the focus. For example, groups have studied the combined effect of primary and quaternary blast effects in what is referred to as a burn-blast model [34]. In any study, methods can be utilized to create a spectrum of isolated and combined injury modes. However, it is important to remember that the animal and restraining system should be less than 20% blockage of ABS in order to maintain optimal shock wave development (*see* **Note 4**).

4 Notes

There are several important aspects of the blast procedures that need further extrapolation in order to help standardize methodology between labs. Four major points are highlighted and discussed below.

1. *Placement of specimen*: "*Inside* vs. *Outside.*" As a first note, it is imperative to expand discussion on the location of animal placement during testing. Some groups have previously reported data from blast neurotrauma studies which have tested animals directly outside a conventional ST and suggested that this method of testing live animals was appropriate for primary blast injury studies [35–43]. When reviewing blast physics principles and shock wave development within a conventional ST, it is clear that placing a specimen directly outside the ST is not the appropriate setting if primary blast injury simulation is intended. The combination of the emerging shock wave and venting gas from the ST causes the formation of a vortex and high flow velocity region called end-jet. Venting jet-flow phenomena from STs are well known from theory, experiment, and computational modeling [44–49]. In fact, an immediate flow transition takes place as the shock exits from the open end (Fig. 7).

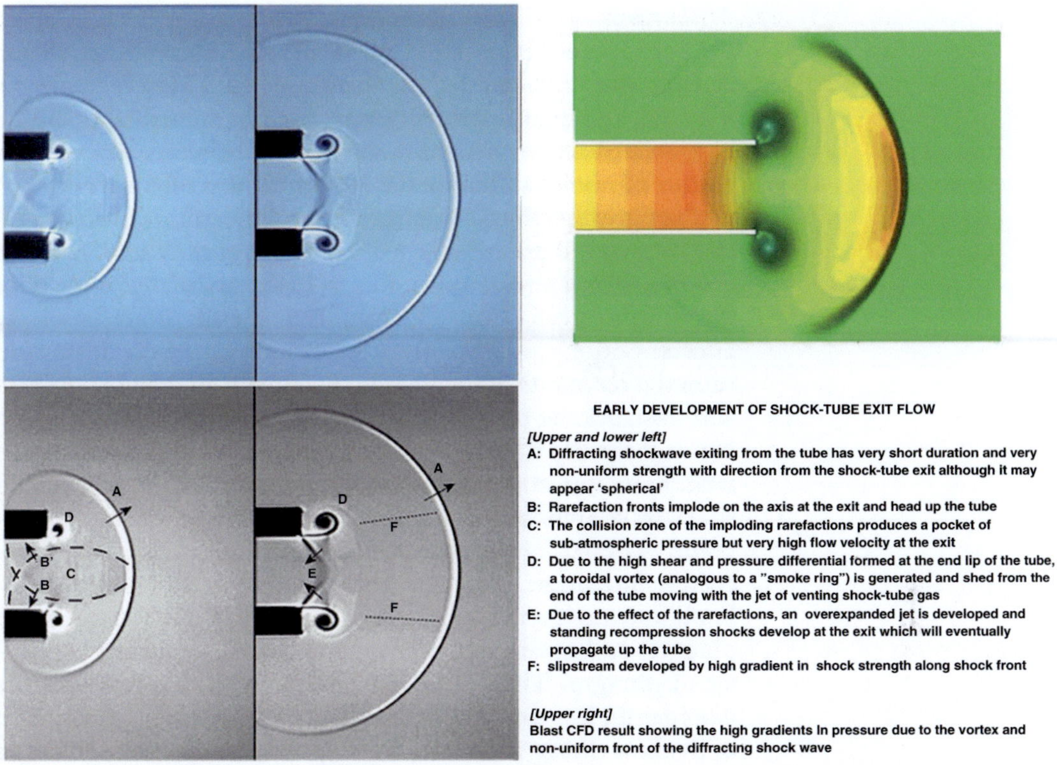

Fig. 7 Combination of the emerging shock wave and venting gas from the ST causes of the formation of a vortex and high flow velocity region called end-jet. Image provided by Dyn-FX Consulting

Intense rarefactions from the open end dramatically drop the static pressure and accelerate the gas outflow generating a collimated jet stream having extreme dynamic pressure. The gas dynamic conditions near or beyond the open end are entirely unlike those of free-field explosive blast particularly with respect to the partitioning of *static* and *dynamic* pressure. Although there is a nonuniform diffracting shock front to the venting flow field, the primary loading conditions for a target within the jet do not represent those of a propagating shock wave at all. Studies that include the testing of specimens directly outside of the ST in-line with the end-jet flow subject the specimen to loading artifacts that are not associated with explosive blast. These artifacts will create exaggerated "throw" forces, causing physical trauma to specimens that does not reproduce the actual mechanisms of injury in blast which is primarily a severe transient "crushing" action. Results from blast TBI studies which involve testing specimens in the end-jet flow must be treated with skepticism. Furthermore, it can be shown that ST are capable of replicating what may appear to be credible static pressure–time records for explosive blast yet are entirely incorrect with regard to the other gas dynamic conditions. This leads to serious misrepresentation of specimen loading and injury interpretations.

2. *Membranes, driving gas, and flow.* A number of factors affect the performance and output of ST and ABS such as size, shape, driving gas, driven gas, temperature, humidity, membrane material, and membrane thickness. Each of these factors should be standardized to maintain consistency between testing. The matter of nonideal membrane rupture is too often overlooked and yet greatly affects the shock wave generation. Even a carefully staged ST experiment will typically produce less than half the theoretical shock levels due to losses associated with poor diaphragm performance, turbulence, and boundary-layer effects. Serious waveform complications and flow losses are caused if remnants of the membrane remain attached at the wall and obstruct the outflow from the driver. The choice of membrane material, thickness, reinforcement, as well as clamping technique and rupture method are important in controlling shock wave amplitudes and waveform. Three widely used membrane materials are Mylar, acetate, and thin metal foils, especially aluminum; each of these behaves differently [50]. Frangible mesh reinforcements can also be used to support diaphragms and greatly extend the shock pressures for any particular diaphragm. They offer less energy expended opening the diaphragm, less diaphragm fragmentation, and less obstruction of the flow due to the mesh. The type and temperature of driver gas used to generate the shock wave is another factor which can affect wave profiles; compressed air, nitrogen, and helium are common. Helium is a more efficient gas than air as a driver gas, and provides a higher peak overpressure and shorter duration for the same driver pressure. The rate of gas filling of the driver, which can be affected by the residual pressure in the supply-gas tank, can also affect diaphragm performance; slow fill rates will often yield poor or irregular diaphragm ruptures. Therefore, it is important to monitor the pressure of the supply gas tank and ideally computer-control the filling of the driver to ensure a rapid steady fill rate. A pressure gauge can be used to ensure that a consistent hydraulic clamping force is applied between the driver and driven sections to ensure repeatable ruptures and minimize potential gas leaks (for manually bolted diaphragm clamping, a torque-wrench can be used to ensure uniform consistent clamping pressure). Furthermore, it is important to note that the shock pressures at the test location will vary with changes in temperature and in some cases the accumulation of driver gas in the test section. Overall, maintaining records on all aspects of the testing environment may seem tedious, but is key to generating reproducible data.
3. *Importance of recognizing sensor limits and reporting pressures.* As described earlier, a blast wave exposure includes prescribed conditions for static and dynamic pressure which should be verified in any simulation. It is important to note that *static*, or side-on, overpressure is easiest to measure in blast field trials

and has been compiled in databases and main references in "energy-scaled" terms as the key descriptor for blast conditions from a designated charge. Unfortunately, an outcome of such references quantifying only the static pressure waveforms for a designated blast threat is that readers new to the area, who have not fully researched the subject, infer that this is the only relevant gas dynamic parameter necessary to replicate in blast simulations. Various pressure sensors can be used to measure static and total (or stagnation) pressures in experiments from which the dynamic pressure conditions can be derived. Static pressure can be measured using a shock-pressure sensor mounted flush to the wall of the ABS. However, traces from some models of wall-mounted gauges can be affected by wall vibration during testing. These fluctuations are not gas dynamic disturbances and hence do not represent anomalous loading of the specimen but confuse the monitoring of exposure conditions. It is recommended to assess shock wave exposure conditions by means of a Pitot-static "pencil" sensor which is stood away from the wall and close to the specimen. This sensor is called a Pitot-static probe since it measures both the total pressure and static pressure via separate ports. Such a probe can be mounted by means of an aerodynamic "sting" to be close to the specimen. An important attribute of using such a probe is that the arrival time of the shock front at each sensor allows tracking of the shock speed from which an analytic solution can be determined for the peak shock levels to complement those measured by the sensors themselves. The analytic solution for peak shock levels is obtained from the Rankine-Hugoniot relations. The speed of the shock front can also be measured by multiple pressure sensors placed along the wall of the test section. After calculating the theoretical equations for static and dynamic pressures, the results should be reported with data.

4. *Blockage of the shock flow*. In free-field blast exposures a uniform shock front fully envelops the specimen, and it is this full "wrap-around" loading of the shock wave reflection/diffraction process which imparts the distinctive stresses and momentum transfer of actual blast/target interactions. Pressures at the front of the specimen continue to be affected by the free-field shock wave as it diffracts around the sides. Proper simulation of the loading dynamics requires having a uniform incident shock front which is several "body widths" on either side of the target. It is not valid blast load conditions to expose only a portion of a target which will produce a "focal" exposure. In addition, if the specimen and holding fixture significantly block the shock wave flow, reflective waves will develop. These reflective waves can be amplified due to convergence on the specimen, imparting complex loading unlike the free-field case. It is

advisable that the specimen and holding fixture should present less than 20 % area blockage in order to replicate free-field conditions for animal testing. If large specimens are required, expansion sections are often incorporated. These sections allow for widening of the test section in order to accommodate larger targets; however, magnitude of shock wave will be affected.

References

1. Owens BD, Kragh JF Jr, Macaitis J, Svoboda SJ, Wenke JC (2007) Characterization of extremity wounds in OIF and OEF. J Orthop Trauma 21:254–257
2. Macgregor AJ, Dougherty AL, Galarneau MR (2011) Injury-specific correlates of combat-related traumatic brain injury in Operation Iraqi Freedom. J Head Trauma Rehabil 26:312–318
3. Murray CK, Reynolds JC, Schroeder JM, Harrison MB, Evans OM, Hospenthal DR (2005) Spectrum of care provided at an echelon II medical unit during Operation Iraqi Freedom. Mil Med 170:516–520
4. Elder GA, Mitsis EM, Ahlers ST, Cristian A (2010) Blast-induced mild traumatic brain injury. Psychiatr Clin North Am 33:757–781
5. Hoffman SW, Harrison CR (2009) The interaction between psychological health and traumatic brain injury: a neuroscience perspective. Clin Neuropsychol 8:1400–1415
6. Hoge CW, McGurk D, Thomas JL, Cox AL, Engel CC, Castro CA (2008) Mild traumatic brain injury in U.S. soldiers returning from Iraq. N Engl J Med 358:453–463
7. Arciniegas DB (2011) Addressing neuropsychiatric disturbances during rehabilitation after traumatic brain injury: current and future methods. Dialogues Clin Neurosci 13:325–345
8. Belanger HG, Kretzmer T, Yoash-Gantz R, Pickett T, Tupler LA (2009) Cognitive sequelae of blast-related versus other mechanisms of brain trauma. J Int Neuropsychol Soc 15:1–8
9. Halbauer JD, Ashford JW, Zeitzer JM, Adamson MM, Lew HL, Yesavage JA (2009) Neuropsychiatric diagnosis and management of chronic sequelae of war-related mild to moderate traumatic brain injury. J Rehabil Res Dev 46:757–796
10. Riggio S (2011) Traumatic brain injury and its neurobehavioral sequelae. Neurol Clin 29:35–47
11. Warden D (2006) Military TBI during the Iraq and Afghanistan wars. J Head Trauma Rehabil 21:398–402
12. Rohling ML, Meyers JE, Millis SR (2003) Neuropsychological impairment following traumatic brain injury: a dose response analysis. Clin Neuropsychol 17:289–302
13. Kochanek PM, Bauman RA, Long JB, Dixon CR, Jenkins LW (2009) A critical problem begging for new insight and new therapies. J Neurotrauma 26:813–814
14. Cawley PJ, Mokadam NA (2010) Delayed complications from exposure to improvised explosive devices. Ann Intern Med 153:278–279
15. Miller G (2011) Neuropathology. A battle no soldier wants to fight. Science 29:517–519
16. Levin HS, Wilde E, Troyanskaya M, Petersen NJ, Scheibel R, Newsome M, Radaideh M, Wu T, Yallampalli R, Chu Z, Li X (2010) Diffusion tensor imaging of mild to moderate blast-related traumatic brain injury and its sequelae. J Neurotrauma 27:683–694
17. Heltemes KJ, Dougherty AL, MacGregor AJ, Galarneau MR (2011) Alcohol abuse disorders among U.S. service members with mild traumatic brain injury. Mil Med 176:147–150
18. Diaz-Arrastia R, Kochanek PM, Bergold P, Kenney K, Marx CE, Grimes CJ, Loh LT, Adam LT, Oskvig D, Curley KC, Salzer W (2014) Pharmacotherapy of traumatic brain injury: state of the science and the road forward: report of the Department of Defense Neurotrauma Pharmacology Workgroup. J Neurotrauma 15:135–158
19. Bethe HA, Fuchs K, Hirshfelder JO, Magee JL, Peierls RE, von Neumann J (1947) Blast wave. Los Alamos Scientific Laboratory, LA-2000, Physics & Mathematics, TID-4500, 13th edn., Suppl
20. Brode HL (1959) Blast wave from a spherical charge. Phys Fluids 2:217
21. Dewey JM (1971) The air velocity in blast waves from T.N.T. explosions. Proc Roy Soc Lond A 324:275–299
22. Sternberg HM, Hurwitz H (1976) Calculated spherical shock waves produced by condensed explosives in air and water. Proc. 6th Symp (Int) on Detonation, ACR-221, Office of Naval Research, Arlington VA, NSWC, White Oak, 528–39
23. Glasstone S, Dolan PJ (1977) The effects of nuclear weapons, 3rd edn., US Dept of Defense

& Energy Research & Development Agency. Available at: http://www.atomicarchive.com/Docs/Effects/index.shtml

24. Kingery CN, Bulmash G (1984) Airblast parameters from TNT spherical air burst and hemispherical surface burst, ARBRL-TR-02555. Ballistic Research Laboratory, Aberdeen Proving Ground, MD
25. Glass II, Hall JG (1959) Handbook of supersonic aerodynamics, Section 18 Shock tubes. NAVORD Report, vol 6. Bureau of Ordnance Publication, Washington, DC, p 1488
26. Iremonger M (1997) Physics of detonations and blast-waves. In: Cooper GJ, Dudley HAF, Gann DS et al (eds) Scientific foundation of trauma, vol 1. Butterworth-Heinemann, Oxford, pp 189–199
27. Celander H, Clemedson CJ, Ericsson U, Hultman H (1955) The use of compressed-air-operated shock tube for physiological blast research. Acta Physiol Scand 32:6–13
28. Celander H, Clemedson CJ, Ericsson UA, Hultman HI (1955) A study on the relation between the duration of the shock wave and the severity of the blast injury produced by it. Acta Physiol Scand 33:14–18
29. Richmond DR, Wetherbe MB, Taborelli RV, Sanchez RT, Sherping F, Goldizen VC, White CS (1959) Shock tube studies of the effects of sharp-rising, long-duration overpressures on biological systems. Progress Report AEC Contract No. AT(29-1)-1242, TID-6056. The Lovelace Foundation for Medical Education and Research, Albuquerque, NM
30. Ritzel DV, Parks SA, Roseveare J, Rude G, Sawyer T (2011) Experimental blast simulation for injury studies. NATO/RTO HFM-207 Symposium, Halifax, Canada
31. Cho HJ, Sajja VSS, VandeVord PJ, Lee YW (2013) Blast induces oxidative stress, inflammation, neuronal loss and subsequent short-term memory impairment in rats. Neuroscience 3:9–20
32. Sajja VSS, Ereifej ES, VandeVord PJ (2014) Hippocampal vulnerability and subacute response following varied blast magnitudes. Neurosci Lett 6:33–37
33. Sajja VSS, Hubbard WB, VandeVord PJ (2015) Subacute oxidative stress and glial reactivity in the amygdala are associated with increased anxiety following blast neurotrauma. Shock 44:71
34. Chai JK, Liu W, Deng HP, Cai JH, Hu QG, Zou XF, Shen CA, Yin HN, Han YF, Zhang XB, Chi YF, Ma L, Sun TJ, Feng R, Lan YT (2013) A novel model of burn-blast combined injury and its phasic changes of blood coagulation in rats. Shock 40:297–302
35. Courtney AC, Andrusiv LP, Courtney MW (2012) Oxy-acetylene driven laboratory scale shock tubes for studying blast wave effects. Rev Sci Instrum 83:045111-1–045111-7
36. Reneer DV, Hisel RD, Hoffman JM, Kryscio RJ, Lusk BT, Geddes JW (2011) A multi-mode shock tube for investigation of blast-induced traumatic brain injury. J Neurotrauma 28:95–104
37. Chavko M, Watanabe T, Adeeb S, Lankasky J, Ahlers ST, McCarron RM (2011) Relationship between orientation to a blast and pressure wave propagation inside the rat brain. J Neurosci Methods 195:61–66
38. Goldstein LE, Fisher AM, Tagge CA, Zhang XL, Velisek L, Sullivan JA, Upreti C, Kracht JM, Ericsson M, Wojnarowicz MW, Goletiani CJ, Maglakelidze GM, Casey N, Moncaster JA, Minaeva O, Moir RD, Nowinski CJ, Stern RA, Cantu RC, Geiling J, Blusztajn JK, Wolozin BL, Ikezu T, Stein TD, Budson AE, Kowall NW, Chargin D, Sharon A, Saman S, Hall GF, Moss WC, Cleveland RO, Tanzi RE, Stanton PK, McKee AC (2012) Chronic traumatic encephalopathy in blast-exposed military veterans and a blast neurotrauma mouse model. Sci Transl Med 4:60
39. Courtney MW, Courtney AC (2011) Working toward exposure thresholds for blast-induced traumatic brain injury: thoracic and acceleration mechanisms. Neuroimage 54(Suppl 1):S55–S61
40. Rafaels K, Bass CR, Salzar RS, Panzer MB, Woods W, Feldman S, Cummings T, Capehart B (2011) Survival risk assessment for primary blast exposures to the head. J Neurotrauma 28:2319–2328
41. Shridharani JK, Wood GW, Panzer MB, Capehart BP, Nyein MK, Radovitzky RA, Bass CR (2012) Porcine head response to blast. Front Neurol 3:1–12
42. Gullotti DM, Beamer M, Panzer MB, Chen YC, Patel TP, Yu A, Jaumard N, Winkelstein B, Bass CR, Morrison B, Meaney DF (2014) Significant head accelerations can influence immediate neurological impairments in a murine model of blast-induced traumatic brain injury. J Biomech Eng 136:091004
43. Panzer MB, Wood GW, Bass CR (2014) Scaling in neurotrauma: how do we apply animal experiments to people? Exp Neurol 14:120–126
44. Golub VV (1994) Development of shock wave and vortex structures in unsteady jets. Shock Waves 3:279–285
45. Minota T (1998) Shock/vortex interaction in a flow field behind a shock wave emitted from

a shock-tube. Proc. of 2nd Int. Workshop on Shock-Wave/Vortex Interaction, 2–11

46. Skews BW (1967) The perturbed region behind a diffracting shock wave. J Fluid Mech 29:705–719

47. Kitajima S, Iwamoto J, Tamura E (2009) A study on the behavior of shock wave and vortex ring dischargeed from a pipe. Proc of 10th Annual Internations Conference of Fluid Control, Measurement and Visualization

48. Kainuma M, Havermann M, Sun M, Takayama K (2005) Effects of the shock tube open-end shape on vortex loops released from it. Shock Waves, 505–10

49. Baird JP (1987) Supersonic vortex rings. Proc R Soc Lond A 409:1836

50. Alphonse V, Sajja VSS, Kemper AR, Ritzel DV, Duma SM, VandeVord PJ (2014) Membrane characteristics for biological blast overpressure testing using blast simulators. Biomed Sci Instrum 50:248

Chapter 8

Cellular Mechanisms and Behavioral Outcomes in Blast-Induced Neurotrauma: Comparing Experimental Setups

Zachary S. Bailey*, W. Brad Hubbard*, and Pamela J. VandeVord

Abstract

Blast-induced neurotrauma (BINT) has increased in incidence over the past decades and can result in cognitive issues that have debilitating consequences. The exact primary and secondary mechanisms of injury have not been elucidated and appearance of cellular injury can vary based on many factors, such as blast overpressure magnitude and duration. Many methodologies to study blast neurotrauma have been employed, ranging from open-field explosives to experimental shock tubes for producing free-field blast waves. While there are benefits to the various methods, certain specifications need to be accounted for in order to properly examine BINT. Primary cell injury mechanisms, occurring as a direct result of the blast wave, have been identified in several studies and include cerebral vascular damage, blood–brain barrier disruption, axonal injury, and cytoskeletal damage. Secondary cell injury mechanisms, triggered subsequent to the initial insult, result in the activation of several molecular cascades and can include, but are not limited to, neuroinflammation and oxidative stress. The collective result of these secondary injuries can lead to functional deficits. Behavioral measures examining motor function, anxiety traits, and cognition/memory problems have been utilized to determine the level of injury severity. While cellular injury mechanisms have been identified following blast exposure, the various experimental models present both concurrent and conflicting results. Furthermore, the temporal response and progression of pathology after blast exposure have yet to be detailed and remain unclear due to limited resemblance of methodologies. This chapter summarizes the current state of blast neuropathology and emphasizes the need for a standardized preclinical model of blast neurotrauma.

Key words Blast, Neurotrauma, Traumatic brain injury, In vivo, Methodology, Preclinical models, Neuropathology, Shock wave, Shock tube

1 Introduction

Blast-induced neurotrauma (BINT) has become the center of military health concern because of the increasing incidence of blast-induced traumatic brain injury (bTBI) in combatants over the past two decades [1]. Cognitive impairments that impede function and performance can be sustained from bTBI. Upon detonation,

*Zachary S. Bailey and W. Brad Hubbard contributed equally as co-primary author.

Firas Kobeissy et al. (eds.), *Injury Models of the Central Nervous System: Methods and Protocols*, Methods in Molecular Biology, vol. 1462, DOI 10.1007/978-1-4939-3816-2_8, © Springer Science+Business Media New York 2016

improvised explosive devices (IEDs) create a blast wave that involves a rapid overpressure change as high-speed energy moves through ambient air, impacting surrounding war fighters. In blast-induced injuries, the resulting insult can be categorized into four different modes: primary, secondary, tertiary, and quaternary. Primary injury is caused directly by the blast overpressure (BOP) wave and any barotrauma to the organs can be attributed to this exposure. Secondary injury is a result of blunt impact from any missiles, such as shrapnel, that are projected towards the body due to the blast. These insults can cause external, penetrating injuries including, but not limited to, bruises and lesions. Tertiary injury is due to the impact of the body with a rigid surface as a result of the blast wind. When soldiers are "thrown" by the blast wind, there is a potential for broken bones or even a concussion, which can compound bTBI. Quaternary injury is a result of any chemical or burn exposure that occurs from the explosion itself. These exposures, which could include infectious agents, can have subsequent systemic implications, such as lung injuries, which then exacerbate the brain injury. For the scope of this chapter, the discussion will focus on primary blast injury to the brain.

The mechanical transmission of blast energy to the brain has not been fully elucidated. However, while several mechanisms of primary blast injury to the brain have been hypothesized, only the "skull dynamics" theory is supported by computational modeling and experimental evidence. This theory is based on the fact that the skull is not perfectly rigid and it must flex/deform, especially when subjected to impulsive loading. Thus, the shock wave invokes different skull dynamic response modes which cause injury [2–9]. While there is an undeniable theory as well as independent experimental validations of the "skull dynamics" mechanism, how the skull dynamic response modes lead to injury at the cellular level still needs to be elucidated. The imparted stress waveforms from the blast energy can lead to localized stress and strains on the cells. Morphological deficits including cytoskeletal and axonal damage can lead to necrosis if injury thresholds are reached. Arguably, more emphasis has been placed on determining the time course of secondary cellular injury mechanisms. These include regulation of neuroinflammation through cytokine signaling, glial activation, and oxidative stress through accumulation of reactive oxygen species (ROS). Delayed apoptosis can result from these cascades and has a potential link to chronic behavioral and functional deficits. Many cellular cues have been explored, and yet the exact mechanism of injury at the cellular level has not been determined.

A significant limitation of evolvement within the blast neurotrauma community is the lack of standardized preclinical models. Many sources of data variability come from variances induced by different experimental setups. This has become such a significant concern that the NIH/NINDS has recently developed a list of common data elements (CDEs) for preclinical TBI research to promote the use of standard reporting and facilitate comparisons across studies [10]. Thus, the goal of this chapter is to present and compare blast neurotrauma methodology and subsequent data from the current literature to stress the importance of implementing standardizing preclinical methods.

1.1 Current Methodologies for Small Animal Preclinical Testing

Several experimental approaches are reported within the current blast neurotrauma literature. Various methods may contribute to inconsistent findings surrounding BINT. For those interested in studying the effects of BOP on the brain, isolation of the primary blast wave is crucial (*see* **Note 1**). However, some studies have examined the effect of composite injuries involving multiple injury modes [11]. Others have studied the combined effects of primary and quaternary blast injuries in what is referred to as a burn-blast model [12]. Knowing the exact blast parameters used to cause injury is crucial for comparison between research models.

Prior to experimentation, understanding blast physics, or working with experts in this field, is critical to designing appropriate environments to simulate a particular blast condition. The previous chapter (Chapter 7) by VandeVord et al. presents a detailed review of blast physics fundamentals. Briefly, BOP, defined as a deviation from atmospheric pressure, can be classified as static or dynamic. Static overpressure is considered the crushing force of the blast wave and can be measured perpendicular to the flow direction such that it does not impede the flow. Dynamic pressure is called the “blast wind” and for certain conditions is the major cause for blast displacement of objects as opposed to the static pressure, which causes crushing. In BINT studies, duration and magnitude of static overpressure are usually reported and most studies aim to represent a simplistic blast wave, the Friedlander waveform (Fig. 1) [13]. A comprehensive review of testing methods found in literature reporting primary blast brain injuries is summarized in Fig. 2. All pressure measurements reported within the review are static overpressures given by the cited article (*see* **Note 2**).

1.2 Detonation of Explosives

Open-field detonations use an explosive to induce BINT. An explosion is a phenomenon that results in a sudden release of energy and creates a blast wave which propagates outward from the explosion. Trinitrotoluene (TNT) is commonly used in varying amounts depending on the magnitude of the detonation required for animal testing. One advantage of conducting open-field testing is that there is no flow hindrance of the blast wave, which allows for testing of either larger animal models or multiple small animal models (Fig. 3A, B) [14–16]. On the other hand, administering anesthesia and controlling for bacterial exposure to survival animals are more difficult in the harsh outside environment (*see* **Note 3**).

While detonations are arguably the most successful in recreating battlefield explosions, a limitation is that repetition of the same

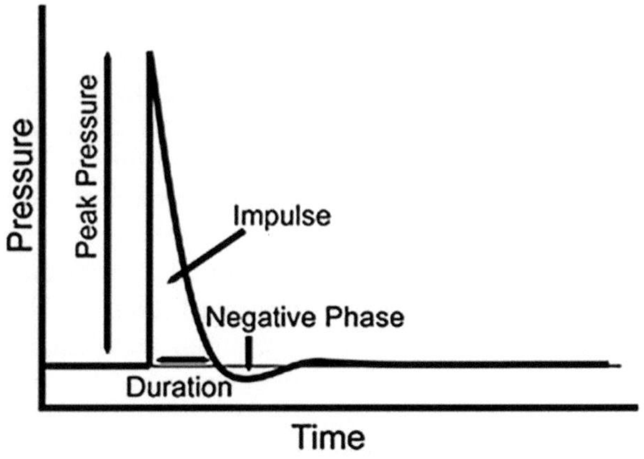

Fig. 1 Classical free-field blast wave, referred to as the "Friedlander waveform," is a simplified representation of the variation through time of static pressure for a classical free-field blast wave as would be measured at a fixed location passed by the blast. It does not account for reflection against the ground or other naturally occurring anomalies. In general, the classical free-field blast wave is characterized by a single high-pressure pulse, or shock front, followed by a rapid exponential decrease of the overpressure (positive phase) and terminating with a period of negative overpressure (negative phase) before returning to ambient conditions

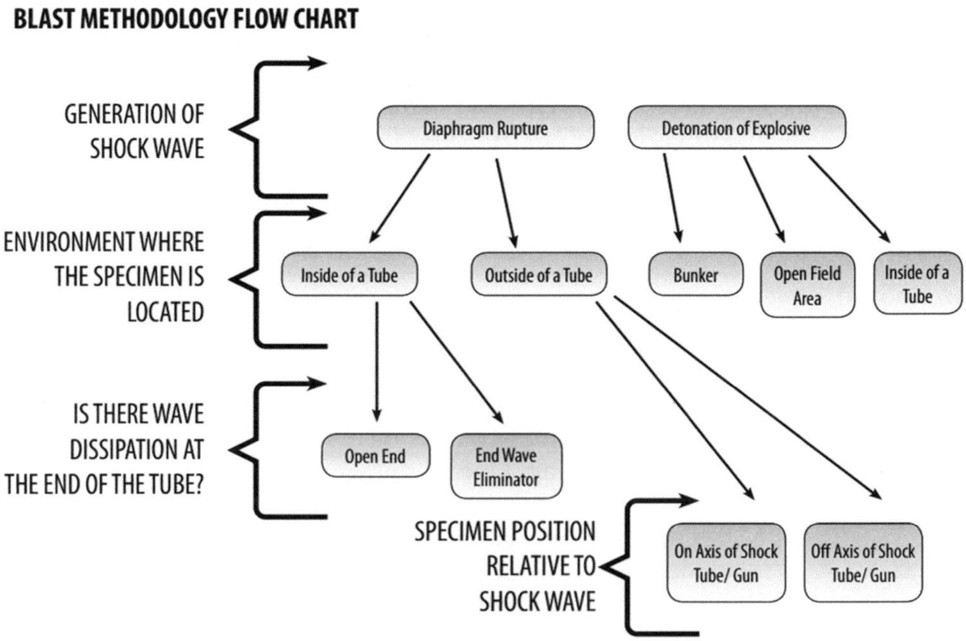

Fig. 2 Schematic of various methodologies which are used for BINT identified from literature. Defining characteristics, such as animal positioning and how the shock wave is generated, serve as distinctive features

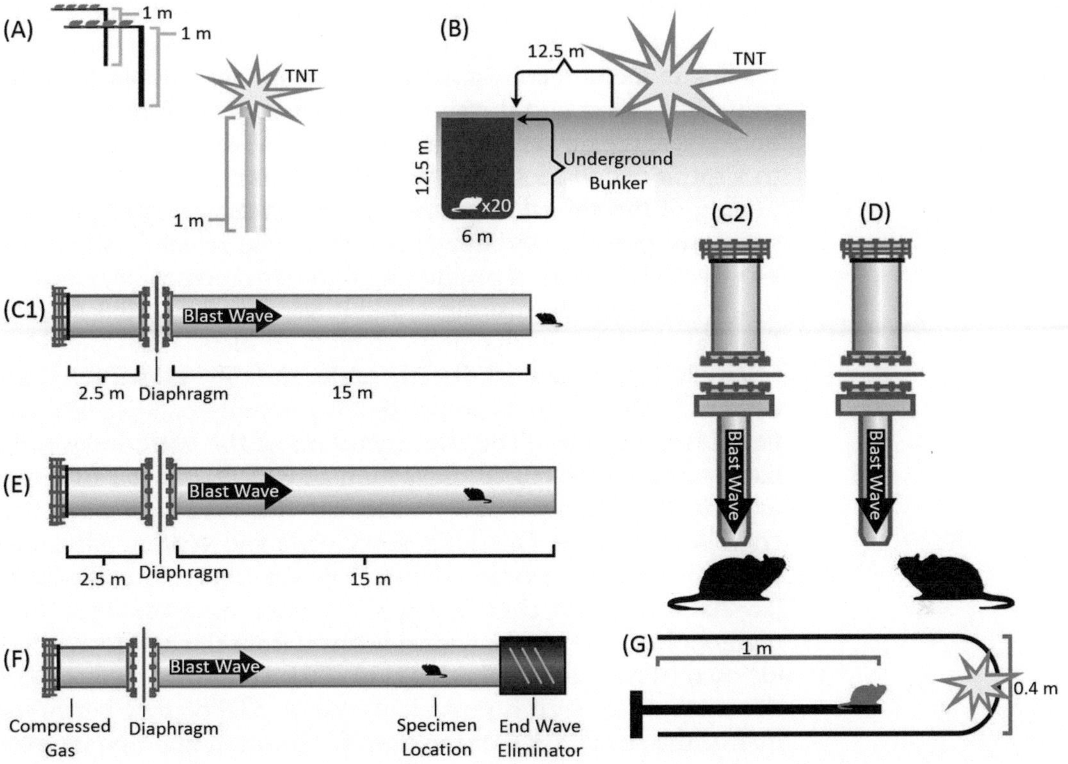

Fig. 3 Examples of different blast experimental setups. (**A**) Open-field testing (Rubovitch et al. [15]) (**B**) Bunker setup (Kaur et al.). (**C1**) Shock tube with animal outside (Long et al. [75]). (**C2**) Shock gun with animal on-axis (Svetlov et al.). (**D**) Shock gun with animal off-axis (Svetlov et al.). (**E**) Shock tube with an open end (Nambier et al.). (**F**) Advanced blast simulator (VandeVord et al. [54]). (**G**) Combined explosive in tube (Saljo et al.)

testing conditions is nearly impossible given variability between explosive devices and reflections of the blast wave. This makes reproducing results and testing set parameters of blast overpressure and duration a formidable task. Though this may mimic IED warfare in theatre, consistent exposure is crucial for obtaining statistical significance within research data. Static overpressure profiles do not resemble the ideal Friedlander waveform and positive durations have been measured up to 18 ms [14]. Obtaining the clearance to perform open-field blast studies in an approved facility is difficult. Yet, some researchers have used explosives in conjunction with blast tubes to replicate battlefield explosions in an experimental environment, eliminating the need for open space (Fig. 3G) [17]. Another drawback is that secondary injuries, such as those sustained from debris, often occur with explosives setups, thus precautions (e.g., animal shielding) are required in order to limit studies to primary injuries (*see* **Note 4**). While it is difficult to isolate primary blast wave, animal protection such as containment within a cage has been utilized for open-field blasts [15, 18].

1.3 Shock Guns

Since live detonation testing is not optimal for research laboratories, several devices have been constructed to recreate shock waves. The shock gun method consists of a narrow tube (usually vertical) that contains a driver and driven tube separated by a diaphragm. The bursting of the diaphragm creates a shock wave that is transmitted to a specimen positioned outside of the tube. In this setup, positioning of the animal is crucial in terms of achieving exposure to static overpressure without exposure to the dynamic winds (*see* **Note 4**). The effects of positioning the animal perpendicular to the shock wave have been studied, and it was found that this orientation causes head accelerations atypical of blast TBI (Fig. 3C2) [19]. Placing the animal directly under the tube causes very high dynamic overpressure exposure that is not representative of open-field blast exposure. The fast expansion of the wave leaving the narrow tube causes rapid dissipation of energy and can therefore make it difficult to produce an accurate free-field blast wave. The combination of the emerging shock wave and venting gas causes the formation of a vortex and high-flow-velocity region called end-jet. As such, researchers using this device have modified their methodology so that the animal is offset from the end-jet and not exposed to reflections (Fig. 3D) [11, 20]. One advantage of using this method is that it is possible to use on a laboratory bench top due to the smaller size of the device. However, appropriate measurements need to be collected to verify that the conditions resemble an appropriate blast environment.

1.4 Conventional Shock Tubes (ST) and Advanced Blast Simulators (ABS)

Historically, conventional shock tubes (ST) have been used to mimic blast conditions within the laboratory setting. This method allows for manipulation of the shock wave within a controlled environment with high repeatability. Most recently, modifications to the ST have led to the design of the advanced blast simulator (ABS), which was designed to intrinsically replicate all the key features of blast wave flow conditions, including the negative phase and secondary shock as described previously by VandeVord et al. (Chapter 7). The ST is composed of two separate chambers: the driver, where the pressure is created by means of an air compressor system or other gas, and the driven, where the shock wave propagates through the test section [21]. Because the wave is produced by compressed gas bursting a membrane instead of an actual chemical explosion, the term shock wave is used instead of blast wave. It is important to understand how the shock wave develops within the tube and how end-of-tube rarefaction leads to an imbalance of high dynamic pressures and yet reduced static pressure conditions, amounting to extremely adverse effects. Thus, experiments staged with a specimen near the end of the tube, where the static pressure decreases and dynamic pressure increases, should be avoided (Fig. 3C1; *see* **Note 4**). In order to create a more accurate blast environment for animal testing, the ABS was designed. There are three chambers in the ABS device: a driver, driven, and end-wave eliminator (EWE) (Fig. 3F). The EWE consists of a dump tank that can contain the expanding gases from entering the lab space and at the same time creates some overpressure reflection that counteracts the rarefaction wave. Baffling is incorporated into the EWE to break up the venting shock front and prevents the waves from traveling back up the device, in contrary to the ST which has an open end causing a reflection of the wave and exposing the animal to multiple extraneous shocks that do not exist in real blast conditions (Fig. 3E). A disadvantage of using either the ST or ABS designs is that shock flow constraints require less than 20 % restriction of specimen in the device in order to recreate the most accurate blast flow conditions. Large animal studies would require a much larger chamber (approximately 16 square foot cross section) or the addition of an expansion section for optimal flow specifications, which ultimately leads to laboratory space concerns.

2 Materials

2.1 Animals

Male adult Sprague Dawley rats (250–300 g) from Harlan Labs (San Diego, CA, USA) are used for these experiments. Animals are acclimated 12-h light/dark cycle with food and water provided ad lib in an Association for Assessment and Accreditation of Laboratory Animal Care-approved facility.

2.2 Anesthesia

The animals are anesthetized with a continuous flow of 3 % isoflurane and 60 % oxygen for 5 min.

2.3 Materials for Testing

1. Compressed gas (helium preferred).
2. ABS (driver section, test section, end-wave eliminator).
3. Membrane (Acetate, Mylar, Vinyl, Metal; Grafix Plastics, Cleveland, OH, USA).
4. Pressure sensors (PCB Piezotronics Inc., Depew, NY, USA).
5. Signal Conditioner (PCB Piezotronics Inc., Depew, NY, USA).
6. Dash 8HF data acquisition system (Astro-Med, Inc, West Warwick, RI, USA).
7. Animal sling holder (Custom made from mesh).
8. High-speed video camera (Phantom Miro eX2, Vision Research, Wayne, NJ, USA).
9. Physiological monitoring system (Nonin PulseSense VET Pulse Oximeter, Henry Schein Inc. Melville, NY, USA).

3 Primary Mechanisms of Blast Neurotrauma

In all of the described methods for inducing blast neurotrauma, the mechanical insult from a blast wave ultimately causes direct damage to cells in terms of morphological defects. While there are several hypotheses regarding which component of blast loading is most closely correlated to injury, it is accepted that brain cells are exposed to high-speed mechanical impulses. bTBI is a diffuse injury due to the global exposure to the entire skull. This widespread injury differentiates bTBI from blunt impact TBI, which is a focal insult accompanied by head rotational acceleration [22]. Even though energy translation leading to injury differs between blunt and blast TBI, cortical, and hippocampal injuries have been observed with both mechanisms, as well as similar clinical symptoms [23–25]. Shearing and stretching of cells occur due to the imparted wave stress at the microscopic level. Since the methodology of inducing BINT may lead to differing injuries, the following sections aim to provide a comparison of the primary injury mechanisms of BINT resulting from the various experimental blast models. Identifying specific biomarkers of primary injury mechanisms can serve to differentiate between the experimental methods of BINT.

3.1 Cytoskeletal Damage

The cytoskeleton is the supporting network of fibers and filaments in the cytoplasm of a cell and has various functions, including deformation resistance through supporting structure, cell signaling, and intracellular transport. Damage to the cytoskeleton has been shown to elicit downstream abnormalities in acute brain injury [26]. Due to cellular shearing during exposure to blast, cytoskeletal breakage can occur, impairing cytoskeletal functions. A study of explosive-induced blast neurotrauma reported an increase of phosphorylated neurofilament proteins (p-NFH) in the cortex and hippocampus following blast exposure at 240 kPa at 18 h post-blast [17]. These changes were found during acute stages and were resolved by 21 days after blast [17]. Cytoskeletal degradation in the hippocampus as well as the cortex was observed in several low-level (11.5 kPa) shock gun studies at 12 h post-exposure with some recovery [27, 28]. The ST model produced a decrease of actin (microfilament protein) in the nucleus accumbens at 3 days post-blast [29], as well as an increased amount of cytoskeletal enzymes, which contribute to the degradation of structural cytoskeletal fibers, in the cortex and cerebellum [30]. These changes were observed at 1 and 7 days following BOP exposure. While there was natural resolution of cytoskeletal impairments in an explosive model at 240 kPa by 21 days, there was faster recovery (by 7 days) in an 11.5 kPa BOP shock gun model. These results show that cytoskeletal degradation is a common outcome occurring throughout the brain; however, the extent of degradation and recovery time differ based on methodology used to induce injury and magnitude of BOP.

3.2 Blood–Brain Barrier Dysfunction

The blood–brain barrier (BBB) is a selective membrane that is essential to creating a controlled environment suitable for brain function. The BBB functions to protect the brain from pathogens and other harmful molecules through selective transport and permeability. The loss of integrity of this barrier may lead to neurological dysfunction and disease. Following TBI, it is known that the BBB loses integrity which enhances subsequent damage through neuroinflammation and increased permeability [31]. Similarly, BBB disruption has been reported following BINT. BBB permeability/integrity is identified by evaluating the quantity of endothelial cells in models of BINT. While secondary mechanisms ensue following injury and can affect BBB integrity, BBB disruption, including acute microlesions, has been reported as the result of the mechanical insult. In a shock gun study which utilized a rifle barrel to generate the shock wave, BBB disruption was observed to be dependent on the magnitude of the blast overpressure from 145 to 323 kPa but independent of the time point of assessment [20]. BBB breakdown, characterized by immunoglobulin (IgG) extravasation, was also reported in open-ended ST studies to be present in the cortex at 3 and 24 h post-blast but not at longer time points [32, 33]. Skotak et al. [34] found BBB degradation from over 190 kPa BOP in an open-ended ST experiment at 24 h post-exposure. However, these changes were found throughout the brain and were not localized to the cortex. BBB disruption appears to occur immediately in high-severity blasts and is possibly the best correlate of cellular damage along with primary blast level [20, 32].

3.3 Neurovascular Disruption

Cerebral vasculature includes arteries and veins that supply the brain with oxygen and nutrients necessary for brain function and transport of blood back to the heart. Any disruption of these vessels could have catastrophic consequences on brain health. In concussive brain injury, traumatic cerebral vascular injury plays a large role in pathophysiology [35]. Abnormalities in vascular structure and impaired vascular integrity are seen in morphological stains and can be due to the shearing effects at density interfaces during injury. Rapid overpressure is reported to cause mechanical damage at the brain-blood interface, which can be identified by vascular damage [36]. In open-field explosive studies, narrowed and permeable vasculature were found in a mouse mode of BINT with pressures ranging from 48 to 77 kPa and recovery up to 4 days post-exposure [14]. Cerebral microvascular lesions and downregulation of type IV collagen (basal lamina component) are commonly reported in studies using a ST with the rodent near the end of the tube, demonstrating a diffuse response throughout the brain [37–40]. Gama Sosa et al. [38] reported shear-related injuries, such as microhemorrhage and degeneration in cerebral microvessels, in cortical vessels in a repeated-exposure (74.5 kPa) ST model at 24 h and 10 months post-exposure. Kamnaksh et al. [40] reported increased plasma levels of vascular endothelial growth factor

(VEGF) in the repeated-exposure group at 2 h post-exposure and in the single-blast group at 22 days post-exposure. Damage of cerebral vasculature was observed with all blast methodologies but the time course of recovery is debatable, depicting the diffuse nature of BINT resulting in lasting damage. Multiple blast exposures can also contribute to early presentation of vascular damage.

3.4 Axonal Injury

The axon is an extension of the neuron that transmits impulses away from the soma to the axonal terminal to relay information. Axons can signal over long distances; thus, having axonal integrity is crucial for optimal brain activity. In TBI, axonal injury, characterized by axonal swelling and cytoskeletal damage, is commonly found due to the susceptibility of axons to deformation [41]. In BINT, axonal injury has been identified in some modalities. Axonal injury is often diagnosed using beta-amyloid precursor protein (β-APP), but gross morphology through histological analysis can also be a predictor. β-APP is an integral membrane protein that gives rise to beta-amyloid following posttranslational modifications. In explosive open-field modalities, morphological axonal abnormalities quantified using diffusion tensor imaging (DTI) at 7 and 30 days post-blast have been observed [15] as well as increased β-APP at 24 h post-blast for 49 and 77 kPa exposures [14]. An abnormal distribution of p-NFH, indicating damage of axons, was seen in a model with an explosive driver at 240 kPa [17]. Axonal injury (increased β-APP) at acute stages has also been observed for experimental tests outside the ST using explosives to generate the shock wave [36]. The reports suggest that nonuniform, uncontrolled shock waves from explosive devices will result in high injury severity that culminates in axonal disruption.

In an open-end ST model, Valiyaveettil, et al. [30] found diffuse axonal injury following repeated blasts on mice, while Koliatsos et al. [42] reported axonal injury with single exposures of 68–183 kPa static overpressure. Damage was localized to the cerebellum and brainstem. In these studies, multiple blast exposures could have an effect on progression of axonal injury throughout the brain. Conversely, using a model with animals placed within the ST [43], β-APP was not found to be elevated within the hippocampus; thus, no axonal injury was identified. While pressure magnitude likely affects the results, one possible explanation is that a reflection wave due to an open end produces an upstream shock that causes multiple or enhanced shock exposures that would cause uncharacteristic shearing of axons in specific regions of the brain. Garman et al. [33] reported that exposure in an ST produced deep axonal injury in the cerebellum and brainstem, with tendencies of diffuse axonal injury. However, the animal was placed in a body shielding device in which the head will undergo rapid accelerations (brainstem damage is an indicator of acceleration-based injury) while the body is shielded, giving an inaccurate representation of true primary blast exposure and the possibility of the tertiary mode of blast injury.

4 Secondary Mechanisms of Blast Neurotrauma

Following the initial mechanical insult, secondary injury pathologies ensue through the activation of various cellular cascades. The cellular response to blast injury encompasses but is not limited to chronic inflammation and oxidative stress. Similar responses have been observed in other neurological disorders and have been linked with cognitive impairments [44, 45]. While the roles of pro-inflammatory and pro-oxidative pathways have been investigated, their temporal appearance and subsequent progression in BINT remains unknown. Since various methods of inducing BINT have led to differing pathologies, the following section aims to provide a comparison of the secondary mechanism of BINT resulting from different experimental blast models.

4.1 Inflammation

Brain injury, including BINT, is followed by a chronic inflammatory response characterized by a sustained activation of glial cells including microglia and astrocytes [46, 47]. The response from glial cells is important to cell survival and neuroprotective efforts; however if not controlled, it can contribute to sustained brain injury. Microglia are the resident immune cells of the central nervous system (CNS) and their activation following injury involves the release of several inflammatory molecules. Several studies have observed microglia activation as a result of BOP exposure both in the acute and chronic stages of recovery [18, 32, 48–50]. Microglia activation has not been investigated as a result of the shock gun method but has been demonstrated in both open-field and ST experimental setups. In the open-field setup, Kaur et al. [49] found lasting microglia activation for 14 days in the pineal gland. Similarly, with animals placed inside an ST, increased activation of microglia was observed for up to 30 days in the hippocampus and brainstem [48]. An increase in the microglia population, observed using Iba-1 and ox-42, has been reported for open-field, ABS, and ST testing [14, 49–51]. These changes were reported in the hippocampus and corpus callosum and have been shown to correlate with injury severity [14, 51]. This indicates a diverse response of microglia with varied pressure magnitudes. Such diversity may imply that microglial response may be suitable for use as a biological pressure sensor in which activation of certain molecular pathways depends on the severity of the blast exposure.

Alterations in astrocyte activation have also been observed. Astrocytes are a key player in brain homeostasis and their activation occurs through a mechanism of hypertrophy coupled with an increase of intermediate filaments such as glial fibrillary acidic protein (GFAP). Widespread changes to astrocyte intensity, through swelling and proliferation, have been observed in several brain regions following BINT with various experimental set-ups. Increases in GFAP following blast have been reported in the hippocampus [43, 48, 50–54], cortex [50, 53–55], amygdala [53, 54], and other brain regions [29, 48] and appear to correlate with elevated levels of GFAP in the blood [11, 40]. Both acute (less than 3 days) and chronic changes (lasting up to 21 days) in GFAP expression and protein levels have been reported following BINT in rodent models. Using a shock gun method, Svetlov et al. [52] found no changes to GFAP levels in the cortex but found increased GFAP levels in the hippocampus 7 days after exposure. More prolonged GFAP changes have been observed in the open-field, ABS, and ST experimental setups. In the open-field setup, GFAP immunoreactivity increased both in the cortex and in the hippocampus for up to 21 days following injury [50]. Similar results were found in the hippocampus using an ST and ABS with the animal within the device [43, 48, 51, 54] and an open ST with the animal located just outside [53].

Inflammation is regulated by pro-inflammatory mediators called cytokines. Cytokines including various interleukins, tumor necrosis factor α (TNF-α), and interferon γ (IFN-γ) have been implicated in the pathology of BINT. The involvement of these molecules has been demonstrated in animal models and multiple experimental setups but their role in the pathology of the injury is not yet fully understood.

Interleukins are a family of cytokines that play a critical role in mediating the inflammatory and immune response, several of which have been found to change following BINT [53, 56, 57]. Interleukin 6 (IL-6) has been the most studied interleukin likely due to its involvement in the inflammatory response following brain injury [58]. Using a rat model, IL-6 protein levels were found to be increased in the hippocampus and amygdala [53]. The injury was administered using an open-ended ST with the animal placed at the end of the tube. Valiyaveettil et al. [57] showed that mRNA expression of IL-6 was increased in the midbrain of mice placed within the ST. These results indicate upregulation of IL-6 in several brain regions in the acute stages following BINT which differ from results observed in serum. Sajja et al. [56] found that IL-6 protein levels were decreased in the blood serum from rats 72 h following injury using an open-end ST with animals placed inside. These results suggest potential local release of IL-6 from glia and/or relocalization of IL-6 from circulation to the site of injury. However, the inflammatory response appears to be delayed.

Other interleukins have been investigated following BOP exposure in animal models. Using an open-ended ST with the animal inside the tube, mRNA expression of various interleukins has been observed to vary within various brain regions. IL-2 was decreased in the hippocampus, IL-28 was decreased in the cerebellum but IL-7 increased in the frontal cortex [57]. Cytokine expression changes have not been limited to the brain. Interleukins have been shown to be altered in the blood serum in the acute stages following BOP exposure [56]. A significant increase in IL-5 was observed in the nucleus accumbens, anterior motor cortex, prefrontal cortex, and anterior striatum [59]. Authors speculate that IL-5 could be a key regulator of inflammation after blast injury since the response was so diffuse in the brain. Other reported cytokines were found elevated within the brain, including GM-CSF, IL-1α, IL-10, and IL-1β [56, 59].

TNF-α and IFN-γ are additional key cytokines found in the brain and are involved in the inflammatory response through transcriptional regulation [60, 61]. Dalle Lucca et al. [62] reported acute increases in TNF-α following injury using an open-ended ST with the animal placed inside the tube. Similarly, IFN-γ upregulation was demonstrated in the acute stages of recovery from BOP exposure. Cho et al. [63] found increased IFN-γ in the hippocampus resulting from the animal being placed within an ABS. Kamnaksh et al. [53] found a similar increase in the hippocampus as well as an increase in the amygdala following exposure from an open-end ST with the animal at the end. Upregulation of each of these cytokines may have a detrimental impact on cell survival. Many of the genes that are regulated by TNF-α are also regulated by IFN-γ and the combination of these cytokines can lead to a large increase in the transcription of these genes [64]. This transcriptional change may directly affect the chronic pathology of BINT.

4.2 Oxidative Stress

Oxidative stress occurs following the aberrant production of ROS, including hydrogen peroxide and superoxide anion. ROS are produced as by-products of normal cell metabolism, and play a role in cell signaling and homeostasis [65]. ROS levels are maintained by a dynamic equilibrium between their production during metabolism and degradation facilitated by antioxidants. Excess accumulation of ROS can cause neurotoxicity and neurodegeneration by prolonged upregulation of pro-inflammatory mediators [66] and has been shown to play a role in the secondary injury process of TBI and BINT. Increased levels of ROS have been found in both acute and chronic stages following blast injury [32, 63, 67–71]. The oxidative stress appears to be widespread following BINT as it has been observed in the hippocampus [63, 68], cortex [67, 69], and hypothalamus [70]. These changes have been found in rodent models using both the ST and shock gun experimental setups. The shock gun method has been shown to produce increased oxidative stress in the hypothalamus 6 h following injury [70], but longer time points have not been investigated. In the ABS setup (animal placed inside), increased oxidative stress was observed in the hippocampus from 4 h to 2 weeks post-injury [63]. Open-ended STs have shown similar increases in the hippocampus [32, 68] as well as increases in the cortex [67, 69]. The prolonged, widespread accumulation of ROS undoubtedly affects brain function, and has been shown to induce changes to the BBB that cause increased neuroinflammation [69]. This secondary effect was observed in an ST rat model [69] and an open-field explosion mouse model where BBB dysfunction was observed 1 month after injury [15]. This can potentially explain the chronic inflammatory response associated with the secondary injury mechanisms.

In order to ameliorate the pro-oxidative environment, superoxide dismutase 1 (SOD1) and superoxide dismutase 2 (SOD2) are natural antioxidants that function to restore the proper balance of superoxide radicals. These antioxidants have been shown to be altered following BINT. In rodent models, SOD1 and SOD2 expression and protein levels were increased following injury. Both open ST and ABS have shown elevated SOD1 in the hippocampus when the animal is placed within the device [51, 68, 72]. In the open-field setup, Rubovitch et al. [15] observed altered SOD2 levels in the areas surrounding vasculature. Cernak et al. [68] found transient SOD2 levels in the hippocampus following blast injury inside an open-end ST. SOD2 was significantly increased 5 days after injury but no change was found after 24 h. In contrast, Huber et al. [72] found increased SOD2 in the hippocampus 24 h after injury which subsided 30 days after injury. These results show that the antioxidant, SOD2, is upregulated in response to injury. This response may be triggered by the aberrant accumulation of ROS. Upregulation of antioxidant enzymes indicates potential neuroprotective efforts of glial cells to reduce the pro-oxidative environment. These efforts appear to decrease in the chronic stages as the antioxidant enzymes return to basal level expression. It has been well established that the accumulation of ROS is critical in BINT pathology. Therefore, natural methods to ameliorate the pro-oxidative environment become very critical to recovery.

5 Functional and Cognitive Outcomes Following Blast Neurotrauma

Cognitive and functional deficits have been observed clinically following blast-induced TBI [24]. Loss of motor function, memory deficits, and increased anxiety are some of the key features associated with BINT pathology. Experimental reproduction of blast injuries in animal models has shown similar outcomes. The changes have been observed predominantly in rodent models in controlled ST models, but other methodologies have been also used to characterize behavior after injury. This section begins to summarize some of the functional, behavioral, and cognitive changes observed following BINT in animal models.

5.1 Motor Function

As previously described, evidence of cellular disturbances in primary and secondary modes of blast injury has been observed within the cortex and can begin to explain the loss of motor function observed by BINT pathology. Loss of motor function in the rodent model has been well established and has been demonstrated in the shock gun, ST, and open-field experimental setups [28, 42, 48, 67, 73, 74]. In the open-field setup, animals showed decreased motor function up to 3 days following injury [74] but not at 7 or 30 days [15]. In a shock gun setup, Park et al. [27] found lasting motor function deficits until 9 days after exposure. Results from ST experiments show differing results. Some studies have shown that motor function deficits followed blast exposure and are apparent in the acute stages [42, 67, 73] but they have also been shown to last for 21 days following the injury [48]. However, other investigations have shown no deficits in motor function [54].

Motor function impairment appears to occur immediately following the blast injury and only be present temporarily [42, 48, 73, 75], which leads to a concern regarding anesthesia effects (*see* **Note 3**). While motor control may be diminished at the acute stage, the timeframe of impairment remains unclear. Using a shock tube paradigm, it has been shown that motor function returns to basal levels at 2 h following the injury according to the balance beam task [73]. While another study found that the motor function was not restored until 21 days later using the Rotarod task [48]. These differences may result from either the varied sensitivities of the performed tasks or the different magnitudes of BOP exposure (*see* **Note 5**).

5.2 Behavior

Anxiety-like behavior has been described as a symptom of BINT through animal models. Several behavioral tests have been performed to assess anxiety including an open-field test, light dark box, and the elevated plus maze. In all behavioral tests, anxiety following BINT has been observed [28, 40, 42, 48, 73]. While the shock gun, ST, and open-field experimental setups have been utilized to assess anxiety, only the shock gun and the ST setups produced anxiety-like behavior. The open-field experiment did not produce anxiety within 7 days following exposure [25]. Anxiety-like behavior has been found to persist in both the acute and chronic stages for up to 30 days following BOP exposure [48]. These changes likely result from primary injuries sustained to the amygdala and the activation of secondary cascades, which were noted and described previously.

5.3 Cognition

Primary and secondary mechanisms of blast neurotrauma have been reported to create biological changes in the hippocampus (which is crucial for various forms of memory). As a result, memory loss is an important outcome of BINT and brain injury in general. Memory deficits have been studied following BINT in the rodent model using a variety of cognitive tests. These tests have been conducted to investigate changes in recognition and recollection, associative memory, and spatial memory following BINT.

In order to investigate recognition memory, novel object recognition (NOR) tests have been performed using a rodent model of BINT [25, 63, 76]. This test capitalizes on the rodent's natural instinct to explore new objects in their environment. Blast exposure has been shown to correlate with a decline in recognition memory observed at 7 days [25, 63] and 14 days [63] following the injury induced by an open-field environment and the ABS. Interestingly, memory changes were not observed at 72 h following injury which may indicate a delayed response and dependence on secondary injury mechanisms [76].

Avoidance tasks use certain cues associated with aversive stimuli to test associative memory, which entails the ability of the animal to associate cues with a stimulus. These tests have demonstrated deficits in associative memory immediately following blast exposure [73] and lasting at least 30 days [48]. These behavioral deficiencies were observed following blast exposure in an open-end ST setup with the animal inside the tube and were not observed in an open-field setup [25]. However, associative memory changes have been demonstrated to be dependent on injury severity and overpressure magnitude [48, 73] which may be the cause of these contradictions.

Behavioral tests like the Y-maze, MWM, and Barnes maze are useful in assessing changes to spatial memory following BINT. Spatial memory is challenged based on the animals' ability to remember and perform in a previously explored environment. While deficits to spatial memory have been observed in rodents, the changes appear to be transient following BINT. Many studies have demonstrated deficits following BINT only at various time points in the acute stages following injury in the ST [42, 54, 73, 75, 77]. The results from these tests are inconsistent with open-field tests as lasting spatial memory deficits have been observed [15]. Still, other studies have found, both in the ST and in an open-field environment, little to no changes in spatial memory [25, 40]. The inconsistency between findings may result from the differences in the BOP exposure and subsequent injury severity which can cause differences in the primary and secondary mechanism of injury. Despite some contradictions, it appears that spatial memory is altered in the acute stages following BINT.

6 Notes

This chapter summarizes the various cellular and behavioral outcomes that are seen in different experimental setups of BINT. Common themes have been established, such as morphological defects, inflammation, and oxidative stress, but the specific time course at which these are presented differs with respect to the method of inducing bTBI. Differences in cognitive and functional outcomes are also seen between the various blast methodologies. There are many inconsistencies between researchers in the methodology of inducing bTBI. These inconsistencies manifest in altered cellular regulation due to the modes of injury and the injury severity. Some methods do not accurately recreate the free-field blast wave or use incorrect animal positioning, which can introduce secondary and tertiary forms of injury and are unrepresentative of BINT. In order to be able to more effectively to compare data between research groups, limitations need to be minimized. Such limitations include the following:

1. Improper animal harnessing which exposes the animal to tertiary injury and head accelerations and thus may lead to different injury pathologies.
2. Differences in pressure sensors and location of static pressure measurements which can lead to inconsistent relationships between peak static overpressure, dynamic pressure, and duration.
3. The use of different anesthetics during testing, some of which are neuroprotectants and have various recovery rates leading to inconsistent results.
4. Improper animal location which results in exposure to multiple waveforms or a dynamic blast wind injuring the live specimen.
5. Biological and behavioral assessments that vary between laboratories and have different sensitivities to pathological outcomes.

References

1. Elder GA, Mitsis EM, Ahlers ST, Cristian A (2010) Blast-induced mild traumatic brain injury. Psychiatr Clin North Am 33:757–781
2. Clemedson C-J, Criborn CO (1955) Mechanical response of different parts of a living body to a high explosive shock wave impact. Am J Physiol 181:471
3. Romba J, Martin P (1961) The propagation of air shock waves on a biophysical model
4. Moss WC, King MJ, Blackman EG (2009) Skull flexure from blast waves: a mechanism for brain injury with implications for helmet design. Phys Rev Lett 103:108702
5. Alley MD, Schimizze BR, Son SF (2011) Experimental modeling of explosive blast-related traumatic brain injuries. Neuroimage 54(Suppl 1):S45–S54
6. Bauman RA, Ling G, Tong L, Januszkiewicz A, Agoston D, Delanerolle N, Kim Y, Ritzel D, Bell R, Ecklund J, Armonda R, Bandak F, Parks S (2009) An introductory characterization of a combat-casualty-care relevant swine model of closed head injury resulting from exposure to explosive blast. J Neurotrauma 26:841–860
7. Leonardi A (2011) An investigation of the biomechanical response from shock wave loading to the head. In: Biomedical engineering. Wayne State University, Detroit, MI
8. Leonardi AD, Bir CA, Ritzel DV, VandeVord PJ (2011) Intracranial pressure increases during exposure to a shock wave. J Neurotrauma 28:85–94
9. Bolander R, Mathie B, Bir C, Ritzel D, VandeVord P (2011) The contribution of skull flexure as a possible mechanism for neurotrauma in the rat when exposed to a shock wave. Ann Biomed Eng 39:2550–2559
10. Hicks R (2014) Common data elements (CDE) for preclinical TBI research, NIH/NINDS
11. Svetlov SI, Prima V, Glushakova O, Svetlov A, Kirk D, Gutierrez H, Serebruany V, Curley K, Wang KKW, Hayes RL (2012) Neuro-glial and systemic mechanisms of pathological responses in rat models of primary blast overpressure compared to 'composite' blast. Front Neurol 3:15
12. Chai JK, Liu W, Deng HP, Cai JH, Hu QG, Zou XF, Shen CA, Yin HN, Han YF, Zhang XB, Chi YF, Ma L, Sun TJ, Feng R, Lan YT

(2013) A novel model of burn-blast combined injury and its phasic changes of blood coagulation in rats. Shock 40:297–302

13. Kinney GF (ed) (1962) Explosive shocks in air. Springer, New York, NY
14. Pun PB, Kan EM, Salim A, Li Z, Ng KC, Moochhala SM, Ling EA, Tan MH, Lu J (2011) Low level primary blast injury in rodent brain. Front Neurol 2:19
15. Rubovitch V, Ten-Bosch M, Zohar O, Harrison CR, Tempel-Brami C, Stein E, Hoffer BJ, Balaban CD, Schreiber S, Chiu WT, Pick CG (2011) A mouse model of blast-induced mild traumatic brain injury. Exp Neurol 232:280–289
16. Axelsson H, Hjelmqvist H, Medin A, Persson JK, Suneson A (2000) Physiological changes in pigs exposed to a blast wave from a detonating high-explosive charge. Mil Med 165:119–126
17. Saljo A, Bao F, Haglid KG, Hansson HA (2000) Blast exposure causes redistribution of phosphorylated neurofilament subunits in neurons of the adult rat brain. J Neurotrauma 17:719–726
18. Kaur C, Singh J, Lim MK, Ng BL, Yap EP, Ling EA (1995) The response of neurons and microglia to blast injury in the rat brain. Neuropathol Appl Neurobiol 21:369–377
19. Gullotti DM, Beamer M, Panzer MB, Chen YC, Patel TP, Yu A, Jaumard N, Winkelstein B, Bass CR, Morrison B, Meaney DF (2014) Significant head accelerations can influence immediate neurological impairments in a murine model of blast-induced traumatic brain injury. J Biomech Eng 136:091004
20. Yeoh S, Bell ED, Monson KL (2013) Distribution of blood-brain barrier disruption in primary blast injury. Ann Biomed Eng 41:2206–2214
21. Celander H, Clemedson C-J, Ericsson UA, Hultman HI (1955) The use of a compressed air operated shock tube for physiological blast research. Acta Physiol Scand 33:6–13
22. Stemper BD, Pintar FA (2014) Biomechanics of concussion. Prog Neurol Surg 28:14–27
23. Belanger HG, Proctor-Weber Z, Kretzmer T, Kim M, French LM, Vanderploeg RD (2011) Symptom complaints following reports of blast versus non-blast mild TBI: does mechanism of injury matter? Clin Neuropsychol 25:702–715
24. Mac Donald CL, Johnson AM, Wierzechowski L, Kassner E, Stewart T, Nelson EC, Werner NJ, Zonies D, Oh J, Fang R, Brody DL (2014) Prospectively assessed clinical outcomes in concussive blast vs nonblast traumatic brain injury among evacuated US military personnel. JAMA Neurol 71:994–1002
25. Tweedie D, Rachmany L, Rubovitch V, Zhang Y, Becker KG, Perez E, Hoffer BJ, Pick CG, Greig NH (2013) Changes in mouse cognition and hippocampal gene expression observed in a mild physical- and blast-traumatic brain injury. Neurobiol Dis 54:1–11
26. Fitzpatrick MO, Dewar D, Teasdale GM, Graham DI (1998) The neuronal cytoskeleton in acute brain injury. Br J Neurosurg 12:313–317
27. Park E, Gottlieb JJ, Cheung B, Shek PN, Baker AJ (2011) A model of low-level primary blast brain trauma results in cytoskeletal proteolysis and chronic functional impairment in the absence of lung barotrauma. J Neurotrauma 28:343–357
28. Park E, Eisen R, Kinio A, Baker AJ (2013) Electrophysiological white matter dysfunction and association with neurobehavioral deficits following low-level primary blast trauma. Neurobiol Dis 52:150–159
29. Sajja VS, Galloway M, Ghoddoussi F, Kepsel A, VandeVord P (2013) Effects of blast-induced neurotrauma on the nucleus accumbens. J Neurosci Res 91:593–601
30. Valiyaveettil M, Alamneh YA, Wang Y, Arun P, Oguntayo S, Wei Y, Long JB, Nambiar MP (2014) Cytoskeletal protein α–II spectrin degradation in the brain of repeated blast exposed mice. Brain Res 1549:32–41
31. Alves JL (2014) Blood-brain barrier and traumatic brain injury. J Neurosci Res 92:141–147
32. Readnower RD, Chavko M, Adeeb S, Conroy MD, Pauly JR, McCarron RM, Sullivan PG (2010) Increase in blood-brain barrier permeability, oxidative stress, and activated microglia in a rat model of blast-induced traumatic brain injury. J Neurosci Res 88:3530–3539
33. Garman RH, Jenkins LW, Switzer RC 3rd, Bauman RA, Tong LC, Swauger PV, Parks SA, Ritzel DV, Dixon CE, Clark RS, Bayir H, Kagan V, Jackson EK, Kochanek PM (2011) Blast exposure in rats with body shielding is characterized primarily by diffuse axonal injury. J Neurotrauma 28:947–959
34. Skotak M, Wang F, Alai A, Holmberg A, Harris S, Switzer RC, Chandra N (2013) Rat injury model under controlled field-relevant primary blast conditions: acute response to a wide range of peak overpressures. J Neurotrauma 30:1147–1160
35. DeWitt DS, Prough DS (2003) Traumatic cerebral vascular injury: the effects of concussive brain injury on the cerebral vasculature. J Neurotrauma 20:795–825
36. Kuehn R, Simard PF, Driscoll I, Keledjian K, Ivanova S, Tosun C, Williams A, Bochicchio G, Gerzanich V, Simard JM (2011) Rodent model of direct cranial blast injury. J Neurotrauma 28:2155–2169
37. Gama Sosa MA, De Gasperi R, Paulino AJ, Pricop PE, Shaughness MC, Maudlin-Jeronimo

E, Hall AA, Janssen WG, Yuk FJ, Dorr NP, Dickstein DL, McCarron RM, Chavko M, Hof PR, Ahlers ST, Elder GA (2013) Blast overpressure induces shear-related injuries in the brain of rats exposed to a mild traumatic brain injury. Acta Neuropathol Commun 1:51

38. Gama Sosa MA, De Gasperi R, Janssen PL, Yuk FJ, Anazodo PC, Pricop PE, Paulino AJ, Wicinski B, Shaughness MC, Maudlin-Jeronimo E, Hall AA, Dickstein DL, McCarron RM, Chavko M, Hof PR, Ahlers ST, Elder GA (2014) Selective vulnerability of the cerebral vasculature to blast injury in a rat model of mild traumatic brain injury. Acta Neuropathol Commun 2:67

39. Bir C, Vandevord P, Shen Y, Raza W, Haacke EM (2012) Effects of variable blast pressures on blood flow and oxygen saturation in rat brain as evidenced using MRI. Magn Reson Imaging 30:527–534

40. Kamnaksh A, Kwon SK, Kovesdi E, Ahmed F, Barry ES, Grunberg NE, Long J, Agoston D (2012) Neurobehavioral, cellular, and molecular consequences of single and multiple mild blast exposure. Electrophoresis 33:3680–3692

41. Smith DH, Meaney DF, Shull WH (2003) Diffuse axonal injury in head trauma. J Head Trauma Rehabil 18:307–316

42. Koliatsos VE, Cernak I, Xu L, Song Y, Savonenko A, Crain BJ, Eberhart CG, Frangakis CE, Melnikova T, Kim H, Lee D (2011) A mouse model of blast injury to brain: initial pathological, neuropathological, and behavioral characterization. J Neuropathol Exp Neurol 70:399–416

43. Sajja VS, Galloway MP, Ghoddoussi F, Thiruthalinathan D, Kepsel A, Hay K, Bir CA, VandeVord PJ (2012) Blast-induced neurotrauma leads to neurochemical changes and neuronal degeneration in the rat hippocampus. NMR Biomed 25:1331–1339

44. Amor S, Puentes F, Baker D, Van Der Valk P (2010) Inflammation in neurodegenerative diseases. Immunology 129:154–169

45. Uttara B, Singh AV, Zamboni P, Mahajan RT (2009) Oxidative stress and neurodegenerative diseases: a review of upstream and downstream antioxidant therapeutic options. Curr Neuropharmacol 7:65–74

46. Kou Z, VandeVord PJ (2014) Traumatic white matter injury and glial activation: from basic science to clinics. Glia 62:1831–1855

47. Kumar A, Loane DJ (2012) Neuroinflammation after traumatic brain injury: opportunities for therapeutic intervention. Brain Behav Immun 26:1191–1201

48. Cernak I, Merkle AC, Koliatsos VE, Bilik JM, Luong QT, Mahota TM, Xu L, Slack N, Windle D, Ahmed FA (2011) The pathobiology of blast injuries and blast-induced neurotrauma as identified using a new experimental model of injury in mice. Neurobiol Dis 41:538–551

49. Kaur C, Singh J, Lim MK, Ng BL, Ling EA (1997) Macrophages/microglia as 'sensors' of injury in the pineal gland of rats following a non-penetrative blast. Neurosci Res 27:317–322

50. Saljo A, Bao F, Hamberger A, Haglid KG, Hansson HA (2001) Exposure to short-lasting impulse noise causes microglial and astroglial cell activation in the adult rat brain. Pathophysiology 8:105–111

51. Sajja VS, Ereifej ES, VandeVord PJ (2014) Hippocampal vulnerability and subacute response following varied blast magnitudes. Neurosci Lett 570:33–37

52. Svetlov SI, Prima V, Kirk DR, Gutierrez H, Curley KC, Hayes RL, Wang KK (2010) Morphologic and biochemical characterization of brain injury in a model of controlled blast overpressure exposure. J Trauma 69:795–804

53. Kamnaksh A, Kovesdi E, Kwon SK, Wingo D, Ahmed F, Grunberg NE, Long J, Agoston DV (2011) Factors affecting blast traumatic brain injury. J Neurotrauma 28:2145–2153

54. VandeVord PJ, Bolander R, Sajja VSS, Hay K, Bir CA (2012) Mild neurotrauma indicates a range-specific pressure response to low level shock wave exposure. Ann Biomed Eng 40:227–236

55. de Lanerolle NC, Bandak F, Kang D, Li AY, Du F, Swauger P, Parks S, Ling G, Kim JH (2011) Characteristics of an explosive blast-induced brain injury in an experimental model. J Neuropathol Exp Neurol 70:1046–1057

56. Sajja VS, Tenn C, McLaws LJ, Vandevord PJ (2012) A temporal evaluation of cytokines in rats after blast exposure. Biomed Sci Instrum 48:374–379

57. Valiyaveettil M, Alamneh YA, Miller SA, Hammamieh R, Arun P, Wang Y, Wei Y, Oguntayo S, Long JB, Nambiar MP (2013) Modulation of cholinergic pathways and inflammatory mediators in blast-induced traumatic brain injury. Chem Biol Interact 203:371–375

58. Lenzlinger P, Morganti-Kossmann M-C, Laurer H, McIntosh T (2001) The duality of the inflammatory response to traumatic brain injury. Mol Neurobiol 24:169–181

59. Sajja VSSS, Tenn C, Mclaws LJ, Vandevord P (2014) IL-5; a diffuse biomarker associated with brain inflammation after blast exposure. Biomed Sci Instrum 50:375

60. Feuerstein GZ, Liu T, Barone FC (1994) Cytokines, inflammation, and brain injury: role of tumor necrosis factor-alpha. Cerebrovasc Brain Metab Rev 6:341–360
61. Mühl H, Pfeilschifter J (2003) Anti-inflammatory properties of pro-inflammatory interferon-γ. Int Immunopharmacol 3:1247–1255
62. Dalle Lucca JJ, Chavko M, Dubick MA, Adeeb S, Falabella MJ, Slack JL, McCarron R, Li Y (2012) Blast-induced moderate neurotrauma (BINT) elicits early complement activation and tumor necrosis factor alpha (TNFalpha) release in a rat brain. J Neurol Sci 318:146–154
63. Cho HJ, Sajja VSSS, VandeVord PJ, Lee YW (2013) Blast induces oxidative stress, inflammation, neuronal loss and subsequent short-term memory impairment in rats. Neuroscience 253:9–20
64. Schroder K, Hertzog PJ, Ravasi T, Hume DA (2004) Interferon-gamma: an overview of signals, mechanisms and functions. J Leukoc Biol 75:163–189
65. Devasagayam TP, Tilak JC, Boloor KK, Sane KS, Ghaskadbi SS, Lele RD (2004) Free radicals and antioxidants in human health: current status and future prospects. J Assoc Physicians India 52:794–804
66. Lull M, Block M (2010) Microglial activation and chronic neurodegeneration. Neurotherapeutics 7:354–365
67. Wang Y, Wei Y, Oguntayo S, Wilkins W, Arun P, Valiyaveettil M, Song J, Long JB, Nambiar MP (2011) Tightly coupled repetitive blast-induced traumatic brain injury: development and characterization in mice. J Neurotrauma 28:2171–2183
68. Cernak I, Wang Z, Jiang J, Bian X, Savic J (2001) Ultrastructural and functional characteristics of blast injury-induced neurotrauma. J Trauma 50:695–706
69. Abdul-Muneer PM, Schuetz H, Wang F, Skotak M, Jones J, Gorantla S, Zimmerman MC, Chandra N, Haorah J (2013) Induction of oxidative and nitrosative damage leads to cerebrovascular inflammation in an animal model of mild traumatic brain injury induced by primary blast. Free Radic Biol Med 60:282–291
70. Tümer N, Svetlov S, Whidden M, Kirichenko N, Prima V, Erdos B, Sherman A, Kobeissy F, Yezierski R, Scarpace PJ, Vierck C, Wang KKW (2013) Overpressure blast-wave induced brain injury elevates oxidative stress in the hypothalamus and catecholamine biosynthesis in the rat adrenal medulla. Neurosci Lett 544:62–67
71. Ahmed FA, Kamnaksh A, Kovesdi E, Long JB, Agoston DV (2013) Long-term consequences of single and multiple mild blast exposure on select physiological parameters and blood-based biomarkers. Electrophoresis 34:2229–2233
72. Huber BR, Meabon JS, Martin TJ, Mourad PD, Bennett R, Kraemer BC, Cernak I, Petrie EC, Emery MJ, Swenson ER, Mayer C, Mehic E, Peskind ER, Cook DG (2013) Blast exposure causes early and persistent aberrant phospho- and cleaved-tau expression in a murine model of mild blast-induced traumatic brain injury. J Alzheimers Dis 37:309–323
73. Ahlers ST, Vasserman-Stokes E, Shaughness MC, Hall AA, Shear DA, Chavko M, McCarron RM, Stone JR (2012) Assessment of the effects of acute and repeated exposure to blast overpressure in rodents: toward a greater understanding of blast and the potential ramifications for injury in humans exposed to blast. Front Neurol 3:32
74. Moochhala SM, Md S, Lu J, Teng CH, Greengrass C (2004) Neuroprotective role of aminoguanidine in behavioral changes after blast injury. J Trauma 56:393–403
75. Long JB, Bentley TL, Wessner KA, Cerone C, Sweeney S, Bauman RA (2009) Blast overpressure in rats: recreating a battlefield injury in the laboratory. J Neurotrauma 26:827–840
76. Sajja VSSS, Perrine SA, Ghoddoussi F, Hall CS, Galloway MP, VandeVord PJ (2014) Blast neurotrauma impairs working memory and disrupts prefrontal myo-inositol levels in rats. Mol Cell Neurosci 59:119–126
77. Saljo A, Svensson B, Mayorga M, Hamberger A, Bolouri H (2009) Low-level blasts raise intracranial pressure and impair cognitive function in rats. J Neurotrauma 26:1345–1352

Chapter 9

Application of Systems Biology to Neuroproteomics: The Path to Enhanced Theranostics in Traumatic Brain Injury

Zaynab Jaber, Patrick Aouad, Mohamad Al Medawar, Hisham Bahmad, Hussein Abou-Abbass, and Firas Kobeissy

Abstract

The application of systems biology tools in analyzing heterogeneous data from multiple sources has become a necessity, especially in biomarker discovery. Such tools were developed with several approaches to address different types of research questions and hypotheses. In the field of neurotrauma and traumatic brain injury (TBI), three distinct approaches have been used so far as systems biology tools, namely functional group categorization, pathway analysis, and protein-protein interaction (PPI) networks. The databases allow for query of the system to identify candidate targets which can be further studied to elucidate potential downstream biomarkers indicative of disease progression, severity, and improvement. The various systems biology tools, databases, and strategies that can be implemented on available TBI data in neuroproteomic studies are discussed in this chapter.

Key words Systems biology, Functional group categorization, Pathway analysis, Neurotrauma, Protein-protein interaction network, TBI

1 Introduction

In the last decade, a wealth of proteomic information has been gathered using high-throughput technologies such as whole-genome sequencing and mass spectrometry. The development and use of these technologies necessitated a shift from the reductionist approach that focuses on features of a single molecule to the holistic approach that considers the system as a whole. This has allowed for the assessment of levels, modifications, and interactions of all macromolecules in different physiological and pathological conditions in a given cell, tissue, or organ. This approach, however, requires sophisticated computer software that can put together known interactions, and predict unknown ones depending on similarities in structure, composition, and placement to other known molecules.

Firas Kobeissy et al. (eds.), *Injury Models of the Central Nervous System: Methods and Protocols*, Methods in Molecular Biology, vol. 1462, DOI 10.1007/978-1-4939-3816-2_9, © Springer Science+Business Media New York 2016

Based on this premise, the discipline of systems biology developed to combine advanced mathematical and computer informatics tools with generated data to synthesize and infer biological hypotheses.

Systems biology utilizes intricate computational and mathematical algorithms, backed by experimentally validated and documented data, to predict the involvement of molecules in processes, pathways, mechanisms, or conditions. In this way, systems biology serves as a great tool for the analysis and prediction of factors involved in metabolic and cell signaling networks that govern cell function and interaction with its microenvironment. It is also associated with newly emerging disciplines that generate a plethora of complex raw data such as genomics, epigenomics, transcriptomics, proteomics, metabolomics, and lipidomics [1].

These "omics" studies investigate the level, modification, and interaction of macromolecules in an entire system. Recently gaining much popularity, proteomics concerns the study of all proteins in cells, tissues, organs, or biofluids. Proteomics studies complement genomics studies by elucidating the effect of gene expression and translation starting from the cell to the entire organism [1]. One of the main goals of proteomics is to interpret protein-protein interactions and functions in both physiological and pathological states for their crucial goal in diagnosis and therapy. In fact, proteomics allows for the discovery of novel proteins and peptides that could be indicators of disease manifestation, severity, or types, consequently acting as potential targets for drug design and development. The increased interest in proteomics is attributed to its high sensitivity and sophistication, in addition to its promise in defining new biomarkers for diagnosis and characterization of pathologies and traumas [2].

Many resources for the discipline of systems biology have been developed to study and categorize proteins in bio-fluids, cells, tissues, and organs as shown in Fig. 1. Tools such as Gene Ontology (GO) [3], PANTHER [4], UniProt [5], and Pfam [6] could be used to categorize comprehensive lists of genes and proteins into functional groups. GO categorizes based on placement of the molecule within the cell, processes it is involved in, and the type of its activity. Another set of tools employed in systems biology is under the umbrella of pathway analysis. Kyoto Encyclopedia of Genes and Genomes (KEGG) pathways database and Ingenuity Pathway Analysis software are extensively used in illustrating the complexity and hierarchy of pathways in a biological process. However, pathway analysis does not cover the entire proteome. Hence, software that characterizes networks of protein-protein interactions, such as Interactome and BioGRID, has been developed to complement pathway analysis and include proteins that are not found in pathways. These databases depend on data from high-throughput technologies for protein-protein interactions such as yeast two-hybrid screening and tandem affinity purification to experimentally document and validate interactions [1].

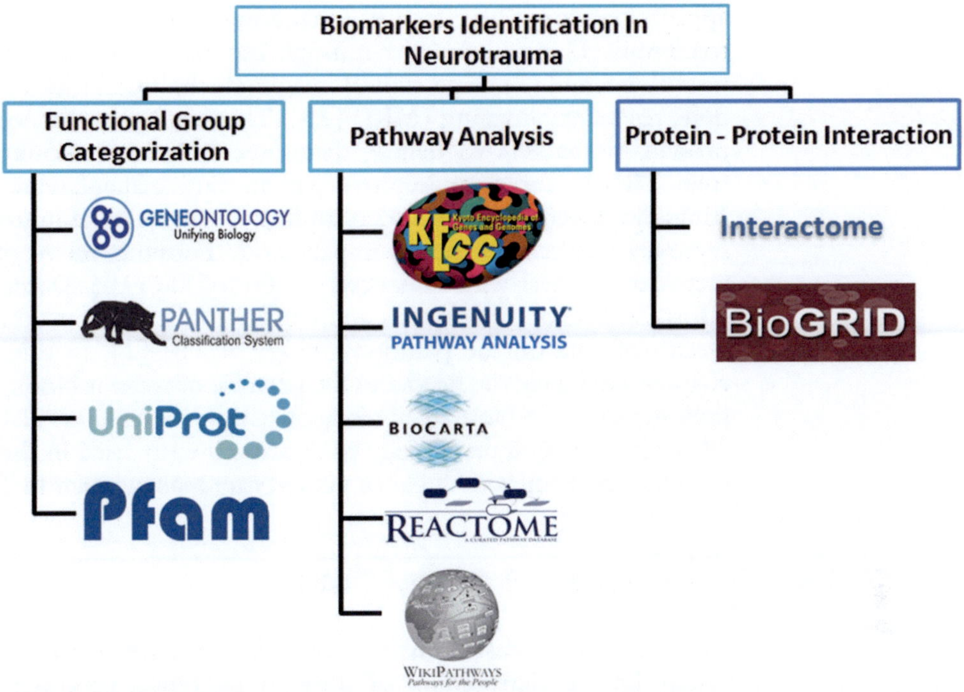

Fig. 1 Different system biology tools used for the identification of biomarkers in the field of traumatic brain injury

Biomarkers are considered one of the fundamental research-based proteomic tools that can be used in personalized medicine currently. They are physiological indicators that can be assessed or measured quantitatively in certain pathological states, whether a disease or an injury, to allow for a better diagnosis and assessment of the disease process and for monitoring response to treatment [7]. This proteomic tool is now so commonly used that its presence as primary endpoints in clinical trials is accepted without question [8]. While proteomics has excelled in furthering our understanding of cancer, injury, and aging, studies on neurological conditions such as traumatic brain injury are still in their infancy.

Traumatic brain injury (TBI), defined as neurotrauma caused by mechanical force applied to the head, is of particular interest, especially since it is a leading cause of death worldwide [7, 9]. TBIs pose a great health concern, with nearly two million people per year afflicted in the USA alone [10]. Of these, nearly 100,000 patients die, another 500,000 are hospitalized, and thousands of others suffer short- or long-term effects [11–13].

Despite the alarming statistics associated with TBI as a leading health epidemic, and that its worldwide prevalence is on the rise [14–16], there are no specific therapeutic treatments approved by the US Food and Drug Administration yet [9, 17]. It remains a challenge to define specific mechanisms for which therapeutic approaches would be effective. Furthermore, TBI may be difficult to identify (1) in patients with multiple traumas, (2) using cognitive tests alone, and (3) using clinical imaging techniques, such as magnetic resonance imaging (MRI) [18]. This highlights the need for protein biomarkers to define, diagnose, categorize, monitor, or treat TBI, in order to improve patient care management [10]. Biomarkers may assist in understanding mechanisms of injury and recovery, and can potentially impact wound healing, recovery, and increased survival with enhanced quality of life [10]. Decades of efforts in animal experiments and clinical studies have produced a wealth of data on the pathophysiology of TBI [1]. In this book chapter we discuss the need and the promise of systems biology and bioinformatics in biomarker research within the area of TBI, and what different systems biology tools are currently used in the field of neuroproteomics to enhance patient care management in TBI.

2 Systems Biology Applications to Neurotrauma Studies

Considered the latest domain in biological science that aims for a system-level understanding of complex biological processes [19–23], systems biology attempts to explain the function of molecules and biological structures in a cellular component under normal or perturbed conditions [24–26]. By evaluating the perturbation at the systems level, and by monitoring the changes as they are linked to the upstream and downstream components, we are able to predict the source of altered functionality in the biological system. To this end, systems biology assists in bridging the gap among the generated high-throughput data and combines experimental, basic science data sets, proteomic and genetic data sets, and literature and text mining and integrates it with computational modeling, bioinformatics, and pathway/interaction mapping methods (*see* Fig. 2). These systems biology components can provide a context or framework for understanding biological responses within physiological networks at the organism level, rather than in isolation [27].

Systems biology can be applied in brain injury, an outcome of complex biological systems responses and a condition that is mediated by several pathways at various time points. In vitro experiments on brain biopsy to simulate TBI or performing clinical trials for developing treatments for TBI is not an option. An ideal biomarker would be a perfect solution, but since such a biomarker has not yet been proven to exist, the best of what we can do is to identify a panel or signature of markers [28] and try to find out accurate networks that can describe and simulate the pathophysiology of TBI [29]. As a result, a neurosystems biological approach would be ideal for holistically studying the altered networks in TBI to better understand the underlying processes that are occurring. In this way, we can better understand the dynamic changes

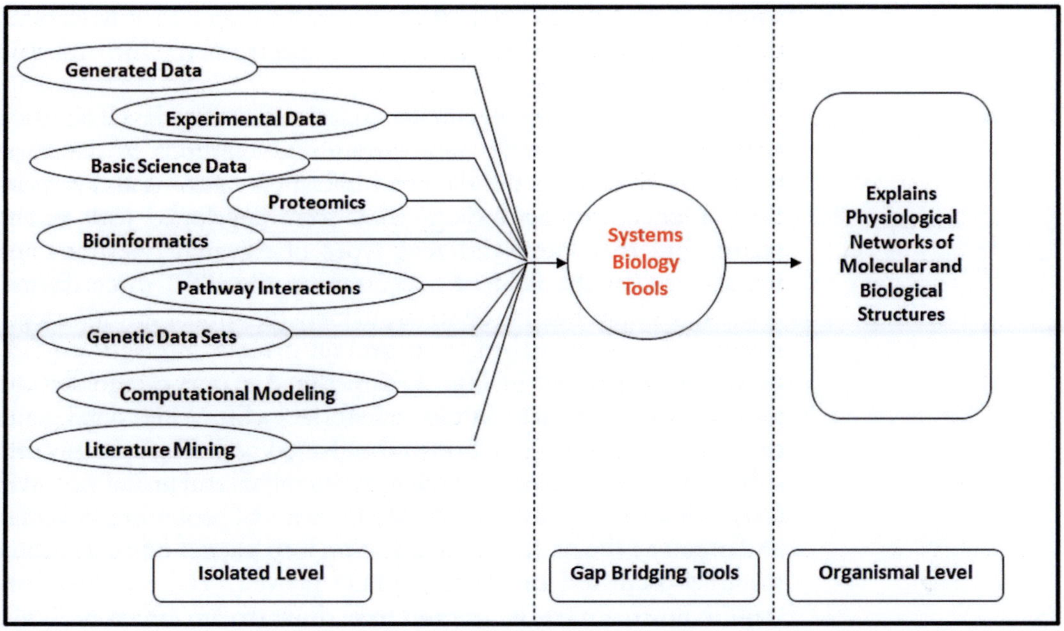

Fig. 2 Schematic representing how systems biology tools bridge the gap between protein analyses at isolated level to infer about protein interactions at organismal level

of mass proteins (potential biomarkers) in a specific condition at a particular time point. This endeavor is crucial towards discovering biomarkers with the capacity of indicating injury severity levels as well as the clinical outcome of patients with a TBI [30]. Additionally, systems biology has made systems modeling and simulation possible towards future development of effective therapeutics. It is now possible to generate visual presentations of network models by incorporating genome, proteome, and metabolome data through such bioinformatics software as Pathway Studio (Ariadne, Rockville, MD), Gene Spring (Agilent Technologies, Santa Clara, CA), and Ingenuity Pathway Analysis (IPA) (Ingenuity Systems, Redwood City, CA).

In fact, different formats have emerged that include quantitative data from biological studies and have found further use in system simulation and analysis, with the ability to construct functional interaction maps of the generated high-throughput data [31]. In this way, systems biology tools can be incorporated to overcome limitations of simple proteomics studies by using protein databases and canonical pathways to map the proteins detected through common proteomics approaches. These functional interaction maps can bring to light altered subsets of genes or proteins describing disturbed cellular functions characteristic to TBI. They can predict and identify certain genes or proteins that have been missed by experimental analysis, as well as provide potential functions of identified proteins with an unknown physiology [9]. For instance,

Boutte et al. (2012) used IPA software to map 321 differentially expressed proteins obtained by mass spectrometry into relevant cellular pathways [29, 32].

It is therefore important to complement the data collection and accumulation using high-throughput methods in databases with tools to analyze the data and infer new patterns and associations. Hence, systems biology tools were developed with several approaches to address different types of research questions and hypotheses. In the field of neurotrauma and TBI, three distinct approaches have been used so far, namely functional group categorization, pathway analysis, and protein-protein interaction (PPI) networks, illustrated in Fig. 3. Functional group categorization, also known as gene set analysis, categorizes a list of identified genes or proteins based on function and activated pathological responses subsequent to a neurotrauma. Pathway analysis and protein-protein interaction (PPI) networks overlap in terms of proteomic coverage and organize the molecules in question into a series of interactions, whether predicted or experimentally proven. These interactions require upstream regulators and have downstream effectors. While pathway analysis offers insights into how the sequence of interactions influence a biological process, PPI networks dissect the network of interactions of a single protein based on experimental data and the motifs it contains that allow prediction of potential activity. Moreover, the two approaches differ in the amount of proteins they cover, while pathway analysis presents useful insights on proteins that have been documented or predicted to be involved in a defined pathway only, whereas PPI networks can cover many more

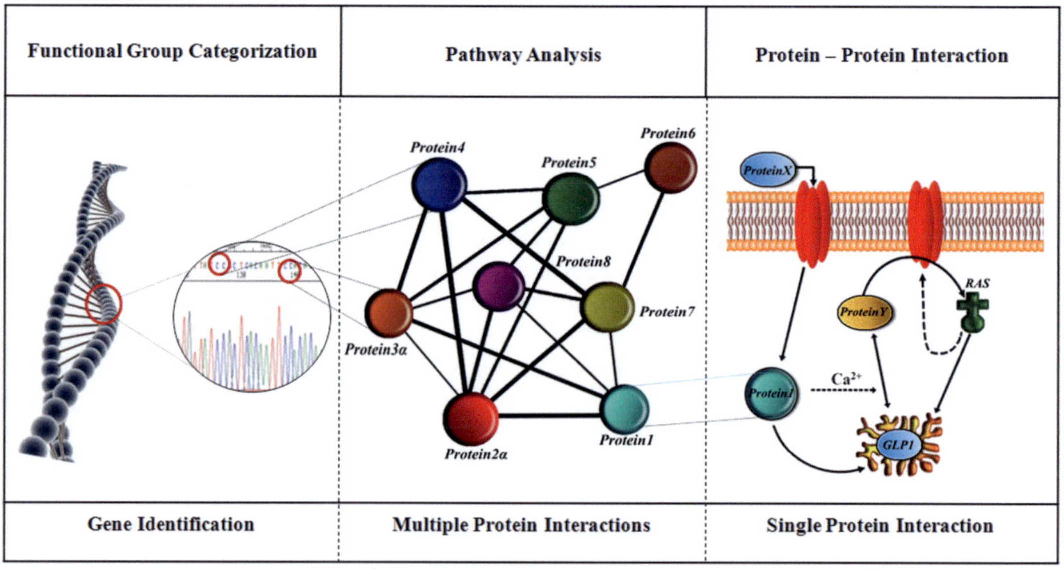

Fig. 3 Three different approaches used in TBI for identification of biomarkers, namely functional categorization, pathway analysis, and protein-protein interaction emphasizing the role of each tool

proteins not included in pathways yet but have been proven experimentally or predicted to interact with other proteins. In each approach, dozens of tools and analysis methods have been developed to serve different purposes, with some being more established and more extensively checked for quality and reliability than others. In this section, we review studies that have used these approaches in identifying biomarkers for TBI diagnosis, prognosis, and monitoring of therapeutics.

2.1 Functional Group Categorization

Before the introduction of clustering and gene set analysis tools, studies generating large quantities of data with genomics and proteomics had to be subject to manual clustering of genes. In 2002, a study by Matzilevich et al. clustered a set of 500 genes that they found had a change in their expression level upon cortical impact injury performed on rats [33]. The genes were manually categorized into 12 functional classes based on extensive literature searches and GenBank annotations of each gene. The following year, Natale et al. utilized the Gene Ontology database to classify differentially expressed genes in TBI mouse models into three functional categories including inflammation, transcriptional regulation, and extracellular formation/cell adhesion [34]. A decade later, in a recent review, Feala et al. gathered a list of high-throughput datasets for TBI [28].

Studies involving bottom-up methods for biomarker discovery lead to the identification of hundreds of genes and proteins. Since it would be impossible to study each one by one, the solution becomes the integration of high-throughput data with biological pathways and networks. Data is then further interpreted and screened through contextual "biological filters" [28]. Function annotations used to categorize genes include Pfam [6], PANTHER [4], and UniProt [5]. However, the most widely used function categorization scheme is Gene Ontology (GO) [3]. The GO project provides a controlled vocabulary of terms to standardize the annotation of genes and gene products. These GO terms are grouped in three main categories: molecular function, biological process, and subcellular localization. Within each category, a number of GO terms describe gene function for easy and flexible categorization of genes into groups with controlled functional granularity. A protein kinase, for example, could be annotated by a set of generic GO terms such as "macromolecule modification," "transferase activity," and more specific GO terms such as "kinase activity" and "protein kinase activity" [1]. In addition to GO, the DAVID Bioinformatics Resources integrate a number of databases, including GO, in one place to provide comprehensive function annotation, categorization, and systems biology analysis for user-uploaded data [1].

Pathway diagrams can be constructed based on experimental data, literature, and established theories or hypotheses. These pathway diagrams illustrate the molecular mechanism of a particular biological process. Databases such as KEGG (Kyoto Encyclopedia of Genes and Genomes) provide pathways that are constructed and regularly updated by experts from the Bioinformatics Center and the Human Genome Center at the University of Tokyo [35]. These pathways are organized into categories including metabolism, genetic and environmental information processing, cellular process, organismal systems, and human diseases. Additionally, KEGG provides pathways for a variety of species, including over 400 human pathways consisting of around 5000 proteins. KEGG pathways for different species could be derived from the same generalized pathway called a canonical pathway that represents common properties of a biological process. Therefore, KEGG pathways may not be suitable in the study of species-specific mechanisms.

2.2 Pathway Analysis

A similar concept, REACTOME is a pathway database constructed by expert biologists in collaboration with REACTOME editorial staff [36, 37]. It is open source, open access, manually created, and peer reviewed. Developed for 21 species, REACTOME pathways include over 1400 human pathways consisting of over 7000 proteins. This is almost twice the total number of proteins covered by the KEGG human pathways. REACTOME pathways differ from KEGG pathways in that the former distinguish protein isoforms whereas the latter use a single protein-coding gene to represent all isoforms. In this way, REACTOME may be more suitable and optimal than KEGG for proteomic studies that require distinguishing protein isoforms and their functional differences.

While KEGG and REACTOME have their own staff to develop pathways, BioCarta and WikiPathways provide an open platform that allows everyone to construct his/her own pathways and publish on their websites [1]. The open platform is an effective strategy to combine intelligence of individuals to quickly update and expand the collection of molecular pathways. The disadvantage, however, is that the quality of these pathways is questionable. To ensure quality, BioCarta has a group of experts in their research fields to curate newly submitted pathways, whereas WikiPathways uses a peer-review strategy to correct false information in its pathways. Essentially, quality of the two databases is heavily dependent on the quantity of human experts participating.

Biological pathways constructed by databases are proving to be important tools in trying to understand how biomolecules and their interactions contribute to biological processes. Currently the pathways are limited: only a fraction of proteins in a whole proteome could be covered by these pathways. In fact, combining KEGG and REACTOME human pathways together only covers about 8000 unique proteins, which is less than 40% of total proteins in a human proteome. The protein-protein interaction (PPI) network (discussed in detail below) is a complementary resource that can cover many more proteins that are not presented in any pathways.

When identifying key genes involved in TBI that are major players in the progression of the disease and that can be used as biomarkers, one should always take into consideration the complexity level of the different intracellular signaling pathways. Molecular pathways are seldom linear because many proteins have multiple interaction sites different protein families. Normally, proteins interact with each other, in a cascading manner, to reach the final outcome: either the activation of the target gene or its suppression. The general method for identifying activated or suppressed molecular pathways is termed "pathway analysis" in which a set of genes is analyzed employing gene set analysis methods to find significant alterations that might be at the core of the disease. Such methods have been scrutinized in recent reviews as being the first choice for gaining insights into molecular pathways, specifically methods that deal with differentially expressed genes (enrichment analysis methods) or the correlation between pathway genes and the class of the samples (functional class scoring methods), shown in Fig. 4 [38]. Divided into three distinct generations, pathway analysis has developed with the evolution of its methods in the last generation to become much more accurate and precise by reducing the complexity of the system which initially includes a set of genes grouped into smaller and related sets. That results in the identification of active signaling pathways that are significantly modified between sample groups [39]. However, the major limitation which comes with these gene set analysis methods is the lack of information about the different interaction between proteins and other molecules termed "pathway's topological information." Ignoring such information results in erroneous data, hence raising the need to convey information about the role of different protein interplays and the location of pathway members instead of simply analyzing all set members equally as in the gene set analysis. Practical pathway analysis treats members according to their locations in the pathway and thus provides a high-throughput molecular measurement with more relevance and accuracy. Pathways analysis approaches have been designed as such to employ topological information by taking into consideration the inter- and intra-pathway connectivity information. The next section will describe the different approaches.

Tarca et al. [38] developed a novel signaling pathway impact analysis (SPIA) that takes into consideration the differentially expressed genes with a novel type of evidence which measures the perturbation factor, that is, the differential expression of upstream genes connected to the pathway. Altogether, SPIA combines data from the classical enrichment analysis methods and accounts for

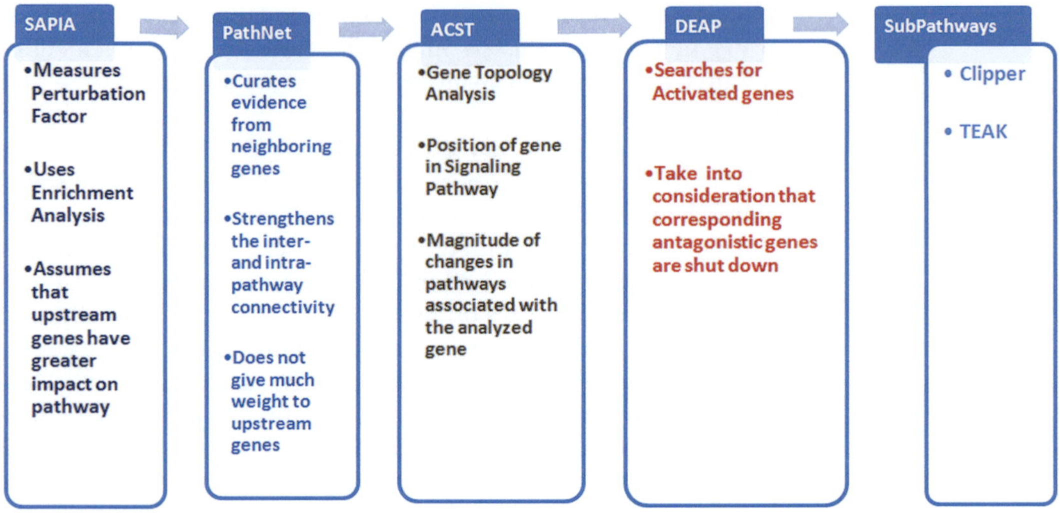

Fig. 4 Pathway analysis approaches used in TBI with the added function that each tool addresses. *SPIA* signaling pathway impact analysis, *ACST* analysis of consistent signaling transduction, *DEAP* differential expression analysis for pathways, *TEAK* topology enrichment analysis framework

some of its limitations by introducing the perturbation factor. That indeed increases sensitivity and specificity of the method, and makes it different from the standard gene set over-expression analysis [1]. Accounting for the evidence provided by enrichment pathway method and perturbation factor, simulations have shown that it is independent of one another. This allows calculating the global significance obtained from combining *P*-values (perturbation *P*-value and enrichment *P*-value). These probabilities can be calculated using a bootstrap approach. To assess the capabilities of SPIA, four datasets were examined using over-representation analysis (ORA), gene set enrichment analysis (GSEA), and SPIA. The latter approach showed increased sensitivity and specificity over the two existing pathway analysis approaches, as it was capable of assessing relevant and activated signaling pathways to each condition when compared to GSEA, with reduced false-positive pathways and a meaningful ranking of the pathways when compared to ORA. It is worth noting here that SPIA uses an assumption that upstream genes have greater impact on the signaling pathway since they will affect the significance of all downstream genes, which cannot be verified by experimental studies. Similar to SPIA, PathNet method uses evidence from differentially expressed genes termed "direct evidence," but instead of computing the perturbation factor, it uses evidence from their neighboring genes termed "indirect evidence." PathNet then combines both evidences and calculates their statistical significance. What is novel about PathNet is that it strengthens the inter- and intra-pathway connectivity by unraveling genes that directly interact with the activated signaling transduction and thus the more significant neighboring genes are, the more significant a direct evidence becomes [1]. While this method detects activated signaling pathways based on differentially expressed genes and their neighbors of genes, it does not give much weight to upstream genes as opposed to SPIA. To evaluate its capabilities in identifying relevant biological pathways relevant to Alzheimer's disease (AD), PathNet was compared to three existing algorithms (hypergeometric test, SPIA, and GSEA). Indeed, PathNet identified the AD pathway and other biologically relevant pathways in two gene microarray datasets when the three other existing methods often failed to do so. SPIA and PathNet determine significant genes based on an arbitrary threshold, which implies that all insignificant genes have no contribution to the activated pathway and only the significant ones, are at the heart of the pathway of interest. Although a gene with a minor change in expression might not greatly or significantly affect a certain pathway, however, multiple genes might collectively have an impact on a certain pathway. To account for that, Fang et al. [40] developed Gene Association Network-based Pathway Analysis (GANPA), which measures gene weights, integrates them into classical gene set analysis, and corrects for the "over-counting" associated with multi-subunit proteins. GANPA computes the weight of each gene involved in an activated pathway, which is proportional to the fraction of interaction of a gene with several other associated genes over the total number of interactions involving that gene. Thus, it takes into account the nonequivalence of genes within pathways and provides weighted means that can be integrated in gene set analysis (GSA) methods.

Recently developed pathway analysis methods aimed at identifying biologically relevant activated or suppressed sub-pathways within a predefined signaling transduction cascade while still conveying topological pathway information. As such, analysis of consistent signaling transduction (ACST) method was developed; it takes into account a pathway's topological information, the position of a gene in a predefined signaling transduction, and the magnitude of changes associated with the analyzed gene [41]. ACST searches for genes that are activated while taking into consideration their corresponding antagonistic genes as being shut down. This web of networking genes constitutes a subgraph that partially or significantly contributes to the activated pathway. For example, upon the binding of a growth factor to a receptor tyrosine kinase (RTK), genes such as JAK1 are activated and in turn activate downstream genes such as PI3K, PDK, and AKT, whereas their antagonistic genes such as PTEN are shut down. This part of the whole pathway constitutes a consistent subgraph. A pathway is scored by the expression of all genes in its maximal consistent subgraph and the subgraph's distance to the downstream target gene of the pathway. Statistical analyses are then run to determine the significance of the pathways and whether its score is significantly higher than the score of a pathway with randomly expressed genes. Similar to ACST, Differential Expression Analysis for Pathways (DEAP) developed by Haynes et al. [42] includes pathway topological information and determines the most differentially expressed gene in a pre-defined pathway. It screens for all linear pathways and assess their significance using particular formulas. Linear paths do not strictly necessitate that their member genes are consistently active, but inconsistent expression of the genes would reduce the score of a path making it less favorable to be identified when there is consistent expression of genes in another paths.

Other pathway analysis approaches that search for sub-pathways include Clipper and Topology Enrichment Analysis framework (TEAK) [43, 44]. The former method converts pathway topological information into Gaussian graphical model and tests differences in mean and in covariance matrices between two experimental conditions, whereas the latter one employs Gaussian Bayesian network to rank and estimate the significance of sub-pathways.

For the past few years, gene set analysis has witnessed great improvements in conveying pathway's topological information, which was totally ignored in several existing approaches. Recent methods have focused on identifying subpaths within biologically relevant signaling pathways; this approach provided more sensitivity and specificity when compared to famous gene set analysis methods. The integration of subpaths remains complicated with more computational resources to be utilized. It is now approved that deregulations in signaling pathways drive many diseases, in particular cancer. Thus, understanding the differential expression of genes in predefined signaling pathways would provide potential new therapeutic targets.

2.3 Protein-Protein Interaction Network

In the previous section, we scrutinized the different pathway analysis methods that are currently being utilized to determine deferentially expressed genes based on experimental data and predefined pathways. Similar to that approach, protein-protein interaction (PPI) network analysis tries to identify network regions, relevant to a study by referring to experimental data and a single protein interaction network. In this section, we focus on the PPI network analysis in unraveling novel molecular mechanisms from high-throughput experimental data.

The first steps of PPI network analysis involve superimposing high throughput experimental data such as differential gene expression obtained from gene microarray data or the proteome abundance from proteomics studies. Numerous modified proteins do not randomly nor uniformly scatter through the network, they are usually expected to condense in specific regions of such network and are termed "network modules" or "subnetworks" and represent the underlying molecular mechanism of a study.

One fundamental component of PPI network analysis is engaging a subnetwork search algorithm in order to find all the subnetworks [1]. Many components of such an algorithm exist, but essentially there is a scoring function through which the subnetwork is evaluated and a searching algorithm where an optimal subnetwork with a maximal score is obtained. Basically, the scoring function is evaluated through a biological hypothesis that includes the characteristics of a subnetwork representing a particular biological pathway, and such evaluation is significant in choosing PPI network analysis for a given study. As for the searching algorithm, theoretically speaking, with any scoring function, an intensive search can find the optimal subnetwork for a particular study. However, practically speaking, it has been proven that finding such optimal score subnetwork is NP-hard meaning that such process is extremely time consuming.

Previous publications have been postulated to find the suboptimal subnetworks. For instance, to do so, Ideker et al. made use of a normalized aggregate z-score (sum of z-score of all genes) and employed a heuristic simulated annealing algorithm [45].

The subnetwork search begins with a working subnetwork that randomly selects a protein and ending after several rounds of expansions and updating of such working network. Since the z-score of a gene measures its differential expression, the scoring and searching functions in this study were designed in a way to increase the score of a subnetwork retrieving all subnetworks that are enriched with differentially expressed genes. The authors who showed that they could indeed identify subnetworks consistent with known regulatory circuits validated this method. Consequently, this method has been cited vastly over the past decade and motivated the finding of many similar methods that aim to acquire biological pathways from a PPI network.

To measure the co-expression of genes in a subnetwork, Guo et al. developed a scoring method, which calculates the final score based on raw score normalized by the mean and the standard deviation [46]. Similar to the latter method, module analysis via topology of interactions and similar sets (MATISSE) calculates the final score by taking into consideration the co-expression of genes in a subnetwork and allowing uncorrelated genes to be present. In addition, MATISSE assigns ranks to different genes in a subnetwork to represent their co-expression in a biologically relevant area. In a predefined signaling pathway and under particular condition, only a fraction of genes is usually coherently regulated; which answers why the method allows the presence of uncorrelated genes while still considering the co-expression of active genes. The score of a given subnetwork is further defined as the edge score (ECF-statistic) and the sum of normalized node score (F-statistics), or can be subtly defined in a specific study. For instance, the DEGAS (DysrEgulated Gene set Analysis via Subnetworks) method defines deregulated genes in several cases of a case–control study of a disease. By comparing their expression value, one can tell if a gene is deregulated in one case when compared to the other, and how significant this difference is. Ultimately, the numbers of such deregulated genes define the subnetwork they are in. Nevertheless, a scoring function should be replaced by a shared distribution of a group of random variables when probabilistic graphic models are to be applied in the identification of specific subnetworks. However, an end user should be aware of the underlying assumption that the connected genes tend to be co-regulated. The method developed by Segal et al., and the most recent method NIPD (network inference with pooling data) [47, 48] are two examples of methods that used probabilistic graphic models. Basically, a scoring function is always used to determine the characteristics of a subnetwork regardless of its gene expression status, but the efficiency in getting the optimal or suboptimal subnetwork is essentially dependent upon the subnetwork-searching algorithm. Several other algorithms have been proposed other than the simulated annealing algorithm and these include variants of greedy search strategies [49, 50], genetic algorithm, and a more efficient algorithm based on integer-linear programming [51]. However, within the neurotrauma community, end users should select PPI network analysis tools with the appropriate scoring functions and this is more significant than tools that are employed with effective searching algorithms. Since neurotrauma represents a serious health problem worldwide, there is an urgent need for identifying marker candidates using high proteomics-based methods that ensure high specificity and sensitivity. Although finding key markers from a big pool of markers from different types of cells remains challenging, applying protein-protein interaction network methods can help attenuate the complexity of the system while focusing on specific biologically relevant pathway that could generate potential markers for better diagnosis and treatment.

3 Conclusion

A disease in which heterogeneous complications are not readily apparent, TBI is often referred to as a "silent epidemic," especially when it comes to the lack of awareness about this disease among the general public [52]. There are currently challenges in confirming diagnosis of brain injury, since the early manifestation of the changes elicited by impact to the head are biochemical and molecular in nature. Therefore it is essential for biochemical and molecular testing to detect and assess the severity of TBI. Various systems biology tools, databases and strategies can be implemented on available TBI data in biomarker discovery. In fact, in the field of neurotrauma and TBI, three distinct approaches have been used so far, namely functional group categorization, pathway analysis, and protein-protein interaction (PPI) networks. The databases allow for query of the system to identify candidate targets which can be further studied to elucidate potential downstream biomarkers indicative of disease progression, severity, and improvement. The integration of proteomic data sets through bioinformatics enables the generation of databases which are powerful tools to aid researchers in understanding the mechanism of neural injury. Thus, the application of neuroproteomics will revolutionize the characterization of protein dynamics, leading to an enhanced understanding of post-injury biochemistry.

References

1. Yu C, Kobeissy F (2015) Frontiers in Neuroengineering Systems Biology Applications to Decipher Mechanisms and Novel Biomarkers in CNS Trauma. In: Kobeissy FH, editor. Brain Neurotrauma: Molecular, Neuropsychological, and Rehabilitation Aspects. Boca Raton (FL): CRC Press/Taylor & Francis (c) 2015 by Taylor & Francis Group, LLC.
2. Baumgartner C, Osl M, Netzer M, Baumgartner D (2011) Bioinformatic-driven search for metabolic biomarkers in disease. J Clin Bioinformatics 1:2
3. Ashburner M, Ball CA, Blake JA, Botstein D, Butler H, Cherry JM, Davis AP, Dolinski K, Dwight SS, Eppig JT, Harris MA, Hill DP, Issel-Tarver L, Kasarskis A, Lewis S, Matese JC, Richardson JE, Ringwald M, Rubin GM, Sherlock G (2000) Gene ontology: tool for the unification of biology. The Gene Ontology Consortium. Nat Genet 25:25–29
4. Thomas PD, Campbell MJ, Kejariwal A, Mi H, Karlak B, Daverman R, Diemer K, Muruganujan A, Narechania A (2003) PANTHER: a library of protein families and subfamilies indexed by function. Genome Res 13:2129–2141
5. Magrane M, Consortium U (2011) UniProt Knowledgebase: a hub of integrated protein data. Database, bar009
6. Punta M, Coggill PC, Eberhardt RY, Mistry J, Tate J, Boursnell C, Pang N, Forslund K, Ceric G, Clements J, Heger A, Holm L, Sonnhammer EL, Eddy SR, Bateman A, Finn RD (2012) The Pfam protein families database. Nucleic Acids Res 40:D290–D301
7. Jeter CB, Hergenroeder GW, Hylin MJ, Redell JB, Moore AN, Dash PK (2013) Biomarkers for the diagnosis and prognosis of mild traumatic brain injury/concussion. J Neurotrauma 30:657–670
8. Strimbu K, Tavel JA (2010) What are biomarkers? Curr Opin HIV AIDS 5:463–466
9. Kobeissy FH, Sadasivan S, Oli MW, Robinson G, Larner SF, Zhang Z, Hayes RL, Wang KK (2008) Neuroproteomics and systems biology-based discovery of protein biomarkers for traumatic brain injury and clinical validation. Proteomics Clin Appl 2:1467–1483
10. Boutte A, Kobeissy F, Wang KK, Zhang Z, Tortella F, Dave JR, Schmid K (2014) Protein biomarkers in traumatic brain injury: an omics approach. In: Biomarkers of brain injury and neurological disorders. CRC Press, Boca Raton, FL, p 42
11. Coronado VG, Xu L, Basavaraju SV, McGuire LC, Wald MM, Faul MD, Guzman BR, Hemphill JD, Centers for Disease, C., and Prevention (2011) Surveillance for traumatic brain injury-related deaths--United States, 1997-2007. Morb Mortal Wkly Rep Surveill Summ 60:1–32
12. Galarneau MR, Woodruff SI, Dye JL, Mohrle CR, Wade AL (2008) Traumatic brain injury during operation Iraqi freedom: findings from the United States Navy-Marine Corps Combat Trauma Registry. J Neurosurg 108:950–957
13. Thompson HJ, McCormick WC, Kagan SH (2006) Traumatic brain injury in older adults: epidemiology, outcomes, and future implications. J Am Geriatr Soc 54:1590–1595
14. Feigin VL, Barker-Collo S, Krishnamurthi R, Theadom A, Starkey N (2010) Epidemiology of ischaemic stroke and traumatic brain injury. Best Pract Res Clin Anaesthesiol 24:485–494
15. Ghajar J (2000) Traumatic brain injury. Lancet 356:923–929
16. Maas AI, Stocchetti N, Bullock R (2008) Moderate and severe traumatic brain injury in adults. Lancet Neurol 7:728–741
17. Guingab-Cagmat JD, Cagmat EB, Hayes RL, Anagli J (2013) Integration of proteomics, bioinformatics, and systems biology in traumatic brain injury biomarker discovery. Front Neurol 4:61
18. Yuh EL, Mukherjee P, Lingsma HF, Yue JK, Ferguson AR, Gordon WA, Valadka AB, Schnyer DM, Okonkwo DO, Maas AI, Manley GT, Investigators T-T (2013) Magnetic resonance imaging improves 3-month outcome prediction in mild traumatic brain injury. Ann Neurol 73:224–235

19. Armstrong JD, Pocklington AJ, Cumiskey MA, Grant SG (2006) Reconstructing protein complexes: from proteomics to systems biology. Proteomics 6:4724–4731
20. Assmus HE, Herwig R, Cho KH, Wolkenhauer O (2006) Dynamics of biological systems: role of systems biology in medical research. Expert Rev Mol Diagn 6:891–902
21. Bard J (2007) Systems developmental biology: the use of ontologies in annotating models and in identifying gene function within and across species. Mamm Genome 18:402–411
22. Barrett CL, Kim TY, Kim HU, Palsson BO, Lee SY (2006) Systems biology as a foundation for genome-scale synthetic biology. Curr Opin Biotechnol 17:488–492
23. Boettler T, Schultheiss M, Blum HE (2007) Systems biology. Dtsch Med Wochenschr 132:2702–2705
24. Kitano H (2002) Computational systems biology. Nature 420:206–210
25. Kitano H (2002) Systems biology: a brief overview. Science 295:1662–1664
26. Davidov E, Holland J, Marple E, Naylor S (2003) Advancing drug discovery through systems biology. Drug Discov Today 8:175–183
27. Chen SS, Haskins WE, Ottens AK, Hayes RL, Denslow N, Wang KKW (2007) Bioinformatics for traumatic brain injury: proteomic data mining. Springer Ser Optim A 7:363–387
28. Feala JD, Abdulhameed MD, Yu C, Dutta B, Yu X, Schmid K, Dave J, Tortella F, Reifman J (2013) Systems biology approaches for discovering biomarkers for traumatic brain injury. J Neurotrauma 30:1101–1116
29. Ali A, Zahraa S, Zhiqun Z, Firas K, Kevin KWW (2014) Neuro-proteomics and neurosystems biology in the quest of TBI biomarker discovery. In: Biomarkers of brain injury and neurological disorders. CRC Press, Boca Raton, FL, pp 3–41
30. Beltrao P, Kiel C, Serrano L (2007) Structures in systems biology. Curr Opin Struct Biol 17:378–384
31. Hucka M, Finney A, Sauro HM, Bolouri H, Doyle JC, Kitano H, Arkin AP, Bornstein BJ, Bray D, Cornish-Bowden A, Cuellar AA, Dronov S, Gilles ED, Ginkel M, Gor V, Goryanin II, Hedley WJ, Hodgman TC, Hofmeyr JH, Hunter PJ, Juty NS, Kasberger JL, Kremling A, Kummer U, Le Novere N, Loew LM, Lucio D, Mendes P, Minch E, Mjolsness ED, Nakayama Y, Nelson MR, Nielsen PF, Sakurada T, Schaff JC, Shapiro BE, Shimizu TS, Spence HD, Stelling J, Takahashi K, Tomita M, Wagner J, Wang J, Forum S (2003) The systems biology markup language (SBML): a medium for representation and exchange of biochemical network models. Bioinformatics 19:524–531
32. Boutte AM, Yao C, Kobeissy F, May Lu XC, Zhang Z, Wang KK, Schmid K, Tortella FC, Dave JR (2012) Proteomic analysis and brain-specific systems biology in a rodent model of penetrating ballistic-like brain injury. Electrophoresis 33:3693–3704
33. Matzilevich DA, Rall JM, Moore AN, Grill RJ, Dash PK (2002) High-density microarray analysis of hippocampal gene expression following experimental brain injury. J Neurosci Res 67:646–663
34. Natale JE, Ahmed F, Cernak I, Stoica B, Faden AI (2003) Gene expression profile changes are commonly modulated across models and species after traumatic brain injury. J Neurotrauma 20:907–927
35. Kanehisa M, Goto S, Furumichi M, Tanabe M, Hirakawa M (2010) KEGG for representation and analysis of molecular networks involving diseases and drugs. Nucleic Acids Res 38:D355–D360
36. Croft D (2013) Building models using Reactome pathways as templates. Methods Mol Biol 1021:273–283
37. D'Eustachio P (2013) Pathway databases: making chemical and biological sense of the genomic data flood. Chem Biol 20:629–635
38. Tarca AL, Draghici S, Khatri P, Hassan SS, Mittal P, Kim JS, Kim CJ, Kusanovic JP, Romero R (2009) A novel signaling pathway impact analysis. Bioinformatics 25:75–82
39. Khatri P, Sirota M, Butte AJ (2012) Ten years of pathway analysis: current approaches and outstanding challenges. PLoS Comput Biol 8:e1002375
40. Fang Z, Tian W, Ji H (2012) A network-based gene-weighting approach for pathway analysis. Cell Res 22:565–580
41. Mieczkowski J, Swiatek-Machado K, Kaminska B (2012) Identification of pathway deregulation--gene expression based analysis of consistent signal transduction. PLoS One 7:e41541
42. Haynes WA, Higdon R, Stanberry L, Collins D, Kolker E (2013) Differential expression analysis for pathways. PLoS Comput Biol 9:e1002967
43. Martini P, Sales G, Massa MS, Chiogna M, Romualdi C (2013) Along signal paths: an empirical gene set approach exploiting pathway topology. Nucleic Acids Res 41:e19
44. Judeh T, Johnson C, Kumar A, Zhu D (2013) TEAK: topology enrichment analysis framework for detecting activated biological subpathways. Nucleic Acids Res 41:1425–1437

45. Ideker T, Ozier O, Schwikowski B, Siegel AF (2002) Discovering regulatory and signalling circuits in molecular interaction networks. Bioinformatics 18(Suppl 1):S233–S240
46. Guo Z, Wang L, Li Y, Gong X, Yao C, Ma W, Wang D, Li Y, Zhu J, Zhang M, Yang D, Rao S, Wang J (2007) Edge-based scoring and searching method for identifying condition-responsive protein-protein interaction sub-network. Bioinformatics 23:2121–2128
47. Stuart JM, Segal E, Koller D, Kim SK (2003) A gene-coexpression network for global discovery of conserved genetic modules. Science 302:249–255
48. Yang J, Roy A, Zhang Y (2013) Protein-ligand binding site recognition using complementary binding-specific substructure comparison and sequence profile alignment. Bioinformatics 29:2588–2595
49. Sohler F, Hanisch D, Zimmer R (2004) New methods for joint analysis of biological networks and expression data. Bioinformatics 20:1517–1521
50. Nacu S, Critchley-Thorne R, Lee P, Holmes S (2007) Gene expression network analysis and applications to immunology. Bioinformatics 23:850–858
51. Dittrich MT, Klau GW, Rosenwald A, Dandekar T, Muller T (2008) Identifying functional modules in protein-protein interaction networks: an integrated exact approach. Bioinformatics 24:i223–i231
52. Faul M, Xu L, Wald MM, Coronado V, Dellinger AM (2010) Traumatic Brain Injury in the United States: National Estimates of Prevalence and Incidence, 2002-2006. Inj Prev 16:A268

Chapter 10

Role of Systems Biology in Brain Injury Biomarker Discovery: Neuroproteomics Application

Zaynab Jaber, Patrick Aouad, Mohamad Al Medawar, Hisham Bahmad, Hussein Abou-Abbass, Hiba Ghandour, Stefania Mondello, and Firas Kobeissy

Abstract

Years of research in the field of neurotrauma have led to the concept of applying systems biology as a tool for biomarker discovery in traumatic brain injury (TBI). Biomarkers may lead to understanding mechanisms of injury and recovery in TBI and can be potential targets for wound healing, recovery, and increased survival with enhanced quality of life. The literature available on neurotrauma studies from both animal and clinical studies has provided rich insight on the molecular pathways and complex networks of TBI, elucidating the proteomics of this disease for the discovery of biomarkers. With such a plethora of information available, the data from the studies require databases with tools to analyze and infer new patterns and associations. The role of different systems biology tools and their use in biomarker discovery in TBI are discussed in this chapter.

Key words Neurotrauma, Traumatic brain injury TBI, Biomarkers, Systems biology tools, UCHL1, SBDPs, NSE, GFAP, S100β

1 Introduction

As systems biology and the field of proteomics continue to rapidly evolve, fundamental changes are being catalyzed toward the future of health care worldwide [1]. Research in these fields holds major implications in medicine, especially in enhancing the ability to improve diagnosis and treatment of diseases. We are currently witnessing an increased interest in personalized medicine; therefore, bridging the gap between basic research and clinical applications becomes imperative. One research-based proteomic tool at the forefront of personalized medicine is biomarkers. Biomarkers are quantitative physiological indicators of a biological disease or injury state that allow for diagnosis and assessment of the disease process and help monitor the response to treatment [2]. In clinical

Firas Kobeissy et al. (eds.), *Injury Models of the Central Nervous System: Methods and Protocols*, Methods in Molecular Biology, vol. 1462, DOI 10.1007/978-1-4939-3816-2_10, © Springer Science+Business Media New York 2016

medicine, biomarkers have uses in diagnosis, prognosis, and determination of physiological status. They can manifest through vital signs, X-rays, and other imaging modalities as well as through laboratory analysis of biological indicators such as ribonucleic acid (RNA), metabolites, lipids, peptides, proteins, or autoantibodies against proteins released from the diseased/injured tissue [3]. Interestingly, much of medical practice involves interpreting and monitoring biomarkers, the diagnostic accuracy of which is quantitatively denoted by sensitivity and specificity.

Traumatic brain injury (TBI) is a neurotrauma caused by mechanical force applied to the head. It is of great concern since it is a leading cause of death worldwide [2, 4]. While traffic accidents and assault are the main causes of TBIs in younger populations, falls are the predominant reason for TBIs in older individuals, followed by traffic accidents [5–7]. A subset of the adult population in the USA, deployed military servicemen and women, are particularly vulnerable and are at high risk for TBI. They are often exposed to a variety of combat traumas. In fact, recent studies report that approximately 20% of Operation Enduring Freedom/Operation Iraqi Freedom veterans have clinical diagnosis of TBI [8]. More than 30,000 military personnel suffered a TBI in 2012. Another 13,000 or more people had a TBI in 2013 [9]. In addition, this population often exhibits comorbidities such as posttraumatic stress disorder (PTSD) or depression that can lead to an increased risk of misdiagnosis [10–14].

TBI does not describe a physical injury to the head, such as laceration, contusion, or fracture, but rather the change in brain function as a result of damage from an external force to the brain. This can be caused by various ways. One example is the case of rapid backward and forward motion caused by rapid acceleration and deceleration, such as that experienced during motor vehicle accidents or shaken-baby syndrome [15]. Another way is through impact due to falling, especially among the elderly, or caused by sporting injuries. TBIs can also result from blunt force trauma such as an assault or from exposure to blasts resulting in rapid changes in pressure. Penetration wounds to the head caused by high-velocity projectiles can also cause TBI [15, 16].

TBI is heterogeneous, as it is highly variable and characterized by several severities (mild, moderate, severe) in addition to multiple injury types (concussive, nonpenetrating, penetrating). It occurs in two phases: first as primary injury which then leads to secondary injury. Upon impact, primary injuries occur when there is deformation of the gray and white matter of the brain, causing a disruption of cell membranes and the release of intracellular contents [15]. Hours and days following the initial insult, secondary injuries occur as a result of brain edema, free radical formation, or the release of inflammatory mediators. These secondary injuries may exacerbate the initial injury through the mediation of cell damage or death resulting in a poor neurological outcome. Brain damage may include excessive neuronal activity caused by unregulated glutamate release, changes in neurotransmitter levels, hemorrhage, changes in cerebral blood flow, damage to axons, and/or disruptions to the blood–brain barrier (BBB) [16]. After the incidence of primary injuries, the focus of TBI patient management becomes prevention or reduction of the extent of secondary injuries.

The transfer of energy that occurs following the insult can cause structural, pathological, and functional changes in the brain that may yield neurological, cognitive, and behavioral symptoms that can be long lasting. Symptoms of TBI may include confusion, concussion or altered levels or loss of consciousness, seizure, coma, focal sensory deficits, or motor neurologic deficits. The long-term effects of TBI may include depression, anxiety, psychiatric disorders, memory loss, reduced motor function, reduced social functioning, impaired vision, insomnia, dizziness, mood disturbances, and deficits in cognition. Moreover, substance abuse was found to be associated with individuals who have experienced a TBI, and for many patients, family life and relationships may be adversely affected [15]. Prominent neurological symptoms include headache, vomiting, nausea, imbalance, vision, dizziness, fatigue, drowsiness, sensitivity to light or noise, and sleep disturbances. Of the cognitive symptoms, problems with attention, concentration, memory, processing speed, and executive functions (e.g., working memory and decision making) are most frequently reported. Existing literature indicates that in the majority of patients, these symptoms will resolve within 10 days to 2 weeks of the injury [17]. In more than 25 % of the cases, however, symptomology can continue long beyond this timeframe [18–20].

In this book chapter, we will tackle the role of systems biology tools, bioinformatics, and biomarker research in the area of TBI. In particular, we will underline the need for biomarker discovery in TBI and how the major advances in the field of proteomics will further aid this quest for enhanced TBI patient care management.

2 Putting It All Together: Data Mining

Enormous amounts of data generated from high-throughput technologies require data mining tools to analyze data and visualize patterns, which are otherwise tedious and sometimes impossible to detect. An example of data mining methods is correspondence analysis which investigates the relation between features and data samples. Feature selection is another method that allows visualization and comprehension of data patterns. The use of these methods in TBI biomarker discovery has been documented in several reports.

A Multiple Correspondence Analysis (MCA) can be used to detect relationship patterns in data collected on multiple variables pertaining to the participants. These data points and variables are projected on graphs known as principal components that help visualize the clustering of data points and account for the highest amount of variance in the data. Points that cluster in proximity are indicated to have similarities while those that cluster further away from each other have more differences. Martinez et al. performed MCA on data collected from chronic TBI patients undergoing either cognitive training or a control program. The analysis was done by grouping the patients based on the type of head injury they suffered and the corresponding patterns in cognitive performance including assessment of memory, attention, and task switching. The analysis yielded 53% of variance detected by the first principal component based on cognitive performance in all assessments. The second principal component detected 8. 79% of variance based on assessment of memory between the different injury types. Moreover, principal component projections for individuals with blast-related injuries were clustered in the low cognitive performance side compared to projections of other injury types that were less clustered and more evenly distributed between high and low cognitive performances. This shows that MCA accurately clustered cognitive deficits detected in individuals suffering from blast-related injuries. This clustering is quite logical given the complex nature of this trauma that includes the initial shockwave followed by acceleration and deceleration shearing forces, and hence the devastating cognitive damage [21].

Recently, Ou et al. analyzed microarray data previously published by Shojo et al. [22] in Gene Expression Omnibus (GEO) database for differential gene expression profiles in rat models of TBI. After normalizing gene expression intensities with a robust multiarray average (RMA) algorithm, differentially expressed genes (DEGs) between control rats and those subjected to moderate fluid percussion of different durations were identified. This was done through implementing a *t*-test to calculate the probability of DEGs between different groups and the respective *p*-values. In turn, the *p*-values were analyzed in R [23] using a *q*-value package [24] to compute the false discovery rate. Significant GEDs were chosen based on a *q*-value < 5 %. In this study, microarray data was obtained on a TBI model from Gene Expression Omnibus (GEO) database and analysis of the altered gene expression profile was conducted. Results suggested that gene expression profiles were significantly altered in the late period after TBI. These altered genes were mainly involved in steroid biosynthesis, cell cycle, metal ion transport, inflammation, and apoptosis [25].

Given the enormity and heterogeneity of raw data generated from basic science research, there is a need to accelerate the translation of preclinical knowledge into clinical therapeutics. Accordingly,

Nielson et al. have recently developed a database for translational neurotrauma research dubbed Visualized Syndromic Information and Outcomes for Neurotrauma-SCI (VISION-SCI) [26]. In this study, syndromic analysis on data from several species published in the last two decades was collected, which allowed the identification of conserved biological mechanisms of recovery that can be used in monitoring of therapy of neurotrauma patients.

3 Deciphering Molecular Mechanisms of Neurotrauma Using Proteomics

Proteins are major effectors driving cell behavior. Accordingly, the field of proteomics was established and devoted entirely to the systemic study of proteins [1]. The goal of proteomics research is to understand the expression and function of proteins on a global level which requires more than simply cataloguing the proteome; it involves the characterization of protein structure, function, and interaction in all its complexities. The ability to capture and compare all of this information between two cellular states is essential for understanding cellular responses [1]. Thus, proteomics is becoming a well-established approach for protein biomarkers discovery with the ability to identify proteome dynamics in response to experimental stimuli [27]. The collective number of published reports and citations utilizing proteomics in brain injuries is steadily increasing [9].

TBI neuroproteomics studies have used biofluids and injured tissue to identify clinical markers that may correlate with injury severity and may be able to determine therapeutic response [28]. In one study, altered differential proteins were evaluated in normal human postmortem cerebrospinal fluid (CSF) [29]. Since postmortem CSF resembles a model of massive brain injury and cell death, its use could allow for identification of protein markers of injury through comparison of the protein profile of postmortem CSF with that of the CSF of individuals with brain injuries. In this study, 172 of the 229 proteins identified were novel and not previously described. Postmortem CSF was thus used to evaluate altered protein levels similarly occurring after traumatic insult. Additionally, differential proteins of intracellular origin were identified in the CSF. This corroborates the suggestion that protein leakage into the CSF occurs following brain injury [30, 31]. Since neuronal-specific proteins leak from injured brain directly to the CSF, this is crucial to identifying protein markers [27].

CSF in a rat model of TBI was also evaluated in another proteomic study by Siman et al. [32] In this study, tau protein fragment of 17 kDa, αII-spectrin breakdown product of 150 kDa, and collapsing response mediated protein-4 were released as a general response to brain insult. The findings from the experiments may suggest surrogate biomarkers for injury severity and may have the potential for increasing our understanding of the mechanism of brain injury by shedding light on the process of how these proteins are observed in the CSF biofluid at specific time points [32]. In another study, Waybright et al. [33] characterized the proteome of human ventricular CSF obtained from hydrocephalic patients. They were able to identify more than 1500 unique proteins which were then compared with the Human Proteome Organization serum proteome database. Human ventricular CSF was then concluded to contain a large array of proteins unique to CSF [33].

Studies undertaking the catalog of cellular elements under various conditions and in various organisms are well underway and becoming increasingly possible with the maturity of global technologies. This is where systems biology should rise to meet the demand of high-throughput data by helping understand how the elements discovered are coordinated to form functional biological systems. Though systems level integration of data is still in its infancy, a number of new concepts have emerged (such as those discussed earlier). The importance of this data integration is twofold: (1) it allows for minimization of noise inherent in data generated through the high-throughput biology and (2) it serves to reveal new biological phenomena not readily apparent from any single analysis [1]. Ultimately the goal is to characterize the information flow through protein networks that reflect the interconnection between the extracellular microenvironment and gene regulatory networks in response to effector functions of development and physiological responses.

Studies conducted by Kobeissy et al. used Pathway Studio to construct a functional interaction map linking 59 proteins significantly increased or decreased post-TBI [4, 34]. The altered pathways were found to be associated with inflammation, cell survival/proliferation, and synaptic plasticity. In another recent study by Feala et al. [35], around 32 TBI biomarker candidates from the literature were analyzed. These biomarkers' associations with four KEGG pathways were found to be statistically significant, three of the four of which (apoptosis pathway, amyotrophic lateral sclerosis pathway, and Alzheimer's disease pathway) were relevant to TBI or the nervous system. By performing a PPI network analysis, they were able to show that the 32 TBI biomarker candidates were tightly connected to each other on a PPI network of over ten thousand proteins.

4 Inferring Molecular Biomarkers in Neurotrauma

Systems biology study of neurotrauma is moving toward revealing the complex molecular processes induced by brain trauma [36]. The field of proteomics serves as a powerful tool in this endeavor, showing great promise in the identification of specific proteins implicated in TBI. Proteomics can lead toward the discovery of many candidate biomarkers to help ascertain the mechanisms of TBI. Already biomarkers have demonstrated great success and reliability in diagnosis of some diseases such as in cardiac injury. For instance, cardiac troponin proteins (T and I) and various forms of brain natriuretic peptide (BNP) are routinely used to facilitate accurate diagnosis of congestive heart failure and myocardial infarction in patients presenting with chest pain.

There is an increased recognition for the need of biomarker discovery which has led to the Biomarkers Consortium launched in October of 2006 as a public–pharmaceutical industry partnership that includes the National Institutes of Health (NIH), the Food and Drug Administration (FDA), the Centers for Medicare and Medicaid Services, in addition to pharmaceutical industry representatives, nonprofit organizations, and advocacy groups [37]. Importantly, an NIH workshop on improving diagnosis of TBI for targeting therapies stressed the need for biomarker identification [38].

However, despite the efforts in brain injury research, there are no clinically validated biomarkers to diagnose TBI. The efforts to identify sensitive, universal, and specific biomarkers are hindered mainly by challenges such as brain tissue complexity and the heterogeneous nature of brain injury models [27, 39]. Even though extensive studies are being pursued to move protein biomarkers to clinical validation, the work is still under development.

Biomarkers can be discovered through traditional strategies such as knowledge-driven or discovery-driven methods, which are also called "top-down" and "bottom-up" methods [36]. While the knowledge-driven strategy infers biomarkers through understanding disease pathology and molecular mechanism, it is restricted by our knowledge of diseases. Due to the lack of understanding of the molecular mechanisms of action of TBI, it is a less effective approach in the search for TBI biomarkers. On the other hand, the discovery-driven strategy employs high-throughput technologies to screen a large number of genes and proteins to determine those whose abundance change could indicate TBI. The limitations to this approach may be inherent noise and the semiquantification nature of high-throughput technologies may lead to false positives passing the screening [36].

In 2006, Kobeissy and colleagues identified 59 proteins 48 h post-TBI using a rat model and they found that proteins that were decreased in abundance included CRMP-2, glyceraldehyde-3-phosphate dehydrogenase, microtubule-associated proteins MAP2A/2B, and hexokinase [34]. Proteins that were upregulated included C-reactive proteins, transferrin, and breakdown products of CRMP-2, synaptotagmin, and αII-spectrin. The changes in these proteins were confirmed by western blotting. This study generated candidate biomarkers that can aid in the evaluation of the severity and progression of injury as well as in the development of possible therapies.

The use of a systems biology-based approach to drug discovery and development for TBI based on the advances in genomics, proteomics, bioinformatic tools, and systems biology software has been shown [28]. In 2012, Boutte and colleagues conducted a proteomic analysis and brain specific systems biology in a rodent model of penetrating ballistic-like brain injury (PBBI) where they used a combination of 2D-gel electrophoresis and Mass Spectrometry (MS) to screen for biomarkers. After identifying 321 upregulated and 65 downregulated proteins 24 h post PBBI compared to sham controls, pathway analysis indicated that these proteins were involved in neurite outgrowth and cell differentiation. Among these proteins that indicated consistent increase in the brain tissue and CSF at several time points post PPBI were UCHL1, tyrosine hydroxylase, and syntaxin-6.

While systems biology is interested in complex biological processes as they are governed by the interactions of multiple genes and proteins, it may seem that the intention to search for a TBI biomarker candidate from TBI-relevant pathways or interaction network is against the principle of systems biology. This is why a panel of biomolecules serving as TBI biomarker profiles should be suggested by systems biology [36]. In fact, GFAP and UCHL1 have been proposed together as TBI biomarkers [40]. There are huge numbers of possible combinations of multiple proteins in which systems biology will prove useful in identifying most effective combinations of proteins for TBI biomarker panels.

Soluble biomarkers ideal for use in the diagnosis of TBI should be absent in the peripheral tissue unless the brain tissue has been injured [10]. The ideal biomarker should be a small molecule that can be rapidly measured in the serum or CSF for a reasonable period after injury. Additionally, it would be ideal for the biomarker to have a level that corresponds to the degree of brain injury.

5 Traumatic Brian Injury Candidate Biomarkers Identified After Applying Systems Biology Concepts to Neuroproteomics

Listed below are examples of the most studied candidate protein biomarkers for TBI and have shown high sensitivity and specificity in independent studies (Table 1). UCHL1, SBDPs, and neuron-specific enolase (NSE) are neuronal and axonal protein biomarkers whereas GFAP and S100β are glial-specific markers [41]. Combining neuroproteomic methods with relevant animal models, systematic assessments have been made to identify additional protein biomarkers for TBI [34, 42–45].

5.1 Ubiquitin Carboxy-Terminal Hydrolase L1 Protein (UCHL1)

UCHL1 is a cysteine protease of relatively small size (around 25 kDa and comprises 1–2 % of the total soluble protein in the brain) that is predominantly expressed in neurons, although it is also expressed in small amounts in neuroendocrine cells. UCHL1 is known to hydrolyze the C-terminal bond of ubiquitin or unfolded polypeptides [10,

Table 1
Putative biomarkers of traumatic brain injury

TBI biomarker	Source of sample	Origin cell type	Function	Use
UCHL1	Blood serum	Neurons	Ubiquitin	BBB disruption
	Blood plasma	Neuroendocrine	Hydrolysis	Injury outcome
	CSF	Tissue		
SBDPs	Blood serum	Neuron axons	Activation of intracellular proteases	Axonal damage
	Blood plasma	Presynaptic terminals		Injury severity
	CSF			
NSE	Blood serum	Neurons	Glycolytic pathway enzyme	Injury severity
	Blood plasma	Oligodendrocyte		
	CSF			
GFAP	Blood serum	Astroglia	Structural filament	Injury severity
	Blood plasma			Outcomes
	CSF			
S100β	Blood serum	Glial cells	Intracellular signaling	BBB disruption
	Blood plasma	Astrocytes	Calcium homeostasis	Injury severity
	CSF			

CSF cerebrospinal fluid, *BBB* blood–brain barrier, *NSE* neuron-specific enolase, *GFAP* glial fibrillary acidic protein, *UCHL-1* ubiquitin carboxy-terminal hydrolase L1, *TBI* traumatic brain injury, *SBDP* Alfa II spectrin breakdown product

41, 46]. Mutations in UCHL1 may be associated with Parkinson's disease and other neurodegenerative disorders [46]. Importantly, UCHL1 has previously been shown to be elevated in patients with severe TBI [10] and several publications have indicated that UCHL1 can be a biomarker for TBI. UCH-L1 CSF and serum levels were found to be elevated in patients with severe TBI correlating with the severity and outcome of injury [15, 47–49].

The elevation of levels of UCH-L1 post-TBI is proposed to be secondary to BBB dysfunction [50]. In addition, several recent studies also demonstrated the detectability of UCH-L1 in blood following mild TBI [51–53].

5.2 α II-Spectrin Breakdown Products (SBDPs)

Among the novel biomarkers studied for their clinical relevance in TBI, alpha II-spectrin is a cytoskeletal protein primarily found in neurons and is concentrated in axons and presynaptic terminals

[41, 54–56]. Though alpha II-spectrin is present in various nucleated cells, and most tissues, its high abundance and enrichment of brain qualifies it as a candidate biomarker, especially if combined with another brain-specific marker [37].

The breakdown products (SBDPs) of alpha II-spectrin is due to activation of intracellular proteases such as calpain and caspase in the brain after TBI, thus reflecting axonal damage [10, 54, 57]. While SBDP150 (molecular weight 150 kDa) and SBDP145 (molecular weight 145 kDa) are characteristics of calpain activation (associated in acute necrotic neuronal cell death), SBDP120 is produced by action of caspase-3 (associated with delayed apoptotic neuronal death) [10, 27]. Elevation levels of SBDPs in CSF were reported as a possible outcome predictor in patients with severe TBI, rather than mild TBI [54, 58–60]. Not only can SBDPs provide important information on severity of brain injury, but also on underlying pathophysiological mechanisms associated with necrotic and apoptotic cell death.

5.3 Neuron-Specific Enolase (NSE)

Highly expressed in neuronal cytoplasm, neuron-specific enolase (NSE) is a glycolytic pathway enzyme of different isoforms [10, 54]. The gamma-gamma homodimer isoform is highly enriched in the neuronal cell body [61], but is present in multiple other cell types, such as erythrocytes, platelets neuroendocrine cells, and oligodendrocyte [62]. NSE has been shown to have the sensitivity and specificity to detect neuronal cell death [63]. Increased CSF and serum levels of NSE have been reported after TBI, with levels that are detectable within six hours postinjury [2, 10]. Studies have also shown that NSE levels in CSF and serum correlate with severity of injury and clinical outcome [10, 41, 54, 64, 65]. However, the specificity and sensitivity of NSE have been reported as unsatisfactory [66–71]. The limitations on NSE as a biomarker of TBI may be due to the high sensitivity of NSE to hemolysis [72]. Therefore, it has been proposed that NSE is not to be used as a standalone screening biomarker for brain injury [71].

5.4 Glial Fibrillary Acidic Protein (GFAP)

Glial fibrillary acidic protein (GFAP) is an intermediate filament protein that forms networks that support the astroglial cells. First reported in 1971, GFAP is found exclusively in the astroglial cytoskeleton [54, 61, 73]. Of the candidate biomarkers available for TBI, GFAP has been assessed in different studies of clinical studies [74–77]. Part of what makes this an ideal biomarker candidate for TBI is that this protein is not found outside the central nervous system [78]. Even if the body is subjected to multiple forms of trauma, GFAP does not increase without brain injury [79, 80]. Thus, GFAP can be considered as a potential biomarker-specific glial injury.

GFAP was studied in both CSF and sera of patients with TBI [56, 66, 81–83]. Upregulation of GFAP follows damage to the astroglial cells (astrogliosis) [10]. Astroglial cells react during injury by generating more GFAP. Evidence points to elevated serum GFAP levels in several types of brain damage, including TBI [79, 82, 84]. GFAP can also predict death or unfavorable outcomes [83, 85] and validation studies in humans are already ongoing [3] according to the proceedings of the military mild TBI diagnostic workshop [10].

5.5 S100β

One of the earliest and most extensively studied biomarkers of brain damage is S100β which belongs to a family of low molecular weight (9–13 kDa) calcium-binding S100 proteins important in intracellular calcium regulation [9, 86]. S100β is mainly found in astroglia and Schwann cells [87, 88]. S100β aids in cell homeostasis and prevents neuronal death by increasing cellular calcium concentrations [89]. It also acts as a neurotrophic factor, promoting neurite outgrowth and astrocytic proliferation [2]. Its potential as a biomarker for TBI is found in its increased concentration in the CSF and serum after injury [90]. This protein is not influenced by hemolysis and has a biological half-life of two hours. Studies have correlated this biomarker with injury and outcome [91–94]. The first study to emphasize the role of serum S100β in TBI patients was done by Ingebrigtsen et al. who showed that elevated serum S100β levels in patients with negative CT results are correlated with occurrence of postconcussive symptoms [95].

Several other studies have investigated the clinical prognostic value of elevated serum S100β levels in TBI patients with conflicting evidence [80, 83, 94, 96–104]. Interestingly, in 2010 Unden and Romner did a meta-analysis of studies on mild head injury in which CT findings and S100β were compared in the acute phase of injury [105]. In the 12 eligible articles (total 2466 patients) they discovered a high sensitivity of low levels of S100β in the prediction of negative CT findings. In fact, Unden and Romner suggested that a low serum S100β level (<0.10 μg/L) in the first three hours after injury has more than 90% negative predictive value of the presence of clinically relevant CT findings. These findings are further confirmed by other studies which also suggest the use of serum S100β as a substitute for CT in assessment of mTBI patients [106, 107]. S100β has also been studied as a useful indicator of patients with intracranial lesion [94].

However, even if those studies demonstrate the sensitivity of the use of S100β, there are several limitations on this biomarker candidate. Since S100β is not specific to the brain, it can show up outside the central nervous system [9, 39, 61, 108, 109]. Therefore, general trauma without brain injury can increase levels of this protein [110]. In fact, S100β can be elevated in bone fractures without head injury [111–113]. Despite the abundance of studies reporting serum S100β elevation, studies of CSF levels of S100β in TBI is still limited [56]. Additionally, elevated S100β occurs after hemorrhagic shock, correlating the concentration to shock severity

[91, 114, 115]. Because of this, S100β cannot be used as a single biomarker for TBI. The ratio of S100β against GFAP has been investigated, instead of S100β alone, and this was used to determine brain damage and prognosis [84].

6 Conclusion

The short-term and long-term effects of TBI, in the absence of any FDA approved treatment [116], highlight the urgency for detection of biomarkers to improve the quality of life and decrease mortality among patients with TBI. Multiple individual soluble biomarkers currently show promise in the diagnosis of brain injury, with the ability to predict degree of injury and clinical outcome. The breakdown products of α-II spectrin and the serum levels of UCH-L1 were found to change in a similar manner to that of S100β and GFAP postinjury. Hence all these putative biomarkers can be used as important predictors of outcome in patients with moderate-to-severe brain injury [55, 117]. Given the limitations in each biomarker, it is likely that no single biomarker will have adequate sensitivity and specificity for accurate diagnosis of TBI. The better approach may be in using bioinformatics to discover and combine biomarkers in order to improve diagnostic accuracy. The field of neuroproteomics is still in the developing stage and its full potential remains to be explored to reveal the integral molecular and cellular mechanisms of gene dynamics involved in brain injury.

References

1. Weston AD, Hood L (2004) Systems biology, proteomics, and the future of health care: toward predictive, preventative, and personalized medicine. J Proteome Res 3:179–196
2. Jeter CB, Hergenroeder GW, Hylin MJ, Redell JB, Moore AN, Dash PK (2013) Biomarkers for the diagnosis and prognosis of mild traumatic brain injury/concussion. J Neurotrauma 30:657–670
3. Marion DW, Curley KC, Schwab K, Hicks RR, mTBI Diagnostics Workshop (2011) Proceedings of the military mTBI Diagnostics Workshop, St. Pete Beach, August 2010. J Neurotrauma 28:517–526
4. Kobeissy FH, Sadasivan S, Oli MW, Robinson G, Larner SF, Zhang Z, Hayes RL, Wang KK (2008) Neuroproteomics and systems biology-based discovery of protein biomarkers for traumatic brain injury and clinical validation. Proteomics Clin Appl 2:1467–1483
5. Faul M, Xu L, Wald MM, Coronado V, Dellinger AM (2010) Traumatic Brain Injury in the United States: National Estimates of Prevalence and Incidence, 2002-2006. Inj Prev 16:A268–A268
6. Hemphill III, J. C., Phan, N., Aminoff, M. J., & Wilterdink, J. L. (2012). Traumatic brain injury: epidemiology, classification, and pathophysiology. UpToDate. UpToDate, 21
7. Koepsell TD, Rivara FP, Vavilala MS, Wang J, Temkin N, Jaffe KM, Durbin DR (2011) Incidence and descriptive epidemiologic features of traumatic brain injury in King County, Washington. Pediatrics 128:946–954
8. Warden D (2006) Military TBI during the Iraq and Afghanistan wars. J Head Trauma Rehabil 21:398–402
9. Boutte A, Kobeissy F, Wang KK, Zhang Z, Tortella F, Dave JR, Schmid K (2014) Protein biomarkers in traumatic brain injury: an omics approach. In: Biomarkers of brain injury and neurological disorders. CRC Press, Boca Raton, FL, p 42
10. Cook GA, Hawley JS (2014) A review of mild traumatic brain injury diagnostics: current perspectives, limitations, and emerging technology. Mil Med 179:1083–1089

11. Bryant RA, Harvey AG (1999) The influence of traumatic brain injury on acute stress disorder and post-traumatic stress disorder following motor vehicle accidents. Brain Inj 13:15–22
12. Hoge CW, McGurk D, Thomas JL, Cox AL, Engel CC, Castro CA (2008) Mild traumatic brain injury in U.S. Soldiers returning from Iraq. N Engl J Med 358:453–463
13. Creamer M, O'Donnell ML, Pattison P (2005) Amnesia, traumatic brain injury, and posttraumatic stress disorder: a methodological inquiry. Behav Res Ther 43:1383–1389
14. Schneiderman AI, Braver ER, Kang HK (2008) Understanding sequelae of injury mechanisms and mild traumatic brain injury incurred during the conflicts in Iraq and Afghanistan: persistent postconcussive symptoms and posttraumatic stress disorder. Am J Epidemiol 167:1446–1452
15. Mundy L (2013) Biomarkers for the diagnosis and management of traumatic brain injury (TBI). In: Technology brief. Health Policy Advisory Committee on Technology, HealthPACT, State of Queensland, Australia
16. Dash PK, Zhao J, Hergenroeder G, Moore AN (2010) Biomarkers for the diagnosis, prognosis, and evaluation of treatment efficacy for traumatic brain injury. Neurotherapeutics 7:100–114
17. d'Hemecourt P (2011) Subacute symptoms of sports-related concussion: outpatient management and return to play. Clin Sports Med 30:63–72, viii
18. Dikmen S, Machamer J, Fann JR, Temkin NR (2010) Rates of symptom reporting following traumatic brain injury. J Int Neuropsychol Soc 16:401–411
19. Lannsjo M, af Geijerstam JL, Johansson U, Bring J, Borg J (2009) Prevalence and structure of symptoms at 3 months after mild traumatic brain injury in a national cohort. Brain Inj 23:213–219
20. Sigurdardottir S, Andelic N, Roe C, Jerstad T, Schanke AK (2009) Post-concussion symptoms after traumatic brain injury at 3 and 12 months post-injury: a prospective study. Brain Inj 23:489–497
21. Martinez D, Krawczyk D, Rodgers BN, Chapman S (2014) Categorizing cognitive performance in traumatic brain injury using multiple correspondence analysis. Arch Phys Med Rehabil 95:e54–e55
22. Shojo H, Kaneko Y, Mabuchi T, Kibayashi K, Adachi N, Borlongan CV (2010) Genetic and histologic evidence implicates role of inflammation in traumatic brain injury-induced apoptosis in the rat cerebral cortex following moderate fluid percussion injury. Neuroscience 171:1273–1282
23. Sullivan SM, Sullivan RK, Miller SM, Ireland Z, Bjorkman ST, Pow DV, Colditz PB (2012) Phosphorylation of GFAP is associated with injury in the neonatal pig hypoxic-ischemic brain. Neurochem Res 37:2364–2378
24. Storey JD, Tibshirani R (2003) Statistical significance for genomewide studies. Proc Natl Acad Sci U S A 100:9440–9445
25. Ou S, Liu GD, Zhou LS, Xia X, Bai SR, Li J, Cui J, Cheng JM, Li YM, Zhang XY, Gu JW (2014) Bioinformatics analysis of gene expression profiles in the rat cerebral cortex following traumatic brain injury. Eur Rev Med Pharmacol Sci 18:101–107
26. Nielson JL, Guandique CF, Liu AW, Burke DA, Lash AT, Moseanko R, Hawbecker S, Strand SC, Zdunowski S, Irvine KA, Brock JH, Nout-Lomas YS, Gensel JC, Anderson KD, Segal MR, Rosenzweig ES, Magnuson DS, Whittemore SR, McTigue DM, Popovich PG, Rabchevsky AG, Scheff SW, Steward O, Courtine G, Edgerton VR, Tuszynski MH, Beattie MS, Bresnahan JC, Ferguson AR (2014) Development of a database for translational spinal cord injury research. J Neurotrauma 31:1789–1799
27. Kobeissy FH, Guingab-Cagmat JD, Razafsha M, O'Steen L, Zhang Z, Hayes RL, Chiu WT, Wang KK (2011) Leveraging biomarker platforms and systems biology for rehabilomics and biologics effectiveness research. PM R 3:S139–147
28. Zhang Z, Larner SF, Kobeissy F, Hayes RL, Wang KK (2010) Systems biology and theranostic approach to drug discovery and development to treat traumatic brain injury. Methods Mol Biol 662:317–329
29. Burgess JA, Lescuyer P, Hainard A, Burkhard PR, Turck N, Michel P, Rossier JS, Reymond F, Hochstrasser DF, Sanchez JC (2006) Identification of brain cell death associated proteins in human post-mortem cerebrospinal fluid. J Proteome Res 5:1674–1681
30. Dumont D, Noben JP, Raus J, Stinissen P, Robben J (2004) Proteomic analysis of cerebrospinal fluid from multiple sclerosis patients. Proteomics 4:2117–2124
31. Hammack BN, Fung KY, Hunsucker SW, Duncan MW, Burgoon MP, Owens GP, Gilden DH (2004) Proteomic analysis of multiple sclerosis cerebrospinal fluid. Mult Scler 10:245–260
32. Siman R, McIntosh TK, Soltesz KM, Chen Z, Neumar RW, Roberts VL (2004) Proteins released from degenerating neurons are surrogate markers for acute brain damage. Neurobiol Dis 16:311–320

33. Waybright T, Avellino AM, Ellenbogen RG, Hollinger BJ, Veenstra TD, Morrison RS (2010) Characterization of the human ventricular cerebrospinal fluid proteome obtained from hydrocephalic patients. J Proteomics 73:1156–1162
34. Kobeissy FH, Ottens AK, Zhang Z, Liu MC, Denslow ND, Dave JR, Tortella FC, Hayes RL, Wang KK (2006) Novel differential neuroproteomics analysis of traumatic brain injury in rats. Mol Cell Proteomics 5:1887–1898
35. Feala JD, Abdulhameed MD, Yu C, Dutta B, Yu X, Schmid K, Dave J, Tortella F, Reifman J (2013) Systems biology approaches for discovering biomarkers for traumatic brain injury. J Neurotrauma 30:1101–1116
36. Yu C, Kobeissy F (2015) Frontiers in Neuroengineering Systems Biology Applications to Decipher Mechanisms and Novel Biomarkers in CNS Trauma. In: Kobeissy FH, editor. Brain Neurotrauma: Molecular, Neuropsychological, and Rehabilitation Aspects. Boca Raton (FL): CRC Press/Taylor & Francis (c) 2015 by Taylor & Francis Group, LLC.
37. Zhang Z, Mondello S, Kobeissy F, Rubenstein R, Streeter J, Hayes RL, Wang KK (2011) Protein biomarkers for traumatic and ischemic brain injury: from bench to bedside. Trans Stroke Res 2:455–462
38. Denslow N, Michel ME, Temple MD, Hsu CY, Saatman K, Hayes RL (2003) Application of proteomics technology to the field of neurotrauma. J Neurotrauma 20:401–407
39. Berger RP (2006) The use of serum biomarkers to predict outcome after traumatic brain injury in adults and children. J Head Trauma Rehabil 21:315–333
40. Mondello S, Jeromin A, Buki A, Bullock R, Czeiter E, Kovacs N, Barzo P, Schmid K, Tortella F, Wang KK, Hayes RL (2012) Glial neuronal ratio: a novel index for differentiating injury type in patients with severe traumatic brain injury. J Neurotrauma 29:1096–1104
41. Guingab-Cagmat JD, Cagmat EB, Hayes RL, Anagli J (2013) Integration of proteomics, bioinformatics, and systems biology in traumatic brain injury biomarker discovery. Front Neurol 4:61
42. Ottens AK, Bustamante L, Golden EC, Yao C, Hayes RL, Wang KK, Tortella FC, Dave JR (2010) Neuroproteomics: a biochemical means to discriminate the extent and modality of brain injury. J Neurotrauma 27:1837–1852
43. Yao C, Williams AJ, Ottens AK, Lu XC, Liu MC, Hayes RL, Wang KK, Tortella FC, Dave JR (2009) P43/pro-EMAPII: a potential biomarker for discriminating traumatic versus ischemic brain injury. J Neurotrauma 26:1295–1305
44. Yao X, Liu J, McCabe JT (2008) Alterations of cerebral cortex and hippocampal proteasome subunit expression and function in a traumatic brain injury rat model. J Neurochem 104:353–363
45. Liu MC, Akle V, Zheng W, Dave JR, Tortella FC, Hayes RL, Wang KK (2006) Comparing calpain- and caspase-3-mediated degradation patterns in traumatic brain injury by differential proteome analysis. Biochem J 394:715–725
46. Setsuie R, Wada K (2007) The functions of UCH-L1 and its relation to neurodegenerative diseases. Neurochem Int 51:105–111
47. Alawieh A, Zaraket FA, Li JL, Mondello S, Nokkari A, Razafsha M, Fadlallah B, Boustany RM, Kobeissy FH (2012) Systems biology, bioinformatics, and biomarkers in neuropsychiatry. Front Neurosci 6:187
48. Mondello S, Linnet A, Buki A, Robicsek S, Gabrielli A, Tepas J, Papa L, Brophy GM, Tortella F, Hayes RL, Wang KK (2012) Clinical utility of serum levels of ubiquitin C-terminal hydrolase as a biomarker for severe traumatic brain injury. Neurosurgery 70:666–675
49. Mondello S, Palmio J, Streeter J, Hayes RL, Peltola J, Jeromin A (2012) Ubiquitin carboxy-terminal hydrolase L1 (UCH-L1) is increased in cerebrospinal fluid and plasma of patients after epileptic seizure. BMC Neurol 12:85
50. Blyth BJ, Farahvar A, He H, Nayak A, Yang C, Shaw G, Bazarian JJ (2011) Elevated serum ubiquitin carboxy-terminal hydrolase L1 is associated with abnormal blood-brain barrier function after traumatic brain injury. J Neurotrauma 28:2453–2462
51. Diaz-Arrastia R, Wang KK, Papa L, Sorani MD, Yue JK, Puccio AM, McMahon PJ, Inoue T, Yuh EL, Lingsma HF, Maas AI, Valadka AB, Okonkwo DO, Manley GT, Investigators T-T (2014) Acute biomarkers of traumatic brain injury: relationship between plasma levels of ubiquitin C-terminal hydrolase-L1 and glial fibrillary acidic protein. J Neurotrauma 31:19–25
52. Papa L, Lewis LM, Silvestri S, Falk JL, Giordano P, Brophy GM, Demery JA, Liu MC, Mo J, Akinyi L, Mondello S, Schmid K, Robertson CS, Tortella FC, Hayes RL, Wang KK (2012) Serum levels of ubiquitin C-terminal hydrolase distinguish mild traumatic brain injury from trauma controls and are elevated in mild and moderate traumatic brain injury patients with intracranial lesions

and neurosurgical intervention. J Trauma Acute Care Surg 72:1335–1344

53. Siman R, Roberts VL, McNeil E, Dang A, Bavaria JE, Ramchandren S, McGarvey M (2008) Biomarker evidence for mild central nervous system injury after surgically-induced circulation arrest. Brain Res 1213:1–11

54. Ali A, Zahraa S, Zhiqun Z, Firas K, Kevin KWW (2014) Neuro-proteomics and neuro-systems biology in the quest of TBI biomarker discovery. In: Biomarkers of brain injury and neurological disorders, Boca Raton, FL: CRC Press, pp 3–41

55. Berger RP, Hayes RL, Richichi R, Beers SR, Wang KK (2012) Serum concentrations of ubiquitin C-terminal hydrolase-L1 and alphaII-spectrin breakdown product 145 kDa correlate with outcome after pediatric TBI. J Neurotrauma 29:162–167

56. Zetterberg H, Smith DH, Blennow K (2013) Biomarkers of mild traumatic brain injury in cerebrospinal fluid and blood. Nat Rev Neurol 9:201–210

57. Pike BR, Flint J, Dave JR, Lu XC, Wang KK, Tortella FC, Hayes RL (2004) Accumulation of calpain and caspase-3 proteolytic fragments of brain-derived alphaII-spectrin in cerebral spinal fluid after middle cerebral artery occlusion in rats. J Cereb Blood Flow Metabol 24:98–106

58. Cardali S, Maugeri R (2006) Detection of alphaII-spectrin and breakdown products in humans after severe traumatic brain injury. J Neurosurg Sci 50:25–31

59. Mondello S, Robicsek SA, Gabrielli A, Brophy GM, Papa L, Tepas J, Robertson C, Buki A, Scharf D, Jixiang M, Akinyi L, Muller U, Wang KK, Hayes RL (2010) alphaII-spectrin breakdown products (SBDPs): diagnosis and outcome in severe traumatic brain injury patients. J Neurotrauma 27:1203–1213

60. Pineda JA, Lewis SB, Valadka AB, Papa L, Hannay HJ, Heaton SC, Demery JA, Liu MC, Aikman JM, Akle V, Brophy GM, Tepas JJ, Wang KK, Robertson CS, Hayes RL (2007) Clinical significance of alphaII-spectrin breakdown products in cerebrospinal fluid after severe traumatic brain injury. J Neurotrauma 24:354–366

61. Olsson B, Zetterberg H, Hampel H, Blennow K (2011) Biomarker-based dissection of neurodegenerative diseases. Prog Neurobiol 95:520–534

62. Kovesdi E, Luckl J, Bukovics P, Farkas O, Pal J, Czeiter E, Szellar D, Doczi T, Komoly S, Buki A (2010) Update on protein biomarkers in traumatic brain injury with emphasis on clinical use in adults and pediatrics. Acta Neurochir 152:1–17

63. Selakovic V, Raicevic R, Radenovic L (2005) The increase of neuron-specific enolase in cerebrospinal fluid and plasma as a marker of neuronal damage in patients with acute brain infarction. J Clin Neurosci 12:542–547

64. Ross SA, Cunningham RT, Johnston CF, Rowlands BJ (1996) Neuron-specific enolase as an aid to outcome prediction in head injury. Br J Neurosurg 10:471–476

65. Ingebrigtsen T, Romner B (2003) Biochemical serum markers for brain damage: a short review with emphasis on clinical utility in mild head injury. Restor Neurol Neurosci 21:171–176

66. Bohmer AE, Oses JP, Schmidt AP, Peron CS, Krebs CL, Oppitz PP, D'Avila TT, Souza DO, Portela LV, Stefani MA (2011) Neuron-specific enolase, S100B, and glial fibrillary acidic protein levels as outcome predictors in patients with severe traumatic brain injury. Neurosurgery 68:1624–1630, discussion 1630-1621

67. Fridriksson T, Kini N, Walsh-Kelly C, Hennes H (2000) Serum neuron-specific enolase as a predictor of intracranial lesions in children with head trauma: a pilot study. Acad Emerg Med 7:816–820

68. Geyer C, Ulrich A, Grafe G, Stach B, Till H (2009) Diagnostic value of S100B and neuron-specific enolase in mild pediatric traumatic brain injury. J Neurosurg Pediatr 4:339–344

69. Meric E, Gunduz A, Turedi S, Cakir E, Yandi M (2010) The prognostic value of neuron-specific enolase in head trauma patients. J Emerg Med 38:297–301

70. Skogseid IM, Nordby HK, Urdal P, Paus E, Lilleaas F (1992) Increased serum creatine kinase BB and neuron specific enolase following head injury indicates brain damage. Acta Neurochir 115:106–111

71. Topolovec-Vranic J, Pollmann-Mudryj MA, Ouchterlony D, Klein D, Spence J, Romaschin A, Rhind S, Tien HC, Baker AJ (2011) The value of serum biomarkers in prediction models of outcome after mild traumatic brain injury. J Trauma 71:S478–486

72. Ramont L, Thoannes H, Volondat A, Chastang F, Millet MC, Maquart FX (2005) Effects of hemolysis and storage condition on neuron-specific enolase (NSE) in cerebrospinal fluid and serum: implications in clinical practice. Clin Chem Lab Med 43:1215–1217

73. Eng LF, Vanderhaeghen JJ, Bignami A, Gerstl B (1971) An acidic protein isolated from fibrous astrocytes. Brain Res 28:351–354

74. Huang XJ, Glushakova O, Mondello S, Van K, Hayes RL, Lyeth BG (2015) Acute temporal

profiles of serum levels of UCH-L1 and GFAP and relationships to neuronal and astroglial pathology following traumatic brain injury in rats. J Neurotrauma 32:1179–1189
75. Papa L, Mittal MK, Ramirez J, Ramia M, Kirby S, Silvestri S, Giordano P, Weber K, Braga CF, Tan CN, Ameli NJ, Lopez M, Zonfrillo MR (2016) In children and youth with mild and moderate traumatic brain injury GFAP out-performs S100beta in detecting traumatic intracranial lesions on CT. J Neurotrauma 33:58
76. Takala RS, Posti JP, Runtti H, Newcombe VF, Outtrim J, Katila AJ, Frantzen J, Ala-Seppala H, Kyllonen A, Maanpaa HR, Tallus J, Hossain MI, Coles JP, Hutchinson P, van Gils M, Menon DK, Tenovuo O (2016) GFAP and UCH-L1 as outcome predictors in traumatic brain injury. World Neurosurg 87:8
77. Wang KK, Yang Z, Yue JK, Zhang Z, Winkler EA, Puccio A, Diaz-Arrastia R, Lingsma H, Yuh EL, Mukherjee P, Valadka A. Plasma Anti-Glial Fibrillary Acidic Protein (GFAP) Autoantibody Levels During the Acute and Chronic Phases of Traumatic Brain Injury-A TRACK-TBI Pilot Study. Journal of neurotrauma. 2015 Nov 11(ja).
78. Galea E, Dupouey P, Feinstein DL (1995) Glial fibrillary acidic protein mRNA isotypes: expression in vitro and in vivo. J Neurosci Res 41:452–461
79. Pelinka LE, Kroepfl A, Schmidhammer R, Krenn M, Buchinger W, Redl H, Raabe A (2004) Glial fibrillary acidic protein in serum after traumatic brain injury and multiple trauma. J Trauma 57:1006–1012
80. Vos PE, Lamers KJ, Hendriks JC, van Haaren M, Beems T, Zimmerman C, van Geel W, de Reus H, Biert J, Verbeek MM (2004) Glial and neuronal proteins in serum predict outcome after severe traumatic brain injury. Neurology 62:1303–1310
81. Honda M, Tsuruta R, Kaneko T, Kasaoka S, Yagi T, Todani M, Fujita M, Izumi T, Maekawa T (2010) Serum glial fibrillary acidic protein is a highly specific biomarker for traumatic brain injury in humans compared with S-100B and neuron-specific enolase. J Trauma 69:104–109
82. Nylen K, Ost M, Csajbok LZ, Nilsson I, Blennow K, Nellgard B, Rosengren L (2006) Increased serum-GFAP in patients with severe traumatic brain injury is related to outcome. J Neurol Sci 240:85–91
83. Vos PE, Jacobs B, Andriessen TM, Lamers KJ, Borm GF, Beems T, Edwards M, Rosmalen CF, Vissers JL (2010) GFAP and S100B are biomarkers of traumatic brain injury: an observational cohort study. Neurology 75:1786–1793
84. Pelinka LE, Kroepfl A, Leixnering M, Buchinger W, Raabe A, Redl H (2004) GFAP versus S100B in serum after traumatic brain injury: relationship to brain damage and outcome. J Neurotrauma 21:1553–1561
85. Zurek J, Fedora M (2012) The usefulness of S100B, NSE, GFAP, NF-H, secretagogin and Hsp70 as a predictive biomarker of outcome in children with traumatic brain injury. Acta Neurochir (Wien) 154:93–103, discussion 103
86. Heizmann CW, Fritz G, Schafer BW (2002) S100 proteins: structure, functions and pathology. Front Biosci 7:d1356–1368
87. Donato R (1986) S-100 proteins. Cell Calcium 7:123–145
88. Donato R, Prestagiovanni B, Zelano G (1986) Identity between cytoplasmic and membrane-bound S-100 proteins purified from bovine and rat brain. J Neurochem 46:1333–1337
89. Barger SW, Van Eldik LJ, Mattson MP (1995) S100 beta protects hippocampal neurons from damage induced by glucose deprivation. Brain Res 677:167–170
90. Townend W, Dibble C, Abid K, Vail A, Sherwood R, Lecky F (2006) Rapid elimination of protein S-100B from serum after minor head trauma. J Neurotrauma 23:149–155
91. Pelinka LE, Bahrami S, Szalay L, Umar F, Redl H (2003) Hemorrhagic shock induces an S 100 B increase associated with shock severity. Shock 19:422–426
92. Berger RP, Adelson PD, Pierce MC, Dulani T, Cassidy LD, Kochanek PM (2005) Serum neuron-specific enolase, S100B, and myelin basic protein concentrations after inflicted and noninflicted traumatic brain injury in children. J Neurosurg 103:61–68
93. Kleindienst A, Hesse F, Bullock MR, Buchfelder M (2007) The neurotrophic protein S100B: value as a marker of brain damage and possible therapeutic implications. Prog Brain Res 161:317–325
94. Egea-Guerrero JJ, Revuelto-Rey J, Murillo-Cabezas F, Munoz-Sanchez MA, Vilches-Arenas A, Sanchez-Linares P, Dominguez-Roldan JM, Leon-Carrion J (2012) Accuracy of the S100beta protein as a marker of brain damage in traumatic brain injury. Brain Inj 26:76–82
95. Ingebrigtsen T, Romner B, Kongstad P, Langbakk B (1995) Increased serum concentrations of protein S-100 after minor head

injury: a biochemical serum marker with prognostic value? J Neurol Neurosurg Psychiatry 59:103–104
96. Bazarian JJ, Zemlan FP, Mookerjee S, Stigbrand T (2006) Serum S-100B and cleaved-tau are poor predictors of long-term outcome after mild traumatic brain injury. Brain Inj 20:759–765
97. de Kruijk JR, Leffers P, Menheere PP, Meerhoff S, Twijnstra A (2001) S-100B and neuron-specific enolase in serum of mild traumatic brain injury patients. A comparison with health controls. Acta Neurol Scand 103:175–179
98. Ingebrigtsen T, Romner B, Marup-Jensen S, Dons M, Lundqvist C, Bellner J, Alling C, Borgesen SE (2000) The clinical value of serum S-100 protein measurements in minor head injury: a Scandinavian multicentre study. Brain Inj 14:1047–1055
99. Ingebrigtsen T, Waterloo K, Jacobsen EA, Langbakk B, Romner B (1999) Traumatic brain damage in minor head injury: relation of serum S-100 protein measurements to magnetic resonance imaging and neurobehavioral outcome. Neurosurgery 45:468–475, discussion 475-466
100. Mercier E, Boutin A, Lauzier F, Fergusson DA, Simard JF, Zarychanski R, Moore L, McIntyre LA, Archambault P, Lamontagne F, Legare F, Randell E, Nadeau L, Rousseau F, Turgeon AF (2013) Predictive value of S-100beta protein for prognosis in patients with moderate and severe traumatic brain injury: systematic review and meta-analysis. BMJ 346:f1757
101. Muller K, Townend W, Biasca N, Unden J, Waterloo K, Romner B, Ingebrigtsen T (2007) S100B serum level predicts computed tomography findings after minor head injury. J Trauma 62:1452–1456
102. Schiavi P, Laccarino C, Servadei F (2012) The value of the calcium binding protein S100 in the management of patients with traumatic brain injury. Acta Biomed 83:5–20
103. Spinella PC, Dominguez T, Drott HR, Huh J, McCormick L, Rajendra A, Argon J, McIntosh T, Helfaer M (2003) S-100beta protein-serum levels in healthy children and its association with outcome in pediatric traumatic brain injury. Crit Care Med 31:939–945
104. Unden J, Astrand R, Waterloo K, Ingebrigtsen T, Bellner J, Reinstrup P, Andsberg G, Romner B (2007) Clinical significance of serum S100B levels in neurointensive care. Neurocrit Care 6:94–99
105. Unden J, Romner B (2010) Can low serum levels of S100B predict normal CT findings after minor head injury in adults?: an evidence-based review and meta-analysis. J Head Trauma Rehabil 25:228–240
106. Biberthaler P, Linsenmeier U, Pfeifer KJ, Kroetz M, Mussack T, Kanz KG, Hoecherl EF, Jonas F, Marzi I, Leucht P, Jochum M, Mutschler W (2006) Serum S-100B concentration provides additional information fot the indication of computed tomography in patients after minor head injury: a prospective multicenter study. Shock 25:446–453
107. Zongo D, Ribereau-Gayon R, Masson F, Laborey M, Contrand B, Salmi LR, Montaudon D, Beaudeux JL, Meurin A, Dousset V, Loiseau H, Lagarde E (2012) S100-B protein as a screening tool for the early assessment of minor head injury. Ann Emerg Med 59:209–218
108. Zimmer DB, Cornwall EH, Landar A, Song W (1995) The S100 protein family: history, function, and expression. Brain Res Bull 37:417–429
109. Donato R (2001) S100: a multigenic family of calcium-modulated proteins of the EF-hand type with intracellular and extracellular functional roles. Int J Biochem Cell Biol 33:637–668
110. Rothoerl RD, Woertgen C (2001) High serum S100B levels for trauma patients without head injuries. Neurosurgery 49:1490–1491, author reply 1492-1493
111. Anderson RE, Hansson LO, Nilsson O, Dijlai-Merzoug R, Settergren G (2001) High serum S100B levels for trauma patients without head injuries. Neurosurgery 48:1255–1258, discussion 1258-1260
112. Routsi C, Stamataki E, Nanas S, Psachoulia C, Stathopoulos A, Koroneos A, Zervou M, Jullien G, Roussos C (2006) Increased levels of serum S100B protein in critically ill patients without brain injury. Shock 26:20–24
113. Unden J, Bellner J, Eneroth M, Alling C, Ingebrigtsen T, Romner B (2005) Raised serum S100B levels after acute bone fractures without cerebral injury. J Trauma 58:59–61
114. Pelinka LE, Szalay L, Jafarmadar M, Schmidhammer R, Redl H, Bahrami S (2003) Circulating S100B is increased after bilateral femur fracture without brain injury in the rat. Br J Anaesth 91:595–597
115. Pelinka LE, Toegel E, Mauritz W, Redl H (2003) Serum S 100 B: a marker of brain damage in traumatic brain injury with and without multiple trauma. Shock 19:195–200
116. Narayan RK, Michel ME, Ansell B, Baethmann A, Biegon A, Bracken MB, Bullock MR, Choi SC, Clifton GL, Contant CF, Coplin WM, Dietrich WD, Ghajar J, Grady SM, Grossman RG, Hall ED, Heetderks W, Hovda DA, Jallo J, Katz RL, Knoller N, Kochanek PM, Maas AI, Majde J,

Marion DW, Marmarou A, Marshall LF, McIntosh TK, Miller E, Mohberg N, Muizelaar JP, Pitts LH, Quinn P, Riesenfeld G, Robertson CS, Strauss KI, Teasdale G, Temkin N, Tuma R, Wade C, Walker MD, Weinrich M, Whyte J, Wilberger J, Young AB, Yurkewicz L (2002) Clinical trials in head injury. J Neurotrauma 19:503–557

117. Mondello S, Muller U, Jeromin A, Streeter J, Hayes RL, Wang KK (2011) Blood-based diagnostics of traumatic brain injuries. Expert Rev Mol Diagn 11:65–78

Part III

Classical TBI Models and Their Link with Pathophysiological Features of CNS Injury—Models

Chapter 11

The Controlled Cortical Impact Model of Experimental Brain Trauma: Overview, Research Applications, and Protocol

Nicole Osier and C. Edward Dixon

Abstract

Controlled cortical impact (CCI) is a commonly used and highly regarded model of brain trauma that uses a pneumatically or electromagnetically controlled piston to induce reproducible and well-controlled injury. The CCI model was originally used in ferrets and it has since been scaled for use in many other species. This chapter will describe the historical development of the CCI model, compare and contrast the pneumatic and electromagnetic models, and summarize key short- and long-term consequences of TBI that have been gleaned using this model. In accordance with the recent efforts to promote high-quality evidence through the reporting of common data elements (CDEs), relevant study details—that should be reported in CCI studies—will be noted.

Key words Traumatic brain injury (TBI), Experimental brain injury, Preclinical, Animal model, Controlled cortical impact (CCI), Common data elements (CDE)

1 Introduction

Animal models have been used to study traumatic brain injury (TBI) for over a century and they remain widely used today to better understand outcomes of brain trauma and test novel therapies [1–4]. Today, preclinical TBI researchers have the choice between several models including: weight drop injury (WDI), fluid percussion injury (FPI), blast-induced TBI (bTBI), and controlled cortical impact (CCI), the focus of this chapter. CCI was originally developed to study TBI in ferrets [5], and its desirable properties (e.g., reproducibility; control over injury parameters) have led researchers to scale the model and apply it to many other species. The original design uses a pneumatically driven piston to induce TBI, while a newer alternative added an element of portability by using an electromagnetically driven piston which is lighter in weight and negates the need for a cylinder filled with compressed N_2 gas.

Firas Kobeissy et al. (eds.), *Injury Models of the Central Nervous System: Methods and Protocols*, Methods in Molecular Biology, vol. 1462, DOI 10.1007/978-1-4939-3816-2_11, © Springer Science+Business Media New York 2016

The purpose of this chapter is to introduce readers to the CCI model so that thoughtful decisions can be made in their own completion of CCI research or in their consumption of the research literature. In doing so, the following will be discussed: (1) historical development, (2) key features, (3) comparison of the pneumatic and electromagnetic devices, (4) research applications, (5) relevant common data elements, as well as (6) factors that influence outcomes and data quality. A standard protocol will be described for pneumatic CCI in rats, along with a list of all required supplies and equipment.

1.1 History of Controlled Cortical Impact and Key Features

Animals have been used to study TBI for over 100 years, with considerable refinement of methodologies within the last three decades [1–3, 6–14]. In comparison to the many experimental TBI models available today, CCI is relatively new. It was originally developed by J.W. Lighthall and colleagues in the late 1980s and early 1990s with the goal of inducing TBI in ferrets [5, 15]. The desirable properties of CCI including the control over important injury parameters and ability to induce reproducible injury led C.E. Dixon and colleagues to scale the CCI model for use in rats during the early 1990s [16]. Since that time, CCI has been further scaled for use in several other species including mice, pigs, and nonhuman primates.

The scalability and other desirable features have resulted in CCI becoming one of the most popular and widely used preclinical TBI models. One noteworthy feature is that CCI provides quantitative control over important biomechanical parameters of TBI, in particular, the velocity, depth, and force of the tip are controlled across a wide range of contact velocities; there are also different options for tip size, geometry, and positioning, as discussed later in this chapter. Taken together, the control and customization of CCI allows researchers to address a multitude of research questions as well as scale the injury as needed to study the histopathological and functional deficits of interest. Reporting the injury parameters is of critical importance when reproducing, interpreting, and comparing published study findings, as described under the Common Data Elements heading.

A detailed protocol for inducing CCI in rats is included later, but to set the stage, a brief summary of CCI will be provided first. Traditionally, CCI is an invasive model whereby the exposed cortex is subjected to trauma following a craniectomy; in invasive CCI studies sham animals are used as controls to ensure that the results seen are not due to anesthesia and craniectomy but rather the CCI itself. Following craniectomy, the CCI device is used to transfer mechanical energy onto the intact dura mater, producing a TBI. Traditionally, the tip is pneumatically driven, though a newer model affords portability by using an electromagnetic device to drive the tip [14]. Both types of CCI allow for control over the tip depth, dwell time, and velocity; additional details about the pneumatic and electromagnetic CCI models are provided in the following section. It is also worth acknowledging that while invasive CCI remains widely used, CCI has been extended to model closed head injury [17–19] as described later in this chapter. It is also notable that even with an invasive CCI model, only one surgical procedure is necessary, as opposed to standard FPI which requires two surgeries.

1.2 Controlled Cortical Impact Types

1.2.1 Pneumatic Controlled Cortical Impact

The pneumatic CCI device (Fig. 1) was the first to be developed and it remains the most commonly used today. For these reasons, the pneumatic model will be emphasized in this chapter with a brief discussion on how it differs from the electromagnetic alternative. A standard pneumatic CCI device features a small bore (19.75 mm) reciprocating a double-acting pneumatic piston with a 50 mm stroke length. The piston is used to drive a tip of a desired size and geometry into the neural tissue (or in some cases the intact skull) to induce brain trauma. The cylinder is held by a crossbar which can be stereotaxically adjusted for variable mounting positions allowing the tip to be either aligned vertically or angled with respect to the brain. The velocity of the piston is monitored by a sensor and can be controlled to promote uniform injury across test animals. Researchers using the CCI model are able to control the depth, duration (a.k.a. dwell time), and velocity of injury as well as choose what size and shape of tip to use.

1.2.2 Electromagnetic Controlled Cortical Impact

The electromagnetic CCI device (Fig. 2) is very similar to the pneumatic model and also uses a stereotaxic frame for adjustability. The electromagnetic CCI devices are considered to be more portable due to their lighter weight. Another similarity is the

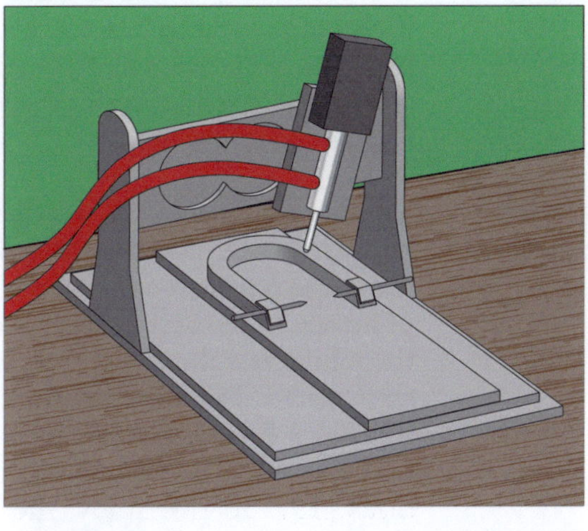

Fig. 1 Diagram depicting a standard pneumatic CCI device (without the associated cylinder of compressed N_2 gas). Depending on the research goals, researchers can choose the ideal tip (e.g., size, geometry) and injury parameters (e.g., depth, dwell time, velocity)

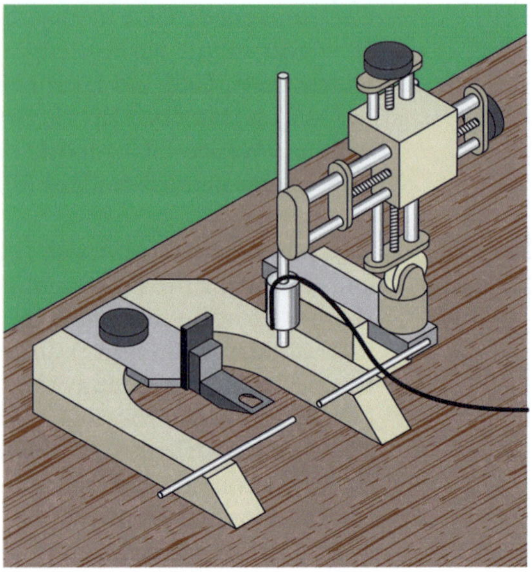

Fig. 2 Diagram depicting a standard electromagnetic CCI device. As with pneumatic CCI, the tip and injury parameters can be adjusted. Unlike with the pneumatic model, the tip is driven by an electromagnetic actuator negating the need for N_2 gas

availability of tips with varying sizes and geometry (e.g., flat, beveled, round). Generally speaking, tip scaling correlates with animal size; for example, 3 mm tips are commonly used for mice, 5–6 mm tips for rats, 10 mm for nonhuman primates, and 15 mm for pigs. Depending on the vendor, the device may come with a variety of tips. Alternatively, there may be additional tips available to purchase separately. Some researchers have also modified tips based on their unique research needs with one group using vulcanized rubber from a lacrosse ball to cover the tip [19]. A list of commercial vendors who sell both electromagnetic and pneumatic CCI devices is included in Table 1.

1.3 Applications of Controlled Cortical Impact

1.3.1 Closed Head Injury

Though originally CCI was developed as an invasive TBI model, it has been more recently adapted to study closed head injury (CHI) including repeated concussions. CHI models have become an area of increased research emphasis as the risk to individuals who are in the military or those involved in various athletic activities is further appreciated. The aforementioned strengths of CCI make it a popular choice for researchers studying CHI [17–19]. Applications of CCI to study CHI include a study modeling sports-related concussions [19]. In this study, the researchers combined elements of CCI with elements of Marmarou's impact acceleration model in an attempt to enhance control over clinically relevant variables. For example, a foam pad was placed under the rodent to limit rotational acceleration and instead promote linear acceleration. In this

Table 1
Commercially available pneumatic- and electromagnetic-CCI devices

CCI Type	Company	Model
Pneumatic	Precision Systems and Instrumentation	LLC TBI-0310 Impactor
	Pittsburgh Precision Instruments	Pneumatic Powered Controlled Cortical Impact Device
	AmScien Instruments	Pneumatic (Cortical) Impact Device (Model: AMS 201)
Electromagnetic	Leica	Impact One Stereotaxic Impactor for CCI
	Hatteras Instruments	Pinpoint PCI3000 Precision Cortical Impactor

Suppliers and models are listed for commonly used pneumatic- and electromagnetic-CCI devices

study, more deficits were observed after repeated TBI than a single TBI, as assessed using a battery of neurobehavioral tests including measures of cognition, memory, and sleep [19].

1.3.2 Long-Term Outcomes

The high survivability of CCI makes it a good choice for studying the long-term changes associated with TBI. Available evidence suggests that CCI results in chronic and progressive changes. For example, one study which assessed the test animals up to 1 year post-TBI reported ongoing deficits including progressive tissue loss and ventricular expansion [20]. In addition to the aforementioned histopathological consequences, there are several chronic behavioral deficits that have been reported. For example, after CCI (vs. sham) memory and learning deficits have been found to persist into the long term as assessed using the MWM [21–35]. Notably, persistent MWM deficits are more rarely reported in other models, including lateral FPI [36, 37] and medial FPI [38]. Long-term deficits in motor function have also been reported, as assessed using the foot fault test [22, 34, 39, 40], whereas the authors of this chapter were only able to find one FPI study reporting long-term motor deficits using this measure [41]. That is to say that CCI is a good choice when researchers are interested in exploring long-term motor and memory symptoms after brain injury. Conversely, no CCI studies were identified where motor deficits on the inclined plane task were reported, though such deficits have been reported after lateral FPI [42]. Similarly, no CCI studies were identified where chronic deficits in reversal learning were reported, although this has been reported in several lateral FPI studies [36, 37, 43]. A detailed review of long-term outcomes for the major experimental TBI models is available to interested readers [44].

1.4 Species Used

One of the assets of the CCI model is that it can be translated to induce experimental TBI in many species. This is accomplished by scaling the injury parameters so as to maintain the percent of brain volume deformed in relation to total brain volume taking into account desired extent of injury to address the research questions. In order to induce CCI in large animals (e.g., pigs), modifications may be necessary to ensure the piston is high enough. Table 2 provides a summary of the various species that CCI has been applied to including injury parameters commonly used in each species.

1.4.1 Rat

Graded TBI can be easily produced in rats using the CCI model. Indeed, the effects of CCI are well categorized in rats where it has been found to result in diverse histopathological and functional changes consistent with what occurs in clinical TBI cases, including but not limited to: blood–brain-barrier disruption, derangements in blood flow and pressure, axonal injury, inflammation, and edema. It is also worth noting that functional symptoms of TBI (e.g., deficits in learning, memory, and motor function) can be assessed using neurobehavioral testing, which is well characterized in rats. For example, memory is readily assessed using the Morris Water Maze (MWM), Barnes Maze, or Novel Object Recognition (NOR) task; whereas motor function can be assessed using the Beam Balance Task, Beam Walking Task, Rotarod Task, and Wire Grip Task.

1.4.2 Mouse

Shortly after translation of CCI from ferrets to rats, the model was further extended to mice [45–47]. Refinement of CCI in mice has paralleled the increasing application of genetically modified mice to TBI research to explore the role of genes and gene products in brain injury recovery [48]. Generally speaking, to scale CCI down for mice entails decreasing the injury depth to adjust for the relatively thinner

Table 2
Commonly used CCI parameters for various species

Species	Injury site	Crani. size (mm)	Tip diameter (mm)	Velocity (m/s)	Dwell time (ms)	Depth (mm)
Mouse	Parietal Cortex	4–5	3	4–6	50–250	0.5–2
Rat	Parietal Cortex; Midline	6–8	5–6	4	50–250	1–3
Primate	Frontal Lobe	11–12	10	3.5	150	7
Pig	Frontal Lobe	15–18	15	2–4	50–400	12

For each species, injury details commonly used are provided including injury site, craniectomy size, tip diameter, velocity, dwell time, and depth

cortex in mice compared to rats. Though slightly less well categorized, a plethora of behavioral testing strategies are available for use in mice including the MWM, NOR, and BBT [31, 33, 49, 50].

1.4.3 Pig

To adequately address some research questions, larger, more human-like brains are needed, necessitating the use of a pig or other large mammal models. The main difference from the rat and mouse models is the considerably larger impactor tip and greater depth to which the neural tissue is deformed (*see* Table 2). In one study, CCI was scaled to induce TBI in piglets; injury parameters were chosen after adjusting for differences in brain morphology (e.g., size and dimensions) in the study's test animals [51]. As with rodent models of CCI, when pigs are used the histopathological changes mirror what is seen in TBI patients, including but not limited to, deranged blood flow and changes to vasculature, ongoing neurodegeneration via a number of mechanisms, and edema. To date, pig models have been used to add to the evidence surrounding TBI biomechanics [52] and as part of an effort to identify clinically relevant biomarkers of underlying brain injury [53]. However, despite these efforts there is a relative dearth of normative data specific to pigs when compared to rodents [54]. Additionally, behavioral testing is less well characterized in pigs and not to mention more challenging to perform due to their larger size and relative intelligence.

1.4.4 Nonhuman Primate

An alternative to the pig model is the nonhuman primate model of CCI, which is typically applied over the frontal cortex [55]. As with the other models, the histopathological changes reported after nonhuman primate CCI mimic what is seen clinically, including but not limited to edema, macrophage accumulation, and neurodegeneration. Nonhuman primates play a critical role in establishing the safety of novel therapies before translation to humans. Notably, due to the increased ethical considerations, care requirements, and cost, a relatively limited number of research institutions have primate research facilities. Consequently, nonhuman primate studies represent only a small fraction of TBI studies. Indeed, the use of nonhuman primate models is only justified when there are major factors that prohibit the use of a less sentient animal.

1.5 Special Considerations, Problems, and Troubleshooting

High-quality CCI research relies on thoughtful study design and careful execution. A few important confounders that have been empirically studied will be addressed as follows. When appropriate, troubleshooting strategies will be noted.

1.5.1 Tip Geometry

Commercially available tips come in round or beveled flat shapes of various sizes. When CCI was developed in ferrets, round tips were used [5, 15]; though still in use, beveled flat tips have become the norm [47, 48, 56–59] and are especially preferred for mouse CCI. Despite convention, little empirical research has tested the effects of tip geometry on outcomes. In one study, Pleasant and colleagues compared flat vs. rounded tips in a mouse (C57BL6J) model; they found more extensive cortical hemorrhage and neuronal loss (proportionally) with flat tips. The rate of neocortical loss was faster with flat tips, with a plateau in neurodegeneration 20 h earlier than rounded tips (4 h vs. 24 h) making the latter a more desirable choice when studying secondary injury cascades in the subacute period [60].

1.5.2 Anesthesia

Overall, careful choice of whether to use anesthesia and what anesthesia regimen to use is critical. In deciding, researchers should consider the study goals, along with established guidelines for the treatment of research animals. Empirical evidence shows that differences in outcomes of TBI are associated with various anesthesia agents; notably, isoflurane results in less hippocampal damage than fentanyl as well as fewer behavioral deficits [61]. Another study found that preinjury isoflurane had neuroprotective effects [62]. It is hypothesized that fentanyl contributes to neural suppression, whereas isoflurane reduces excitotoxicity and promotes blood flow [54]. Ketamine also demonstrates neuroprotective properties via antagonism of *N*-methyl-D-aspartate (NMDA) receptors [63]. Halothane has also been reported to have neuroprotective properties after contusion injuries [64]. Use of neuroprotective anesthesia can obscure deficits in performance in all anesthetized groups. For fentanyl, the concern is that the resulting neural suppression could worsen performance and may obscure treatment effects in a drug study.

Despite these concerns, the overwhelming majority of CCI studies use anesthesia, commonly isoflurane. Volatile gases (e.g., isoflurane, halothane) are often preferred due to their relatively short half-life compared to long-acting options (e.g., pentobarbital), facilitating evaluation of righting reflex shortly after anesthesia discontinuation [16]. Typically a high dose (e.g., 4%) is used to induce anesthesia followed by a reduced maintenance dose (*see* Subheading 4). Anesthetized animals should be monitored to ensure consciousness is not regained during surgery. Assessments like the toe pinch can be used to assess sufficiency of anesthesia in accordance with institutional and national policies. One recent study of repetitive closed head injury used unanaesthetized mice that were instead comfortably restrained to avoid the confounding effects of anesthesia on TBI outcomes and promote clinical relevance [19].

Commonly encountered problems surrounding anesthesia are summarized later. First, if animals are intubated and the animal is fighting the ventilator, changes in tube placement may alleviate the problem. Also, if consciousness is regained, then the anesthesia induction system should be checked to ensure that the anesthesia is set to the appropriate level and there are no leaks in the tubing. It is also important to consider that if gaseous anesthesia is used, specialized laboratory equipment is required including appropriate ventilation and scavenging systems to ensure the safety of personnel; isoflurane detection systems are available to monitor exposure.

1.5.3 Craniectomy and Sham Procedure

Researchers typically produce craniectomy using a pneumatic or electric drill, although a handheld trephine is sometimes used. Efforts to reduce heat production during the craniectomy can help reduce potential confounders, and sterile saline can be applied to the craniectomy site to reduce the temperature if the procedure is prolonged. Since anesthesia and craniectomy can result in behavioral and functional changes it is important that when invasive CCI is used control animals be exposed to sham. Empirical evidence shows that craniectomy results in inflammation and lesions regardless of whether a trephine or electric drill is used, as compared to naïve rats (anesthesia only). However, lesions were largest and behavioral deficits were most severe in animals that received their craniectomies using a drill [65]. Craniectomy location also affects outcome with midline craniectomy leading to more sagittal bleeding [16] than parasagittal craniectomy [66, 67]. Bilateral craniectomies have been used to enhance lateral movement of tissue and produce subsequent bilateral cortical contusions [68, 69]; producing bilateral craniectomies is also a good way to train individuals on the procedure. Details regarding the control group should be provided in publications [65], including any bleeding, mortality, inconsistency across animals, etc.

2 Materials

2.1 Animals

A strength of the CCI model is that it can be used in many species. In much of our work, adult male Sprague Dawley rats (280–320 g; Charles River Labs, Raleigh, VA, USA) are used; thus, this protocol is specific to Sprague Dawley rats. Our animals are routinely housed in a climate-controlled room with a 12 h light/dark cycle and are regularly monitored by the Department of Laboratory Animal Research.

2.2 Anesthesia

- Induction Dose: 4.0 % isoflurane in 2:1 N_2O:O_2.
- Maintenance Dose: 2 % delivered in 2:1 N_2O:O_2.

2.3 Supplies and Equipment

1. Homeothermic heating system (Harvard Apparatus, MA, USA).
2. Stereotaxic frame.
3. Isoflurane.
4. Cylinders of compressed N_2O and O_2 for isoflurane delivery.
5. Cylinder of compressed N_2 to drive pneumatic tip.
6. Anesthesia chamber.
7. Gas scavenging system.
8. Cannula for intubation.

9. Laryngoscope to assist with intubation, if necessary.
10. Animal trimmers.
11. Pneumatic drill with drill bits.
12. Compressed air for drill.
13. Betadine.
14. Sterile drape.
15. Cotton-tipped applicators.
16. Gauze.
17. Saline-filled syringe.
18. Sterile surgical instruments (scalpel; scissors; periosteal elevator; microdissecting forceps; rongeurs; bulldog clips, etc.).
19. Temperature probe and associated readout (Harvard Apparatus, MA, USA).
20. MouseOx Plus, blood oxygenation monitoring system, and associated readout (Starr Life Sciences Corp., PA, USA).
21. Suture kit.
22. Pneumatic CCI device (Pittsburgh Precision Instruments, PA, USA).
23. Tip of desired size and shape.

3 Methods

1. Before starting the surgeries, ensure the CCI device is in good working order. Does the piston fire freely? Is the impact velocity and dwell time consistent with what is set?
2. Place the rat in an anesthesia chamber and induce anesthesia with 4% isoflurane in a 2:1 mixture of $N_2O{:}O_2$; ensure the animal is sufficiently anesthetized using a toe pinch test.
3. Intubate the rat.
4. Place the anesthetized animal in a stereotaxic frame and secure the incisor and ear bars to keep the animal secure throughout the surgery.
5. Adjust the anesthesia to the maintenance dose of 2% isoflurane (*see* **Note 1**).
6. Assess the animal's level of alertness using the toe pinch test for suppression of pedal response (or another similar test) to ensure sufficient anesthesia is being delivered.
7. Use the hair trimmers to shave the rat's scalp moving both with and against the grain.
8. Use a sterile drape to cover the animal such that the only opening in the drape is directly over the exposed scalp.

9. Use gauze and antiseptic solution (e.g., Betadine) to scrub the scalp and prepare the surgical site.
10. Use a scalpel to make a midline incision (*see* **Note 2**).
11. Separate the muscle from the skull using the periosteal elevator and microdissecting forceps.
12. Reflect the skin and fascia to expose the underlying skull and scrub the surface of the skull with a cotton-tipped applicator.
13. Use pneumatic drill (hooked up to a compressed air cylinder) to create a craniectomy. Center the craniectomy between the sagittal suture and coronal ridge with the borders near the lambda and bregma for unobstructed tip clearance (*see* **Note 3**).
14. If necessary, use rongeurs to elongate the craniectomy until it is large enough to accommodate the impactor tip; carefully lift away the resulting bone flap, so as to avoid dura breech (*see* **Note 4**).
15. To ensure that the tip is centered over the craniectomy site, manually extend the shaft on the CCI device and gently lower the impactor tip so that it lightly and briefly touches the exposed dura mater.
16. With the piston statically pressurized and in the full stroke position, zero the tip to the cortical surface.
17. Carefully withdraw the tip and adjust the piston assembly to the desired impact depth based on the research goals and study protocol (*see* **Note 5**).
18. Induce injury by actuating the CCI device; discontinue the anesthesia (*see* **Note 6**).
19. Close the surgical site using sutures or another method. Apply topical anesthetic (e.g., lidocaine) to the surgical site to minimize discomfort.
20. Remove the rat from the stereotaxic frame and extubate.
21. Complete any assessments desired in the immediate postinjury period (e.g., righting reflex latency) and postsurgical monitoring.
22. Keep the test animal in a holding cage until it is able to fully recover from anesthesia, as evidenced by the return of spontaneous locomotion.
23. Once the animal has fully recovered, return it to the animal housing room and resume normal husbandry.
24. Continue to administer analgesic in accordance with institutional and government guidelines for pain management in laboratory animals.

4 Common Data Elements

The National Institute for Neurological Diseases and Stroke (NINDS) has published a set of common data elements (CDEs) for experimental TBI research including details surrounding the animals used (e.g., species, strain, commercial supplier), demographics (e.g., age, sex), metrics (e.g., weight), animal husbandry, and outcome assessment(s) used (e.g., timing of assessment, measures). Beyond the CDE's generic to all experimental TBI models, the NINDS recognizes a set of CDEs specific to CCI research. The CCI-specific CDEs include, but are not limited to craniectomy size, tip (size, shape, angle, rigidity), and injury parameter settings (depth, dwell time, velocity). Researchers are encouraged to review the current list of CDEs during study planning, grant writing, and dissemination of findings to promote comparison of studies and the conduct of high-quality research [70].

5 Conclusion

CCI is one of the best characterized models of experimental TBI and it remains a popular choice for studying the physiologic and functional deficits that occur acutely and chronically following TBI. Traditionally, CCI is an invasive model that is preceded by craniectomy, but recently the model has been applied to study concussion and other types of closed head injury. The original CCI model was pneumatically driven but more recently, an electromagnetic alternative has been introduced which provides increased portability.

Researchers employing the CCI model should give care and attention to study design and selection of injury parameters. Control over important confounding variables (e.g., hypothermia, hyperthermia) is critical to adequately address the research questions. The first step is to thoroughly explore the literature to consider how various injury parameters have panned out with respect to histopathological and functional consequences in the past. Researchers are also encouraged to conduct pilot work in order to tailor the experimental design (e.g., injury parameters, tip size, anesthesia type) and subsequently facilitate addressing the research goals. Pilot research also provides a valuable opportunity to ensure that the device is in proper working order and is calibrated.

This chapter introduced the CCI model including a brief overview of its development and extension to various species and research applications. A list of required supplies and equipment was provided as well as a detailed protocol for pneumatic CCI in rats. Discussion of confounding factors and troubleshooting methods were briefly discussed. Lastly, the importance of CDEs was extolled and exemplars of CDEs specific to CCI were noted. This introductory chapter will enable readers to be thoughtful consumers of publications describing CCI research and have the requisite knowledge needed to design and conduct a CCI study.

6 Notes

1. Traditionally, the authors of this chapter use a maintenance dose of 2% isoflurane in a 2:1 mixture of $N_2O:O_2$ titrating up the dose if the animal is showing signs of regaining consciousness.
2. The incision made in our lab is approximately 20 mm long for rats (shorter for mice).
3. The authors strive to make consistent craniectomies that are approximately 6 mm in diameter to facilitate clearance of a 5 mm diameter tip.
4. It is common practice to discard the bone flap rather than attempt to reattach it, as this can lead to secondary injury (e.g., increased intracranial pressure).
5. In our lab, we induce moderate TBI using a 5 mm tip to deform the neural tissue of a rat to a depth of 2.8 mm at a velocity of 4 m/s.
6. Depending on the preference of the researchers and the method used to close the surgical site, anesthesia can be discontinued before or after wound site closure.

Acknowledgements

Support for this chapter comes from the following government funding sources: Department of Veterans Affairs grant RR&D B1127-I, NIH-NINDS grant R01-NS079061, and NIH-NINR grants 1F31NR014957 and T32NR009759. Additional support for this chapter comes from the following foundations and professional societies: The Pittsburgh Foundation, Sigma Theta Tau International Eta Chapter, the International Society for Nurses in Genetics, and the American Association of Neuroscience Nursing/Neuroscience Nursing Foundation. We would also like to acknowledge Mr. Michael D. Farmer for his time in generating the figures and Mrs. Marilyn K. Farmer for her continued editorial support.

References

1. Kramer SP (1896) A contribution to the theory of cerebral concussion. Anim Surg 23:163–173
2. Rinder L, Olsson Y (1968) Studies on vascular permeability changes in experimental brain concussion. I. Distribution of circulating fluorescent indicators in brain and cervical cord after sudden mechanical loading of the brain. Acta Neuropathol 11:183–200
3. Denny-Brown D, Russell W (1941) Experimental cerebral concussion. Brain 64:93
4. Lindgren S, Rinder L (1965) Experimental studies in head injury. I. Some factors influencing results of model experiments. Biophysik 2:320–329
5. Lighthall JW (1988) Controlled cortical impact: a new experimental brain injury model. J Neurotrauma 5:1–15

6. Gennarelli TA, Thibault LE, Adams JH, Graham DI, Thompson CJ, Marcincin RP (1982) Diffuse axonal injury and traumatic coma in the primate. Ann Neurol 12:564–574
7. Govons SR, Govons RB, VanHuss WD, Heusner WW (1972) Brain concussion in the rat. Exp Neurol 34:121–128
8. Nilsson B, Pontén U, Voigt G (1977) Experimental head injury in the rat. Part 1: Mechanics, pathophysiology, and morphology in an impact acceleration trauma model. J Neurosurg 47:241–251
9. Ommaya AK, Geller A, Parsons LC (1971) The effect of experimental head injury on one-trial learning in rats. Int J Neurosci 1:371–378
10. Ommaya AK, Gennarelli TA (1974) Cerebral concussion and traumatic unconsciousness. Correlation of experimental and clinical observations of blunt head injuries. Brain 97:633–654
11. Sullivan HG, Martinez J, Becker DP, Miller JD, Griffith R, Wist AO (1976) Fluid-percussion model of mechanical brain injury in the cat. J Neurosurg 45:521–534
12. Cannon WB (1901) Cerebral pressure following trauma. Am J Physiol 6:91–121
13. Parkinson D, West M, Pathiraja T (1978) Concussion: comparison of humans and rats. Neurosurgery 3:176–180
14. Onyszchuk G, Al-Hafez B, He Y-Y, Bilgen M, Berman NEJ, Brooks WM (2007) A mouse model of sensorimotor controlled cortical impact: characterization using longitudinal magnetic resonance imaging, behavioral assessments and histology. J Neurosci Methods 160:187–196
15. Lighthall JW, Goshgarian HG, Pinderski CR (1990) Characterization of axonal injury produced by controlled cortical impact. J Neurotrauma 7:65–76
16. Dixon C, Clifton G, Lighthall J, Yaghmai A, Hayes R (1991) A controlled cortical impact model of traumatic brain injury in the rat. J Neurosci Methods 39:253–262
17. Shitaka Y, Tran HT, Bennett RE, Sanchez L, Levy MA, Dikranian K, Brody DL (2011) Repetitive closed-skull traumatic brain injury in mice causes persistent multifocal axonal injury and microglial reactivity. J Neuropathol Exp Neurol 70:551–567
18. Klemenhagen KC, O'Brien SP, Brody DL (2013) Repetitive concussive traumatic brain injury interacts with post-injury foot shock stress to worsen social and depression-like behavior in mice. PLoS One 8:e74510
19. Petraglia AL, Plog BA, Dayawansa S, Chen M, Dashnaw ML, Czerniecka K, Walker CT, Viterise T, Hyrien O, Iliff JJ, Deane R, Nedergaard M, Huang JH (2014) The spectrum of neurobehavioral sequelae after repetitive mild traumatic brain injury: a novel mouse model of chronic traumatic encephalopathy. J Neurotrauma 31:1211–1224
20. Dixon C, Kochanek P, Yan H, Schiding J, Griffith R, Baum E, Marion D, DeKosky S (1999) One-year study of spatial memory performance, brain morphology, and cholinergic markers after moderate controlled cortical impact in rats. J Neurotrauma 16:109–122
21. Xiong Y, Zhang Y, Mahmood A, Meng Y, Zhang ZG, Morris DC, Chopp M (2012) Neuroprotective and neurorestorative effects of thymosin β4 treatment initiated 6 hours after traumatic brain injury in rats. J Neurosurg 116:1081–1092
22. Meng Y, Xiong Y, Mahmood A, Zhang Y, Qu C, Chopp M (2011) Dose-dependent neurorestorative effects of delayed treatment of traumatic brain injury with recombinant human erythropoietin in rats. J Neurosurg 115:550–560
23. Longhi L, Watson DJ, Saatman KE, Thompson HJ, Zhang C, Fujimoto S, Royo N, Castelbuono D, Raghupathi R, Trojanowski JQ, Lee VM-Y, Wolfe JH, Stocchetti N, McIntosh TK (2004) Ex vivo gene therapy using targeted engraftment of NGF-expressing human NT2N neurons attenuates cognitive deficits following traumatic brain injury in mice. J Neurotrauma 21:1723–1736
24. Longhi L, Gesuete R, Perego C, Ortolano F, Sacchi N, Villa P, Stocchetti N, De Simoni M-G (2011) Long-lasting protection in brain trauma by endotoxin preconditioning. J Cereb Blood Flow Metab 31:1919–1929
25. Cheng JP, Shaw KE, Monaco CM, Hoffman AN, Sozda CN, Olsen AS, Kline AE (2012) A relatively brief exposure to environmental enrichment after experimental traumatic brain injury confers long-term cognitive benefits. J Neurotrauma 29:2684–2688
26. Fox GB, Faden AI (1998) Traumatic brain injury causes delayed motor and cognitive impairment in a mutant mouse strain known to exhibit delayed Wallerian degeneration. J Neurosci Res 53:718–727
27. Dixon CE, Hamm RJ, Taft WC, Hayes RL (1994) Increased anticholinergic sensitivity following closed skull impact and controlled cortical impact traumatic brain injury in the rat. J Neurotrauma 11:275–287
28. Marklund N, Morales D, Clausen F, Hånell A, Kiwanuka O, Pitkänen A, Gimbel DA, Philipson O, Lannfelt L, Hillered L, Strittmatter SM, McIntosh TK (2009) Functional outcome

is impaired following traumatic brain injury in aging Nogo-A/B-deficient mice. Neuroscience 163:540–551
29. Chauhan NB, Gatto R (2010) Synergistic benefits of erythropoietin and simvastatin after traumatic brain injury. Brain Res 1360:177–192
30. Chauhan NB, Gatto R (2011) Restoration of cognitive deficits after statin feeding in TBI. Restor Neurol Neurosci 29:23–34
31. Byrnes KR, Loane DJ, Stoica BA, Zhang J, Faden AI (2012) Delayed mGluR5 activation limits neuroinflammation and neurodegeneration after traumatic brain injury. J Neuroinflammation 9:43
32. Zhang Y, Chopp M, Mahmood A, Meng Y, Qu C, Xiong Y (2012) Impact of inhibition of erythropoietin treatment-mediated neurogenesis in the dentate gyrus of the hippocampus on restoration of spatial learning after traumatic brain injury. Exp Neurol 235:336–344
33. Tomasevic G, Laurer HL, Mattiasson G, van Steeg H, Wieloch T, McIntosh TK (2012) Delayed neuromotor recovery and increased memory acquisition dysfunction following experimental brain trauma in mice lacking the DNA repair gene XPA. J Neurosurg 116:1368–1378
34. Xiong Y, Zhang Y, Mahmood A, Meng Y, Qu C, Chopp M (2011) Erythropoietin mediates neurobehavioral recovery and neurovascular remodeling following traumatic brain injury in rats by increasing expression of vascular endothelial growth factor. Transl Stroke Res 2:619–632
35. Han R-Z, Hu J-J, Weng Y-C, Li D-F, Huang Y (2009) NMDA receptor antagonist MK-801 reduces neuronal damage and preserves learning and memory in a rat model of traumatic brain injury. Neurosci Bull 25:367–375
36. Shultz SR, Bao F, Omana V, Chiu C, Brown A, Cain DP (2012) Repeated mild lateral fluid percussion brain injury in the rat causes cumulative long-term behavioral impairments, neuroinflammation, and cortical loss in an animal model of repeated concussion. J Neurotrauma 29:281–294
37. Shultz SR, Bao F, Weaver LC, Cain DP, Brown A (2013) Treatment with an anti-CD11d integrin antibody reduces neuroinflammation and improves outcome in a rat model of repeated concussion. J Neuroinflammation 10:26
38. Hamm RJ, Pike BR, Temple MD, O'Dell DM, Lyeth BG (1995) The effect of postinjury kindled seizures on cognitive performance of traumatically brain-injured rats. Exp Neurol 136:143–148
39. Hoane MR (2004) Magnesium therapy and recovery of function in experimental models of brain injury and neurodegenerative disease. Clin Calcium 14:65–70
40. Xiong Y, Mahmood A, Zhang Y, Meng Y, Zhang ZG, Qu C, Sager TN, Chopp M (2011) Effects of posttraumatic carbamylated erythropoietin therapy on reducing lesion volume and hippocampal cell loss, enhancing angiogenesis and neurogenesis, and improving functional outcome in rats following traumatic brain injury. J Neurosurg 114:549–559
41. Rau TF, Kothiwal AS, Rova AR, Brooks DM, Poulsen DJ (2012) Treatment with low-dose methamphetamine improves behavioral and cognitive function after severe traumatic brain injury. J Trauma Acute Care Surg 73:S165–S172
42. Hallam TM, Floyd CL, Folkerts MM, Lee LL, Gong Q-Z, Lyeth BG, Muizelaar JP, Berman RF (2004) Comparison of behavioral deficits and acute neuronal degeneration in rat lateral fluid percussion and weight-drop brain injury models. J Neurotrauma 21:521–539
43. Thompson HJ, LeBold DG, Marklund N, Morales DM, Hagner AP, McIntosh TK (2006) Cognitive evaluation of traumatically brain-injured rats using serial testing in the Morris water maze. Restor Neurol Neurosci 24:109–114
44. Osier ND, Carlson SW, DeSana A, Dixon CE (2015) Chronic histopathological and behavioral outcomes of experimental traumatic brain injury in adult male animals. J Neurotrauma 32:1861. doi:10.1089/neu.2014.3680
45. Fox GB, LeVasseur RA, Faden AI (1999) Behavioral responses of C57BL/6, FVB/N, and 129/SvEMS mouse strains to traumatic brain injury: implications for gene targeting approaches to neurotrauma. J Neurotrauma 16:377–389
46. Hannay HJ, Feldman Z, Phan P, Keyani A, Panwar N, Goodman JC, Robertson CS (1999) Validation of a controlled cortical impact model of head injury in mice. J Neurotrauma 16:1103–1114
47. Fox GB, Fan L, LeVasseur RA, Faden AI (1998) Sustained sensory/motor and cognitive deficits with neuronal apoptosis following controlled cortical impact brain injury in the mouse. J Neurotrauma 15:599–614
48. Smith DH, Soares HD, Pierce JS, Perlman KG, Saatman KE, Meaney DF, Dixon CE, McIntosh TK (1995) A model of parasagittal controlled cortical impact in the mouse: cognitive and histopathologic effects. J Neurotrauma 12:169–178
49. Han X, Tong J, Zhang J, Farahvar A, Wang E, Yang J, Samadani U, Smith DH, Huang JH (2011) Imipramine treatment improves cognitive outcome associated with enhanced hippocampal neurogenesis after traumatic brain injury in mice. J Neurotrauma 28:995–1007
50. Scafidi S, Racz J, Hazelton J, McKenna MC, Fiskum G (2010) Neuroprotection by acetyl-

L-carnitine after traumatic injury to the immature rat brain. Dev Neurosci 32:480–487
51. Duhaime AC, Margulies SS, Durham SR, O'Rourke MM, Golden JA, Marwaha S, Raghupathi R (2000) Maturation-dependent response of the piglet brain to scaled cortical impact. J Neurosurg 93:455–462
52. Manley GT, Rosenthal G, Lam M, Morabito D, Yan D, Derugin N, Bollen A, Knudson MM, Panter SS (2006) Controlled cortical impact in swine: pathophysiology and biomechanics. J Neurotrauma 23:128–139
53. Costine BA, Quebeda-Clerkin PB, Dodge CP, Harris BT, Hillier SC, Duhaime A-C (2012) Neuron-specific enolase, but not S100B or myelin basic protein, increases in peripheral blood corresponding to lesion volume after cortical impact in piglets. J Neurotrauma 29:2689–2695
54. Kline AE, Dixon CE (2001) Contemporary in vivo models of brain trauma and a comparison of injury responses. In: Miller LP, Hayes RL (eds) Head trauma: basic, preclinical, and clinical directions. John Wiley & Sons, New York, NY, pp 65–84
55. King C, Robinson T, Dixon CE, Rao GR, Larnard D, Nemoto CEM (2010) Brain temperature profiles during epidural cooling with the ChillerPad in a monkey model of traumatic brain injury. J Neurotrauma 27:1895–1903
56. Dennis AM, Haselkorn ML, Vagni VA, Garman RH, Janesko-Feldman K, Bayir H, Clark RSB, Jenkins LW, Dixon CE, Kochanek PM (2009) Hemorrhagic shock after experimental traumatic brain injury in mice: effect on neuronal death. J Neurotrauma 26:889–899
57. Sandhir R, Berman NEJ (2010) Age-dependent response of CCAAT/enhancer binding proteins following traumatic brain injury in mice. Neurochem Int 56:188–193
58. Hemerka JN, Wu X, Dixon CE, Garman RH, Exo JL, Shellington DK, Blasiole B, Vagni VA, Janesko-Feldman K, Xu M, Wisniewski SR, Bayır H, Jenkins LW, Clark RSB, Tisherman SA, Kochanek PM (2012) Severe brief pressure-controlled hemorrhagic shock after traumatic brain injury exacerbates functional deficits and long-term neuropathological damage in mice. J Neurotrauma 29:2192–2208
59. Monaco CM, Mattiola VV, Folweiler KA, Tay JK, Yelleswarapu NK, Curatolo LM, Matter AM, Cheng JP, Kline AE (2013) Environmental enrichment promotes robust functional and histological benefits in female rats after controlled cortical impact injury. Exp Neurol 247:410–418
60. Pleasant JM, Carlson SW, Mao H, Scheff SW, Yang KH, Saatman KE (2011) Rate of neurodegeneration in the mouse controlled cortical impact model is influenced by impactor tip shape: implications for mechanistic and therapeutic studies. J Neurotrauma 28:2245–2262
61. Statler KD, Kochanek PM, Dixon CE, Alexander HL, Warner DS, Clark RS, Wisniewski SR, Graham SH, Jenkins LW, Marion DW, Safar PJ (2000) Isoflurane improves long-term neurologic outcome versus fentanyl after traumatic brain injury in rats. J Neurotrauma 17:1179–1189
62. Statler KD, Alexander H, Vagni V, Holubkov R, Dixon CE, Clark R, Jenkins L, Kochanek PM (2006) Isoflurane exerts neuroprotective actions at or near the time of severe traumatic brain injury. Brain Res 1076:216–224
63. McDonald JW, Roeser NF, Silverstein FS, Johnston MV (1989) Quantitative assessment of neuroprotection against NMDA-induced brain injury. Exp Neurol 106:289–296
64. McPherson RW, Kirsch JR, Salzman SK, Traystman RJ (1994) The neurobiology of central nervous system trauma. Oxford University Press, New York, NY
65. Cole JT, Yarnell A, Kean WS, Gold E, Lewis B, Ren M, McMullen DC, Jacobowitz DM, Pollard HB, O'Neill JT, Grunberg NE, Dalgard CL, Frank JA, Watson WD (2011) Craniotomy: true sham for traumatic brain injury, or a sham of a sham? J Neurotrauma 28:359–369
66. Shin SS, Bray ER, Dixon CE (2012) Effects of nicotine administration on striatal dopamine signaling after traumatic brain injury in rats. J Neurotrauma 29:843–850
67. Shin SS, Bales JW, Yan HQ, Kline AE, Wagner AK, Lyons-Weiler J, Dixon CE (2013) The effect of environmental enrichment on substantia nigra gene expression after traumatic brain injury in rats. J Neurotrauma 30:259–270
68. Meaney DF, Ross DT, Winkelstein BA, Brasko J, Goldstein D, Bilston LB, Thibault LE, Gennarelli TA (1994) Modification of the cortical impact model to produce axonal injury in the rat cerebral cortex. J Neurotrauma 11:599–612
69. He J, Evans C-O, Hoffman SW, Oyesiku NM, Stein DG (2004) Progesterone and allopregnanolone reduce inflammatory cytokines after traumatic brain injury. Exp Neurol 189:404–412
70. RIGOR Improving the quality of NINDS-supported pre-clinical and clinical research through rigorous study design and transparent reporting

Chapter 12

Weight Drop Models in Traumatic Brain Injury

Brian T. Kalish and Michael J. Whalen

Abstract

Weight drop models in rodents have been used for several decades to advance our understanding of the pathophysiology of traumatic brain injury. Weight drop models have been used to replicate focal cerebral contusion as well as diffuse brain injury characterized by axonal damage. More recently, closed head injury models with free head rotation have been developed to model sports concussions, which feature functional disturbances in the absence of overt brain damage assessed by conventional imaging techniques. Here, we describe the history of development of closed head injury models in the first part of the chapter. In the second part, we describe the development of our own weight drop closed head injury model that features impact plus rapid downward head rotation, no structural brain injury, and long-term cognitive deficits in the case of multiple injuries. This rodent model was developed to reproduce key aspects of sports concussion so that a mechanistic understanding of how long-term cognitive deficits might develop will eventually follow. Such knowledge is hoped to impact athletes and war fighters and others who suffer concussive head injuries by leading to targeted therapies aimed at preventing cognitive and other neurological sequelae in these high-risk groups.

Key words Mice, Closed head injury, Weight drop, Concussion, Cognitive deficits, Morris water maze, Diffuse injury, Mild traumatic brain injury

1 Introduction

The creation of clinically relevant animal models of traumatic brain injury (TBI) has proven difficult given the biomechanical and pathophysiologic complexity of the injury process. Most models cannot simulate the entire spectrum of human TBI or replicate common mechanisms of injury. All models are confounded by the inherent variability in injury severity and neurologic outcome. These limitations make molecular and translational studies challenging.

Weight drop models are a relatively nascent area of investigation, but the models are gaining momentum given their similarities to human TBI. Weight drop models can simulate the full spectrum of TBI, ranging from mild concussion to severe TBI. Common models of TBI, such as fluid percussion and controlled cortical impact produce a focal brain contusion with little axonal injury. Conversely, weight drop models aim to reproduce diffuse brain

Firas Kobeissy et al. (eds.), *Injury Models of the Central Nervous System: Methods and Protocols*, Methods in Molecular Biology, vol. 1462, DOI 10.1007/978-1-4939-3816-2_12, © Springer Science+Business Media New York 2016

injury. In the first portion of this chapter, we summarize the characteristics of some of the well-known weight drop models. In the second part, we describe the development of our own weight drop models of repetitive mild traumatic brain injury.

1.1 Review of Weight Drop TBI Models

Dr. Anthony Marmarou proposed the first weight drop model of closed skull TBI in rats [1]. This model produces shear stress and diffuse axonal injury, as opposed to prior TBI models that produced focal brain contusion. The scalp of anesthetized mice is shaved, and an incision is made to expose the periosteum. A stainless steel helmet is fixed to the skull with dental acrylic. The helmet distributes kinetic energy over the brain, thereby preventing focal injury. The head injury device consists of a column of brass weights that fall freely through a plexiglass column. The falling weight ranges from 50 to 500 g. The rat is placed on a foam bed of known spring constant. The weights are dropped from a set height to induce a reproducible injury. Data from the initial description of the model demonstrates a mortality rate of 44 % and a skull fracture rate of 12.5 % when a 450-g weight is dropped from 2 m. The injury is accompanied by seizures, apnea, and hypertension. The use of mechanical ventilation significantly improved survival [2]. Postmortem analysis demonstrated brain edema; ventriculomegaly; and widespread damage to neurons, axons, and microvasculature. Diffuse axonal injury was found in the brain stem, optic tracts, corpus callosum, internal capsule, and the cerebral and cerebellar peduncles [2]. A biomechanical analysis of impact dynamics estimated that the 450-g weight dropped from 2 m produced a peak acceleration of $900 \times g$.

Adelson et al. described a modification of the Marmarou model that produces diffuse cortical swelling [3]. The authors found that the 450-g weight used by Marmarou produced too great a mortality, and therefore studied a 75- and 100-g weight dropped from 2 m. The 75- and 100-g injury severity levels produced mortality rates of 18.2 % and 38 %, respectively. Pathological examination of the brains from the severely injured animals demonstrated neuronal death, vascular disruption, and diffuse cerebral edema. No gross contusions were noted. Of Note, this model produced a brain stem injury that may be responsible for at least some of the observed mortality.

Shohami and colleagues developed a weight drop model in the rat that utilizes a free-falling rod rather than weights [4]. This injury produces blood–brain barrier disruption, cerebral edema, and neurological deficits as well as focal contusion and cell death. In this model, the scalp is incised, and a free falling, silicone-tipped rod delivers a cranial impact over the left hemisphere (1–2 mm lateral to midline). Blood–brain barrier permeability peaks in 4 h and is present up to 4 days [4, 5]. Brain edema peaks at 18 h postinjury [5]. The investigators developed a neurological severity score (NSS) to assess motor function and cognitive deficits after injury. The NSS correlates closely with the pathologic severity of brain damage [5]. The Shohami laboratory has extensively characterized the functional and biochemical response to injury in their model. Closed head injury is accompanied by the rapid production of eicosanoids (5-HETE and prostaglandin E2) [6] and cytokines (IL-6 and TNF-alpha) [7]. Several therapeutics, including endothelin antagonists [8], acetylcholinesterase inhibitors [9, 10], cannabinoids [11, 12], and TNF-alpha modulators [13], have proven protective in this model.

Given the emergence of genetically modified mice, the Shohami group adapted their rat model to produce a similar mechanism of injury in mice [14]. Similar to the rat model, the severity of brain injury could be modulated by the falling height and the mass of the weight. Using a weight of the falling rod between 333 and 1600 g, a 2-cm drop height is associated with mild injury, while a 3-cm drop height is associated with severe injury. There is an increased probability of skull fractures with increased injury severity. This model consistently produces disruption in the blood–brain barrier, cerebral edema, and neuronal cell death below the contusion site and remotely in the hippocampus. There is 13% mortality in the immediate postinjury period, with an additional 13% mortality in the subsequent 24 h [14].

Feeney et al. used a weight drop device to generate graded, focal cortical contusions [15]. The contusing apparatus consisted of a 40 cm stainless steel tube attached to a circular footplate that was positioned over the exposed dura of rats. Contusions were created in the hindpaw region of the brain that has both motor and somatosensory functions. Behavioral deficits were observed in the contralateral hindlimb. Mild forces (50 g/cm) did not produce surface hemorrhaging, but more severe forces (200–1000 g/cm) produced surface hemorrhaging and cortical disorganization in some cases. Necrotic cavitation with subcortical cell loss was observed 24 h postinjury with 200 and 600 g/cm forces. After 2 weeks, these cavitary lesions were lined with macrophages and stained positive for acid phosphatase. Animal performance on a balance beam demonstrated trauma–dose relationship, and a persistent deficit was observed at 90 days postinjury.

1.2 Review of Other Closed Head TBI Models

Nedergaard and colleagues developed a "Hit and Run" model of closed head injury that did not require stereotactic fixation of the head or preparation of a cranial window, as in other models [16]. This model induces a closed head injury, which creates cerebral edema and intracranial hypertension commonly observed in human TBI. The device used in this model is a modification of the pneumatic cortical impact device. The instrument is rotated and mounted 90° such that the metal impact rod is oriented horizontally. After anesthesia, mice are hung vertically from their incisor teeth from a metal ring. The impactor rod strikes the mouse skull laterally between the eye and ear. The mouse head is free to move in response to the impact. The velocity of the rod can be varied to create different severities of injury. "Mild injuries" (impact speed 4.8 m/s) were characterized histologically by diffuse cortical reactive gliosis without gross tissue disruption. "Moderate injuries" (impact speed 5.2 m/s) were associated with cortical disruption and the formation of a glial scar. The impact depth and contact times were held constant between groups. Both levels of severity were associated with a loss of myelination; this effect was more prolonged in the "moderate injury" group. Axon degeneration and loss of white matter were also observed in both mild and moderate injury groups. Neither injury severity level was associated with early death, and delayed death was very rare. No mice suffered skull fractures. Of Note, the authors only described a single-hit model; the effect of repetitive injuries has not been reported.

The Wayne State model reported by Kane et al. uses a 95 g weight dropped down a guide tube from a height of 1 m onto the head of a lightly anesthetized mouse suspended on aluminum foil [17]. The foil allows for unrestricted movement of the animal at the time of impact, and a string tied to the weight prevents it from bouncing and hitting the mouse a second time. After a single hit, mice do not experience seizures or paralysis. The authors have studied mice after repeated hits (1 hit per day for 5 days) and demonstrated a reproducible cognitive deficit compared to control mice. In particular, the mice had deficits in motor coordination that recovered over time. In injured mice 30 days after injury, there is evidence of gliosis, astrocytosis, and elevated phosphorylated Tau. There was no microglial activation, disruption of the blood–brain barrier, or extensive loss of cortical white matter. Preliminary studies did not find evidence of β-amyloid deposition.

The Wayne State model shares several similarities with the Harvard weight drop model described later: both models allow for unconstrained movement of the head and body after impact, and both models can be adapted for multiple impacts. Neither model induces skull fractures, cerebral edema, or intracranial hemorrhage. In the section that follows, we describe our own model of closed head concussive brain injury.

2 Development of a Mild Closed Head Injury Model of Repetitive Sports Concussions

2.1 Considerations for Development of a Rodent Model of Sports Concussions

2.1.1 Choice of Mouse as the Species for Model Development

In considering which species to use for development of a new sports concussion model, we considered rats versus mice. Advantages of the rat include its excellent performance in the Morris water maze, a gold standard test of learning and memory applied to almost all TBI models; its relatively cheap cost and ease of maintenance; ease of line placement for intravenous drug injections; and availability of antibodies and PCR reagents to detect a wide range of rat proteins and RNA species. The large size of the rat head is a distinct advantage when considering positron emission tomography (PET) imaging, as well as functional magnetic resonance imaging (fMRI) and magnetic resonance spectroscopy, as most magnets can successfully incorporate coils to fit the head of a rat. Mice are more challenging, and spatial resolution of most imaging studies is not as good as rats because of the smaller brain size. On the other hand, mice are much cheaper than rats, many strains perform well in behavioral tests including the Morris water maze, and the existence of genetically modified mice allows for investigation of mechanisms of secondary brain injury associated with concussion. Based mostly on the latter consideration, we chose to develop our weight drop concussion model in mice.

2.1.2 Anesthetic Considerations

A number of anesthetic agents have been used in experimental TBI models including inhaled agents (isoflurane, sevoflurane, halothane), intravenous agents (ketamine/xylazine, barbiturates, propofol), and others such as chloral hydrate and avertin. Almost all anesthetics have neuroprotective effects in models of brain injury including antiapoptosis, anti-inflammatory, and energy sparing effects [18]. Several studies have shown dose- and exposure time-dependent neurodegenerative effects in developing rodent brains, which may influence anesthetic choice in pediatric concussion models; however, no anesthetic agent has a clear advantage over others in terms of potential for neuroprotection that might confound model development [18]. Another important issue is how a particular anesthetic agent might interact with the secondary response to concussive brain trauma; for example, sevoflurane anesthesia was associated with a decrease in reduction of loss of consciousness after multiple concussions compared to a single injury in mice, an effect that was thought to be an interaction between sevoflurane and repetitive injury [19]. For our purposes, we chose to use isoflurane/nitrous oxide to facilitate quick recovery from anesthesia and thereby accurately measure loss of consciousness time. The decision to use a given anesthetic is not trivial, as any agent that interferes with the natural history of concussive injury in the mouse will necessarily limit translation to humans. For this reason, at least one group has chosen to avoid the use of anesthesia all together in a multiple hit concussion model in mice [20].

2.1.3 Choice of Injury Mechanics

To best model sports concussions, we wanted a weight drop model in which the head was free to move downward after impact, to mimic both impact and rapid head acceleration forces experienced by athletes. Studies of impacts experienced by football players suggest that angular and rotational forces may significantly influence whether or not a collision produces a concussion [21–27] (and reviewed by [19]). In primates, loss of consciousness is more efficiently produced by impact plus acceleration forces rather than by whiplash injury alone [28, 29] or striking a head fixed in place [30]. These considerations ruled out controlled cortical impact and fluid percussion injury models in which the head is held fixed, as well as rotational acceleration models that lack impact which often result in severe brain injury [31].

2.1.4 Histopathology Requirements

Human concussion TBI is defined as a complex functional alteration in brain activity rather than structural damage detectable by routinely used computerized tomography or structural magnetic resonance imaging sequences [32]. No doubt this definition will change as more sensitive noninvasive methods are developed to detect potential subtle histopathological features of concussion, and when functional imaging becomes more commonplace. Our goals for histopathology in a concussion model were lack of structural brain damage such as contusion, hemorrhage, blood–brain barrier disruption, edema, and acute cell death. A schematic of the gross pathology of the experimental closed head injury is shown in Fig. 1. Diffusion tensor imaging studies have suggested an association between concussion TBI and white matter injury [33], and we expected to see some evidence for axonal injury in a concussion model. Other histopathological features that we sought were diffuse gliosis [34] and increase in phosphorylated tau species, as well as increased beta amyloid protein and plaque formation, brain atrophy [35] and ventricular enlargement, consistent with development of chronic traumatic encephalopathy (CTE) seen in younger athletes with a prior history of sports concussions [36–44]. On the other hand, it is possible that debilitating neurological symptoms of concussion can be produced independently of tau and beta amyloid pathology. The histopathology of a concussion model is a critical issue because clinical case series and autopsy studies alone cannot be used to prove that concussions eventually lead to development of CTE in athletes [45]. Animal models are needed to directly test this hypothesis and provide a direct link [19].

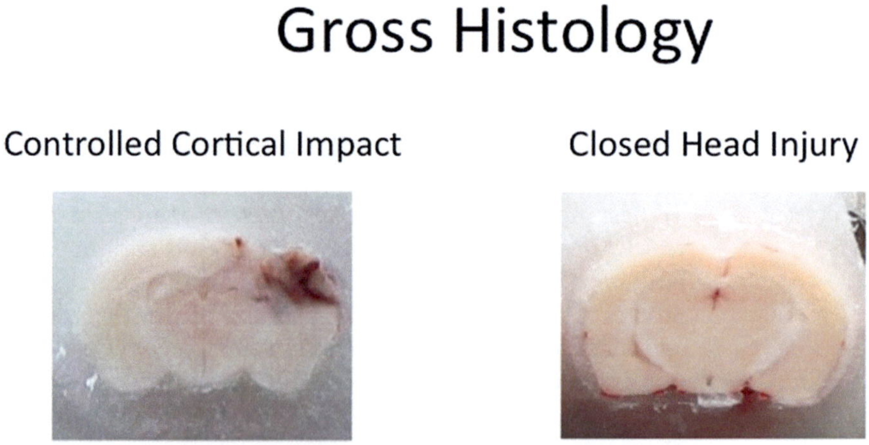

Fig. 1 Gross pathology of closed head injury. Compared to contusion injury generated by controlled cortical impact (**a**), the brain is structurally normal after closed head injury (**b**)

2.1.5 Single Versus Repetitive Injuries

Human concussion presents with cognitive dysfunction early after injury, and repetitive concussions in humans are thought to increase neurological deficits over time. Therefore, we hoped that we could develop a mouse concussion model that would exhibit early functional deficits after a single injury, and more pronounced deficits and perhaps histopathology after multiple injuries [46–52]. If so, the model would allow us to test several clinically relevant hypotheses regarding human concussion, such as whether the number of concussions directly correlate with cognitive deficits, whether an injury-free time interval exists between concussions that mitigates permanent cognitive dysfunction, and whether increasing the level of injury might lengthen the vulnerable period between concussions, defined as that period of time in which additional injury may lead to increased neurological and psychological deficits associated with human concussion such as depression and anxiety [53–57]. As will be discussed later, in a multiple hit concussion model one might not want cognitive deficits after one or even the first few injuries in order to model milder forms of injury that synergize to produce cognitive and other deficits over time. Finally, by modulating the injury level (changing either the bolt weight or the drop height, see below) we hoped that we could extend an adult concussion model to immature mice [19], since a high percentage of sports-related concussions occur in children and adolescents [58, 59].

3 Materials

3.1 Development of a Single Hit Concussion Model in Mice

The materials needed for execution of our concussion model are the following:

1. A metal guide tube (this can be made of other materials as well, such as plastic, fiberglass, etc.) 66 in. long.
2. Tape to attach the guide tube to a wall or other solid structure.
3. Metal bolts of various weights (54, 83 g) that fit through the guide tube. Our initial goal was to develop a single hit weight drop concussion model in mice that resulted in rapid onset cognitive deficits and no structural brain injury. Male C57Bl/6 mice (2–3 months of age, 25–30 g) were anesthetized in 4.5 % isoflurane/70 % nitrous oxide/balance oxygen for 45 s and placed face down on a Kimwipe napkin with the experimenter grasping the mouse by the tail on the Kimwipe (Fig. 2).
4. Conceptually, the head and upper body of the mouse would be free to accelerate downward through a tear in the Kimwipe after the bolt struck the head, thus providing a whiplash component of injury relevant to sports concussion.

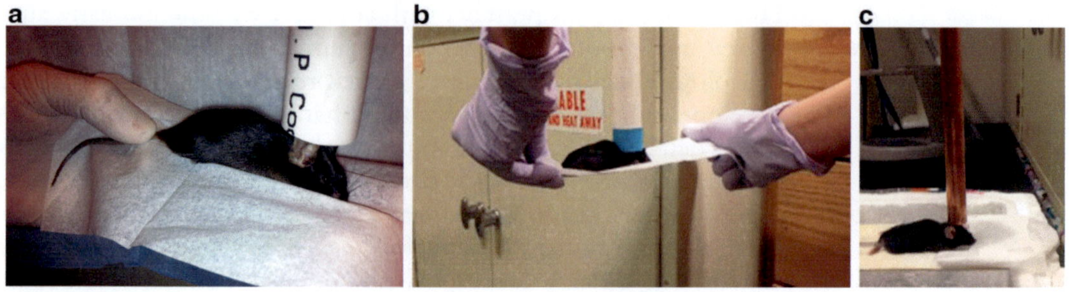

Fig. 2 Description of the weight drop closed head injury model. (**a**, **b**) Mice are gripped by the tail on a Kimwipe napkin and the head is placed under a guide tube. (**c**) Alternatively, a platform can be used to hold the mouse in place, allowing for a single operator to perform the model

3.2 Methods

1. We used a hollow guide tube and a metal bolt of various weights (54, 83 g) dropped from various heights (ranging from 28 to 66 in.) to modulate the injury level (*see* **Note 1**).
2. Mice were grasped on the Kimwipe by the tail and the head was centered underneath the guide tube (10 mm diameter). The bolt was dropped through the guide tube onto the head by one investigator while another held the Kimwipe and mouse under the opening (*see* **Note 2**).
3. In the beginning, it took a fair amount of practice to overcome the natural tendency to pull the mouse away from the tube opening just prior to impact with the bolt.
4. Using a bolt weight of 83 g and drop height 66 in., we experienced high mortality rates from apnea. Interestingly, if we rotated the head slightly left or right so that the bolt struck over the right or left ear, death from apnea was virtually eliminated. Moreover, we found that mortality rates seemed to depend in part on how snugly the mouse head was placed within the guide tube: if the head was held up to the guide tube with upward pressure, mortality was higher than it was when the mouse was held a few millimeters under the guide tube opening without upward pressure.
5. It was noticed that in order to achieve consistency between operators, it would be ideal to strike the mouse on the center of the head, so we reduced the bolt weight to 54 g. Using a drop height of 66 in. and bolt weight 54 g, loss of consciousness (defined as return to spontaneous ambulation in our initial studies, but later changed to return of righting reflex in later studies) was 450 ± 20 s in injured mice compared to 36 ± 1.8 s in sham-injured mice (mice subjected to anesthesia without weight drop, $p < 0.0001$), suggesting a robust injury [60].
6. Mortality, mainly from apnea, in this model was approximately 20%. In a subset of five mice we placed femoral arterial lines using p10 tubing and measured blood pressure and blood gases

before and after injury. We found a transient, modest increase in blood pressure at 2 min that returned to baseline values by 4 min [60]. As expected with apnea, closed head injury (CHI) caused a transient drop in PaO_2 and a modest increase in $PaCO_2$ that resolved by 4 min. However, no mice were hypoxic and all mice recovered blood pressure and blood gases similar to preinjury values within a few minutes after CHI [60].

4 Results

We next examined whether CHI caused gross structural brain injury using 53 g and 66 in. injury parameters. We were pleased that none of the mice had skull fractures or intraparenchymal hemorrhages and no gross structural brain damage. We assessed brain edema using the wet-dry/wet weight method (brains were weighed and dried in an oven at 90 °C for 48 h then weighed dry) and found no increase in brain water content in either hemisphere at 24 or 48 h compared to sham-injured mice. We assessed blood–brain barrier permeability to Evans blue albumin over the first 24 h after injury and found no increase in injured mice compared to shams. For this experiment, 2% Evans blue in PBS was injected intravenously (5 ml/kg) before CHI and 24 h later mice were transcardially perfused with PBS and brains were placed in *N,N* dimethylformamide for 3 days. Evans blue extracted by *N,N* dimethylformamide was measured spectrophotometrically. Alternatively, we performed immunohistochemistry to detect mouse IgG in brain tissue sections [61]. The finding of lack of edema, BBB damage, and structural brain damage satisfied many of our histological criteria for a mouse concussion model (*see* **Note 3**).

Most of the existing concussive TBI models induce some degree of acute cell death. One of our primary goals was to produce a model lacking acute cell death in order to study mechanisms of cognitive dysfunction without this confounding factor. Moreover, it is generally thought that human concussion is a "mild" form of TBI that probably does not induce significant neuronal death, although this is impossible to prove in clinical studies that do not include autopsy results. To determine whether our model involved acute cell death, mice were injured and brain sections obtained at 24, 48, or 72 h after injury and subjected to fluoro-jade B staining (a marker of acute neuronal degeneration), in vivo propidium iodide staining (a marker for loss of membrane integrity), TUNEL (a marker of double strand DNA damage and cell death), hematoxylin and eosin (H&E) staining, GFAP staining (astrocytosis) and IBA-1 staining (microgliosis) [60]. Amyloid precursor protein immunohistochemistry and electron microscopy were used to assess axonal injury. At later times (60 days) after injury, hippocampal cell loss and brain atrophy were assessed using image analysis of H&E stained brain sections. No acute cell death or cell loss was observed at any of the time points assessed. Axonal injury was only occasionally detected by electron microscopy, and there was no brain atrophy at 60 days [60]. Strikingly, there was robust astrocytosis and microgliosis at 48–72 h in cortex and hippocampus of injured (but not sham-injured) mice, demonstrating an acute inflammatory response to concussion [60]. Similar findings have been reported in autopsy studies of humans with TBI at acute and chronic stages, suggesting that our model recapitulates at least some of the features of concussive TBI reported in clinical case studies [34, 43, 62].

To examine possible biochemical mechanisms of concussion-induced inflammation, we performed reverse transcriptase polymerase chain reaction (RTPCR) and nuclear factor kappa B activity assays on brain tissue obtained within 24 h of CHI. We found biochemical evidence of acute inflammation with early increases in TNF alpha and Fas mRNA, and TNF alpha protein followed by increased activation of NFkB [60]. Importantly, CHI produced deficits in motor performance (wire grip test) and cognitive deficits (Morris water maze performance) within 1–3 days after injury. Mice deficient in TNF and Fas had increased cognitive deficits suggesting that TNF/Fas induction in the concussion model is a protective response to injury, although a limitation of the experimental design (use of knockout mice) did not allow us to dissect early versus later effects of TNF/Fas inactivation. Nonetheless, this was an important result because TNF/Fas antagonism was protective in a contusion TBI model [63], and the results in our concussion model provided evidence for the concept that individual pathways activated in different pathoanatomic subtypes of TBI (e.g., focal contusion vs. diffuse concussion) may influence outcome in opposite ways. We believe these findings have significant implications for treatment trials of patients with focal vs. concussive TBI [63].

Taken together, the aforementioned data suggested that we had developed a mouse model of human concussion that featured an early cognitive deficit in the absence of gross and microscopic brain damage that was associated with a neuroinflammatory response and manipulable by TNF/Fas antagonism. However, a significant weakness of the model was that cognitive outcome was highly operator dependent and cognitive deficits were not always apparent from one operator to the next (also *see* **Note 4**). This was a difficult problem that was not solved by increasing the bolt weight to 83 g, which resulted in higher mortality but not always increased cognitive deficits (*see* **Note 5**). We tried recovering the mice at 37 °C to maintain normothermia after recovery from anesthesia, but that did not make a difference with respect to consistent cognitive outcome. Another caveat with our findings is that Morris water maze testing was performed at a time when motor deficits were present (between 1 and 3 days after CHI). Although motor function has not been shown to correlate with MWM performance in TBI models, ideally we would have waited until motor deficits had resolved to perform MWM testing. However, we wanted to model the clinical scenario in which cognitive deficits appear early (within 24 h) after concussion so we began MWM testing the day after injury. Lastly, we initially started model development using CD1 mice; however, uninjured mice from this outbred strain did not perform consistently in the MWM so we developed the model using C57Bl/6J mice from Jackson laboratories (Bar Harbor, ME).

5 Development of a Multiple Hit Mouse Concussion Model

After establishing a single hit concussion model, we next set about developing a multiple hit concussion model of sports concussions. Factors that we considered in the planning stages included all of those identified for the single hit model (such as lack of overt brain damage and measurable cognitive deficits). Additional features that we wanted in a multiple hit model included (1) no detectable cognitive deficit after a single hit (and no structural or microscopic brain injury); (2) a reproducible increase in cognitive deficits with increasing numbers of injuries; (3) presence of psychiatric manifestations of concussion such as anxiety and depression [53, 57, 64]; (4) a measurable vulnerable period within which additional concussions lead to worse cognitive deficits, and conversely injuries outside the vulnerable period do not lead to long-term cognitive dysfunction; (5) an increase in phosphorylated tau species and emergence of beta amyloid plaques and tau tangles as mice age, recapitulating features of chronic traumatic encephalopathy as reported in the brains of athletes diagnosed via autopsy findings [44, 64].

Because the injury parameters of the aforementioned single hit model (66 in. drop height) produced a cognitive deficit after 1 hit, we lowered the drop height to 38 or 42 in. and injured mice either once, three times daily (1 hit per day for 3 days), five times daily, or 10 daily hits. MWM testing was begun 24 h, 1 month, and/or 1 year after the final injury; repeat MWM testing was done by placing the goal platform in a different quadrant for each repeated test. Data obtained in repeat MWM tests are somewhat limited because procedural learning persists and is already present for subsequent tests. Nonetheless, repeat MWM testing can be used to assess new spatial learning in injured mice. The 38-in. drop height model produced no cognitive deficit after 1 hit, modest deficits after 5 hits, and severe cognitive deficits after 10 hits at either 38 or 42 in. drop heights. Moreover, hidden platform deficits persisted at 1 month and 1 year after injury. When we tested additional groups of mice injured five times with injuries separated by a week or a month, mice injured weekly (but not biweekly or monthly) still performed worse than sham-injured mice, suggesting a safe rest interval in this model between 1 week and 1 month.

Repetitive injury did not lead to structural brain damage, acute cell death, brain edema, or blood–brain barrier damage assessed by IgG immunostaining [61]. Thus, the 38-in. drop height repetitive CHI model satisfied nearly all of our criteria for modeling human concussion. However, we noticed that approximately 25 % of mice had convulsions after injury. We initially reported these events as seizures [61], but in retrospect these limb movements, which occur several seconds after impact and last for less than a minute, may also be due to electrocerebral silence induced by concussion with transient loss of cortical inhibition of spinal cord tracts resulting in limb movement. We are planning EEG studies to distinguish between these two possibilities, but the occurrence of seizure-like activity prompted us to test even lower drop heights to produce a model of repetitive injury devoid of convulsions.

To determine a drop height that eliminated convulsions but still led to cognitive deficits we tested several heights between 38- and 18-in.. Pilot studies suggested that a 28-in. drop height would satisfy both conditions, but lower drop heights did not produce cognitive deficits using a 54 g weight. Mice (3 months old males, Jackson labs) were anesthetized for 45 s in isoflurane as earlier and subjected to sham injury or CHI by dropping the bolt on the vertex of the head (*see* **Note 6**). This model did not produce loss of consciousness in injured compared to sham-injured mice (*see* **Note 7**), nor did it produce convulsions, similar to the majority of athletes who suffer concussions [65].

We next determined that a 5 or 7 hit model (1 hit daily for 5 or 7 days) did not produce acute cell death or overt brain damage, and we were now poised to examine cognitive deficits in a truly mild multiple concussion model. Using a 5 hit daily injury model, we performed experiments with 1 hit CHI and shams, 5 hit daily CHI and shams, 5 hit weekly CHI and shams, 5 hit biweekly CHI and shams, and 5 hit monthly CHI and their respective shams. All injured groups were compared only to their respective sham-injured mice because the groups were injured at different times, making direct comparisons among injured groups problematic.

We tested the hypothesis that a safe rest interval exists such that repeated injuries would not lead to long-term cognitive deficits, and found that indeed mice injured biweekly or monthly had no long-term cognitive deficits whereas mice injured daily or weekly had deficits compared to sham at 6 months (*see* **Note 8**). Because these were long-term experiments that could not be repeated in a reasonable time frame, we powered most of our studies with $n = 12$–16 mice per group and obtained tightly clustered data that allowed confident determination of statistically significant results. With these studies, we confirmed the important concept that a safe rest interval existed in our repetitive concussion model, thus in part validating the concept that rest between concussions may mitigate development of permanent cognitive deficits [32].

Interestingly, we found a chronic astrocytosis at 6 months in mice injured daily for 5 days, but no overt neuronal loss. Contrary to published studies in adults with TBI, mice expressing human APOE4 did not have worse cognitive deficits compared to WT suggesting no contribution of astrocytic APOE4 in this model. Somewhat disappointingly, we did not see evidence of CTE in terms of phosphorylated tau species or beta amyloid assessed by ELISA and immunohistochemistry. However, it may require more than 6 months for these processes to become manifest in mice, and more sensitive immunohistochemical reagents may yield positive findings in future studies. Alternatively, it is possible, and even likely, that mechanisms of short-term neurological deficits differ from those associated with development of CTE, which may represent the most severe end of the spectrum for patients with repetitive concussions and severe neurodegeneration [36].

Another finding in this model that may be incongruent with human studies is that cognitive deficits are detectable within a few days after a threshold number of injuries, and these deficits remain even at 6 months after the last injury; thus, rather than develop over the course of time, permanent cognitive deficits produced by our repetitive CHI model are present very early on. It might be more interesting for a concussion model to produce increasing cognitive deficits over the course of several months (or longer with aging)—and this might be the case if we were to test these mice over longer periods of time. Very long-term studies may not be feasible however, as 4/4 mice subjected to 7 daily concussions (28 in., 54 g) died at 14 months after injury compared to 0/4 shams, suggesting that multiple concussions might shorten the lifespan of injured mice. Although this hypothesis requires formal testing with larger numbers, it is an interesting observation that might be applicable to humans with repetitive concussions as well. Alternatively, it may be a shortcoming of the mouse model.

6 Future Considerations

Arguably the development of our CHI model, and similar weight drop models [16, 17] that have been published subsequent to Khuman et al. [60], represent forward steps in modeling human concussion that will facilitate discovery of relevant disease mechanisms. A major question for us and others in the field [19] is whether inhaled anesthetics interact with injury models to reduce neurological deficits, as LOC times decrease with increasing number of injuries in our model (unpublished observations) and in others [19]. One group has managed to avoid the issue all together by subjecting unanesthetized mice to repetitive TBI [20]. Studies examining effects of noninhaled anesthetics are currently underway in our laboratory. Another caveat of our repetitive hit models is that similar to the single hit model, cognitive outcome and mortality in the 5 hit daily model is highly operator dependent for reasons that remain incompletely defined. In the future, we would like to develop a holding device that is operator independent to ensure consistency in the injury level produced by CHI. Another future direction for our laboratory is development of an adolescent mouse concussion model that recapitulates symptoms of sports concussions in high school and college athletes. Finally, measurements of inflammation, cerebral blood flow, and brain metabolism will be important components of concussion model development as these mechanisms are hypothesized to underlie long-term neurological deficits in patients with neurological degeneration and history of TBI [41–43, 50].

7 Notes

1. There are two ways to increase the level of injury—raising the drop height or increasing the dropped weight. Each lab should experiment with both approaches, calibrating to no structural brain damage and a robust cognitive deficit in the Morris water maze or other behavior test of choice.
2. Although we started with a two person operator system (one to hold the mouse and the other to drop the bolt), the apparatus can be standardized better by using a platform to hold the mouse, thus making injury independent of the second operator. This approach will ensure greater consistency of injury over time and among different operators.
3. The model lends itself to cerebral blood flow testing because of the lack of structural brain injury. We have used diffuse correlation spectroscopy in the past for this, and it is also possible to use laser speckle and laser Doppler flowmetry as well.
4. There may be marked gender effects in the closed head injury model but before examining this issue it is important to use age- and weight-matched male and female mice, as the lighter female mice may sustain less inertial injury after impact.
5. Use of a guide tube that is approximately the diameter of the mouse head will result in more accurate hits with the dropped weight, which ideally should be cylindrical in shape and should easily pass through the guide tube but fit snugly within it. Some groups use metal, others use plastic, or other transparent materials in order to follow the course of the dropped weight.
6. Younger (adolescent) mice may have increased mortality from apnea that can be prevented by injuring over one side of the head or the other.
7. Because repeated injuries may alter the loss of consciousness time for subsequent injuries, it is recommended to measure loss

of consciousness time (defined as righting reflex) at very least after the first in a series of repetitive injuries rather than wait until several injuries to begin measurement of awakening.

8. It is critical to use a mouse strain that performs well in cognitive tests such as the Morris water maze. We have found that C57Bl/6 mice work well (Jackson Laboratories, Bar Harbor, ME). CD1 and other outbred strains may not perform consistently.

References

1. Marmarou A, Foda MA, van den Brink W, Campbell J, Kita H, Demetriadou K (1994) A new model of diffuse brain injury in rats. Part I: Pathophysiology and biomechanics. J Neurosurg 80:291–300
2. Foda MA, Marmarou A (1994) A new model of diffuse brain injury in rats. Part II: Morphological characterization. J Neurosurg 80:301–313
3. Adelson PD, Robichaud P, Hamilton RL, Kochanek PM (1996) A model of diffuse traumatic brain injury in the immature rat. J Neurosurg 85:877–884
4. Shapira Y, Shohami E, Sidi A, Soffer D, Freeman S, Cotev S (1988) Experimental closed head injury in rats: mechanical, pathophysiologic, and neurologic properties. Crit Care Med 16:258–265
5. Shapira Y, Setton D, Artru AA, Shohami E (1993) Blood-brain barrier permeability, cerebral edema, and neurologic function after closed head injury in rats. Anesth Analg 77:141–148
6. Shohami E, Shapira Y, Cotev S (1988) Experimental closed head injury in rats: prostaglandin production in a noninjured zone. Neurosurgery 22:859–863
7. Shohami E, Novikov M, Bass R, Yamin A, Gallily R (1994) Closed head injury triggers early production of TNF alpha and IL-6 by brain tissue. J Cereb Blood Flow Metab 14:615–619
8. Barone FC, Ohlstein EH, Hunter AJ, Campbell CA, Hadingham SH, Parsons AA, Yang Y, Shohami E (2000) Selective antagonism of endothelin-A-receptors improves outcome in both head trauma and focal stroke in rat. J Cardiovasc Pharmacol 36:S357–S361
9. Chen Y, Shohami E, Bass R, Weinstock M (1998) Cerebro-protective effects of ENA713, a novel acetylcholinesterase inhibitor, in closed head injury in the rat. Brain Res 784:18–24
10. Chen Y, Shohami E, Constantini S, Weinstock M (1998) Rivastigmine, a brain-selective acetylcholinesterase inhibitor, ameliorates cognitive and motor deficits induced by closed-head injury in the mouse. J Neurotrauma 15:231–237
11. Shohami E, Novikov M, Bass R (1995) Long-term effect of HU-211, a novel non-competitive NMDA antagonist, on motor and memory functions after closed head injury in the rat. Brain Res 674:55–62
12. Shohami E, Novikov M, Mechoulam R (1993) A nonpsychotropic cannabinoid, HU-211, has cerebroprotective effects after closed head injury in the rat. J Neurotrauma 10:109–119
13. Shohami E, Gallily R, Mechoulam R, Bass R, Ben-Hur T (1997) Cytokine production in the brain following closed head injury: dexanabinol (HU-211) is a novel TNF-alpha inhibitor and an effective neuroprotectant. J Neuroimmunol 72:169–177
14. Flierl MA, Stahel PF, Beauchamp KM, Morgan SJ, Smith WR, Shohami E (2009) Mouse closed head injury model induced by a weight-drop device. Nat Protoc 4:1328–1337
15. Feeney DM, Boyeson MG, Linn RT, Murray HM, Dail WG (1981) Responses to cortical injury: I. Methodology and local effects of contusions in the rat. Brain Res 211:67–77
16. Ren Z, Iliff JJ, Yang L, Yang J, Chen X, Chen MJ, Giese RN, Wang B, Shi X, Nedergaard M (2013) 'Hit & Run' model of closed-skull traumatic brain injury (TBI) reveals complex patterns of post-traumatic AQP4 dysregulation. J Cereb Blood Flow Metab 33:834–845
17. Kane MJ, Angoa-Perez M, Briggs DI, Viano DC, Kreipke CW, Kuhn DM (2011) A mouse model of human repetitive mild traumatic brain injury. J Neurosci Methods 203:41–49
18. Schifilliti D, Grasso G, Conti A, Fodale V (2010) Anaesthetic-related neuroprotection: intravenous or inhalational agents? CNS Drugs 24:893–907
19. Angoa-Perez M, Kane MJ, Briggs DI, Herrera-Mundo N, Viano DC, Kuhn DM (2014) Animal models of sports-related head injury: bridging the gap between preclinical research and clinical reality. J Neurochem 129:916–931
20. Petraglia AL, Plog BA, Dayawansa S, Chen M, Dashnaw ML, Czerniecka K, Walker CT, Viterise T, Hyrien O, Iliff JJ, Deane R,

Nedergaard M, Huang JH (2014) The spectrum of neurobehavioral sequelae after repetitive mild traumatic brain injury: a novel mouse model of chronic traumatic encephalopathy. J Neurotrauma 31:1211–1224

21. Guskiewicz KM, Mihalik JP, Shankar V, Marshall SW, Crowell DH, Oliaro SM, Ciocca MF, Hooker DN (2007) Measurement of head impacts in collegiate football players: relationship between head impact biomechanics and acute clinical outcome after concussion. Neurosurgery 61:1244–1252, discussion 1252-1243
22. Meaney DF, Smith DH (2011) Biomechanics of concussion. Clin Sports Med 30(19–31):vii
23. Pellman EJ, Viano DC, Tucker AM, Casson IR, Committee on Mild Traumatic Brain Injury, N. F. L. (2003) Concussion in professional football: location and direction of helmet impacts-Part 2. Neurosurgery 53:1328–1340, discussion 1340-1321
24. Pellman EJ, Viano DC, Tucker AM, Casson IR, Waeckerle JF (2003) Concussion in professional football: reconstruction of game impacts and injuries. Neurosurgery 53:799–812, discussion 812-794
25. Urban JE, Davenport EM, Golman AJ, Maldjian JA, Whitlow CT, Powers AK, Stitzel JD (2013) Head impact exposure in youth football: high school ages 14 to 18 years and cumulative impact analysis. Ann Biomed Eng 41:2474–2487
26. Viano DC, Casson IR, Pellman EJ (2007) Concussion in professional football: biomechanics of the struck player--part 14. Neurosurgery 61:313–327, discussion 327-318
27. Viano DC, Pellman EJ (2005) Concussion in professional football: biomechanics of the striking player--part 8. Neurosurgery 56:266–280, discussion 266-280
28. Ommaya AK, Gennarelli TA (1974) Cerebral concussion and traumatic unconsciousness. Correlation of experimental and clinical observations of blunt head injuries. Brain 97:633–654
29. Ommaya AK, Hirsch AE (1971) Tolerances for cerebral concussion from head impact and whiplash in primates. J Biomech 4:13–21
30. Denny-Brown D, Russell R (1941) Experimental Cerebral Concussion. Brain 64:93–164
31. Raghupathi R, Mehr MF, Helfaer MA, Margulies SS (2004) Traumatic axonal injury is exacerbated following repetitive closed head injury in the neonatal pig. J Neurotrauma 21:307–316
32. McCrory P, Meeuwisse W, Aubry M, Cantu B, Dvorak J, Echemendia R, Engebretsen L, Johnston K, Kutcher J, Raftery M, Sills A, Benson B, Davis G, Ellenbogen R, Guskiewicz K, Herring SA, Iverson G, Jordan B, Kissick J, McCrea M, McIntosh A, Maddocks D, Makdissi M, Purcell L, Putukian M, Schneider K, Tator C, Turner M (2013) Consensus statement on Concussion in Sport - the 4th International Conference on Concussion in Sport held in Zurich, November 2012. Phys Ther Sport 14:e1–e13
33. Gardner A, Kay-Lambkin F, Stanwell P, Donnelly J, Williams WH, Hiles A, Schofield P, Levi C, Jones DK (2012) A systematic review of diffusion tensor imaging findings in sports-related concussion. J Neurotrauma 29:2521–2538
34. Johnson VE, Stewart JE, Begbie FD, Trojanowski JQ, Smith DH, Stewart W (2013) Inflammation and white matter degeneration persist for years after a single traumatic brain injury. Brain 136:28–42
35. Ross DE, Ochs AL, Seabaugh JM, Demark MF, Shrader CR, Marwitz JH, Havranek MD (2012) Progressive brain atrophy in patients with chronic neuropsychiatric symptoms after mild traumatic brain injury: a preliminary study. Brain Inj 26:1500–1509
36. DeKosky ST, Ikonomovic MD, Gandy S (2010) Traumatic brain injury: football, warfare, and long-term effects. Minn Med 93:46–47
37. Guskiewicz KM, Marshall SW, Bailes J, McCrea M, Cantu RC, Randolph C, Jordan BD (2005) Association between recurrent concussion and late-life cognitive impairment in retired professional football players. Neurosurgery 57:719–726, discussion 719-726
38. Guskiewicz KM, McCrea M, Marshall SW, Cantu RC, Randolph C, Barr W, Onate JA, Kelly JP (2003) Cumulative effects associated with recurrent concussion in collegiate football players: the NCAA Concussion Study. JAMA 290:2549–2555
39. Iverson GL (2014) Chronic traumatic encephalopathy and risk of suicide in former athletes. Br J Sports Med 48:162–165
40. Jordan BD (2013) The clinical spectrum of sport-related traumatic brain injury. Nat Rev Neurol 9:222–230
41. McKee AC, Cantu RC, Nowinski CJ, Hedley-Whyte ET, Gavett BE, Budson AE, Santini VE, Lee HS, Kubilus CA, Stern RA (2009) Chronic traumatic encephalopathy in athletes: progressive tauopathy after repetitive head injury. J Neuropathol Exp Neurol 68:709–735
42. McKee AC, Stern RA, Nowinski CJ, Stein TD, Alvarez VE, Daneshvar DH, Lee HS, Wojtowicz SM, Hall G, Baugh CM, Riley DO, Kubilus CA, Cormier KA, Jacobs MA, Martin BR, Abraham CR, Ikezu T, Reichard RR, Wolozin BL, Budson AE, Goldstein LE, Kowall NW, Cantu RC (2013) The spectrum of disease in chronic traumatic encephalopathy. Brain 136:43–64

43. Smith C, Gentleman SM, Leclercq PD, Murray LS, Griffin WS, Graham DI, Nicoll JA (2013) The neuroinflammatory response in humans after traumatic brain injury. Neuropathol Appl Neurobiol 39:654–666
44. Smith DH, Johnson VE, Stewart W (2013) Chronic neuropathologies of single and repetitive TBI: substrates of dementia? *Nature reviews*. Neurology 9:211–221
45. McCrory P, Meeuwisse WH, Kutcher JS, Jordan BD, Gardner A (2013) What is the evidence for chronic concussion-related changes in retired athletes: behavioural, pathological and clinical outcomes? Br J Sports Med 47:327–330
46. Gronwall D, Wrightson P (1975) Cumulative effect of concussion. Lancet 2:995–997
47. Hamberger A, Viano DC, Saljo A, Bolouri H (2009) Concussion in professional football: morphology of brain injuries in the NFL concussion model--part 16. Neurosurgery 64:1174–1182
48. Laurer HL, Bareyre FM, Lee VM, Trojanowski JQ, Longhi L, Hoover R, Saatman KE, Raghupathi R, Hoshino S, Grady MS, McIntosh TK (2001) Mild head injury increasing the brain's vulnerability to a second concussive impact. J Neurosurg 95:859–870
49. Longhi L, Saatman KE, Fujimoto S, Raghupathi R, Meaney DF, Davis J, McMillan BSA, Conte V, Laurer HL, Stein S, Stocchetti N, McIntosh TK (2005) Temporal window of vulnerability to repetitive experimental concussive brain injury. Neurosurgery 56:364–374, discussion 364-374
50. Prins ML, Alexander D, Giza CC, Hovda DA (2013) Repeated mild traumatic brain injury: mechanisms of cerebral vulnerability. J Neurotrauma 30:30–38
51. Prins ML, Hales A, Reger M, Giza CC, Hovda DA (2010) Repeat traumatic brain injury in the juvenile rat is associated with increased axonal injury and cognitive impairments. Dev Neurosci 32:510–518
52. Uryu K, Laurer H, McIntosh T, Pratico D, Martinez D, Leight S, Lee VM, Trojanowski JQ (2002) Repetitive mild brain trauma accelerates Abeta deposition, lipid peroxidation, and cognitive impairment in a transgenic mouse model of Alzheimer amyloidosis. J Neurosci 22:446–454
53. Covassin T, Elbin RJ 3rd, Larson E, Kontos AP (2012) Sex and age differences in depression and baseline sport-related concussion neurocognitive performance and symptoms. Clin J Sport Med 22:98–104
54. Didehbani N, Munro Cullum C, Mansinghani S, Conover H, Hart J Jr (2013) Depressive symptoms and concussions in aging retired NFL players. Arch Clin Neuropsychol 28:418–424
55. Guskiewicz KM, Marshall SW, Bailes J, McCrea M, Harding HP Jr, Matthews A, Mihalik JR, Cantu RC (2007) Recurrent concussion and risk of depression in retired professional football players. Med Sci Sports Exerc 39:903–909
56. Kerr ZY, Marshall SW, Harding HP Jr, Guskiewicz KM (2012) Nine-year risk of depression diagnosis increases with increasing self-reported concussions in retired professional football players. Am J Sports Med 40:2206–2212
57. Kontos AP, Covassin T, Elbin RJ, Parker T (2012) Depression and neurocognitive performance after concussion among male and female high school and collegiate athletes. Arch Phys Med Rehabil 93:1751–1756
58. Koepsell TD, Rivara FP, Vavilala MS, Wang J, Temkin N, Jaffe KM, Durbin DR (2011) Incidence and descriptive epidemiologic features of traumatic brain injury in King County, Washington. Pediatrics 128:946–954
59. Toledo E, Lebel A, Becerra L, Minster A, Linnman C, Maleki N, Dodick DW, Borsook D (2012) The young brain and concussion: imaging as a biomarker for diagnosis and prognosis. Neurosci Biobehav Rev 36:1510–1531
60. Khuman J, Meehan WP 3rd, Zhu X, Qiu J, Hoffmann U, Zhang J, Giovannone E, Lo EH, Whalen MJ (2011) Tumor necrosis factor alpha and Fas receptor contribute to cognitive deficits independent of cell death after concussive traumatic brain injury in mice. J Cereb Blood Flow Metab 31:778–789
61. Meehan WP 3rd, Zhang J, Mannix R, Whalen MJ (2012) Increasing recovery time between injuries improves cognitive outcome after repetitive mild concussive brain injuries in mice. Neurosurgery 71:885–891
62. Hernandez-Ontiveros DG, Tajiri N, Acosta S, Giunta B, Tan J, Borlongan CV (2013) Microglia activation as a biomarker for traumatic brain injury. Front Neurol 4:30
63. Bermpohl D, You Z, Lo EH, Kim HH, Whalen MJ (2007) TNF alpha and Fas mediate tissue damage and functional outcome after traumatic brain injury in mice. J Cereb Blood Flow Metab 27:1806–1818
64. Stern RA, Daneshvar DH, Baugh CM, Seichepine DR, Montenigro PH, Riley DO, Fritts NG, Stamm JM, Robbins CA, McHale L, Simkin I, Stein TD, Alvarez VE, Goldstein LE, Budson AE, Kowall NW, Nowinski CJ, Cantu RC, McKee AC (2013) Clinical presentation of chronic traumatic encephalopathy. Neurology 81:1122–1129
65. Meehan WP 3rd, Bachur RG (2009) Sport-related concussion. Pediatrics 123:114–123

Chapter 13

Midline (Central) Fluid Percussion Model of Traumatic Brain Injury

Rachel K. Rowe, Daniel R. Griffiths, and Jonathan Lifshitz

Abstract

Research models of traumatic brain injury (TBI) hold significant validity towards the human condition, with each model replicating a subset of clinical features and symptoms. After 30 years of characterization and implementation, fluid percussion injury (FPI) is firmly recognized as a clinically relevant model of TBI, encompassing concussion through severe injury. The midline variation of FPI may best represent mild and diffuse clinical brain injury, because of the acute behavioral deficits, the late onset of subtle behavioral morbidities, and the absence of gross histopathology. This chapter outlines the procedures for midline (diffuse) FPI in adult male rats and mice. With these procedures, it becomes possible to generate brain-injured laboratory animals for studies of injury-induced pathophysiology and behavioral deficits, for which rational therapeutic interventions can be implemented.

Key words Traumatic brain injury (TBI), Concussion, Diffuse, Fluid percussion, Trauma, Rodent, Rat, Mouse, Experimental model, Righting reflex, Fencing response, Postoperative care

1 Introduction

1.1 Model Selection

Midline fluid percussion permits the study of experimental traumatic brain injury (TBI) in a model that is reproducible, clinically relevant, and scalable between species and injury severities. Brain injury is induced by a rapid (~20 ms) fluid pulse through a craniectomy onto the intact dura that follows the inner curvature of the skull and creates an elastic decompression of the brain [1, 2]. While fluid percussion injury (FPI) necessitates breaching the cranial vault, the skull is sealed to the injury device, recreating a closed system, which approximates a closed head injury with decompressive craniectomy. The mechanical forces disrupt cell membranes, blood vessels, and neuronal processes. By increasing the angle from which the pendulum hammer falls, greater pressures can be generated to travel through the fluid-filled cylinder and impact the brain. This model can be implemented to evaluate pathophysiological

Firas Kobeissy et al. (eds.), *Injury Models of the Central Nervous System: Methods and Protocols*, Methods in Molecular Biology, vol. 1462, DOI 10.1007/978-1-4939-3816-2_13, © Springer Science+Business Media New York 2016

mechanisms underlying histological and behavioral deficits, and therapeutic interventions to mitigate degeneration and promote recovery of function.

2 Materials

2.1 Animals

Fluid percussion brain injury has been successfully performed on various species, including cats, rabbits, pigs, rats, and mice. The adaptation of fluid percussion to rats [3–5] was followed by its implementation in mice [6]. The procedures outlined in this chapter focus on adult male Sprague-Dawley rats (approximately 300–400 g) and 8-week-old adult male C57BL/6 mice (approximately 20–30 g). To maximize the success of brain injury, examine all animals for any signs of ill health (e.g., rough coat, bleeding or dirty eyes, runny or bleeding nose, and scratched around eyes or nose area). Weigh all animals prior to surgery in order to track injury-induced weight loss.

2.2 Equipment

2.2.1 Injury Device

1. Fluid percussion injury device (Fig. 1).
   (a) Custom Design and Fabrication.
   (b) Virginia Commonwealth University.

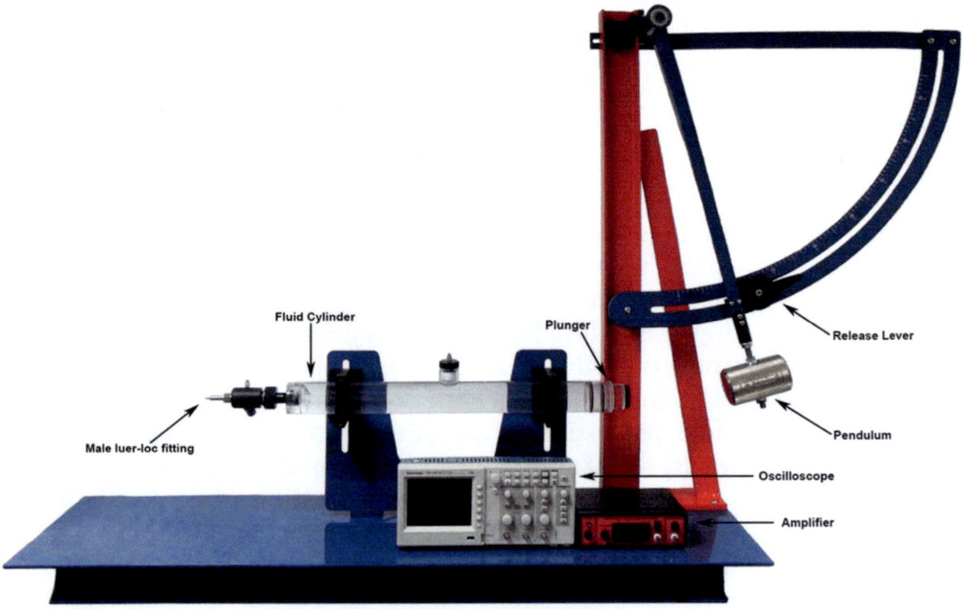

Fig. 1 Fluid percussion injury device. Injury is induced by a 20-ms fluid pulse delivered onto the intact dura via a craniectomy and surgically implanted injury hub. The fluid pulse is generated by the pressure wave produced when the weighted end of the pendulum arm strikes the end of fluid cylinder. The force of the pulse is detected by a transducer and the signal is amplified before being sent to the oscilloscope which outputs the millivolts. The millivolts can then be converted to atmospheres of pressure

(c) http://www.radiology.vcu.edu/research/customdesign/fpi.html.

(d) Product information including assembly manual, operation manual, and product brochure are provided on the website, for cleaning instructions (*see* **Note 1**).

2. Recording oscilloscope (recommended: Tektronix, Model 1001B).
3. Industrial Velcro to secure the device to the bench to prevent movement.
4. High-vacuum grease (e.g., Fisher Scientific, #14-635-5D).
5. Dishwashing solution to clean fluid cylinder.
6. Jet Dry finishing rinse to minimize air bubbles in the cylinder upon filling.

2.2.2 Anesthesia

1. Vaporizer for delivery of inhaled anesthesia (*see* **Note 2** for safety tips).
2. Tubing/petcocks.
3. Induction chamber.
4. Isoflurane.
5. Oxygen.
6. Rodent nose cone for inhaled anesthetic that is compatible with the stereotaxic frame.

2.2.3 Surgical Supplies

1. Gauze sponges.
2. Cotton tip applicators.
3. Heating pad (recommended: Deltaphase isothermal heating pad-BrainTree Scientific, #39DP).
4. 20-gauge needles (recommended: 1″ length).
5. 1 mL syringes.
6. ≥10 mL syringes, Luer-lock tip.
7. Small animal trimmer for fur removal (e.g., Wahl, Mini Arco Animal Trimmer).
8. Ophthalmic ointment to prevent drying of eyes during surgery.
9. 4 % Chlorhexidine solution (or Betadine scrub) for preparation of the incision.
10. 70 % Ethanol (or alcohol pads).
11. Cyanoacrylate (e.g., Super Glue).
12. Perm Reline and Repair resin, liquid, and powder (All for Dentist, #H00327).
13. Antibiotic ointment.
14. Saline-filled syringe, blunted needle bent 90°.

2.2.4 Surgical Instruments

1. Small animal stereotaxic frame.
2. Scalpel handle and blade.
3. Delicate bone scraper (Fine Science Tools, #10075-16).
4. Chisels Wedelstaedt ¾ DE (Henry Schein, #600-4972).
5. Bull Dog clips (Fine Science Tools, #18050-28, #18051-28).
6. Needle holder and scissors.

2.2.5 Rat Surgical Instruments

1. Dremel tool with engraving cutter #106.
2. Trephine (4.7 mm) (Miltex, #26-140).
3. Fingernail drill with 5/64″ drill bit (Miltex, #33-232).
4. Stainless steel skull screws (2–56 × 3/16″) (Small parts Inc., #MX-0256-03B-25).

2.2.6 Mouse Surgical Instruments

1. Custom trephine (3 mm) (Machine Shop, Arizona State University, Tempe, AZ).
   (a) Contact Rachel Rowe, rkro222@email.arizona.edu.
2. Weed whacker line for cranial disc (1.7 mm diameter).
3. Side-grasping forceps (7 × 7) (Henry Schein, #6-124XL).
4. 3 M Vetbond tissue adhesive (Henry Schein, #700-3449).

2.2.7 Injury Hub (Fig. 2)

1. 1½″ needle (20 gauge) (Becton Dickinson, #305176).
2. Syringe (1 cm^3).
3. Razor blades.
4. Tissue forceps (Henry Schein, #6-114).

2.2.8 Rat Injury Hub

1. Cosmetic pencil sharpener.

2.2.9 Mouse Injury Hub

1. Luer-loc extension tubing (Baxter, #2C5643).

3 Methods

3.1 Record Keeping

A standard surgery sheet should be used to record information pertaining to the surgical procedure, injury, and both immediate and long-term postoperative care (*see* Appendix). Postoperative observation and treatment of each animal should be maintained and include Notes about the general condition of the animal and any supportive care the animal received (e.g., saline injections).

3.2 Preoperative Preparation

Appropriate personal protective equipment should be worn: clean lab coat or scrubs, gloves, face mask, hair covering, and protective eyewear. Assess the animal for signs of pain, distress, or disease and record this information on the data sheet (e.g., abnormal posture,

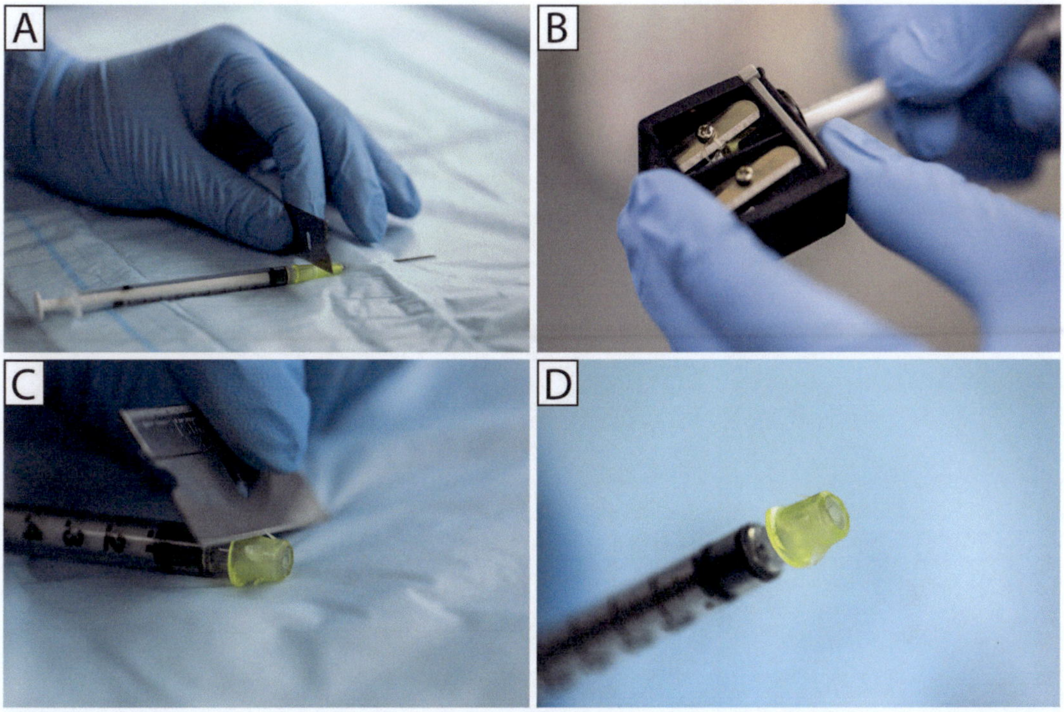

Fig. 2 Injury hub construction. Firmly attach a 20-gauge needle to a 1 cm^3 syringe and insert the needle into a laboratory bench pad to prevent the needle from becoming projected after it is cut (**a**). Use a razor blade to cut off the tip of the needle (**a**). For the rat, the injury hub is beveled using a cosmetic pencil sharpener (**b**). Using a razor blade, score the exterior of the hub making burrs at even intervals around the hub (**c**). When finished, the cut end should be flat and even, and parallel to the Luer-Loc plane (**d**)

movement, poor grooming, and evidence of porphyrin accumulation on eyes, nose, or fur).

3.3 Administer Anesthesia and Secure in Head Holder

1. Anesthetize the animal with 5% isoflurane for 5 min in an induction chamber.
2. Shave or remove hair from scalp, as appropriate (Fig. 3a).
3. Secure animal in a stereotaxic frame equipped with a nose cone for continuous inhalation of isoflurane (2.0–2.5%) (Fig. 3a). The back of the front incisors should be flush with the bite bar, without tension applied to the teeth. If you observe mouth breathing, check the positioning of the teeth over the bite bar and/or reposition the nose cone to allow for normal respiration.
4. Apply ophthalmic ointment to the eyes to keep them moist during the surgery.
5. Prepare the surgical area with 70% alcohol and betadine solution (antiseptic).

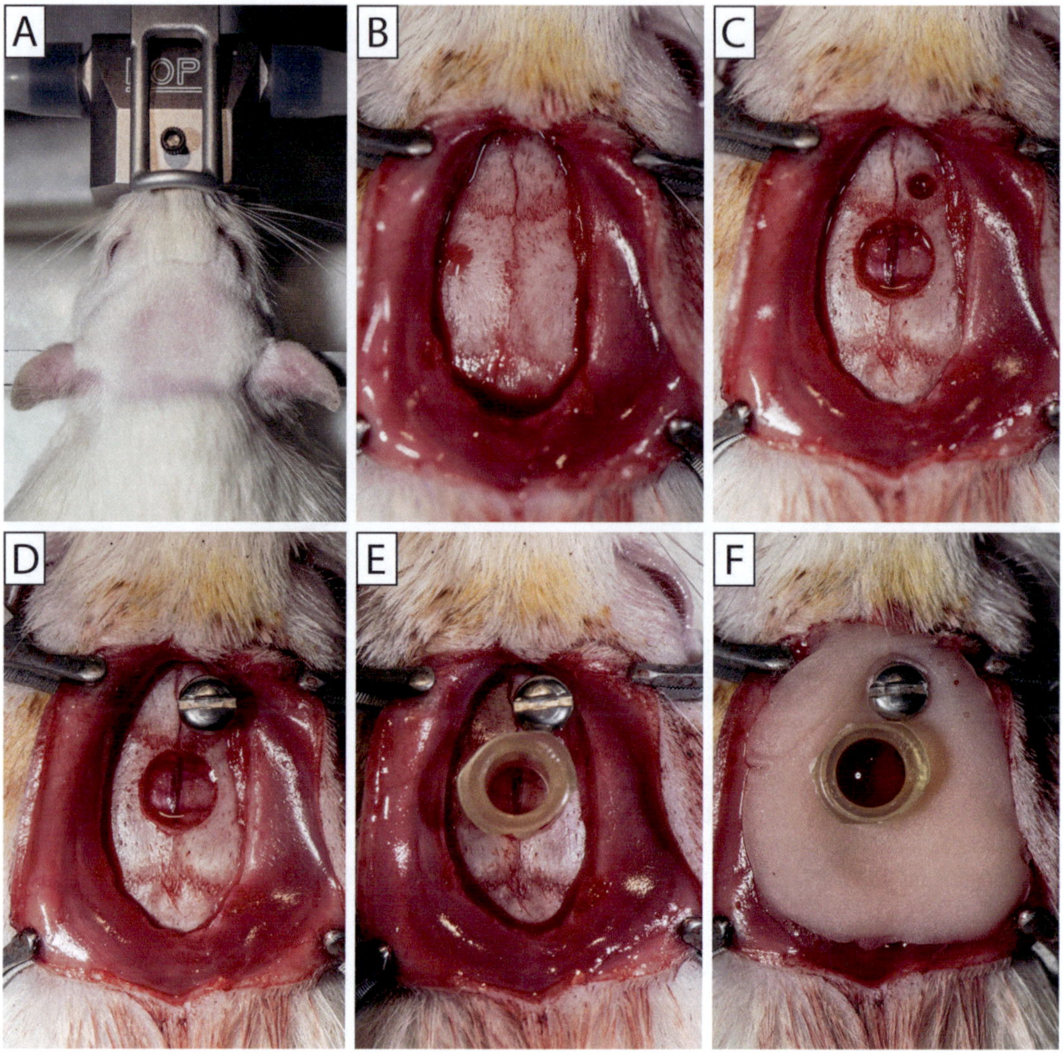

Fig. 3 Cranial surgery for hub placement in the rat. The rat's head is shaved and the animal is secured in a stereotaxic frame with a continuous flow of isoflurane via a nose cone (**a**). A midline incision is made to expose the skull and the overlying fascia is removed (**b**). A Dremel tool is used to make two pilot holes. The screw hole is expanded with a finger nail drill and 5/64″ drill bit (**c**). A 4.7 mm diameter trephine is used to create a cranial disc that is removed to expose the underlying dura (**c**). A stainless steel screw is secured into the screw hole (**d**). Small drops of cyanoacrylate gel are placed on the outside of the constructed injury hub, and the hub placed inside the craniectomy (**e**). After the cyanoacrylate gel dries, the injury hub and screw are covered in methyl methacrylate cement and the injury hub is filled with saline (**f**)

6. Monitor anesthesia by observing muscle relaxation, in addition to assessing the toe pinch reflex. Animals under appropriate anesthesia will have a steady respiration rate.

3.4 Cranial Surgery for Hub Placement

1. Make a midline sagittal incision extending from between the eyes, to the base of the skull, just past the ears. To avoid excessive bleeding, avoid cutting the muscle at the base of the skull (Figs. 3b and 4a).

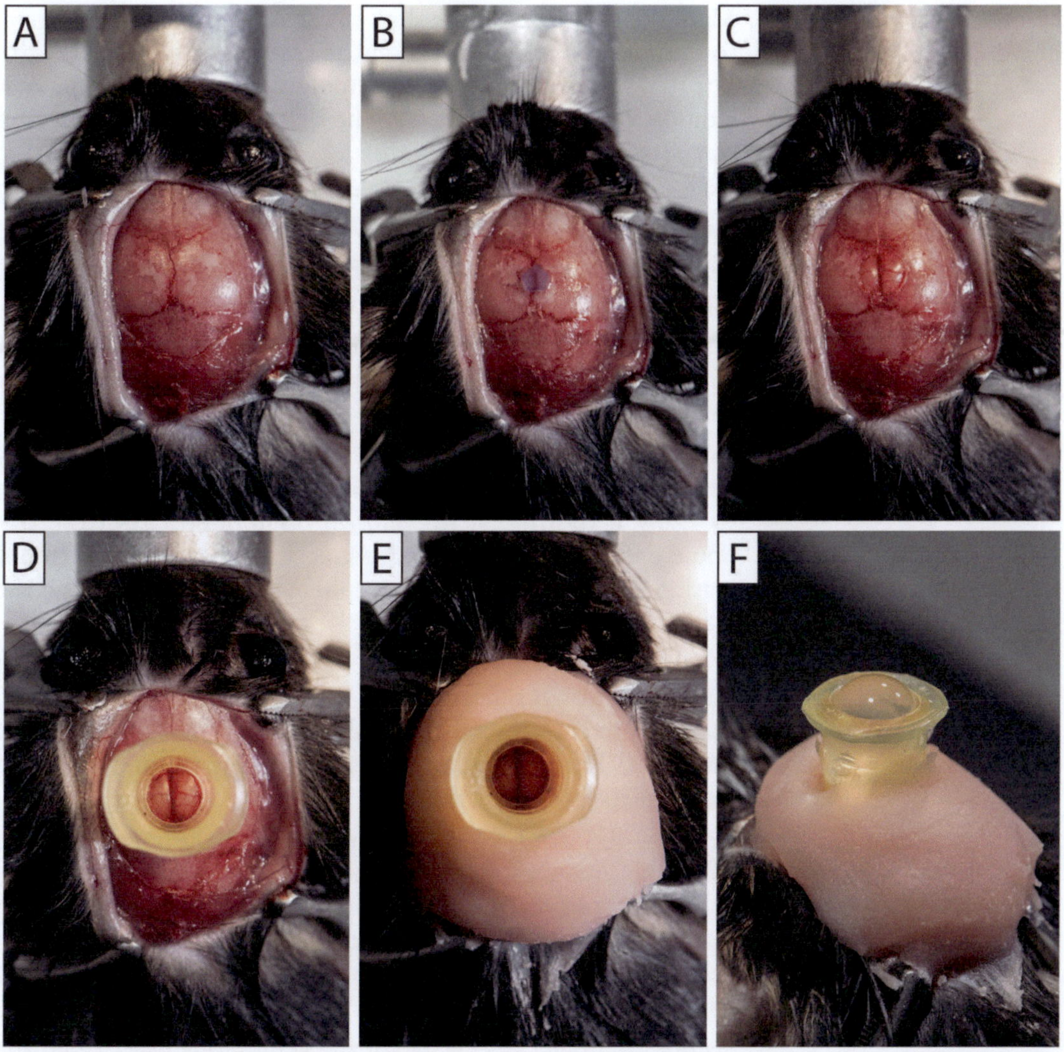

Fig. 4 Cranial surgery for hub placement in the mouse. A midline incision is made to expose the skull and the overlying fascia is removed (**a**). Vetbond tissue adhesive is used to secure a disc shaved from weed whacker line at the location of the craniectomy (**b**). A 3 mm diameter trephine is used to create a cranial disc that removed to expose the underlying dura (**c**). Small drops of cyanoacrylate gel are placed on the outside of the constructed injury hub, and the hub placed outside the craniectomy (**d**). After the cyanoacrylate gel dries, the injury is covered in methyl methacrylate cement (**e**) and the injury hub is filled with saline (**f**)

2. Expose the skull and scrape the fascia from the skull using a delicate bone scraper, cotton swabs, and gauze. Clear away temporal muscle as necessary. If greater exposure is needed, stretch the skin by applying pressure with the fingers (Figs. 3b and 4a).
3. Attach Bull Dog clips to the edges of the incision (two anterior, two posterior) to expose the surgical site. When the Bull Dog clips fall down, the weight will hold the incision open.

3.5 Cranial Surgery for Hub Placement-Rat

1. Mark the locations on the skull of the screw hole (1 mm lateral to bregma and 1 mm rostral to the coronal suture on the right side) and craniectomy center (midway between bregma and lambda, over sagittal suture). The skull screw is used to secure the injury hub in place.
2. Drill pilot holes at both markings using the Dremel tool and burr bit.
3. Expand the screw hole with a finger nail drill and 5/64″ drill bit (Fig. 3c).
4. Place the centering pin inside the 4.7 mm diameter trephine. Anchor the centering pin in the pilot hole at the craniectomy center.
5. Continually turn and spin the trephine to make a craniectomy without disrupting the underlying dura. Keep trephine clean by using a toothbrush to remove bone debris from the trephine teeth. Apply saline to moisten the bone and aid in trephination. As needed, angle the trephine to evenly cut around the craniectomy.
6. Frequently check the progress of the craniectomy by applying mild pressure to the center of the craniectomy. As the skull thins, the craniectomy will be able to move independently of the skull. The craniectomy is complete when the bone can move freely in all directions.
7. Remove the bone piece working around the circumference using the Wedelstaedt and scalpel, or two Wedelstaedt instruments without disrupting the dura. When the bone has been removed, gently clear any blood from the craniectomy site (Fig. 3c). If the surgery site continues to bleed when skull is removed (*see* **Note 3**).
8. Secure a stainless steel screw in the skull screw hole. Hold the screw with forceps and advance the screw with a screwdriver (Fig. 3d).

3.6 Cranial Surgery for Hub Placement—Mouse

1. Shave weed whacker line with a razor blade as thin as possible to make a circular disc that is an equal thickness on all sides. Disc should be level when placed on the skull.
2. Pick up the disc with side-grasping forceps. Dip the cranial disc into a drop of Vetbond tissue adhesive placed on a nonabsorbent surface.
3. Place the disc at the location of the craniectomy (midway between bregma and lamda on the sagittal suture). To drop the disc, release the forceps and use a wooden applicator stick to properly position the disc. Once in position, use a Kimwipe tissue to wick away any excess Vetbond (Fig. 4a, b). Allow the Vetbond to fully dry before beginning to trephine.

4. Place the 3.0 mm trephine over the disc and perform the craniectomy by continually turning and spinning the trephine without disrupting the underlying dura. Keep trephine clean by using a toothbrush to remove bone debris from the trephine. Apply saline to moisten the bone and aid in trephination. As needed, angle the trephine to evenly cut around the craniectomy. If the disc comes off while trephining (*see* **Note 4**).
5. Frequently check the progress of the craniectomy by applying mild pressure to the center of the craniectomy. As the skull thins, the craniectomy will be able to move independently of the skull. The craniectomy is complete when the bone can move freely in all directions.
   (a) Under magnification, remove the bone piece working around the circumference using the Wedelstaedt and scalpel, or two Wedelstaedt instruments without disrupting the dura (Fig. 4c). When the bone has been removed, gently clear any blood from the craniectomy site.

3.7 Injury Hub

3.7.1 Injury Hub Construction

1. Attach a 22-gauge, 1½″ needle to a 1 cm^3 syringe. Place the needle into a laboratory bench pad (Fig. 2a).
2. Cut the female Luer-Loc hub from the needle using a razor blade (Fig. 2a). The cut is made parallel to the Luer-loc with an outer diameter of ~4.7 mm for the rat, and ~3.0 mm for the mouse. *See* **Note 5** for tips.
3. Inspect the cut edge of the injury hub and trim to size and level as necessary.
4. For the rat, bevel the cut edge of the injury hub with a cosmetic pencil sharpener (Fig. 2b).
5. Shave thin burrs around the injury hub starting at the Luer-Loc edge in the direction of the cut edge using a razor blade (Fig. 2c).

3.7.2 Injury Hub Placement

1. Hold the hub in tissue forceps (behind the teeth). Apply small drops of cyanoacrylate gel on the outside of the hub, just above the cut end.
2. Position the hub over the craniectomy (using magnification for the mouse). For the rat, the injury hub fits inside the craniectomy (Fig. 3e). For the mouse, the injury hub fits outside the craniectomy (Fig. 4d).
3. Using a wooden applicator stick (cut a sharp angle) gently scrape the cyanoacrylate gel down the injury hub onto the skull. Apply more cyanoacrylate gel if needed to the junction between the injury hub and the skull to firmly adhere the injury hub to the skull in addition to creating a seal.

4. After the cyanoacrylate gel dries, cover the injury hub (and screw) in methyl methacrylate cement (Figs. 3f and 4e). Apply the methyl methacrylate cement from a 1 cm^3 syringe when it is thick enough to hold shape.
5. When the methacrylate cement has dried, fill the injury hub with saline (Figs. 3f and 4f).
6. Place a suture at both the anterior and posterior edges of the incision.
7. Remove the animal from the stereotaxic frame and anesthesia. Place the animal in a recovery cage on a heating pad until the animal is awake and alert. Monitor animals for outward signs of pain or distress.

3.8 Injury

Before using the injury device, check that when the weighted pendulum arm is hanging in a neutral position (at 0°) that it is flush and centered on the foam pad at the end of the plunger. Adjust as needed. Drop the pendulum hammer several times to prime the device. Remove any air bubbles (*see* **Notes 6–8**).

1. Reanesthetize the animal after an approximately 60-min recovery period from surgery.
2. Visually inspect inside the injury hub for debris, blood or dried dental acrylic. Clean out the injury hub using a small cotton tip applicator or irrigate with saline if necessary, *see* **Note 9**.
3. Fill the injury hub with sterile saline until a bead of fluid is formed by surface tension (Figs. 5a and 6a). Remove any air bubbles from inside the hub.
4. To avoid air between the hub and device, press the plunger so that a drop of fluid is produced at the end of the injury device. Connect the female Luer-Loc injury hub on the animal to the male Luer-Loc fitting on the injury device (Figs. 5b and 6b). Create continuity between the fluid of the cylinder and the fluid in the injury hub. For handling and placement during injury, *see* **Note 10**.
5. Check the animal for a toe pinch response. Once a normal breathing pattern returns (1–2 breaths per second) and the animal has a positive toe pinch response, release the pendulum to injure the animal. Secure the pendulum after it strikes the plunger and return it to the catch.
6. Immediately after the injury, start a timer to measure the duration of the suppression of the righting reflex.
7. Remove the injury hub by pressing on the bridge of the nose for leverage. Visually inspect the hub for obstructions.
8. Observe and record the duration and extent of apnea or seizure. Note the condition/appearance of the surgical site and

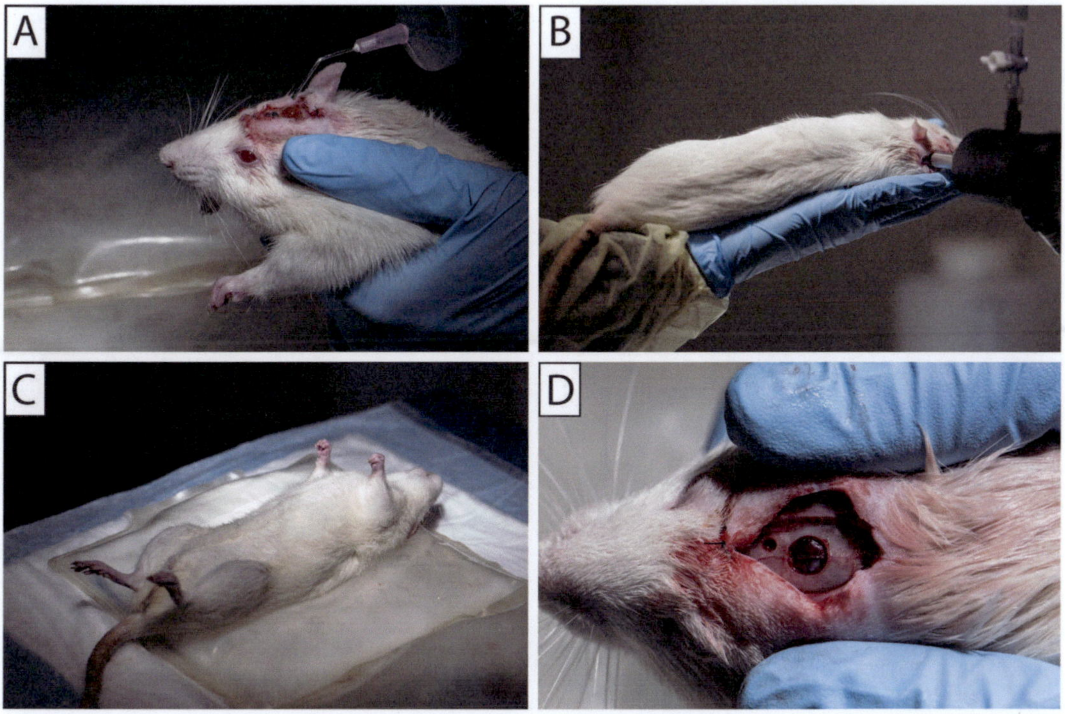

Fig. 5 Midline fluid percussion injury in the rat. The injury hub is filled with sterile saline until a bead of fluid is formed (**a**). The plunger is pressed to produce a drop of water at the end of the injury device. Creating continuity between the fluid of the cylinder and the fluid in the injury hub, the female Luer-Loc of the injury hub is connected to the male Luer-Loc fitting on the injury device by laying the rat on their right side and holding it with your left hand (**b**). Following a toe pinch response, the pendulum is released to injure the animal. The animal is placed in a supine position on a heating pad until the animal spontaneously rights itself (**c**). The condition/appearance of the surgical site and brain tissue beneath the injury site is observed and recorded (**d**)

brain tissue beneath the injury site and record brain herniation and hemorrhage. If the dura is breached, the animal should be euthanized and not included in the study, *see* **Note 11** for details.

9. Control bleeding if necessary. Leave the craniectomy open. Close the wound (i.e., suture or staple) and apply topical lidocaine and antibiotic ointment.
10. Place the animal in a supine position on a heating pad (Fig. 5c). The time elapsed until the animal spontaneously rights is recorded as the righting reflex time.
11. Once the animal has righted, place it in a designated recovery area equipped with a heating pad.
12. When the animal regains normal ambulatory behavior, it can be returned to its home cage.

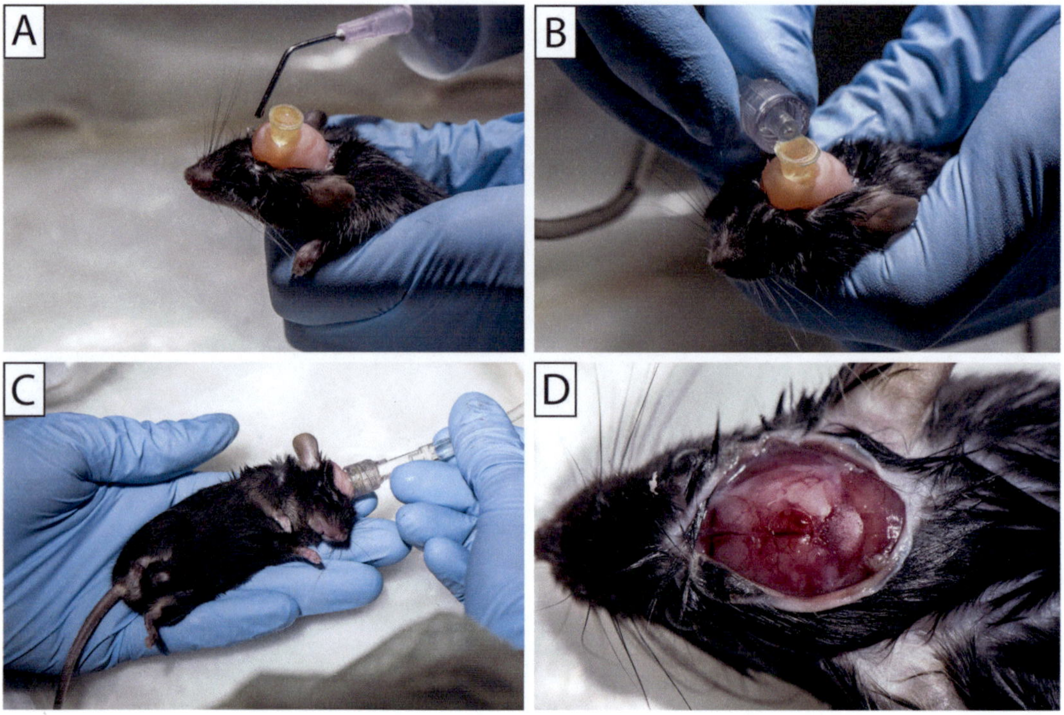

Fig. 6 Midline fluid percussion injury in the mouse. The injury hub is filled with sterile saline until a bead of water is formed (**a**). Creating continuity between the fluid of the Luer-Loc extension tube and the fluid in the injury hub, the female Luer-Loc of the injury hub is connected to the male Luer-Loc fitting on extension tube which is connected to injury device (**b**). Following a toe pinch response, the pendulum is released to injure the animal while it is lying on its side (**c**). Immediately after injury the animal is placed on its side until the animal spontaneously rights itself (**c**). The condition/appearance of the surgical site and brain tissue beneath the injury site is observed and recorded (**d**)

3.9 Postoperative Care

3.9.1 Postoperative Evaluations

1. Following injury, animals should be visually monitored for continued recovery every 10 min post-injury (for the first hour). Within 15–20 min after injury, surviving animals should be alert. Within 1 h after injury, animals should be ambulatory. Brain-injured and uninjured control animals typically show no outward effects once they have recovered from anesthesia, and resume normal eating, drinking, and grooming patterns. Typically animals return to sleep, as the injury occurs during their sleep cycle.
2. Postoperative evaluations should be done daily (for a minimum of 3 days). Follow the postoperative evaluation sheet to record the external examination, physical examination, suture site, and a pain evaluation. Typically, animals require no special supportive care after surgery. This injury does not produce overt signs of postoperative pain, and do not call for pain monitoring or drugs to manage pain. Caution should be taken in administering such compounds, as they can influence outcome (for review *see* [7]).

3.9.2 Postoperative Weight

1. Weigh animals daily. Record weights on the evaluation sheet.
2. Animals can lose up to 20 % of their body weight after surgery and injury. It is beneficial to prophylactically provide mash (chow + water) and/or place normal rat chow on the floor of the cage to facilitate weight gain.
3. If by the second day post-injury, there is continued weight loss, the animals will likely require fluid injections (0.9 % sterile saline) to prevent dehydration. Consult a local veterinarian for advice.
4. Weight loss exceeding 20 % of their body weight indicates significant injuries that require intensive postoperative care or euthanasia.

4 Advantages, Limitations, Complications

4.1 Advantages

1. Midline fluid percussion injury (mFPI) is scalable to induce a highly reproducible brain injury that models the clinical sequelae of concussion. Using this model, brain injury is induced by a 20 ms fluid pulse delivered onto the intact dura through a craniectomy [8], defining this technique as a model of traumatic brain injury rather than a head injury. While this model necessitates breaching the cranial vault, the skull is sealed to the injury device, recreating a closed system, which approximates a closed head injury with a decompressive craniectomy. This model allows for the injury to be induced after a recovery from anesthesia which returns the animals to a condition that resembles the human condition, thus enhancing the face validity. The bilateral diffuse pathology in the absence of a cavitation also resembles the human condition of concussion, in which cavitation is rare.

4.2 Limitations

1. In contrast to the tissue destruction caused by other brain injury models, mFPI results in tissue disruption. There is variability in the extent of damaged tissue and range of physiological responses among animals because tissue disruption, unlike tissue destruction, does not have a ceiling effect. The injury to brain occurs within a range which necessitates large group sizes to detect significant effects.

5 Notes

1. The plunger impact pad on the fluid cylinder should be replaced every 8–12 months. Information and instructions for the setup, cleaning, and maintenance of the FPI device can be

found in the FPI Operation Manual: http://www.radiology.vcu.edu/docs/FPIOperationManual.pdf.

2. When using an inhaled anesthetic, it is recommended that all procedures are performed in a well-ventilated area, on a down-draft or similar table, or in a type II biosafety cabinet to minimize anesthesia exposure to the surgeon (current OSHA recommendation for halogenated gasses is <2 ppm).
3. If the surgery site continues to bleed when the skull is removed, lightly remove blood with gauze. Adding saline can create hydrostatic pressure that will reduce bleeding. If the site continues to bleed, control the bleeding with Gelfoam. Excessive wiping or dabbing at the craniectomy site will prevent blood clotting and worsen the bleed.
4. If the disc comes off while trephining during a mouse surgery, clean excess dried glue from the area and apply a new disc using Vetbond. However, if the bone can move independently of the skull in an area, use a small dot of superglue to attach a new disc. Vetbond will run and may touch the surface of the dura compromising the surgery.
5. When constructing the injury hub for mice, to confirm the proper diameter you can place the trephine through the hub and confirm a tight fit.
6. Air bubbles in the FPI device can prevent an accurate measurement of the injury magnitude. When air is present in the device, the oscilloscope reading will have many jagged peaks instead of a smooth curve with one peak. The syringe ports can be used to remove any air that enters the device. One way for air to become trapped in the fluid cylinder is after cleaning of the cylinder. This can be minimized by rinsing with a spot remover solution for the dishwasher (e.g., Jet Dry).
7. Air bubbles can also enter the device during the impact. To prevent air bubbles it is necessary to use two 10 mL syringes during preparation of the device between rat injuries. After injury, remove the rat from the male Luer-loc fitting on the end of the device (Fig. 1). Next, a 10 mL syringe with a female Luer-loc fitting should be attached to the device. Pull up on the syringe to remove fluid contaminated with blood or air from the device. Lastly, a second 10 mL syringe containing clean deionized water should be attached to the device. Pull up on the syringe to remove any air bubbles from the device. Check that when the weighted pendulum arm is hanging in a neutral position (at 0°) that it is flush and centered on the foam pad at the end of the plunger. Adjust as needed. Drop the pendulum hammer several times to prime the device. Between every rat injury a syringe should be attached and "dirty" water removed, then a second "clean" syringe should be attached to prime the device.

8. It is important to make sure that the Luer-loc extension tubing is free of air bubbles before each mouse injury. Hold the end of the tubing higher than the connection point on the device. Lightly tapping the extension tubing will force air bubbles to the end of the tubing where they can easily be removed. Between each mouse injury, remove all air bubbles from the tubing. Check that when the weighted pendulum arm is hanging in a neutral position (at 0°) that it is flush and centered on the foam pad at the end of the plunger. Adjust as needed. Drop the pendulum hammer several times to prime the device.
9. When placing the injury hub over the craniectomy, cyanoacrylate gel can spread on to the dura. If the cyanoacrylate is not thoroughly dry forming a seal, the methyl methacrylate can also spread under the injury hub and onto the dura. These substances on the dura will change mechanical properties and alter the injury. Visual inspection is necessary to identify cyanoacrylate gel or methyl methacrylate on the dura, as well as any other obstruction over the injury site, such as a blood clot.
10. For the injury, rats should be held in your left hand lying on their right side. Attach the rat directly to the device (Fig. 5b). For mice, attach them to the device using a Luer-Loc extension tube (Fig. 6b).
11. During the cranial surgery, the dura can be compromised by the trephination or removal of the bone. When the injury is induced, pressure from the fluid pulse will cause the dura to tear and the brain will herniate through the craniectomy. If the dura is compromised, the injury becomes inconsistent and should be classified as a technical failure. A dura breach will extend the opening of the blood–brain barrier and displace neural tissue. Animals with a dura breach should be excluded from any study.

6 Appendix: Fluid Percussion Brain Injury

Animal ID#: ____________ Species: rat mouse Animal Marking: ___________

Protocol #_____________________

Investigator: ________________ Strain: ____________________

Barcode #: _______________ Emergency Ph. #:___(___)___-______

Contact Person:___________ Lab Phone#:__(___)___-______

Surgery Date:______________ Injury Date:_____/_____/_______

Experiment:____

Date	Clinical Notes

Euthanized: Date: ____/____/______

Substance Used: ____________ Amount: ________ Route: ____________

Perfusion

Perfusion Date: ______________ Surgeon: ______________

Sodium Pentobarbital: Route: __________ Amount: ______________

Perfusion Technique: Intracardial ____________

Saline: ______________ __________ml

Fixative: ______________ __________ml

Perfusion Quality: Good Moderate Marginal Poor

Post-Fix: Brain / SkullDays: ________ Fixative: ______________

Brain Sectioning

Sectioning Date: ____/____/_____

Cutting Scheme:____

____

SURGERY

Animal ID#: ______________

Pre-Surgical Observation*

Good Poor

**Animals assessed for normal activity, inquisitiveness, condition of hair coat, eating, drinking, defecation, urination, both eyes are present, appearance of eyes, breathing rate, gait, and bodyweight.*

Surgical Preparation

Weight: _________g

Anesthesia Type:____________ __________________

Percentage: _______% Duration: _____________ Time: ___:____

Surgical Procedure

Anesthesia Type:______Isoflurane______ ______________

Percentage: _______ Duration: _____________ Time: ___:____

- Secured in stereotaxic frame ___
  Set stereotaxic frame: Earbar: **5.4** Nose cone unit: **-7.0**
- Lacri-Lube ointment ___
- Betadine & Alcohol Scrub ___
- Sterile Gloves___
- Midline incision and skull exposure ___
- Midline craniotomy between bregma and lambda ___
- Cut needle hub at lower white line & fashion injury hub___
- Needle hub placed over midline craniotomy with cyanoacrylate ___
- Dental cement poured around hub and over screws ___
- Fill hub with saline ___
- Wound closure with suture ___
- Lidocaine ___
- Bacitracin ___
- Animal Returned to Holding Cage ___ Time:___:____

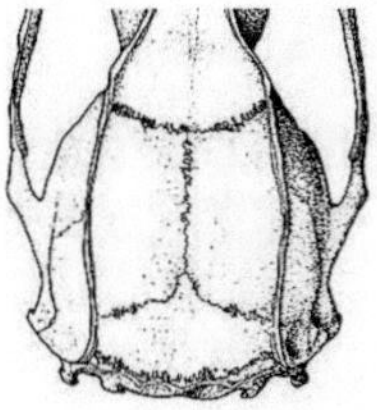

Surgical notes:____
____
____
____
____

Procedure was completed: ❑ As expected ❑ Better than expected ❑ Worse than expected

Surgeon Signature: ____________________________________ Date: _____/_____/________

FLUID PERCUSSION INJURY

Animal ID#: ______________

INJURY **SHAM**

Pre-Injury Observation*

Good Poor

**Animals assessed for normal activity, inquisitiveness, condition of hair coat, eating, drinking, defecation, urination, both eyes are present, appearance of eyes, breathing rate, gait, and bodyweight.*

Anesthesia Type: Isoflurane ______________

Percentage: ______% Duration: ____________ Time: ___:___

Pre-Injury Inspection

- Visual inspection of injury hub ______________________________
- Visual inspection of the dura ____ ______________________________

Hammer Angle: __________ Injury Time: ___:___

Magnitude of injury: __________ ATM __________mV (147.28 mV = 1 ATM)

MAX: __________

MIN: __________

Rise Time: __________

Fall Time: __________

Positive Width: __________

Oscilloscope Filename:______________________________

Animal Status Post-Injury

Apnea: __________ sec. ❑ massage of thorax ❑ no intervention

Seizure: __________ sec.

Herniation: ❑ none ❑ small ❑ moderate ❑ large

Hematoma: ❑ none ❑ small ❑ moderate ❑ large ❑ bloody

Dura Breach: ❑ none ❑ small ❑ extensive

Breathing: ❑ normal ❑ deep ❑ regular ❑ rapid ❑ labored ❑ shallow ❑ irregular ❑ slow

Righting Reflex Time: ______________

Notes following injury / sham: ____

____

____

- Wound closure with suture ___ Staples ___
- Lidocaine ___
- Bacitracin ___

Time returned to incubator:________

Time returned to colony:__________

Suture/Staple Removal Date: __________________

Procedure was completed: ❑ As expected ❑ Better than expected ❑ Worse than expected

Surgeon Signature: ________________________________ Date: ______/______/________

POST-OPERATIVE EVALUATION

Animal ID#: ______________

Date	/ /	/ /	/ /	/ /	/ /
Days post-procedure					
Time	:	:	:	:	:
Observer's Initials					

EXTERNAL EXAMINATION

Active?	Y N	Y N	Y N	Y N	Y N
Inquisitive?	Y N	Y N	Y N	Y N	Y N
Crusty eyes?	Y N	Y N	Y N	Y N	Y N
Eating?	Y N	Y N	Y N	Y N	Y N
Drinking?	Y N	Y N	Y N	Y N	Y N
Feces?	Y N	Y N	Y N	Y N	Y N
Urine?	Y N	Y N	Y N	Y N	Y N

PHYSICAL EXAMINATION

Type of Breathing?					
Gait?					
Coat Appearance?					
Dehydration?*	Y N	Y N	Y N	Y N	Y N
Body Appearance?					
Body Weight?					

SUTURE SITE

Wound edges red?	Y N	Y N	Y N	Y N	Y N
Swelling around incision?	Y N	Y N	Y N	Y N	Y N
Swelling under incision?	Y N	Y N	Y N	Y N	Y N
Sutures missing?	Y N	Y N	Y N	Y N	Y N
Exudate from incision?	Y N	Y N	Y N	Y N	Y N

PAIN EVALUATION**

Behavior					
Locomotion					
Appearance of incision					
Pain on palpation of surgery site					
Total					

N = normal L = labored R = rapid S = shallow Y = yes N = no
G = good F = fair P = poor
*Gently pinch up a fold of skin. Skin of dehydrated animals will stay pinched up.
**Scale 0-3

References

1. Dixon CE, Lighthall JW, Anderson TE (1988) Physiologic, histopathologic, and cineradiographic characterization of a new fluid-percussion model of experimental brain injury in the rat. J Neurotrauma 5:91–104
2. Thibault LE, Meaney DF, Anderson BJ, Marmarou A (1992) Biomechanical aspects of a fluid percussion model of brain injury. J Neurotrauma 9:311–322
3. Dixon CE, Lyeth BG, Povlishock JT, Findling RL, Hamm RJ, Marmarou A, Young HF, Hayes RL (1987) A fluid percussion model of experimental brain injury in the rat. J Neurosurg 67:110–119
4. McIntosh TK, Noble L, Andrews B, Faden AI (1987) Traumatic brain injury in the rat: characterization of a midline fluid-percussion model. Cent Nerv Syst Trauma 4:119–134
5. McIntosh TK, Vink R, Noble L, Yamakami I, Fernyak S, Soares H, Faden AL (1989) Traumatic brain injury in the rat: characterization of a lateral fluid-percussion model. Neuroscience 28:233–244
6. Carbonell WS, Maris DO, McCall T, Grady MS (1998) Adaptation of the fluid percussion injury model to the mouse. J Neurotrauma 15:217–229
7. Rowe RK, Harrison JL, Thomas TC, Pauly JR, Adelson PD, and Lifshitz J (2013) Using anesthetics and analgeics in experimental traumatic brain injury. Lab Anim (NY) 42:286–291 doi:10.2038/laban.257
8. Lifshitz J (2008) Fluid percussion injury. In: Chen J, Xu X-M, Xu Z, Zhang CJ (eds) Animal models of acute neurological injuries. The Humana Press Inc., Totowa, NJ

Chapter 14

Lateral (Parasagittal) Fluid Percussion Model of Traumatic Brain Injury

Ken C. Van and Bruce G. Lyeth

Abstract

Fluid percussion was first conceptualized in the 1940s and has evolved into one of the leading laboratory methods for studying experimental traumatic brain injury (TBI). Over the decades, fluid percussion has been used in numerous species and today is predominantly applied to the rat. The fluid percussion technique rapidly injects a small volume of fluid, such as isotonic saline, through a circular craniotomy onto the intact dura overlying the brain cortex. In brief, the methods involve surgical production of a circular craniotomy, attachment of a fluid-filled conduit between the dura overlying the cortex and the outlet port of the fluid percussion device. A fluid pulse is then generated by the free-fall of a pendulum striking a piston on the fluid-filled cylinder of the device. The fluid enters the cranium, producing a compression and displacement of the brain parenchyma resulting in a sharp, high magnitude elevation of intracranial pressure that is propagated diffusely through the brain. This results in an immediate and transient period of traumatic unconsciousness as well as a combination of focal and diffuse damage to the brain, which is evident upon histological and behavioral analysis. Numerous studies have demonstrated that the rat fluid percussion model reproduces a wide range of pathological features associated with human TBI.

Key words Fluid percussion, Craniectomy, Endotracheal intubation, Trephination, Mechanical ventilation

1 Introduction

Several excellent reviews have been written documenting the pathophysiology of fluid percussion traumatic brain injury (TBI) in laboratory animals and the high degree of relevance to human TBI [1]. The fluid percussion technique has been applied to a number of species including mouse [2, 3], cat [4], rabbit [5], dog, sheep [6], and pig [7], with the overwhelming majority of applications to the rat (for review *see* [1]). This chapter provides a brief historical background of the fluid percussion model from its inception in larger animals to the current use predominantly in the rat. A major goal is to provide the new user with a practical guide for successful application of the model to the laboratory rat. A secondary goal is to share

Firas Kobeissy et al. (eds.), *Injury Models of the Central Nervous System: Methods and Protocols*, Methods in Molecular Biology, vol. 1462, DOI 10.1007/978-1-4939-3816-2_14, © Springer Science+Business Media New York 2016

tips and nuances that the authors have accumulated over several decades of use that may be helpful even for the seasoned user.

The earliest fluid percussion model procedures involved striking a fluid-filled column secured to the exposed dura of an animal. Denny-Brown and Russell produced a localized pressure pulse to the exposed dura of cats in order to produce a generalized loading to the brain rather than a focal disturbance [8]. They achieved this by rapidly applying extradural injections of fluid and termed this a "percussion concussion" to distinguish the injury from an acceleration concussion.

Gurdjian used a similar technique in dogs, but used compressed air (rather than a fluid) to rapidly and transiently raise intracranial pressure to produce a concussion [9]. Later studies produced concussive injuries by rapidly injecting fluid into a column of water attached to the rabbit vertex [10] or by dropping a weight onto a column of water attached directly to the cerebral cortex of dogs [11].

The evolution of the fluid percussion procedure continued with modifications to control for different amplitudes and durations of the fluid pulse to the brain of rabbits by Stalhammar and colleagues in Sweden [12–14]. The Richmond group, led by Povlishock and Becker, modified the Stalhammar device and applied fluid percussion to the midline of the cat [15]. The next major advance was Dixon's characterization of midline fluid percussion in the rat [16] followed by McIntosh's characterization of the lateral (parasagittal) orientation [17]. The transition to the rat model greatly expanded the applications for cognitive and motor behavioral analysis and testing of pharmacological interventions. The lateral (parasagittal) approach is currently the most commonly used orientation and is the focus of the chapter.

2 Material

2.1 Fluid Percussion Device Setup (Fig. 1)

1. Fluid percussion device (Custom Design and Fabrication model 01-B, Richmond, VA).
2. Extracranial transducer (Sensym ICT model SPTmV0100PG5W02, Milpitas, CA).
3. Digital storage oscilloscope (Tektronix Inc. model TDS 1002, Beaverton, OR).
4. Pressure transducer amplifier (Custom Design and Fabrication, Richmond, VA).

2.2 Animals

1. Male Sprague–Dawley rats (300–325 g; Harlan Laboratories, Hayward, CA).

2.3 Anesthesia Induction and Maintenance

1. Plexiglas® anesthesia induction chamber.
2. Isoflurane, USP.

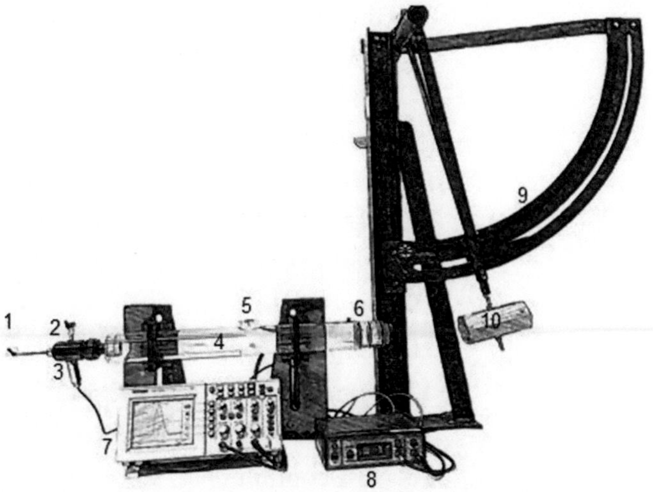

Fig. 1 Schematic diagram of fluid percussion injury device (Model 01-B Custom Design & Fabrication). Luer-lock outlet port (*1*), Luer-lock bubble removal port (*2*), transducer housing (*3*), fluid cylinder (*4*), fill port (*5*), Piston and O-rings (*6*), oscilloscope (*7*), transducer amplifier (*8*), protractor (*9*), and impact hammer (*10*)

3. Isoflurane vaporizer (Highland Medical Equipment, Temecula, CA).
4. Laryngoscope with fiber optics handles (American Diagnostics Corporation).
5. Plexiglas® frame for intubation (Fig. 3).
6. Y-tube anesthesia setup (Fig. 4b).
7. Endotracheal tube (Fig. 4c).
8. Rodent volume ventilator (Harvard Apparatus model 683, Holliston, MA).
9. Blunt end forceps.
10. Polyethylene tubing-PE50 (Becton Dickinson, Franklin Lakes, NJ).

2.4 Animal Preparation and Temperature Measurements

1. Lubricating ophthalmic ointment.
2. Hair clippers.
3. Heating lamp.
4. Temperature controller pad (CWE model TC-1000, Ardmore, PA).
5. Thermalert monitoring thermometer (Physitemp model TH-5, Clifton, NJ).
6. Needle temperature probe (Physitemp model MT-29/2, Clifton, NJ).

2.5 Surgery

1. Stereotaxic frame (Kopf Instruments, Tujunga, CA, USA).
2. Alcohol and betadine swabs.

3. Bupivacaine HCl 0.5% (5 mg/mL) (Hospira Inc., Lake Forest, IL).
4. Scalpel blade holder and blade No. 10 (Henry Schein Inc., Melville, NY).
5. Cotton-tipped applicators.
6. Gauze sponge, non-woven.
7. Lactated Ringer's solution or Saline.
8. Syringes (1, 6, 12 mL) and needles (18–20 G, 22–23 G).
9. Jewelers forceps.
10. Trephine (Miltex model 26-140, York, PA) (Fig. 8a, b).
11. Bone curette (Miltex model 21-322, York, PA) (Fig. 8c).
12. Pin vice with drill bit (No. 47) (Fig. 8d).
13. TBI conduit (Fig. 9).
14. Fluid percussion injury connector tube (Fig. 11b).
15. Anchor screws (round-head machine 2-56, 2.1 mm diameter, 6.0 mm length, Grainger, West Sacramento).
16. Super glue gel (Loctite, West Lake, Ohio).
17. Crosslinked flash acrylic liquid (The Motloid Company, Chicago, IL).
18. Crosslinked flash acrylic powder (Yates & Bird, Chicago, IL).
19. Suture needles and 4.0 braided silk sutures.
20. Needle holder with suture cutters.

3 Methods

3.1 Fluid Percussion Device

The first fluid percussion device was originally developed over several decades ago by Sullivan et al. studying brain trauma in cats [15]. The design and manufacture of the percussion device was later adapted for TBI in rodents and it has become a mainstay in the field of neurotrauma research. There are several manufacturers of the percussion devices, including AmScien Instruments, Dragonfly Inc., and Custom Design and Fabrication (formerly VCU Biomedical Engineering). The Custom Design and Fabrication (Model 01-B) consists of a Plexiglass® cylindrical chamber filled with deionized distilled water or normal saline (Fig. 1). The fluid chamber has a solid Plexiglass® piston with O-rings on one end and a pressure transducer housing with male Luer-lock at the other end. An impact hammer attached to a pendulum can be adjusted on a fixed protractor to allow for varying heights of free-fall to produce different severities of the injury. To induce TBI, the female Luer injury conduit on the acrylic assembly on the rat is connected to the male Luer output port of the pressure transducer housing. Once the experimenter releases the pendulum, the impact hammer strikes the piston, producing a transient pressure that is transmitted through the fluid-filled cylinder to the Luer-lock outlet at the opposite end of the device. The fluid pressure comes into contact with the dura surface of the animal.

The changes in the transient pressure are detected by the pressure transducer (model SPTmV0100PG5W02; Sensym ICT, Milpitas, CA), relayed to the transducer amplifier and recorded on the digital storage oscilloscope (model TDS 1002; Tektronix Inc., Beaverton, OR). The voltage changes are converted into atmospheres of pressure according to a calibration equation. It is important to accurately calibrate the fluid percussion transducer/oscilloscope system output voltage to a range of precise known pressures. This will ensure accuracy and consistency of pressure recorded during the injury process. Extreme care should be taken to remove all air bubbles throughout the entire fluid system, as air bubbles will greatly alter the pressure dynamics during injury. If air bubbles are present in the system, they will be detected on the oscilloscope tracing as erratic peaks or multiple peaks (be sure to disengage any smoothing filters on the transducer amplifier) (Fig. 2) (*see* **Note 1**). Refer to the manufacturer's website for additional detailed information on the assembly, operations, and troubleshooting of the fluid percussion device.

3.2 Anesthetics

There has been a long-standing debate over the potential neuroprotective effects of anesthetics after brain injury. Since the 1960s, specific properties of certain volatile anesthetics and sedative agents have been recognized and reported as potential neuroprotective agents [18–23]. Hence, the effect of anesthesia in experimental TBI is noteworthy and should require a careful evaluation. Two broad categories of anesthetics are inhalable anesthetics and injectable anesthetics. Halothane and isoflurane are examples of inhalable anesthetics while barbiturates and dissociative agents are examples

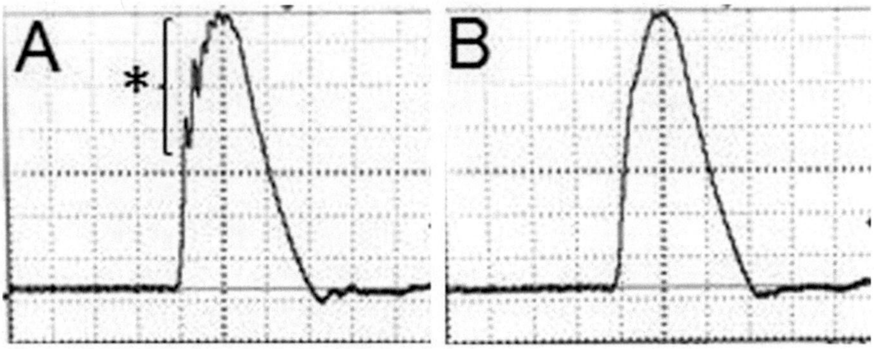

Fig. 2 Fluid percussion pressure traces on the digital oscilloscope for a moderate TBI (3.18 V = 2.16 ATM). The multiple peaks on the pressure curve (see * bracket) are due to the air bubbles inside the fluid cylinder (**a**). The pressure curve is smooth without the air bubbles (**b**)

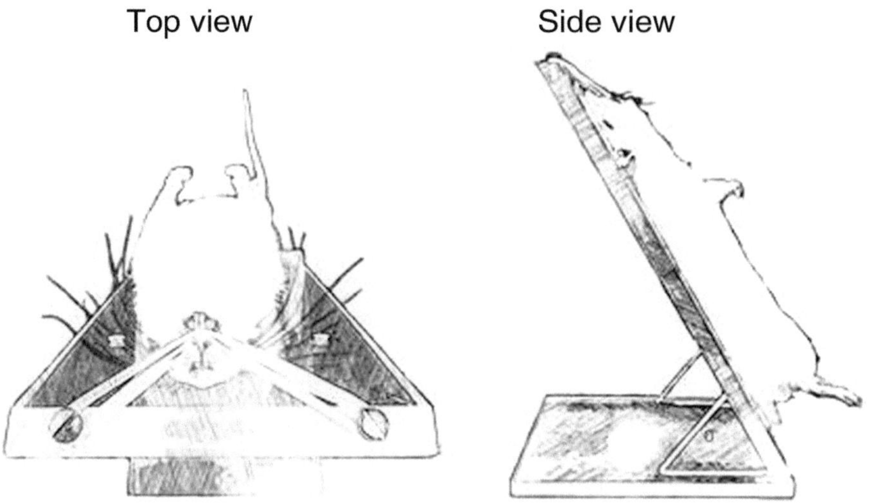

Fig. 3 Endotracheal intubation in an animal. After anesthesia induction, the animal is positioned on the Plexiglas® frame tilted at 60°

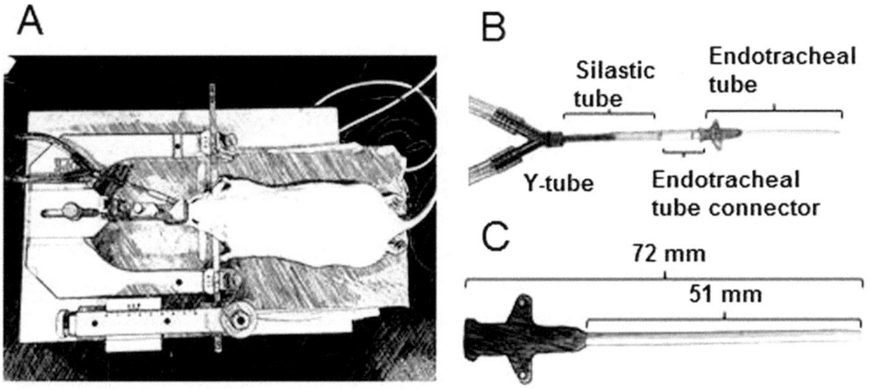

Fig. 4 Stereotaxic setup of an animal under isoflurane anesthesia prior to the craniectomy surgery (**a**), Y-tube and endotracheal intubation unit (**b**), and an example of an endotracheal intubation tube (**c**)

of injectable anesthetics. Injectable anesthetics are commonly used in animal studies, however, there are two disadvantages associated with them: (1) the need for repeated administration if the surgical procedure is longer than anticipated and (2) delayed emergence from anesthesia if the surgical procedure is completed early. These disadvantages make injectable anesthetics difficult for application in TBI studies. In contrast, characteristics of inhalable anesthetics that make them ideal for TBI studies include: (1) rapid induction, (2) control over depth and duration of general anesthesia by the experimenter, and (3) rapid emergence and recovery time.

Halothane and isoflurane are the most common inhalation anesthetics for laboratory animal use. However, evidence showing positive effects of inhalable anesthetics, especially in neuroprotection studies, has prompted researchers to weigh their options carefully and design their experiments accordingly. TBI often produces alterations of cerebral blood flow, cerebrovascular autoregulation dysfunction, cerebral metabolic impairment, and elevated intracranial pressure [24–28]. Halothane is infrequently used in cerebral ischemia and TBI studies because it is a very potent cerebral vasodilator. Even at a low concentration, halothane significantly increases cerebral blood flow [29] and raises intracranial pressure when mixed with nitrous oxide [30–32].

Anesthetics that alter the cerebral blood flow and intracranial pressure can have a significant effect on experimental outcomes [33]. As a result, isoflurane has become the preferred anesthetic for veterinary medicine and experimental TBI studies. The appropriate isoflurane concentration used for induction and surgical maintenance is 4 % and 2 %, respectively. Nitrous oxide (NO) is generally used as a carrier gas to reduce the dose of isoflurane required during anesthesia maintenance. NO is mixed with oxygen at 70 % and 30 %, respectively. The following steps should be taken to minimize total anesthesia time and reduce experimental variability: (1) Anesthesia should be completely switched off immediately after injury. Isoflurane is only required during the TBI surgery and is no longer needed post-injury. Usually, moderate to severe TBI would render the animal unconscious for longer than 10 min so additional anesthesia is not necessary when closing the surgical incision. (2) Total time spent under anesthesia should be the same for each animal. For sham-injured animals, there is no TBI so the animal should remain on anesthesia until the surgical incision is closed.

3.3 Mechanical Ventilation and Endotracheal Intubation

Severe clinical TBI patients often require intubation and mechanical ventilation to control respiration and prevent further insults to the brain. The use of a mechanical ventilator is becoming commonplace in animal research [34], especially in experimental animal TBI studies. TBI impact is usually followed immediately by apnea which, if prolonged causes hypoxia, hypercapnia, and cerebral edema leading to poorer outcomes [35]. Secondary insults, such as hypotension and hypoxia, markedly exacerbate sensorimotor dysfunction, cognitive deficits, neuronal degeneration [36–38], and increase mortality and morbidity [39]. In experimental TBI, important considerations for the proper settings of a mechanical ventilator should be addressed to prevent hypoventilation or hyperventilation of the animal. Adequate sedation and oxygenation are usually maintained at the tidal volume of 2.0–2.5 mL and 75–85 strokes per min for adult rats (~300 g). Ventilator settings to achieve normal blood gases should be confirmed by arterial blood gases analysis.

When used in conjunction with a mechanical ventilator, endotracheal intubation can be used for the precise delivery of different concentrations of anesthetics. It also keeps the airway patent after injury, an important factor for reducing complications after TBI. Inhalable anesthetics are also often delivered via a nose cone but there are limitations to this method in TBI studies. The nose cone anesthesia setup does not allow the experimenter to have control over the animal's respiration and apnea post-TBI. In contrast, endotracheal intubation with mechanical ventilation enables the experimenter to do the following: (1) maintain precision control over the rate of respiration and (2) reduce the incidences of hyperventilation or hypoventilation. It is important to maintain normoventilation to minimize complications associated with TBI and reduce experimental variability.

Endotracheal intubation involves insertion of a plastic cannula between the vocal chords into the trachea. There are several variations on how to perform an endotracheal intubation [40, 41]. The technique is challenging and requires considerable practice. Intravenous over-the-needle catheters (size 14) are ideal for use as an endotracheal intubation tube for rats (Fig. 4) (*see* **Note 2**).

After anesthesia induction, the unconscious animal is suspended by its superior incisors on a 60° tilted Plexiglas® frame in a supine position (Fig. 3). Blunt end forceps are used to gently pull the tongue to the side and then a small size laryngoscope—with a light source attached to the retractor—is used to hold the tongue against the base of the mouth to visualize the vocal folds and glottis. A blunt-end polyethylene tube (PE50) is used as a stylet to guide the endotracheal tube (*see* **Note 3**). A needle stylet should not be used because of the high risk of bleeding caused by the sharp, rigid metal piece. To facilitate the insertion of the endotracheal tube into the trachea, insert the tip of the polyethylene tube stylet between the vocal chords during inspiration. The endotracheal tube is then advanced into the larynx. The guide stylet is withdrawn while securing the endotracheal tube. Any observations of gurgling and bubbling sounds should be addressed immediately by withdrawing the fluids with a suctioning catheter.

After anesthesia induction, a successful intubation should take place in less than 30 s to avoid the need to re-anesthetize the animal. The number of intubation attempts should be limited. Continue to perform the intubation attempts only if there are no signs of tracheal bleeding or oropharyngeal swelling or damage. Verify the proper placement of the endotracheal tube before proceeding to the next step of the surgery (*see* **Note 4**).

3.4 Body Weight Measurements

The appropriate weight range for the specific study should be maintained within a narrow range to ensure experimental consistency. On the day of surgery, pre-injury body weight of the animals is recorded as a baseline measurement. Fasting before surgery is optimal, but generally not necessary. It is important that the animal's body weight be recorded daily from the baseline date until the experimental endpoint. Generally, TBI will cause appetite loss and fatigue during the first few days post-injury. Frequently, changes in the animal body weight are due primarily to dehydration, therefore, fluid management is critical for the animal's well-being and to avoid introducing variability into the experimental design. Severe dehydration can also cause other complications that can confound the experimental results.

Room temperature lactated Ringer's or normal saline (6 mL) can be administered daily (subcutaneous) for balanced fluid replacement. The variation in body weight loss post-injury and the interval when the animal begins to regain weight are usually dependent on the injury severity. As generally observed in the author's experiments, the steady rebound to baseline body weight for mild, moderate, and severe TBI-injured rats is between 1 and 3 days, 5–7 days, and greater than 10 days, respectively. Animals that lose up to 20 % of their body weight should be euthanized and excluded from the study. For sham-TBI animals, the craniectomy and surgical anesthesia should have no significant effect on their grooming and appetite. Sham-injured animals should regain baseline body-weight within 1 day post-surgery.

3.5 Temperature

Surgical anesthesia as well as TBI can alter the animal's ability to regulate and maintain normal body and brain temperature [42, 43]. Raised body and brain temperature are common occurrences in both the clinical and experimental TBI studies [44–46] and even a 1–2 °C elevation in body and brain temperature is detrimental to recovery [44]. On the other end of the spectrum, post-traumatic brain hypothermia can improve outcomes in both clinical [47–49] and experimental TBI [50–53]. For animal TBI studies, a temperature-controlled heating pad (CWE model TC-1000, Ardmore, PA) and heat lamp are useful to help maintain the body temperature within the normal range during surgery and post-injury. The brain temperature can be monitored with a 29-gauge needle temperature probe (Physitemp unit TH-5, probe MT-29/2, Clifton, NJ) placed in the temporalis muscle and is a dependable indirect measure of brain temperature [43, 54]. It is important to monitor and maintain appropriate body and temporalis muscle temperatures at 37 ± 0.5 °C and 36 ± 0.5 °C, respectively.

3.6 The Lateral Fluid Percussion Surgery

3.6.1 Anesthesia Induction

The survival surgery procedures described herein should be performed in accordance with aseptic rodent surgery guidelines. For the purpose of endotracheal intubation, the animal is placed in an induction chamber and lightly anesthetized with 4 % isoflurane in 100 % air for 4–5 min. Afterwards, the head is shaved prior to intubation. The muscle relaxation effect of isoflurane should allow a relatively easy access for oral endotracheal intubation. After successfully performing the intubation, the animal is connected to a mechanical ventilator (Harvard Apparatus model 683, Holliston, MA) and isoflurane vaporizer (Highland Medical Equipment) setup that allows inspiration and expiration via a y-shaped tube (Fig. 4) (*see* **Note 5**). The surgical level of anesthesia is maintained with 2 % isoflurane in a 70 % nitrous oxide and 30 % oxygen gas mixture.

Always check and ensure that the animal is anesthetized before proceeding to the next step of the procedure. Use the hindpaw withdrawal reflex pinch to monitor for adequate depth of anesthesia. Also, ensure there is adequate oxygenation by observing skin color of paws and ears. Adjust the anesthesia level accordingly if the animal is responsive to a toe pinch. If necessary, temporarily and slightly increase the anesthesia to sedate the animal; however, remember to reduce to an appropriate level to avoid anesthesia overdosing. As part of standard preoperating procedures, the following steps should be taken prior to any incision: (1) Apply an ophthalmic ointment to the animal's eyes to prevent dryness of the corneas. (2) Apply and clean the surgical site with alcohol and betadine swabs.

3.6.2 Skull Exposure

The application of a local anesthetic is recommended to minimize any potential residual postoperative pain and discomfort from the surgical procedures. Prior to an incision, a nonsteroidal local analgesic, bupivacaine (0.025 % diluted in saline), is applied to the subcutaneous space along the incision site. A 2-cm midline scalp incision extending from the eyes to the tip of ear is made with the scalpel (#10 blade) to expose the bone for trephining the craniectomy (*see* Fig. 5 for a step-by-step illustration of the craniectomy surgery). The incision should not be made past the ears to avoid major bleeding of the neck muscles. Reflect and retract the skin laterally. Use sterile cotton-tipped swabs to scrape the periosteal connective tissue and fascia from the skull. There will be an increase in bleeding from the

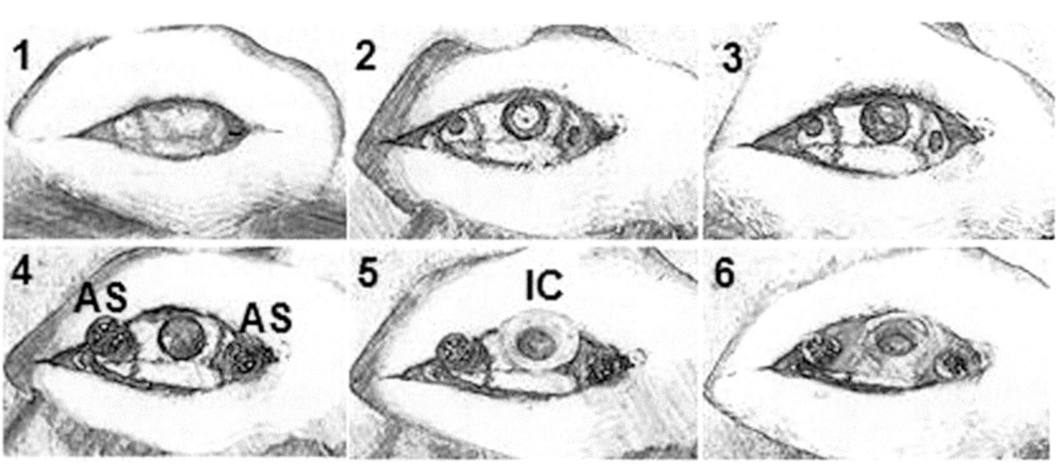

Fig. 5 Step-by-step procedure of the craniectomy surgery. Midline scalp incision (step 1). The craniectomy is created by trephination of the skull over the right parietal bone, two burr holes for anchor screws (step 2). The bone flap is removed with the dura intact (step 3). The two steel anchor screws (AS) are secured in place (step 4). The injury conduit (IC) is positioned over the intact dura (step 5). Dental acrylic TBI injury assembly (step 6)

skull surface if using a scalpel to scrape away the periosteum. A syringe containing room temperature Ringer's solution or normal saline is used to rinse the skull when cleaning out the fascia.

3.6.3 Craniectomy: Size and Location

The animal's head is mounted and secured in a stereotaxic frame (Kopf Instruments, Tujunga, CA, USA) with the head level at the interaural axis. The lateral fluid percussion injury involves creating a right-side craniectomy to access the dura overlying the brain parenchyma by removal of a specific size bone flap from the parietal cortex. The bone flap will not be replaced after the injury. The authors perform a 4.8 mm diameter craniectomy. However, other variations in the size of the craniectomy are reported in the literature [55–57]. For adult rats, the distance from bregma to lambda is 9 mm. Bregma is the intersection of the sagittal and coronal sutures. Lambda is the invisible point where the lambdoid and the sagittal suture meet. The animal's head should be aligned to a skull flat position so that bregma and lambda are in the same horizontal plane (Fig. 6). Adjust the incisor bar on the stereotaxic instrument so that the dorsal–ventral readings on the Vernier scale are similar between the bregma and lambda landmarks. Studies using animals of the same age do not require re-leveling of the incisor bar between animals and experiments. Placement of the center of the craniectomy is midpoint between bregma and lambda (–4.5 mm A-P) and midpoint between the sagittal suture and the lateral ridge (+3.0 mm M-L) (Fig. 7). The craniectomy's position should be exact to reduce experimental variability between animals. This is especially important in TBI behavioral and pathological studies because small shifts can affect behavioral outcome and differential lesion development [58, 59].

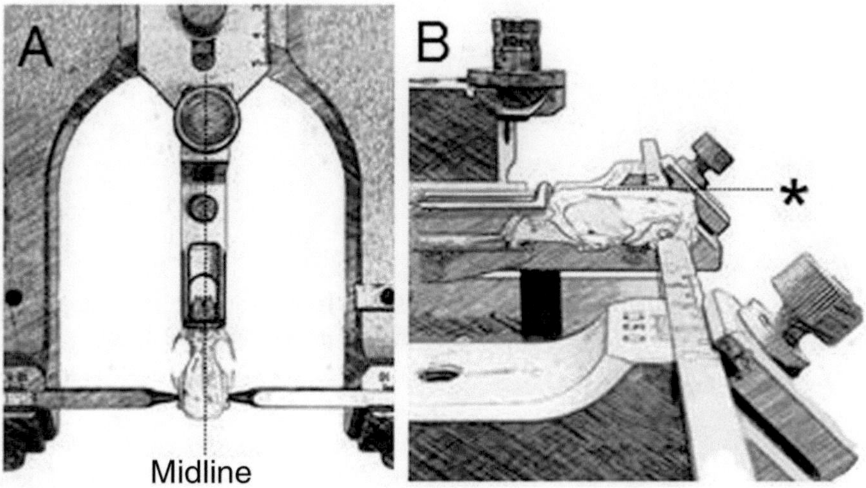

Fig. 6 Stereotaxic headholder (Kopf instruments) with a rat's skull (Sprague–Dawley) immobilized by the ear bars (**a**). The incisor bar is adjusted so that bregma and lambda lie in the same horizontal plane (*) for the accurate positioning of the craniectomy (**b**)

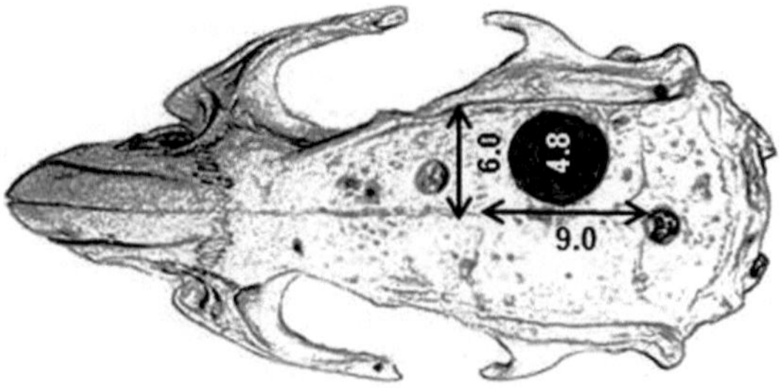

Fig. 7 Stereotaxic coordinates for the craniectomy and anchor screws. The center of the 4.8 mm craniectomy is positioned at 4.5 mm A-P and 3.0 mm M-L from bregma. Burr holes are made approximately 2 mm anterior to bregma and 1 mm posterior to lambda. Dimensions are in millimeters

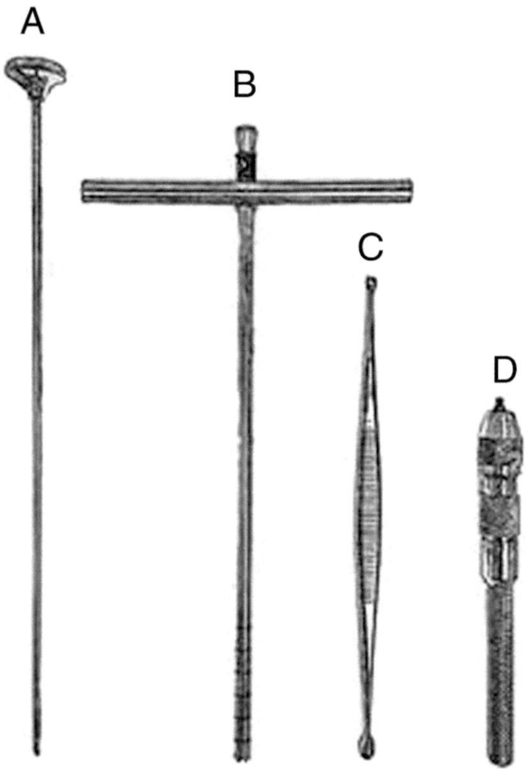

Fig. 8 Surgical instruments used for lateral fluid percussion. Trephine guide stylet (**a**), T-Trephine (**b**), bone curette (**c**), pin vise with drill bit (**d**)

3.6.4 Trephination

A craniectomy can be created by using an electric dental drill or a handheld T-style trephine (Miltex 26-140) (Fig. 8). The use of an electric drill may cause heat-induced cellular death in the cortex directly near the drilling site that is not due to the TBI impact [unpublished data]. Hence, the small T-style handheld trephine is the recommended drilling instrument (*see* **Note 6**). Unlike the

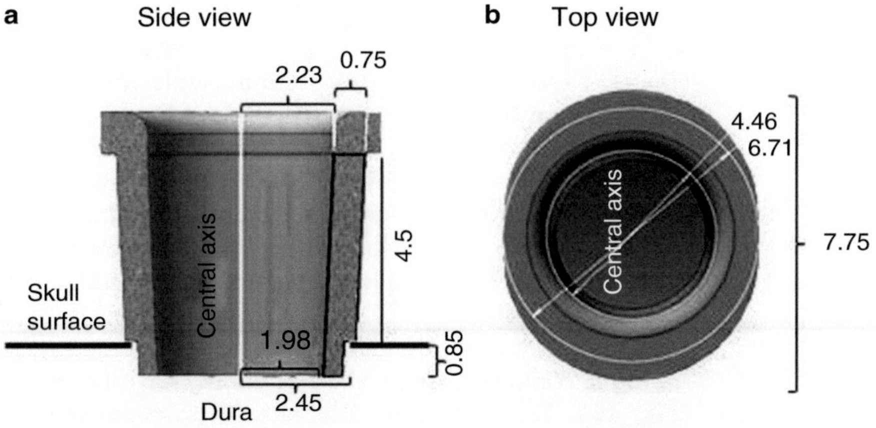

Fig. 9 The TBI injury conduit from a side view and top view rotated about the central axis. The injury conduit is made with the exact dimensions to fit the craniectomy and connector tube. The inside dimensions are the female Luer-lock taper. Dimensions are in millimeters

electric drill technique, the T-style handheld trephine can drill and generate a perfect circular craniectomy that is usually free of broken bone chips (*see* **Note 7**). The most common surgical mistake when drilling a craniectomy is the use of excessive force on the trephine. It is easy to mistakenly break through the dura and damage the brain. It is both necessary and important to use the surgical microscope to monitor the drilling process. Bleeding should not occur from underneath the craniectomy site since no major blood vessels are located in that area. Fine forceps can be used to gently tap on the bone flap for movement. The drilling should come to a halt once the skull bone becomes loosened. Carefully and gently slide the forceps underneath the skull bone flap to lift and remove it while at the same time avoiding any contact with the dura. A bone curette is used to remove any bone fragments. By keeping the dura intact, the brain tissues will not be in direct contact with the fluid bolus during TBI. Any animals showing signs of damage to the dura should be excluded from the study.

3.6.5 Anchor Screws

The components of a TBI acrylic assembly consist of two anchoring screws, an injury conduit and a hardened dental cement, to secure the conduit to the craniotomy. The placement of the anchor screws (round-head machine 2-56, 2.1 mm diameter, 6.0 mm length) is essential to prevent the dislodging of the acrylic assembly during the impact. Two small burr holes for the anchor screws are positioned away from the site of impact approximately 2 mm rostral to Bregma and 1 mm caudal to Lambda. Gently drill at an angle using drill bit (size 47) in a pin vise tool to create a burr hole that is shallow enough to allow the screws to catch without creating a deep penetration into the skull (*see* **Note 8**).

3.6.6 Injury Conduit and TBI Acrylic Assembly

The rigid injury conduit (Fig. 9) is made with a 3D printer from UV curing resin (Stratasys) to specific dimensions which allow the conduit to securely fit into the 4.8 mm craniectomy. An alternative option is to modify a Luer-Loc needle conduit (22 G) with similar dimensions (*see* **Note 9**). In preparation for TBI, the conduit is placed into the craniectomy over the exposed intact dura (*see* **Notes 10** and **11**). A perfectly secured conduit will tilt at an angle from the skull surface (Fig. 10). Cyanoacrylate adhesive acrylic (Plastics One, Roanoke, VA) is applied to secure the conduit and anchoring screws (*see* **Note 12**). The acrylic drying process usually takes 5 min depending on the dilution of the acrylic mixture. Once the drying is complete, attach the FPI connector tube to the conduit (Fig. 11). The connector tube should be completely filled with normal sterile saline to prevent any bubbles from forming in the connector tube. This volume of the sterile saline will also prevent the non-sterile fluid in the fluid percussion device cylinder from entering into the cranium during the TBI impact. The connector tube is attached to the fluid percussion device for delivery of the fluid pressure to the dura (Fig. 12).

3.6.7 Lateral Fluid Percussion TBI

Ensure that the percussion device is working properly and adjusted for the desired magnitude of injury before inducing the TBI. The isoflurane anesthesia is turned off just prior to injury. The animal is

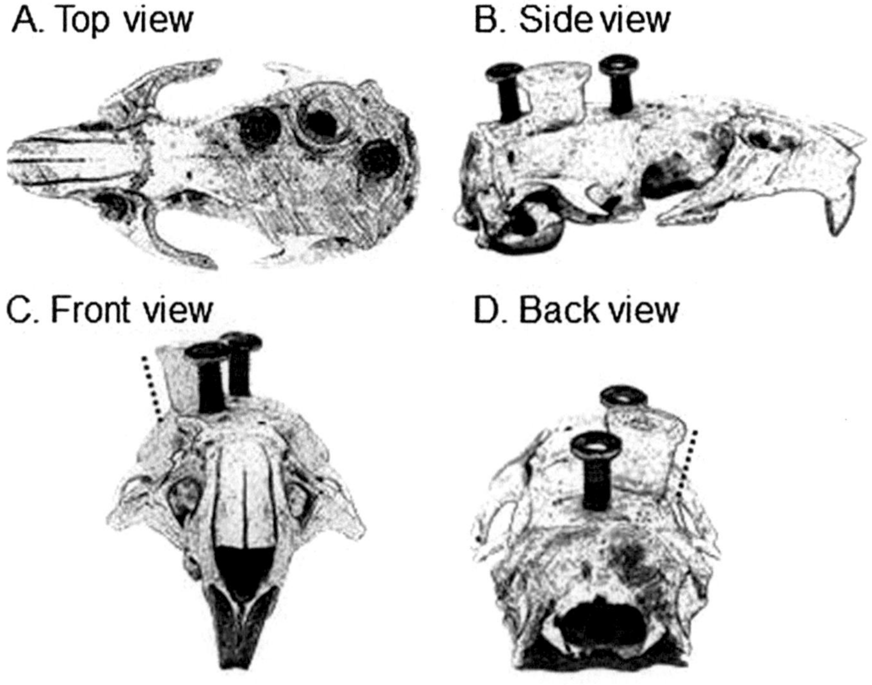

Fig. 10 Multiple views of the animal's skull with the anchor screws and injury conduit attached to the craniectomy. When properly fit, the injury conduit is slightly tilted to the side due to the natural slope of the skull surface (*dashed lines*)

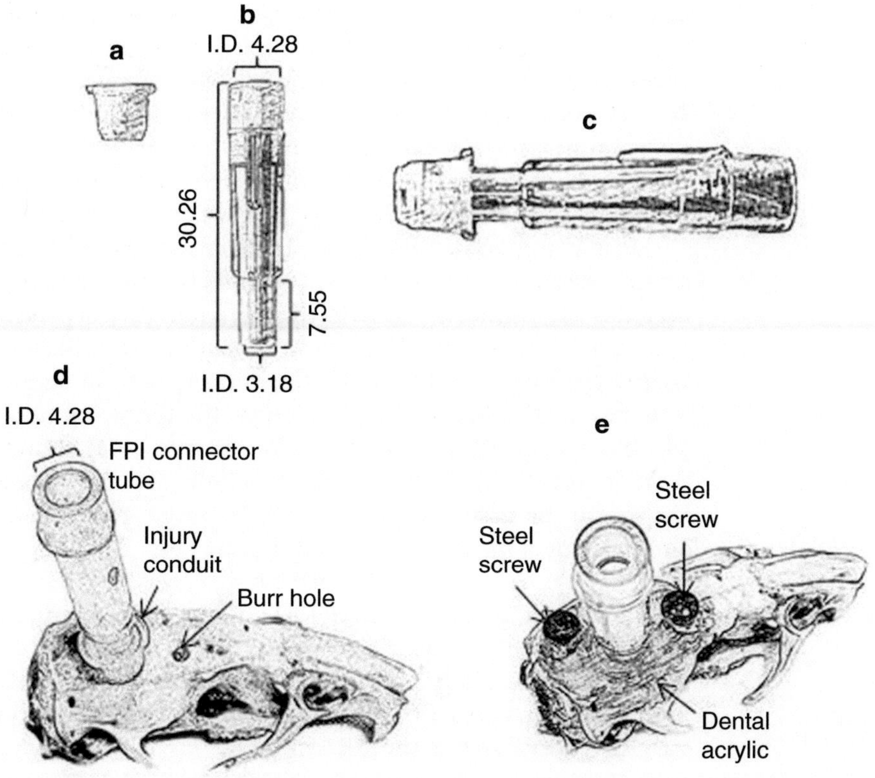

Fig. 11 The specific components of a TBI injury assembly. The TBI injury conduit (**a**) and the fluid percussion device connector tube (**b**) are connected together (**c**) and are attached to the craniectomy (**d**). The dental acrylic and anchor screws secured the injury conduit and connector tube (**e**). Dimensions are in millimeters

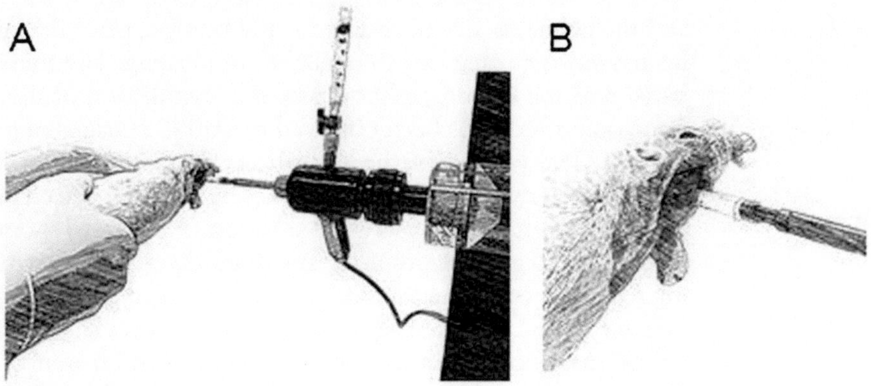

Fig. 12 Lateral orientation of the animal for delivery of a pressurized fluid pulse to the exposed intact dura. The animal is connected to the fluid percussion device by an injury connector tube that is prefilled with sterile saline

disconnected from the mechanical ventilator and immediately connected to the fluid percussion device. The LFP injury is a quick process that takes place in less than 10 s while the animal is still sedated. It is important to not introduce air bubbles into the FPI device connector tube to avoid dampening of the pressure pulse. A weakened pressure pulse will diminish the anticipated TBI pressure resulting in less injury to the brain.

It is better to have two people carry out the TBI stage of the experiment. The first experimenter holds the animal on its side and attaches the connector tube to the device. Injury is produced with the lateral orientation. The second experimenter releases the pendulum to induce the TBI. After the impact, one can determine that a successful TBI has occurred based on the following observations: (1) No fluid leaks from the acrylic assembly, (2) the column of fluid in the connector tube remains full, (3) the pressure pulse shown on the oscilloscope is within the target range and the pressure tracing is smooth with a single peak (Fig. 2).

3.6.8 Sham-TBI

When designing experimental TBI studies, it is essential to have a sham-TBI control group to properly assess the effects of surgical anesthesia or the craniectomy on the animals. Sham-TBI animals are not subjected to TBI induction but they do undergo the same surgical procedures as the TBI animals.

3.6.9 Postoperative Care

After inducing TBI, disconnect the injury tube from the device and promptly reconnect the animal to the mechanical ventilator. The animal is ventilated with a 2:1 nitrous oxide/oxygen mixture in the absence of isoflurane until its spontaneous breathing resumes. A spatula is used to remove the acrylic cap from the skull. Warm lactated Ringer's or normal saline is used to rinse and clean the incision. Perform the hindpaw withdrawal reflex pinch to ensure that the animal is unresponsive before suturing the scalp incision. Once the incision is closed and the animal is able to spontaneously breathe, place the animal in a heated recovery cage for postoperative monitoring. The animal is extubated and the righting reflex is assessed. Extubation of the animal is done only after it has been observed to exhibit no signs of respiratory distress. Gently withdraw the intubation tube from the animal to avoid introducing any trauma to its trachea. The animal is placed in a supine position at regular intervals (~20 s) to test its ability to spontaneously revert back to a prone position. The duration of the suppressed righting reflex is used as an additional indicator of injury severity. The experimenter should observe the animal until it becomes ambulatory. If the animal does not regain its consciousness within 30 min of the TBI induction, it should be euthanized. This is typically an indication of hemorrhage around the brainstem. Any residual effects of isoflurane anesthesia typically do not alter consciousness for longer than 2 min. TBI animals remain unconscious for 10–15 min due to injured brain functions rather than anesthetic effects.

3.7 TBI: Complications, Morbidity, and Mortality

The lateral fluid percussion model reproduces morbidity and mortality that are often comparable to those of humans. Similar to clinical patients, TBI-injured animals also exhibited brain injury hallmarks such as cerebral contusions, subdural hematomas, and intracranial hemorrhages [60]. The complications and mortality associated with experimental TBI are usually indicative of the injury severity. Common complications included prolonged apnea, respiratory distress [61, 62], acute pulmonary edema [63, 64], and dysautonomia [65] (decorticate and decerebrate postures). Mortality after moderate to severe TBI usually occurs within the first 30 min after injury and is associated with excessive intracranial hemorrhage. The authors have observed that mild, moderate, and severe LFP injuries produce a mortality rate of approximately 0, 25 %, and greater than 50 %, respectively.

4 Notes

1. Removal of air bubbles from the fluid percussion device: Filling the apparatus with degassed water or saline will reduce the formation of air bubbles. Degassing can be achieved prior to filling the apparatus by applying a vacuum to the distilled water in an Erlenmeyer flask (using a magnetic stir bar aids in the process).
2. An endotracheal tube can be constructed with PE 200 polyethylene tube (I.D. 1.40 mm; O.D. 1.9 mm).
3. A soft material guide stylet should be utilized to minimize trauma during endotracheal intubation. The polyethylene guide stylet (PE50) is inserted inside and approximately 4 mm past the distal end of the endotracheal tube.
4. Since direct visualization of the tracheal intubation is not possible in the animal, the correct placement of the tracheal tube position is confirmed by observing the exhaled breath on a smooth black surface (e.g., lab bench).
5. *See* Fig. 4b for the different parts used to connect the anesthesia Y-tube to the endotracheal tube. A short piece of Silastic tube is attached to the Y-tube and to the endotracheal tube. The Silastic tubing can be connected directly to the PE 200 endotracheal tube. If using the IV over-the-needle endotracheal tube, the Silastic tubing needs to be secured to a hard plastic connector with a male Luer outlet for insertion into the IV over-the-needle endotracheal tube. It is important to perform routine internal cleaning of the anesthesia Y-tube unit and endotracheal tubing. A clogged tube will restrict an adequate amount of anesthesia and oxygen to the animal. If not closely monitored, the animal will become hypoxic and possibly die during surgery.

6. Producing a sharp point on the tip of the trephine stylet with a file will prevent the trephine from "walking" from the precise skull location during the trephining process.
7. Normal saline or lactated Ringer's solution at room temperature can be used to irrigate the skull to facilitate the drilling process. The trephine can penetrate moistened bone more freely and provide a cleaner cut.
8. The placement of the anchor screws into the burr holes should be done prior to removing the craniectomy bone flap. The intact bone flap keeps the brain protected from accidental trauma from a slip of the surgical screwdriver into the craniectomy during the placement of the screws.
9. The main advantage of producing a conduit with a 3D printer is a superior fit without the uneven edges that can potentially breach the dura and compromise the experiment.
10. A 1 mL syringe can be temporarily attached to the conduit to help guide and facilitate the steady placement of the conduit into the craniectomy. Once the conduit is locked into the craniectomy, gently rotate the syringe in a counterclockwise motion to detach it from the hub. It is difficult to handle and properly guide the small size conduit into the craniectomy with the use of the fingers.
11. Use the wooden end of a cotton-tipped swab to spread a very small amount of superglue around the conduit's stepped-shoulder bottom to ensure a sealed, tighter fit in the craniectomy. It is important to use the glue sparingly to prevent it from contacting the dura surface. Dried glue on top of the dura will form a thick barrier, impeding and diminishing the fluid pulse pressure.
12. It is recommended to use a 1 mL syringe to slowly inject a small amount of the acrylic adhesive to create the acrylic assembly. The acrylic-filled syringe allows the experimenter to have control over the flow rate of the injection, preventing any accidental excessive overflowing. Avoid excessive thickness of the acrylic adhesive since it is exothermic upon curing and could cause a thermal lesion to the brain. Prior to curing, separate the acrylic adhesive from the skin while the acrylic is still soft.

References

1. Thompson HJ, Lifshitz J, Marklund N, Grady MS, Graham DI, Hovda DA, McIntosh TK (2005) Lateral fluid percussion brain injury: a 15-year review and evaluation. J Neurotrauma 22:42–75
2. Alder J, Fujioka W, Lifshitz J, Crockett DP, Thakker-Varia S (2011) Lateral fluid percussion: model of traumatic brain injury in mice. J Vis Exp (54)
3. Carbonell WS, Maris DO, McCall T, Grady MS (1998) Adaptation of the fluid percussion injury model to the mouse. J Neurotrauma 15:217–229
4. Povlishock JT, Becker DP, Cheng CL, Vaughan GW (1983) Axonal change in minor head injury. J Neuropathol Exp Neurol 42:225–242
5. Hartl R, Medary M, Ruge M, Arfors KE, Ghajar J (1997) Blood-brain barrier breakdown

occurs early after traumatic brain injury and is not related to white blood cell adherence. Acta Neurochir Suppl 70:240–242

6. Millen JE, Glauser FL, Fairman RP (1985) A comparison of physiological responses to percussive brain trauma in dogs and sheep. J Neurosurg 62:587–591
7. Pfenninger EG, Reith A, Breitig D, Grunert A, Ahnefeld FW (1989) Early changes of intracranial pressure, perfusion pressure, and blood flow after acute head injury. Part 1: An experimental study of the underlying pathophysiology. J Neurosurg 70:774–779
8. Denny-Brown D, Russell WR (1940) Experimental cerebral concussion. J Physiol 99:153
9. Gurdjian ES, Lissner HR, Webster JE, Latimer FR, Haddad BF (1954) Studies on experimental concussion: relation of physiologic effect to time duration of intracranial pressure increase at impact. Neurology 4:674–681
10. Lindgren S, Rinder L (1966) Experimental studies in head injury. II. Pressure propagation in "percussion concussion". Biophysik 3:174–180
11. Metz B (1971) Acetylcholine and experimental brain injury. J Neurosurg 35:523–528
12. Stalhammar D (1975) Experimental brain damage from fluid pressures due to impact acceleration. 1. Design of experimental procedure. Acta Neurol Scand 52:7–26
13. Stalhammar D (1975) Experimental brain damage from fluid pressures due to impact acceleration. 2. Pathophysiological observations. Acta Neurol Scand 52:27–37
14. Stalhammar D, Olsson Y (1975) Experimental brain damage from fluid pressures due to impact acceleration. 3. Morphological observations. Acta Neurol Scand 52:38–55
15. Sullivan HG, Martinez J, Becker DP, Miller JD, Griffith R, Wist AO (1976) Fluid-percussion model of mechanical brain injury in the cat. J Neurosurg 45:521–534
16. Dixon CE, Lyeth BG, Povlishock JT, Findling RL, Hamm RJ, Marmarou A, Young HF, Hayes RL (1987) A fluid percussion model of experimental brain injury in the rat. J Neurosurg 67:110–119
17. McIntosh TK, Vink R, Noble L, Yamakami I, Fernyak S, Soares H, Faden AL (1989) Traumatic brain injury in the rat: characterization of a lateral fluid-percussion model. Neuroscience 28:233–244
18. Statler KD, Alexander H, Vagni V, Holubkov R, Dixon CE, Clark RS, Jenkins L, Kochanek PM (2006) Isoflurane exerts neuroprotective actions at or near the time of severe traumatic brain injury. Brain Res 1076:216–224
19. Kawaguchi M, Furuya H, Patel PM (2005) Neuroprotective effects of anesthetic agents. J Anesth 19:150–156
20. Koerner IP, Brambrink AM (2006) Brain protection by anesthetic agents. Curr Opin Anaesthesiol 19:481–486
21. Matchett GA, Allard MW, Martin RD, Zhang JH (2009) Neuroprotective effect of volatile anesthetic agents: molecular mechanisms. Neurol Res 31:128–134
22. Statler KD, Alexander H, Vagni V, Dixon CE, Clark RS, Jenkins L, Kochanek PM (2006) Comparison of seven anesthetic agents on outcome after experimental traumatic brain injury in adult, male rats. J Neurotrauma 23:97–108
23. Statler KD, Kochanek PM, Dixon CE, Alexander HL, Warner DS, Clark RS, Wisniewski SR, Graham SH, Jenkins LW, Marion DW, Safar PJ (2000) Isoflurane improves long-term neurologic outcome versus fentanyl after traumatic brain injury in rats. J Neurotrauma 17:1179–1189
24. Udomphorn Y, Armstead WM, Vavilala MS (2008) Cerebral blood flow and autoregulation after pediatric traumatic brain injury. Pediatr Neurol 38:225–234
25. Bouma GJ, Muizelaar JP (1992) Cerebral blood flow, cerebral blood volume, and cerebrovascular reactivity after severe head injury. J Neurotrauma 9(Suppl 1):S333–S348
26. Bouma GJ, Muizelaar JP (1995) Cerebral blood flow in severe clinical head injury. New Horiz 3:384–394
27. Brown JI, Moulton RJ, Konasiewicz SJ, Baker AJ (1998) Cerebral oxidative metabolism and evoked potential deterioration after severe brain injury: new evidence of early posttraumatic ischemia. Neurosurgery 42:1057–1063, discussion 1063-1054
28. Philip S, Udomphorn Y, Kirkham FJ, Vavilala MS (2009) Cerebrovascular pathophysiology in pediatric traumatic brain injury. J Trauma 67:S128–S134
29. Christensen MS, Hoedt-Rasmussen K, Lassen NA (1967) Cerebral vasodilatation by halothane anaesthesia in man and its potentiation by hypotension and hypercapnia. Br J Anaesth 39:927–934
30. Bain JA, Catton DV, Cox JM, Spoerel WE (1967) The effect of general anaesthesia on the tolerance of cerebral ischaemia in rabbits. Can Anaesth Soc J 14:69–78
31. Jennett WB, Barker J, Fitch W, McDowall DG (1969) Effect of anaesthesia on intracranial pressure in patients with space-occupying lesions. Lancet 1:61–64

32. McDowall DG, Barker J, Jennett WB (1966) Cerebro-spinal fluid pressure measurements during anaesthesia. Anaesthesia 21:189–201
33. Smith AL, Wollman H (1972) Cerebral blood flow and metabolism: effects of anesthetic drugs and techniques. Anesthesiology 36:378–400
34. Rivard AL, Simura KJ, Mohammed S, Magembe AJ, Pearson HM, Hallman MR, Barnett SJ, Gatlin DL, Gallegos RP, Bianco RW (2006) Rat intubation and ventilation for surgical research. J Invest Surg 19:267–274
35. Chesnut RM, Marshall LF, Klauber MR, Blunt BA, Baldwin N, Eisenberg HM, Jane JA, Marmarou A, Foulkes MA (1993) The role of secondary brain injury in determining outcome from severe head injury. J Trauma 34:216–222
36. Bramlett HM, Dietrich WD, Green EJ (1999) Secondary hypoxia following moderate fluid percussion brain injury in rats exacerbates sensorimotor and cognitive deficits. J Neurotrauma 16:1035–1047
37. Bramlett HM, Green EJ, Dietrich WD (1999) Exacerbation of cortical and hippocampal CA1 damage due to posttraumatic hypoxia following moderate fluid-percussion brain injury in rats. J Neurosurg 91:653–659
38. Feng JF, Zhao X, Gurkoff GG, Van KC, Shahlaie K, Lyeth BG (2012) Post-traumatic hypoxia exacerbates neuronal cell death in the hippocampus. J Neurotrauma 29:1167–1179
39. Werner C, Engelhard K (2007) Pathophysiology of traumatic brain injury. Br J Anaesth 99:4–9
40. Costa DL, Lehmann JR, Harold WM, Drew RT (1986) Transoral tracheal intubation of rodents using a fiberoptic laryngoscope. Lab Anim Sci 36:256–261
41. Thet LA (1983) A simple method of intubating rats under direct vision. Lab Anim Sci 33:368–369
42. Mrozek S, Vardon F, Geeraerts T (2012) Brain temperature: physiology and pathophysiology after brain injury. Anesthesiol Res Pract 2012:989487
43. Jiang JY, Lyeth BG, Clifton GL, Jenkins LW, Hamm RJ, Hayes RL (1991) Relationship between body and brain temperature in traumatically brain-injured rodents. J Neurosurg 74:492–496
44. Childs C (2008) Human brain temperature: regulation, measurement and relationship with cerebral trauma: part 1. Br J Neurosurg 22:486–496
45. Sacho RH, Childs C (2008) The significance of altered temperature after traumatic brain injury: an analysis of investigations in experimental and human studies: part 2. Br J Neurosurg 22:497–507
46. Thompson HJ, Tkacs NC, Saatman KE, Raghupathi R, McIntosh TK (2003) Hyperthermia following traumatic brain injury: a critical evaluation. Neurobiol Dis 12:163–173
47. Marion DW, Penrod LE, Kelsey SF, Obrist WD, Kochanek PM, Palmer AM, Wisniewski SR, DeKosky ST (1997) Treatment of traumatic brain injury with moderate hypothermia. N Engl J Med 336:540–546
48. Clifton GL, Allen S, Barrodale P, Plenger P, Berry J, Koch S, Fletcher J, Hayes RL, Choi SC (1993) A phase II study of moderate hypothermia in severe brain injury. J Neurotrauma 10:263–271, discussion 273
49. Marion DW, Obrist WD, Carlier PM, Penrod LE, Darby JM (1993) The use of moderate therapeutic hypothermia for patients with severe head injuries: a preliminary report. J Neurosurg 79:354–362
50. Dietrich WD, Alonso O, Busto R, Globus MY, Ginsberg MD (1994) Post-traumatic brain hypothermia reduces histopathological damage following concussive brain injury in the rat. Acta Neuropathol 87:250–258
51. Clifton GL, Jiang JY, Lyeth BG, Jenkins LW, Hamm RJ, Hayes RL (1991) Marked protection by moderate hypothermia after experimental traumatic brain injury. J Cereb Blood Flow Metab 11:114–121
52. Jiang JY, Lyeth BG, Kapasi MZ, Jenkins LW, Povlishock JT (1992) Moderate hypothermia reduces blood-brain barrier disruption following traumatic brain injury in the rat. Acta Neuropathol 84:495–500
53. Lyeth BG, Jiang JY, Liu S (1993) Behavioral protection by moderate hypothermia initiated after experimental traumatic brain injury. J Neurotrauma 10:57–64
54. Hasegawa Y, Latour LL, Sotak CH, Dardzinski BJ, Fisher M (1994) Temperature dependent change of apparent diffusion coefficient of water in normal and ischemic brain of rats. J Cereb Blood Flow Metab 14:383–390
55. Rink A, Fung KM, Trojanowski JQ, Lee VM, Neugebauer E, McIntosh TK (1995) Evidence of apoptotic cell death after experimental traumatic brain injury in the rat. Am J Pathol 147:1575–1583
56. Hicks RR, Zhang L, Atkinson A, Stevenon M, Veneracion M, Seroogy KB (2002) Environmental enrichment attenuates cognitive deficits, but does not alter neurotrophin gene expression in the hippocampus following lateral fluid percussion brain injury. Neuroscience 112:631–637
57. Sinson G, Voddi M, McIntosh TK (1995) Nerve growth factor administration attenuates cognitive but not neurobehavioral motor dysfunction or

hippocampal cell loss following fluid-percussion brain injury in rats. J Neurochem 65:2209–2216
58. Vink R, Mullins PG, Temple MD, Bao W, Faden AI (2001) Small shifts in craniotomy position in the lateral fluid percussion injury model are associated with differential lesion development. J Neurotrauma 18:839–847
59. Floyd CL, Golden KM, Black RT, Hamm RJ, Lyeth BG (2002) Craniectomy position affects morris water maze performance and hippocampal cell loss after parasagittal fluid percussion. J Neurotrauma 19:303–316
60. Maas AI, Stocchetti N, Bullock R (2008) Moderate and severe traumatic brain injury in adults. Lancet Neurol 7:728–741
61. Holland MC, Mackersie RC, Morabito D, Campbell AR, Kivett VA, Patel R, Erickson VR, Pittet JF (2003) The development of acute lung injury is associated with worse neurologic outcome in patients with severe traumatic brain injury. J Trauma 55:106–111
62. Rincon F, Ghosh S, Dey S, Maltenfort M, Vibbert M, Urtecho J, McBride W, Moussouttas M, Bell R, Ratliff JK, Jallo J (2012) Impact of acute lung injury and acute respiratory distress syndrome after traumatic brain injury in the United States. Neurosurgery 71:795–803
63. Bahloul M, Chaari AN, Kallel H, Khabir A, Ayadi A, Charfeddine H, Hergafi L, Chaari AD, Chelly HE, Ben Hamida C, Rekik N, Bouaziz M (2006) Neurogenic pulmonary edema due to traumatic brain injury: evidence of cardiac dysfunction. Am J Crit Care 15:462–470
64. Baumann A, Audibert G, McDonnell J, Mertes PM (2007) Neurogenic pulmonary edema. Acta Anaesthesiol Scand 51:447–455
65. Hendricks HT, Heeren AH, Vos PE (2010) Dysautonomia after severe traumatic brain injury. Eur J Neurol 17:1172–1177

Chapter 15

Impact Acceleration Model of Diffuse Traumatic Brain Injury

Sarah C. Hellewell*, Jenna M. Ziebell*, Jonathan Lifshitz, and M. Cristina Morganti-Kossmann

Abstract

The impact acceleration (I/A) model of traumatic brain injury (TBI) was developed to reliably induce diffuse traumatic axonal injury in rats in the absence of skull fractures and parenchymal focal lesions. This model replicates a pathophysiology that is commonly observed in humans with diffuse axonal injury (DAI) caused by acceleration–deceleration forces. Such injuries are typical consequences of motor vehicle accidents and falls, which do not necessarily require a direct impact to the closed skull. There are several desirable characteristics of the I/A model, including the extensive axonal injury produced in the absence of a focal contusion, the suitability for secondary insult modeling, and the adaptability for mild/moderate injury through alteration of height and/or weight. Furthermore, the trauma device is inexpensive and readily manufactured in any laboratory, and the induction of injury is rapid (~45 min per animal from weighing to post-injury recovery) allowing multiple animal experiments per day. In this chapter, we describe in detail the methodology and materials required to produce the rat model of I/A in the laboratory. We also review current adaptations to the model to alter injury severity, discuss frequent complications and technical issues encountered using this model, and provide recommendations to ensure technically sound injury induction.

Key words Diffuse traumatic brain injury, Traumatic axonal injury, Neuronal pathology, Rat model

1 Introduction

The rodent I/A model was designed by Marmarou and co-workers in 1994 to replicate the characteristics of DAI, a pathology commonly observed in humans following diffuse brain trauma [1]. The I/A model (also commonly known as the Marmarou model) has been adopted by laboratories around the world, generating more than 150 publications in just over 20 years. Together with the midline fluid percussion injury model, it remains one of the most validated experimental paradigms to study the diffuse

*Authors contributed equally in this chapter

Firas Kobeissy et al. (eds.), *Injury Models of the Central Nervous System: Methods and Protocols,* Methods in Molecular Biology, vol. 1462, DOI 10.1007/978-1-4939-3816-2_15, © Springer Science+Business Media New York 2016

traumatic axonal pathology (TAI, the experimental counterpart of human DAI) intrinsic to diffuse brain trauma. The I/A model is also one of the few models of TBI resulting in post-traumatic apnea in rodents, thus faithfully reproducing an immediate post-traumatic loss of consciousness of TBI. The model is scalable to produce multiple injury severities; it is suited to the addition of secondary insults (e.g. hypoxia); and compatible with a number of quantifiable behavioral tests ranging from motor, sensorimotor, and cognitive/memory function to determine the extent of initial neurological deficits, ongoing recovery post-trauma, and therapeutic efficacy on functional outcome.

The I/A model is produced using a weight-drop device consisting of a cylindrical brass weight of 250–450 g that falls through a Plexiglas tube from 1 to 2 m in height. The falling weight impacts a stainless steel disk that is fixed to the rat skull prior to trauma with dental acrylic, while the animal lies prone on a foam cushion of known density (14 kg/m^3). The steel disk prevents the skull from fracturing by distributing the impact force uniformly across the skull and into the brain. Following the impact of the weight, the head of the animal is accelerated into the foam pad, and then decelerated, mimicking the forces transferred to TBI patients during motor vehicle accidents (e.g. whiplash injury). When using the highest severity combination of a 450 g weight from 2 m, a velocity of 6.06 m/s is achieved, resulting in a brain acceleration of 900 G and a brain compression gradient of 0.28 mm [1]. This force translates into an immediate suppression of neurological reflexes and apnea [2], as well as widespread axonal injury with a specific target on the corpus callosum and brain stem. The severity of injury can be reduced by varying the falling weight (<450 g, commonly 250 g) or the fall height (1, 1.5 m).

The I/A model has been particularly useful in elucidating the pathology underlying DAI [3], and to characterize other cellular and molecular mechanisms leading to secondary brain damage, including BBB dysfunction, edema, inflammation, glial activation, blood leukocyte infiltration, and neuronal degeneration [4–9]. Furthermore, this model is useful to study the impact of secondary post-traumatic hypoxia in exacerbating neurological deficits caused by TAI [4, 10]. This combined insult model is of particular relevance in the clinical setting since post-traumatic hypoxia is known to double mortality and worsen outcomes after severe TBI [11].

Behavioral impairment has been widely reported after TAI in the rat, specifically in relation to neurological reflexes, beam walk, beam balance, inclined plane, forelimb placement, and the Rotarod [4, 12, 13]. These deficits have been shown to persist for at least 4 weeks post-trauma [4, 14], with the Rotarod being the most sensitive amongst the multiple motor assessments. Long-lasting cognitive changes and memory loss have also been reported, primarily using the Morris Water Maze [14].

2 Materials

2.1 Animals

This model was established using adult male Sprague–Dawley rats, with a weight range of 350–400 g. Rats of this strain and weight have provided consistent replication of outcome measures (*see* **Note 1**).

2.2 Trauma device

The components of the I/A device are described briefly below. For full description of parts and construction, refer to the original manuscript describing the model [1]. The I/A device consists of:

1. Wood/metal base (77 × 60 × 2, width × length × height in cm).
2. Wood/metal stand to support the tube (4.5 × 4.5 × 200, $W \times L \times H$ in cm).
3. Plexiglas tube (height of 220 cm, diameter of 2.5 cm) (*see* **Note 2**).
4. Small U-shaped metal brace to provide external stability for the tube (2 × 2.5 × 230 cm).
5. Roll of 100% cotton twill tape, 12 mm wide (Medline, Mountainside Medical Equipment, Marcy, NY; *see* **Note 3**).
6. Cylindrical brass column of desired weight (for 450 g: height of 25 cm, diameter of 1.59 cm; *see* **Note 4**).

2.3 Equipment

2.3.1 Anesthesia

Endotracheal intubation and mechanical ventilation are required to adequately anesthetize the rat. Control of ventilation is also advantageous where investigators wish to superimpose a secondary insult such as hypoxia (*see* **Notes 5–7**). For this setup, you will require:

1. Small animal ventilator with pressure and volume control (Inspira ASVV, Harvard Apparatus, Holliston, MA).
2. Isoflurane Anesthesia Machine with oxygen and nitrogen inputs capable of gas mixture (e.g. CDS 9000 small animal anesthesia machine, Smiths Medical, Dublin, OH).
3. Oxygen (piped, or in a cylinder).
4. Nitrogen (piped, or in a cylinder).
5. Gas regulators (if using cylinders).
6. 4× lots of corrugated heavy-duty plastic tubing with collared ends, cut into lengths of: 64 cm, 2 × 58 cm, and 15 cm (222 mm diameter, Implox Healthcare, Adelaide, SA, Australia).
7. 1 large "Y" piece connector, plastic or metal (Implox Healthcare).
8. 1 small barbed "Y" piece connector, plastic (Implox Healthcare).
9. 1 male luer lock adapter (B. Braun, Bethlehem, PA).
10. 4× lots of clear PVC tubing cut into lengths of 60 cm and 3 × 40 cm (4 mm internal diameter, ANPROS, Bayswater, VIC, Australia).
11. Surgical tape, to secure tubing in place.

12. Induction chamber.
13. Laryngoscope handle and blade (e.g. Harvard Apparatus, #586592, #596774).
14. Intubation rack.
15. 16 G IV catheter for intubation, needle cut down approximately 2 cm and blunted (Introcan Safety® catheter, B. Braun, Bethlehem, PA).
16. Yarn or suture, cut into 20 cm lengths to secure endotracheal catheter to snout.

2.3.2 Surgical Supplies and Instruments

1. Clinical record sheets and surgery data sheets.
2. Two heat pads (one for placement under the animal during surgery, the other for the recovery box) (e.g. Harvard Apparatus, #340925).
3. Bench pads.
4. Acetone (for glue removal).
5. 70 % alcohol.
6. Local anesthetic (e.g. Lignocaine or Lidocaine).
7. Povidone-iodine solution.
8. Dental cement or Superglue.
9. Cotton-tipped applicators.
10. Gauze.
11. Small animal clippers (e.g. Wahl cordless pocket pro, #9961-2801, Sterling, IL).
12. Scalpel blade and holder.
13. Hemostats (e.g. Aesculap, #MB893R, Center Valley, PA).
14. Forceps (e.g. Aesculap #MB780R).
15. Surgical scissors (e.g. Aesculap, #MB925R).
16. Staples and stapler applicator or suture.
17. Helmet: stainless steel disk, 10 mm in diameter and 2 mm in thickness (scored on one side if adhering using superglue; *see* **Notes 8** and **9**).
18. Tape/hook and loop fastener.
19. Foam (type E bed foam, density of 14 kg/m^3 and stiffness of 2500 N/m, measuring 13 × 43 × 11.5 cm, Foam to Size, Ashland, VA; *see* **Notes 10** and **11**).
20. Wooden or Plexiglass box, with the dimensions 14 × 44 × 11.5 cm ($W \times L \times H$).
21. Step ladder (If unable to reach 2 m).
22. Injury device (as described above).
23. Warmed clean cage for animals to recover from surgery (recovery box).

24. Sodium pentobarbital-based anesthetic (e.g. Lethabarb or Euthosolv) for animal euthanasia in the event of skull fracture.

3 Methods

Firstly, the rat should be weighed and inspected for any signs of sickness. If healthy and within the weight range, the rat can be included in the study.

3.1 Experimental Set Up

1. Place the trauma device on a level floor close to where the surgical procedures will be undertaken, and put a stepladder on the side closest to the weight.
2. Position anesthesia machine at the end of the surgical table (Fig. 1a). Connect oxygen and nitrogen sources to anesthetic device.
3. Prepare tubing: One 64 cm corrugated tube with collared ends is required to deliver the anesthesia/gas mixture between the intubation chamber and the rat. Connect one end to the anesthesia machine, and attach the y-connector to the other (*see* Fig. 1a). Attach one of the 58 cm corrugated tubes and the 15 cm corrugated tube to each of the "y" arms of the y-connector.

   (a) Connect the 58 cm tube to the anesthetic induction chamber. To the smaller 15 cm tube, attach one 60 cm clear PVC tube using surgical tape, and connect this PVC tube to the ventilator "gas input" (Fig. 1d). Attach the remaining corrugated tube to the opposite end of the induction chamber to the input tube. This last corrugated tube is required for waste gas removal.
4. Turn on gas inputs, and balance O_2 and N_2 to 22%/78% respectively, to achieve a "room-air" ventilation.
5. Ensure the lid is closed on the induction chamber. Turn isoflurane to 5% and allow anesthetic gas to mix and fill the chamber.

   (a) Check for leaks in tubing, which will be evident by strong odor and whistling noises. Turn isoflurane off until ready to commence anesthesia.
6. Attach two of the 40 cm clear PVC tubes to the rodent ventilator to maintain anesthesia and gas delivery. These tubes are required for gas input and exhaust—connect them to the ventilator's input/output attachments (Fig. 1b, d). Join the end of each PVC tube not attached to the ventilator to the "y" arms of the small barbed y connector. Attach the opposite end of the y connector to the luer lock adapter, with the male hub facing outwards. This will connect to the female catheter port once the rat is intubated.

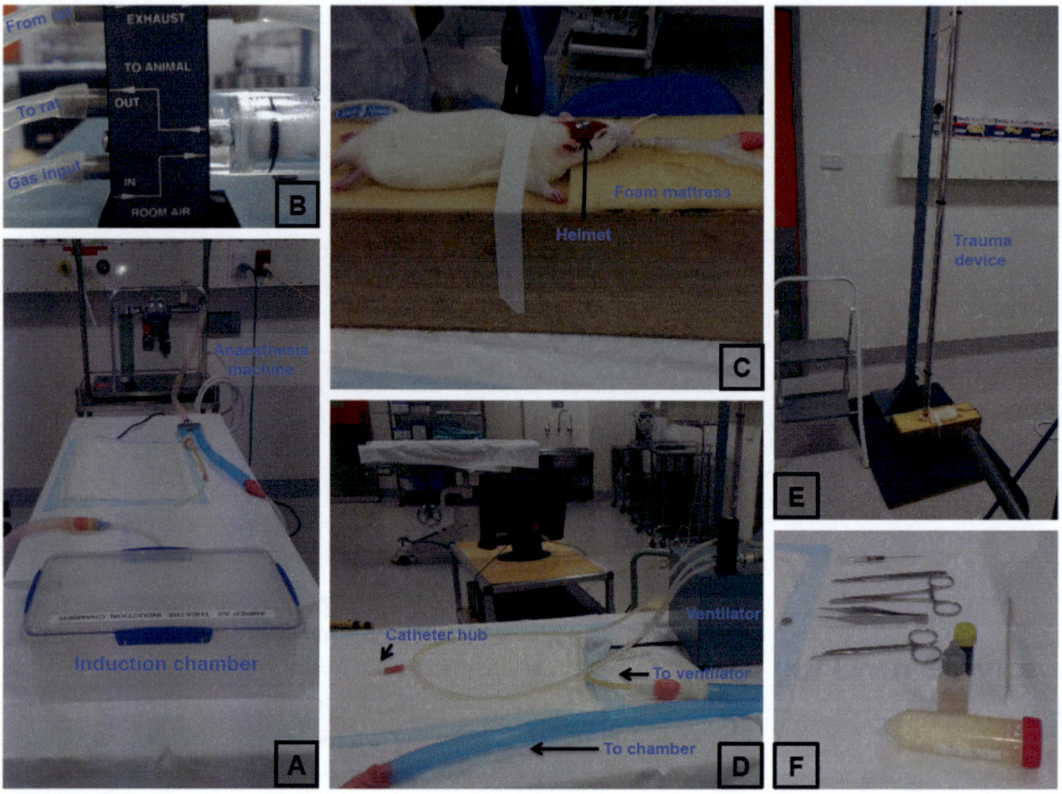

Fig. 1 Experimental set-up for impact acceleration traumatic axonal injury. (**a**) Set up anesthetic machine close to the surgical table, with tubing arranged as described to provide anesthesia to the induction chamber and ventilator. (**b**) Arrange the ventilator tubes so that anesthesia and gas are both passing through the ventilator to the rat. (**c**) The helmet should sit flat on the skull, and the rat should be placed prone on the foam mattress. Keep the male luer lock adapter hub from the ventilator (**d**) attached to the female endotracheal tube port until ready to induce trauma. (**e**) One researcher releases the weight after climbing the stepladder to reach it, while the other positions the rat beneath the device. (**f**) Surgical equipment should be kept close at hand

7. Adjust the settings on the ventilator to 80 breaths/min and 2.5 ml/breath.
8. Place one heatpad near the ventilator and cover with a bench pad. Place all surgical equipment, saline, Betadine, lignocaine, superglue, etc. close at hand (Fig. 1f).
9. Cut yarn into lengths of approximately 25 cm and tie tightly around catheter. These will be used to secure the catheter around the snout.

3.2 Anesthesia Induction

1. Place rat in the anesthesia induction chamber and leave animal for 5–7 min, or until deep anesthesia is achieved. This is observed by slow, regular breathing, decreased muscle tone ("floppy" posture), and absence of a pedal withdrawal reflex.
2. Once deep anesthesia has been achieved, remove rat and hang by teeth on intubation rack. Using forceps, move tongue to one

side and insert laryngoscope. Visualize the vocal cords and gently insert the 16 G catheter between vocal cords and into the trachea. If placed correctly, the catheter should feel like passing over small corrugations. Place the rat in a supine position and attach the male hub from the ventilator to the female port of the catheter to check whether endotracheal intubation has been achieved. This is confirmed with the rib cage moving in slow rhythmic breaths synchronized to the ventilator movement. If expansion and contraction are seen in the abdomen, or breathing is out of time with the ventilator, it is likely that the tube has been placed in the esophagus. If this has occurred, remove the catheter and place the rat back in the intubation chamber before attempting re-catheterization (*see* **Notes 12** and **13**).

3. Once endotracheal intubation has been achieved, tie the catheter firmly in place with the yarn, looping around the snout to ensure the catheter does not come loose or slip out.
4. Turn the isoflurane concentration down to 2–3 % to maintain the rat under anesthesia, and turn the rat carefully into a prone position, adjusting the endotracheal catheter as you do so to prevent twisting.

3.3 Surgical Preparation

1. Using clippers, shave the rat's head.
2. Place Betadine onto gauze and wipe shaved scalp.
3. Inject a small amount of local anesthetic into scalp, wait 3–5 min.
4. Check once again that pedal withdrawal reflex is absent. Using a scalpel, make a 1.5–2 cm incision in the scalp from a rostral to a caudal direction, starting above the level of the ears.
5. Expose the skull by finely cutting the overlying periosteum with the scalpel in a crosshatch pattern out to the most lateral edges of the incision.
6. Using gauze, apply gentle pressure to absorb any blood and dry the skull thoroughly.
7. If using dental acrylic: mix the powder with the liquid and work into a wet paste. Apply about half a teaspoon evenly to the skull, and then place the steel disk ("helmet") over the parietal bone midway between bregma and lambda, directly over the midline suture (*see* **Note 14**).
8. If using superglue: Apply the superglue to the underside of the helmet, and then place the disk on the skull midway between bregma and lambda, directly over the midline suture (Fig. 1c; *see* **Note 9**).
9. Allow the acrylic/glue to dry completely. Superglue will take approximately 3–5 min, while acrylic may take up to 10 min.
10. While waiting, prepare the foam mattress. It should be placed snugly inside the wooden or Plexiglas box, so that the head

support portion sits level with the top of the box, and the body support portion sits approximately 1 cm below the box. Have it close to the surgical bench so you are able to swiftly move the rat from bench to mattress (*see* **Note 11**).

11. One person should monitor the mattress, whilst another transports the rat, supporting the head and neck as well as the body to ensure the endotracheal catheter stays in place. Put the rat onto the mattress and place it prone on the foam bed, ensuring the head is positioned straight so the disk is flat (Fig. 1c).
12. Disconnect the rat from the ventilator and secure to the box using surgical or masking tape over the dorsal surface below the armpit, attaching it to either side of the box (Fig. 1c).
13. Rapidly move the rat and the mattress underneath the trauma device, and align the rat so the weight will hit directly onto the middle of the steel helmet. While one person maintains the animal in position, the other should be ready to release the weight, climbing the stepladder if necessary (Fig. 1e).
14. Induce injury by releasing the weight from the desired height through the Plexiglas tube. Ensure the rat is rapidly moved away from the device to avoid a "second-hit" of the rebounding weight.
15. Rapidly reconnect the rat to mechanical ventilation and turn off anesthesia, so that rat is ventilated only with supporting room air.
16. Remove helmet using hemostats and inspect for skull fracture, cleaning off any remaining blood with gauze (*see* **Note 15**).
17. Suture or staple the scalp incision closed.
18. Once showing signs of wakefulness (independent breathing, recovery of righting reflex), wean the rat from mechanical ventilation. The time may be variable, though most rats should be awake 10–20 min after the impact. Place the rat in a warmed recovery box on a heatpad for 60 min and monitor for recovery.

4 Discussion

The I/A model originally described by Marmarou and colleagues in 1994 [1, 2] reliably recreates the pathophysiology of human DAI, including diffuse white matter damage, neurological and cognitive deficits, neuroinflammation, and neuronal damage and dysfunction [3, 5, 7, 10, 15, 16].

This model is advantageous in that it requires minimal material and is easily set up, with a low mortality rate provided rats are adequately ventilated after injury. Typically, animal numbers as low as 5 per treatment group are sufficient to reach statistical significance, minimizing the number of animals to be used. The wide range of histological and biochemical outcomes, as well as long-term behavioral and cognitive consequences of the I/A model make it particularly suitable for pharmacological intervention studies. Research thus far has focused largely on attenuation of axonal pathology directly through mitigation of the cysteine protease calpain, which causes axonal proteolysis [17, 18] or attenuation of mitochondrial dysfunction [19, 20], or indirectly via multifunctional therapies such as erythropoietin, which prevents early deleterious signaling cascades and subsequent edema formation and behavioral dysfunction [21–23].

However, the I/A model has several drawbacks, including difficulty in translation to other species such as mice due to variations in skull size and thickness, and a failure of the model to translate to immature rats, where the graded cognitive dysfunction seen in adult rats over differing height/weight combinations is absent. A severe injury of 100 g/2 m is sufficient to cause diffuse edema and early motor deficits in postnatal day 17 rats, lasting 3–4 days in the absence of cognitive deficit [24], which is only observed when this injury is performed with the "ultra severe" combination of 150 g/2 m [25, 26]. Few studies have examined the effect of I/A injury in aged rats, however one important factor that has been reproduced is persistent cognitive dysfunction, with decreased capacity for recovery when compared to young adult rats [27].

To achieve success with this I/A model, researchers must pay careful attention to the fine procedural details, which will largely dictate whether standardized reproducible injuries are achieved. For example, incorrect or multiple attempts at placing the endotracheal catheter will cause airway swelling and mucous buildup, which will compromise airway recovery when animals are extubated. However, careful and correct placement will eliminate this consequence. Placement of the steel disk on the rat's skull is also of critical importance; if the disk is not adhered flat onto the skull, the impacting weight will strike the highest point, increasing the likelihood of unilateral damage and skull fracture. Cooperation between the researcher releasing the weight and the researcher below with the animal is also needed to reach agreement on the exact moment the weight is released, whereby the rat must be moved directly after the first hit of the weight and prior to the weight rebounding, while ensuring the rat is not moved prior to a full first hit of the weight. This timing of movement of the rat requires keen attention, and is likely the most important factor of success in the I/A model.

5 Notes

1. The body weight of the rat is a strong determinant of outcome, with larger animals having little apnea and high survival rates even when non-ventilated, as opposed to smaller animals that frequently experience long apnea, higher incidence of skull fracture, and greater mortality. Using rats heavier than indicated (400 g) will also reduce the likelihood of significant

axonal injury, and increase variability between rats, so it is important to maintain a consistent weight at trauma throughout the study group.

2. The Plexiglas tube ought to be higher than 200 cm to enclose the weight prior to its release. Drill a hole at 1 or 2 m from the bottom of the Plexiglas tube and place a bolt through the hole to support the weight. The weight falls when the bolt is removed.
3. Tie the weight securely using the twill tape or string and attach the opposite end to the supporting arms of the injury stand. Ensure excess tape is present to allow the weight to reach through to the Plexiglas tube without becoming taut. While twill tape could be substituted for another high-strength woven tape, the use of fishing wire is not advisable as it may lead to abrasion of the top of the Plexiglas tube over time.
4. Reducing the height of the falling weight will minimize adverse events, with graded pathological response ranging from most severe at 2 m, to mildest when the weight is released from a height of 1 m [4, 28, 29]. Additionally, the weight may be adjusted rather than the height, with a graded response also produced, in which weights of 250–300 g from 2 m resulted in mild injury [30].
5. Isoflurane is a transient inhalation anesthetic with advantages including rapid induction, titratable dose, and swift recovery. However, isoflurane may have adverse side effects to the researcher. The use of this anesthetic should be conducted in a well-ventilated room, with active scavenging systems. As for the animals, isoflurane anesthesia may also result in respiratory depression, and play a role in neuroprotection [31]. Barbiturates such as sodium pentobarbital and sodium thiopental are occasionally used in acute recording or monitoring studies in which rodents are not planned for recovery. This is due to long-lasting anesthesia without depression of cortical evoked responses. Barbiturates are administered via i.p. injection, eliminating the need for anesthesia machinery. However, barbiturates may cause respiratory depression, hypotension with decreased ICP, and elevated mortality if not administered at the correct dose.
6. When originally described, the investigators assessed incidence of apnea and mortality in this model using rats with or without ventilation [1]. They found that while all rats experienced a period of apnea, unventilated rats had a mortality rate of 58.6%, compared to those that were ventilated, having a mortality of 8.7%. Aside from survival, ventilation has the added advantage of administering secondary post-traumatic insult such as hypoxia or hyperoxia, with gas concentrations readily adjusted.
7. Additional (secondary) insults to TBI are commonplace clinically, where isolated brain injury is a rare occurrence. The most frequent insults are hypoxia and hypotension, which

substantially worsen outcomes [11]. While hypoxia may occur inherently in this model due to apnea when rats are not ventilated, depth and duration of the insult cannot be controlled, and so administration of hypoxia can be achieved via exchange of normoxic gas (22 % O_2/78 % N_2) to a hypoxic state. Decreasing the oxygen to 12 % O_2 (balance N_2) in this model for 30 min reliably produces a severe hypoxic insult in which PO_2 is reduced to approximately 40–50 mmHg [10, 32]. Hypoxia/ischemia may also be induced in this model via bilateral carotid artery occlusion after injury [33]. Additionally, this model produces a substantial hypertensive response, with mean arterial pressure significantly elevated above sham levels at least 45 min post-injury [10].

8. The use of other steel materials for the helmet can lead to variations in the force transferred to the brain. It is not recommended to alter the material of the helmet within a study.
9. Scoring allows for better adherence of the superglue to the helmet and reduces the risk of it moving during the impact. Placing the helmet in acetone between each animal may aide in removing superglue.
10. The mattress height should be 11.5 cm in the first 15.5 cm of the mattress' length to support the rats' head, and then should be cut down 1 cm to be 10.5 cm for the remainder of the mattress length, to support the body.
11. The density of the foam bed is critical and needs to be checked regularly to ensure consistency of results. From first use, the foam may soften up to 30 %, with a stress reduction of 5 % thereafter [34]. However, any stress reduction after first use is recoverable, with a recommended period of 1 h between uses ensuring proper foam recovery. There will also be a gradual loss with repeated use, so the foam will need regular replacement. As a guideline, the foam should be replaced after every 40–50 weight drops or 3 months.
12. There are several options for rodent intubation, with the main difference being investigator preference. Proper lighting is essential to visualize the vocal cords, which can be achieved by tracheoscope or surgical lights. Rather than hanging the rat from the intubation rack, some prefer to use a bench top and align the animal to the edge of the counter. A string is placed over the front teeth and then attached to hemostats that are allowed to hang toward the floor. This technique opens the mouth and allows visualization of the vocal cords from another angle, with intubation proceeding as per the protocol for an intubation rack. The placement of an endotracheal catheter is multifold, including direct inflation of the lungs if apnea occurs, protection of the airway from fluids or secretions, and direct delivery of anesthetic.

13. This procedure can be repeated a maximum of two times as swelling of the trachea can impair further intubation attempts and injure the animal.
14. Make sure the cement is evenly distributed on the skull and no residues are left on its surface. Ensure that the helmet is placed in flat position. Given the requirement for a relatively flat dorsal skull surface, the model is difficult to use in any species other than rats. The helmet needs to be cleaned well after each experiment to ensure all glue is removed to allow maximal adherence in the next injury.
15. If a skull fracture is noted, immediately euthanize the animal with sodium pentobarbital.

5.1 Conclusion

The I/A rat model of diffuse TBI replicates the unique pathophysiological consequences of human DAI, with a distinct experimental benefit of inducing apnea directly after injury. The I/A model has now been well characterized and utilized in more than 150 publications, with this model proving to be suitable for examination of both short- and long-term histopathological consequences as well as behavior and cognition. These factors, combined with the ease of use of the apparatus, make the I/A model a suitable choice for the experimental study of diffuse TBI, and a strong candidate to examine emerging neuroprotective strategies.

References

1. Marmarou A, Foda MA, van den Brink W, Campbell J, Kita H, Demetriadou K (1994) A new model of diffuse brain injury in rats. Part I: Pathophysiology and biomechanics. J Neurosurg 80:291–300
2. Foda MA, Marmarou A (1994) A new model of diffuse brain injury in rats. Part II: Morphological characterization. J Neurosurg 80:301–313
3. Povlishock JT, Marmarou A, McIntosh T, Trojanowski JQ, Moroi J (1997) Impact acceleration injury in the rat: evidence for focal axolemmal change and related neurofilament sidearm alteration. J Neuropathol Exp Neurol 56:347–359
4. Beaumont A, Marmarou A, Czigner A, Yamamoto M, Demetriadou K, Shirotani T, Marmarou C, Dunbar J (1999) The impact-acceleration model of head injury: injury severity predicts motor and cognitive performance after trauma. Neurol Res 21:742–754
5. Csuka E, Hans VH, Ammann E, Trentz O, Kossmann T, Morganti-Kossmann MC (2000) Cell activation and inflammatory response following traumatic axonal injury in the rat. Neuroreport 11:2587–2590
6. Donkin JJ, Nimmo AJ, Cernak I, Blumbergs PC, Vink R (2009) Substance P is associated with the development of brain edema and functional deficits after traumatic brain injury. J Cereb Blood Flow Metab 29:1388–1398
7. Hans VH, Kossmann T, Lenzlinger PM, Probstmeier R, Imhof HG, Trentz O, Morganti-Kossmann MC (1999) Experimental axonal injury triggers interleukin-6 mRNA, protein synthesis and release into cerebrospinal fluid. J Cereb Blood Flow Metab 19:184–194
8. Rancan M, Otto VI, Hans VH, Gerlach I, Jork R, Trentz O, Kossmann T, Morganti-Kossmann MC (2001) Upregulation of ICAM-1 and MCP-1 but not of MIP-2 and sensorimotor deficit in response to traumatic axonal injury in rats. J Neurosci Res 63:438–446
9. Stahel PF, Kossmann T, Morganti-Kossmann MC, Hans VH, Barnum SR (1997) Experimental diffuse axonal injury induces enhanced neuronal C5a receptor mRNA expression in rats. Brain Res Mol Brain Res 50:205–212
10. Hellewell SC, Yan EB, Agyapomaa DA, Bye N, Morganti-Kossmann MC (2010) Post-traumatic hypoxia exacerbates brain tissue damage: analysis of axonal injury and glial responses. J Neurotrauma 27:1997–2010

11. Chesnut RM, Marshall LF, Klauber MR, Blunt BA, Baldwin N, Eisenberg HM, Jane JA, Marmarou A, Foulkes MA (1993) The role of secondary brain injury in determining outcome from severe head injury. J Trauma 34:216–222
12. Heath DL, Vink R (1995) Impact acceleration-induced severe diffuse axonal injury in rats: characterization of phosphate metabolism and neurologic outcome. J Neurotrauma 12:1027–1034
13. Stibick DL, Feeney DM (2001) Enduring vulnerability to transient reinstatement of hemiplegia by prazosin after traumatic brain injury. J Neurotrauma 18:303–312
14. O'Connor C, Heath DL, Cernak I, Nimmo AJ, Vink R (2003) Effects of daily versus weekly testing and pre-training on the assessment of neurologic impairment following diffuse traumatic brain Injury in rats. J Neurotrauma 20:985–993
15. Nimmo AJ, Cernak I, Heath DL, Hu X, Bennett CJ, Vink R (2004) Neurogenic inflammation is associated with development of edema and functional deficits following traumatic brain injury in rats. Neuropeptides 38:40–47
16. Yan EB, Hellewell SC, Bellander BM, Agyapomaa DA, Morganti-Kossmann MC (2011) Post-traumatic hypoxia exacerbates neurological deficit, neuroinflammation and cerebral metabolism in rats with diffuse traumatic brain injury. J Neuroinflammation 8:147
17. Buki A, Farkas O, Doczi T, Povlishock JT (2003) Preinjury administration of the calpain inhibitor MDL-28170 attenuates traumatically induced axonal injury. J Neurotrauma 20:261–268
18. Saatman KE, Murai H, Bartus RT, Smith DH, Hayward NJ, Perri BR, McIntosh TK (1996) Calpain inhibitor AK295 attenuates motor and cognitive deficits following experimental brain injury in the rat. Proc Natl Acad Sci U S A 93:3428–3433
19. Buki A, Okonkwo DO, Povlishock JT (1999) Postinjury cyclosporin A administration limits axonal damage and disconnection in traumatic brain injury. J Neurotrauma 16:511–521
20. Okonkwo DO, Melon DE, Pellicane AJ, Mutlu LK, Rubin DG, Stone JR, Helm GA (2003) Dose-response of cyclosporin A in attenuating traumatic axonal injury in rat. Neuroreport 14:463–466
21. Bouzat P, Francony G, Thomas S, Valable S, Mauconduit F, Fevre MC, Barbier EL, Bernaudin M, Lahrech H, Payen JF (2011) Reduced brain edema and functional deficits after treatment of diffuse traumatic brain injury by carbamylated erythropoietin derivative. Crit Care Med 39:2099–2105
22. Valable S, Francony G, Bouzat P, Fevre MC, Mahious N, Bouet V, Farion R, Barbier E, Lahrech H, Remy C, Petit E, Segebarth C, Bernaudin M, Payen JF (2010) The impact of erythropoietin on short-term changes in phosphorylation of brain protein kinases in a rat model of traumatic brain injury. J Cereb Blood Flow Metab 30:361–369
23. Hellewell SC, Yan EB, Alwis DS, Bye N, Morganti-Kossmann MC (2013) Erythropoietin improves motor and cognitive deficit, axonal pathology, and neuroinflammation in a combined model of diffuse traumatic brain injury and hypoxia, in association with upregulation of the erythropoietin receptor. J Neuroinflammation 10:156
24. Adelson PD, Robichaud P, Hamilton RL, Kochanek PM (1996) A model of diffuse traumatic brain injury in the immature rat. J Neurosurg 85:877–884
25. Adelson PD, Dixon CE, Robichaud P, Kochanek PM (1997) Motor and cognitive functional deficits following diffuse traumatic brain injury in the immature rat. J Neurotrauma 14:99–108
26. Adelson PD, Jenkins LW, Hamilton RL, Robichaud P, Tran MP, Kochanek PM (2001) Histopathologic response of the immature rat to diffuse traumatic brain injury. J Neurotrauma 18:967–976
27. Maughan PH, Scholten KJ, Schmidt RH (2000) Recovery of water maze performance in aged versus young rats after brain injury with the impact acceleration model. J Neurotrauma 17:1141–1153
28. Kallakuri S, Cavanaugh JM, Ozaktay AC, Takebayashi T (2003) The effect of varying impact energy on diffuse axonal injury in the rat brain: a preliminary study. Exp Brain Res 148:419–424
29. Signoretti S, Di Pietro V, Vagnozzi R, Lazzarino G, Amorini AM, Belli A, D'Urso S, Tavazzi B (2010) Transient alterations of creatine, creatine phosphate, N-acetylaspartate and high-energy phosphates after mild traumatic brain injury in the rat. Mol Cell Biochem 333:269–277
30. Ucar T, Tanriover G, Gurer I, Onal MZ, Kazan S (2006) Modified experimental mild traumatic brain injury model. J Trauma 60:558–565
31. Statler KD, Alexander H, Vagni V, Holubkov R, Dixon CE, Clark RS, Jenkins L, Kochanek PM (2006) Isoflurane exerts neuroprotective actions at or near the time of severe traumatic brain injury. Brain Res 1076:216–224

32. Gao G, Oda Y, Wei EP, Povlishock JT (2010) The adverse pial arteriolar and axonal consequences of traumatic brain injury complicated by hypoxia and their therapeutic modulation with hypothermia in rat. J Cereb Blood Flow Metab 30:628–637
33. Stiefel MF, Tomita Y, Marmarou A (2005) Secondary ischemia impairing the restoration of ion homeostasis following traumatic brain injury. J Neurosurg 103:707–714
34. Zhang L, Gurao M, Yang KH, King AI (2011) Material characterization and computer model simulation of low density polyurethane foam used in a rodent traumatic brain injury model. J Neurosci Methods 198:93–98

Chapter 16

Experimental Models for Neurotrauma Research

Johan Davidsson and Mårten Risling

Abstract

Physical trauma in the central nervous system (CNS) is usually the result of a number of forces in different directions and dimensions. A large number of experimental models have been developed to improve the possibilities to understand the outcome of CNS trauma. In this chapter, we will describe the need for a variety of experimental models for research on traumatic brain injury (TBI) and spinal cord injury (SCI). Models can serve different needs, such as: to test new treatments for injuries, to reveal thresholds for injuries, to provide a better understanding of injury mechanisms, or to test tools and methods for translation between experiments and clinical data. In this chapter, we will discuss on the validation of models and translation between experimental and clinical studies.

Key words Traumatic brain injury (TBI), Diffuse axonal injury (DAI), Mechanisms, Blast trauma, White matter

1 Introduction

The central nervous system (CNS) is the most delicate and complex part of the body. The definition for the CNS is based on the glial cells that surround and support neurons. Both the brain and the spinal cord are included in the CNS, since these parts of the nervous system comprise astrocytes and oligodendrocytes. Spinal nerve roots and peripheral nerves contain Schwann cells and are therefore regarded as parts of the peripheral nervous system (PNS).

Physical trauma in the central nervous system is often the result of a number of forces in different directions and dimensions. The timeline for the interaction between the acting forces and the nervous tissue is usually very narrow and it can therefore be very difficult to assess the influence of each contributing force. In addition, human patients, which come in different ages and sizes, may have a number of additional injuries or diseases as well a genetic background that may influence the outcome of the injury.

Firas Kobeissy et al. (eds.), *Injury Models of the Central Nervous System: Methods and Protocols*, Methods in Molecular Biology, vol. 1462, DOI 10.1007/978-1-4939-3816-2_16, © Springer Science+Business Media New York 2016

1.1 Spinal Cord Injury (SCI)

It has been known since ancient civilizations, such as the Pharaonic Egypt, that spinal cord injury (SCI) leads to chronic loss of functions, such as paralysis. The complex cellular response and lack of successful regenerative growth in CNS was described in great detail by Ramón y Cajal [1]. Although a substantial amount of research on SCI has been conducted, there has been only limited success in the identification of new strategies to improve the regenerative response after SCI. Models for SCI usually have a more regenerative focus than the models for brain injury.

1.2 Traumatic Brain Injury (TBI)

TBI is a leading cause for death and disability in both civilian life and at the battlefield. The signature TBI has evidently changed from penetrating TBI to blast-induced TBI in recent military conflicts. All types of TBI may occur in both the civilian and military setting. However, there are some important considerations for military TBI, mostly relating to the extreme energy transfer.

1. TBI can be graded from mild to severe and primary or secondary. TBI can be classified as diffuse or focal, and very often in a complex mixture.
   (a) *Mild TBI*: A TBI is often classified as mild (concussion or commotion) if loss of consciousness or confusion is shorter than 30 min. MRI (magnetic resonance imaging) and CT (computer tomography) scans are usually normal but the patient may have headache, cognitive problems (memory problems, mood disturbance, attention deficits), and the effect on the patient can be devastating. The majority of blast-induced TBI fall into this category, although the pathophysiology is largely unknown. Cerebral concussion is often associated with other types of brain injury.
   (b) *Moderate and Severe TBI*: These injuries can be divided into closed head injuries and penetrating injuries. Closed head injuries may be both diffuse and focal. The penetrating TBI will always induce a focal injury and often diffuse secondary injuries.
   (c) *Diffuse TBI*: The most common diffuse injury is the diffuse axonal injury (DAI), which is defined as the presence of diffuse damage to axons in the cerebral subcortical parasagittal white matter, corpus callosum, brain stem, and cerebellum. This is usually the result of an acceleration/deceleration trauma. Different parts of the brain move at different speeds because of their relative density. If this is a rotational trauma, the positions of the axis of rotation will be an important factor in the injury mechanism and areas at a greater distance from this axis will sustain larger forces. This can lead to shearing injury and DAI. Beta-amyloid precursor protein (APP) has been proven to be an excellent

marker for axonal injury in histology. Modern imaging techniques, such as MRI with DTI (diffusion tensor imaging), have provided improved possibilities to detect DAI.

(d) *Focal TBI*: Focal brain injuries include contusions and penetrating TBI. Focal TBI is usually the result of bullet and shrapnel entrance in the skull cavity or a contact between the head and a sharp or semi-blunt object that results in skull fracture. Also acceleration/deceleration trauma can cause focal injuries to the brain.

(e) *Secondary traumatic brain damage* occurs as a complication of the different types of primary TBI and includes ischemic and hypoxic damage, swelling, raised intracranial pressure, and infection. The onset of these complications varies from minutes to hours. The secondary TBI is potentially reversible with adequate treatment. Axonal damage in both DAI and focal injuries interferes with axoplasmic transport. Severe traumatic injury results in primary axonal disruption or transection termed "primary axotomy" or sets in train a series of ill-understood events culminating in secondary axonal degeneration or secondary axotomy. Thus, adequate treatment could probably limit the axonal damage.

2. Experimental models for studies of CNS injuries
   Experimental research is a strategy to take control over confounding factors that may obscure data from real life clinical CNS injuries.
3. Experimental models may serve many different purposes, such as
   (a) To test new treatments for injuries.
   (b) To provide a better understanding of injury mechanisms.
   (c) To reveal thresholds for injuries.
   (d) To develop test tools and methods for clinical settings. For example to develop protocols for MRI, new biomarkers, protocols for EEG.

1.3 Model requirements

It is difficult to provide a complete list of all different models that are employed in neuroscience. We will limit the description to examples of models for traumatic injuries and therefore exclude models for stroke, neurodegenerative diseases, infections, etc. It is usually important to test if the data from the employed model can be translated and provide any conclusions for the condition that is supposed to mimic. It is also important to enable verification of any data that can be harvested from the model. Thus, a good monitoring and documentation of physical data should be provided from all new models. The model needs to be both validated and verified. Thus, to show that the model can replicate exactly what it is suggested to replicate and that the experiments can be repeated by both the scientists that have developed the model and by any other scientist that has the proper tools. This is a highly complex process, since the methods for measurement also need to be validated and verified. It can be supposed that many of the failures in translating new treatments from experimental setups to clinical practice may be due to mistakes in verification and validation of the model or the experimental setup. It has also been said that a lot of published experimental data cannot be repeated, possibly due to improper validation and description of the experimental setup.

1.4 Obstacles for Translation

1. There are a number of more or less obvious factors that may impede translation between an experimental model and the clinical situation—even if the model has been properly characterized:
   (a) Trauma in the clinical situation is the result of many complex forces, delivered during a short time. It is therefore extremely difficult to produce a model that can in all details simulate a clinical situation.
   (b) Severe trauma to the head or spine is usually combined with other injuries, thus a multitrauma situation that can modify the outcome of the TBI or spinal cord injury by for example a systemic inflammation, infection, and hypoxia. The importance of such contributing factors can be tested in the experimental situation, but will still provide an obstacle in translation between experiments and the clinical populations.
   (c) The time scale for biological response in experimental animals, such as rodents, is probably very different than that in humans.
   (d) The genetic variation in rodent strains is usually very limited compared to clinical populations. It is therefore possible that some results are over-amplified in the experimental situation and may not at all be relevant in a clinical situation.

2 Experimental Models of Brain Neurotrauma

Below are a set of different experimental models of brain trauma. For each model, we will try to describe the rationale (why the model was developed), the development, results obtained, and limitations with the model. We will also describe the work with validation, verification, and translation. In each section, we also give examples of alternative useful and validated models.

2.1 Spinal Injury, with Focus on Motor Axons

The pioneering studies of Ramón y Cajal demonstrated that injured neurons in the central nervous system (CNS) have a strong capacity to extend new axons into the peripheral nervous system (PNS)

[1], he showed that motor axons that had been severed due to spinal cord injury had the capacity to regrow to neighboring ventral roots. The observation that neurons in the CNS could extend axons into PNS tissue was reinvestigated by Richardson and colleagues [2].

2.1.1 A Lesion in the Ventral Funiculus of the Spinal Cord

With background from the observations that motoneurons in adult cats survived after peripheral nerve injuries to a much better degree than sensory neurons, it was decided to analyze if motoneurons would survive also after very proximal injuries. The most proximal location would then be to cut the axons inside the spinal cord, but outside the motor nuclei in the grey matter. The selected site was the ventral funiculus, i.e. the white matter between the ventral horn and the transition to the spinal ventral nerve root. The experiments were performed in adult cats that were subjected to a lumbar laminectomy. The denticulte ligament at the side of the lumbosacral enlargement was divided and used to gently twist the spinal cord so that the junction between the spinal cord and the ventral root was visible. A thin fragment of a razor blade (or a specialized knife for micro-surgery) was pushed into the lateral aspect of the spinal cord segment L7 and moved longitudinally to produce a selective lesion in the ventral funiculus. The aim was to cut the motor axons at this location, without damage to the grey matter (where the cell bodies reside) or the connection between the spinal cord and the nerve root. It was observed that, although the animals were deeply anesthetized, the muscles in the hindlimb corresponding to the L7 segment (the L7 myotome) responded with contractions during this procedure. The erector spinae muscle, fascia, and skin were closed with sutures in separate layers. The animals were sacrificed after survival times from 1 to around 300 days and fixed by intravascular fixation with glutaraldehyde. Using electron microscopy, it was revealed that the lesion was located in the white matter and had spared both the grey matter and the nerve root exit zone. A large proportion of the motoneurons actually had survived the lesion (Fig. 1).

It was shown with electron microscopy and intracellular labeling with horseradish peroxidase that axons from such surviving motoneurons penetrated the scar tissue in the ventral funiculus of the spinal cord and entered the ventral roots [3–5]. The regrowing axons penetrate through a highly unusual CNS environment which lacks a blood–brain barrier (BBB) function [6] but possesses a high content of cells bearing neurotrophin receptors [7–9] and matrix molecules [10, 11].

2.1.2 Avulsion Injuries: The Clinical Use of the Observed Regenerative Response

A clinical counterpart to this injury is a ventral root avulsion at the border between the CNS and PNS, typically caused by a high-energy trauma such as a motorbike accident causing excessive trauma to the shoulder and head resulting in stretching and rup-

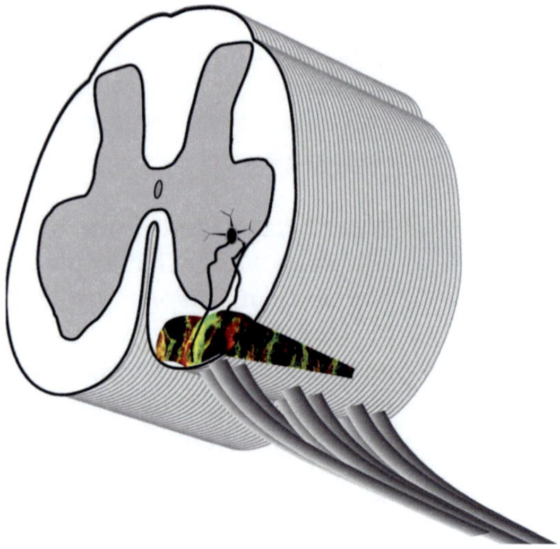

Fig. 1 A schematic representation of the ventral funiculus lesion model. The lesion area in the ventral funiculus of the spinal cord is penetrated by regrowing motor axons, surrounded by a thin sheet or glial tissue. The extracellular room is expanded and contains large amounts of matrix molecules such as laminin and collagen, the blood–brain barrier has a more or less permanent defect

turing of ventral roots. Ventral root avulsion is not followed by spontaneous regrowth, since the avulsed roots are widely separated from the spinal cord inside the subarachnoid space or even pulled to a position outside the vertebral channel. Replantation of avulsed spinal ventral roots into the spinal cord (Fig. 2) has been shown to enable significant and useful regrowth of motor axons in both experimental animals and human clinical cases [12–15]. Gene expression array studies show that replantation of the ventral root can initiate a rapid upregulation in the expression of genes associated with neurite growth, in contrast to neurons subjected to avulsion only [16]. Thus, the replantation can influence gene programs in the spinal cord.

The results from such treatment in humans are less successful in older patients and good reinnervation in distal muscles like in the hand is seldom possible. In a recent case report, Carlstedt et al. [14] described a preadolescent boy with complete brachial plexus avulsion injury that was treated by replantation of five ventral roots. Shoulder muscle recovery started 8–10 months after the spinal cord operation. At 12–15 months, elbow function began to recover, followed 2 years postoperatively by forearm, wrist, and intrinsic hand muscle activity. This led to recovery of hand motor function without restoration of sensation. Bilateral motor cortex activity and activation of the sensory cortex on use of the affected hand was demonstrated by fMRI. The findings of that case study

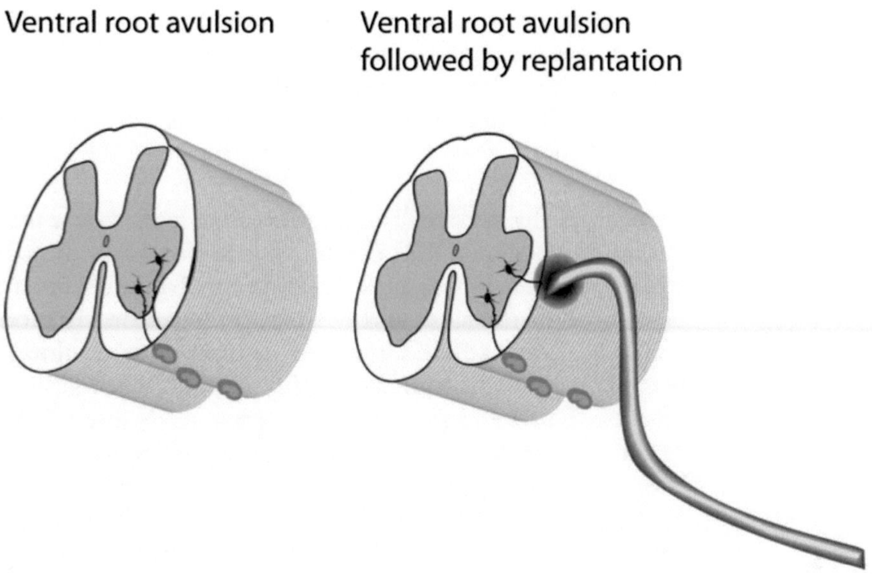

Fig. 2 A schematic representation of replantation of avulsed ventral roots. The ventral roots are avulsed at the border between CNS and PNS (spinal cord surface). The rootlets are then replanted into the ventro-lateral surface of the spinal cord. Regrowth into the ventral root can occur both directly within the spinal cord and by growth in the pia mater. After reaching the ventral root, motor axons can continue and reinnervate denervated muscle

suggested that the restored hand function might rely on cortical sensory programs established before the injury. Although the time needed for recovery seems very long, it cannot be excluded that the final result is dependent on several biological programs that are elicited in the early acute stage.

The method for producing nerve root injuries has a significant role for the outcome of the injury. Rhizotomy (division of the root) leads to a milder representation of the spinal cord trauma that occurs after "true" avulsion injury [17]. Shortland and colleagues employ a model for extra-vertebral avulsion (without need for laminectomy) [18]. However, for replantation surgery laminectomy is necessary. One limitation with most animal studies for replantation is that the injury and the replantation surgery are performed at the same time, whereas delayed surgery is the typical clinical situation.

This is to our knowledge, the only example of an experimental model that has been translated into practical clinical use for treatment of a traumatic injury that involves axonal growth in the spinal cord. Many models for SCI seem to simplifications of the usual clinical type of SCI. Maybe it would be of benefit to develop models for spinal cord contusion that could represent a better representation of actual clinical situations, with high-energy transfer, rapid deformation of vertebra and incomplete injury to the spinal cord.

2.2 TBI in the Military Setting: Modeling Primary Blast TBI

The use of Improvised Explosive Devices (IED) in contemporary warfare has changed the scene and spectrum of TBI at the battlefield. At the same time, new equipment for body protection has increased the survival rate after TBI at the battlefield. It may be assumed that several mechanisms contribute to the injury. TBI has been identified as major health problem in military personnel returning from service. The injuries range from severe multitrauma to a number of mild TBI that still has to be settled. Propagation of blast waves is very complex. It could involve both direct propagation through the skull and indirect propagation via blood vessels and it is obvious that a systemic response comes along with the blast TBI. It is difficult to identify a reliable borderline between mild blast TBI and posttraumatic stress syndrome (PTSD). Many of the symptoms are similar and many patients might suffer from both TBI and PTSD. The energy transfer during an explosion is extremely complex and it is difficult to predict how energy is propagated into the body and absorbed. Exposure data from clinical situations at the battlefield are usually lacking. One way to generate a better understanding of the mechanisms of TBI after a blast exposure is to perform controlled experiments in animals. Many clinical TBI cases have a complex mixture of diffuse and focal injuries, which are complicated by secondary injury events. Individuals exposed to a blast often suffer from multiple injuries, i.e., pulmonary lesions or amputations. Severe blast-related TBI with brain edema and vascular spasm [19] can be assumed to be the result of a combination of more than one injury mechanism (*see* Fig. 3).

1. One way to understand the effects of a blast wave is to divide the mechanism into
   (a) *Primary blast*: Effects of the primary blast wave, thus the propagation of a supersonic pressure transient with short duration. The threshold for injuries is determined by factors such as peak pressure, duration, and shape of the wave.
   (b) *Secondary effects of blast*, i.e., due to the impact of flying objects, such as shrapnel fragments, which can generate penetrating injuries. The proportion of such injuries was larger in previous conflicts and seems to have been reduced by improvements in helmet construction.
   (c) *Tertiary effects of blast*, i.e., the result of acceleration movements, which may result in tissue shearing and diffuse injuries, such as diffuse axonal injuries (DAI).
   (d) *Quaternary effects of blast*, the result of heat, smoke, or emission of electromagnetic pulses (EMP).
2. Open field exposure

   Large-scale experiments during the 1950s in desert areas and ponds generated fundamental data for effects of blast with simple wave forms. Large numbers of animals of different species

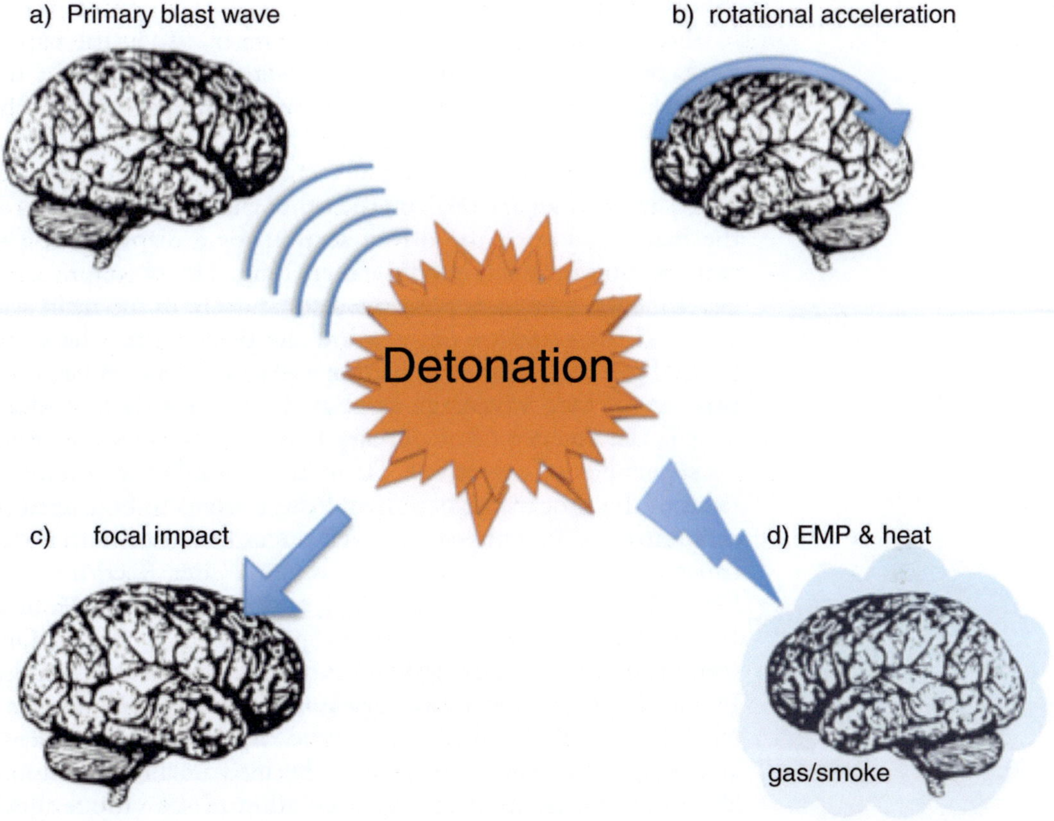

Fig. 3 A schematic representation of different mechanisms for blast-induced neurotrauma. Previously published in [94]

were subjected to open field exposure to blast. These experiments determined thresholds for mortality and injuries such as bleeding in air-filled organs such as the lungs and intestines. The potential effects on the central nervous system were generally not assessed. These experiments provided the Friedländer type of wave and dose–response curves (the Bowen curves) were determined [20–22]. Large amounts of explosives are needed and dosimetry can be difficult. Good control of the physiology of the experimental animals and proper tissue collection is usually not possible in outdoor conditions in combination with a large number of animals in a single experiment. However, open field experiments may allow for more realistic experiments with large animals that are more similar in size to humans. It also makes it possible to use waveforms relevant for simulation of IED, for example reflection from the ground or vehicles. Examples of new models with open field exposures aimed to produce mild TBI include the Combat Zone-like blast scenery for mice [23] and a primate model [24].

3. Shock tubes with compressed air or gas

Shock tubes employ compressed air or gas, rather than explosives. Systems with compressed air were used already in the 1950s [25]. Most systems comprise two chambers, separated by a membrane (Fig. 4). Compressed gas is loaded into one of these chambers (the overpressure chamber or the driver section), which is separated from the other chamber, referred to as the main section or the driven section, by a diaphragm. The main section is usually several meters long. The object, i.e., the experimental animal, is positioned somewhere in the main section and the operator can rupture the diaphragm. The compressed gas enters the main section and simulates a propagating blast wave. One advantage associated with this type of shock tube is the absence of quaternary blast effects as well as other disadvantages of explosives. There are a number of modifications of the shock tube design and there seems to be a need to calibrate the different systems. Well-documented modern shock tubes can for instance be found at the Walter Reed institute [26], Wayne State University [27], and the US Naval Medical Research Center (Maryland, Silver Spring, USA) [28, 29]. One sophisticated shock tube system has been installed at the Applied Physics Laboratory at Johns Hopkins University [30]. This is a modular, multi-chamber shock tube capable of reproducing complex shock wave signatures. The instrumentation should allow direct measurement and calculation of the various shock loading characteristics, including static pressure, total pressure, and overpressure impulse.

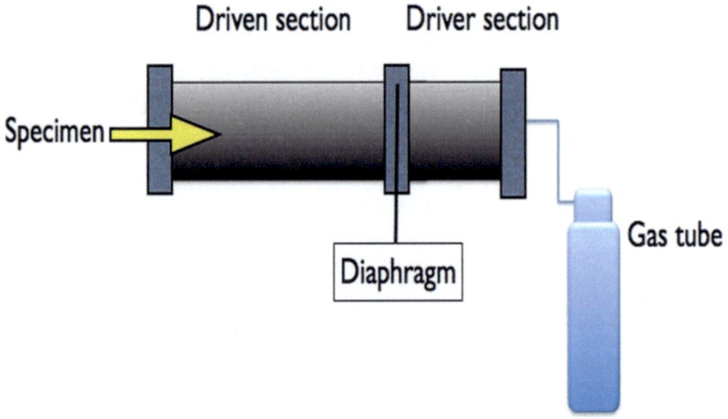

Fig. 4 A schematic representation of a typical over pressure shock tube composed of two chambers separated by a membrane. Compressed air/gas in the driver section is released as a pressure wave to the other (driven) compartment by the controlled puncture of the membrane. Previously published in [94]

4. A Swedish Blast tube for studies on primary blast

In 1949, Carl-Johan Clemedson published a thesis titled "*An experimental study on air blast injuries*" in 1949 at the university in Uppsala [31]. He continued his research at FOA (Swedish Defence Research Establishment) using compressed air [25] and finally a newly constructed blast tube [32] in which a charge of plastic explosive (pentaerythritol tetranitrate PETN) was used. The system was composed of a cylindrical 400 mm wide and 565 mm long cast iron tube, with a cone shaped tip where the charge is placed. The wall thickness at the tip is 100 mm. At the other end, the iron tube is elongated to a total length by an extension tube of steel (wall thickness 10 mm). The open end of the extension tube was by a steel disc connected to a pendulum in Clemedson's initial experiments. At the conical end of the tube, there is a threaded screw plug for the insertion of the charge that is mounted on the detonator. The charge is fixed at a distance of 100 mm from the inner tip. Clemedson and his coworkers published a number of studies on muscle tissue [33] as well as vascular and respiratory effects of blast in the rabbit [34]. After some time, this work was extended to include the central nervous system [35] and the cerebral vasculature [36]. The blast tube (without the pendulum door—which results in a more complex waveform) was modified for work with rodents by Anders Suneson. Annette Säljö used this system in her thesis work [37–39]. The ignition system was later changed to a non-electric ignition (NONEL from Nobel). The anesthetized rat is mounted in the blast tube at a distance of 1 m from the charge. The PETN generates a pulse with rapid raise time and very short duration (<0.3 ms). An increase in the charge has a direct and proportional effect on the peak pressure, but only a small effect on the duration. Secondary reflections are limited. By moving the object to the orifice of the tube (1.5 m from the charge) the peak pressure drops to about half, whereas the duration is increased somewhat. All rat experiments have been performed at the 1 m distance from the charge (Fig. 5), which has been varied from 0.5 to 5 g PETN (Spherical charges of Swedish army plastic explosive *m*/46, containing 86 % pentaerythritol tetranitrate PETN and mineral oil, where used. In the following text, the weights of the plastic explosive charges are given in gram). We have previously reported that the peak pressure during detonation of 2.5 g PETN would be 260 kPa and nearly 600 kPa with a 5 g charge. To get representative recordings of the peak pressure during a detonation of explosives is connected to substantial methodological difficulties due to the extreme requirement for good dynamics. Recent recordings with a set of open silicon piezoresistive sub-miniature pressure transducers (Entran Sensors & Electronics) indi-

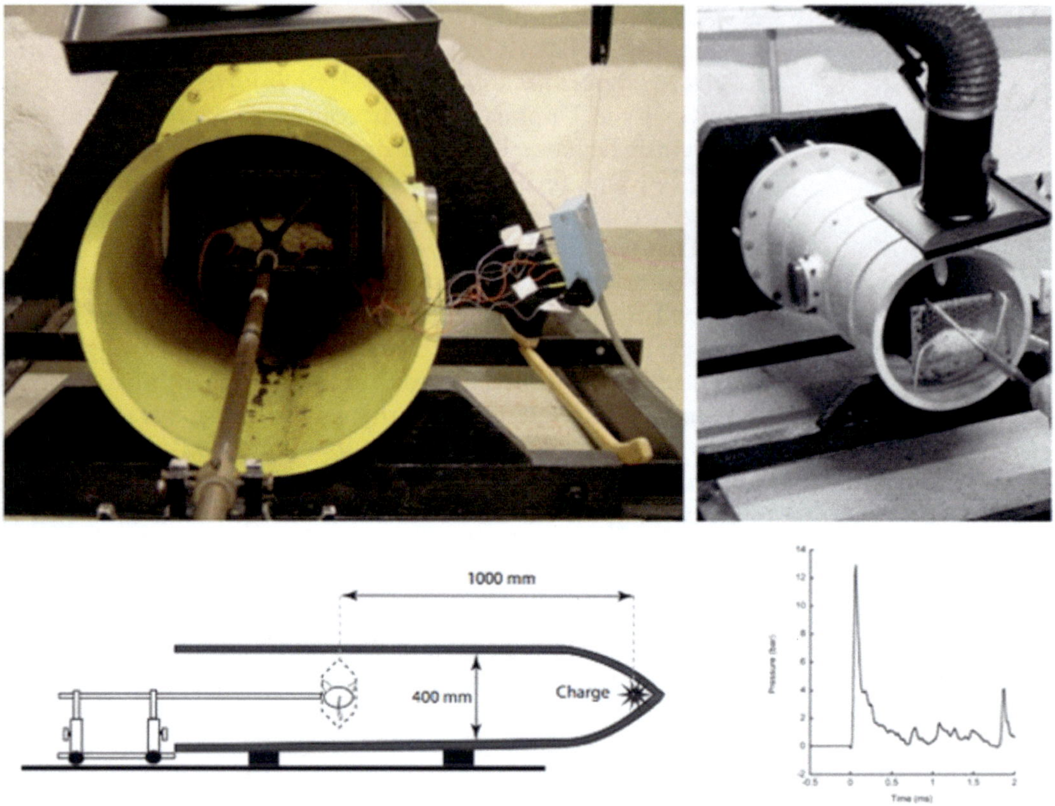

Fig. 5 Pictures showing the blast tube that was constructed by the Swedish scientist Clemedson in the 1950s. This system may be one of the oldest systems that still are in use. Previously published in [94]

cate a peak pressure exceeding 10 bar during detonation of 5 g PETN, as recorded in front of a rat dummy at the 1 m position in the tube. A charge of 2.5 g in an unprotected animal (side-on at 1 m from the charge) results in about 50% lethality and at 3 g the majority of the exposed animals die from pulmonary bleedings. On the other hand, the lethality is less than 10% at 2 g PETN. In recent experiments we have used a full protection (rigid steel tube) of the body, except for the head and had no lethality with 5 g charges.

In spite of the substantial lethality with this system (in unprotected animals), structural changes in the brain seem very limited or absent. The cell death in the brain that was reported by Säljö and coworkers [39], employing a similar protocol, has not been possible to verify. Instead, there is a significant cell death and gene expression changes [40, 41] in the inner ear. The hippocampus of the brain of rats exposed to a 2 g charge has been subjected to examination of the gene expression by use of Affymetrix Rat Gene Arrays. In such experiments, the expression of all known genes can be evaluated at the same time. It was found that about 100 genes had a significant expression change 24 h survival after the blast. The majority (76 genes) had a decrease in their expression whereas an increased expression was found in 39 genes [42]. Affected systems include synaptic transmission and neurogenesis. Both unprotected rats exposed to detonation of 2 g PETN and rats with the torso protected by a rigid steel tube during exposure to a 5 g blast have been examined. Taken together, the lack of structural damage combined with some functional changes indicates that the primary blast in this model generates a mild TBI. Future work with the blast tube will include repeated exposures and variations in body position. The blast tube in the current setting creates a very simple and short pulse that may imitate the situation at short distance with an open field charge. This pulse form is probably not at all similar to the situation in a protected vehicle that is hit by a road bomb. It should therefore be of interest to modify the length of the blast tube and introduce reflecting objects that could create a more complex pulse form. One aspect with this type of blast tube is the heat and gas emission that is a result of the use of explosives instead of compressed air. Thus, quaternary blast is added to the primary blast. Acceleration movements are however limited, by the montage of the animal, and impacts of flying objects do not occur. Secondary and tertiary blast is instead imitated and better controlled in other models.

5. Breacher studies

   "Breachers" are military or law enforcement personnel that are routinely exposed to low-level blast during training. Such repeated exposure has been associated with symptoms similar to that of sports concussion. The physical parameters of the blast exposure can be monitored during training in a much better way than in battle-field situations. This creates a situation that is similar to an experimental model and collection of samples, such as biomarkers [43], can therefore be compared to animal experiments for primary blast.

3 Models for Acceleration TBI

Although TBI can be associated with skull fractures, it commonly occurs without fractures [44]. About 40% of all TBI patients admitted to hospitals are non-focal injuries [45] and are usually referred to as distributed brain injuries (DBI).

At least four categories of DBI can be identified: diffuse axonal injury (DAI); diffuse hypoxic, anoxic, or ischemic injury; diffuse swelling; and diffuse vascular injury. DAI is the most common type of DBI and commonly results in unconsciousness or death [44, 46]. The DAI pathology, which is characterized by perturbations to the axoplasmic transport along the length of axons [47], is likely to cause axonal swelling or degeneration which can reduce the functionality or disconnect the axons from their existing networks [48]. It has been reported that DAI commonly are localized in the subcortical white matter, grey–white matter interface and corpus callosum [44, 49], and at points of attachment, such as cranial nerves [50].

DBI is commonly a result of inertial induced loads; intracranial motions arise when the skull is accelerated and the brain mass, due to its inertia, lags behind or continues its motion relative to the skull. These inertia-induced loads are most common in rapid head rotations [51] which often occur in fall accidents, traffic accidents, and military assaults. Mechanical and mathematical models have been used to show that these inertia-induced loads produce strains in the brain tissue and that these strains cause the following neurological deficiencies [48, 52–58].

An attractive approach to study DBI pathology and its associated injury mechanism and threshold would be to reconstruct well-documented accident cases in which the patient is slightly injured. The real life accidents are however commonly rather complex, the injuries are regularly sever, and the patient suffers from a multitude of injuries. Therefore, anesthetized animals have been used in the past, and are in use, to study DBI and DAI.

The rotational weight drop model that [59] was developed by Marmarou and coworkers [60, 61] has generated very important data on development of diffuse brain injuries, including an improved understanding of diffuse axonal injury [47]. One of the advantages/disadvantages with this model is that it combines DAI with a contusion injury. Andersen [62] developed a model in which a stunner was used to accelerate the head. Both sever local brain damage and axonal injury was produced by this model. Both the forth-mentioned models have been found to be less useful for threshold studies on DAI and when DAI is to be studied separately.

Several models were developed in the past to study DAI using primates [44, 63], miniature swine [59], and rabbits [64, 65]. Few of these are in use today due to ethical considerations, or a lack of appropriate methods to assess the effects of the trauma, or due to excessive meningeal bleedings.

Models developed for rats provide several advantages over models using large animals. Ellingson et al. [66] and Fijalkowski et al. [67] exposed rats at low to medium severity rotational accelerations in the coronal plane. Despite the higher accelerations, the rats suffered from classical concussion injuries with minimal histological abnormalities.

3.1 A Rotational Acceleration Model

A rotational acceleration model aimed for studies into diffuse brain injuries and mTBI is described in detail by Davidsson [68]. It is designed to produce brain injuries by sagittal plane rearward rotational acceleration. This model allows for the production of graded injury and studies into injury thresholds.

In brief, the skull of an anesthetized adult Sprague–Dawley rat, weighing from 280 to 450 g, is tightly secured to a rotating bar. The bar is impacted by a striker that causes the bar and the animal head to rotate rearward; the acceleration phase lasts 0.4 ms and is followed by a rotation at constant speed and a gentle deceleration when the bar makes contact with a padded stop (Fig. 6). By adjusting the air pressure in the rifle used to accelerate the striker, rotational acceleration between 0.3 and 2.1 Mrad/s^2 can be produced.

Numerous combinations of trauma levels, posttrauma survival times, brain and serum retrieval, and tissue preparation techniques were adopted to characterize this model. The trauma cause subdu-

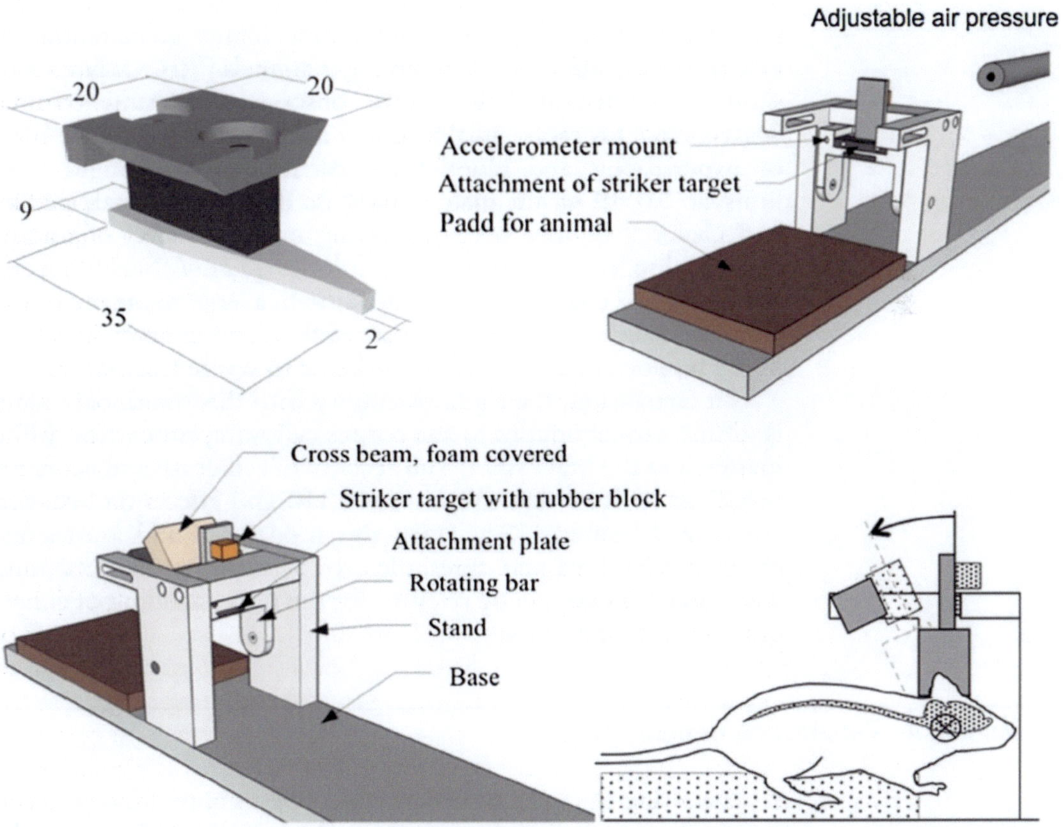

Fig. 6 Diagrams. Top row: skull cap and attachments plate and oblique view of test device (crossbeam removed for visibility). Bottom row: oblique view of test device and side view of the head with central nervous system schematically depicted (screws and accelerometer not depicted) (Previously published in [94])

ral bleedings in animals exposed to severe trauma although such hematoma can occasionally be found at rather low acceleration levels [69]. Staining brain tissue with amyloid precursor protein (APP) antibodies, which expose axons with reduced axioplasmic transport, revealed widespread axonal injuries (AI) in the frontal region of the corpus callosum, the upper and lower borders of the corpus callosum, and sometimes stretched into the caudate putamen. In the mid brain region APP-stained axons were found in the same regions as well as in the thalamus and hippocampus regions. Further, APP-stained axons were found in some of the tracts in the brain stem. Staining brain tissue with FD Neurosilver that detects degenerating axons confirmed the injury pattern [69]. For older subjects, those weighing on average 700 g, fewer positive axons appeared in the corpus callosum in animals while injured axons were found in clusters elsewhere: in the internal capsules, inside and in the vicinity of the anterior commissure, and in the structures between the lateral ventricles [70]. In addition to these areas, positive axons were spread out and frequent in structures close to the centerlines and the skull bases. The observed AI was apparent only when the rotational acceleration level was moderate (1.1 Mrad/s^2) and above [71]. Older animals required higher accelerations to exhibit similar injury levels as younger animals [70]. Only limited signs of contusion injury were observed following trauma. Macrophage invasions, glial fibrillary acidic protein redistribution or hypertrophy, and blood–brain barrier (BBB) changes were unusual. S100B serum analyses indicate that blood vessel, axonal, and glial cell injuries occur following moderate level of trauma despite absence of obvious BBB injuries [72]. Affymetrix gene arrays showed changes in the expression in a large number of gene families including cell death, inflammation, and neurotransmitters in the hippocampus 24 h after moderate to severe trauma [42].

In conclusion, the signature injury with this rotational model is diffuse axonal injuries in the corpus callosum, subcortical white matter, and the brain stem. The absence of cell death and excessive bleedings indicate that this is a mild TBI and effects on behavior are indeed limited [72]. Thus, this model can add knowledge about mechanisms and thresholds for acceleration-induced mild TBI and such data can be relevant for the understanding of consequences of tertiary blast.

4 Models for Penetrating Ballistic TBI

A penetrating ballistic TBI occurs when objects impact the head and penetrate skin, skull and meninges and cause injury directly to the brain tissue. In contrast to closed head injury, penetrating ballistic TBI involves direct laceration of brain tissue, often complicated by secondary effects such as hemorrhage, edema, inflammation, and higher risk of coagulopathy [73]. The Vietnam head injury study has generated a lot of extremely valuable data on the long-term effects of penetrating TBI [74–76]. The Vietnam head injury study included baseline (pre-injury) data and a fairly homogenous group of patients (in terms of age and type of injury). Thus, interesting possibilities for translation emerge if this type of injury can be properly represented in experimental models.

A number of models have been developed to study penetrating ballistic TBI using cat [77], dog [78], monkey [79], and sheep [80]. None of these models are currently in routine use. To date, there are only two models in use and these both use rodents. In one of these a designed probe is inserted in the brain at the desired location and rapid inflation of an attached balloon is used to mimic the temporary cavity caused by energy dissipation from the bullet [81]. In the other model, a probe penetrates the brain at high velocity [82]. When shooting this probe into gel samples large temporary cavities are formed that are correlated to both speed of penetration and the shape of the penetrating object. The latter model will be presented in greater detail below.

4.1 Penetration TBI at High Velocity

A midline incision was made through the skin and periosteum of the anesthetized rat head, and a burr hole of 2.75 mm in diameter was drilled with its center slightly lateral and posterior to bregma. The rat was thereafter placed in a stereotactic frame and positioned so that a probe, commonly 2 mm in diameter with a spherical tip, was positioned directly above the dura exposed by the burr hole. This probe was fitted into a holder and guided by a narrow tube (Fig. 7). A lead pellet was accelerated by air pressure from a specially designed air rifle and impacted the probe. The probe penetrates 5–6 mm into the brain at high speed; 90 m/s at the time it starts to enter the brain.

The injury caused severe damage to the lateral and medial parietal cortecies, corpus callosum, hippocampus, and several parts of the posterior thalamus [82]. After a few days after trauma, a large cavity was formed in the brain. The penetration trauma also causes hemorrhages, blood–brain barrier breakdown, neurodegenerations, gliosis, transient changes in several behavior tests, and persistent deficiency of reference memory. There are significant changes in gene expression, both in the cerebral cortex that surrounds the penetration and in the hippocampus [42] and leakage of S-100B into serum [83]. The inflammatory response in the cortex includes an activation of the terminal pathway of the complement system [84]. This model has also been used to characterize changes in BDNF and neurotrophin receptors after penetrating injury [85] in order to provide a correlate to data on the importance of BDNF gene polymorphism for the outcome of penetrating TBI in the Vietnam Head Injury study [86].

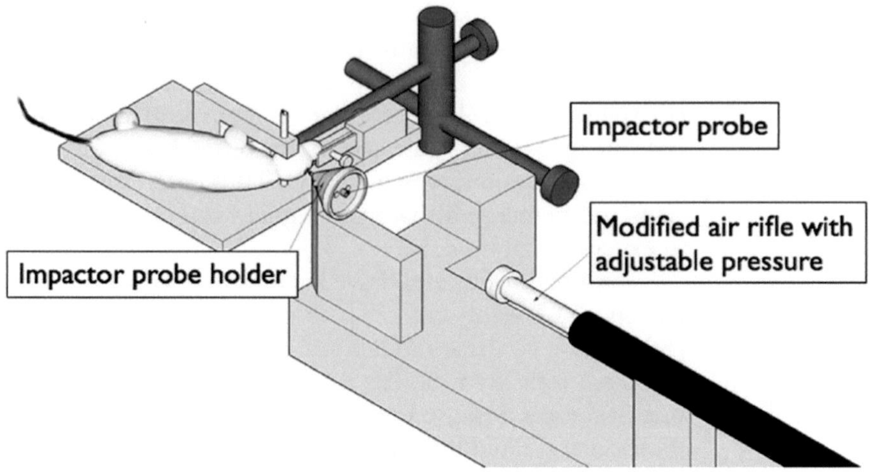

Fig. 7 The penetration rig (Previously published in [94])

5 Concluding Remarks

Far to many experimental studies end up with data that cannot be translated into clinical work. Translation can fail if the models are not properly validated and provide a good representation of the clinical injury they are supposed to model. Translation can also be impeded by large differences in the methods for outcome measure [87]. In order to facilitate the translation of experimental data, it may be important to include outcome recordings that could be used also in clinical work, such as serological biomarkers and MRI [88]. One additional strategy is to develop and use Finite Element Models of the animal and the model that is used to load the animal during the experiment. One model for the rat head has recently been employed to model rotational acceleration injury [89]. Such work can enable scaling and comparison with human clinical cases of TBI [56, 90, 91]. The choice of the animal species or strain can obviously have a significant impact on the outcome of the injury. Differences in body size and the geometry of the skull can be assumed to represent critical factors in experimental design. For example, experiments with rotational acceleration are very dependent on the distance to the axis of rotation, thus a larger brain may be far less resistant to rotational injury. Different rat strains may exhibit different inflammatory responses and reactions to TBI.

Validated and verified experimental models are necessary for successful identification of new drug candidates for treatment of TBI or SCI [92]. Differences in time scales between experimental animals, such as rodents, and humans are difficult to represent and compensate for in experimental work [93].

References

1. Ramón y Cajal S (1928) Degeneration and regeneration of the nervous system. In: DeFelipe J, Jones EG (eds) History of neuroscience, vol 5. Oxford University Press, New York, Reprinted 1991 ed
2. Richardson PM, McGuinness UM, Aguayo AJ (1980) Axons from CNS neurons regenerate into PNS grafts. Nature 284:264–265
3. Risling M, Cullheim S, Hildebrand C (1983) Reinnervation of the ventral root L7 from ventral horn neurons following intramedullary axotomy in adult cats. Brain Res 280(1):15–23
4. Lindå H, Risling M, Cullheim S (1985) 'Dendraxons' in regenerating motoneurons in the cat: do dendrites generate new axons after central axotomy? Brain Res 358(1-2):329–333
5. Lindå H, Cullheim S, Risling M (1992) A light and electron microscopic study of intracellularly HRP-labeled lumbar motoneurons after intramedullary axotomy in the adult cat. J Comp Neurol 318(2):188–208
6. Risling M, Linda H, Cullheim S, Franson P (1989) A persistent defect in the blood-brain barrier after ventral funiculus lesion in adult cats: implications for CNS regeneration? Brain Res 494(1):13–21
7. Risling M, Fried K, Lindå H, Cullheim S, Meier M (1992) Changes in nerve growth factor receptor-like immunoreactivity in the spinal cord after ventral funiculus lesion in adult cats. J Neurocytol 21(2):79–93
8. Frisén J, Verge VM, Cullheim S, Persson H, Fried K, Middlemas DS, Hunter T, Hokfelt T, Risling M (1992) Increased levels of trkB mRNA and trkB protein-like immunoreactivity in the injured rat and cat spinal cord. Proc Natl Acad Sci U S A 89(23):11282–11286
9. Frisén J, Risling M, Korhonen L, Zirrgiebel U, Johansson CB, Cullheim S, Lindholm D (1998) Nerve growth factor induces process formation in meningeal cells: implications for scar formation in the injured CNS. J Neurosci 18(15):5714–5722
10. Risling M, Fried K, Lindå H, Carlstedt T, Cullheim S (1993) Regrowth of motor axons following spinal cord lesions: distribution of laminin and collagen in the CNS scar tissue. Brain Res Bull 30(3-4):405–414
11. Deckner M, Lindholm T, Cullheim S, Risling M (2000) Differential expression of tenascin-C, tenascin-R, tenascin/J1, and tenascin-X in spinal cord scar tissue and in the olfactory system. Exp Neurol 166(2):350–362
12. Carlstedt T, Linda H, Cullheim S, Risling M (1986) Reinnervation of hind limb muscles after ventral root avulsion and implantation in the lumbar spinal cord of the adult rat. Acta Physiol Scand 128(4):645–646
13. Carlstedt T, Grane P, Hallin RG, Noren G (1995) Return of function after spinal cord implantation of avulsed spinal nerve roots. Lancet 346(8986):1323–1325
14. Carlstedt T, Hultgren T, Nyman T, Hansson T (2009) Cortical activity and hand function restoration in a patient after spinal cord surgery. Nat Rev Neurol 5(10):571–574
15. Cullheim S, Carlstedt T, Linda H, Risling M, Ulfhake B (1989) Motoneurons reinnervate skeletal muscle after ventral root implantation into the spinal cord of the cat. Neuroscience 29(3):725–733
16. Risling M, Ochsman T, Carlstedt T, Linda H, Plantman S, Rostami E, Angeria M, Skold MK (2011) On acute gene expression changes after ventral root replantation. Front Neurol 1:159
17. Chew DJ, Carlstedt T, Shortland PJ (2011) A comparative histological analysis of two models of nerve root avulsion injury in the adult rat. Neuropathol Appl Neurobiol 37(6):613–632
18. Chew DJ, Murrell K, Carlstedt T, Shortland PJ (2013) Segmental spinal root avulsion in the adult rat: a model to study avulsion injury pain. J Neurotrauma 30(3):160–172
19. Armonda RA, Bell RS, Vo AH, Ling G, DeGraba TJ, Crandall B, Ecklund J, Campbell WW (2006) Wartime traumatic cerebral vasospasm: recent review of combat casualties. Neurosurgery 59(6):1215–1225, discussion 1225
20. White CS, IG Bowen, Richmond DR (1965) Biological tolerance to air blast and related biomedical criteria. CEX-65.4. CEX [reports]; civil effects exercise. U.S. Atomic Energy Commission, 1–239
21. Richmond DR, Damon EG, Bowen IG, Fletcher ER, White CS (1967) Air-blast studies with eight species of mammals. Techn Progr Rep DASA 1854. Fission product inhalation project [technical progress report]. Lovelace Foundation for Medical Education and Research, 1–44
22. Richmond DR, Damon EG, Fletcher ER, Bowen IG, White CS (1967) The relationship between selected blast-wave parameters and the response of mammals exposed to air blast. Techn Progr Rep DASA 1860. Fission product inhalation project [technical progress report]. Lovelace Foundation for Medical Education and Research, 1–36
23. Rubovitch V, Ten-Bosch M, Zohar O, Harrison CR, Tempel-Brami C, Stein E, Hoffer BJ,

Balaban CD, Schreiber S, Chiu WT, Pick CG (2011) A mouse model of blast-induced mild traumatic brain injury. Exp Neurol 232(2): 280–289

24. Lu J, Ng KC, Ling GS, Wu J, Poon JF, Kan EM, Tan MH, Wu YJ, Li P, Moochhala S, Yap E, Lee LK, Teo AL, Yeh IB, Sergio DM, Chua F, Kumar SD, Ling EA (2012) Effect of blast exposure on the brain structure and cognition in the Macaca fascicularis. J Neurotrauma 29:1434
25. Celander H, Clemedson CJ, Ericsson UA, Hultman HI (1955) The use of a compressed air operated shock tube for physiological blast research. Acta Physiol Scand 33(1):6–13
26. Long JB, Bentley TL, Wessner KA, Cerone C, Sweeney S, Bauman RA (2009) Blast overpressure in rats: recreating a battlefield injury in the laboratory. J Neurotrauma 26(6):827–840
27. Bolander R, Mathie B, Bir C, Ritzel D, VandeVord P (2011) Skull flexure as a contributing factor in the mechanism of injury in the rat when exposed to a shock wave. Ann Biomed Eng 39(10):2550–2559
28. Chavko M, Koller WA, Prusaczyk WK, McCarron RM (2007) Measurement of blast wave by a miniature fiber optic pressure transducer in the rat brain. J Neurosci Methods 159(2):277–281
29. Chavko M, Watanabe T, Adeeb S, Lankasky J, Ahlers ST, McCarron RM (2011) Relationship between orientation to a blast and pressure wave propagation inside the rat brain. J Neurosci Methods 195(1):61–66
30. Cernak I, Merkle AC, Koliatsos VE, Bilik JM, Luong QT, Mahota TM, Xu L, Slack N, Windle D, Ahmed FA (2011) The pathobiology of blast injuries and blast-induced neurotrauma as identified using a new experimental model of injury in mice. Neurobiol Dis 41(2):538–551
31. Clemedson CJ (1949) An experimental study on air blast injuries. Acta Physiol Scand 18(Suppl LXI):7
32. Clemedson CJ, Criborn CO (1955) A detonation chamber for physiological blast research. J Aviat Med 26(5):373–381
33. Clemedson CJ, Jonsson A, Pettersson H (1956) Propagation of an air-transmitted shock wave in muscular tissue. Nature 177(4504):380–381
34. Clemedson CJ, Hultman H (1958) Cardiac output in early phase of blast injury in rabbits. Am J Physiol 194(3):601–606
35. Clemedson CJ (1956) Shock wave transmission to the central nervous system. Acta Physiol Scand 37(2-3):204–214
36. Clemedson CJ, Hartelius H, Holmberg G (1957) The effect of high explosive blast on the cerebral vascular permeability. Acta Pathol Microbiol Scand 40(2):89–95
37. Säljo A, Bao F, Haglid KG, Hansson HA (2000) Blast exposure causes redistribution of phosphorylated neurofilament subunits in neurons of the adult rat brain. J Neurotrauma 17(8):719–726
38. Säljö A, Bao F, Hamberger A, Haglid KG, Hansson HA (2001) Exposure to short-lasting impulse noise causes microglial and astroglial cell activation in the adult rat brain. Pathophysiology 8(2):105–111
39. Säljö A, Bao F, Jingshan S, Hamberger A, Hansson HA, Haglid KG (2002) Exposure to short-lasting impulse noise causes neuronal c-Jun expression and induction of apoptosis in the adult rat brain. J Neurotrauma 19(8):985–991
40. Kirkegaard M, Murai N, Risling M, Suneson A, Jarlebark L, Ulfendahl M (2006) Differential gene expression in the rat cochlea after exposure to impulse noise. Neuroscience 142(2):425–435
41. Murai N, Kirkegaard M, Jarlebark L, Risling M, Suneson A, Ulfendahl M (2008) Activation of JNK in the inner ear following impulse noise exposure. J Neurotrauma 25(1):72–77
42. Risling M, Plantman S, Angeria M, Rostami E, Bellander BM, Kirkegaard M, Arborelius U, Davidsson J (2011) Mechanisms of blast induced brain injuries, experimental studies in rats. Neuroimage 54(Suppl 1):S89–S97
43. Tate CM, Wang KK, Eonta S, Zhang Y, Carr W, Tortella FC, Hayes RL, Kamimori GH (2013) Serum brain biomarker level, neurocognitive performance, and self-reported symptom changes in soldiers repeatedly exposed to low-level blast: a breacher pilot study. J Neurotrauma 30(19):1620–1630
44. Gennarelli TA, Thibault LE, Adams JH, Graham DI, Thompson CJ, Marcincin RP (1982) Diffuse axonal injury and traumatic coma in the primate. Ann Neurol 12(6): 564–574
45. Wismans J, Janssen E, Beusenberg M, Bovendeerd P (2000) Injury bio-mechanics. Course book. Eindhoven Technical University
46. Melvin J, Lighthall WJ, Ueno K (1993) Accidental injury, biomechanics and prevention. Springer, New York, NY
47. Povlishock JT, Jenkins LW (1995) Are the pathobiological changes evoked by traumatic brain injury immediate and irreversible? Brain Pathol 5(4):415–426
48. Povlishock JT (1992) Traumatically induced axonal injury: pathogenesis and pathobiological implications. Brain Pathol 2(1):1–12

49. Smith DH, Meaney DF (2000) Axonal damage in traumatic brain injury. Neuroscientist 6(6):483–495
50. Viano DC (197) Brain injury biomechanics in closed-head impact: studies on injury epidemiology, tolerance criteria, biomechanics, and traffic injury prevention, Karolinska institutet
51. Holbourn AHS (1943) The mechanics of head injuries. Lancet ii:438–441
52. Adams JH, Doyle D, Ford I, Gennarelli TA, Graham DI, McLellan DR (1989) Diffuse axonal injury in head injury: definition, diagnosis and grading. Histopathology 15(1):49–59
53. Margulies SS, Thibault LE, Gennarelli TA (1990) Physical model simulations of brain injury in the primate. J Biomech 23(8):823–836
54. Margulies SS, Thibault LE (1992) A proposed tolerance criterion for diffuse axonal injury in man. J Biomech 25(8):917–923
55. Zhang L, Yang KH, King AI (2004) A proposed injury threshold for mild traumatic brain injury. J Biomech Eng 126(2):226–236
56. Kleiven S (2007) Predictors for traumatic brain injuries evaluated through accident reconstructions. Stapp Car Crash J 51:81–114
57. Antona-Makoshi J, Davidsson J, Ejima S, Ono K, Brolin K, Anata K (2013) Correlation of global head and brain tissue injury criteria to experimental concussion derived from monkey head trauma experiments. Proceeding of the 2013 International IRCOBI Conference on the Biomechanics of Impact
58. Strich SJ (1961) Shearing of nerve fibres as a cause of brain damage due to head injury. A pathological study of twenty cases. Lancet ii:443–448
59. Ross DT, Meaney DF, Sabol MK, Smith DH, Gennarelli TA (1994) Distribution of forebrain diffuse axonal injury following inertial closed head injury in miniature swine. Exp Neurol 126(2):291–299
60. Marmarou A, Foda MA, van den Brink W, Campbell J, Kita H, Demetriadou K (1994) A new model of diffuse brain injury in rats. Part I: Pathophysiology and biomechanics. J Neurosurg 80(2):291–300
61. Foda MA, Marmarou A (1994) A new model of diffuse brain injury in rats. Part II: Morphological characterization. J Neurosurg 80(2):301–313
62. Andersson RWG (2000) A study on the biomechanics of axonal injury, in University of Adelaide
63. Ono K, Kikuchi A, Nakamura M, Kobayashi H, Nakamura N (1980) Human head tolerance to sagittal impact reliable estimation deduced from experimental head injury using subhuman primates and human cadaver skulls. Proc. of the 24th Stapp Car Crash Conf., 101–160
64. Runnerstam M, Bao F, Huang Y, Shi J, Gutierrez E, Hamberger A, Hansson HA, Viano D, Haglid K (2001) A new model for diffuse brain injury by rotational acceleration: II. Effects on extracellular glutamate, intracranial pressure, and neuronal apoptosis. J Neurotrauma 18(3):259–273
65. Gutierrez E, Huang Y, Haglid K, Bao F, Hansson HA, Hamberger A, Viano D (2001) A new model for diffuse brain injury by rotational acceleration: I model, gross appearance, and astrocytosis. J Neurotrauma 18(3):247–257
66. Ellingson BM, Fijalkowski RJ, Pintar FA, Yoganandan N, Gennarelli TA (2005) New mechanism for inducing closed head injury in the rat. Biomed Sci Instrum 41:86–91
67. Fijalkowski RJ, Stemper BD, Pintar FA, Yoganandan N, Crowe MJ, Gennarelli TA (2007) New rat model for diffuse brain injury using coronal plane angular acceleration. J Neurotrauma 24(8):1387–1398
68. Davidsson J (2008) A new model, experiments and injury threshold for sagittal plane rotational induced diffuse brain injuries in 6th Frame work EU-project APROSYS2008
69. Davidsson J, Risling M (2011) A new model to produce sagittal plane rotational induced diffuse axonal injuries. Front Neurol 2(41):1–11
70. Davidsson J, Angeria M, Risling M (2013) Effect of age on amount and distribution of diffuse axonal injury after rotational trauma. Proceeding of JSAE Annual Congress, Yokohama, Japan, May 22–24, 2013
71. Davidsson J, Angeria M, Risling M (2009) Injury threshold for sagittal plane rotational induced diffuse axonal injuries. International IRCOBI conference on the biomechanics of injury. York, UK: IRCOBI Conference
72. Rostami E, Davidsson J, Ng KC, Lu J, Gyorgy A, Walker J, Wingo D, Plantman S, Bellander BM, Agoston DV, Risling M (2012) A model for mild traumatic brain injury that induces limited transient memory impairment and increased levels of axon related serum biomarkers. Front Neurol 3:115
73. Talving P, Benfield R, Hadjizacharia P, Inaba K, Chan LS, Demetriades D (2009) Coagulopathy in severe traumatic brain injury: a prospective study. J Trauma 66(1):55–61, discussion 61-2
74. Raymont V, Salazar AM, Krueger F, Grafman J (2011) "Studying injured minds" - the Vietnam head injury study and 40 years of brain injury research. Front Neurol 2:15

75. Grafman J, Schwab K, Warden D, Pridgen A, Brown HR, Salazar AM (1996) Frontal lobe injuries, violence, and aggression: a report of the Vietnam Head Injury Study. Neurology 46(5):1231–1238
76. Salazar AM, Schwab K, Grafman JH (1995) Penetrating injuries in the Vietnam war. Traumatic unconsciousness, epilepsy, and psychosocial outcome. Neurosurg Clin N Am 6(4):715–726
77. Carey ME, Sarna GS, Farrell JB, Happel LT (1989) Experimental missile wound to the brain. J Neurosurg 71(5 Pt 1):754–764
78. Tan Y, Zhou S, Liu Y, Li Z (1998) A gross and microscopic study of cerebral injuries accompanying maxillofacial high-velocity projectile wounding in dogs. J Oral Maxillofac Surg 56(3):345–348
79. Crockard HA, Brown FD, Johns LM, Mullan S (1977) An experimental cerebral missile injury model in primates. J Neurosurg 46(6):776–783
80. Finnie JW (1993) Pathology of experimental traumatic craniocerebral missile injury. J Comp Pathol 108(1):93–101
81. Williams AJ, Hartings JA, Lu XC, Rolli ML, Dave JR, Tortella FC (2005) Characterization of a new rat model of penetrating ballistic brain injury. J Neurotrauma 22(2):313–331
82. Plantman S, Ng KC, Lu J, Davidsson J, Risling M (2012) Characterization of a novel rat model of penetrating traumatic brain injury. J Neurotrauma 29(6):1219–1232
83. Risling M, Sköld M, Larsson IL, Angeria M, Davidsson J (2004) Leakage of S-100 protein after high velocity penetration injury to the brain. 7th international neurotrauma symposium. Adelaide: Medimond
84. Rostami E, Davidsson J, Gyorgy A, Agoston DV, Risling M, Bellander BM (2013) The terminal pathway of the complement system is activated in focal penetrating but not in mild diffuse traumatic brain injury. J Neurotrauma 30(23):1954–1965
85. Rostami E, Krueger F, Plantman S, Davidsson J, Agoston D, Grafman J, Risling M (2014) Alteration in BDNF and its receptors, full-length and truncated TrkB and p75(NTR) following penetrating traumatic brain injury. Brain Res 1542:195–205
86. Rostami E, Krueger F, Zoubak S, Dal Monte O, Raymont V, Pardini M, Hodgkinson CA, Goldman D, Risling M, Grafman J (2011) BDNF polymorphism predicts general intelligence after penetrating traumatic brain injury. PLoS One 6(11):e27389
87. Agoston DV, Risling M, Bellander BM (2012) Bench-to-bedside and bedside back to the bench; coordinating clinical and experimental traumatic brain injury studies. Front Neurol 3:3
88. Kamnaksh A, Budde MD, Kovesdi E, Long JB, Frank JA, Agoston DV (2014) Diffusion tensor imaging reveals acute subcortical changes after mild blast-induced traumatic brain injury. Sci Rep 4:4809
89. Antona-Makoshi J, Davidsson J, Risling M, Ejima S, Ono K (2014) Validation of local brain kinematics of a novel rat brain finite element model under rotational acceleration. Int J Automot Eng 5:31–37
90. Kleiven S, Hardy WN (2002) Correlation of an FE model of the human head with local brain motion--consequences for injury prediction. Stapp Car Crash J 46:123–144
91. Kleiven S (2013) Why most traumatic brain injuries are not caused by linear acceleration but skull fractures are. Front Bioeng Biotechnol 1:15
92. Agoston DV, Risling M (2012) Where will the (new) drugs for traumatic brain injury treatment be coming from? Front Neurol 3:27
93. Agoston DV (2013) Of timescales, animal models, and human disease: the 50th anniversary of C. elegans as a biological model. Front Neurol 4:129
94. Risling M, Davidsson J (2012) Experimental animal models for studies on the mechanisms of blast induced neurotrauma. Front Neurol 3(30):1–9

Chapter 17

A Porcine Model of Traumatic Brain Injury via Head Rotational Acceleration

D. Kacy Cullen, James P. Harris, Kevin D. Browne, John A. Wolf, John E. Duda, David F. Meaney, Susan S. Margulies, and Douglas H. Smith

Abstract

Unique from other brain disorders, traumatic brain injury (TBI) generally results from a discrete biomechanical event that induces rapid head movement. The large size and high organization of the human brain makes it particularly vulnerable to traumatic injury from rotational accelerations that can cause dynamic deformation of the brain tissue. Therefore, replicating the injury biomechanics of human TBI in animal models presents a substantial challenge, particularly with regard to addressing brain size and injury parameters. Here we present the historical development and use of a porcine model of head rotational acceleration. By scaling up the rotational forces to account for difference in brain mass between swine and humans, this model has been shown to produce the same tissue deformations and identical neuropathologies found in human TBI. The parameters of scaled rapid angular accelerations applied for the model reproduce inertial forces generated when the human head suddenly accelerates or decelerates in falls, collisions, or blunt impacts. The model uses custom-built linkage assemblies and a powerful linear actuator designed to produce purely impulsive non-impact head rotation in different angular planes at controlled rotational acceleration levels. Through a range of head rotational kinematics, this model can produce functional and neuropathological changes across the spectrum from concussion to severe TBI. Notably, however, the model is very difficult to employ, requiring a highly skilled team for medical management, biomechanics, neurological recovery, and specialized outcome measures including neuromonitoring, neurophysiology, neuroimaging, and neuropathology. Nonetheless, while challenging, this clinically relevant model has proven valuable for identifying mechanisms of acute and progressive neuropathologies as well as for the evaluation of noninvasive diagnostic techniques and potential neuroprotective treatments following TBI.

Key words Traumatic brain injury (TBI), Biomechanics, Neuropathology, Diffuse brain injury (DBI), Axonal injury, Modeling, Degeneration, Concussion

1 Introduction

1.1 Traumatic Brain Injury (TBI)

TBI represents a major health and socioeconomic problem, as annually in the USA alone there are over 80,000 deaths with over five million exhibiting chronic neurological deficits [1–6]. The so-called "mild" TBI, otherwise known as concussion, is astonishingly

Firas Kobeissy et al. (eds.), *Injury Models of the Central Nervous System: Methods and Protocols*, Methods in Molecular Biology, vol. 1462, DOI 10.1007/978-1-4939-3816-2_17, © Springer Science+Business Media New York 2016

prevalent as it is estimated that 1.6–3.8 million sports-related concussions occur in the USA each year [2, 7–11]. Moreover, recent military conflicts have seen a dramatic increase in the prevalence of TBI compared to twentieth century wars, causing TBI to be described as the "signature injury" of the modern warfighter [12]. The persisting and even progressive neuropathology and neurological dysfunction triggered by this mechanical injury represents a particularly unique challenge, as noted in preclinical models and humans [13–20].

As a heterogeneous disorder, long-term outcome following TBI is dependent on the type and severity of the initial physical event (primary injury) compounded by multifaceted pathophysiological consequences (secondary injuries) [21–27]. The primary injury represents physical damage on the macro- (tissue tears, vascular disruption), micro- (cell shearing), or nano-scale (cytoskeletal breakage, plasmalemmal damage). Complex secondary pathophysiological cascades include inflammation and reactive gliosis, edema, metabolic deficits, loss of ionic homeostasis, aberrant enzymatic activation, increase in reactive oxygen species, excitotoxicity, hypoxia, and altered cell signaling [16, 26, 28–32]. These deleterious cascades may lead to prolonged cellular dysfunction, axonal degeneration, and cell death [32–36]. Collectively, the initial injury and evolving pathophysiology often lead to neurodegeneration and other pathologies progressing over weeks, months, years, or even decades [16, 20, 24, 31, 32, 37–41]. Due to chronic and progressive mechanisms of neurophysiological dysfunction and neuronal/axonal degeneration, TBI can be considered an acute biophysical trauma that can lead to a neurodegenerative disease state in some cases.

Across the severity spectrum of TBI, outcomes vary from temporary, mild cognitive deficits to permanent, severely debilitating changes affecting motor function, emotion, and cognition [41, 42]. Even concussion may lead to cognitive disruptions immediately post-injury as well as persistent neurological deficits [43–46]. Moreover, functional impairment following TBI may be prolonged due to complex degenerative cascades, the limited regenerative ability of the brain, and lack of effective treatments. Unfortunately, despite reports of hundreds of treatments that have show efficacy in rodent models of TBI, none have translated to clinical use despite over 30 clinical trials based on the preclinical data [47, 48]. While the lack of positive findings in clinical studies may reflect the complexity and heterogeneity of human TBI and challenges in clinical trial design [28], it may also serve as a cautionary tale of the inability of rodent models to replicate the pathophysiology and neurodegenerative sequelae of clinical TBI [49].

1.2 TBI Biomechanics: The Importance of Head Rotational Acceleration

TBI is unique from any other neurological affliction in that it is induced by a discrete physical event. The vast majority of clinical TBIs are closed-head (i.e., non-penetrating) diffuse brain injuries caused by inertial loading to the head [2, 17, 30, 50–52]. The cause of the inertial loading is the transfer of kinetic energy, typically based on the body/head having momentum and impacting a larger object (e.g., concrete in the case of a fall or a dashboard in the case of a motor vehicle collision) to cause rapid deceleration; and/or the head being impacted by a object having significant momentum (e.g., tackler in American football, head collisions in soccer) to rapidly accelerate the head (with deceleration often occurring due to another impact and/or due to anatomical limitations, e.g., the chin impacting the chest). Of note, the head may be loaded without an impact, as the transfer of kinetic energy can occur through the body (e.g., restrained occupant in a motor vehicle collision). In moderate-to-severe TBI, the impact loading itself commonly exerts focal effects such as overt bleeding and contusion on the brain surface. However, approximately 90% of clinical TBIs are classified as mild, which by definition are closed-head injuries and not typically associated with bleeds or contusions, although there is debate in this area [53, 54]. In concussion, although impact and inertial loading of the head generally act in combination, the linear forces associated with head impact primarily serve to rapidly accelerate or decelerate the head relative to the body, thus generating rapid angular acceleration–deceleration of the head.

Such rapid rotational loading of the head generates complex stress–strain fields throughout brain tissue, which was suggested over 70 years ago to be the principal cause of diffuse brain injury [55, 56] (Fig. 1). This relationship is exemplified with concussion in particular, where loss-of-consciousness or other neurological deficits occur absent a focal (impact) contusion, suggesting that a blow to the head causes diffuse strain fields associated with rapid head rotation (inertial loading). Indeed, the importance of head movement/rotation to the etiology of concussion was established in seminal studies by Drs. Denny-Brown and Russell, who found that transient loss-of-consciousness was readily induced when the head was free to rotate, but not when forces were applied to a fixed head [55]. This work was extended in a series of seminal studies conducted by Dr. Ommaya and colleagues at the National Institutes of Health in the 1960s and 1970s establishing the importance of head rotational versus linear (translational) accelerations in loss-of-consciousness and associated neuropathology using nonhuman primates (NHP) [57–62]. Here, neurological endpoints such as loss-of-consciousness and coma were rarely obtained with impacts causing linear head motion—and when present, only at extremely high "g" levels—but rather occurred with much lower impact thresholds when the head was free to rotate causing angular acceleration. This was also reflected in the presence and distribution of diffuse axonal injury (DAI)—the hallmark pathology of closed-head TBI—which was suggested to be biomechanically induced based on the generation of diffuse strain patterns in the brain following rotation, but was not present with linear head motion.

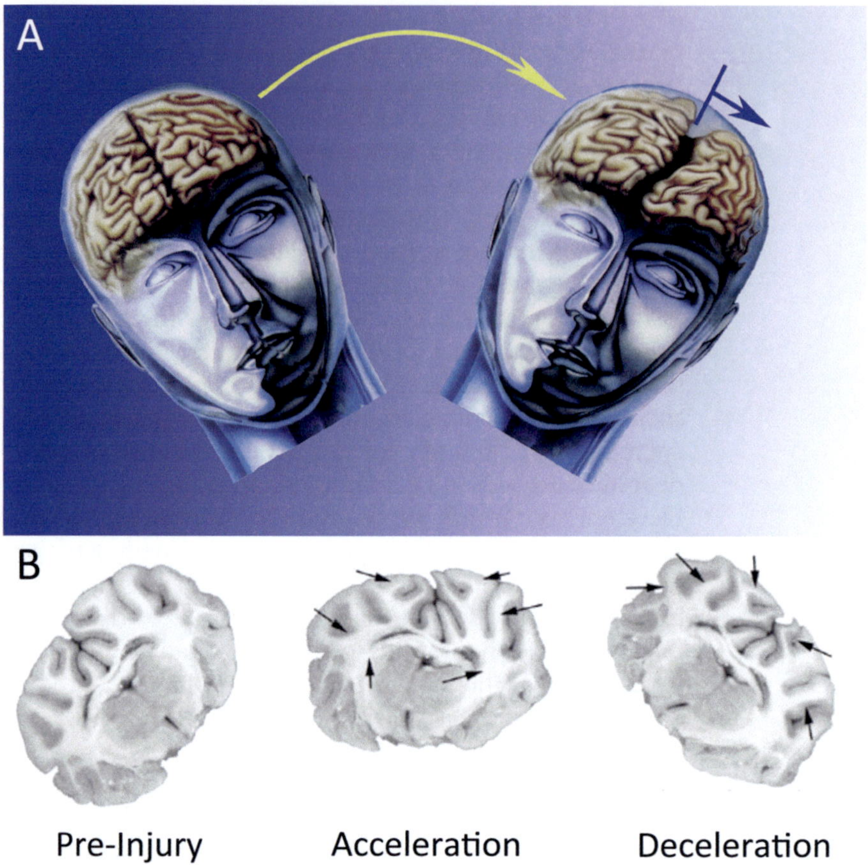

Fig. 1 Modeling closed-head diffuse brain injury in swine. (**a**) Conceptual schematic of diffuse TBI in humans, which is most often caused by rapid rotational acceleration/deceleration of the head. Such inertial loading due to angular acceleration/velocity generates diffuse strain patterns in the brain. (**b**) Mock-up showing predicted tissue deformation during head rotational acceleration–deceleration loading of the pig brain during rapid head rotation in the coronal plane. Head rotational parameters, in particular angular acceleration, may be scaled from humans to gyrencephalic mammals based on brain mass. Figure adapted with permission from [30]

After numerous studies rigorously testing linear versus rotational head acceleration in loss-of-consciousness, coma, intracerebral hemorrhage, and DAI, Ommaya and Gennarelli concluded that:

> At equivalent levels of input acceleration, rotation of the head appears to be necessary for loss of consciousness as well as productive of diffuse and focal lesions in the brain, the main damage distribution being at brain surfaces and at zones of changes in density of the intracranial tissues. Translation of the head in the horizontal plane on the other hand produces essentially focal effects only, resulting in well-circumscribed cerebral contusions and intracerebral hematomas; such focal effects do not appear adequate for the production of cerebral concussion or other evidence of diffuse effects on the brain. [50]

Collectively, this body of work over the 1960–1970s established a causal link between the physical and physiological consequences of TBI, and in particular the importance of head rotational acceleration in the etiology of diffuse brain injury. Building on these seminal studies, subsequent work applied analytical and physical models to determine the relevant tissue-level strain fields resulting in the previously observed neuropathology, suggesting the need for high strains (10–50 %) at the tissue- and cell-level, delivered at rapid strain rates of 10–50 s^{-1} (i.e., over tens of milliseconds) [63–67]. Of note, many biophysical responses in neural cells have been shown to be strain rate-dependent owing to the viscoelastic nature of cells where at high strain rates—characteristic of TBI—the cellular structures may behave in a brittle manner whereas at low strain rates structures may be compliant even for large strains [68–74]. At or beyond these empirically derived biomechanical thresholds, there will be an element of physical tissue, cellular, and/or axonal damage that may be focal, multifocal, or completely diffuse, resulting in varied manifestation from overt disruption of intra- and extra-cerebral vasculature (e.g., acute subdural hematoma) to subtle cellular and/or subcellular damage (e.g., diffuse axonal injury in subcortical white matter); with particularly vulnerable brain regions depending on factors such as the local neuroanatomy, micro-structural discontinuities (i.e., interfaces), cell (tract) orientation, and plane of head rotation [63, 64, 75–79]. Moreover, in closed-head diffuse brain injury, the dominant mode of brain tissue deformation is shear strain, owing to the shear modulus being at least several orders of magnitude lower than the bulk modulus of brain tissue [56, 70, 80, 81]. From a tissue mechanics standpoint, this makes the strain fields in the brain highly dependent on rotational loading, but relatively impervious to linear translation—thus furthering the links between head rotational acceleration, tissue strain levels, and resulting macro- to micro-neuropathology.

Remarkably, the collective experimental observation over the past 50 years have borne out predictions made by Holbourn in the 1940s, who postulated that rotational forces were necessary to generate shear strain patterns in viscoelastic (and virtually incompressible) soft tissue, whereas significant tissue deformation fields would not be achieved by linear forces [56]. Holbourn's own work on the mechanics of head injury coupled with the physiological work of his contemporaries led him to conclude that "concussion is a rotational injury" [56]. Thus, it is the consensus view in the field that this rapid rotational loading of the head and neck about the craniocervical junction and torso (i.e., acceleration–deceleration inertial loading) is the proximal cause of diffuse brain injury in general and concussion in particular. Therefore, attention to the tissue and cellular biomechanics of injury—based on tissue- and cellular-level strain fields—is critically important to fully describe clinical TBI across the spectrum of severities as well as to validate experimental preclinical models.

1.3 The Importance of Large Animal Models of TBI and Human Relevance

The complex neuroanatomy and neurophysiology of humans and other large mammals such as pigs, denoted by gyrencephalic brains, substantial white matter domains and specific pathophysiological features, may be key factors in the development of specific features of trauma-induced neuropathology [30, 82]. Although rodent models provide a powerful platform to elucidate mechanisms of trauma-induced neurodegeneration, they may not be suitable to mimic all aspects of clinical TBI. Therefore, it is important that biomechanically and neuroanatomically dependent phenomena elucidated in rodent models be subsequently evaluated in large animal models of TBI prior to extrapolation to humans. In particular, there are crucial differences between the rat (or mouse) and pig brain that must be considered in modeling closed-head TBI. First, brain mass is an important consideration, and is a key parameter in closed head inertial brain injury where mass-mass effects dominate [56, 61, 62, 75]. Large-animal models of closed-head TBI may be uniquely capable of replicating the tissue-level biomechanics of inertial brain injury marked by diffuse strain patterns in the brain. It is a significant challenge to achieve sufficient head rotational acceleration in small rodents to mimic the tissue-level forces without inducing compression effects or rupture of the vasculature, although some have proposed models for this purpose [83, 84]. Another key similarity between the human and pig brain is gross neuroanatomy, as humans and most large mammals possess gyrencephalic (3D gyri and sulci) brains with substantial white matter domains whereas rats and mice have lissencephalic brains with a paucity of white matter. Specifically, human and porcine brains exhibit a similar 60:40 ratio of white matter to gray matter, whereas that ratio is 14:86 in rats and 10:90 in mice [85–87]. This discrepancy in white matter geometry and volume between large mammals versus rodents is a crucial component for the fidelity of modeling the mechanisms and distribution of DAI, the hallmark pathology of closed-head diffuse brain injury across a range of severities in humans [15, 30, 51, 52, 88–91]. Also, cortical neuronal degeneration immediately post-trauma and in chronic traumatic encephalopathy follows a distinct pattern with respect to the macro neuroanatomy [38]—which would be impossible to mimic using lissencephalic rodents. Pathophysiological components are also important. Rodents (absent genetic modifications) normally do not acquire specific neurodegenerative pathologies such as Aβ plaques; however, these pathologies are found in swine post-TBI [19, 92]. Moreover, the majority of rodent TBI studies utilize open skull techniques and are dominated by focal/impact injuries, which poorly replicate diffuse brain injury and inherently possess craniectomy and/or bleeding, which confound any attempt at "mild" levels of TBI [93]. Overall, models employing lissencephalic animals may fail to capture the mechanisms and distribution of acute pathophysiological responses and neuropathological manifestation such as neuronal and axonal degeneration.

The relatively low mass rodent brain make attempts to properly scale closed-head inertial forces prohibitively challenging, and DAI is a pathology difficult to replicate in lissencephalic rodents with a paucity of white matter. Thus, a large animal model with biomechanical fidelity to clinical TBI is valuable to identify acute and chronic pathophysiological and neurodegenerative changes following closed-head TBI and to validate pathophysiological mechanisms found using rodents.

1.4 Overview of Large Animal TBI Studies at the University of Pennsylvania

After Ommaya's seminal studies in the 1960s and 1970s, researchers focused on gaining an improved understanding of the biomechanics, injury etiology, and treatment of diffuse traumatic white matter injury using several species, ages, assessments, and post-injury intervals ranging from hours to months [64, 75–77, 79, 94–110]. The initial studies in the 1970s through the early 1980s investigated a spectrum of brain injuries ranging from mild cerebral concussion through severe injuries such as DAI with prolonged coma and/or acute subdural hematoma in the NHP [75–77, 94]. Of note, while mild brain injuries (e.g., concussion) are the most common types of brain injuries, DAI and subdural hematoma are responsible for approximately 70 % of the mortality and morbidity associated with brain injury. These landmark studies included a broad range of rotational loading directions, acceleration amplitudes, as well as repeated and single loads, and revealed thresholds for concussion, coma, and subdural hematoma as a function of rotational acceleration levels across different planes of rotation [75–77, 94]. Moreover, this work revealed patterns of prominent neuropathology including DAI similar to that seen in humans postmortem [75]. These initial studies were highly impactful for the TBI field, greatly increasing our understanding of the biomechanical etiology and thresholds for a range of neurological and neuropathological consequences of TBI. Importantly, these studies built on Ommaya's work establishing that diffuse brain injuries were predominantly caused by rotational accelerations of the head.

1.5 Overview of the Injury Apparatus for Rapid Head Rotation Without Impact

In the early 1990s, Gennarelli began working with Meaney and Smith to develop a model of non-impact closed-head rotational TBI in adult swine [64, 79, 92, 95–99, 111–114]. This model was later adapted by Margulies and colleagues as a pediatric model of TBI using neonatal to adolescent swine [100–109, 115–124]. For these studies, the injury device subjects the porcine head to non-impact, rapid angular acceleration to induce inertial forces common in human TBI resulting from falls, impacts, or collisions [64, 79, 125]. This porcine model has been shown to induce reproducible neurological and neuropathological deficits ranging from short-term neurological abnormalities to prolonged loss-of-consciousness/coma, mild edema to profound increases in intracranial pressure, vascular abnormalities with or without the presence of overt hemorrhages, and, upon postmortem examination, astrogliosis, neuro-inflammation, perikaryal degeneration and multifocal DAI [79, 92, 96, 98] (*see* section 3 *Outcome Measurements* below). Due in large part to neuroanatomical similarities between humans and pigs and biomechanical inputs representative of diffuse rotational loading in humans, this model is the most clinically relevant model of closed-head diffuse brain injury in use today. Of note, outcomes vary based on plane of head rotation, escalating head rotational acceleration/velocity kinematics, and time post-injury. Importantly, the emergence and distribution of these features—with proper biomechanical scaling—mirror that seen in humans across the spectrum from "mild" to "severe" TBI. Indeed, it has been well established that the porcine model is suitable to model clinical TBI as it satisfies key considerations related to the human condition, including the injury biomechanics, cellular biophysical responses, pathophysiological progression, and neurodegenerative sequelae of clinical TBI.

2 Application of the HYGE Device as a Large Animal Model of Closed-Head TBI

Using a pneumatic device (HYGE, Inc., Kittanning, PA; formerly BENDIX, Corp.), the injury paradigm was developed to establish a preclinical model of closed-head diffuse brain injury with biomechanical and neuropathological fidelity to inertial TBI in humans and not confounded by impact or focal contusion effects. The HYGE model is based on pure impulsive head rotational acceleration and deceleration using mammals with large brain mass and complex gyrencephalic neuroanatomy. This well-characterized model subjects the head to rapid angular acceleration using custom-built linkage assemblies coupled to a pneumatic actuator to convert linear motion to angular motion. In this fashion, this model produces pure impulsive non-impact head rotation in different planes at controlled rotational acceleration levels [64, 79, 98, 126] (Fig. 2). Multiple strains of pigs have been used with the HYGE, including Hanford and Yucatan miniature swine and Yorkshire swine (standard North American farm strain) over a range of immature ages to the adult. In the past, all tests were conducted using female pigs; however, male pigs have recently been incorporated into studies. In all these strains, brain masses range from 35–80 g for neonatal to adolescent pigs and 80–150 g for adult pigs, and total body weight at the time of injury ranges from approximately 2–50 kg.

2.1 Induction of Head Rotational Diffuse Brain Injury Using the HYGE Pneumatic Actuator

All procedures with the model are carried out in accordance with the University of Pennsylvania's Institutional Animal Care and Use Committee and adhere to policies set forth in the *Guide for the Care and Use of Laboratory Animal, Eighth Edition*. During the procedure, the animals are fully anesthetized and physiological

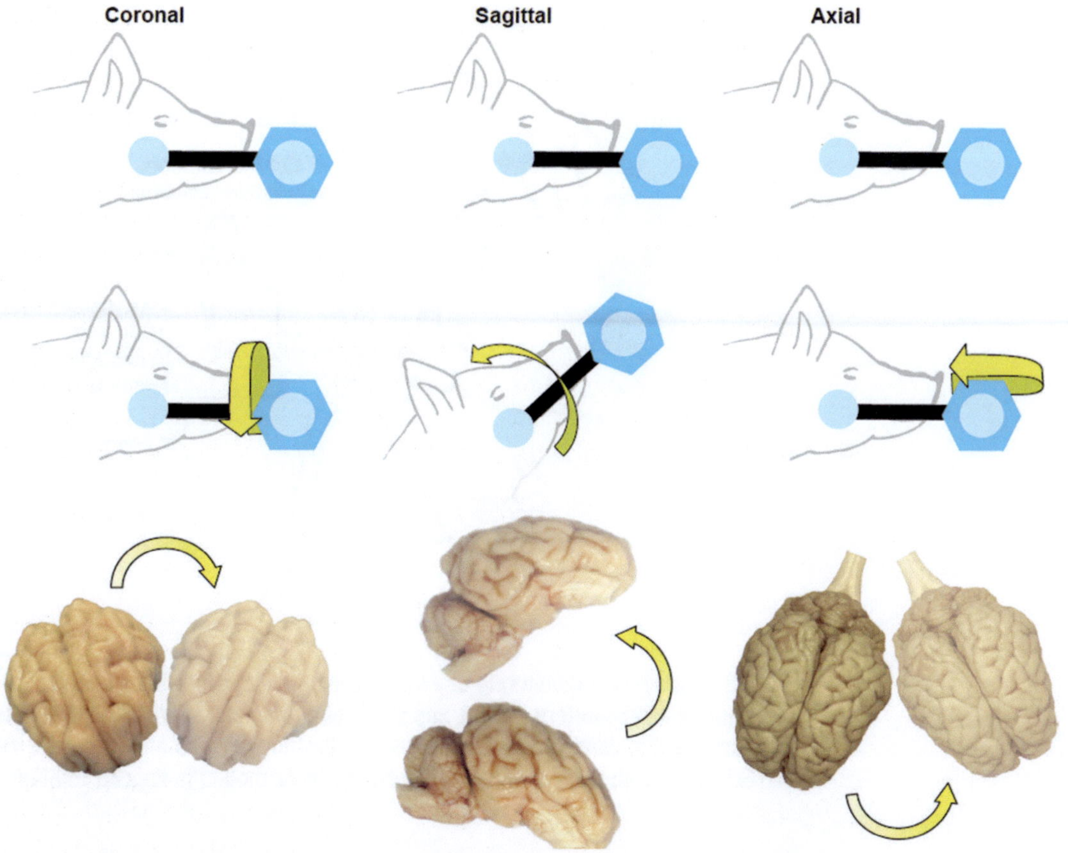

Fig. 2 Methodology for closed-head rotational acceleration TBI in swine. Closed-head diffuse brain injury was induced using rotational acceleration–deceleration of the head/brain in the coronal, sagittal, or axial plane. This provides control over the afflicted anatomical substrates and the extent of injury

parameters are monitored continuously throughout the procedure, including SpO_2, heart rate, respiratory rate, and temperature (Fig. 3). For coupling to the linkage assembly, the animals are secured to a custom-built bite plate designed to accommodate the jaw and snout.

2.2 HYGE Device Operation

The HYGE device uses compressed gas to accelerate an internal piston that moves a thrust column in a programmable linear fashion. This shaft is then coupled externally to a custom-built external linkage assembly to produce the desired kinematics (Fig. 4). Specifically, the kinematic linkage assembly is directly coupled to the thrust column of the HYGE actuator, and converts the linear action of the thrust column to angular (rotational) motion. The actuator consists of a pneumatic cylinder 6 in. in diameter, and the internal piston is surrounded by a hydraulic fluid (i.e., transmission fluid) within the shaft. The piston is driven by a preset differential pressure using compressed nitrogen and is capable of generating 40,000 lb of thrust (over 18,000 kg)

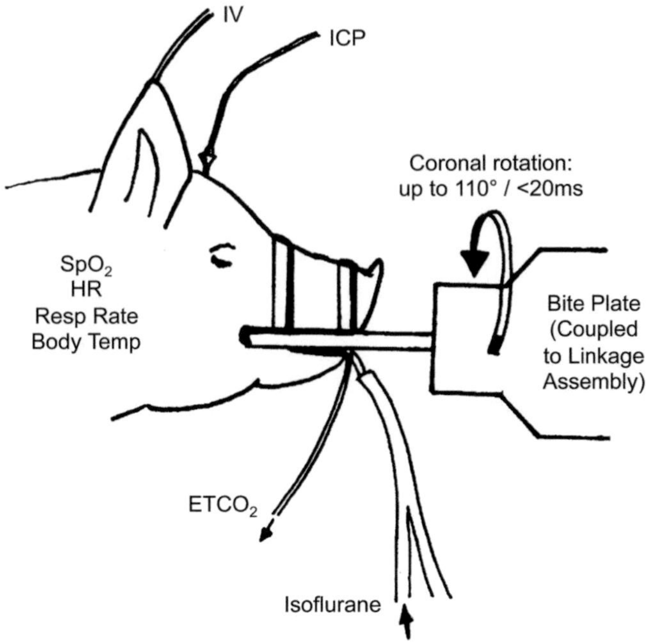

Fig. 3 Physiological monitoring of experimental subjects. Physiological parameters are measured before, during, and after head rotational injury using the HYGE device. The bite plate demonstrates the way that the pig was mounted onto the HYGE device with custom linkages to convert linear motion to angular rotation

in less than 6 ms. The linkage incorporates the ability to independently control the center of rotation, degree of angular excursion, and direction of the motion relative to the anatomy. All of the aspects of acceleration waveform and magnitude are controlled by custom designed metering pins located internal to the HYGE device.

The side arms and bite plate (and thus the animals' head) are transduced rotationally by the linkage assembly upon activation of the HYGE piston (Fig. 4). Importantly, the HYGE device permits head rotations in the sagittal, coronal, horizontal, and oblique planes, with a center of rotation about the cervical spine. The force generated by the piston is determined by differential pressure levels inputted into load and set chambers, thus providing direct control of the magnitude of the rotational acceleration transduced by the linkage assembly. Two metering pins are used to create a biphasic, acceleration–deceleration load time history. The magnitude of these components can be adjusted by changing the metering pin profiles to produce a predominant deceleration phase if desired. Thus, the HYGE device has the capability to independently modify angular acceleration/deceleration and angular velocity by using alternative acceleration and deceleration metering pins, respectively (Fig. 4). Although the duration of the inertial load is comparable to impact loading conditions (3–20 ms), a limitation of the

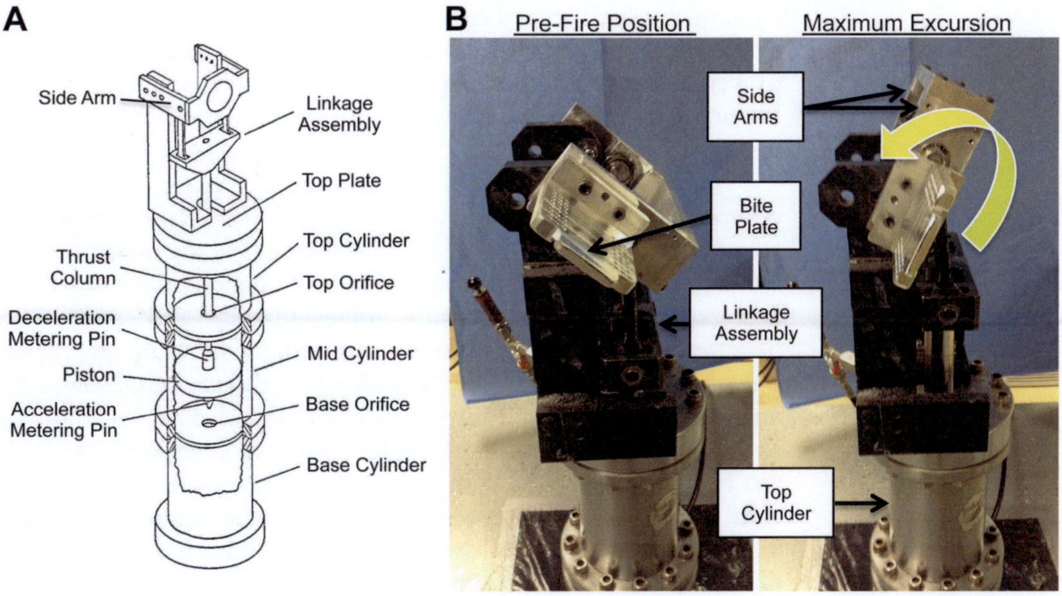

Fig. 4 HYGE device and custom-built linkage assembly for head rotation in the coronal plane. (**a**) Schematic of the HYGE pneumatic actuator and custom linkage assembly to convert the linear motion of the HYGE piston to angular motion. (**b**) Pictures of the HYGE device used to deliver head rotational acceleration in swine, demonstrating the pre-loading and maximum rotation position

HYGE device is that it may not model the sudden deceleration (or acceleration) that can occur with severe blunt impact. Long-term usage of the system has provided calibration curves to enable the HYGE to be set to a specific angular velocity based on the input pressures and the particular side arms used. As a result, one can reproducibly control the angular velocity and acceleration of the head, and thus the severity of the injury.

2.3 Measurement of Injury Kinematics

The injury kinematics are measured for each study. Specifically, angular velocity is measured by a magnetohydrodynamic sensor (custom-built ARS-06 from Applied Technology Associates, Albuquerque, NM) mounted to the linkage assembly sidearm. Since the linkage arm and the head are rigidly affixed to each other, tracking the linkage arm angular velocity equates with tracking the angular velocity of the head without the confounder of direct attachment to the scalp (angular velocity is the same anywhere along the same lever arm). The sensor transduces angular velocity into an electric field (voltage) that is generated by the movement of a conducting fluid in relation to a permanent magnet. Each sensor has been calibrated by Applied Technology Associates to allow conversion of voltage to angular velocity. The voltage is measured by a National Instruments data acquisition system running custom written LabView software to acquire voltage samples at 10 kHz (one sample every 0.1 ms), and then these measurements are converted to angular velocity based on the calibration of each individual sensor. The angular movement can be visualized by plotting angular velocity versus time. Angular displacement is calculated by integrating the angular velocity, whereas angular acceleration is calculated by taking the derivative of the angular velocity as described [107].

A second derivative of the filtered angular velocity can also be computed to calculate the angular jerk. From these operations, traces for the angular position, velocity, and acceleration may be attained for a given injury (*see* Table 1). For each injury, traces are analyzed to compute maximums, minimums, and averages for each parameter (Fig. 5). The start and end points are manually picked out with assistance of custom software that shows each instance the angular velocity crossed zero (negative to positive value or positive to negative value). The peak velocity point is a key component in the HYGE movement. The point not only provides a straightforward metric for injury severity, but this point also provides a dividing mark between a positive phase of acceleration and a negative phase of acceleration (deceleration). Analysis is performed separately on both phases. From all the traces and start/end points, motion parameters of the injury may be extracted and compared from trial to trial. Under these parameters, the HYGE device is capable of excursions up to 110° in <20 ms (generally <12 ms), generating angular velocities of up to 350 rad/s at angular accelerations up to 300,000 rad/s.

Table 1
HYGE kinematic parameters

Angular velocity (rad/s)
Maximum and minimum velocity Average positive velocity (mean, median)
Time: Start of movement to max velocity Time: Maximum velocity to zero velocity Time: Total movement time
Angular acceleration (rad/s^2)
Maximum acceleration Minimum acceleration Maximum – minimum acceleration Average positive acceleration (mean, median) Average negative acceleration (mean, median)
Time: Duration of positive acceleration Time: Duration of negative acceleration Time: Max velocity (zero acceleration) to minimum acceleration
Position (radians and degrees)
Distance (radians) to maximum velocity point Total distance (radians)

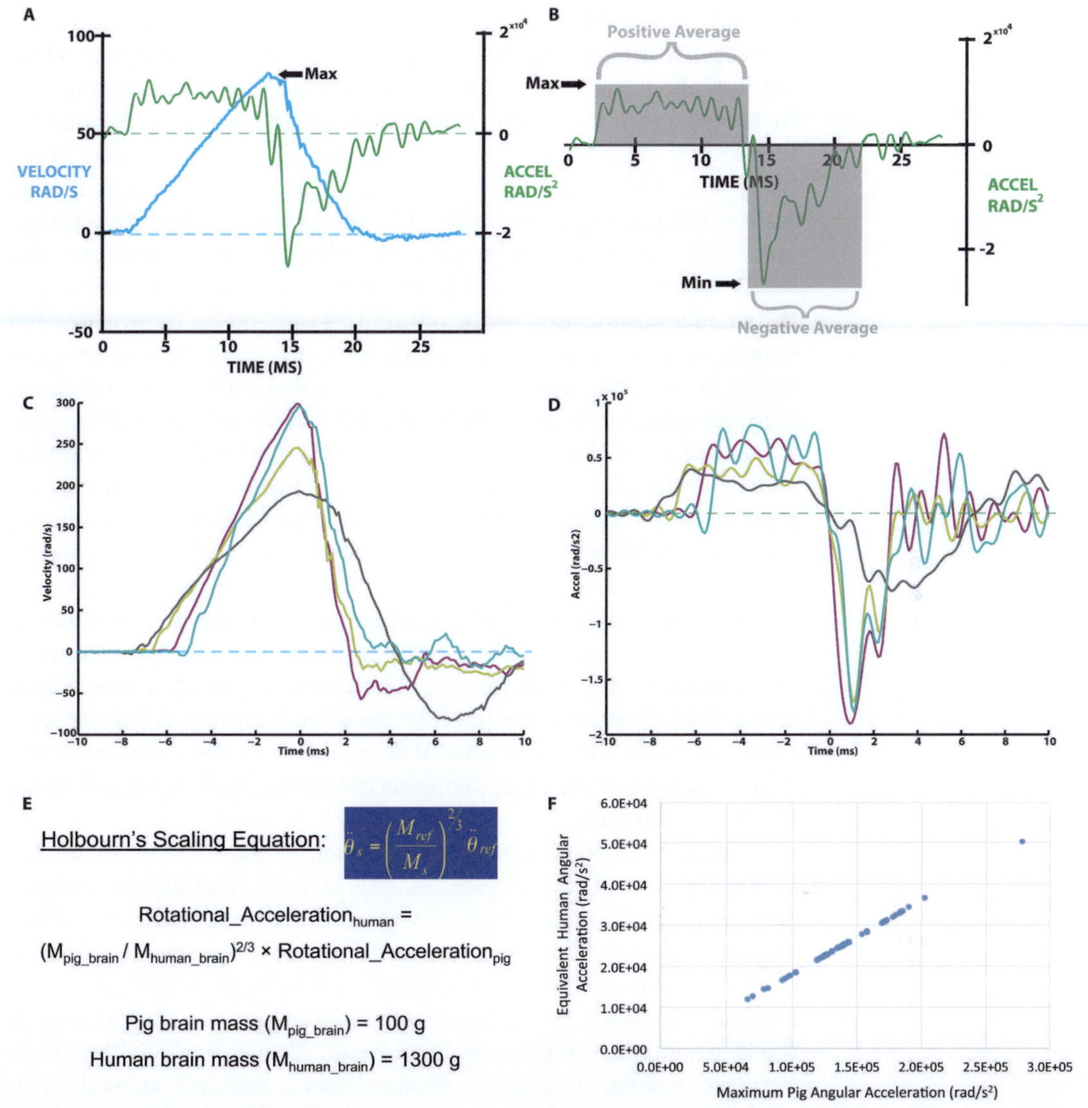

Fig. 5 HYGE kinematics and scaling to human inertial TBI. (**a**) Representative concurrent angular velocity and angular acceleration traces. (**b**) Angular acceleration trace highlighting the acceleration and deceleration phases. (**c-d**) Overlay of multiple color-coded (**c**) angular velocity and (**d**) angular acceleration traces to show changes in waveform slopes and duration for various rotational scenarios. (**e**) Holbourn's Scaling Equation and relevant assumptions to scale head rotational loading from humans to pigs. (**f**) For a fixed ratio of brain masses, angular acceleration scales linearly between humans and pigs

2.4 Injury Biomechanics: Human Scaling, Physical Models, and Anatomical Considerations

The loading conditions generated by this device closely approximate the conditions of inertial brain injury in humans based on brain mass scaling. Traditionally, the rotational accelerations necessary to scale tissue-level forces from human to pigs has been accomplished based on Holbourn's Scaling Equation with subsequent refinements (A.H.S. Holbourn [1956]; private communication to Dr. Sabina Stritch, October 13, 1956; [127]). Thus, the smaller porcine brain mass (<150 g) requires higher levels of rotational acceleration to produce injuries that mimic those seen in the adult human (brain mass approximately 1000–1500 g) (Fig. 5). Porcine to human kinematic scaling has been improved to account for additional factors such as age and brain tissue mechanical properties [104]. An alternative scaling relationship based on rotational velocity has also previously been proposed [127]. Recently, we have used these simplified scaling relationships to transfer the rotational accelerations associated with concussion in humans, described in recent studies to be 5600–8000 rad/s [128–132], to equivalent rotational motions in the smaller young adult swine brain [114]. Based on these estimated scaling relationships and the range of brain mass for humans and young adult swine, we calculated that coronal plane rotational accelerations ranging from 28,000–59,000 rad/s (corresponding with rotational velocities of approximately 110–150 rad/s) were associated with these concussion thresholds in humans. In addition, we previously determined that axial plane accelerations in pigs caused increased localized strains in the brainstem region [133] compared to coronal plane accelerations, so we therefore calculated a proportionally lower level of peak rotational acceleration in this plane of 14,000–30,000 rad/s (corresponding with rotational velocities of approximately 95–120 rad/s). Therefore, equivalent tissue-level strain fields between the porcine brain and human brain are predicted to occur during rotational acceleration at rates approximately 4–6 times greater for pigs, owing to the reduced brain mass of swine relative to humans.

There is a broad range of rotational acceleration levels attainable using the HYGE device, as well as the capability for angular acceleration and deceleration to be modified independently of the angular velocity. This unique capability may allow the elucidation of the relative contributions of rotational velocity, rotational acceleration, and acceleration duration to determine the importance of these kinematic variables on injury risk. To further establish links between macro- and micro-biomechanical features as well as to advance scaling to human injuries, the HYGE device has been used to subject physical models of the skull - brain structures to identical loading conditions used to produce specific brain injuries in animals [63, 65, 66, 104, 126]. The data from these physical model experiments together with an analytical approximation of the deformations of the tissues of the brain allowed the development of early correlations between the specific brain injuries and the loading conditions. Thus, the HYGE device is ideally suited to identify biomechanical thresholds for concussion and various neuropathologies based on targeted kinematic parameters, and relate these outcomes to predicted and measured tissue-level strain fields.

A consideration in extrapolating findings from this porcine model to human TBI is based on whole-organism neuroanatomical differences. As quadrupeds, pigs alter the dynamics of tissue strains across distinct structures based on brain anatomy, likely affecting specific regions such as cerebellar versus cerebral strain fields. Also, the relationship between the center-of-mass and the center-of-rotation varies across injury planes for human versus pigs. In humans, the center of mass and center of rotation are approximately the same for rotation in the horizontal/axial plane, but different for head rotation in the sagittal and coronal planes. Alternatively, in quadrupeds, the center of mass and center of rotation are the same for rotation in the coronal plane, but differ in the sagittal and horizontal/axial planes (see Fig. 2). Notably, sagittal is the only plane where the relationship between the location of the center of mass and center of rotation is similar between humans and quadrupeds (proportionally). These notable differences between human and porcine anatomy are known parameters that can be accounted for experimentally and in mathematical simulations of strain fields.

3 Outcome Measurements

Over the last 20 years, this unique porcine model of TBI has been instrumental in seminal discoveries linking the biomechanics, pathobiology, physiological, and cognitive/behavioral outcomes of closed-head TBI. Moreover, these studies have improved our ability to monitor and noninvasively assess overt and subtle pathological features of injury. Thus far, studies using this model have been reported in numerous publications [64, 79, 92, 95–109, 111–124].

3.1 Neurological Recovery

This porcine model of non-impact closed-head rotational-acceleration induced TBI results in different neurological outcomes which are dependent on the plane of head rotation and the level of rotational acceleration/velocity [67, 98, 114, 126, 134]. In general, adult swine undergoing rotational injury in the coronal plane at the levels tested (typically 120–300 rad/s) experience brief or no apnea and do not present a measurable loss-of-consciousness. These coronal-rotated animals generally recover quickly and without overt signs of injury. Specifically, the animals regain consciousness within 15–30 min of removing anesthesia (i.e., indistinguishable from sham animals), require little to no oversight during recovery, and become ambulatory, regain balance, and self-feed within a few hours of the procedure. In contrast, animals are more vulnerable to head rotation in the sagittal or axial planes. At head rotational velocity levels above 110 rad/s in these planes, the animals generally exhibit some degree of loss-of-consciousness (i.e., transient or prolonged/coma), typically on the order of hours.

These animals often need continuous oversight during recovery, with many requiring ventilation due to vascular compromise, brain swelling and marked increases intracranial pressure (ICP) as described below. Head rotational velocity beyond 130 rad/s generally results in persistent coma and prolonged neurointensive care and ventilation. Moreover, acute changes in EEG activity have been noted in injured animals [98]. Changes included slowing of alpha rhythms in the frontal and parietal areas and intermittent rhythmic high amplitude theta and delta activity. Thus the full spectrum of acute neurological outcomes may be attained based on rotational levels, including no overt changes, transient loss-of-consciousness, prolonged coma, and even death. Overall, based on loss-of-consciousness and neurological recovery, head rotation in the coronal plane at levels generated by the HYGE are considered to induce a "mild" TBI phenotype, whereas sagittal or axial plane rotation is considered to induce a "mild," "moderate," or "severe" TBI phenotype depending on the head rotational levels employed [98, 114, 126, 134]. Of note, these observations have been consistent across sexes and strains, and are specifically based on the ranges of angular velocities/acceleration tested to date.

To semiquantitatively assess the acute neurological recovery and depth of unconsciousness in brain-injured pigs, we have developed a numerical coma scale scoring system based on the following categories: corneal reflex (0 = absent, 1 = unilateral, 2 = bilateral); response to pain (0 = absent, 1 = movement without any sign of intention, 2 = movement with intention); spontaneous eye opening (0 = negative, 2 = positive); and righting reflex (2 = positive). The severity of coma was determined by the sum of the scores: 0–1 represented severe coma, 2 or 3 moderate coma, 4 or 5 mild coma, and 6–8 emergence from coma [98]. A coma scale score is determined for each animal at 30-min intervals beginning immediately following the injury. Further evaluation was based on gross neurosensory examination including normal startle reflexes, gait, rooting behavior, eating, and drinking; however, these are only qualitatively assessed.

3.2 ICP Changes and Edema

A common cause of death and long-term disability following severe TBI is devastating elevations in ICP caused by vascular compromise and/or secondary sequelae causing edema. Indeed, the control of increased ICP is a major therapeutic goal in neurointensive care and neurosurgical settings. As such, ICP measurements before and after TBI in a completely closed-head environment have a significant research value in order to better understand the mechanisms of TBI-induced ICP increases and ultimately to optimize effective ICP-management therapies for patients. The porcine model of closed-head rotational TBI is ideally suited to acquire such dynamic measurements using human-scale devices (Fig. 6). Accordingly, in recent studies we developed custom-built, small, fully implantable wireless devices (both analog and digital) capable of continuously measuring ICP prior to, during, and after head rotational acceleration in swine [135–137]. Across various studies, the mean baseline ICP ranged from 9.5 ± 3.4 to 16.7 ± 4.6 mmHg (mean ± standard deviation; typically measured over a ~24 h period before injury), and varied based on activity level and body position. Following rapid head rotation, device integrity and positioning

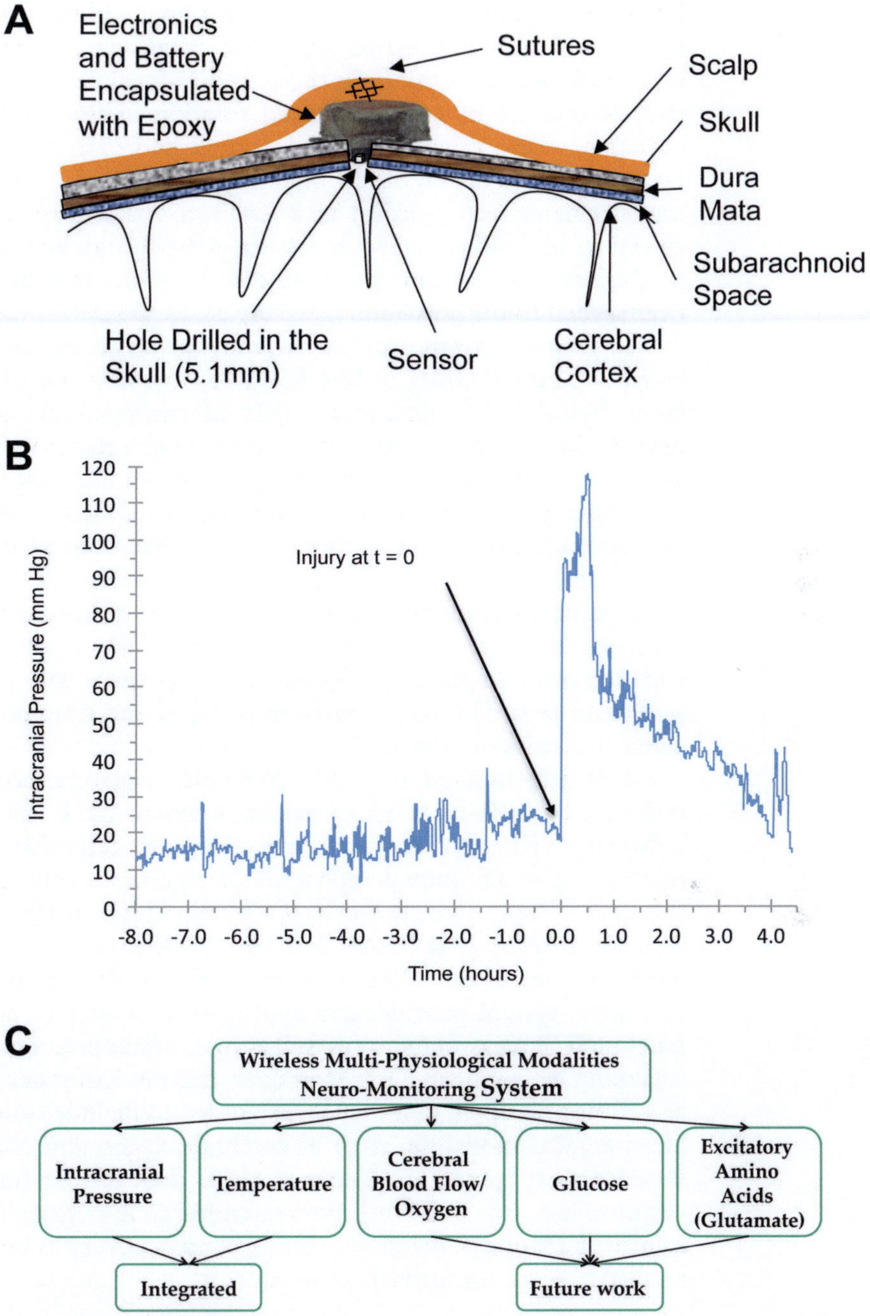

Fig. 6 Telemetry-based neuromonitoring system in swine model of closed-head rotational injury. (**a**) Schematic of implanted custom-built device contained above the skull with a burr hole for sensor access to CSF. (**b**) Example ICP trace from wireless device before and after closed-head TBI in swine. In this study, baseline ICP readings were relatively stable over the 8 h prior to injury at 16.7 ± 4.6 mmHg (mean ± standard deviation). We found that closed-head rotation TBI induced a rapid and extreme ICP spike occurring directly upon injury. The acute elevation in ICP generally lasted for 40–60 min, followed by a gradual decline to maintain a persistently elevated level over several hours post-injury. (**c**) Current and future capabilities of this fully implantable, wireless neuromonitoring system

remained suitable for dynamic ICP reading within 2 min post-injury, which is impressive given the rotational forces used to induce diffuse brain injury in these studies (peak angular acceleration of over 50,000 rad/s). Head rotation in the sagittal plane inducing severe injury produced a rapid and extreme ICP spike occurring immediately upon injury, generally 5–7 times baseline measurements and peaking at 8–10 min post-injury. The acute elevation in ICP generally lasted for 40–60 min, followed by a gradual decline to maintain an elevated level of 2–4 times baseline over several hours post-injury.

To confirm our measurements, the gold standard Camino ICP monitor (1104B, Integra Life Sciences) was introduced into the parenchyma 1–3 h post-injury (placed contralateral to wireless device). Over multiple trials, Camino measurements were within 10% of concurrent measurements with custom implanted devices, with discrepancies potentially attributed to different placement (intraparenchyma versus subdural). Gross pathology revealed subdural hematoma in animals experiencing immediate ICP changes, whereas persistently elevated ICP was likely influenced by both cytotoxic and vasogenic edema. Moreover, the results attained with our novel implantable devices were consistent with previously published post-TBI ICP trends obtained by the Camino catheter using this swine model [79].

This fully implantable, telemetry-based neuromonitoring system may be utilized as a tool to diagnose and track ICP changes following TBI for a range of severities with diminished risk of infection. Our findings demonstrated a significant spike in ICP at the time of head acceleration and a sustained increase in ICP over a period of time post-injury. Moreover, different peak ICP levels were observed at the different injury levels. While not unexpected, this novel system provides the opportunity to acquire per-animal baseline ICP measurements as well as to continuously measure ICP following closed-head TBI. Moreover, this miniature device serves as a robust platform that may be expanded to include other critical physiological modalities such as cerebral oxygen and blood flow. Based on experimental objectives, these data can be transmitted continuously (up to 17 m) over extended time periods following injures at a range of severities, during acute recovery as well as later in awake, behaving animals post-injury.

3.3 Neuroimaging

A major goal for TBI diagnosis and treatment is the noninvasive detection of the acute and evolving neuropathological consequences of the injury. Such information would be invaluable in assessing the extent and distribution of subtle pathology following mild TBI—often difficult to diagnose—and in identifying patients most likely to require therapy or other interventions. Moreover, these techniques will provide a means to improve the efficiency and sensitivity of studies evaluating the efficacy of therapeutic intervention. Here, large animal studies using brains that more closely resemble human anatomy and employing conventional clinical imaging equipment, such as magnetic resonance imaging (MRI), are particularly valuable. Early in the history of the porcine TBI model, Kimura et al. employed magnetization transfer ratio imaging (MTR) to correlate changes in MRI scans to histological evidence of damage, opening the door to a range of imaging modalities [99, 138]. Subsequent studies utilized proton magnetic resonance spectroscopy (MRS) to detect a decline in N-acetylaspartate (NAA) in areas of confirmed axonal damage and a decrease in intracellular magnesium [139]. The introduction of newer radiological techniques, such as susceptibility weighted imaging (SWI), diffusion tensor imaging (DTI), and diffusion kurtosis imaging (DKI), along with stronger magnets, have provided more detailed analysis of changes that occur after TBI. By combining the images generated with histological evidence, this model provides the ideal vehicle to study the chronological progression of TBI, and thus facilitate diagnosis and treatment.

3.4 Gross Pathology and Neuropathology

A crucial component of the porcine-HYGE model is that it provides the opportunity to directly evaluate the gross and histopathological consequences of TBI. Typically, there are no overt changes in the gross appearance of the brain and often no evidence of bleeds in animals that have undergone rotation in the coronal plane (over the range of head rotational levels evaluated to date), and in rare cases where bleeds are present they are localized and modest. In contrast, brains injured in the sagittal or axial planes may display signs of edema (moderate level) and/or subarachnoid/subdural hemorrhage (moderate and severe). When present, blood is usually found in the tentorium and around the base of the brain (Fig. 7). The varied neurological recovery (described previously) and level of vascular involvement underscores the multifaceted nature and complexity of head rotational TBI in this swine model, as is the case across the severity spectrum of TBI in humans.

The porcine model has been used to increase our understanding of the distribution, progression, and mechanisms of neuropathology following closed-head diffuse brain injury. Prominent changes have varied based on neuroanatomical locale and severity of head rotation, and include astrogliosis, neuro-inflammation, perikaryal degeneration and multifocal DAI (Fig. 8) [79, 92, 96, 98]. In particular, DAI has been demonstrated to be one of the most common and important pathologic features following closed-head TBI in humans and in animal models [50–52, 75, 79]. DAI is a major feature of closed-head diffuse brain injury in swine, and manifests as accumulation of axonal transport proteins such as amyloid precursor protein (APP) in swollen regions of axons with eventual degeneration in stereotypical distribution suggestive of a biomechanical etiology [30, 140] (Fig. 9).

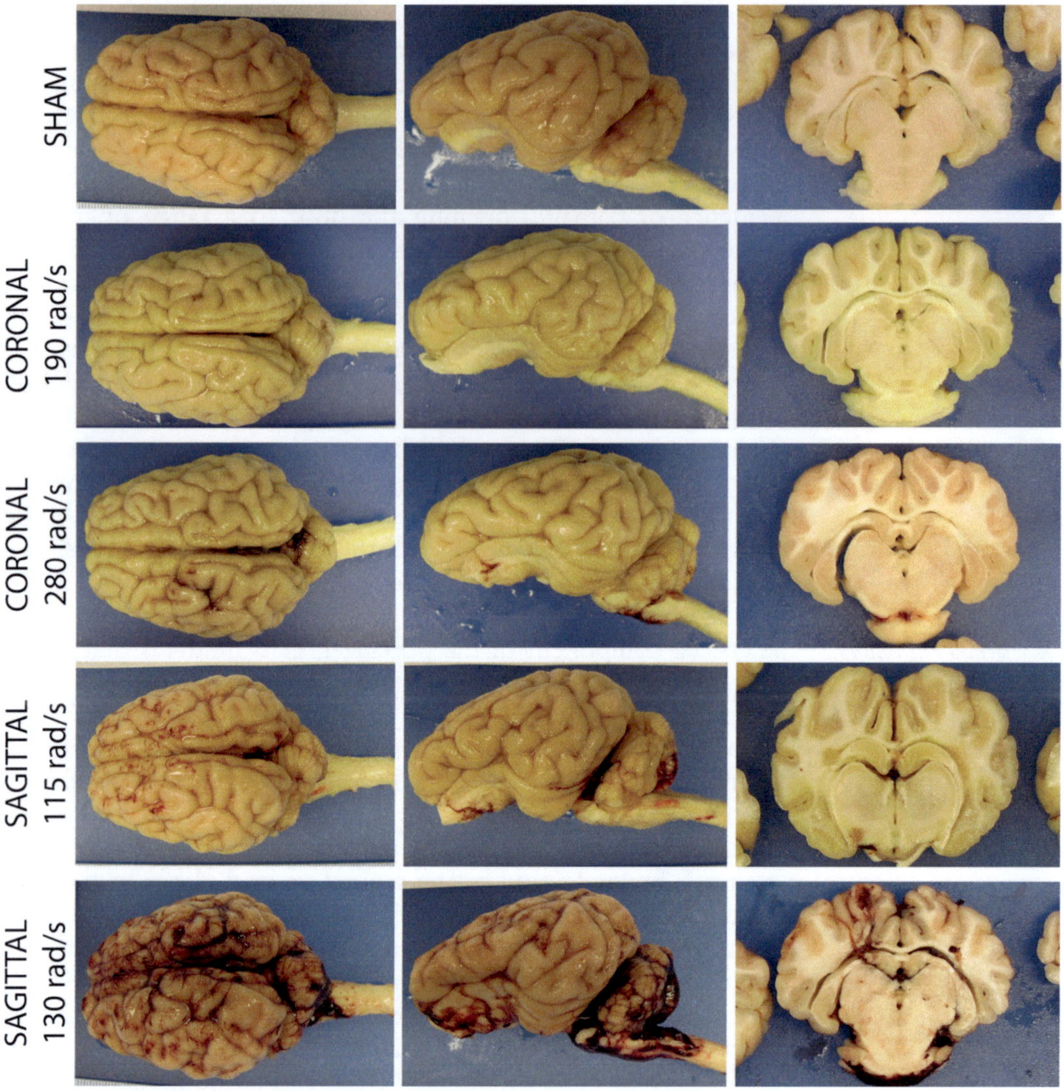

Fig. 7 Gross pathology. Thresholds for vascular compromise as a function of rotational plane and peak angular velocity. Gross pathological examinations were performed to assess the severity level and complexity of the injuries. In general, brains appeared grossly normal following coronal head rotation, and in the rare case of bleeding, such was localized and minor. Following high sagittal injuries, subdural hematoma, extensive bleeding on the brain stem and spinal cord, as well as blood accumulation within the ventricles and within cortical sulci may be observed

Moreover, the morphology of degenerating axons closely resembles that seen in human brains post-TBI (Fig. 10). In both animal models and human postmortem studies of TBI, multiple notable proteins have been shown to accumulate in degenerating axons, including Aβ, neurofilament (NF) proteins, and α-synuclein [140, 142–144] (Fig. 11). Numerous additional proteins—both structural and enzymatic—have been shown to aggregate in degenerating

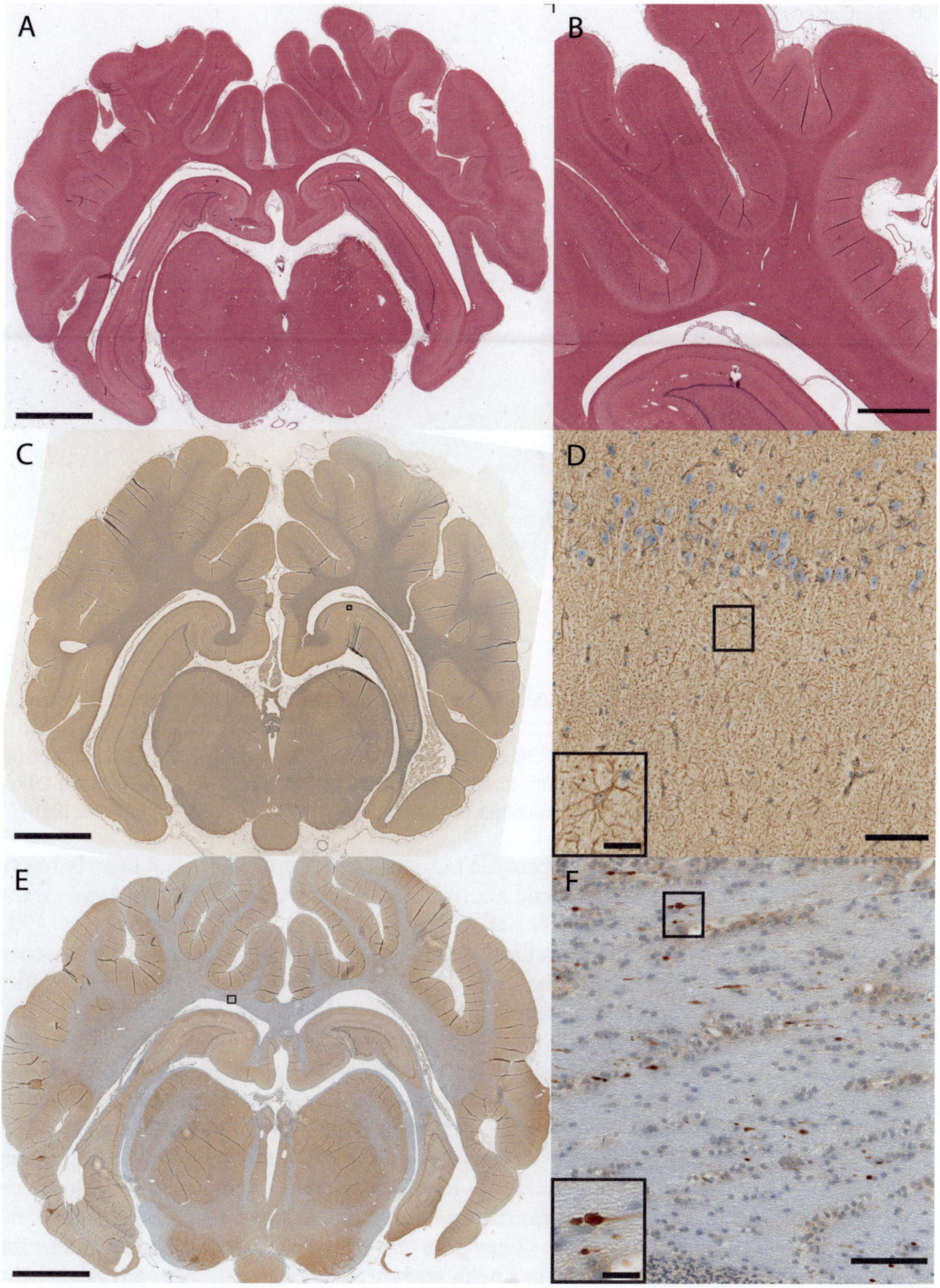

Fig. 8 Routine histopathology examination of the swine brain. (**a**, **b**) H&E examination for histological assessment of tissue/cellular structure, including detection of cell infiltration, edema, and pyknosis. (**c**, **d**) Immunohistochemistry for reactive astrogliosis based on glial fibrillary acidic protein (GFAP) immunoreactivity showing astrocyte hypertrophy (AB5804 polyclonal antibody; 1:500; Millipore, Billerica, MA). (**e**, **f**) DAI detection based on amyloid precursor protein (APP) immunoreactive axons in the subcortical white matter displaying the classic morphological appearance of traumatic axonal injury, including terminally disconnected swollen axonal bulbs (monoclonal antibody specific for the N-terminal amino acids 66-81 of APP; 1:50,000; Millipore)

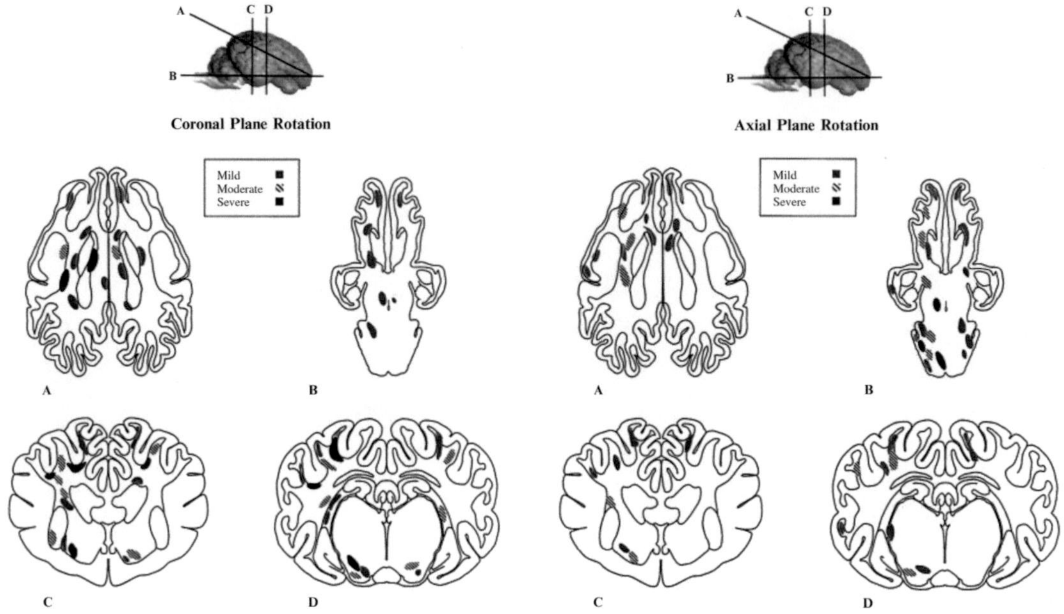

Fig. 9 Distribution of axonal pathology in pig model. Schematic representation of the distribution and severity of axonal injury following head coronal plane (*left*) and axial plane (*right*) rotation. *Lines* through the brain shown at the *top* of the figure demarcate anatomical regions of interest: frontal lobe, basal ganglia, and occipital lobe (**a**), brainstem through brain base (**b**), rostral thalamic level (**c**), and dorsal hippocampal level (**d**). Regions of axonal injury are *shaded* according to severity (mild, moderate, or severe). Reprinted with permission from ref. [98]

axons, which provide tantalizing pathophysiological links between TBI and chronic neurodegenerative sequelae, and underscore the important role of progressive DAI in these processes [92] (Fig. 12). Specifically, many of these proteins are the primary constituents of the pathologic inclusions found in several neurodegenerative diseases [19, 31, 142, 145]. It is suspected that axonal degeneration plays a critical role in chronic neurodegeneration post-TBI, whereby axonal transport is progressively blocked by the accumulation of pathological proteins, creating conditions for the propagation of such pathology gradually over time [19, 31]. Additionally, perikaryal degeneration has been observed in this model, generally following severe loading conditions and in specified neuroanatomical regions, including the cerebral cortex and hippocampus (Fig. 13). In particular, hippocampal neuronal degeneration has been shown only following relatively high head rotational levels (often complicated by hematoma) but not lower levels associated with concussion [79, 95, 114, 134].

3.5 Future Directions

Ongoing studies are expanding the use of this model to include novel measurements as well as to translate outcomes utilized in pediatric swine, including behavioral assessment [100, 109, 146],

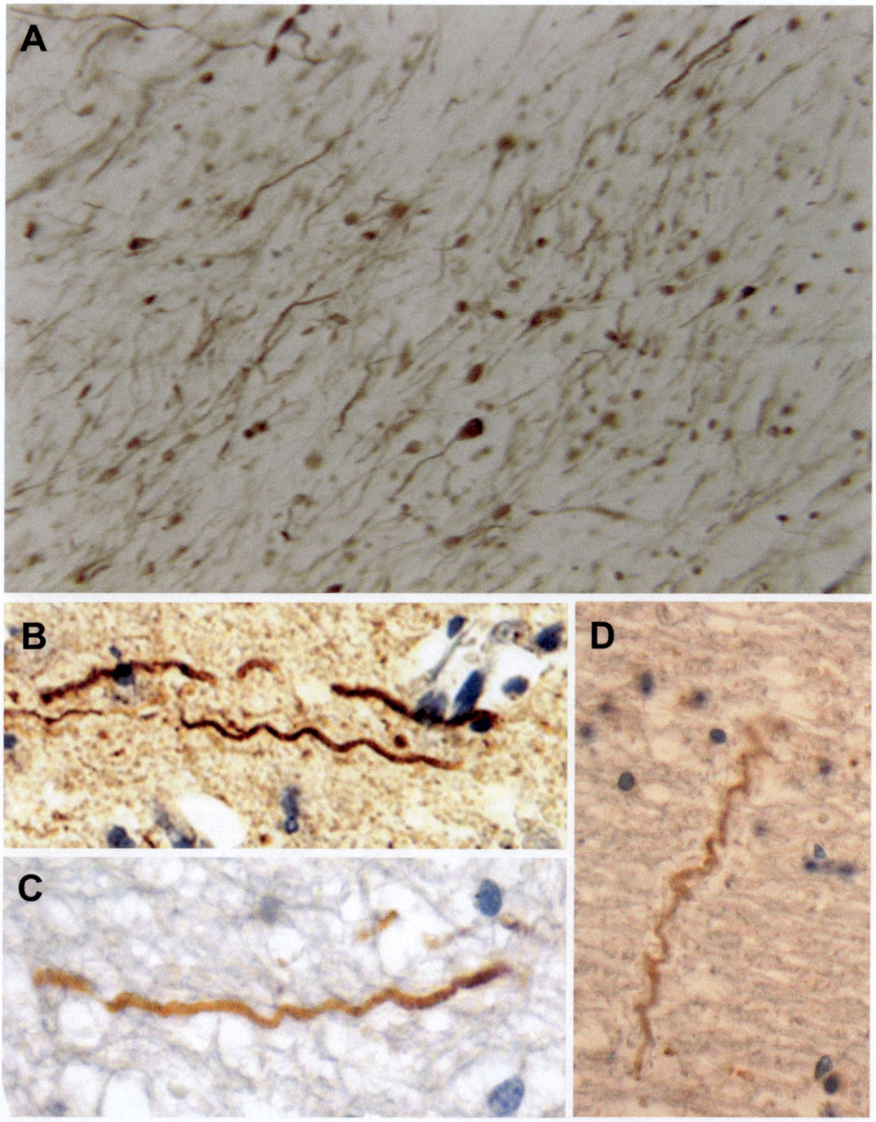

Fig. 10 Examples of axonal pathology in swine model in comparison to humans post-TBI [141]. Immunohistochemistry using specific antibodies for neurofilament and APP to identify intra-axonal accumulations and other morphological changes. (**a**) A multitude of axonal varicosities and axonal bulbs, demonstrating widespread traumatic axonal injury in pigs following head rotational acceleration. (**b–d**) Examples of trauma-induced axonal undulations in pigs and humans: (**b**) Pig 3 h post-TBI. (**c**) Human TBI: 18-year-old male, deceased 10 h following assault. (**d**) Human TBI: 18-year-old female, deceased 22 h following a motor vehicle collision. The HYGE model of TBI in swine produces diffuse axonal injury that mirrors that detected in humans post-TBI. Adapted with permission from ref. [141]

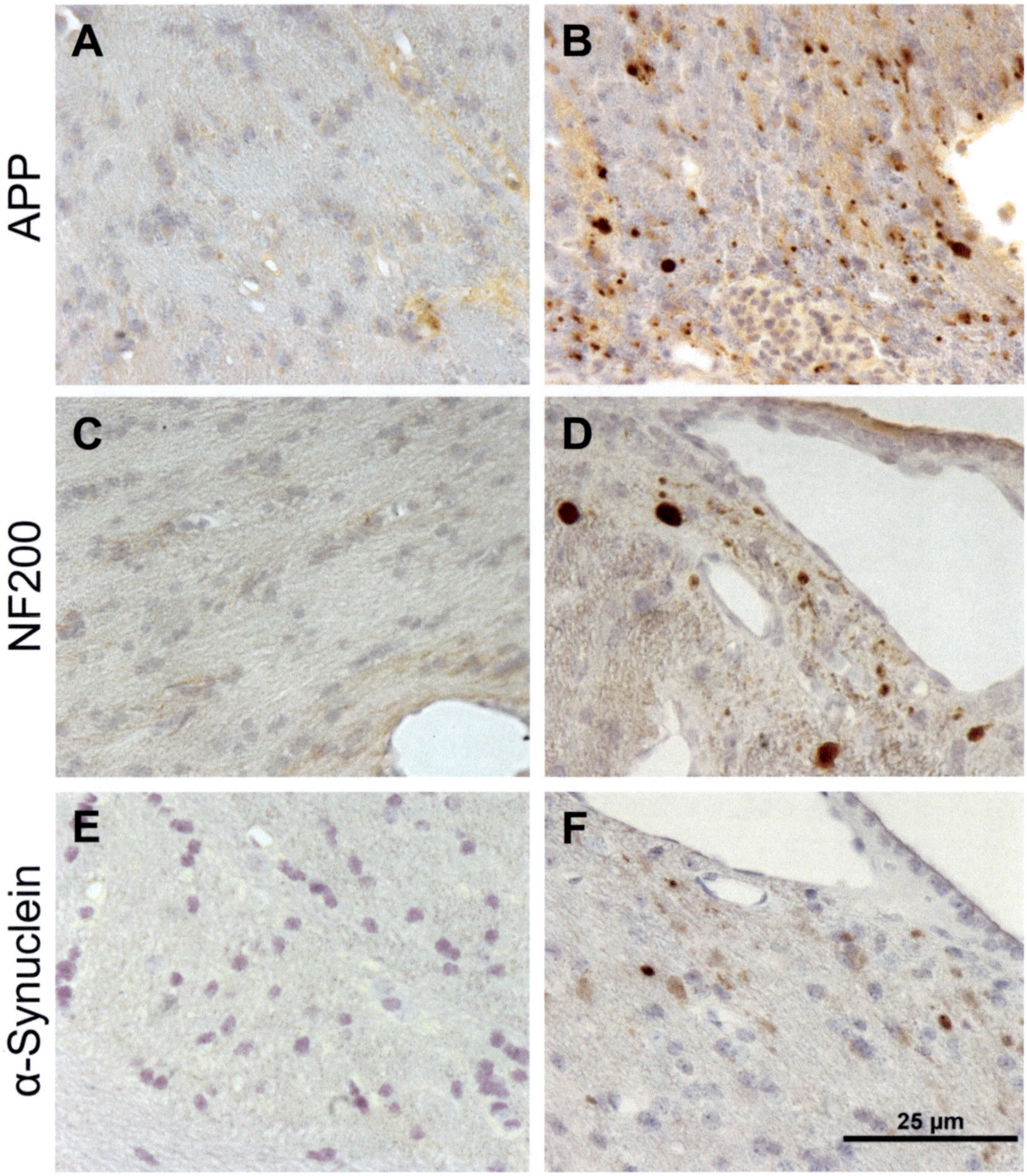

Fig. 11 Pathological accumulation of multiple proteins in axons in swine post-TBI. Corpus callosum from (**a**, **c**, **e**) sham pigs compared to (**b**, **d**, **f**) pigs at 7 days following rotational acceleration induced TBI using the HYGE device. Immunohistochemistry revealed protein accumulations of (**b**) APP, (**d**) neurofilament (NF200), and (**f**) α-synuclein (Syn303)

Fig. 12 (continued) (22C11/Red) in (*e*), and PS-1 (PS-1/Red) in (*h*). Co-accumulation of BACE (*Green*) was found with APP (*Red*) in (*c*) and (*i*), kinesin (L1/Red) in (*d*) and (*k*), and CCA (*Red*) in (*j*). Co-accumulation of APP (*Red*) was found with PS-1 (*Green*) in (*l*). In neurons, A (*Green*) co-accumulated with APP (*Red*) in (*m*) and CCA (*Red*) in (*n*). Macrophages demonstrated co-immunoreactivity of A (13335/Green) with OX42 (CD11b/Red) in (*o*) and (*p*). Scale bar = 25 μm. Reprinted with permission from ref. [92]

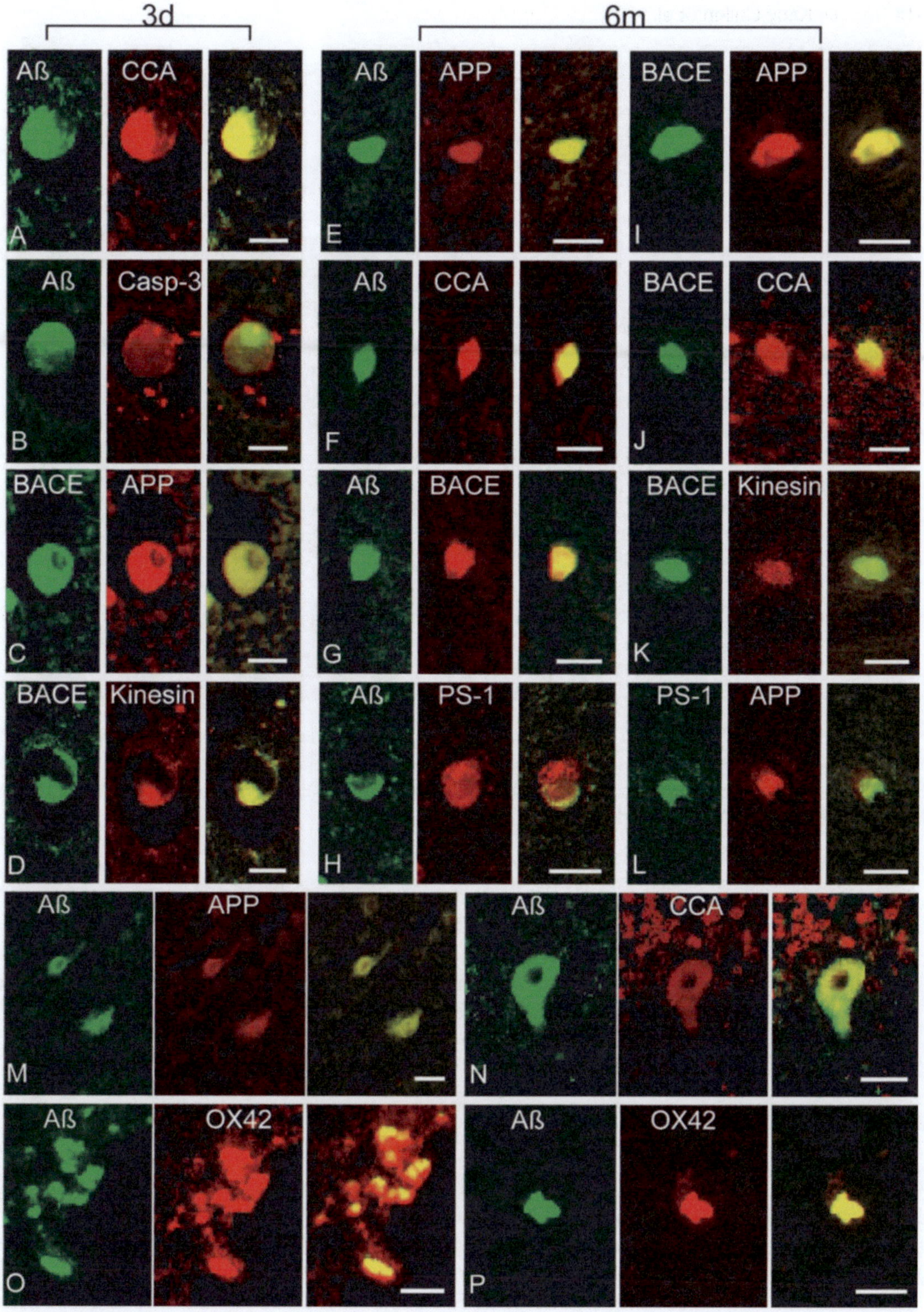

Fig. 12 Co-localization of multiple proteins in cells and axons in swine post-TBI. Representative double-immunofluorescence photomicrographs demonstrating co-accumulations of proteins in damaged (*a–l*) axons, (*m–n*) neurons and (*o–p*) macrophages at 3 days and 6 months post-injury. Merged green and red fluorescence shown in *yellow*. In axon bulbs in the white matter, co-accumulation A (antibodies 6F3D and 13335/Green) was found with CCA (249/Red) in (*a*) and (*f*), caspase-3 (P20/Red) in (*b*), BACE (BACE-2/Red) in (*g*), APP

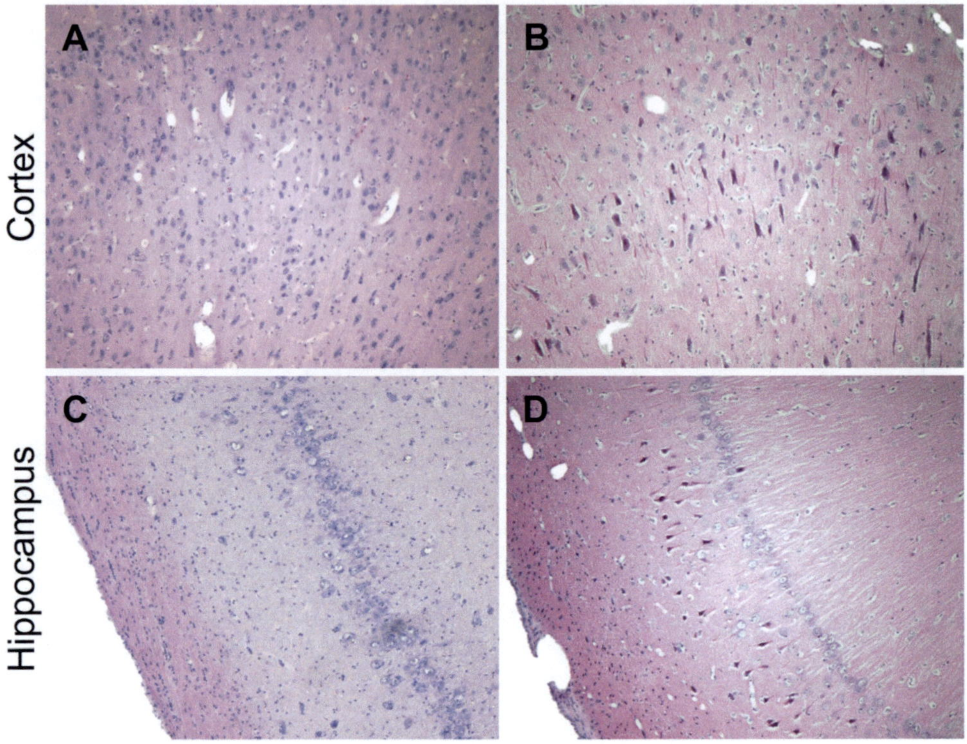

Fig. 13 Neuronal degeneration following moderate-to-severe TBI in swine. H&E staining of the cerebral cortex and hippocampus in (**a**, **c**) sham pigs and (**b**, **d**) pigs subjected to closed-head rotational acceleration using the HYGE device. Neuronal degeneration, as shown by neuronal pyknosis, was observed at 7 days following sagittal plan rotation in the (**b**) cortex and (**d**) hippocampus

neurophysiological changes [134, 146, 147], and multi-model neuromonitoring [101–103, 137]. These lines of study will further advance the capabilities, clinical relevance, and impact of this important model for improving our understanding of TBI sequelae and enhancing our ability to restore neurological function following TBI across a range of severities.

4 Overview: Why Model TBI in Pigs?

4.1 Relevance to Human TBI: Why Input Biomechanics Matters

This swine model of closed-head rotational acceleration induced TBI is a well-characterized model with biomechanical and neuroanatomical fidelity to human TBI. It is clearly important to improve our understanding of the links between the physical and physiological consequences of TBI to guide the development of targeted therapeutics to address the predominantly afflicted cell populations based on the mechanisms of injury. As in a drug study, where it is important to test the effects of a compound at a "physiologically relevant" concentration (i.e., the concentration in which it is present in the brain), when studying TBI it is crucial that the injury levels applied be "biomechanically relevant" to human TBI. Failure to be within the "relevant" regime may lead to measurement of confounding responses that do not represent pathways in the human condition. Thus, models with sufficient biomechanical fidelity to human TBI are critical to advancing our understanding of the cellular, tissue, and whole-organism responses to neurotrauma.

4.2 Biomechanical Thresholds and Injury Risk Criteria

Biomechanical thresholds and risk criteria for TBI are vigorously being investigated due to the high incidence of sports- and military-related TBI and an increased understanding of the long-term neurological and neurodegenerative consequences. It has been demonstrated that diffuse brain injury thresholds and outcomes depend on the direction of head motion as well as on the magnitude of rotational kinematics [94, 126]. Moreover, the time over which head rotation occurs is an important component of injury thresholds, as animal studies have indicated that the incidence of concussion increases when the duration of rotational acceleration is increased [57]. Furthermore, animal studies have demonstrated that the location of brain deformation may affect the resulting injury, suggesting that even a concussion-specific brain deformation threshold may vary with region [148–151].

Moreover, as described previously, it has been demonstrated in human and animal studies that higher rotational velocities and accelerations—rather than linear accelerations—are associated with larger diffuse brain deformations and worsened neurological and neuropathological outcomes [152–154]. In addition, animal studies have shown that purely linear motions produce little brain deformation or distortion and no concussion [50, 154, 155]. Of note, whereas impact forces have recently been shown to correlate with concussion thresholds and injury severity, linear accelerations are not the proximal cause of injury, rather brain tissue strain fields—and the resulting TBI—are primarily caused by the resulting head rotational acceleration levels. Although, for rare instances when head impact is in line with the center of mass of the head, linear acceleration correlates with head rotational acceleration. In these situations, linear acceleration is a reasonable surrogate for the average deformation response in the brain. However, the majority of TBIs are due to combined rotation and linear head motion; for these cases, computational simulations have predicted the relationship between the location of head impact, the kinematic response of the head (linear and rotational accelerations), and the predicted diffuse strain fields in the brain [156, 157]. In these more common non-centroidal head impacts, linear and rotational accelerations are not significantly correlated, and the rotational acceleration component of the head correlates most strongly with brain tissue deformations. As such, for the most common head impact scenarios, the linear acceleration component of the head will not adequately describe the brain's deformation response, and therefore will not be a robust predictor of TBI risk when used alone. Consequently, recent efforts to define macro-biomechanical thresholds based on linear forces will undoubtedly possess limited predictive utility for loss-of-consciousness, neuropathology, and later neurological outcomes. Although TBI risk metrics based on kinetics and kinematics of head motion are complex and multifaceted, the porcine-HYGE model is an ideal platform to relate input biomechanics to neurological and neuropathological outcomes due to fidelity to closed-head TBI in humans.

4.3 Summary

The porcine-HYGE model is biomechanically representative of human TBI based on many considerations. Closed-head diffuse brain injury in humans generally results from rapid head rotation, occurring with or without head impact, the severity of which is dependent upon the acceleration and/or deceleration of the head (non-impact or impact-induced) [56, 63–65, 75, 158–161]. There are key similarities between pig and human brain that are crucial in replicating TBI pathology. For instance, this model may uniquely represent the relevant injury biomechanics based on rapid head acceleration/rotation. This is due primarily to the anatomical advantage of the pigs possessing gyrencephalic architecture with substantial white matter domains. This complex brain architecture in swine allows replication of the diffuse tissue/cell-level strain fields responsible for cell injury in TBI in humans. As such, relevant pathophysiological consequences are also reproduced, including DAI—the predominant pathology in closed-head TBI in humans—in patterns and extent similar to humans.

Importantly, this model is biomechanically well characterized to produce primarily diffuse damage, while previous physical model and computational studies allow detailed analysis of the resulting strain fields. Indeed, when angular acceleration is scaled based on the relative masses of human and pig brains, the loading conditions generated closely approximate the conditions of inertial brain injury in humans [64, 114]. In addition, the acute neurological and gross consequences closely mirror the human condition, as this is the only model to produce the full range of acute neurological effects, ranging from no/transient loss-of-consciousness ("mild" TBI) to prolonged coma ("severe" TBI). Furthermore, vascular compromise (e.g., subdural hematoma) thresholds are scalable to human TBI, and are only present at "moderate-to-severe" injury levels. Moreover, unlike rodents, non-genetically modified pigs have been shown to develop hallmark neurodegenerative pathologies in TBI studies. Finally, outcome parameters are more relevant to humans, and include neurobehavioral, neuroimaging, neurophysiological, and neuropathological outcomes.

While this porcine model is cumbersome and labor-intensive, it is the most clinically relevant animal model of closed-head TBI in use today. Unfortunately, due to the complex nature of procedures and apparatus, our HYGE device is currently the only one in the world that is utilized as a preclinical model of TBI. The combination of high cost, low throughput, and in-depth expertise are formidable impediments to this model being adopted by more labs. However, due to the unparalleled clinical relevance it would be beneficial for this model to be employed more widely provided sufficient expertise is present. Indeed, biomechanical input parameters and animal selection should be carefully considered, and therapeutic advancements made in rodent models that are to be applied to human TBI should be considered first for confirmation in a large animal model such as the one presented here.

Acknowledgements

Financial support for this work was provided by the Department of Veterans Affairs/Rehabilitation Research & Development (Merit Review #B1097-I), the National Institutes of Health/NINDS (R01-NS-038104, R01-NS-050598 & T32-NS-043126), and University of Pennsylvania's University Research Foundation. The authors wish to thank Dr. William Stewart of the Dept. of Neuropathology and Glasgow TBI Archive, Southern General Hospital, Glasgow, UK for consultation on immunohistochemical protocols. We also thank Victoria E. Johnson, Daniel P. Brown, Michael R. Grovola, Laura A. Struzyna, and Constance J. Mietus for technical contributions.

Conflict of interest: The authors have no conflicts of interest related to this work to disclose.

References

1. Langlois JA, Rutland-Brown W, Thomas KE (2004) Traumatic brain injury in the United States: emergency department visits, hospitalizations, and deaths. U.S. Department of Health and Human Services, Washington, DC
2. Langlois JA, Rutland-Brown W, Wald MM (2006) The epidemiology and impact of traumatic brain injury: a brief overview. J Head Trauma Rehabil 21:375–378
3. Hyder AA, Wunderlich CA, Puvanachandra P, Gururaj G, Kobusingye OC (2007) The impact of traumatic brain injuries: a global perspective. NeuroRehabilitation 22:341–353
4. Thornhill S, Teasdale GM, Murray GD, McEwen J, Roy CW, Penny KI (2000) Disability in young people and adults one year after head injury: prospective cohort study. BMJ 320:1631–1635
5. Humphreys I, Wood RL, Phillips CJ, Macey S (2013) The costs of traumatic brain injury: a literature review. Clinicoecon Outcomes Res 5:281–287
6. Thurman DJ, Alverson C, Dunn KA, Guerrero J, Sniezek JE (1999) Traumatic brain injury in the United States: a public health perspective. J Head Trauma Rehabil 14:602–615
7. Graham R, Rivara FP, Ford MA, Spicer CM (2014) Sports-related concussions in youth: improving the science, changing the culture. The National Academies Press, Washington, DC
8. Jordan BD (2013) The clinical spectrum of sport-related traumatic brain injury. Nature reviews. Neurology 9:222–230
9. Prevention, C. f. D. C. a., and Control, N. C. f. I. P. a. (2003) Report to congress on mild traumatic brain injury in the United States: steps to prevent a serious public health problem. Centers for Disease Control and Prevention, Atlanta, GA

10. Prevention, C. f. D. C. a. (2011) Nonfatal traumatic brain injuries related to sports and recreation activities among persons aged ≤19 Years—United States, 2001–2009. MMWR 60:1337–1342
11. Faul M, Xu L, Wald MM, Coronado VG (2010) Traumatic brain injury in the United States: emergency department visits, hospitalizations, and deaths. Centers for Disease Control and Prevention, National Center for Injury Prevention and Control, Atlanta, GA
12. Hoge CW, McGurk D, Thomas JL, Cox AL, Engel CC, Castro CA (2008) Mild traumatic brain injury in U.S. Soldiers returning from Iraq. N Engl J Med 358:453–463
13. Smith DH, Lowenstein DH, Gennarelli TA, McIntosh TK (1994) Persistent memory dysfunction is associated with bilateral hippocampal damage following experimental brain injury. Neurosci Lett 168:151–154
14. Adelson PD, Dixon CE, Kochanek PM (2000) Long-term dysfunction following diffuse traumatic brain injury in the immature rat. J Neurotrauma 17:273–282
15. Povlishock JT, Katz DI (2005) Update of neuropathology and neurological recovery after traumatic brain injury. J Head Trauma Rehabil 20:76–94
16. Smith DH, Chen XH, Pierce JE, Wolf JA, Trojanowski JQ, Graham DI, McIntosh TK (1997) Progressive atrophy and neuron death for one year following brain trauma in the rat. J Neurotrauma 14:715–727
17. Coronado VG, McGuire LC, Sarmiento K, Bell J, Lionbarger MR, Jones CD, Geller AI, Khoury N, Xu L (2012) Trends in traumatic brain injury in the U.S. and the public health response: 1995–2009. J Safety Res 43:299–307
18. Bigler ED (2013) Traumatic brain injury, neuroimaging, and neurodegeneration. Front Hum Neurosci 7:395
19. Johnson VE, Stewart W, Smith DH (2010) Traumatic brain injury and amyloid-beta pathology: a link to Alzheimer's disease? Nat Rev Neurosci 11:361–370
20. Smith DH, Johnson VE, Stewart W (2013) Chronic neuropathologies of single and repetitive TBI: substrates of dementia? Nat Rev Neurol 9:211–221
21. McIntosh TK, Smith DH, Meaney DF, Kotapka MJ, Gennarelli TA, Graham DI (1996) Neuropathological sequelae of traumatic brain injury: relationship to neurochemical and biomechanical mechanisms. Lab Invest 74:315–342
22. Gennarelli TA (1997) The pathobiology of traumatic brain injury. Neuroscientist 3:73–81
23. Gennarelli TA (1993) Mechanisms of brain injury. J Emerg Med 11(Suppl 1):5–11
24. McIntosh TK, Saatman KE, Raghupathi R, Graham DI, Smith DH, Lee VM, Trojanowski JQ (1998) The Dorothy Russell memorial lecture. The molecular and cellular sequelae of experimental traumatic brain injury: pathogenetic mechanisms. Neuropathol Appl Neurobiol 24:251–267
25. McIntosh TK, Juhler M, Raghupathi R, Saatman KE, Smith DH (1999) Secondary brain injury: neurochemical and cellular mediators. In: Marion W (ed) Traumatic brain injury. Thieme Medical Publishers, New York, NY, pp 39–54
26. Loane DJ, Faden AI (2010) Neuroprotection for traumatic brain injury: translational challenges and emerging therapeutic strategies. Trends Pharmacol Sci 31:596–604
27. Smith DH, Meaney DF, Shull WH (2003) Diffuse axonal injury in head trauma. J Head Trauma Rehabil 18:307–316
28. Saatman KE, Duhaime AC, Bullock R, Maas AI, Valadka A, Manley GT (2008) Classification of traumatic brain injury for targeted therapies. J Neurotrauma 25:719–738
29. Stein SC, Spettell C, Young G, Ross SE (1993) Delayed and progressive brain injury in closed-head trauma: radiological demonstration. Neurosurgery 32:25–30, discussion 30-21
30. Smith DH, Meaney DF (2000) Axonal damage in traumatic brain injury. Neuroscientist 6:483–495
31. Johnson VE, Stewart W, Smith DH (2013) Axonal pathology in traumatic brain injury. Exp Neurol 246:35–43
32. Johnson VE, Stewart JE, Begbie FD, Trojanowski JQ, Smith DH, Stewart W (2013) Inflammation and white matter degeneration persist for years after a single traumatic brain injury. Brain 136:28–42
33. Choi DW (1994) Calcium and excitotoxic neuronal injury. Ann N Y Acad Sci 747:162–171
34. Goforth PB, Ellis EF, Satin LS (1999) Enhancement of AMPA-mediated current after traumatic injury in cortical neurons. J Neurosci 19:7367–7374
35. Sattler R, Tymianski M (2000) Molecular mechanisms of calcium-dependent excitotoxicity. J Mol Med 78:3–13
36. Weber JT, Rzigalinski BA, Willoughby KA, Moore SF, Ellis EF (1999) Alterations in calcium-mediated signal transduction after traumatic injury of cortical neurons. Cell Calcium 26:289–299
37. Mouzon B, Bachmeier C, Ferro A, Ojo J-O, Crynen G, Acker C, Davies P, Mullan M,

Stewart W, Crawford F (2013) Chronic neuropathological and neurobehavioral changes in a repetitive mTBI model. Ann Neurol 75(2):241–254

38. McKee AC, Cantu RC, Nowinski CJ, Hedley-Whyte ET, Gavett BE, Budson AE, Santini VE, Lee HS, Kubilus CA, Stern RA (2009) Chronic traumatic encephalopathy in athletes: progressive tauopathy after repetitive head injury. J Neuropathol Exp Neurol 68:709–735
39. McIntosh TK (1994) Neurochemical sequelae of traumatic brain injury: therapeutic implications. Cerebrovasc Brain Metab Rev 6:109–162
40. Raghupathi R (2004) Cell death mechanisms following traumatic brain injury. Brain Pathol 14:215–222
41. Monti JM, Voss MW, Pence A, McAuley E, Kramer AF, Cohen NJ (2013) History of mild traumatic brain injury is associated with deficits in relational memory, reduced hippocampal volume, and less neural activity later in life. Front Aging Neurosci 5:41
42. Ryan LM, Warden DL (2003) Post concussion syndrome. Int Rev Psychiatry 15:310–316
43. Vanderploeg RD, Crowell TA, Curtiss G (2001) Verbal learning and memory deficits in traumatic brain injury: encoding, consolidation, and retrieval. J Clin Exp Neuropsychol 23:185–195
44. De Kruijk JR, Twijnstra A, Leffers P (2001) Diagnostic criteria and differential diagnosis of mild traumatic brain injury. Brain Inj 15:99–106
45. De Monte VE, Geffen GM, Massavelli BM (2006) The effects of post-traumatic amnesia on information processing following mild traumatic brain injury. Brain Inj 20:1345–1354
46. Leininger BE, Gramling SE, Farrell AD, Kreutzer JS, Peck EA (1990) Neuropsychological deficits in symptomatic minor head injury patients after concussion and mild concussion. J Neurol Neurosurg Psychiatry 53:293–296
47. Stein DG (2015) Embracing failure: what the phase III progesterone studies can teach about TBI clinical trials. Brain Inj 29:1259–1272
48. Kabadi SV, Faden AI (2014) Neuroprotective strategies for traumatic brain injury: improving clinical translation. Int J Mol Sci 15:1216–1236
49. Smith DH, Hicks RR, Johnson VE, Bergstrom DA, Cummings DM, Noble LJ, Hovda D, Whalen M, Ahlers ST, LaPlaca M, Tortella FC, Duhaime AC, Dixon CE (2015) Pre-clinical traumatic brain injury common data elements: toward a common language across laboratories. J Neurotrauma 32(22):1725–1735
50. Ommaya AK, Gennarelli TA (1974) Cerebral concussion and traumatic unconsciousness. Correlation of experimental and clinical observations of blunt head injuries. Brain 97:633–654
51. Adams JH, Doyle D, Ford I, Gennarelli TA, Graham DI, McClellan DR (1989) Diffuse axonal injury in head injury: definition, diagnosis, and grading. Histopathology 15:49–59
52. Povlishock JT (1992) Traumatically induced axonal injury: pathogenesis and pathobiological implications. Brain Pathol 2:1–12
53. Santiago LA, Oh BC, Dash PK, Holcomb JB, Wade CE (2012) A clinical comparison of penetrating and blunt traumatic brain injuries. Brain Inj 26:107–125
54. Demetriades D, Kuncir E, Murray J, Velmahos GC, Rhee P, Chan L (2004) Mortality prediction of head abbreviated injury score and Glasgow coma scale: analysis of 7,764 head injuries. J Am Coll Surg 199:216–222
55. Denny-Brown DE, Russell WR (1941) Experimental concussion: (section of neurology). Proc R Soc Med 34:691–692
56. Holbourn AHS (1943) Mechanics of head injury. Lancet 2:438–441
57. Ommaya A, Hirsch A, Flamm E, Mahone R (1966) Cerebral concussion in monkey – an experimental model. Science 153:211
58. Yarnell P, Ommaya AK (1969) Experimental cerebral concussion in the rhesus monkey. Bull N Y Acad Med 45:39–45
59. Letcher FS, Corrao PG, Ommaya AK (1973) Head injury in the chimpanzee. 2. Spontaneous and evoked epidural potentials as indices of injury severity. J Neurosurg 39:167–177
60. Ommaya AK, Corrao P, Letcher FS (1973) Head injury in the chimpanzee. 1. Biodynamics of traumatic unconsciousness. J Neurosurg 39:152–166
61. Gennarelli TA, Thibault LE, Ommaya AK (1971) Comparison of linear and rotational acceleration in experimental cerebral concussion. Proceedings of the 15th Stapp car crash conference, New York: society of automotive engineers, pp 797–803
62. Gennarelli TA, Thibault LE, Ommaya AK (1972) Pathophysiologic responses to rotational and translational accelerations of the head. Proceedings of the 16th Stapp car crash conference, New York: society of automotive engineers, pp 296–308
63. Margulies SS, Thibault LE (1992) A proposed tolerance criterion for diffuse axonal injury in man. J Biomech 25:917–923

64. Meaney DF, Smith DH, Shreiber DI, Bain AC, Miller RT, Ross DT, Gennarelli TA (1995) Biomechanical analysis of experimental diffuse axonal injury. J Neurotrauma 12:689–694
65. Margulies SS, Thibault LE, Gennarelli TA (1990) Physical model simulations of brain injury in the primate. J Biomech 23:823–836
66. Margulies SS, Thibault LE (1989) An analytical model of traumatic diffuse brain injury. J Biomech Eng 111:241–249
67. Sullivan S, Eucker SA, Gabrieli D, Bradfield C, Coats B, Maltese MR, Lee J, Smith C, Margulies SS (2015) White matter tract-oriented deformation predicts traumatic axonal brain injury and reveals rotational direction-specific vulnerabilities. Biomech Model Mechanobiol 14:877–896
68. Galbraith JA, Thibault LE, Matteson DR (1993) Mechanical and electrical responses of the squid giant axon to simple elongation. J Biomech Eng 115:13–22
69. Cullen DK, Simon CM, LaPlaca MC (2007) Strain rate-dependent induction of reactive astrogliosis and cell death in three-dimensional neuronal-astrocytic co-cultures. Brain Res 1158:103–115
70. Cullen DK, Vernekar VN, LaPlaca MC (2011) Trauma-induced plasmalemma disruptions in three-dimensional neural cultures are dependent on strain modality and rate. J Neurotrauma 28:2219–2233
71. LaPlaca MC, Cullen DK, McLoughlin JJ, Cargill RS 2nd (2005) High rate shear strain of three-dimensional neural cell cultures: a new in vitro traumatic brain injury model. J Biomech 38:1093–1105
72. Prado GR, Ross JD, DeWeerth SP, LaPlaca MC (2005) Mechanical trauma induces immediate changes in neuronal network activity. J Neural Eng 2:148–158
73. Geddes DM, Cargill RS 2nd, LaPlaca MC (2003) Mechanical stretch to neurons results in a strain rate and magnitude-dependent increase in plasma membrane permeability. J Neurotrauma 20:1039–1049
74. LaPlaca MC, Lee VM, Thibault LE (1997) An in vitro model of traumatic neuronal injury: loading rate-dependent changes in acute cytosolic calcium and lactate dehydrogenase release. J Neurotrauma 14:355–368
75. Gennarelli TA, Thibault LE, Adams JH, Graham DI, Thompson CJ, Marcincin RP (1982) Diffuse axonal injury and traumatic coma in the primate. Ann Neurol 12:564–574
76. Adams JH, Graham DI, Gennarelli TA (1981) Acceleration induced head injury in the monkey. II. Neuropathology. Acta Neuropathol Suppl 7:26–28
77. Kotapka MJ, Gennarelli TA, Graham DI, Adams JH, Thibault LE, Ross DT, Ford I (1991) Selective vulnerability of hippocampal neurons in acceleration-induced experimental head injury. J Neurotrauma 8:247–258
78. Cullen DK, LaPlaca MC (2006) Neuronal response to high rate shear deformation depends on heterogeneity of the local strain field. J Neurotrauma 23:1304–1319
79. Smith DH, Chen XH, Xu BN, McIntosh TK, Gennarelli TA, Meaney DF (1997) Characterization of diffuse axonal pathology and selective hippocampal damage following inertial brain trauma in the pig. J Neuropathol Exp Neurol 56:822–834
80. Sahay KB, Mehrotra R, Sachdeva U, Banerji AK (1992) Elastomechanical characterization of brain tissues. J Biomech 25:319–326
81. Kleiven S (2013) Why most traumatic brain injuries are not caused by linear acceleration but skull fractures are. Front Bioeng Biotechnol 1:15
82. Duhaime AC (2006) Large animal models of traumatic injury to the immature brain. Dev Neurosci 28:380–387
83. Fijalkowski RJ, Stemper BD, Pintar FA, Yoganandan N, Crowe MJ, Gennarelli TA (2007) New rat model for diffuse brain injury using coronal plane angular acceleration. J Neurotrauma 24:1387–1398
84. Gutierrez E, Huang Y, Haglid K, Bao F, Hansson HA, Hamberger A, Viano D (2001) A new model for diffuse brain injury by rotational acceleration: I model, gross appearance, and astrocytosis. J Neurotrauma 18:247–257
85. Howells DW, Porritt MJ, Rewell SS, O'Collins V, Sena ES, van der Worp HB, Traystman RJ, Macleod MR (2010) Different strokes for different folks: the rich diversity of animal models of focal cerebral ischemia. J Cereb Blood Flow Metab 30:1412–1431
86. Bailey EL, McCulloch J, Sudlow C, Wardlaw JM (2009) Potential animal models of lacunar stroke: a systematic review. Stroke 40:e451–e458
87. Zhang K, Sejnowski TJ (2000) A universal scaling law between gray matter and white matter of cerebral cortex. Proc Natl Acad Sci U S A 97:5621–5626
88. Adams JH, Graham DI, Murray LS, Scott G (1982) Diffuse axonal injury due to nonmissile head injury in humans: an analysis of 45 cases. Ann Neurol 12:557–563
89. Graham DI, Adams JH, Gennarelli TA (1988) Mechanisms of non-penetrating head injury. Prog Clin Biol Res 234:159–168

90. Povlishock JT, Becker DP, Cheng CLY, Vaughan GW (1983) Axonal change in minor head injury. J Neuropathol Exp Neurol 42:225–242
91. Povlishock JT, Becker DP (1985) Fate of reactive axonal swellings induced by head injury. Lab Invest 52:540–552
92. Chen XH, Siman R, Iwata A, Meaney DF, Trojanowski JQ, Smith DH (2004) Long-term accumulation of amyloid-beta, beta-secretase, presenilin-1, and caspase-3 in damaged axons following brain trauma. Am J Pathol 165:357–371
93. Xiong Y, Mahmood A, Chopp M (2013) Animal models of traumatic brain injury. Nat Rev Neurosci 14:128–142
94. Gennarelli TA, Adams JH, Graham DI (1981) Acceleration induced head injury in the monkey. I. The model, its mechanical and physiological correlates. Acta Neuropathol Suppl 7:23–25
95. Ross DT, Meaney DF, Sabol MK, Smith DH, Gennarelli TA (1994) Distribution of forebrain diffuse axonal injury following inertial closed head injury in miniature swine. Exp Neurol 126:291–299
96. Chen XH, Meaney DF, Xu BN, Nonaka M, McIntosh TK, Wolf JA, Saatman KE, Smith DH (1999) Evolution of neurofilament subtype accumulation in axons following diffuse brain injury in the pig. J Neuropathol Exp Neurol 58:588–596
97. Smith DH, Chen XH, Nonaka M, Trojanowski JQ, Lee VM, Saatman KE, Leoni MJ, Xu BN, Wolf JA, Meaney DF (1999) Accumulation of amyloid beta and tau and the formation of neurofilament inclusions following diffuse brain injury in the pig. J Neuropathol Exp Neurol 58:982–992
98. Smith DH, Nonaka M, Miller R, Leoni M, Chen XH, Alsop D, Meaney DF (2000) Immediate coma following inertial brain injury dependent on axonal damage in the brainstem. J Neurosurg 93:315–322
99. Kimura H, Meaney DF, McGowan JC, Grossman RI, Lenkinski RE, Ross DT, McIntosh TK, Gennarelli TA, Smith DH (1996) Magnetization transfer imaging of diffuse axonal injury following experimental brain injury in the pig: characterization by magnetization transfer ratio with histopathologic correlation. J Comput Assist Tomogr 20:540–546
100. Friess SH, Ichord RN, Owens K, Ralston J, Rizol R, Overall KL, Smith C, Helfaer MA, Margulies SS (2007) Neurobehavioral functional deficits following closed head injury in the neonatal pig. Exp Neurol 204:234–243
101. Friess SH, Bruins B, Kilbaugh TJ, Smith C, Margulies SS (2015) Differing effects when using phenylephrine and norepinephrine to augment cerebral blood flow after traumatic brain injury in the immature brain. J Neurotrauma 32:237–243
102. Friess SH, Ralston J, Eucker SA, Helfaer MA, Smith C, Margulies SS (2011) Neurocritical care monitoring correlates with neuropathology in a swine model of pediatric traumatic brain injury. Neurosurgery 69:1139–1147, discussion 1147
103. Friess SH, Smith C, Kilbaugh TJ, Frangos SG, Ralston J, Helfaer MA, Margulies SS (2012) Early cerebral perfusion pressure augmentation with phenylephrine after traumatic brain injury may be neuroprotective in a pediatric swine model. Crit Care Med 40:2400–2406
104. Ibrahim NG, Natesh R, Szczesny SE, Ryall K, Eucker SA, Coats B, Margulies SS (2010) In situ deformations in the immature brain during rapid rotations. J Biomech Eng 132:044501
105. Jaber SM, Sullivan S, Margulies SS (2015) Noninvasive metrics for identification of brain injury deficits in piglets. Dev Neuropsychol 40:34–39
106. Zhou C, Eucker SA, Durduran T, Yu G, Ralston J, Friess SH, Ichord RN, Margulies SS, Yodh AG (2009) Diffuse optical monitoring of hemodynamic changes in piglet brain with closed head injury. J Biomed Opt 14:034015
107. Raghupathi R, Margulies SS (2002) Traumatic axonal injury after closed head injury in the neonatal pig. J Neurotrauma 19:843–853
108. Raghupathi R, Mehr MF, Helfaer MA, Margulies SS (2004) Traumatic axonal injury is exacerbated following repetitive closed head injury in the neonatal pig. J Neurotrauma 21:307–316
109. Friess SH, Ichord RN, Ralston J, Ryall K, Helfaer MA, Smith C, Margulies SS (2009) Repeated traumatic brain injury affects composite cognitive function in piglets. J Neurotrauma 26:1111–1121
110. Zhu Q, Prange M, Margulies S (2006) Predicting unconsciousness from a pediatric brain injury threshold. Dev Neurosci 28:388–395
111. Cecil KM, Lenkinski RE, Meaney DF, McIntosh TK, Smith DH (1998) High-field proton magnetic resonance spectroscopy of a swine model for axonal injury. J Neurochem 70:2038–2044
112. Stein SC, Chen XH, Sinson GP, Smith DH (2002) Intravascular coagulation: a major secondary insult in nonfatal traumatic brain injury. J Neurosurg 97:1373–1377

113. Zhang J, Groff RF, Chen XH, Browne KD, Huang J, Schwartz ED, Meaney DF, Johnson VE, Stein SC, Rojkjaer R, Smith DH (2008) Hemostatic and neuroprotective effects of human recombinant activated factor VII therapy after traumatic brain injury in pigs. Exp Neurol 210:645–655
114. Browne KD, Chen XH, Meaney DF, Smith DH (2011) Mild traumatic brain injury and diffuse axonal injury in swine. J Neurotrauma 28:1747–1755
115. Naim MY, Friess S, Smith C, Ralston J, Ryall K, Helfaer MA, Margulies SS (2010) Folic acid enhances early functional recovery in a piglet model of pediatric head injury. Dev Neurosci 32:466–479
116. Ibrahim NG, Ralston J, Smith C, Margulies SS (2010) Physiological and pathological responses to head rotations in toddler piglets. J Neurotrauma 27:1021–1035
117. Coats B, Binenbaum G, Peiffer RL, Forbes BJ, Margulies SS (2010) Ocular hemorrhages in neonatal porcine eyes from single, rapid rotational events. Invest Ophthalmol Vis Sci 51:4792–4797
118. Kilbaugh TJ, Bhandare S, Lorom DH, Saraswati M, Robertson CL, Margulies SS (2011) Cyclosporin A preserves mitochondrial function after traumatic brain injury in the immature rat and piglet. J Neurotrauma 28:763–774
119. Coats B, Eucker SA, Sullivan S, Margulies SS (2012) Finite element model predictions of intracranial hemorrhage from non-impact, rapid head rotations in the piglet. Int J Dev Neurosci 30:191–200
120. Friess SH, Naim MY, Kilbaugh TJ, Ralston J, Margulies SS (2012) Premedication with meloxicam exacerbates intracranial haemorrhage in an immature swine model of non-impact inertial head injury. Lab Anim 46:164–166
121. Sullivan S, Friess SH, Ralston J, Smith C, Propert KJ, Rapp PE, Margulies SS (2013) Improved behavior, motor, and cognition assessments in neonatal piglets. J Neurotrauma 30:1770–1779
122. Weeks D, Sullivan S, Kilbaugh T, Smith C, Margulies SS (2014) Influences of developmental age on the resolution of diffuse traumatic intracranial hemorrhage and axonal injury. J Neurotrauma 31:206–214
123. Margulies SS, Kilbaugh T, Sullivan S, Smith C, Propert K, Byro M, Saliga K, Costine BA, Duhaime AC (2015) Establishing a clinically relevant large animal model platform for TBI therapy development: using cyclosporin A as a case study. Brain Pathol 25:289–303
124. Clevenger AC, Kilbaugh T, Margulies SS (2015) Carotid artery blood flow decreases after rapid head rotation in piglets. J Neurotrauma 32:120–126
125. Thibault LE, Gennarelli TA (1989) Biomechanics of diffuse brain injuries. *Proceedings of the 10th international technical conference on experimental safety vehicles*. DOT, NHTSA
126. Eucker SA, Smith C, Ralston J, Friess SH, Margulies SS (2011) Physiological and histopathological responses following closed rotational head injury depend on direction of head motion. Exp Neurol 227:79–88
127. Ommaya AK, Yarnell P, Hirsch AE, Harris EH. Scaling of experimental data on cerebral concussion in sub-human primates to concussive thresholds in man. SAE Technical Paper, Proceedings of the 11th Stapp Car Crash Conference, Warrendale, PA 1967:73–80
128. Duma SM, Manoogian SJ, Bussone WR, Brolinson PG, Goforth MW, Donnenwerth JJ, Greenwald RM, Chu JJ, Crisco JJ (2005) Analysis of real-time head accelerations in collegiate football players. Clin J Sport Med 15:3–8
129. Frechede B, McIntosh AS (2009) Numerical reconstruction of real-life concussive football impacts. Med Sci Sports Exerc 41:390–398
130. Greenwald RM, Gwin JT, Chu JJ, Crisco JJ (2008) Head impact severity measures for evaluating mild traumatic brain injury risk exposure. Neurosurgery 62:789–798, discussion 798
131. Newman JA, Shewchenko N, Welbourne E (2000) A proposed new biomechanical head injury assessment function – the maximum power index. Stapp Car Crash J 44:215–247
132. Pellman EJ, Viano DC, Tucker AM, Casson IR, Waeckerle JF (2003) Concussion in professional football: reconstruction of game impacts and injuries. Neurosurgery 53:799–812, discussion 812-794
133. Miller RT, Margulies SS, Leoni M, Nonaka M, Chen XH, Smith DH, Meaney DF (1998) Finite element modeling approaches for predicting injury in an experimental model of severe diffuse axonal injury. 42nd Stapp car crash conference proceedings
134. Wolf JA, Johnson BN, Johnson VE, Browne KD, Mietus CJ, Smith DH, Grady MS, Cohen A, Cullen DK (in review) Concussion induces hippocampal circuitry disruption in swine
135. Meng X, Browne KD, Huang SM, Cullen DK, Tofighi MR, Rosen A (2012) Dynamic study of wireless intracranial pressure monitoring of rotational head injury in a swine model. Electron Lett 48:363–364

136. Meng X, Browne KD, Huang SM, Mietus CJ, Cullen DK, Tofighi MR, Rosen A (2012) Dynamic evaluation of a digital wireless intracranial pressure sensor for the assessment of traumatic brain injury in a swine model. IEEE Trans Microw Theory Tech 61(1):316–325
137. Meng X, Mietus CJ, Browne KD, Tofighi MR, Rosen A, Cullen DK (2013) A telemetry-based neuromonitoring system: validation in a swine model of closed-head rotational acceleration-induced TBI. National neurotrauma society annual meeting
138. McGowan JC, McCormack TM, Grossman RI, Mendonca R, Chen XH, Berlin JA, Meaney DF, Xu BN, Cecil KM, McIntosh TK, Smith DH (1999) Diffuse axonal pathology detected with magnetization transfer imaging following brain injury in the pig. Magn Reson Med 41:727–733
139. Smith DH, Cecil KM, Meaney DF, Chen XH, McIntosh TK, Gennarelli TA, Lenkinski RE (1998) Magnetic resonance spectroscopy of diffuse brain trauma in the pig. J Neurotrauma 15:665–674
140. Smith DH, Chen XH, Iwata A, Graham DI (2003) Amyloid beta accumulation in axons after traumatic brain injury in humans. J Neurosurg 98:1072–1077
141. Tang-Schomer MD, Johnson VE, Baas PW, Stewart W, Smith DH (2012) Partial interruption of axonal transport due to microtubule breakage accounts for the formation of periodic varicosities after traumatic axonal injury. Exp Neurol 233:364–372
142. Uryu K, Chen XH, Martinez D, Browne KD, Johnson VE, Graham DI, Lee VM, Trojanowski JQ, Smith DH (2007) Multiple proteins implicated in neurodegenerative diseases accumulate in axons after brain trauma in humans. Exp Neurol 208:185–192
143. Hutson CB, Lazo CR, Mortazavi F, Giza CC, Hovda D, Chesselet MF (2011) Traumatic brain injury in adult rats causes progressive nigrostriatal dopaminergic cell loss and enhanced vulnerability to the pesticide paraquat. J Neurotrauma 28:1783–1801
144. Uryu K, Giasson BI, Longhi L, Martinez D, Murray I, Conte V, Nakamura M, Saatman K, Talbot K, Horiguchi T, McIntosh T, Lee VM, Trojanowski JQ (2003) Age-dependent synuclein pathology following traumatic brain injury in mice. Exp Neurol 184:214–224
145. Johnson VE, Stewart W, Smith DH (2012) Widespread tau and amyloid-beta pathology many years after a single traumatic brain injury in humans. Brain Pathol 22:142–149
146. Koch ., Tekriwal A, Ulyanova AV, Grovola MR, Cullen DK, Wolf JA (2015) Chronic neurophysiological recording of the hippocampus in awake behaving swine after diffuse brain injury. National neurotrauma society annual meeting
147. Wolf JA, Ulyanova AV, Browne KD, Koch P, Grovola MR, Johnson VE, Cullen DK (2014) Hippocampal network disruptions after diffuse brain injury in swine. Winter conference on brain research
148. Elkin BS, Morrison B 3rd (2007) Region-specific tolerance criteria for the living brain. Stapp Car Crash J 51:127–138
149. Cater HL, Sundstrom LE, Morrison B 3rd (2006) Temporal development of hippocampal cell death is dependent on tissue strain but not strain rate. J Biomech 39:2810–2818
150. Vink R, Mullins PG, Temple MD, Bao W, Faden AI (2001) Small shifts in craniotomy position in the lateral fluid percussion injury model are associated with differential lesion development. J Neurotrauma 18:839–847
151. Yoshino A, Hovda DA, Kawamata T, Katayama Y, Becker DP (1991) Dynamic changes in local cerebral glucose utilization following cerebral conclusion in rats: evidence of a hyper- and subsequent hypometabolic state. Brain Res 561:106–119
152. Gennarelli TA, Pintar FA, Yoganandan N (2003) Biomechanical tolerances for diffuse brain injury and a hypothesis for genotypic variability in response to trauma. Annu Proc Assoc Adv Automot Med 47:624–628
153. Kimpara H, Iwamoto M (2012) Mild traumatic brain injury predictors based on angular accelerations during impacts. Ann Biomed Eng 40:114–126
154. Ommaya AK, Grubb RL Jr, Naumann RA (1971) Coup and contre-coup injury: observations on the mechanics of visible brain injuries in the rhesus monkey. J Neurosurg 35:503–516
155. Hardy WN, Foster CD, Mason MJ, Yang KH, King AI, Tashman S (2001) Investigation of head injury mechanisms using neutral density technology and high-speed biplanar X-ray. Stapp Car Crash J 45:337–368
156. Kleiven S (2007) Predictors for traumatic brain injuries evaluated through accident reconstructions. Stapp Car Crash J 51:81–114
157. Aare M, Kleiven S, Halldin P (2004) Injury tolerances for oblique impact helmet testing. Int J Crashworthines 9:15–23
158. Holbourn AHS (1945) Mechanics of brain injuries. Br Med Bull 3:147–149

159. LaPlaca MC, Simon CM, Prado GR, Cullen DK (2007) CNS injury biomechanics and experimental models. Prog Brain Res 161:13–26

160. Bayly PV, Black EE, Pedersen RC, Leister EP, Genin GM (2006) In vivo imaging of rapid deformation and strain in an animal model of traumatic brain injury. J Biomech 39:1086–1095

161. Hardy WN, Mason MJ, Foster CD, Shah CS, Kopacz JM, Yang KH, King AI, Bishop J, Bey M, Anderst W, Tashman S (2007) A study of the response of the human cadaver head to impact. Stapp Car Crash J 51:17–80

Chapter 18

Pediatric Rodent Models of Traumatic Brain Injury

Bridgette D. Semple, Jaclyn Carlson, and Linda J. Noble-Haeusslein

Abstract

Due to a high incidence of traumatic brain injury (TBI) in children and adolescents, age-specific studies are necessary to fully understand the long-term consequences of injuries to the immature brain. Preclinical and translational research can help elucidate the vulnerabilities of the developing brain to insult, and provide model systems to formulate and evaluate potential treatments aimed at minimizing the adverse effects of TBI. Several experimental TBI models have therefore been scaled down from adult rodents for use in juvenile animals. The following chapter discusses these adapted models for pediatric TBI, and the importance of age equivalence across species during model development and interpretation. Many neurodevelopmental processes are ongoing throughout childhood and adolescence, such that neuropathological mechanisms secondary to a brain insult, including oxidative stress, metabolic dysfunction and inflammation, may be influenced by the age at the time of insult. The long-term evaluation of clinically relevant functional outcomes is imperative to better understand the persistence and evolution of behavioral deficits over time after injury to the developing brain. Strategies to modify or protect against the chronic consequences of pediatric TBI, by supporting the trajectory of normal brain development, have the potential to improve quality of life for brain-injured children.

Key words Traumatic brain injury, Pediatric, Juvenile, Children, Behavior, Mice, Rodents

1 Introduction

Traumatic brain injury (TBI) is a leading cause of death and disability in children including adolescents [1]. For individuals under 20 years of age, a recent review of studies from North America, Europe, Australia, and New Zealand found an estimated annual incidence of 691 per 100,000 population treated in Emergency Departments, 74 per 100,000 hospitalizations, and 9 per 100,000 resulting in mortality [2]. Patients under the age of 5 years typically have a higher risk of sustaining a TBI [3], with the leading cause being falls, which accounts for approximately two-thirds of injuries at this age [2]. Even during childhood, males have an approximately twofold higher risk of injury compared to females [2].

Historically, there has been a widespread dogma that the developing brain is more refractory to injury than the adult brain, based

Firas Kobeissy et al. (eds.), *Injury Models of the Central Nervous System: Methods and Protocols*, Methods in Molecular Biology, vol. 1462, DOI 10.1007/978-1-4939-3816-2_18, © Springer Science+Business Media New York 2016

upon the premise that its capacity for plasticity results in enhanced regeneration and reorganization. While this may be the case for specific developmental periods, there is now considerable evidence from both clinical and basic research indicating that severe brain injuries, acquired early in life, result in worse functional consequences compared to those sustained later in childhood or adulthood [4–7]. In other words, the age-at-insult is a key determinant of unfavorable outcomes, whereby TBI under the age of 4 years significantly impacts subsequent brain development to influence physical, cognitive, and behavioral sequelae [4, 8]. In particular, psychological and social problems may emerge over time post-injury, likely resulting from a disruption of normal brain maturation or the impaired acquisition of new skills [9, 10]. Cognitive and psychosocial deficits can profoundly impact quality of life through adolescence to adulthood, impairing one's ability to function effectively at home, school, work, and socially [8]. Such deficits are often associated with widespread reductions in both cortical and sub-cortical volumes including the hippocampus and deep limbic regions, with evidence of progressive atrophy over time [11–17]. Further, TBI in the pediatric population has been associated with a greater risk of cerebral hemodynamic dysfunction, hypoxic–ischemic injury, and diffuse cerebral edema compared to adults [18]. Together, these findings emphasize the need for relevant preclinical and translational research to better understand age-specific vulnerabilities to TBI and minimize the adverse effects of brain injury during development.

The consequences of a TBI during childhood depends upon a range of factors, including the nature of the injury itself (e.g., severity, type, location) as well as environmental factors (e.g., comorbidities, family support, socioeconomic demographics, interventions) [9]. Age has a considerable influence on the biomechanics of injury, related in part to the composition of the brain across development (i.e., the extent of myelination and brain water content) as well as extra-cranial factors such as the head-to-body ratio, neck strength and stability, and skull rigidity [19]. To gain a better understanding of how each of these factors contributes to outcomes after childhood brain injuries, and begin to delineate the mechanisms of intrinsic vulnerability of the developing brain to insult, most of the previously described models of experimental TBI (see preceding chapters) have been adapted for use in the immature rodent.

In this chapter, we discuss these models in the context of pediatric TBI and their seminal findings to date. We consider experimental design for comparing between the effects of TBI at different ages, including the identification of key milestones of cerebral maturation across postnatal development. Lastly, we note challenges associated with rodent models of pediatric brain injury with a focus on behavioral outcome measures, and highlight areas in need of future research.

2 When Is "Childhood" for a Rodent?

The definition of childhood equivalence is often vague and poorly defined in preclinical literature, and readers will note the interchangeable use of terminology "pediatric," "juvenile," "adolescent," "immature," "neonatal," and "developing" to describe a rodent between birth and adulthood. Attempting to equate the developmental trajectory of laboratory animals relative to humans is a commonly used experimental design in brain injury research, although it relies upon presumptions of appropriate age equivalence across species, and/or scaling of injury severities across different ages. Precise comparisons across species can be made by focusing exclusively on one specific process of interest—for example, an equivalent amount of the enzyme glutamate decarboxylase (16.2% of adult levels) was found to be present in humans at 40 weeks gestation, and rats at postnatal day 7 (PND7) [20]. In contrast, the relative levels of the choline acetyltransferase enzyme was reached considerably later in rats (PND20), indicating that the exact interspecies age match may shift with each outcome measure. This creates a conundrum whereby equating ages based upon a single developmental milestone or event may lead to erroneous interpretations [21].

To address this dilemma, we and others have considered the timing of neurogenesis, synaptogenesis, gliogenesis, oligodendrocyte maturation and age-dependent behaviors, as well as developmentally regulated molecular and biochemical changes, to delineate an approximate timeline in which rodent brain maturation parallels that of a human [22]. Non-cerebral development—including sexual maturation, organ growth and other behavioral changes—should also be considered [21]. Broadly, rodents at PND7–13 are considered equivalent to a neonatal human, PND17–25 is used to approximate early childhood, PND35–42 equates to adolescence, and early adulthood is typically from 2 months of age onwards [22] (*see* Table 1). Importantly, comparison and investigation of injury at different ages across development has revealed distinct periods of regional and age-dependent vulnerability to brain injuries, as described further below.

3 Modeling TBI in Young Rodents

Each of the main approaches to generate experimental TBI in rodents, previously established in the adult, have been adapted and scaled to a younger rodent, including the fluid percussion injury (FPI), impact acceleration weight drop (WD) (open or closed skull), and controlled cortical impact (CCI). The parameters of these models can be modified to deliver a range of varying injury severities. In addition, multiple mild impacts can be delivered to model repeated concussive-like insults.

Table 1
Summary of key developmental processes and behavioral phenotypes across comparable ages in humans and rodents

Human	Rodent	Developmental milestones	Reference(s)
23–32 week gestation (preterm infant)	PND 1–3	Oligodendrocyte maturation state changes—predominance of mitotically active pre-OLs[a].	Craig et al. (2003), Lodygensky et al. (2010), Dean et al. (2011a, b)
		Immune system development.	Holsapple et al. (2003)
		Establishment of the blood–brain barrier.	Engelhardt (2003), Daneman et al. (2010)
36–40 week gestation (term infant)	PND 7–10	Peak brain growth spurt.	Dobbing and Sands (1979), Bockhorst et al. (2008)
		Peak in gliogenesis.	Catalani et al. (2002), Kriegstein and Alvarez-Buylla (2009)
		Increasing axonal and dendritic density.	Cowan (1979), Bockhorst et al. (2008), Baloch et al. (2009)
		Oligodendrocyte maturation state changes—switch to a predominance of immature OLs.	Craig et al. (2003), Dean et al. (2011a, b)
		Consolidation of the immune system.	Holsapple et al. (2003)
2–3 year old	PND 20–21	Brain reaches 90–95 % of adult weight.	Dobbing and Sands (1973, 1979), Dekaban et al. (1987), Giedd et al. (1999)
		Peak in synaptic density at 50 % >adult levels.	Huttenlocher (1979), Micheva and Beaulieu (1996)
		Peak in myelination rate.	Keshavan et al. (2002)
		Neurotransmitter and receptor changes.	Hedner et al. (1986), Romijn et al. (1991)
		Increased activity levels and sociability. Weaning.	Wills et al. (1983), Wood et al. (2003), Terranova and Laviola (2005), Blakemore (2008)
		Increasing working memory.	Herschkowitz et al. (1997)
4–11 year old	PND 25–35	Fractionation/specialization of prefrontal cortex neural networks (structural maturation).	Tsujimoto (2008)
		Maximum volume of grey matter and cortical thickness.	Sowell et al. (1999), Bansal et al. (2008)
		Increased sociability.	Terranova and Laviola (2005)
		Further development of working memory and inhibitory control.	Tsujimoto (2008)
12–18 year old	PND 35–49	Reduced synapse density, reaching a plateau at adult levels. Refinement of cognitive-dependent circuitry. Ongoing myelination; increasing white matter volume and fractional anisotropy.	Huttenlocher (1979), Lidow et al. (1991) Giedd et al. (1999), Chahboune et al. (2007), Baloch et al. (2009), Brouwer et al. (2012)
		Adolescent-type behaviors including sociability, risk-taking, impulsivity.	Spear (2000), Sturman and Muoghaddam (2011)
		Onset of sexual maturity.	Lambert (2009)
		Increased cognitive control capacities.	Laviola et al. (2003), Raznahan et al. (2011)
20 years +	PND 60+	Adult levels of neurotransmitters.	Romijn et al. (1991)
		Adult levels of synaptic density.	Huttenlocher (1979)
		Ongoing myelination and declining grey matter.	Lebel and Beaulieu (2011), Lebel et al. (2012)
		Emergence of adult-type behaviors including reduced risk-taking, reduced impulsivity and increased parental tendencies.	Laviola et al. (2003)

Adapted from ref. [22], reprinted with permission from Elsevier

[a]*OL* oligodendrocyte

Induced by the injection of fluid into the epidural space following a craniotomy, FPI results in a spectrum of severities ranging from mild pathology through to more severe contusions, subarachnoid hemorrhage and diffuse axonal injury, depending upon the volume of fluid and force of the pulse [23–25]. A seminal study using this model characterized different injury severities in PND17, PND28 and adult rats, and found that the youngest animals showed a longer period of apnea, greater mortality and hemodynamic changes including sustained hypotension after injury compared to the older age groups [21, 26].

Weight drop models of TBI typically produce diffuse injuries by impacting the whole intact head, which is either restrained or allowed to move freely upon impact. These models often incorporate the fixation of a helmet or impact plate to the skull to prevent penetration, a particular problem with smaller and younger animals in which skull fractures and convulsions may occur with a greater likelihood after injury [27, 28]. At severe injury levels, WD impacts mimic the diffuse swelling, neuropathology and cognitive dysfunction often seen in brain-injured children [29–32]. For example, impact-acceleration injury at PND17 results in transient vestibulomotor deficits (up to 10 days post-injury) and persistent deficits in the Morris Water Maze test enduring to 3 months post-injury, associated with pronounced astrogliosis and axonal injury [33, 34].

CCI models are most relevant to brain injuries resulting in a focal contusion. This model allows for precise control over mechanical parameters including velocity and depth, and results in reproducible neuropathology to the cortex and hippocampus in the immature rodent [35]. Fixation of the head in a stereotaxic frame, while ensuring high reproducibility across animals, also limits the injury mechanism to one that does not involve acceleration/deceleration or rotational forces. As a consequence, CCI predominantly results in a focal lesion cavity with limited diffuse axonal injury.

Cognitive and sensorimotor deficits often result, and anxiety-like behaviors may persist long-term through 2 months after injury to PND17 rats, alongside evidence of changes in white matter connectivity and neuronal reorganization [36, 37]. In a CCI model in PND21 mice, characteristic pathology includes acute neurodegeneration and inflammation, leading to progressive atrophy of the cortex and hippocampus, and the coincidental emergence of behavioral abnormalities including hyperactivity, spatial memory deficits and social dysfunction [38–41] (Fig. 1). A more anterior impact site in PND21 mice and PND17 rats is characterized by persistent sensorimotor dysfunction [36, 42], as well as the evolution of cognitive deficits up to 6 months post-injury [43].

Several studies have recently examined the consequences of repeated concussive-like injuries to juvenile rodents, using a range of models, number of impacts and the interval between impacts. Although a single concussive injury is typically not associated with

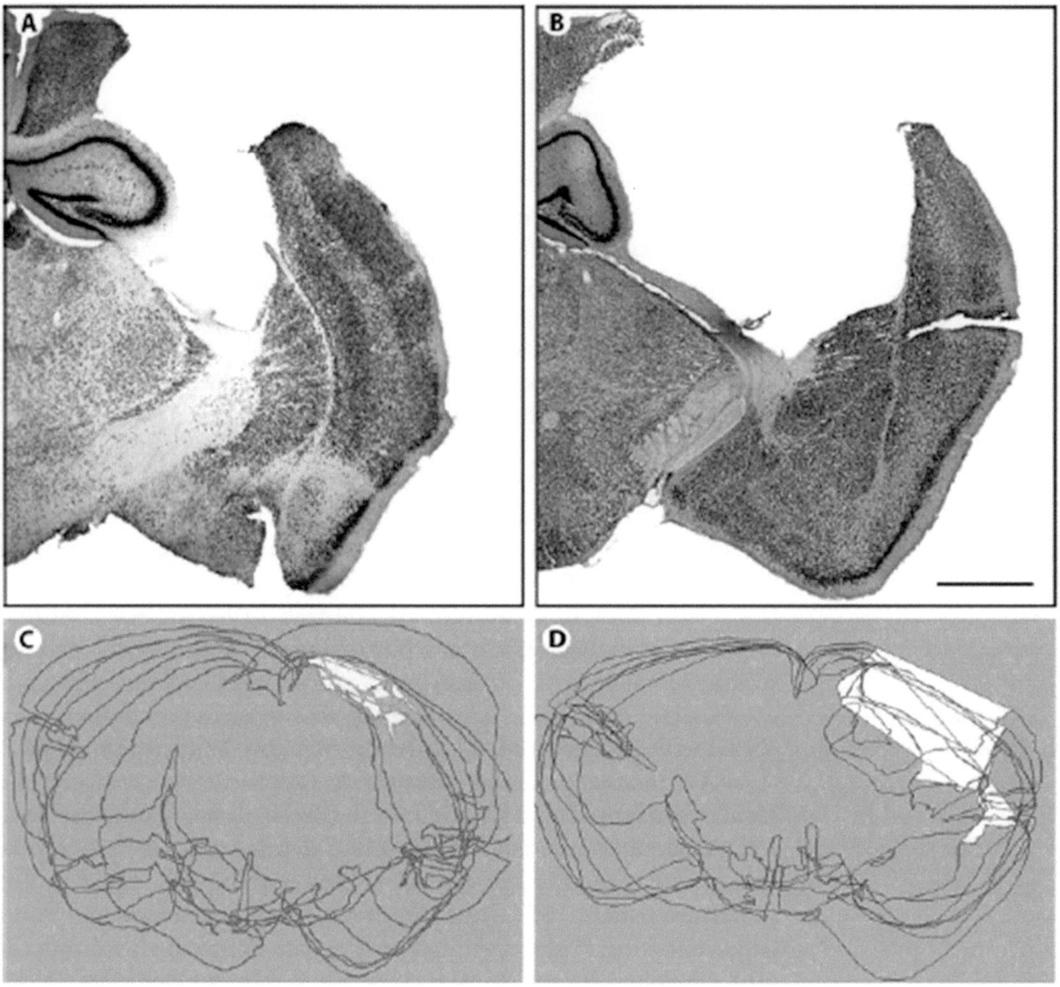

Fig. 1 Progressive neurodegeneration is evident over time after a mouse model of TBI at PND21. A cortical cavity involving the frontal and parietal grey matter and subcortical white matter is observed 5 weeks after injury (**a**). By 4.5 months after injury (**b**), a marked expansion of this cavity is evident. Delineation of total brain volume and lesion volumes by stereological tracings from injured mice at 5 weeks (**c**) and 4.5 months post-injury (**d**) allow for quantification of tissue loss and progressive atrophy. Scale bar = 1.0 mm. From ref. [41], reprinted with permission from S. Karger AG Publishers, Basel

significant neuropathology or long-term consequences, there is increasing evidence that repetitive injuries result in more severe symptoms, a longer recovery time, and an increased risk of adverse chronic outcomes. A CCI impactor has been utilized to generate repeated closed head injuries to PND35 rats, representing concussive-like insults during adolescence. This model is characterized by acute gliosis and axonal injury with transient cognitive deficits and a distinct window of glucose metabolic dysfunction [44, 45]. Another group, although limiting their focus to a single mild brain injury induced by a modified WD paradigm to juvenile (PND30) rats, has detected both transient and persistent behavioral impairments and morphological changes after injury [46–49]. In younger animals, a WD device has also been employed to generate five daily mild impacts to PND20 rats. This results in ventriculomegaly and cortical thinning [50], which was not observed following a similar injury paradigm in adult rats [51], but has been reported in clinical cases of chronic traumatic encephalopathy [52]. Of note, optimal modeling of 'concussions' in rodents remains somewhat controversial, with a lack of consensus regarding the appropriate injury severity, mechanism of impact and clinically relevant interval between repeated impacts. Many models of concussive-like injuries may in fact be modeling a more moderate TBI, as mortality, macroscopic damage and/or hemorrhage have been reported after even a single insult [45, 53]. This is in contrast to the clinical manifestation of a single concussion, which is typically not associated with neuropathology by computed tomography imaging. Furthermore, although loss of consciousness is seen in less than 10% of concussed patients [54], the rodent equivalent of prolonged righting reflex remains a salient characteristic of many concussive-like injury models [55].

4 Rodent Models of Pediatric TBI Reveal Age-Dependent Injury Responses

From birth to early adulthood, the mammalian brain undergoes a multitude of maturational processes including synaptogenesis and synaptic pruning, region-specific neurogenesis, and ongoing myelination. Brain injuries that occur at different ages may have varying neuropathological and functional consequences, depending upon the stage of brain maturation and impact of the insult on particular developmental processes. Indeed, rodent models of pediatric TBI have revealed age-dependent injury responses, which may underlie the vulnerability to poorer outcomes after early life injury seen in the clinical setting.

The peak in neurogenesis throughout the brain occurs during gestation, and by adulthood, the generation of new neurons is largely restricted to the dentate gyrus of the hippocampus and the subgranular zone of the lateral ventricles [56]. However, the postnatal brain maintains a limited capacity to produce neocortical neurons after injury [57], in an age-dependent manner. A more robust proliferative response is seen after brain injuries during early postnatal life compared to injuries in older animals [58, 59], although this regenerative response does not result in significant neuronal replacement [60]. Recent studies have provided support for a role of neurogenesis as a mechanism of functional recovery after injury. When cell proliferation is experimentally ablated after adult TBI, by the administration of ganciclovir in a nestin-HSV-TK transgenic model or by administration of an antimitotic agent, normal cognitive improvement over time post-injury is abolished [61, 62]. However, it is unclear whether similar age-dependent changes in the neurogenic response occur after injury during human brain development, and the prospect of harnessing this proliferative response to improve outcomes remains theoretical.

Following their genesis, neurons undergo a period of synaptogenesis involving an increase in arborization and synaptic contacts, followed by a period of activity-dependent synaptic pruning and refinement. This process is thought to contribute to neural plasticity and the reorganization of circuitry, and the refinement of cognitive processing abilities across development [63]. Synaptic density increases dramatically during early postnatal life, peaking above adult levels by approximately 2 years of age in humans [64, 65], and 3 weeks of age in rodents [66]. A CCI injury that has little effect on PND7 rats results in considerable region-specific changes in dendritic morphology and complexity when the injury occurs at PND17 or PND30, suggesting that the timing of synaptic changes may contribute to age-dependent vulnerability to injury at this time [49, 67, 68]. An enhanced capacity of the developing brain for plasticity is supported by the observation that PND35 rats exhibit synaptic sprouting at a rate fourfold higher than adult animals (PND90) [69, 70], as well as evidence of circuitry reorganization in both the injured cortex and hippocampus with the preservation of somewhat normal topography [37].

The functional consequences of such changes has been detected as enhanced neuronal excitability in the hippocampus at 2 weeks after CCI to PND17 rats [68], which may indicate and underlie a process of post-traumatic epileptogenesis, leading to abnormal electroencephalographic (EEG) spiking seen by 4–11 months post-injury [71]. In contrast, others have identified a reduction in neurophysiological responses in cortical tissue adjacent and remote from the impact site by 2–3 weeks after CCI to PND17 rats [72]. Progressive lesion expansion and transient or late-emerging behavioral phenotypes seen over time after experimental TBI provides evidence in support of a degree of circuitry reorganization [73]. Such findings also suggest that post-TBI plasticity results in abnormal neuronal networks, which may contribute to negative long-term outcomes including seizure susceptibility [72].

The myelination of axons is a prolonged process, which continues throughout childhood and adolescence in a conserved, spatiotemporal pattern. Changes in myelination over time underlie a progressive increase in total white matter in the brain across development [74]. Diffuse axonal injury after TBI implicates perturbations to the axonal membrane as well as the likely loss of myelin, and white matter atrophy in the adult brain may be attributed to acute tissue loss and/or delayed secondary deafferentation. In children, this scenario is super-imposed upon a still-developing system whereby atrophy may reflect a disruption of normal myelination and growth, as well as decreased organization [75, 76]. Traumatic injuries to the brains of children and adolescents (as well as adults) are notoriously heterogeneous in nature, such that the CCI model, for example, will mimic TBI in a subset of pediatric patients with a range of pathological changes including reduced corpus callosal volume and ventricular enlargement [77]. As CCI produces a focal legion with limited diffuse axonal injury, the WD or FPI models which result in a greater degree of axonal pathology are more appropriate for modeling children with widespread white matter damage.

After the primary mechanical insult of a moderate or severe TBI, a plethora of secondary pathological events are initiated in the brain including hemorrhage, necrotic and apoptotic cell death, oxidative stress, edema, disruption of the blood–brain barrier (BBB) and neuroinflammation. The propensity for aberrant cell death after TBI is considerably age-dependent, with younger rodents (PND3) showing a greater degree of apoptotic cell death compared to PND7 and older animals, identifying an early-life window of vulnerability [78, 79].

The developing brain also appears to exhibit a differential cerebral inflammatory response. Compared to the adult, a 3-week-old rodent brain shows enhanced breakdown of the BBB and neutrophil recruitment in response to the inflammatory mediator interleukin-1β [80], potentially via a CXC chemokine recruitment mechanism [81]. This finding of age-dependent susceptibility to an inflammatory response holds true in the PND21 mouse model of pTBI, in which the magnitude and time course of CD45+ Gr1+ leukocyte infiltration is greater than after an equivalent injury in adults [82].

Oxidative stress/injury is another key determinant of recovery after pTBI [83, 84]. The brain is particularly vulnerable to oxidative damage because of its high rate of oxidative activity and relatively low antioxidant capacity. Oxidative stress/injury arises from exposure to reactive oxygen and nitrogen species, glutamate-mediated excitotoxicity, and the degradation products of heme including free iron, that collectively overwhelm developmentally regulated antioxidants (*see* Fig. 2) [83]. Importantly, there is enhanced vulnerability to injury resulting from oxidative stress in the developing brain, as injury may occur at an age when antioxidants are not yet fully established. For example, heme oxygenase, an enzyme that is responsible for degrading the pro-oxidant heme, is low in the rodent brain at birth and during early brain development [84]. From PND7 onwards there is a rise in this enzyme, reaching adult levels at PND21 in rodents. Thus, trauma to the brain during early neonatal and postnatal development may result in prolonged exposure to heme and subsequent enhanced secondary tissue damage [83]. Antioxidant activity in response to an insult may also show age-dependent variability. This is exemplified in studies that have compared the activity of the antioxidant glutathione peroxidase

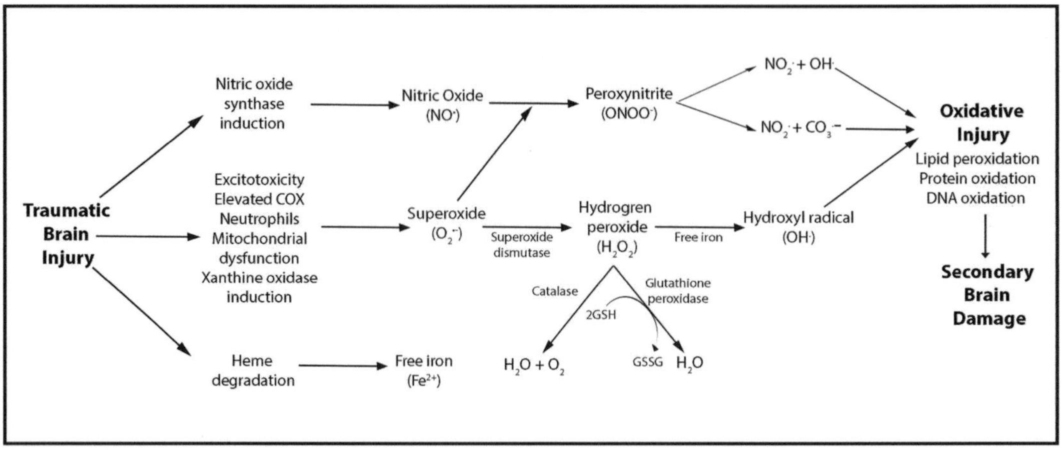

Fig. 2 Key oxidative stress pathways in the injured brain. Adapter from ref. [83] reprinted with permission from Springer Science and Business Media

(GPx) in the rodent brain after TBI at both PND21 and at adulthood. GPx catalyzes the reduction of the pro-oxidant hydrogen peroxidase to water through the oxidation of glutathione (*see* Fig. 2). The baseline activity of GPx is similar in the adult brain as compared to the developing brain at PND21. However, in response to TBI, there is an increase in activity of this antioxidant in the injured adult brain, whereas no such increase is observed in the brain injured at PND21 [85]. This unique age-dependent response likely contributes to the vulnerability of the brain to injuries at this early age, and subsequent long-term cognitive deficits [86].

Increased attention has been focused in recent years on the mechanisms by which TBI influences cerebral metabolism and blood flow, particularly after early-life insult. In young patients, abnormalities in metabolites have been associated with poorer neurological outcomes in infants and children after closed head injury [87, 88]. In young rodents, a decrease or delay in cerebral glucose metabolism has been reported after CCI to PND17–21 rats [89–91]. At the cellular level, abnormalities in mitochondrial oxidative metabolism are also evident in both neurons and astrocytes within hours after TBI to immature rodents, and persisting for at least 7 days [92]. Interestingly, an age-dependent switch in the brain's preferred metabolic substrates, from ketones to glucose as the primary fuel source, may provide the developing brain with some resilience to metabolic disruption. Around weaning (PND15–23 in rats), the immature brain has a sixfold greater capacity to take up and process ketones compared to the adult brain [93, 94]. Following energy challenges, even after this peak (e.g., during adolescence at ~PND35), a younger animal exhibits a greater ability for ketone uptake and metabolism compared to the adult brain, resulting in better recovery of cerebral metabolic and neurochemical balance after injury [94, 95].

Together, these data highlight the vulnerability of the young brain during particular developmental windows, an often underappreciated insight in light of the perception that a juvenile brain shows resilience due to a capacity for plasticity.

5 Age and Time-Appropriate Outcome Measures

There are a wide range of functional outcome measures suitable for use in young rodents, including tasks to assess sensory, motor, cognitive and emotional behaviors (*see* Fig. 3). Of note, preclinical TBI research has traditionally been focused on understanding acute mechanisms of cell death and injury, and functional assessments largely concerned with short-term outcomes. In fact, a recent analysis of published TBI literature found that the majority (68%) of reviewed papers did not evaluate functional outcomes past 1 month post-TBI, with only 10% of studies looking

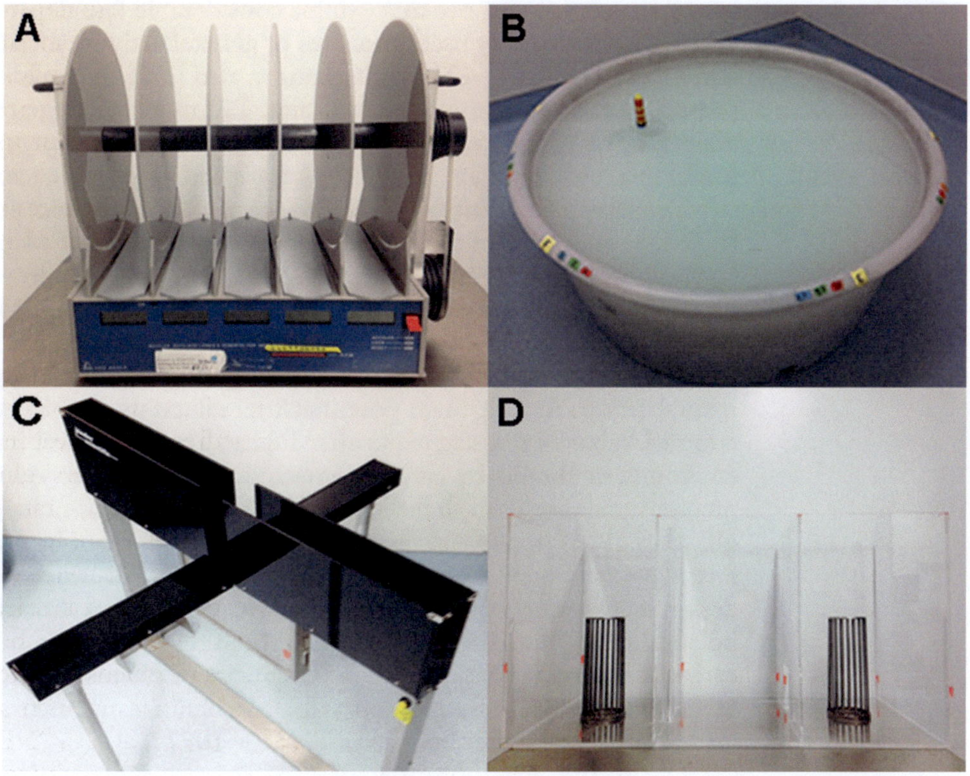

Fig. 3 Common behavioral assays used to evaluate functional outcomes after pTBI in rodents. The accelerating rotarod (**a**) provides a measure of sensorimotor and vestibulomotor function. The Morris water maze (**b**) allows for the evaluation of spatial learning and memory. Anxiety-like behaviors can be detected by time spent in the open and closed arms of the Elevated Plus Maze (**c**). Sociability and social recognition are readily quantified in the three-chamber social approach test (**d**)

long-term (>2 months post-injury) [96]. This is particularly problematic in the context of TBI to the young brain, as clinical findings provide evidence of long-term deficits persisting for months to years after childhood injuries, well into adulthood [97–101]. The evaluation of long-term dysfunction after pediatric TBI is complicated by normal developmental changes due to brain and body maturation during this period. The result is a scenario where recovery after pediatric TBI requires advancement to appropriate developmental milestones for the patient's age, not merely a return to pre-injury baseline [102, 103].

Long-term behavioral testing is of great value after experimental TBI to the developing brain, to track the persistence of functional deficits, as well as identify symptoms, which may resolve or emerge over time as the brain matures. The identification and evaluation of potential therapeutics or interventions in rodent models must also be considered in the context of ongoing behavioral changes with both age and time post-injury. Age-matched sham-operated or naive control animals are necessary to account for variation in the baseline behavioral phenotype, as we and others have found that uninjured rodents exhibit considerable age-dependent changes in commonly used measures of general activity, sociability and anxiety [38, 41, 42]. Such changes also limit the validity of pre-injury testing or training in pediatric TBI models, compared to adult models where pre-injury baseline performance can provide a useful indicator of injury consequences for individual animals.

The National Institute of Neurological Disorders and Stroke (NINDS) and others have placed emphasis on the use of adequate controls, sample sizes and statistical methods, as well as appropriate blinding and randomization during preclinical study design [104]. In line with these recommendations, in the context of pediatric TBI, we recommend that littermates be randomly allocated across experimental groups to control for potential litter effects during the evaluation of behavioral assessments after injury. The behavioral testing environment should be carefully controlled to minimize environmental variability (e.g., lighting, audible and ultrasonic noise, and olfactory distractions), and testing consistently performed by trained investigators blinded to experimental groups. How the experimental design might interfere with normal behavior phenotypes should also be considered. For example, individual housing of animals after an early-life injury can impair the development of normal social and sexual interactions, effect emotional and cognitive measures, and influence recovery after brain injuries [105–107].

Lastly, the age-appropriateness of specific functional assays should be considered. A propensity for social and play-like interactions increases dramatically after weaning in rodents, and the evaluation of such behaviors may detect aberrant psychosocial and emotional function after pediatric TBI [38, 39, 108, 109]. Spatial navigation skills, required for successful task acquisition in the

Morris Water Maze, also show a maturational time course whereby the rate of learning improves with age. The ability to locate the platform in relation to external room cues is reportedly established by approximately PND35 in rats [110], coinciding with a period of ongoing synaptic refinement in key hippocampal structures. In another study, investigators noted that PND17 rats required more trials to reach criterion and took fewer direct paths to the platform compared to older rats (PND28 or adult) [111], suggesting that this common test for cognitive ability is of limited value for assessing preweaning animals. Consideration of body weight is also important, and measures of sensorimotor performance, such as the rotarod or balance beam, may require scaling of the equipment to an age-appropriate size [73].

6 Work to Be Done

Despite a recent increase in preclinical research in pediatric TBI, much of our understanding regarding the long-term consequences remains in its infancy. The underlying mechanisms of comorbidities including post-traumatic epilepsy, neuroendocrine dysfunction and the disruption of sleep-wake behavior remain unclear, and such factors are likely to negatively impact chronic quality of life. In particular, the relationship between pediatric TBI and chronic psychosocial outcomes has received little attention in the preclinical arena, and future studies may reveal age-dependent windows of opportunity for behavioral interventions. How injury to the developing brain differs from injury during adulthood also requires further investigation; for example, a full profile of peripheral and central immune responses after injury at different ages may elucidate the mechanisms underlying age-specific susceptibility to inflammation.

Another oft-neglected consideration in preclinical TBI research is that of sex and gender, which appears to influence outcomes in both patient populations and experimental models [112–114]. The importance of circulating sex hormones compared to innate sexual dimorphisms of the male versus female brain remains undefined, although there is evidence to suggest that even prior to sexual maturation, an individual's sex is an important determinant of long-term recovery [115]. Studies examining sex differences after TBI at different developmental ages will provide a better understanding of these mechanisms.

7 Conclusions

Several adaptations of experimental models commonly used in adult rodents have been successfully employed to investigate the mechanisms and consequences of TBI during childhood and adolescence. A key consideration of study design is the appropriate age in a rodent to model vulnerable pediatric populations, a decision that should be made based upon a multifaceted understanding of normal brain development with age. A range of clinically relevant, functional measures can be evaluated in brain-injured rodents, including sensory, motor and psychosocial. The importance of considering long-term consequences cannot be overstated, particularly in light of abundant clinical evidence of long-term problems persisting to adulthood after pediatric TBI, and the potential for early-life injury to negatively influence brain maturation across development. Over recent years, preclinical TBI research in the pediatric arena has unveiled key mechanisms of oxidative stress, inflammation and cerebral metabolic changes, which may underlie the particular vulnerability of the young brain to poor outcomes after TBI. Future research aimed at modulating or protecting against such processes to support a trajectory of normal brain development has the potential to improve long-term outcomes after brain injury in children.

Acknowledgement

This review was supported by NIH/NINDS R01 NS050159 and NS077767.

References

1. Faul M, Xu L, Wald MM, Coronado VG (2010) Traumatic brain injury in the United States: Emergency department visits, hospitalizations and deaths 2002-2006. Centers for Disease Control and Prevention, National Center for Injury Prevention and Control Atlanta, GA
2. Thurman DJ (2014) The epidemiology of traumatic brain injury in children and youths: a review of research since 1990. J Child Neurol 31:20–27
3. Langlois JA (2000) Traumatic brain injury in the United States: assessing outcomes in children: summary and recommendations from the Expert Working Group, October 26-27. Division of Acute Care, Rehabilitation Research and Disability Prevention, National Center for Injury Prevention and Control, Centers for Disease Control and Prevention, Department of Health and Human Services. Atlanta, GA
4. Anderson V, Catroppa C, Morse S, Haritou F, Rosenfeld J (2005) Functional plasticity or vulnerability after early brain injury? Pediatrics 116:1374–1382
5. Anderson V, Jacobs R, Spencer-Smith M, Coleman L, Anderson P, Williams J, Greenham M, Leventer R (2010) Does early age at brain insult predict worse outcome? Neuropsychological implications. J Pediatr Psychol 35:716–727
6. Robertson CM, Joffe AR, Moore AJ, Watt JM (2002) Neurodevelopmental outcome of young pediatric intensive care survivors of serious brain injury. Pediatr Crit Care Med 3:345–350
7. Luerssen TG, Klauber MR, Marshall LF (1988) Outcome from head injury related to patient's age. A longitudinal prospective study of adult and pediatric head injury. J Neurosurg 68:409–416
8. Beauchamp MH, Anderson V (2013) Cognitive and psychopathological sequelae of pediatric traumatic brain injury. Handb Clin Neurol 112:913–920
9. Anderson VA, Spencer-Smith MM, Coleman L, Anderson PJ, Greenham M, Jacobs R, Lee KJ, Leventer RJ (2014) Predicting neurocognitive and behavioural outcome after early

brain insult. Dev Med Child Neurol 56:329–336

10. Choe MC, Valino H, Fischer J, Zeiger M, Breault J, McArthur DL, Leung M, Madikians A, Yudovin S, Lerner JT, Giza CC (2015) Targeting the epidemic: interventions and follow-up are necessary in the pediatric traumatic brain injury clinic. J Child Neurol 31:109–115
11. Verger K, Junque C, Levin HS, Jurado MA, Perez-Gomez M, Bartres-Faz D, Barrios M, Alvarez A, Bartumeus F, Mercader JM (2001) Correlation of atrophy measures on MRI with neuropsychological sequelae in children and adolescents with traumatic brain injury. Brain Inj 15:211–221
12. Serra-Grabulosa JM, Junque C, Verger K, Salgado-Pineda P, Maneru C, Mercader JM (2005) Cerebral correlates of declarative memory dysfunctions in early traumatic brain injury. J Neurol Neurosurg Psychiatry 76:129–131
13. Wilde EA, Hunter JV, Newsome MR, Scheibel RS, Bigler ED, Johnson JL, Fearing MA, Cleavinger HB, Li X, Swank PR, Pedroza C, Roberson GS, Bachevalier J, Levin HS (2005) Frontal and temporal morphometric findings on MRI in children after moderate to severe traumatic brain injury. J Neurotrauma 22:333–344
14. Wilde EA, Bigler ED, Hunter JV, Fearing MA, Scheibel RS, Newsome MR, Johnson JL, Bachevalier J, Li X, Levin HS (2007) Hippocampus, amygdala, and basal ganglia morphometrics in children after moderate-to-severe traumatic brain injury. Dev Med Child Neurol 49:294–299
15. Wilde EA, Merkley TL, Bigler ED, Max JE, Schmidt AT, Ayoub KW, McCauley SR, Hunter JV, Hanten G, Li X, Chu ZD, Levin HS (2012) Longitudinal changes in cortical thickness in children after traumatic brain injury and their relation to behavioral regulation and emotional control. Int J Dev Neurosci 30:267–276
16. Beauchamp MH, Ditchfield M, Maller JJ, Catroppa C, Godfrey C, Rosenfeld JV, Kean MJ, Anderson VA (2011) Hippocampus, amygdala and global brain changes 10 years after childhood traumatic brain injury. Int J Dev Neurosci 29:137–143
17. Keightley ML, Sinopoli KJ, Davis KD, Mikulis DJ, Wennberg R, Tartaglia MC, Chen JK, Tator CH (2014) Is there evidence for neurodegenerative change following traumatic brain injury in children and youth? A scoping review. Front Hum Neurosci 8:139
18. Giza CC, Mink RB, Madikians A (2007) Pediatric traumatic brain injury: not just little adults. Curr Opin Crit Care 13:143–152
19. Pinto PS, Poretti A, Meoded A, Tekes A, Huisman TA (2012) The unique features of traumatic brain injury in children. Review of the characteristics of the pediatric skull and brain, mechanisms of trauma, patterns of injury, complications and their imaging findings—part 1. J Neuroimag 22:e1–e17
20. Romijn HJ, Hofman MA, Gramsbergen A (1991) At what age is the developing cerebral cortex of the rat comparable to that of the full-term newborn human baby? Early Hum Dev 26:61–67
21. Prins ML, Hovda DA (2003) Developing experimental models to address traumatic brain injury in children. J Neurotrauma 20:123–137
22. Semple BD, Blomgren K, Gimlin K, Ferriero DM, Noble-Haeusslein LJ (2013) Brain development in rodents and humans: identifying benchmarks of maturation and vulnerability to injury across species. Prog Neurobiol 106–107:1–16
23. Dixon CE, Lyeth BG, Povlishock JT, Findling RL, Hamm RJ, Marmarou A, Young HF, Hayes RL (1987) A fluid percussion model of experimental brain injury in the rat. J Neurosurg 67:110–119
24. Thompson HJ, Lifshitz J, Marklund N, Grady MS, Graham DI, Hovda DA, McIntosh TK (2005) Lateral fluid percussion brain injury: a 15-year review and evaluation. J Neurotrauma 22:42–75
25. McIntosh TK, Vink R, Noble L, Yamakami I, Fernyak S, Soares H, Faden AL (1989) Traumatic brain injury in the rat: characterization of a lateral fluid-percussion model. Neuroscience 28:233–244
26. Prins ML, Lee SM, Cheng CL, Becker DP, Hovda DA (1996) Fluid percussion brain injury in the developing and adult rat: a comparative study of mortality, morphology, intracranial pressure and mean arterial blood pressure. Brain Res Dev Brain Res 95:272–282
27. Marmarou A, Foda MA, van den Brink W, Campbell J, Kita H, Demetriadou K (1994) A new model of diffuse brain injury in rats. Part I: pathophysiology and biomechanics. J Neurosurg 80:291–300
28. Feeney DM, Boyeson MG, Linn RT, Murray HM, Dail WG (1981) Responses to cortical injury: I. Methodology and local effects of contusions in the rat. Brain Res 211:67–77
29. Adelson PD, Dixon CE, Kochanek PM (2000) Long-term dysfunction following diffuse traumatic brain injury in the immature rat. J Neurotrauma 17:273–282
30. Adelson PD, Whalen MJ, Kochanek PM, Robichaud P, Carlos TM (1998) Blood brain

barrier permeability and acute inflammation in two models of traumatic brain injury in the immature rat: a preliminary report. Acta Neurochir Suppl 71:104–106

31. Huh JW, Widing AG, Raghupathi R (2008) Midline brain injury in the immature rat induces sustained cognitive deficits, bihemispheric axonal injury and neurodegeneration. Exp Neurol 213:84–92
32. Huh JW, Widing AG, Raghupathi R (2011) Differential effects of injury severity on cognition and cellular pathology after contusive brain trauma in the immature rat. J Neurotrauma 28:245–257
33. Adelson PD, Jenkins LW, Hamilton RL, Robichaud P, Tran MP, Kochanek PM (2001) Histopathologic response of the immature rat to diffuse traumatic brain injury. J Neurotrauma 18:967–976
34. Adelson PD, Fellows-Mayle W, Kochanek PM, Dixon CE (2013) Morris water maze function and histologic characterization of two age-at-injury experimental models of controlled cortical impact in the immature rat. Child's Nerv Syst 29:43–53
35. Osier, N. D., Korpon, J. R., & Dixon, C. E. (2015) Controlled cortical impact model. In: Kobeissy, F. H. ed. Brain neurotrauma: molecular, neuropsychological, and rehabilitation aspects, Boca Raton (FL), pp 421–428
36. Ajao DO, Pop V, Kamper JE, Adami A, Rudobeck E, Huang L, Vlkolinsky R, Hartman RE, Ashwal S, Obenaus A, Badaut J (2012) Traumatic brain injury in young rats leads to progressive behavioral deficits coincident with altered tissue properties in adulthood. J Neurotrauma 29:2060–2074
37. Card JP, Santone DJ Jr, Gluhovsky MY, Adelson PD (2005) Plastic reorganization of hippocampal and neocortical circuitry in experimental traumatic brain injury in the immature rat. J Neurotrauma 22:989–1002
38. Semple BD, Canchola SA, Noble-Haeusslein L (2012) Deficits in social behavior emerge during development after pediatric traumatic brain injury in mice. J Neurotrauma 29:2672–2683
39. Semple BD, Noble-Haeusslein LJ, Kwon YJ, Sam PN, Gibson AM, Grissom S, Brown S, Adahman Z, Hollingsworth CA, Kwakye A, Gimlin K, Wilde EA, Hanten G, Levin HS, Schenk AK (2014) Sociosexual and communication deficits after traumatic injury to the developing murine brain. PLoS One 9(8):e103386
40. Tong W, Igarashi T, Ferriero DM, Noble LJ (2002) Traumatic brain injury in the immature mouse brain: characterization of regional vulnerability. Exp Neurol 176:105–116
41. Pullela R, Raber J, Pfankuch T, Ferriero DM, Claus CP, Koh S-E, Yamauchi T, Rola R, Fike JR, Noble-Haeusslein LJ (2006) Traumatic injury to the immature brain results in progressive neuronal loss, hyperactivity and delayed cognitive impairments. Dev Neurosci 28:396–409
42. Chen CY, Noble-Haeusslein LJ, Ferriero D, Semple BD (2013) Traumatic injury to the immature frontal lobe: a new murine model of long-term motor impairment in the absence of psychosocial or cognitive deficits. Dev Neurosci 35:474–490
43. Kamper JE, Pop V, Fukuda AM, Ajao DO, Hartman RE, Babaut J (2013) Juvenile traumatic brain injury evolves into a chronic brain disorder: behavioral and histological changes over 6 months. Exp Neurol 250:8–19
44. Prins ML, Alexander D, Giza CC, Hovda DA (2013) Repeated mild traumatic brain injury: mechanisms of cerebral vulnerability. J Neurotrauma 30:30–38
45. Prins ML, Hales A, Reger M, Giza CC, Hovda DA (2010) Repeat traumatic brain injury in the juvenile rat is associated with increased axonal injury and cognitive impairments. Dev Neurosci 32:510–518
46. Mychasiuk R, Farran A, Angoa-Perez M, Briggs D, Kuhn D, Esser MJ (2014) A novel model of mild traumatic brain injury for juvenile rats. J Vis Exp (94)
47. Mychasiuk R, Farran A, Esser MJ (2014) Assessment of an experimental rodent model of pediatric mild traumatic brain injury. J Neurotrauma 31:749–757
48. Mychasiuk R, Hehar H, Esser MJ (2015) A mild traumatic brain injury (mTBI) induces secondary attention-deficit hyperactivity disorder-like symptomology in young rats. Behav Brain Res 286:285–292
49. Mychasiuk R, Hehar H, Ma I, Kolb B, Esser MJ (2015) The development of lasting impairments: a mild pediatric brain injury alters gene expression, dendritic morphology, and synaptic connectivity in the prefrontal cortex of rats. Neuroscience 288:145–155
50. Goddeyne C, Nichols J, Wu C, Anderson T (2015) Repetitive mild traumatic brain injury induces ventriculomegaly and cortical thinning in juvenile rats. J Neurophysiol 113:3268–3280. doi:10.1152/jn.00970.2014
51. Kane MJ, Angoa-Pérez M, Briggs DI, Viano DC, Kreipke CW, Kuhn DM (2012) A mouse model of human repetitive mild traumatic brain injury. J Neurosci Methods 203:41–49

52. Albaugh MD, Orr C, Nickerson JP, Zweber C, Slauterbeck JR, Hipko S, Gonyea J, Andrews T, Brackenbury JC, Watts R, Hudziak JJ (2015) Postconcussion symptoms are associated with cerebral cortical thickness in healthy collegiate and preparatory school ice hockey players. J Pediatr 166:394–400.e1
53. Fijalkowski RJ, Stemper BD, Pintar FA, Yoganandan N, Crowe MJ, Gennarelli TA (2007) New rat model for diffuse brain injury using coronal plane angular acceleration. J Neurotrauma 24:1387–1398
54. Guskiewicz KM, McCrea M, Marshall SW, Cantu RC, Randolph C, Barr W, Onate JA, Kelly JP (2003) Cumulative effects associated with recurrent concussion in collegiate football players: the NCAA Concussion Study. JAMA 290:2549–2555
55. DeWitt DS, Perez-Polo R, Hulsebosch CE, Dash PK, Robertson CS (2013) Challenges in the development of rodent models of mild traumatic brain injury. J Neurotrauma 30:688–701
56. Rice D, Barone SJ (2000) Critical periods of vulnerability for the developing nervous system: evidence from humans and animal models. Environ Health Perspect 108:511–533
57. Kernie SG, Parent JM (2010) Forebrain neurogenesis after focal ischemic and traumatic brain injury. Neurobiol Dis 37:267–274
58. Covey MV, Jiang Y, Alli VV, Yang Z, Levison SW (2010) Defining the critical period for neocortical neurogenesis after pediatric brain injury. Dev Neurosci 32:488–498
59. Sun D, Colello RJ, Daugherty WP, Kwon TH, McGinn MJ, Harvey HB, Bullock MR (2005) Cell proliferation and neuronal differentiation in the dentate gyrus in juvenile and adult rats following traumatic brain injury. J Neurotrauma 22:95–105
60. Goodus MT, Guzman AM, Calderon F, Jiang Y, Levison SW (2015) Neural stem cells in the immature, but not the mature subventricular zone respond robustly to traumatic brain injury. Dev Neurosci 37:29–42
61. Blaiss CA, Yu TS, Zhang G, Chen J, Dimchev G, Parada LF, Powell CM, Kernie SG (2011) Temporally specified genetic ablation of neurogenesis impairs cognitive recovery after traumatic brain injury. J Neurosci 31:4906–4916
62. Sun D, Daniels TE, Rolfe A, Waters M, Hamm R (2015) Inhibition of injury-induced cell proliferation in the dentate gyrus of the hippocampus impairs spontaneous cognitive recovery after traumatic brain injury. J Neurotrauma 32:495–505
63. Low LK, Cheng HJ (2006) Axon pruning: an essential step underlying the developmental plasticity of neuronal connections. Philos Trans R Soc Lond B Biol Sci 361:1531–1544
64. Herschkowitz N, Kagan J, Zilles K (1997) Neurobiological bases of behavioral development in the first year. Neuropediatrics 28:296–306
65. Huttenlocher PR (1979) Synaptic density in human frontal cortex—developmental changes and effects of aging. Brain Res 163:195–205
66. Crain B, Cotman C, Taylor D, Lynch G (1973) A quantitative electron microscopic study of synaptogenesis in the dentate gyrus of the rat. Brain Res 63:195–204
67. Casella EM, Thomas TC, Vanino DL, Fellows-Mayle W, Lifshitz J, Card JP, Adelson PD (2014) Traumatic brain injury alters long-term hippocampal neuron morphology in juvenile, but not immature, rats. Child's Nerv Syst 30:1333–1342
68. Nichols J, Perez RS, Wu C, Adelson PD, Anderson T (2014) Traumatic brain injury induces rapid enhancement of cortical excitability in juvenile rats. CNS Neurosci Ther 21:193–203
69. Scheff SW, Benardo LS, Cotman CW (1980) Decline in reactive fiber growth in the dentate gyrus of aged rats compared to young adult rats following entorhinal cortex removal. Brain Res 199:21–38
70. McWilliams JR, Lynch G (1983) Rate of synaptic replacement in denervated rat hippocampus declines precipitously from the juvenile period to adulthood. Science 221:572–574
71. Statler KD, Scheerlinck P, Pouliot W, Hamilton M, White HS, Dudek FE (2009) A potential model of pediatric posttraumatic epilepsy. Epilepsy Res 86:221–223
72. Li N, Yang Y, Glover DP, Zhang J, Saraswati M, Robertson C, Pelled G (2014) Evidence for impaired plasticity after traumatic brain injury in the developing brain. J Neurotrauma 31:395–403
73. Ellis TWJ, Ziebell JM, Adelson PD, Lifshitz J (2014) Commentary on Kamper et al., juvenile traumatic brain injury evolves into a chronic brain disorder: the challenges in longitudinal studies of juvenile traumatic brain injury. Exp Neurol 261:434–439
74. Giedd JN, Blumenthal J, Jeffries NO, Castellanos FX, Liu H, Zijdenbos A, Paus T, Evans AC, Rapoport JL (1999) Brain development during childhood and adolescence: a longitudinal MRI study. Nat Neurosci 2:861–863
75. Ewing-Cobbs L, Prasad MR, Swank P, Kramer L, Cox CS Jr, Fletcher JM, Barnes M, Zhang X, Hasan KM (2008) Arrested development and disrupted callosal microstructure following pediatric traumatic brain injury: relation to neurobehavioral outcomes. Neuroimage 42:1305–1315

76. Tasker RC (2006) Changes in white matter late after severe traumatic brain injury in childhood. Dev Neurosci 28:302–308
77. Bigler ED, Abildskov TJ, Petrie J, Farrer TJ, Dennis M, Simic N, Taylor HG, Rubin KH, Vannatta K, Gerhardt CA, Stancin T, Owen Yeates K (2013) Heterogeneity of brain lesions in pediatric traumatic brain injury. Neuropsychology 27:438–451
78. Bittigau P, Sifringer M, Felderhoff-Mueser U, Ikonomidou C (2004) Apoptotic neurodegeneration in the context of traumatic injury to the developing brain. Exp Toxicol Pathol 56:83–89
79. Ikonomidou C, Qin Y, Labruyere J, Kirby C, Olney JW (1996) Prevention of trauma-induced neurodegeneration in infant rat brain. Pediatr Res 39:1020–1027
80. Anthony DC, Bolton SJ, Fearn S, Perry VH (1997) Age-related effects of interleukin-1 beta on polymorphonuclear neutrophil-dependent increases in blood-brain barrier permeability in rats. Brain 120:435–444
81. Anthony D, Dempster R, Fearn S, Clements J, Wells G, Perry VH, Walker K (1998) CXC chemokines generate age-related increases in neutrophil-mediated brain inflammation and blood-brain barrier breakdown. Curr Biol 8:923–926
82. Claus CP, Tsuru-Aoyagi K, Adwanikar H, Walker B, Whetstone W, Noble-Haeusslein LJ (2010) Age is a determinant of the inflammatory response and loss of cortical volume after traumatic brain injury. Dev Neurosci 32:454–465
83. Potts M, Koh S-E, Whetstone W, Walker B, Yoneyama T, Claus C, Manvelyan H, Noble-Haeusslein L (2006) Traumatic injury to the immature brain: inflammation, oxidative injury, and iron-mediated damage as potential therapeutic targets. Neuroreport 3:143–153
84. Chang EF, Claus CP, Vreman HJ, Wong RJ, Noble-Haeusslein LJ (2005) Heme regulation in traumatic brain injury: relevance to the adult and developing brain. J Cereb Blood Flow Metab 25:1401–1417
85. Fan P, Yamauchi T, Noble L, Ferriero D (2003) Age-dependent differences in glutathione peroxidase activity after traumatic brain injury. J Neurotrauma 20:437–445
86. Tsuru-Aoyagi K, Potts M, Trivedi A, Pfankuch T, Raber J, Wendland M, Claus C, Koh S-E, Ferriero D, Noble-Haeusslein L (2009) Glutathione peroxidase activity modulates recovery in the injured immature brain. Ann Neurol 65:540–549
87. Ashwal S, Holshouser BA, Shu SK, Simmons PL, Perkin RM, Tomasi LG, Knierim DS, Sheridan C, Craig K, Andrews GH, Hinshaw DB (2000) Predictive value of proton magnetic resonance spectroscopy in pediatric closed head injury. Pediatr Neurol 23:114–125
88. Holshouser BA, Ashwal S, Luh GY, Shu S, Kahlon S, Auld KL, Tomasi LG, Perkin RM, Hinshaw DB Jr (1997) Proton MR spectroscopy after acute central nervous system injury: outcome prediction in neonates, infants, and children. Radiology 202:487–496
89. Robertson CL, Saraswati M, Scafidi S, Fiskum G, Casey P, McKenna MC (2013) Cerebral glucose metabolism in an immature rat model of pediatric traumatic brain injury. J Neurotrauma 30:2066–2072
90. Casey PA, McKenna MC, Fiskum G, Saraswati M, Robertson CL (2008) Early and sustained alterations in cerebral metabolism after traumatic brain injury in immature rats. J Neurotrauma 25:603–614
91. Scafidi S, O'Brien J, Hopkins I, Robertson C, Fiskum G, McKenna M (2009) Delayed cerebral oxidative glucose metabolism after traumatic brain injury in young rats. J Neurochem 109(Suppl 1):189–197
92. Robertson CL, Saraswati M, Fiskum G (2007) Mitochondrial dysfunction early after traumatic brain injury in immature rats. J Neurochem 101:1248–1257
93. Nehlig A, Boyet S, Pereira de Vasconcelos A (1991) Autoradiographic measurement of local cerebral beta-hydroxybutyrate uptake in the rat during postnatal development. Neuroscience 40:871–878
94. Prins ML, Matsumoto J (2014) Metabolic response of pediatric traumatic brain injury. J Child Neurol 31:28–34
95. Deng-Bryant Y, Prins ML, Hovda DA, Harris NG (2011) Ketogenic diet prevents alterations in brain metabolism in young but not adult rats after traumatic brain injury. J Neurotrauma 28:1813–1825
96. Gold EM, Su D, López-Velázquez L, Haus DL, Perez H, Lacuesta GA, Anderson AJ, Cummings BJ (2013) Functional assessment of long-term deficits in rodent models of traumatic brain injury. Regen Med 8:483–516
97. Karver CL, Wade SL, Cassedy A, Taylor HG, Stancin T, Yeates KO, Walz NC (2012) Age at injury and long-term behavior problems after traumatic brain injury in young children. Rehabil Psychol 57:256–265

98. Kieslich M, Marquardt G, Galow G, Lorenz R, Jacobi G (2001) Neurological and mental outcome after severe head injury in childhood: a long-term follow-up of 318 children. Disabil Rehabil 23:665–669
99. Koskiniemi M, Kyykkä T, Nybo T, Jarho L (1995) Long-term outcome after severe brain injury in preschoolers is worse than expected. Arch Pediatr Adolesc Med 149:249–254
100. McKinlay A, Grace RC, Horwood LJ, Fergusson DM, Macfarlane MR (2010) Long-term behavioural outcomes of preschool mild traumatic brain injury. Child Care Health Dev 36:22–30
101. Ryan NP, Anderson V, Godfrey C, Beauchamp MH, Coleman L, Eren S, Rosema S, Taylor K, Catroppa C (2013) Predictors of very long-term socio-cognitive function after pediatric traumatic brain injury: support for the vulnerability of the immature 'social brain'. J Neurotrauma 31:649–657
102. Babikian T, Asarnow R (2009) Neurocognitive outcomes and recovery after pediatric TBI: meta-analytic review of the literature. Neuropsychology 23:283–296
103. Giza CC, Kolb B, Harris NG, Asarnow RF, Prins ML (2009) Hitting a moving target: basic mechanisms of recovery from acquired developmental brain injury. Dev Neurorehabil 12:255–268
104. Lapchak PA, Zhang JH, Noble-Haeusslein LJ (2013) RIGOR Guidelines: escalating STAIR and STEPS for effective translational research. Transl Stroke Res 4:279–285
105. Kercmar J, Tobet SA, Majdic G (2014) Social isolation during puberty affects female sexual behavior in mice. Front Behav Neurosci 8:337
106. Venna VR, Xu Y, Doran SJ, Patrizz A, McCullough LD (2014) Social interaction plays a critical role in neurogenesis and recovery after stroke. Transl Psychiatry 4:e351
107. Chabout J, Serreau P, Ey E, Bellier L, Aubin T, Bourgeron T, Granon S (2012) Adult male mice emit context-specific ultrasonic vocalizations that are modulated by prior isolation or group-rearing environment. PLoS One 7:e29401
108. Kaidanovich-Beilin O, Lipina T, Vukobradovic I, Roder J, Woodgett JR (2011) Assessment of social interaction behaviors. J Vis Exp (48):2473. doi:10.3791/2473
109. Terranova ML, Laviola G (2005) Scoring of social interactions and play in mice during adolescence. Curr Prot Toxicol Chapter 13:13.10.1–13.10.11
110. Schenk F (1985) Development of place navigation in rats from weaning to puberty. Behav Neural Biol 43:69–85
111. Prins ML, Hovda DA (1998) Traumatic brain injury in the developing rat: effects of maturation on Morris water maze acquisition. J Neurotrauma 15:799–811
112. Mychasiuk R, Hehar H, Farran A, Esser MJ (2014) Mean girls: sex differences in the effects of mild traumatic brain injury on the social dynamics of juvenile rat play behaviour. Behav Brain Res 259:284–291
113. Russell KL, Kutchko KM, Fowler SC, Berman NEJ, Levant B (2011) Sensorimotor behavioral tests for use in a juvenile rat model of traumatic brain injury: assessment of sex differences. J Neurosci Methods 199:214–222
114. Slewa-Younan S, van den Berg S, Baguley IJ, Nott M, Cameron ID (2008) Towards an understanding of sex differences in functional outcome following moderate to severe traumatic brain injury: a systematic review. J Neurol Neurosurg Psychiatry 79:1197–1201
115. Guevara R, Gianotti M, Oliver J, Roca P (2011) Age and sex-related changes in rat brain mitochondrial oxidative status. Exp Gerontol 46:923–928

Chapter 19

Modeling Pediatric Brain Trauma: Piglet Model of Controlled Cortical Impact

Jennifer C. Munoz Pareja, Kristen Keeley, Ann-Christine Duhaime, and Carter P. Dodge

Abstract

The brain has different responses to traumatic injury as a function of its developmental stage. As a model of injury to the immature brain, the piglet shares numerous similarities in regards to morphology and neurodevelopmental sequence compared to humans. This chapter describes a piglet scaled focal contusion model of traumatic brain injury that accounts for the changes in mass and morphology of the brain as it matures, facilitating the study of age-dependent differences in response to a comparable mechanical trauma.

Key words Controlled cortical impact, CCI, Animal model, Traumatic brain injury, Piglet, Swine, Development

1 Introduction

When it comes to traumatic brain injury research, it is important to understand the differences between species and among different ages of subject. Ideally, the investigator should attempt to match the mechanism, morphology, and/or maturation stages when choosing a model to answer a specific question about traumatic brain injury during immaturity.

Large animals, especially piglets, have human like morphology and physiology; there is evidence to suggest that piglets most closely resemble the human postnatal developmental sequence and thus confer significant advantages for modeling the effects of early insult, including cerebrovascular development [1–4]. The cerebral blood flow in conscious, normal neonatal piglets has been found to be similar to that in conscious human neonates [5]. With respect to brain electrical activity as measured by electroencephalography, piglets have a similar development to humans. They have been studied in the awake and conscious state as well as under conditions of cerebral insult and have electroencephalography patterns

Firas Kobeissy et al. (eds.), *Injury Models of the Central Nervous System: Methods and Protocols*, Methods in Molecular Biology, vol. 1462, DOI 10.1007/978-1-4939-3816-2_19, © Springer Science+Business Media New York 2016

similar to those seen in immature humans [6–8]. For these reasons, piglets have been used in a variety of models of perinatal brain insult, including hypoxia-ischemia, and trauma [9–16]. It is also crucial to identify important anatomical variations between large and small animal models; one of the biggest differences among rodents and piglets is the presence of white matter and gyri. Piglets also share developmental similarities and imaging that can be comparable to humans; some of the disadvantages are the genetic heterogeneity, complex anesthetic considerations, fewer molecular techniques, fewer outcome scales, and the significantly higher cost when compared to small animal models.

1.1 Piglet Model of Controlled Cortical Impact Model

The response of the immature brain to mechanical trauma differs among ages and developmental stages [17]. This model was developed by Drs. Duhaime and Margulies at the University of Pennsylvania, has been described in detail in other publications, and has been used at a number of institutions since that time to create a strictly comparable mechanical input across subjects at different ages in order to identify specific differences in injury response due to maturational changes alone [9]. The controlled cortical impact (CCI) model (Fig. 1) utilizes a portable, skull-mounted, stainless steel spring-loaded blunt indentation device with three point screw fixation, thereby allowing visible confirmation of the initial position of the indentor tip, ensuring a directly perpendicular indentation, and eliminating relative motion of the head and indentor.

The device was tested using a laser transducer and was shown to deliver highly reproducible firing traces with a time course of 4 ms and an indentation velocity of 1.7 m/s. When the device is

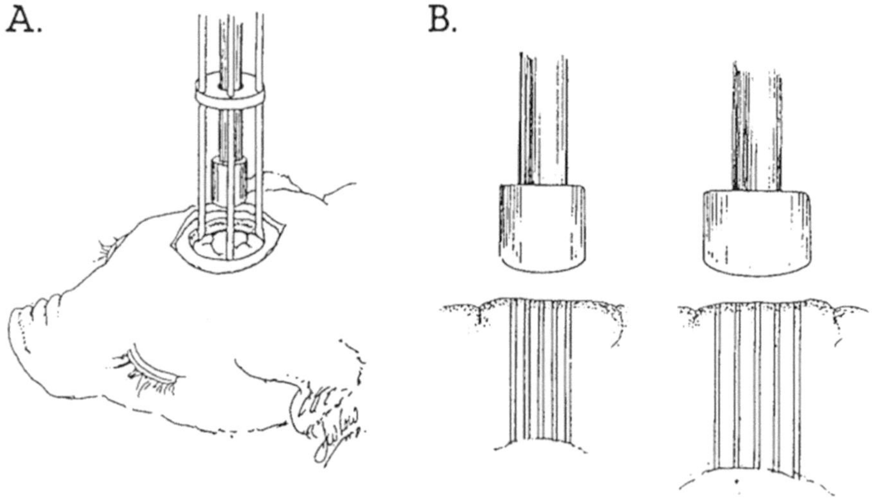

Fig. 1 (**a**) Schematic representation of the cortical injury device. (**b**) Illustration of the scaled indentor tips and their use in histological sampling. The tip diameter and indentation depth increase in proportion to increases in brain dimension with age, from youngest (*left*) to oldest (*right*), as described in the text. Samples are taken at the 0, 25, 50, 75, and 100 % regions of tissue under the indentor tip for each age

fired it creates a rapid displacement of the cortical surface, producing a focal cortical contusion, with underlying white matter damage, decreased cerebral blood flow and somatosensory dysfunction. Lesions are centered at the rostral gyrus of the frontal lobe to facilitate functional outcome comparisons among ages [9, 17, 18]. Indentation volumes are scaled in three dimensions to the dimensions of the growing brain, to displace 1 % of the total brain volume for that age subject, and the dimensions of the indentor tip and depth of indentation are scaled such that comparable anatomical structures are displaced in all ages.

Because of the rapid and forceful indentation of the brain surface, minor differences in mechanical properties of the brain are overcome in this model, which holds strain constant through all ages [9]. This type of scaling has been utilized because of evidence that strain is the best predictor of resultant injury in biomechanical in-vivo models of brain trauma [19, 20]. Because the displacement of the cortex is coupled to the skull, this model produces no inertial motion of the head itself, and in this way embodies a purely focal mechanism. The model creates a well-defined cortical and subcortical contusion with variable hemorrhage which evolves over time, is associated with regional brain swelling and changes in cerebral blood flow, and can be measured by histology or imaging.

1.2 Anesthetic Considerations

Anesthetics for use of this model can be varied depending on the specific outcomes investigated. For example, there are seizure-permissive anesthetics, protocols for experiments performed with subjects on room air, those specific for detailed imaging requiring minimal movements, and other special considerations [20, 21]. Below is one specific anesthetic protocol, though others may be used for specific purposes. In all instances, a rescue medication protocol should be developed, as young animals are particularly prone to hypotension and other systemic perturbations from surgical-level anesthetics, particularly if room air conditions are utilized. Finally, the use of muscle relaxants must be undertaken only with extreme caution and following strict protocols to ensure that subjects remain completely anesthetized and comfortable during any procedure, and that any chance of distress is minimized, as it would be for human children. Ongoing collaboration with a pediatric anesthesiologist or critical care clinician is extremely useful when using immature large animal models to ensure effective and humane anesthetic and analgesic care.

2 Materials

2.1 General Supplies

1. Operating room approved for use with large animals.
2. Medical tape (3M, USA).
3. Scissors (OR sharp/blunt scissors 5.5″).

4. Gauze (Kendall Curity® Non-Sterile Cotton Gauze Sponges: 2″×2″ and 4″×4″).
5. Sterile draping (Kimberly-Clark Sterile Examination Drape).
6. 4-0 Monocryl or other absorbable sutures (Ethicon, USA).
7. Chux pads (Mckesson, USA).
8. Bair hugger (3M, USA).
9. Gel pad water blanket (Norm-O-Temp®, CZS, USA).
10. Appropriate personal protective equipment including sterile gowns and gloves, hair covers, booties, and masks.
11. Bone wax (Ethicon, USA).
12. Chlorhexidine surgical scrub solution 4% (Applicare, USA).
13. Tegaderm (3M, USA).
14. Sterile alcohol pred pad (KENDALL 6818 WEBCOL, USA).
15. Size 11 and size 15 disposable sterile surgical blades (Healthaw, China).
16. Skin glue (Vetbond, 3M, USA).
17. Animal clippers.
18. Eye lubricant.

2.2 Anesthesia Equipment

1. Anesthesia machine equipped with an isoflurane vaporizer.
2. Nose cone (Parkland Scientific, 3–7/16″ OD×1–3/16″ ID×2–7/8″ *L*, Item 93815026).
3. Oxygen, air, and vacuum sources.
4. Argyle rigid plastic Yankauer suction tubing (Vitality medical, USA).
5. Miller laryngoscope, size 2. Other sizes and blades if preferred should be available.
6. Endotracheal tubes 3.0 and 3.5 and 4.0 cuffed (Mallinckrodt™, USA).
7. Intubating stylet (Mallinckrodt™, USA).

2.3 Intravenous Line

1. IV catheters 22, 24, and 26 gauge (26 gauge useful for 5 day piglets).
2. IV tubing such as microbore extension sets and stopcocks.
3. 0.9 NaCl 500 mL or 1 L bags.
4. IV stand.
5. Syringes and needles of different sizes.
6. Infusion syringe pump (Smiths medical, USA).

2.4 Monitors

1. Hemodynamic monitor and connections (Edwards Lifesciences, USA).

2. End tidal CO_2 monitor and special tubing (Edwards Lifesciences, USA).
3. Oxygen saturation monitor and connections (Edwards Lifesciences, USA).
4. Rectal temperature probe.
5. Monitoring electrodes (3M, USA).
6. Neonatal blood pressure cuff.

2.5 Sterile Surgical Pack

1. Scaled cortical impact device and indentor.
2. Medium scissors (Operating Scissors 4.5″ Straight Sharp/Sharp).
3. Halsey needle holder.
4. Halsted mosquito hemostats.
5. Scalpel handlers #3, #4.
6. Standard tissue forceps.
7. Hudson drill with D'Errico burr.
8. Curettes of different sizes.
9. Bone rongeurs.
10. Dural separator.
11. Blue OR towels.
12. Gauze (Kendall Curity® Non-Sterile Cotton Gauze Sponges: 2″ × 2″ and 4″ × 4″).
13. Bowl for chlorhexidine soak.

2.6 Drugs and Supplements

1. Ketamine.
2. Xylazine.
3. Atropine.
4. Buprenorphine.
5. Bupivacaine.
6. Methohexital (*see* Subheading 1.2 above).
7. Rocuronium (*see* Subheading 1.2 above).
8. Ophthalmic lubricant ointment (Puralube® Vet Ointment).
9. Electrolyte water mixture (Bluelite, TechMix, Stewart, MN).
10. Euthasol (for terminal experiments).

3 Methods

3.1 Preoperative Preparation

Prior to surgery, 1-month-old and 4-month-old piglets need to be restricted from food overnight, 5-day-old piglets need to have access to an electrolyte water mixture (Bluelite, TechMix, Stewart, MN) for at least 4 h prior to surgery. *See* **Notes 1** and **2** for special considerations.

3.2 Anesthetic Procedures and Surgical Prep (For Alternatives, Please See Subheading 1.2 Above)

Piglets are deeply anesthetized for all procedures. Induce general anesthesia with injections of ketamine (20 mg/kg IM), xylazine (2 mg/kg IM), and atropine (0.03 mg/kg IM) followed by 5% isoflurane delivered via nose cone mask. Once the animal is anesthetized and no longer needs restraint, decrease the isoflurane to 2–3% (*see* **Notes 3** and **4**). Place an intravenous catheter in a limb or ear vein, administer buprenorphine (0.002 mg/kg IM for 5 day piglets and 0.02 mg/kg IM for 1- and 4-month-old piglets), moisten eyes with lubricating antibacterial eye gel, and secure eyelids down using Tegaderm prior to endotracheal intubation. After intubation, place the piglet on the anesthesia machine and mechanically ventilate with room air, adjust minute ventilation to maintain an end-tidal CO_2 between 35 and 45 mmHg. Isoflurane should be decreased accordingly depending on the level of sedation needed at the time; normally it should be between 1 and 2%. Monitor and record end-tidal CO_2, oxygen saturation, blood pressure, heart rate, and core body temperature at baseline, immediately post-injury, and at 5, 10, and 15 min post injury. Core body temperature can be measured via a rectal probe and needs to be maintained between 37 and 39 °C by adjusting the use of a heating pad (place chux between the pad and the piglet) and Bair Hugger blanket.

Prior to opening sterile instruments, use clippers to shear hair from the snout to the crown of the head. Use tape such as Durapore surgical tape to remove any loose strands after clipping. In the event of hypotension, defined as a mean arterial pressure (MAP) of less than 30, a bolus of normal saline (10–40 mL/kg) should be administered (*see* **Note 5**). Hypotension should prompt complete evaluation of all possible causes, including over-sedation, hypovolemia, and others according to a checklist, and rescue procedures and medications should be immediately available and reviewed before each procedure, please *see* Subheading 4 regarding resuscitation.

3.3 Creation of Scaled Cortical Impact

After routine skin sterilization place sterile drapes around surgical site, inject and secure in place with a sterile surgical stapler. Administer bupivacaine (up to 1 mg/kg) subcutaneously along the intended incision site. Using a 15 blade, perform a midline vertex incision to expose the coronal and sagittal sutures. Expose the right lateral vertex by continuing the incision rostral of the coronal suture such that the incision curves laterally just anterior to the right. Tissue forceps and the size 15 scalpel blade are used to gently detach the skin from the periosteum to expose the skull. The skin flap created can be held out of the way by attaching curved mosquito hemostats to the fascia on the underside of the skin.

To perform the craniectomy, begin by creating a burr hole through the skull just posterior to the junction of the right coronal and sagittal sutures in the right parietal bone using a Hudson drill. Enlarge the burr hole to a diameter 1 cm larger than the diameter of the indentor tip with bone rongeurs in the posterior and lateral direction; take special care while performing this procedure to avoid trauma to the cortical surface. While expanding the burr hole, ensure that the edges of the bone are not sharp or pointed as may result from taking too large of "bites" with the rongeurs. When the burr hole is complete, gently remove any remaining bone bits, irrigate the dural surface, and press bone wax into the skull to stem bleeding. Once the site is clean and clear of any debris, carefully separate the dura from the skull in the area surrounding the burr hole. At this point, it is wise to ensure proper fitting of injury device before opening the dura, with an unimpeded trajectory of the indentor tip (Figs. 1 and 2).

After confirming adequate burr hole size, carefully open the dura using a size 11 scalpel blade. Once a small incision is made, use a dural separator to protect the brain from the scalpel as the dura is continued to be opened in a stellate fashion to expose the cortical surface. Secure the sterilized cortical indentation device

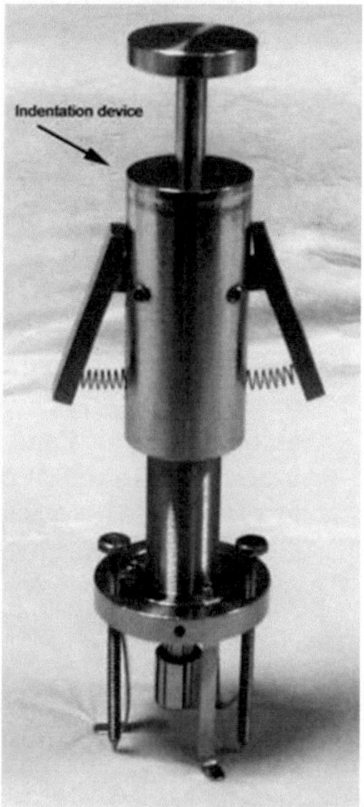

Fig. 2 Photograph of the scaled cortical impact device. After craniectomy, the dura mater is opened widely, the device is firmly attached to the skull, and indentation occurs directly to the exposed cortical surface. The diameter and depth of indentation are scaled to be proportional to brain growth, displacing approximately 1 % of the total brain volume over 400 ms

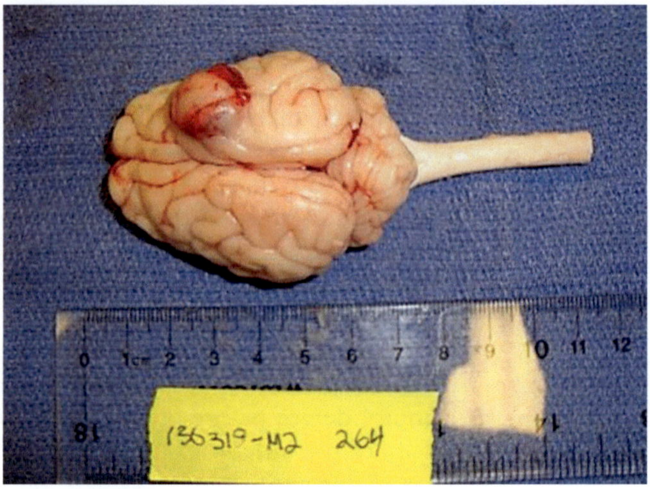

Fig. 3 Postmortem image of a piglet brain demonstrating the controlled cortical impact model

firmly to the skull by placing feet under the skull around the edge of the burr hole and tightening the screw fixation. Screw the indentor tip into the secured apparatus and lower until the tip just makes contact with the cortical surface. Confirm with at least one other person, who is able to look from a different angle, that the tip is just barely touching the cortical surface. In some instances the tip is raised a specific distance from the surface to vary the desired indentation depth. Before injury, ensure core body temperature is within the 37–39 °C range. Utilize the appropriate size indentor tip for each age group; perform scaled cortical displacement to the appropriate indentation depth by releasing the spring-loaded indentor tip. The device described utilized rounded, interchangeable indentor tips and indents as follows: 1.04 cm in diameter and indenting to a depth of 4.8 mm for the 5 day old piglets; 1.07 cm in diameter with 5.9 mm indention for the 1 month old piglets; and 1.267 cm in diameter with 7 mm indention for the 4 month old piglets. The spring-loaded indentor tip will fire at a time course of 4 ms; remove the device from the skull. Gently irrigate the cortical surface and reapproximate the dura. Record the level of hemorrhaging observed, and if pial hemorrhage is present, apply gentle pressure and irrigation and wait for active bleeding to cease before closing the scalp and suturing with 4-0 Monocryl (Fig. 3).

3.4 Postoperative Care

Lighten isoflurane and decrease ventilation rate (increase CO_2) until the piglet begins to breathe over the machine then switch off ventilator. Wait to extubate until the animal is beginning to wake up and move legs. Someone should be with the animal until it is alert and able to ambulate without risk of injuring itself further, this usually takes approximately 15 min from the time of extubation (*see* **Note 6**). Observe the piglet frequently for an hour after surgery to assess recovery. If animal, is acting normal and is bright, alert, and responsive (BAR), follow routinely as per animal care protocols, but recheck at least every few hours. This injury model is specifically designed to avoid obvious neurologic deficit or distress, so most animals will behave normally; alterations should be discussed with the veterinary staff. For the 5-day piglets, it is especially important to monitor weight and eating habits daily (both pre- and post-surgery).

4 Notes

1. Tips to Encourage Socialization and Comfort

   Since pigs are social creatures, it is recommended that if possible they be housed more than one per pen. This is especially important for the younger age groups. Piglets less than 2 weeks old should be kept together with littermates. One month pigs should be kept 2–3 per pen. It is also recommended that researchers who will be handling the animals the most while they are awake (anesthetic induction and recovery, blood draws, etc.) familiarize themselves with the animal by interacting with them prior to any procedures. One useful way to familiarize research personnel with the behavior, intelligence, and sociability of these animals is to feed the piglet's yogurt or other treats as recommended by the veterinary and animal care staff. All attempts to minimize stress and discomfort and to understand and respect the behavioral repertoire of the subjects should be made.

2. Special Considerations for 5 Day Piglets

   Piglets less than 2 weeks must be closely monitored for their ability to thrive. This age group must be shown how to drink warm milk from a dish and may experience diarrhea. Weights should be taken every day, from the day they arrive at the facility until the last day of the study, to be sure they do not lose more than 20% of their body weight. These piglets are also susceptible to apnea while anesthetized and during post-anesthetic recovery. For this reason, these piglets should be induced and allowed to recover from anesthesia while still in the operating room and closely observed.

3. Creating an Anesthesia Cart for Transport

   Depending on the logistics of where the large animal operating room is compared to the animal facility, it may be necessary to transport the animal long distances. In this situation, consider creating an anesthesia cart to aid in transport. A heavy-duty plastic cart with two oxygen tanks, an isoflurane vaporizer, tubing, nose cone, pulse oximeter, covering blankets, and relevant medical supplies results in a safe and secure mode of transport.

4. Alternative Method of Delivering Anesthesia

   Depending on study design, it may be necessary to use alternative methods for sedation. For example, if the goal is to image the brain before or after injury with sequences that are sensitive to motion, it may be important to induce a neuromuscular blockade and use a compatible anesthetic agent. For this situation, we have used Methohexital (1.5–2.5 mg/kg IV bolus for induction; 0.05–0.15 mg/kg/min IV continuous infusion for anesthetic maintenance) combined with Rocuronium (0.8 mg/kg IV). Rocuronium is given just prior to imaging and every 10 min as needed. We have also found that placing the subject supine for MRI instead of prone minimizes the effect of respiratory motion on the head position.

5. Intraoperative Tips

   Have someone be specifically in charge of monitoring and maintaining vital signs and other ongoing data collection such as medications given during the surgery. Develop a clear plan before beginning a study as to what interventions should be taken in certain situations and what should constitute exclusion criteria. An example guideline for excluding animals from a study:

   (a) Failed intubation.
   (b) Airway trauma such that intubation is not possible or tracheal injury is suspected.
   (c) Prolonged oxygen desaturation <90% for greater than 5 min.
   (d) Hypotension unresponsive to lowering isoflurane or up to three boluses of 10 ml/kg saline.
   (e) Cardiac arrest.
   (f) Core temperature below 35.0 or above 41.0 °C on presentation.
   (g) Core temperature cannot be maintained between 37.0 and 39.0 °C for the 15 min prior to injury (It is also advised to have a plan for resuscitation measures which should be taken as described below).
   (h) Atropine (0.015 mg/kg IV) for treatment of bradycardia.
   (i) Ephedrine (0.5 mg/kg IV) for treatment of hypotension.

(j) Epinephrine (0.005 mg/kg IV) for treatment of bradycardia and/or hypotension.

(k) 50% Glucose (0.75 mg/kg IV) for treatment of hypoglycemia.

(l) Lidocaine (1.25 mg/kg IV) for treatment of ventricular arrhythmia.

(m) Naloxone (0.001 mg/kg IV) for treatment of respiratory depression.

6. Recovering from Anesthesia

(a) Typical Recovery Notes

When the pigs are recovering from anesthesia, it is normal for them to squeal and move the limbs forcefully, similar to human children during emergence from anesthesia. Because squealing can be loud, especially in confined spaces, noise-reducing earpieces can be helpful and properly restraining the animal is essential for a smooth and safe recovery. Special care must be taken to ensure that the pig does not accidentally bite its tongue, impact the head, or harm itself in any way during this period. Giving the piglet yogurt is a great way to distract it if it seems upset and also to encourage it to take a few steps when trying to determine whether it is safe to leave the animal.

(b) Emergence Delirium

In a very small subset of animals we have observed the phenomenon of emergence delirium. When this occurs, upon emergence from anesthesia the animal is vocal and thrashing to an extreme degree. They will not be able to eat, track objects with their eyes, or respond to environmental changes. Administer buprenorphine (0.025 mg/kg IV) to alleviate any possible pain. If the delirium persists, give 0.5–1.0 mg/kg propofol (IV).

5 Conclusion

The scaled cortical impact model has proven useful in investigating age-dependent response to focal cortical mechanical deformation, because it causes a measureable lesion without causing significant disability or distress. Piglets have human-like brains, and this is extremely useful for translational research that may improve understanding of traumatic brain injury in higher-order animals, including humans. The use of immature large animals requires attention to many details and carries a responsibility to do all possible to minimize discomfort and distress to meet both scientific and humane goals.

References

1. Dobbing J, Sands J (1979) Comparative aspects of the brain growth spurt. Early Hum Dev 311:79–83
2. Dickerson JWT, Dobbing J (1967) Prenatal and postnatal growth and development of the central nervous system of the pig. Proc R Soc London 166:384–395
3. Flynn TJ (1984) Developmental changes of myelin related lipids in brain of miniature swine. Neurochem Res 9:935–945
4. Buckley NM (1986) Maturation of circulatory system in three mammalian models of human development. Comp Biochem Physiol 83A: 1–7
5. Wootton R, Flecknell PA, John M (1982) Accurate measurement of cerebral metabolism in the conscious, unrestrained neonatal piglet. Blood flow. Biol Neonate 41:209–220
6. Saito T, Watanabe Y, Nemoto T, Kasuya E, Sakumoto R (2005) Radiotelemetry recording of electroencephalogram in piglets during rest. Physiol Behav 84:725–731
7. Gavilanes AWD, Vles JSH, von Siebenthal K, Reulen JP, Neiman FH, van Sprundel R, Blanco CE (2000) Electrocortical brain activity, cerebral hemodynamics and oxygen saturation during progressive hypotension in newborn piglets. Clin Neurophysiol 112:52–59
8. Ioroi T, Peeters-Scholte C, Post I, Leusink C, Groenendaal F, van Bel F (2002) Changes in cerebral hemodynamics, regional oxygen saturation and amplitude integrated continuous EEG during hypoxia-ischemia and reperfusion in newborn piglets. Exp Brain Res 144: 172–177
9. Duhaime AC, Margulies SS, Durham SR, O'Rourke MM, Golden JA, Marwaha S, Raghupathi R (2000) Maturation-dependent response of the piglet brain to scaled cortical impact. J Neurosurg 93:455–462
10. Armstead WM, Kurth CD (1994) Different cerebral hemodynamic responses following fluid percussion brain injury in the newborn pig. J Neurotrauma 11:487–497
11. Barks JDE, Silverstein FS (1992) Excitatory amino acids contribute to the pathogenesis of perinatal hypoxic-ischemic brain injury. Brain Pathol 2:235–243
12. Shaver E, Duhaime AC, Curtis M, Gennarelli LM, Barret R (1996) Experimental acute subdural hematoma in infant piglets. Pediatr Neurosurg 25:123–129
13. Bauer R, Walter B, Torossian A, Fritz H, Schonlonski O, Jochum T, Hoyer D, Reinhart K, Zwiener U (1999) A piglet model for evaluation of cerebral blood flow and brain oxidative metabolism during gradual cerebral perfusion pressure decrease. Pediatr Neurosurg 30:62–69
14. Raghupathi R, Margulies SS (2002) Traumatic axonal injury after closed head injury in the neonatal pig. J Neurotauma 19:843–853
15. Kurth CD, Levy WJ, McCann J (2002) Near-infrared spectroscopy cerebral oxygen saturation thresholds for hypoxia-ischemia in piglets. J Cereb Blood Flow Metab 22:335–341
16. Loepke AW, Golden JA, McCann JC, Kurth CD (2005) Injury pattern of the neonatal brain after hypothermic low flow cardiopulmonary bypass in a piglet model. Anesth Analg 101:340–348
17. Duhaime AC (2006) Large animal models of traumatic injury to the immature brain. Dev Neurosci 28:380–387
18. Margulies S, Thibault L, Gennarelli T (1990) Physical model simulations of brain injury in the primate. J Biomech 23:17–19
19. Duhaime AC, Hunter JV, Grate LL, Kim A, Golden J, Demidenko E, Harris C (2003) Magnetic resonance imaging studies of age-dependent responses to scaled focal brain injury in the piglet. J Neurosurg 99:542–548
20. Duhaime AC, Saykin AJ, McDonald BC, Dodge CP, Eskey CJ, Darcey TM et al (2006) Functional magnetic resonance imaging of the primary somatosensory cortex in piglets. J Neurosurg 104(Suppl 4):259–264
21. Costine BA, Missios S, Taylor SR, McGuone D, Smith CM, Dodge CP, Harris BT, Duhaime AC (2015) The subventricular zone in the immature piglet brain: anatomy and exodus of neuroblasts into white matter after traumatic brain injury. Dev Neurosci 37(2):115–130

Chapter 20

Thromboembolic Model of Cerebral Ischemia and Reperfusion in Mice

Ali Alawieh, Wenxue Wang, Aarti Narang, and Stephen Tomlinson

Abstract

Ischemic stroke is the fourth leading cause of death in the USA and a prominent cause of death globally. Besides thrombolytic therapy used in a small subset of patients, no alternative therapeutic strategy has been shown to improve the outcome of stroke patients. Preclinical models of ischemic stroke are an essential tool for investigating pathogenic processes that happen after the ischemic insult, as well as to screen for candidate therapeutic interventions. There are several models of rodent ischemic stroke including mechanical occlusion, thromboembolic stroke, and photothrombotic stroke. However, models that permit studying stroke in the context of thrombolytic therapy, such as thromboembolic models, are becoming of increasing interest to the research community. In this chapter, we describe a thromboembolic model of ischemic stroke with and without tissue-plasminogen activator-induced reperfusion. We describe protocols for microemboli preparation, surgical procedure, and post-stroke assessment of animals.

Key words Cerebral ischemia reperfusion, Stroke, Thromboembolic stroke, Microemboli, Thrombolytic therapy, Tissue-plasminogen activator

1 Introduction

Ischemic stroke is among the leading causes of death and mortality in the USA and globally [1, 2]. The only currently approved therapy for ischemic stroke is recombinant tissue-plasminogen activator (t-PA) that dissolves the clot and restores perfusion to the ischemic brain [3–5]. However, t-PA has a very limited window of efficacy of up to 4.5 h after stroke onset and carries the risk of fatal intra-cranial hemorrhage [3–5]. There is thus a significant need for new stroke therapeutics.

Ischemic stroke is characterized by two phases of injury: a primary injury ensuing from ischemic insult to the brain parenchyma resulting in cell death within the ischemic core, and secondary injury that occurs in the ischemic penumbra and results from activation of pro-inflammatory and pro-apoptotic cascades. Secondary injury is exacerbated after restoration of blood flow to the ischemic

Firas Kobeissy et al. (eds.), *Injury Models of the Central Nervous System: Methods and Protocols*, Methods in Molecular Biology, vol. 1462, DOI 10.1007/978-1-4939-3816-2_20, © Springer Science+Business Media New York 2016

penumbra by several components of the innate and adaptive immune system that gain access to the injury site after injury-induced permeabilization of the blood–brain barrier [6]. Previous studies have identified a core of immune recognition and response elements that are involved in the amplification of secondary injury after stroke [7, 8]. To better understand the pathophysiology of ischemic stroke, as well as to screen for potential therapeutic interventions, preclinical models of ischemic stroke represent an indispensable tool.

The focus of this chapter is the thromboembolic model of ischemic stroke, although there are several models of ischemic stroke currently being used. The types of rodent model of focal cerebral ischemia include:

1. Mechanical models of middle cerebral artery occlusion (MCAO), such as permanent and transient MCAO [9, 10] or distal MCAO [11].
2. Thromboembolic models that include administration of blood clots into the cerebral circulation to induce focal cerebral ischemia [12].
3. Chemical models inducing focal ischemic lesions by intracerebral endothelin-1 injection [13, 14] or photo-thrombosis [15].

Although mechanical models of MCAO are by far the most commonly used models of ischemic stroke, a major limitation of this type of model is that it does not adequately allow for the study of cerebral ischemia in the context of t-PA therapy. In addition, models of proximal MCAO suffer from other limitations including the onset of massive infarcts that would often be deadly in humans and that involve a high mortality rate in animals. On the other hand, thromboembolic models provide certain advantages such as allowing titration of the extent of injury and studying reperfusion in the context of thrombolytic therapy (t-PA). With regard to the latter point, a recent STAIR recommendation states "Given the expansion of the t-PA time window to 4.5 h, it will be increasingly difficult to test a neuroprotective agent alone, so most neuroprotective agents will be tested in combination with reperfusion therapies, that is, on a background of t-PA as a standard of care" [16]. The FDA also requires demonstration that an investigational agent does not interfere with the fibrinolytic activity of t-PA. Thus, although in the past thromboembolic models have not been frequently utilized due to a more challenging surgical procedure, greater variability, and potential difficulty in outcome assessment strategies especially when mild injuries are induced, they represent an important preclinical model for investigating new stroke therapeutics.

In this chapter, we provide detailed instructions on performing a thromboembolic stroke model in mice based on the model described by Atochin et al. [12]. The model we describe utilizes human plasma to form microemboli that are injected into the internal carotid artery to block the middle cerebral artery. We also describe induction of reperfusion by t-PA administration.

2 Materials

2.1 Microemboli Preparation

Microemboli preparation requires the following materials and devices:

1. Thrombin from human plasma (T7009-100UN; Sigma-Aldrich) stored at −20 °C.
2. Human fibrinogen with plasminogen depleted (FIB 1; enzyme research laboratories) stored at −80 °C (stock solution of 44.44 mg/mL divided into 0.45 ml aliquots).
3. Human plasma treated with sodium citrate and filtered at 2 μm stored at −80 °C.
4. Krebs-Ringer buffer (KRB) prepared by dissolving into 1 l of distilled H_2O: 6.95 g of NaCl, 0.35 g of KCl, 0.29 g of $MgSO_4$, 0.14 g of $CaCl_2$, 2.1 g of $NaHCO_3$, and 0.16 g of KH_2PO_4.
5. $CaCl_2$ stock solution at 400 mM.
6. Z2 Coulter particle count and size analyzer (Beckman Coulter, CA) or similar.
7. Tissue homogenizer.

2.2 Surgery

For successful performance of the surgical procedure, the following surgical instruments, devices, and tools are required, with sterilization as applicable:

1. Dissecting microscope and a light source.
2. Surgical board that allows for stabilizing the mice during the procedure along with a thermoregulatory heating and monitoring device to maintain body temperature during anesthesia.
3. Surgical retractor to allow ease of access to the vessels within the incision area.
4. Two serrated Moloney forceps, two angular Nugent forceps, and two Dumont #1 forceps (Roboz Surgical Instrument, MD).
5. Silk sutures 6-0 for vessel ligation (Ethicon Inc., CA).
6. Microvascular clips (85 psi) for temporary vessel occlusion (Harvard Apparatus, MA).
7. Wound clips and a clip applier if clips are to be used for closing the wound. Alternatively, wounds can be sutured using a 5-0 silk suture with needle.

8. Surgical scissors for skin incisions, and microsurgical scissors to cut through small vessels (Roboz Surgical Instrument, MD).
9. A polyethylene catheter with inner diameter of 0.02 mm (BrainTree Scientific Inc, MA).
10. Cathflo® Activase® (Alteplase, Genentech, CA) tissue-plasminogen activator protein reconstituted in sterile water.
11. Mouse tail illuminator and restrainer.
12. Infusion pump with Hamilton syringe (Harvard Apparatus, MA).
13. Rodent tail vein catheter (BrainTree Scientific, MA).

2.3 Laser Doppler Flowmetry

1. Laser doppler flowmeter (LDF) including a power house, a mouse probe, and the corresponding acquisition software (moorVMS-LDF1 Laser Doppler Blood Flow and Temperature Monitor, Moor Instruments, DE).
2. Stereotactic platform for mounting the animal head with a probe holder (Kopf Instruments, CA).

2.4 Infarct Volume Estimation

1. 2,3,5-Triphenyltetrazolium chloride powder (Sigma, MO) stored at 4 °C.
2. Adult mouse brain slicer, 1.0 mm coronal slice intervals (Harvard Apparatus, MA).
3. Razor blades.
4. Solution of 0.9 % saline or phosphate-buffered saline (PBS).
5. Dissecting microscope with camera and light source to take images of stained brains.
6. NIH ImageJ software for image analysis.

3 Methods

3.1 Microemboli Preparation

Microemboli are prepared according to the following procedures 2 days before surgery:

1. The following reagents mixed into a final volume of 2 ml of KRB in a glass tube:
   - Human plasma: 1.4 ml.
   - Fibrinogen to a final concentration of 10 mg/ml (0.45 ml of stock).
   - Thrombin to a final concentration 0.2 U/ml.
   - $CaCl_2$ to a final concentration of 20 mM (0.1 ml of stock).
   - Supplement with KRB to final volume of 2 ml.

   Precautions: Human plasma should be thawed on ice; however, fibrinogen should be thawed at 37°.

2. Incubate the mixture at room temperature for at least 1 h, after which the solution should form a gel-like, slightly opaque mesh indicating successful formation of the clot. Minimal fluidity should be seen in the glass tube at this stage.
3. Place the coagulated mixture at 4 °C overnight.
4. Prepare cutting razor blades pre-cleaned with 70 % alcohol followed by KRB.
5. Decant the clot from the glass tubes into a petri dish with 1–2 ml of KRB, use the razor blades to cut the clot into smaller pieces, and then transfer the microemboli into polypropylene tubes.
6. Homogenize the microemboli on ice using a clean tissue homogenizer (Fisher) at ½ to ¾ maximum speed for 15–20 s.
7. Centrifuge the homogenized microemboli at 2000 × *g* for 15 min.
8. Discard the supernatant and resuspend the microemboli in 2 ml of KRB.
9. Repeat **steps 6–8** three times, and at the final step resuspend the microemboli in 5–8 ml KRB supplemented with 3 mg/ml of bovine serum albumin (BSA) (*see* **Notes 1** and **2**).
10. The outcome of this process is a large number of microemboli of variable sizes. Allow the solution to sediment for 5 min at 4 °C to remove the large clots, and then remove the supernatant into a clean glass tube (Fig. 1).
11. Prior to their use in surgeries, microemboli need to be sized and counted to record their size distribution and concentration. Two random aliquots from the clot solution should be used for analysis by a Z2 Coulter particle count and size analyzer (Beckman Coulter) or similar. An ideal size of the microemboli is 2–4 μm in diameter. The gain of the particle counter should be set to count particles within a range of 1.9–7.7 μm. Particles more than 10 μm in diameter need to be less than 5 % of the total particle count.

The particle counter will return a distribution curve for the microemboli size as shown in Fig. 2. Titration of the dose of emboli administered to the animal can be performed to simulate different degrees of injury. Based on the current protocol, we use a dose of 5×10^5 particles for mild injury, 1×10^6 particles for moderate injury, and 1.5×10^7 particles for severe injury (*see* **Note 3**).

3.2 Surgical Procedure

All surgical procedures should be approved by the Institutional Animal Care and Use Committee in accordance with the National Institutes of Health Guide for the Care and Use of Laboratory Animals. Surgeries performed for this chapter were approved by the Institutional Animal Care and Use Committee at the Medical

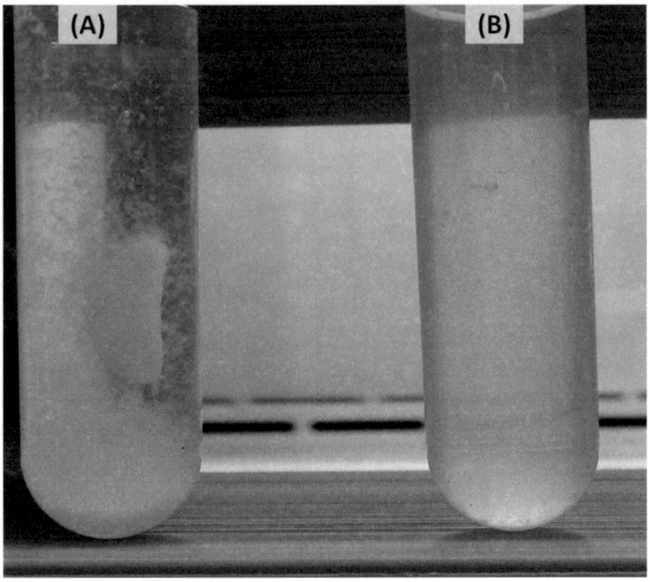

Fig. 1 Images of homogenized microemboli prior to sizing and counting using the particle counter. (**a**) Image of microemboli after the last suspension showing large and clustered microemboli during sedimentation. (**b**) Microemboli solution after sedimentation and prior to sizing. After repetitive homogenization cycles, microemboli of large size (above 10 μm in diameters) may still be present. These emboli need to be excluded prior to sizing and subsequent injection to the animal. If a high concentration of large clots is still present, use longer durations of homogenization or additional homogenization steps

University of South Carolina. The procedure below describes the experimental protocol used to induce microembolic stroke in the right hemisphere along the territory of the middle cerebral artery (MCA).

3.2.1 Preparation for Surgery

1. Depending on the experimental design, animals may be trained on different motor and cognitive tasks prior to the surgical procedures. Once the mice are ready for surgery, prepare and size the microemboli according to the required dosing (*see* **Note 2**). This chapter describes procedures performed in young C57bL/6J mice from Jacksons Laboratories that are 10–12 weeks of age. However, older mice can be also used but with higher expected mortality after the procedure.
2. Prepare the polyethylene catheter by attaching a syringe filled with 100–200 μl of PBS, and push PBS through the catheter to ensure absence of air bubbles.
3. Weigh mice and anesthetize by intraperitoneal injection of 80–100 mg/kg of ketamine and 10 mg/kg of xylazine mixed in saline. (Gas anesthesia using isoflurane is an alternative option.)
4. After reaching surgical stage of anesthesia, shave fur from both the neck area (where the incision is to be made) and the head

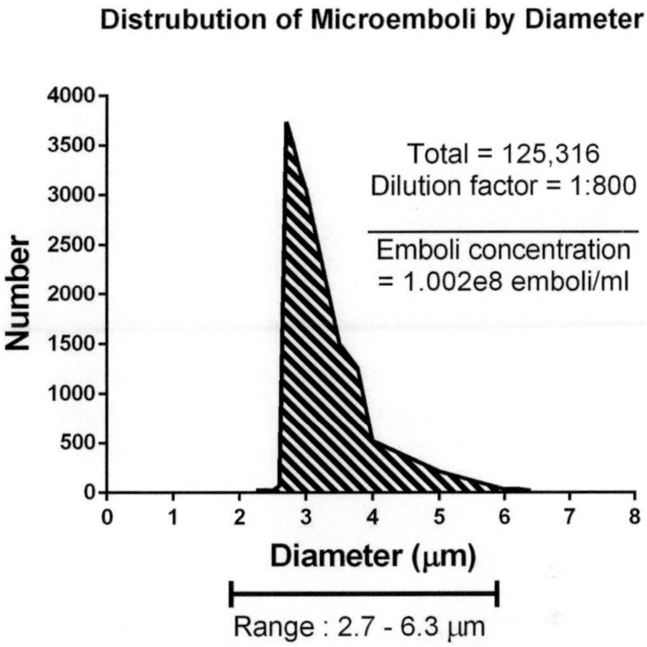

Fig. 2 Example of microemboli diameter distribution. Ideally, the majority of microemboli should fall within the range of 2–4 μm in diameter, with less than 5 % of microemboli exceeding 10 μm in diameter. Shown in the graph is the distribution of microemboli by diameter after sizing using the Z2 Coulter counter. A total of 1.002×10^8 emboli/ml was prepared that can then be diluted to the dose required for injection

area above the calvarium (in preparation for laser Doppler flowmetry). Throughout the following procedures, maintain mice on a thermoregulatory pad. Apply ophthalmic ointments to the eyes and close the eyelids to prevent eye desiccation.

5. Laser Doppler flowmetry is used to confirm the onset of ischemia and to enhance the reproducibility of the model. Mount mice on a stereotactic setup to which the LDF probe is attached. A presurgical baseline of blood flow flux is measured through the LDF probe and compared to post-embolization flux to detect the extent of ischemia.
6. A 1 cm midline skin incision is cut over the calvarium from superior nuchal line to the nasion and the skin is pulled laterally. The LDF probe is placed on the calvarium 1 mm posterior and 5 mm lateral to bregma. Baseline flux is then measured (*see* **Note 4**).

3.2.2 Microembolic Stroke Surgery

7. After baseline flow measurement, place animal in supine position on the surgical board with forepaws, hindpaws, and tail fixed by rubber bands or tape.
8. Under the dissecting microscope, disinfect the shaved neck with ethanol and Betadine, and make a midline incision between the manubrium and the jaw using forceps and scissors.

9. Using forceps, separate the muscles and glands around the midline by gently pulling them apart lateral to midline.
10. Insert the retractor by first placing it under the sternocleidomastoid muscle on one side, gently pulling it laterally, and then inserting the other side of the retractor under the opposite muscle starting from the bottom end and sliding upwards. Once the retractor is in place, extend the opening within the incision area to allow clear field of view.
11. Identify the right common carotid artery (CCA), and separate it from the accompanying vein and the vagus nerve to avoid injury to these structures.
12. Lift the CCA using Nugent forceps, and clamp it using a microvascular clip proximal to its bifurcation (Figs. 3a and 4a).

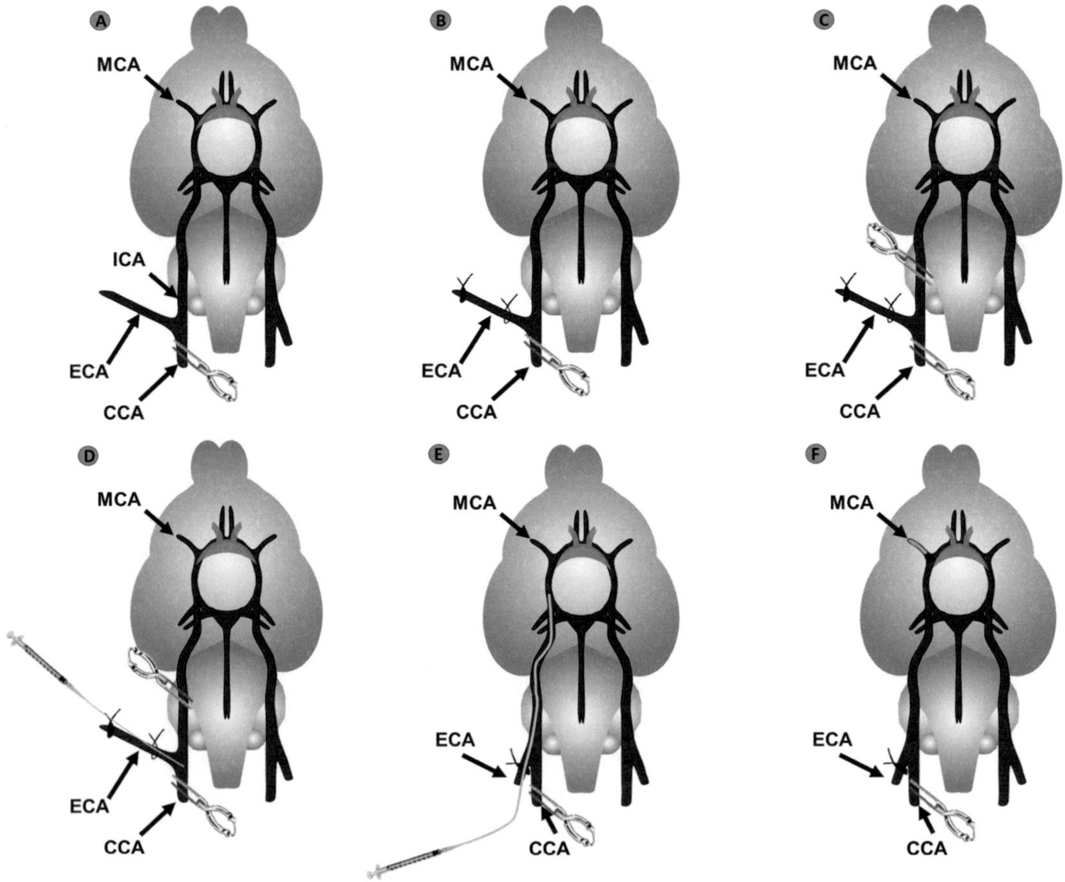

Fig. 3 Illustration of key steps in the surgical procedure. (**a**) Identification and clamping of the CCA. (**b**) Drawing a tight and a loose suture around the ECA. (**c**) Identification and clamping of the ICA. (**d**) Cutting a hole in the ECA between the two sutures and inserting the catheter into the CCA. (**e**) Completely cutting the ECA, removing the clamp from the ICA, and inserting the catheter into the ICA until resistance is felt. (**f**) Administration of the clots and retraction of the catheter prior to closing the wound

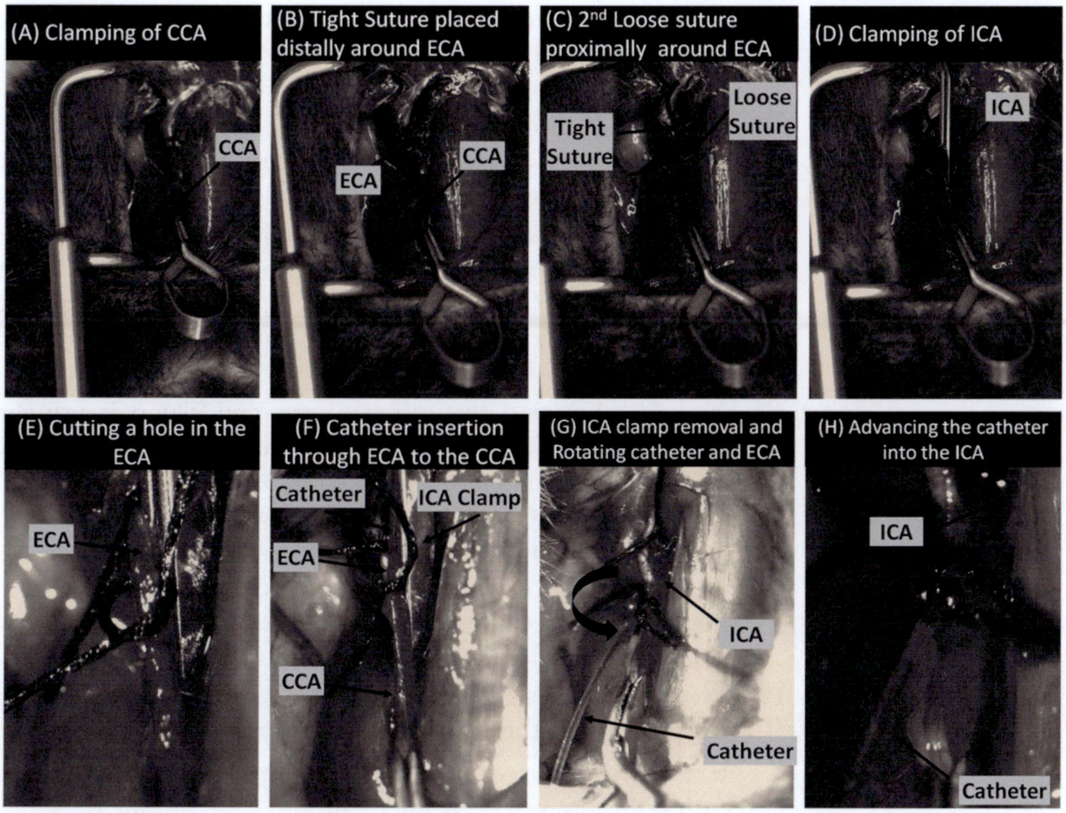

Fig. 4 Images captured during surgical procedures on a dissecting microscope

13. In order to visualize the external carotid artery (ECA), the omohyoid muscle needs to be divided with careful attention so as not to sever the artery. Once the ECA is isolated and separated from the vagus nerve, the ECA is elevated gently using a Nugent forceps and the underlying connective tissue is cleared. A black 6-0 suture ~1 cm in length is inserted under the ECA, pushed distally away from the bifurcation, and tied into a tight knot around the ECA. Another knot is similarly made around the ECA closer to the bifurcation; however, this knot should be kept loose to allow the insertion of the catheter afterwards (Figs. 3b and 4b, c).
14. Prior to cutting the ECA and inserting the catheter, isolate the ICA and clamp to prevent bleeding. The loose suture around the ECA can be used to gently pull the ECA to visualize the pterygopalatine artery, with removal of the lymph nodes and connective tissue covering the vessels. Once both the pterygopalatine artery and the internal carotid artery (ICA) are visualized, apply a microvascular clip to clamp both vessels. Note that the pterygopalatine artery can be coagulated to prevent possible bleeding (Figs. 3c and 4d).

15. After clamping both the CCA and ICA, use the microscissors to make a small hole (less than 1/3 of the vessel diameter) in the ECA between the two sutures and closer to the tight end (Figs. 3d and 4e).

    Precaution: Do not cut the ECA completely since this will make the insertion of catheter harder.

16. Using the Dumont forceps, insert the tip of the catheter into the small hole within the ECA and advance the catheter down into the CCA. Once in the CCA, tighten the loose suture around the ECA and the catheter. Cut the ECA close to the first suture (Fig. 4f).
17. Prior to advancing the catheter into the ICA, remove the microvascular clip on the ICA and then slightly retract the catheter backward by holding the tight knot with one hand and retracting the catheter with the other.
18. Flip the ECA with the catheter so that the angle between the ECA and ICA is now 180°. You may notice some backflow of blood into the catheter at this stage which is anticipated (Figs. 3e and 4g).
19. Advance the catheter into the ICA until resistance is felt. At this stage the catheter is just proximal to the opening of the MCA (Fig. 4h).
20. Vortex the pre-prepared microemboli solution and withdraw the required volume (based on concentration and dosing) using a new syringe. Replace the PBS syringe attached to the catheter with the new syringe loaded with microemboli. Make sure not to pull the catheter from the vessel (*see* **Notes 4** and **5**).
21. Inject the clot solution as a bolus injection, and then retract and withdraw the catheter (Fig. 3f).
22. Tighten and trim the sutures, and then remove the clip on the CCA. Clean any bleeding using a sterile cotton tip.
23. Bring the muscles and skin together and close the wound using a wound clip or by suturing.
24. To assess post-embolic ischemia, the animal is mounted again on the stereotactic platform and using the same coordinates 10–15 min after emboli injection, blood flow flux is measured over the ischemic territory and compared to pre-ischemic levels. Reduction in blood flow through the MCA is dependent on the dose of emboli. High-dose emboli (1.4×10^7 microemboli) will result in more than 50% reduction in MCA blood flow after embolization.
25. After wound closing, the animal is monitored on a heating pad until recovery and animals are allowed free access to water and soft food while being monitored for signs of neurological deficits and recovery.

3.3 Reperfusion Using Recombinant Tissue-Plasminogen Activator

One advantage of microembolic stroke model is that it can be used to induce ischemic stroke with or without reperfusion. Depending on the aims of the experiment, reperfusion may be induced by administration of t-PA protein by tail-vein infusion of 10 mg/kg of Cathflo Activase reconstituted in sterile water (*see* **Note 7**). The timing of t-PA administration will depend on the aims of the experiment; however, a range of 15 min to 5 h after ischemic embolization has been used (*see* **Note 8**) [17–20].

Precaution: t-PA should not be administered as a single bolus injection. Administration requires a 10% bolus injection followed by infusion over at least 15 min (optimally 30 min).

1. To administer t-PA, place the animal in a tail illuminator restrainer to visualize the tail vein. (If the tail vein is hard to detect, placing the tail in warm water may facilitate visualization.)
2. Attach a Hamilton syringe to the rodent tail vein catheter and load the syringe with the required dose of t-PA (10 mg/kg).
3. Insert the tip of the catheter into the tail vein and inject 10% of the t-PA dose as a single bolus, and using the pump infuse the remaining dose over 15–30 min. Ensure that the animal is adequately restrained and the catheter is fixed in place to prevent the dislocation of the catheter during the infusion process.

3.4 Infarct Volume Estimation

Several methods can be used to calculate infarct volume after microembolic stroke including TTC (2,3,5-triphenyltetrazolium chloride) staining, magnetic resonance imaging (MRI), and histological staining (H&E or Nissel). The TTC provides the fastest assessment for infarct volume compared to the other techniques; however, it may not be sensitive for small infarcts occurring after injection of low doses of emboli. If low doses of emboli are to be used, histological staining approaches with Nissl or H&E are more sensitive. MRI imaging, in turn, is less sensitive compared to histological measures; however, it allows for measuring evolution of infarct within the same animal and for relating changes in infarct volume with performance on different motor and cognitive tasks.

TTC staining is performed after perfusing the animal as described below:

1. Prior to animal sacrifice, prepare a 1% TTC solution by dissolving 0.01 g of TTC into 100 ml of PBS or 0.9% saline. Store the solution in the dark prior to use.
2. At the designated time point of sacrifice, euthanize the animal in an isoflurane chamber, and then intra-cardially perfuse with PBS to clear the blood from the brain.
3. After perfusion, gently extract the brain from the skull after pealing the skull bone and dura, and then place brain on the

slicing matrix with the ventral side up. Using razor blades pre-cleaned with PBS or 0.9% saline, slice the brain into 1 mm thick coronal sections, dip into 1% TTC solution, and incubate for 5–10 min in the dark. To achieve optimal staining, flip the sections over after 5 min of incubation.

4. After incubation, the interaction of TTC with cellular enzymes will result in a red discoloration in areas of live tissue and a white discoloration in areas of infarction. To stop the TTC reaction, replace the TTC solution with a fixative (4% paraformaldehyde or 10% formalin solution).
5. To image the brain slices, place sections on a glass slide and visualize under a dissecting microscope. Image the TTC-stained sections using a digital camera attached to the dissecting microscope, and use NIH ImageJ software to measure the infarct area (Fig. 5).
6. To calculate % infarct volume, the area of live tissue in the ipsilateral hemisphere (IpsL) and the area of the contralateral

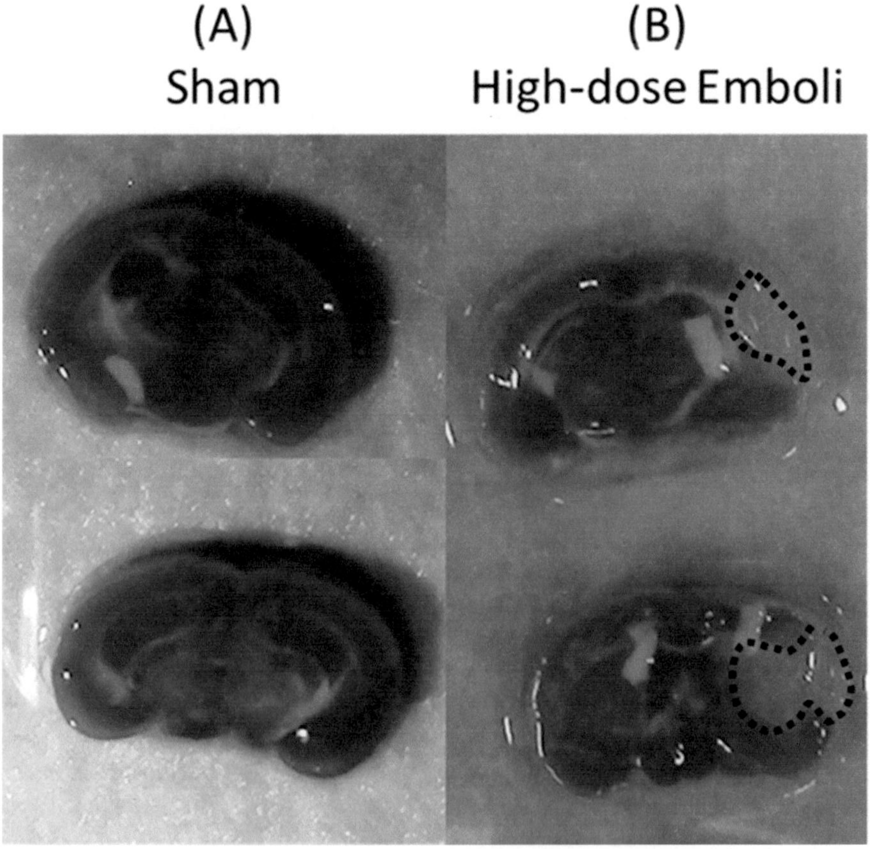

Fig. 5 Representative images of TTC-stained brain sections from (**a**) sham-treated animal, and (**b**) animal treated with a high dose of microemboli. *Dashed lines* outline the infarct region

hemisphere (ContL) are measured using ImageJ for each slice. Then,

$$\%\ \text{Infarct volume} = \frac{V_{\text{ContL}} - V_{\text{ipsi}}}{V_{\text{ContL}}} \times 100$$

3.5 Post-stroke Assessment

Several measures can be used to assess signs of recovery after micro-embolic stroke. These measures are not the focus of this chapter, but some examples are as follows:

- Motor recovery can be assessed by the four-digit neurological scoring system described in [21], the corner test described in [22], grid walking test described in [23], adhesive tape removal test described in [24], or pasta-handling tasks described in [25]. Open-field locomotor activity test can also be used.
- Cognitive recovery can be assessed by Barnes maze task described in [26], passive avoidance task described in [27], or Morris water maze described in [28].

4 Notes

1. The protocol describes three homogenization-centrifugation cycles; however, if the results of sizing indicate high emboli diameters, additional homogenization cycles may be added.
2. Final resuspension volume prior to sizing is dependent on experience. Start by resuspending the emboli in a lower volume (5 ml) of KRB to avoid excessively diluting the sample. Depending on the target dose of emboli, the concentration needs to be adjusted so that the injected volume is 50–100 μl.
3. Using a high dose of emboli is associated with a high mortality rate in this model. In our hands, 60–70 % of mice receiving high-dose emboli will die within 24 h of emboli injection, with low- and medium-dose emboli usually resulting in less than 25 % mortality at 24 h. These outcomes are based on surgeries performed without administration of t-PA.
4. There is an alternative option for performing laser Doppler flowmetry that involves fixing a flexible LDF probe to the skull prior to and throughout the procedure to capture the drop in cerebral blood flow after emboli injection.
5. Because the surgical procedure prior to injection of the emboli may take several minutes, it is preferred not to initially load the emboli syringe onto the catheter. Rather, attach a PBS-loaded syringe to the catheter during the surgery, and once the catheter is in place, replace the syringe with a microemboli-loaded syringe and inject into the ICA. Failure to do so may result in the small emboli clustering together or adhering to the lower ends of the syringe and may affect the reproducibility of injury.

6. Not described in the methods is the option to radiolabel the emboli prior to injection in order to detect the location and dosage of emboli in the ipsilateral hemisphere. If this is required, investigators need to be trained on handling radioactive material, procedures, and waste disposal. Radiolabeling emboli can be performed by incubating the emboli with radioactive iodine as previously described [12, 29, 30].
7. After Alteplase (t-PA) is reconstituted, it should be used within 8–12 h. Therefore, it is recommended to reconstitute immediately before use and to protect from light. If another t-PA product is used, storage and reconstitution should be according to the manufacturer's instructions.
8. In human stroke, t-PA is ideally administered within 3 h of stroke onset. However, a longer window of 4.5 h has also been used, although with an increased risk of adverse events [3–5]. Although there is no obvious and direct mapping of biological time between mice and humans, administering t-PA within 15–60 min after embolic stroke in mice would appear to be a clinically relevant strategy. In rats, the administration of t-PA beyond 60 min of embolic stroke did not consistently show a decrease in infarct volume [17–20].

Acknowledgements

This work is funded by a Department of Veterans Affairs Merit Award (RX00114) and the NIH (P20GM109040) to S.T. and an American Heart Association Pre-doctoral Fellowship Award (15PRE25250009) to A.A.

References

1. Naghavi M, Wang H, Lozano R, Davis A, Liang X, Zhou M, Vollset SE, Ozgoren AA, Abdalla S, Abd-Allah F (2015) Global, regional, and national age-sex specific all-cause and cause-specific mortality for 240 causes of death, 1990-2013: a systematic analysis for the Global Burden of Disease Study 2013. Lancet 385:117–171
2. Mozaffarian D, Benjamin EJ, Go AS, Arnett DK, Blaha MJ, Cushman M, de Ferranti S, Despres JP, Fullerton HJ, Howard VJ, Huffman MD, Judd SE, Kissela BM, Lackland DT, Lichtman JH, Lisabeth LD, Liu S, Mackey RH, Matchar DB, McGuire DK, Mohler ER III, Moy CS, Muntner P, Mussolino ME, Nasir K, Neumar RW, Nichol G, Palaniappan L, Pandey DK, Reeves MJ, Rodriguez CJ, Sorlie PD, Stein J, Towfighi A, Turan TN, Virani SS, Willey JZ, Woo D, Yeh RW, Turner MB, American Heart Association Statistics C, Stroke Statistics S (2015) Heart disease and stroke statistics—2015 update: a report from the American Heart Association. Circulation 131:29–322
3. Wardlaw JM, Koumellis P, Liu M (2013) Thrombolysis (different doses, routes of administration and agents) for acute ischaemic stroke. Cochrane Database System Rev 5, CD000514
4. Broderick JP, Palesch YY, Demchuk AM, Yeatts SD, Khatri P, Hill MD, Jauch EC, Jovin TG, Yan B, Silver FL, von Kummer R, Molina CA, Demaerschalk BM, Budzik R, Clark WM, Zaidat OO, Malisch TW, Goyal M, Schonewille WJ, Mazighi M, Engelter ST, Anderson C, Spilker J, Carrozzella J, Ryckborst KJ, Janis LS, Martin RH, Foster LD, Tomsick TA, Interventional Management of Stroke IIII (2013) Endovascular therapy after intravenous

t-PA versus t-PA alone for stroke. N Engl J Med 368:893–903

5. Hacke W, Donnan G, Fieschi C, Kaste M, von Kummer R, Broderick JP, Brott T, Frankel M, Grotta JC, Haley EC Jr, Kwiatkowski T, Levine SR, Lewandowski C, Lu M, Lyden P, Marler JR, Patel S, Tilley BC, Albers G, Bluhmki E, Wilhelm M, Hamilton S, Investigators AT, Investigators ET, Investigators Nr-PSG (2004) Association of outcome with early stroke treatment: pooled analysis of ATLANTIS, ECASS, and NINDS rt-PA stroke trials. Lancet 363:768–774
6. Alawieh A, Elvington A, Tomlinson S (2015) Complement in the homeostatic and ischemic brain. Front Immunol 6:417
7. Chamorro A, Meisel A, Planas AM, Urra X, van de Beek D, Veltkamp R (2012) The immunology of acute stroke. Nat Rev Neurol 8:401–410
8. Alawieh A, Sabra Z, Sabra M, Tomlinson S, Zaraket FA (2015) A rich-club organization in brain ischemia protein interaction network. Sci Rep 5:13513
9. Hata R, Mies G, Wiessner C, Fritze K, Hesselbarth D, Brinker G, Hossmann KA (1998) A reproducible model of middle cerebral artery occlusion in mice: hemodynamic, biochemical, and magnetic resonance imaging. J Cereb Blood Flow Metab 18:367–375
10. Longa EZ, Weinstein PR, Carlson S, Cummins R (1989) Reversible middle cerebral artery occlusion without craniectomy in rats. Stroke 20:84–91
11. Chen S, Hsu C, Hogan E, Maricq H, Balentine J (1986) A model of focal ischemic stroke in the rat: reproducible extensive cortical infarction. Stroke 17:738–743
12. Atochin DN, Murciano JC, Gursoy-Ozdemir Y, Krasik T, Noda F, Ayata C, Dunn AK, Moskowitz MA, Huang PL, Muzykantov VR (2004) Mouse model of microembolic stroke and reperfusion. Stroke 35:2177–2182
13. Fuxe K, Cintra A, Andbjer B, Änggärds E, Goldstein M, Agnati J (1989) Centrally administered endothelin-1 produces lesions in the brain of the male rat. Acta Physiol Scand 137:155–156
14. Horie N, Maag A-L, Hamilton SA, Shichinohe H, Bliss TM, Steinberg GK (2008) Mouse model of focal cerebral ischemia using endothelin-1. J Neurosci Methods 173:286–290
15. Watson BD, Dietrich WD, Busto R, Wachtel MS, Ginsberg MD (1985) Induction of reproducible brain infarction by photochemically initiated thrombosis. Ann Neurol 17:497–504
16. Albers GW, Goldstein LB, Hess DC, Wechsler LR, Furie KL, Gorelick PB, Hurn P, Liebeskind DS, Nogueira RG, Saver JL (2011) Stroke Treatment Academic Industry Roundtable (STAIR) recommendations for maximizing the use of intravenous thrombolytics and expanding treatment options with intra-arterial and neuroprotective therapies. Stroke 42:2645–2650
17. Brinker G, Franke C, Hoehn M, Uhlenkuken U, Hossmann KA (1999) Thrombolysis of cerebral clot embolism in rat: effect of treatment delay. Neuroreport 10:3269–3272
18. Overgaard K, Sereghy T, Pedersen H, Boysen G (1994) Effect of delayed thrombolysis with rt-PA in a rat embolic stroke model. J Cereb Blood Flow Metab 14:472–477
19. Jiang Q, Zhang RL, Zhang ZG, Ewing JR, Divine GW, Chopp M (1998) Diffusion-, T2-, and perfusion-weighted nuclear magnetic resonance imaging of middle cerebral artery embolic stroke and recombinant tissue plasminogen activator intervention in the rat. J Cereb Blood Flow Metab 18:758–767
20. Busch E, Kruger K, Allegrini PR, Kerskens CM, Gyngell ML, Hoehn-Berlage M, Hossmann KA (1998) Reperfusion after thrombolytic therapy of embolic stroke in the rat: magnetic resonance and biochemical imaging. J Cereb Blood Flow Metab 18:407–418
21. Elvington A, Atkinson C, Kulik L, Zhu H, Yu J, Kindy MS, Holers VM, Tomlinson S (2012) Pathogenic natural antibodies propagate cerebral injury following ischemic stroke in mice. J Immunol 188:1460–1468
22. Zhang L, Schallert T, Zhang ZG, Jiang Q, Arniego P, Li Q, Lu M, Chopp M (2002) A test for detecting long-term sensorimotor dysfunction in the mouse after focal cerebral ischemia. J Neurosci Methods 117:207–214
23. Hunter AJ, Hatcher J, Virley D, Nelson P, Irving E, Hadingham SJ, Parsons AA (2000) Functional assessments in mice and rats after focal stroke. Neuropharmacology 39:806–816
24. Bouet V, Boulouard M, Toutain J, Divoux D, Bernaudin M, Schumann-Bard P, Freret T (2009) The adhesive removal test: a sensitive method to assess sensorimotor deficits in mice. Nat Protoc 4:1560–1564
25. Tennant KA, Asay AL, Allred RP, Ozburn AR, Kleim JA, Jones TA (2010) The vermicelli and capellini handling tests: simple quantitative measures of dexterous forepaw function in rats and mice. J Vis Exp (41):pii: 2076

26. Rosenfeld CS, Ferguson SA (2014) Barnes maze testing strategies with small and large rodent models. J Vis Exp (84):e51194
27. Borlongan CV, Cahill DW, Sanberg PR (1995) Locomotor and passive avoidance deficits following occlusion of the middle cerebral artery. Physiol Behav 58:909–917
28. Bromley-Brits K, Deng Y, Song W (2011) Morris water maze test for learning and memory deficits in Alzheimer's disease model mice. J Vis Exp (53):pii: 2920
29. Murciano J-C, Harshaw D, Neschis DG, Koniaris L, Bdeir K, Medinilla S, Fisher AB, Golden MA, Cines DB, Nakada MT (2002) Platelets inhibit the lysis of pulmonary microemboli. Am J Physiol Lung Cell Mol Physiol 282:L529–L539
30. Bdeir K, Murciano J-C, Tomaszewski J, Koniaris L, Martinez J, Cines DB, Muzykantov VR, Higazi AA-R (2000) Urokinase mediates fibrinolysis in the pulmonary microvasculature. Blood 96:1820–1826

Chapter 21

Animal Stroke Model: Ischemia–Reperfusion and Intracerebral Hemorrhage

Changhong Ren, Christopher Sy, Jinhuan Gao, Yuchuan Ding, and Xunming Ji

Abstract

Stroke is a major health issue worldwide—one with serious financial and public health implications. As a result, ongoing clinical research on novel and improved stroke therapies is not only pertinent but also paramount. Due to the complexity of a stroke-like event and its many sequelae, devising usable methods and experimental models are necessary to study and better understand the pathophysiological processes that ensue. As it stands, animal models that simulate stroke-like events have proven to be the most logical and effective options in regards to experimental studies. A number of animal stroke models exist and have been demonstrated in previous studies on ischemic as well as hemorrhagic stroke. Considering the efficiency and reproducibility of animal models, here, we introduce an ischemic stroke model induced by middle cerebral artery occlusion (MCAO) and an intracerebral hemorrhagic stroke model induced by collagenase injection. The models outlined here have been proven to demonstrate the clinical relevance desired for use in continued research on stroke pathophysiology and the study of future therapeutic options.

Key words Animal stroke model, Ischemia–reperfusion, Middle cerebral artery occlusion, Intraluminal filament, Intracerebral hemorrhage, Collagenase injection

1 Introduction

Stroke places a large financial burden on healthcare systems. In the USA, it is the fourth leading cause of death and is the leading cause of long-term severe disability [1, 2]. Due to the seriousness of a stroke-like event and its many sequelae, devising methods and models to study and better understand the pathophysiological processes that occur are paramount. Several clinical trials and retrospective studies are inherently limited in their usefulness and clinical applicability when it comes to stroke [3]. As a result, animal models that simulate stroke-like events have proven to be the most logical and effective options for experimental studies. In addition to addressing ethical concerns, use of animal models in stroke research

Firas Kobeissy et al. (eds.), *Injury Models of the Central Nervous System: Methods and Protocols*, Methods in Molecular Biology, vol. 1462, DOI 10.1007/978-1-4939-3816-2_21, © Springer Science+Business Media New York 2016

has led to methods that are highly reproducible and outcomes that are both informative and clinically relevant.

Although animal stroke models have been developed using mice, rabbits, cats, dogs, pigs, baboons, and other nonhuman primates, those involving rats are used most commonly for a number of reasons [3–6]. Rats, in general, closely resemble the cerebrovascular anatomy and physiology of humans. They are moderate in size and, as such, are ideal for the monitoring of vital and physical signs—including the subsequent analysis of brain tissue for physical and molecular changes [7–9]. It has been shown that rats exhibit a relative homogeneity within each other and among different strains [9].

Stroke itself can be broadly divided into two major categories based on etiology: hemorrhagic (~20% of cases) and ischemic (~80% of cases) (Bennett et al. 2012). Common to both types of stroke is the cessation of blood circulation to one or more areas of the brain—leading to hypoxia, malnutrition, and the buildup of toxic substances. The end result is neuronal cell death and subsequent neurobehavioral deficits in survivors. Hemorrhagic stroke is characterized by intracranial bleeding within the brain parenchyma, caused either through trauma or spontaneously as a result of an underlying medical condition [10]. Regardless of the cause, cerebrovasculature is damaged, leading to disruption of blood supply within a wide area. Continuous bleeding in hemorrhagic stroke also leads to the formation of a growing hematoma that can cause further mechanical damage in addition to that already caused by a dysfunctional circulation. Ischemic stroke is the more common of the two types and is instead characterized by the focal occlusion of a blood vessel that supplies the brain. Clinically, in humans, it is often the middle cerebral artery (MCA) that is occluded due to its status as the downstream continuation of the internal carotid artery (ICA) [11–13]. As large arteries, these major vessels are where cardiac and systemic emboli are likely to enter. Furthermore, the anatomy of the distal MCA is characterized by the presence of a major branch point where emboli can become easily lodged. Unlike hemorrhagic stroke, the focal occlusion of ischemic stroke leads to a more localized and demarcated lesion consisting of brain parenchyma formerly perfused by the occluded vessel. It is important to *Note* that ischemic stroke also differs from hemorrhagic stroke in that a "penumbra" region of injured, yet still viable, tissue often surrounds the core lesion produced by the lack of blood supply [14, 15]. This penumbra region, preserved by collateral circulation, can potentially be saved but is susceptible to further damage from reperfusion-related injuries. It is the case that rapid reestablishment of circulation has the potential to exacerbate damage to vessels and tissue that have been previously weakened. Regardless of the exact cause and progression on stroke, serious complications can result. All in all, there are a number of rat models which have been developed to simulate both types of stroke pathophysiology. The models most commonly used, due largely to their reproducibility and clinical applicability, will be briefly introduced and reviewed.

Compared to ischemic stroke, models of hemorrhagic stroke are inherently less focal and controlled, owing to the presence of hematomas that cause damage not only through shear force and mass effect but also through direct toxicity from blood components and inflammatory responses. Given the focus of this section, only the most common models will be discussed here for reference. The two models most commonly used are the direct injection of autologous blood and the use of bacterial collagenase [4, 16]. In the model involving injection of autologous blood, the growth and progression of an actual hematoma can be very closely simulated. Control of injection rate (simulating hematoma growth rate) using blood belonging to the original specimen allows preservation of many physiological parameters that would occur with a spontaneous hemorrhagic stroke. Despite this, the use of bacterial collagenase has more recently been found to better simulate what occurs in the clinical setting [4, 17]. Its action relies on the degradation of Type IV collagen located within the basement membrane of the blood–brain barrier (BBB)—leading to disruption and leakage surrounding sites of injection delivery. The concept was first described by Rosenberg et al. in 1990 from observations of elevated collagenase levels released from injured cells [18]. A marked advantage of the collagenase model over autologous injection is that, again due to direct disruption of the BBB, spontaneous hemorrhages occur directly upon delivery of collagenase to the desired site [17, 19]. These spontaneous hemorrhages mimic the clinical setting and cannot be simulated with the autologous injection model. The efficacy of this model, along with its flexibility and ease of use, has allowed successive improvements and modifications. Location, amount, and rate are easily customizable. Choudhri et al., for example, experimented with different infusion rates while targeting the basal ganglia [20].

As far as ischemic stroke models, the two models most often used are the thromboembolic model and the intraluminal suture model [3, 6]. Both rely on the production of focal ischemia through targeted occlusion of cerebral vasculature. The vessel most often targeted is the middle cerebral artery (MCA), which is the cerebral vessel most often occluded within humans in the clinical setting. In the thromboembolic model, clotted material is injected into the internal carotid artery (ICA), which is accessed via the external carotid artery (ECA). Many materials have been used to produce injectable clots though the most often used are human or autologous blood and fibrin-rich blood clots due to their physiological relevance. In the intraluminal suture model access to the MCA is similar—through the ICA via the ECA. The difference involves the insertion of a filamentous suture, instead of clotted material, to physically occlude the MCA at its major branch point. Both of these models hold great value in the experimental setting. Of *Note* are other models of focal cerebral ischemia including photothrombosis, which uses a photoactive dye and irradiation to induce focal endothelial damage and endothelin-1 induced stroke where the eponymous peptide is applied to cause prolonged vasoconstriction [3, 6, 9]. These latter two models have their niches in experimental research but are not often used in clinical studies and therapeutic testing—especially in light of the effectiveness of the thromboembolic and intraluminal suture models.

The thromboembolic model most closely mimics the clinical setting and allows the use of thrombolytic therapies as part of the experimental regimen. Although the intraluminal model cannot accurately simulate the action of thrombolytics it does allow controlled timing of reperfusion through withdrawal of the suture at a desired time. This allows greater flexibility with the intraluminal model, as it is not reliant on thrombolytic therapy for the onset of reperfusion. The intraluminal model has also proven to be more reliable with extremely reproducible results, largely because the technique has become more and more refined. Koizumi et al. first described the concept of using an intraluminal suture in 1986 where, most notably, a silicone-coated suture was used [25]. Both ischemia and reperfusion were achieved but results were variable and inconsistent. The technique was refined in 1989 by Zea Longa et al. where an uncoated suture was used and where a flame was used to blunt the suture tip [26]. Successive improvements, including the use of poly-l-lysine coating [27] and increased diameters [28, 29], for example, have only made the intraluminal model more suitable, adaptable, and effective for clinical stroke studies. As it stands, larger ischemic lesions have been generated with the intraluminal model, which has proven to be of value in studies of neuroprotection. Reproducible transient ischemia ranging from 60 min to 2 h has been well documented. Most notable, perhaps, is the option for reliable permanent ischemia using the intraluminal model. Lastly, the intraluminal model is comparatively less invasive and less complex to perform—lending itself further to the stringent reproducibility, precision, and efficiency expected from clinical studies.

When selecting a particular animal model for experimental studies it is important to *Note* that despite the excellent reproducibility of these models, they can not fully replicate the heterogeneity of stroke pathophysiology. Furthermore, model selection and the success of any study will depend on what mechanisms are of interest, as some models are better suited for certain pursuits and not as well-suited for others. Regardless, the models outlined here have, up to this point, demonstrated the high reproducibility and clinical relevance desired for continued research in stroke pathophysiology and the study of future therapeutic options.

2 Material

The experimental procedures are in accordance with the Animal Care Guidelines of the Animal Experimental Committee of Xuanwu Hospital, Capital Medical University and the National Institutes of Health. Male adult Sprague Dawley rats (280–320 g, Charles River Labs, Beijing, China) were used in these experiments.

2.1 List of Equipment and Supplies

2.1.1 General Supplies

Disposable masks, latex gloves, canvas gloves, mouse board, medical tape, cotton swabs, marker pen.

2.1.2 Animal Handling Equipment and Supplies

1. Rat temperature control, including: 20 × 30 mm heating pad, temperature controller, and rectal probe (Harvard Apparatus).
2. Lubricant ophthalmic ointment (Akorn, Buffalo Grove, Illinois).
3. Rat anesthesia:
   (a) Anesthesia induction box. Oxygen (O_2) and nitrous oxide (N_2O) gas tanks with manifold (Bickford veterinary anesthesia equipment model no. 61010; AM Bickford Inc., Wales Center, NY, USA).
   (b) Vacuum recapture system (Rodent Ventilator Model 683; Harvard Apparatus Inc., Holliston, MA, USA).
   (c) Oxygen (O_2), nitrous oxide (N_2O).
   (d) Isoflurane.
   (e) Endotracheal cannulas (hollow tube) and tube core for endotracheal intubation;
4. Surgical microscope (Carl Zeiss).
5. Electrocoagulator.
6. Trimmer.
7. Betadine.

2.2 Equipment and Supplies for Cerebral Blood Flow Measurement

1. Laser-Doppler Cerebral blood flow detector (PF5001, Perimed).
2. Stereotaxis frame with integrated tooth bar (Kopf Instruments, CA).
3. Surgical tools: fine forceps, blunt-pointed scissors, homemade retractors (two), acutenaculum.
4. Microdrill (MD-1200, Braintree Scientific).
5. 1 mm carbide ball bit (1 mm).
6. Hydrogen peroxide.

7. Supper Bonder medical device adhesive and accelerator.
8. Bone wax (Shanghai Sanyou Co. Ltd.).

2.3 Equipment and Supplies for MCAO

1. Surgical tools: iridectomy scissors, fine forceps, blunt-pointed scissors, hemostats, homemade retractors, acutenaculum (Fig. 1).
2. Bipolar electrocoagulation (ACC100, DEVEL).
3. Suture needle and skill sutures (4-0).
4. Microvessel clip: S&T vascular clamp, 3.5 × 1 mm jaw, 7 mm long (Harvard Apparatus, MA).
5. Heparinized saline (25 U/mL). To make heparinized saline: combine 0.9 % sodium chloride solution with heparin.
6. Commercially available sutures.

2.4 Equipment and Supplies for Collagenase Injection

1. Stereotaxic frame with integrated tooth bar.
2. Surgical tools: fine forceps, blunt-pointed scissors, two homemade retractors, acutenaculum.
3. Microdrill (MD-1200, Braintree Scientific).

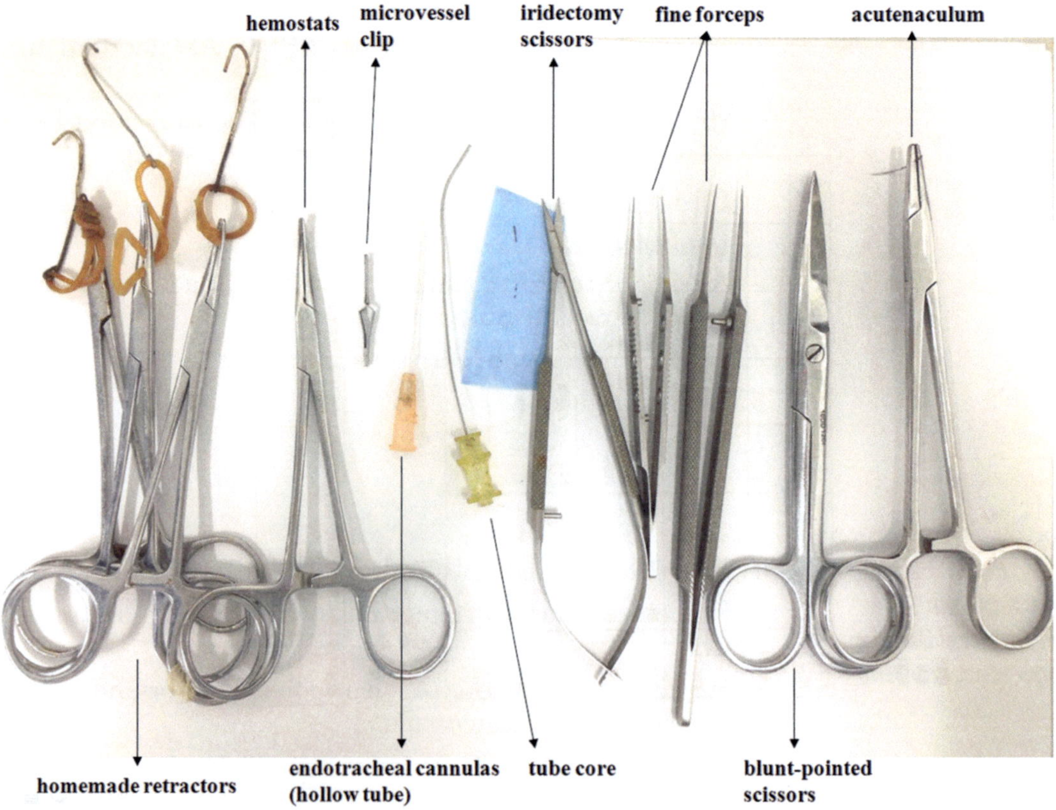

Fig. 1 Surgical instruments used

4. Hydrogen peroxide.
5. Bone wax.
6. Microsyringe with 22-gauge needle.
7. Collagenase VII-S (Sigma, St Louis, MO, USA). To make collagenase solution for injection, combine 0.5 U collagenase VII-S in 1 μL saline. Prepare just before using.

3 Methods

3.1 MCA Occlusion Model

For conducting focal ischemic stroke, middle cerebral artery occlusion (MCAO) is used followed by reperfusion. MCAO is induced using the intraluminal filament model described by Zea Longa et al. [26].

The MCAO model, involving insertion of an intraluminal suture, is performed to simulate focal ischemic stroke. Reperfusion can be established by removal of the suture while non-removal of the suture allows for a model of permanent ischemia. MCAO was induced using the intraluminal filament model described by Zea Longa et al. [26]. Successful MCAO is confirmed, in vivo, using DRT4 laser Doppler flowmetry (PERIMED 5000, Sweden) [30]. Neurological deficits in rats are examined after ischemia and compared to baseline. The deficits were scored using a modified scoring system [26].

3.1.1 Preparation of Suture

All sutures used were commercially available, though the diameter of suture needed is dependent on the weight of rat and should be determined accordingly. In this study, the suture used is suitable for rats weighing 280–320 g. The diameter of the tip is (0.28 mm), whereas the diameter of the remaining suture is (0.36 ± 0.02 mm). Under microscope, mark a 1 mm line at the tip of the suture using black marker pen. Mark a point at 18 mm, 19 mm and 20 mm away from tip, respectively. Store prepared sutures within a container immersed in heparinized saline (25 U/mL).

3.1.2 Surgical Preparation

1. Adult male Sprague-Dawley rat were weighed and prepared (280–320 g range only).
2. Animals were fasted overnight but were allowed free access to water.
3. Induce anesthesia of the animal with 4 % isoflurane and a 30:70 mixture of oxygen and nitrous oxide. Insure that the animal is fully sedated using tail and/or toe pinches.
4. Secure the animal on the board with medical tape.
5. Under a microscope, insert endotracheal (ET) cannula with tube core (tube core exposed 1 cm from cannula). The glottis

usually opens and closes with respiration. While the glottis is open, insert the exposed tube core past the glottis. Next, insert the remaining cannula past the glottis as well. Withdraw the tube core while leaving the ET cannula in place within the glottis. Successful ET intubation is determined by positive air flow from the cannula and can be assessed with a cotton wisp held near the device to observe for air movement.

6. Isoflurane is subsequently lowered from 4% to 1.5–2% for endovascular access, and administered continuously throughout the time-course of these experiments. Care should be taken to minimize the level of isoflurane while maintaining full anesthesia during surgery—this can be periodically assessed using a paw pinch test.

3.1.3 Measurement of Local Cerebral Blood Flow (CBF)

To verify MCAO and reperfusion and to examine changes of the microcirculation after reperfusion, local CBF in cortex supplied by the right MCA should be measured using Laser-Doppler Cerebral blood flow detector LDF before (baseline) and during occlusion.

Skull Preparation

1. Place the rat's head between the ear bars of the stereotaxic frame and gently secure the head to ensure the stability of the rat. Place the heated pad under the animal body. The rectal probe is lubricated with sterile mineral oil, inserted, and secured with adhesive tape to the tail. The core body temperature of rat should be continuously monitored and maintained at 37–37.5 °C (*see* **Note 1**).
2. Shave the scalp from neck region to the frontal area of the rat using a rodent trimmer. Sterilize the operating area with Betadine according to aseptic techniques.
3. Cut a longitudinal incision along the animal's midline, starting from between the ears to between the eyes (Fig. 2a).
4. Retract skin to expose subcutaneous tissue. Wipe the scalp and associated connective tissue with a cotton swab dipped in 30% hydrogen peroxide to complete the skull exposure (Fig. 2b) (*see* **Note 2**).
5. Carefully clean the exposed skull with a cotton swab dipped in 30% hydrogen peroxide. This step is important for cleaning up residual blood and fascia.

Thinning the Skull

1. A micro drill and 1 mm carbide ball bit are used to gently thin the skull over the region to be excised (3 mm posterior and 5 mm lateral to bregma).
2. Once the outer layer of the skull is removed, the spongy bone layer is removed through gentle scraping with dull forceps. Any remaining layers are thinned with the carbide bit, taking great care in not breaking through to the brain parenchyma. Be sure to clear away any bone fragments from the skull as these will interfere with CBF signaling.

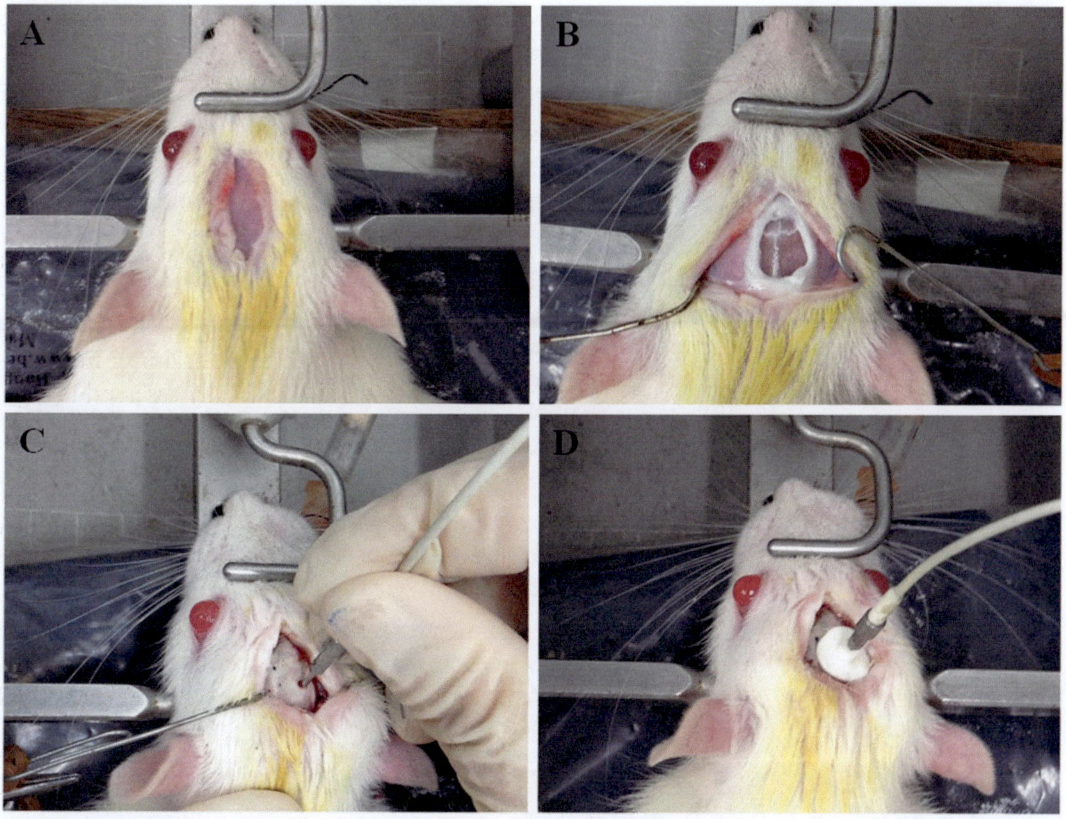

Fig. 2 Schematic diagram of CBF measurement. (**a**) Schematic diagram of the longitudinal incision. (**b**) Schematic diagram of the skull exposure. (**c**) Schematic diagram of skull window for positioning of laser indicator. (**d**) Probe holder is fixed on the skull

3. In order to protect the pad from moisture and any secretions, cotton is stuffed between the scalp and skull to absorb body fluids.

CBF Measurement

1. Adjust PeriCam PSI head to make sure that the red cross dot of (660 nm) laser indicator is positioned in the center of measurement area (Fig. 2c).
2. Insert the probe into the PH07-4 Probe Holder, and then stereotaxically place the Probe Holder (with the probe) on the exposed polished hole (Fig. 2d).
3. Adjust the position of the Probe Holder to collect perfusion signals. When signaling of adequate quality (i.e., stable and normal CBF range) is acquired, apply adhesive around the Probe Holder to fix it onto the skull. The probe holder should be held by hand until the glue is dry. Coagulation accelerator spray can be used to rapidly solidify the glue.

3.1.4 MCAO Surgery

1. For CBF measurement, the animal is placed in a prone position, whereas for MCAO surgery, the animal is supine. Carefully change the position so as to not disturb the placement of the laser probe.

2. Re-secure the animal to the board by applying medical tape to both upper limbs.
3. Shave the hair from the animal's neck region up to the lower jaw using a rodent trimmer and sterilize the operating area with Betadine according to aseptic techniques (Fig. 3a).
4. Cut a longitudinal incision along the midline from the neck (sternal area) up to about 2 cm below the animal's lower jaw (Fig. 3a).
5. After cutting the skin, submandibular glands and sublingual major glands are exposed (Fig. 3b). A longitudinal shallow depression between the left and right glands is visible; this is indicated by a dotted line in Fig. 3b. Carefully cut the fascia along that shallow depression, taking care to avoid shearing of the submandibular gland and the sublingual major gland.
6. The sternocleidomastoid muscle and sternohyoideus muscle are then exposed under the submandibular gland. Separate the right sternocleidomastoid muscle from the sternohyoideus muscle by blunt dissection with hemostats.
7. Retract the skin and fascia towards the upper left portion of the surgical field with retractor #1 (Fig. 3e), retract the sternocleidomastoid muscle to the lower left portion with retractor #2

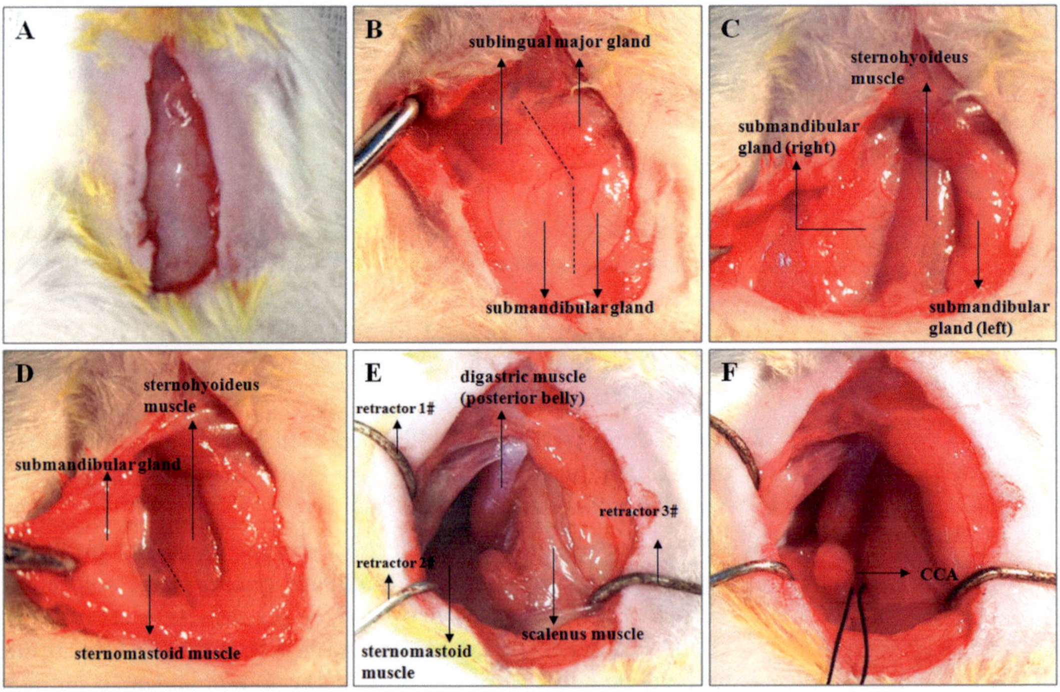

Fig. 3 Schematic diagram of MCAO surgery process. Muscle tissues are separated layer by layer. (**a**) Schematic diagram of the longitudinal incision. (**b**) Schematic diagram of the glands under skin, *dotted line* indicate the position for cutting fascia. (**c, d**) Schematic diagram of the muscle under gland, *dotted line* indicate the position for blunt dissection. (**e**) Schematic diagram of exposed visual field. (**f**) Schematic diagram of exposed CCA

(Fig. 3e), and then retract the scalenus muscle to the lower right portion with retractor #3 to keep the incision open and accessible (Fig. 3e). This step is important for adequate exposure of the surgical field (*see* **Note 3**).

8. With the surgical field exposed, the pulsation of the common carotid artery (CCA) should be visible (a bright red and thick-walled vessel). Under the surgical microscope (magnification: 6× or 10×), carefully separate the fascia around the CCA with fine forceps and avoid tearing the veins that accompany the CCA. Of *Note*, to avoid stressing the animal, there is no need to separate CCA along the entire length. The CCA should be gently separated about 1 mm, just enough so that a 4-0 silk suture can be threaded underneath and around the CCA.
9. A 4-0 silk suture is threaded under the CCA (Fig. 3f).
10. The external carotid artery (ECA), a branch of the CCA, is then separated from its surrounding fascia. The ECA as well as the internal carotid artery (ICA) are the eventual bifurcations of the CCA. When looking at the surgical field, the ECA appears to be the direct, linear, extension of the CCA extending upward towards the brain (Fig. 4). At the bifurcation of

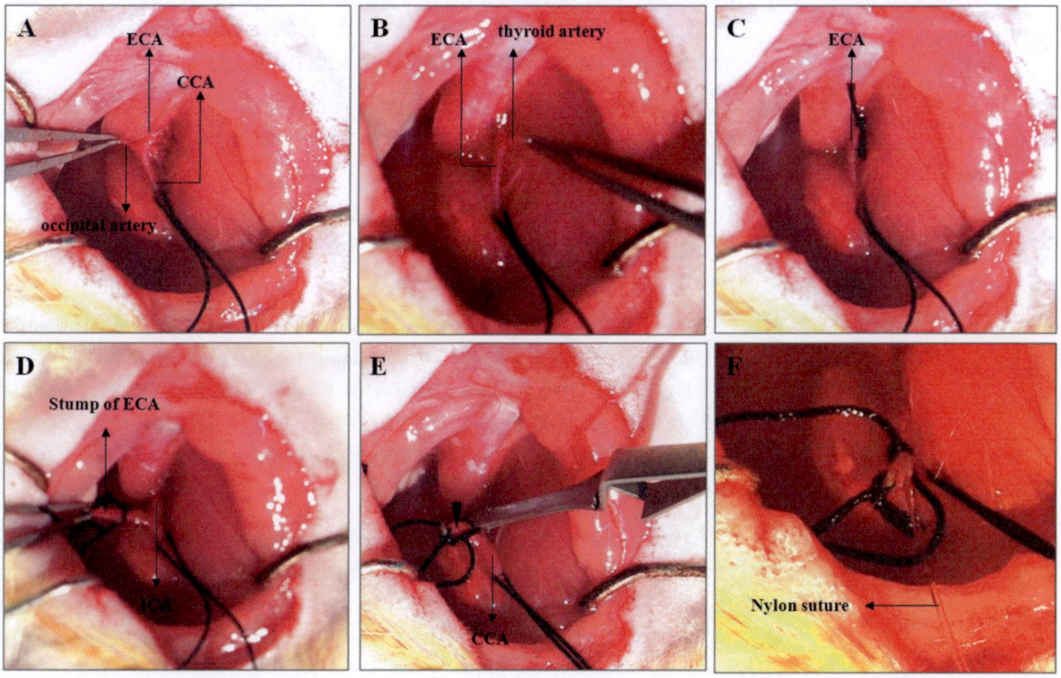

Fig. 4 Schematic diagram of MCAO surgery process. Blood vessels are separated one by one. (**a**) Schematic diagram of ECA and occipital artery. (**b**) Schematic diagram of ECA and thyroid artery. (**c**) Schematic diagram of ECA ligated by suture. The occipital artery and the thyroid artery are cut by electrocoagulation. (**d**) Schematic diagram of stump of ECA and separated ICA. (**e**) Schematic diagram of the ICA clamped by microvessel clamp. *Arrowhead* indicate the position of incision for inserting suture. (**f**) Schematic diagram of insertion of nylon suture

the CCA, the ICA is located under the ECA. Of *Note*, the vagus nerve (appearing like a white ribbon) accompanies the CCA and should be avoided.

11. The occipital artery and thyroid artery are branches of the ECA that should be isolated and coagulated. When the ECA is fully exposed, the occipital artery is visible to the left of ECA (Fig. 4a), near the bifurcation of the CCA. Again, the occipital artery should be identified and separated from surrounding fascia, then ligated by electrocoagulation.
12. The thyroid artery is visible to the right of ECA (Fig. 4b). Similar to the occipital artery, the thyroid artery is also separated from surrounding fascia and ligated by electrocoagulation.
13. The ECA is dissected further distally, ligated, and coagulated along with the terminal lingual and maxillary artery branches, which are then divided (Fig. 4c, d). The resulting stump of the ECA will be used as the insertion point for the intraluminal suture.
14. Loosely tie a 4-0 silk suture around the mobilized ECA stump (Fig. 4d), which is used for securing the intraluminal suture upon insertion and for preventing blood loss.
15. Separate the ICA from its surrounding fascia and from the adjacent vagus nerve at the ICA and ECA branch (Fig. 4d). When isolating ICA, a tiny round muscle close to the branch is visible and must be separated from the ICA.
16. Make sure the 4-0 suture threaded under the CCA is sufficiently tractioned by hemostats to prevent blood flow into the CCA at this time.
17. A microvessel clamp is placed at the bottom edge of the CCA (Fig. 4e). It is best to place the microvessel clamp slightly away from the ICA and ECA branch.
18. A tiny incision is made in the stump of ECA stump with iridectomy scissors at an angle of incidence (45° from parallel), while taking great care to not completely sever the artery. It is best to make this incision close to the end of the ECA stump. Once the artery is completely severed, another incision can be cut under that incision.

Troubleshooting: Massive hemorrhage from the incision = CCA and/or ICA is not completely clamped. If the ICA is not completely clamped, quickly clamp the stump of ECA under the incision with fine forceps, and adjust the microvessel clip to ensure the clip completely blocks arterial blood flow. Sometimes, the ICA is not fully isolated from connected fascia; as a result the clip can not fully block blood flow. As such, it is important to carefully separate the ICA from the connective fascia as well as the muscle. If the CCA is not completely clamped, verify that the tied suture is placing enough tension on the vessel to occlude blood flow.

19. Insert intraluminal nylon suture through the proximal ECA. When the suture is gently inserted up to the microvessel clip, the silk suture around the ECA stump is tightened around the intraluminal nylon suture to prevent bleeding, and the microvascular clip can then be removed.
20. The nylon suture is then gently advanced upward towards the ICA lumen. The position of the suture within the ICA lumen can be seen. Pull the stump of ECA towards the operator so that the ECA and ICA are manipulated into a straight line (Fig. 4f). This creates the easiest possible angle for suture insertion.
21. Gently insert suture, following the path of the ICA towards the upper right. When the insertion length approaches 18 mm (the first marked dot), insert the suture more slowly (*see* **Note 4**). Resistance is felt and a slight curving of the suture or stretching of the ICA is observed, indicating that the blunted tip of the suture has passed the MCA origin and has reached the proximal segment of the anterior cerebral artery (ACA); which has a smaller diameter. At this point, the intraluminal suture has blocked the origin of the MCA. If the MCA is occluded, the CBF can drop to 20 % of baseline. The silk suture around the ECA stump is further tightened around the intraluminal nylon suture to secure its placement and to prevent any potential bleeding.
22. Cut the protruding nylon intraluminal suture at about 0.5 cm away from the incision. Bury the stump of the suture into the tissue parallel with surrounding muscle fiber, be sure the remaining suture lying outside of the lumen does not lie perpendicular to the muscle. Suture the skin.
23. Soften the bone wax by heat. The bone window is then sealed with the soft bone wax.
24. Suture the scalp.
25. Turn off the inhalational anesthesia. When the rat can spontaneous breathe, turn off breath machine. Place the animal in housing for recovery.

3.1.5 Reperfusion

1. Induce anesthesia within the animal with 4 % enflurane and a 30:70 mixture of oxygen and nitrous oxide. A nose cone is used to continuously deliver the anesthetic gases.
2. Open the incision to expose previously manipulated blood vessels. Clamp the CCA with microvessel clip. Pull out the nylon suture very slowly. When three marker points are taken out of the vessel, the nylon suture can be pulled back quickly. When the most distally marked point reaches the bifurcation of the CCA, clamp the ICA with microvessel clip, and proceed in pulling out the suture in its entirety. Coagulate the stump of the ECA. At this point the two microvessel clips can be removed.

3. Suture the skin. Turn off the inhalational anesthesia. Place the animal in housing for recovery.

3.1.6 Behavioral Testing

Neurological deficits in animals are examined after any procedures. The deficits are scored using a five-score system [26].

1. After reperfusion, the animal is held by the middle part of the tail and elevated from the tabletop. Observe the forelimb. 0, no observable deficits; 1, difficulty in fully extending the contralateral forelimb; 2, unable to extend the contralateral forelimb.
2. Put the animal on open ground, if the rat cannot walk in a straight line, but instead moves in a circular course towards the direction contralateral to the infarcted side it is scored as a 3; Severe circling is scored as 4; falling to the contralateral side is scored as a 5 (*see* **Note 5**).

3.1.7 Important Tips

1. Muscle tissue should be separated layer by layer. After cutting the skin, glands are visible. Cut the fascia along the midline of these glands (Fig. 3b). When anterior cervical muscles are exposed (Fig. 3c, d), do not cut muscle to avoid bleeding. Bluntly separate the two muscles with hemostats (separate muscles along the dotted line as indicated in Fig. 3b, d).
2. When separating blood vessels, try to thoroughly clean the connective fascia from the blood vessel as much as possible. If the vessel is isolated clearly, it is easy to conduct subsequent steps including: electrocoagulation, placing microvessel clips, making incisions, and the eventual insertion of the intraluminal suture. This is especially important when making the incision at the stump of ECA using iridectomy scissors, so that the operator will not mistakenly cut completely though the outer membrane of the ECA (If mistakenly cut, repeated operation will result in cutting a bigger incision or severing the stump of ECA. For the purposes of this procedure the smaller this ECA incision is, the better). Because an oblique incision makes it easier for the initial insertion of the suture, it is better to make the incision at a 45° with the vascular wall. Be careful to not stretch the vessel too much, which could potentially cause vasospasm and, thus, insertion failure.
3. If resistance is felt before full length of suture (18 mm) is inserted, this suggests that the suture could have been inserted into the pterygopalatine artery. Pull the suture out of the ICA up until the bifurcation of the CCA. Make another attempt to insert the intraluminal suture again, taking *Note* that insertion direction is important for correct placement. If the direction is straight or left, it is relatively easy to place the suture into the pterygopalatine artery. For beginners, the pterygopalatine artery can be ligated as described by Enrique Zea Longa et al.

to avoid confusion. In brief, identify and dissect the ansa of the glossopharyngeal nerve at the origin of the pterygopalatine artery; this posteriorly directed extracranial branch of the ICA is ligated with 7-0 nylon suture close to its origin. At this point, the ICA is the only remaining extracranial branch of the CCA, allowing less room for error in placement.

4. Try to avoid damage to the vagus nerve as it passes under the bifurcation of the CCA.
5. Operator should be careful to insert the suture gently and smoothly. Despite this, insertion should be as fast as possible as prolongation of this procedure could lead to intravascular thrombosis.
6. In order to reduce the possibility of subarachnoid hemorrhage and vasospasm, repeated suture inserts (over four times) into blood vessels should be avoided.

3.1.8 Exclusion Criteria (Failed Operation)

1. If the depth of the suture insert is less than 18 mm (see above), and there is no obvious neurological deficits, the Zea Longa score is 0.
2. Subarachnoid hemorrhage (SAH). Because hemorrhage also induces brain injury, it would be problematic to distinguish unintended brain damage from the intentional ischemia–reperfusion induced injury.

3.1.9 Cause of Death

Operator must first find out the cause of death, dissect rat brain, and look at whether there is SAH. SAH indicates that the suture is inserted too deep. Operator must pay attention to the length of insertion. If there is no bleeding, operator should observe whether there is a serious hemisphere edema. Any serious edema indicates that the death was caused by prolonged ischemia. As such, a short ischemic time period is required. If bleeding and/or serous cerebral edema is not apparent, operator should pay attention to animal status or whether the living environment was inadequate for any reason.

3.1.10 MCAO Complications

1. Subarachnoid hemorrhage, which can be avoided by gentle operation.
2. ICA thrombosis, which can be avoided by using heparinized sutures.

3.2 Intracerebral Hemorrhagic Models

Injection of collagenase VII is used to induce intracerebral hemorrhage (ICH) for hemorrhagic stroke evaluation [31].

3.2.1 Surgical Preparation

General surgical preparation is the same as described for MCAO preparation (Subheading 2.2, **item 2**).

3.2.2 Skull Preparation

Skull Preparation is the same as described in the part of CBF measurement (Subheading 2.2, **item 3**).

3.2.3 Thinning the Skull

1. Set up microsyringe with 22-gauge needle on the stereotaxis frame.
2. Collagenase VII is injected into the right striatum (from bregma: 1.2 mm posterior, 3.5 mm lateral, and 3.5 mm depth). Needle is located aiming the bregma. And then locate skull window by stereotaxic frame and mark a puncture site using a marker pen (stereotactic coordinates from bregma: 1.2 mm posterior, 3.5 mm lateral). Take the microsyringe away.
3. A micro drill and 1 mm carbide ball bit is used to gently thin the skull over the marked region. Once the outer layer of the skull is removed, the spongy bone layer can be removed with gentle scraping using a dull forceps. The remaining layer can be thinned further and as needed with the carbide bit, with great care so as to not completely break through.

3.2.4 Collagenase VII Injection

1. Draw up 1 μL Collagenase VII into the microsyringe. Ensure that there are no air bubbles in the syringe.
2. Move the needle back to the point just over the skull window. Screw the needle gently onto the surface of the skull window. Screw the needle gently into the brain to about a 4 mm depth and then withdraw 0.5 mm using stereotactic coordinates (Fig. 4).
3. Very slowly screw the stylet of the syringe (at speed 0.1–0.2 μL/min). The entire injection procedure should last 5–10 min.
4. The needle is left in place for 5 min to allow collagenase infusion and action.
5. Slowly withdraw the needle at a speed of about 1 mm/min (*see* **Note 6**).
6. Soften the bone wax by heat. Bone window is sealed with the soft bone wax.
7. Suture the scalp.
8. Turn off the inhalational anesthesia. When the animal can spontaneously breathe, then turn off the breath machine. Place the animal in housing for recovery.

4 Notes

1. Depending on the ambient temperature, place a ceramic heat lamp over the rat to prevent a drop in body temperature while the heating pad stabilizes.

2. In order to reduce the possibility of bleeding from subcutaneous tissue, do not cut subcutaneous tissue with scissors and use hydrogen peroxide.
3. Because trachea is to the right of the scalenus muscle, make sure that the trachea is not compressed by retractor #3.
4. The insertion length depends on the animal's body weight. If the insert length is over 20 mm (the third marked dot), and the suture can still be inserted, please do not insert any more.
5. If the animal does not walk, operator can pinch the tip of the tail. Do not stand beside the animal so as to avoid disturbing it.
6. Increasing the delay in needle withdrawal and infusion course is important for successful generation of the ICH model.

Acknowledgments

This work was partially supported by the program of National Natural Science Foundation of China (81573867), and Scientific Special Funding of Capital Health Development (No. 2011-1001-03).

References

1. Go AS, Mozaffarian D, Roger VL, Benjamin EJ, Berry JD, Borden WB, Bravata DM, Dai S, Ford ES, Fox CS, Franco S, Fullerton HJ, Gillespie C, Hailpern SM, Heit JA, Howard VJ, Huffman MD, Kissela BM, Kittner SJ, Lackland DT, Lichtman JH, Lisabeth LD, Magid D, Marcus GM, Marelli A, Matchar DB, McGuire DK, Mohler ER, Moy CS, Mussolino ME, Nichol G, Paynter NP, Schreiner PJ, Sorlie PD, Stein J, Turan TN, Virani SS, Wong ND, Woo D, Turner MB (2013) Executive summary: heart disease and stroke statistics—2013 update: a report from the American Heart Association. Circulation 127:143–152
2. Wang G, Zhang Z, Ayala C, Dunet DO, Fang J, George MG (2014) Costs of hospitalization for stroke patients aged 18-64 years in the United States. J Stroke Cerebrovasc Dis 23:861–868
3. Krafft PR, Bailey EL, Lekic T, Rolland WB, Altay O, Tang J, Wardlaw JM, Zhang JH, Sudlow CL (2012) Etiology of stroke and choice of models. Int J Stroke 7:398–406
4. James ML, Warner DS, Laskowitz DT (2008) Preclinical models of intracerebral hemorrhage: a translational perspective. Neurocrit Care 9:139–152
5. Ma Q, Khatibi NH, Chen H, Tang J, Zhang JH (2011) History of preclinical models of intracerebral hemorrhage. Acta Neurochir Suppl 111:3–8
6. Bacigaluppi M, Comi G, Hermann DM (2010) Animal models of ischemic stroke. Part two: modeling cerebral ischemia. Open Neurol J 4:34–38
7. Andaluz N, Zuccarello M, Wagner KR (2002) Experimental animal models of intracerebral hemorrhage. Neurosurg Clin N Am 13:385–393
8. Carmichael ST (2005) Rodent models of focal stroke: size, mechanism, and purpose. NeuroRx 2:396–409
9. Durukan A, Tatlisumak T (2007) Acute ischemic stroke: overview of major experimental rodent models, pathophysiology, and therapy of focal cerebral ischemia. Pharmacol Biochem Behav 87:179–197
10. Emiru T, Bershad EM, Zantek ND, Datta YH, Rao GH, Hartley EW, Divani AA (2013) Intracerebral hemorrhage: a review of coagulation function. Clin Appl Thromb Hemost 19:652–662
11. Bogousslavsky J, Van Melle G, Regli F (1988) The Lausanne Stroke Registry: analysis of 1,000 consecutive patients with first stroke. Stroke 19:1083–1092

12. Fischer U, Arnold M, Nedeltchev K, Brekenfeld C, Ballinari P, Remonda L, Schroth G, Mattle HP (2005) NIHSS score and arteriographic findings in acute ischemic stroke. Stroke 36:2121–2125
13. Saqqur M, Uchino K, Demchuk AM, Molina CA, Garami Z, Calleja S, Akhtar N, Orouk FO, Salam A, Shuaib A, Alexandrov AV (2007) Site of arterial occlusion identified by transcranial Doppler predicts the response to intravenous thrombolysis for stroke. Stroke 38:948–954
14. Rordorf G, Koroshetz WJ, Copen WA, Cramer SC, Schaefer PW, Budzik RF Jr, Schwamm LH, Buonanno F, Sorensen AG, Gonzalez G (1998) Regional ischemia and ischemic injury in patients with acute middle cerebral artery stroke as defined by early diffusion-weighted and perfusion-weighted MRI. Stroke 29:939–943
15. Bivard A, Levi C, Spratt N, Parsons M (2013) Perfusion CT in acute stroke: a comprehensive analysis of infarct and penumbra. Radiology 267:543–550
16. Strbian D, Durukan A, Tatlisumak T (2008) Rodent models of hemorrhagic stroke. Curr Pharm Des 14:352–358
17. MacLellan CL, Silasi G, Auriat AM, Colbourne F (2010) Rodent models of intracerebral hemorrhage. Stroke 41:S95–S98
18. Rosenberg GA, Mun-Bryce S, Wesley M, Kornfeld M (1990) Collagenase-induced intracerebral hemorrhage in rats. Stroke 21:801–807
19. Manaenko A, Chen H, Zhang JH, Tang J (2011) Comparison of different preclinical models of intracerebral hemorrhage. Acta Neurochir Suppl 111:9–14
20. Choudhri TF, Hoh BL, Solomon RA, Connolly ES Jr, Pinsky DJ (1997) Use of a spectrophotometric hemoglobin assay to objectively quantify intracerebral hemorrhage in mice. Stroke 28:2296–2302
21. Clark W, Gunion-Rinker L, Lessov N, Hazel K (1998) Citicoline treatment for experimental intracerebral hemorrhage in mice. Stroke 29:2136–2140
22. MacLellan CL, Gyawali S, Colbourne F (2006) Skilled reaching impairments follow intrastriatal hemorrhagic stroke in rats. Behav Brain Res 175:82–89
23. Beray-Berthat V, Delifer C, Besson VC, Girgis H, Coqueran B, Plotkine M, Marchand-Leroux C, Margaill I (2010) Long-term histological and behavioural characterisation of a collagenase-induced model of intracerebral haemorrhage in rats. J Neurosci Methods 191:180–190
24. Takamatsu Y, Ishida A, Hamakawa M, Tamakoshi K, Jung CG, Ishida K (2010) Treadmill running improves motor function and alters dendritic morphology in the striatum after collagenase-induced intracerebral hemorrhage in rats. Brain Res 1355:165–173
25. Koizumi J, Yoshida Y, Nakazawa T, Ooneda G (1986) Experimental studies of ischemic brain edema: a new experimental model of cerebral embolism in rats in which recirculation can be introduced in the ischemic area. Jpn Stroke J 8:1–8
26. Longa EZ, Weinstein PR, Carlson S, Cummins R (1989) Reversible middle cerebral artery occlusion without craniectomy in rats. Stroke 20:84–91
27. Belayev L, Alonso OF, Busto R, Zhao W, Ginsberg MD (1996) Middle cerebral artery occlusion in the rat by intraluminal suture. Neurological and pathological evaluation of an improved model. Stroke 27:1616–1622, discussion 1623
28. Laing RJ, Jakubowski J, Laing RW (1993) Middle cerebral artery occlusion without craniectomy in rats. Which method works best? Stroke 24:294–297, discussion 297–298
29. Shimamura N, Matchett G, Tsubokawa T, Ohkuma H, Zhang J (2006) Comparison of silicon-coated nylon suture to plain nylon suture in the rat middle cerebral artery occlusion model. J Neurosci Methods 156:161–165
30. Ding Y, Li J, Rafols JA, Phillis JW, Diaz FG (2002) Prereperfusion saline infusion into ischemic territory reduces inflammatory injury after transient middle cerebral artery occlusion in rats. Stroke 33:2492–2498
31. Matsushita K, Meng W, Wang X, Asahi M, Asahi K, Moskowitz MA, Lo EH (2000) Evidence for apoptosis after intercerebral hemorrhage in rat striatum. J Cereb Blood Flow Metab 20:396–404

Part IV

Special Topics in CNS Trauma: Comorbid Conditions in CNS Injury

Chapter 22

Combined Neurotrauma Models: Experimental Models Combining Traumatic Brain Injury and Secondary Insults

Dennis W. Simon, Vincent M. Vagni, Patrick M. Kochanek, and Robert S.B. Clark

Abstract

Patients with severe traumatic brain injury (TBI) frequently present with concomitant injuries that may cause secondary brain injury and impact outcomes. Animal models have been developed that combine contemporary models of TBI with a secondary neurologic insult such as hypoxia, shock, long bone fracture, and radiation exposure. Combined injury models may be particularly useful when modeling treatment strategies and in efforts to map basic research to a heterogeneous patient population. Here, we review these models and their collective contribution to the literature on TBI. In addition, we provide protocols and notes for two well-characterized models of TBI plus hemorrhagic shock.

Key words Traumatic brain injury, Secondary brain injury, Combined injury model, Controlled cortical impact, Hemorrhagic shock, Hypoxia

1 Introduction

Combined injury models of traumatic brain injury (TBI) incorporate one or more clinically relevant factors of secondary brain injury into the model, thereby expanding the scope and applicability of these experiments to scenarios commonly encountered in humans. The necessity of developing combined injury models to enhance contemporary animal models of TBI is apparent from recent clinical studies. It is clear that patients with severe TBI often present to the hospital with serious concomitant injuries that have been shown to increase risk of poor neurologic outcome and death. Chestnut et al. prospectively evaluated 717 adults with severe TBI (Glasgow Coma Scale score ≤8) and found hypotension (systolic blood pressure <90 mmHg) and/or hypoxemia ($PaO_2 \leq 60$ mmHg or apnea/cyanosis) in over a third of the cases [1]. Moreover, hypotension and hypoxemia were independently associated with increased morbidity and mortality; a patient with a single episode of hypotension had a 150% greater chance of death from

Firas Kobeissy et al. (eds.), *Injury Models of the Central Nervous System: Methods and Protocols*, Methods in Molecular Biology, vol. 1462, DOI 10.1007/978-1-4939-3816-2_22,

TBI. Similar findings have been noted in pediatric patients with severe TBI. In a study of 58 patients younger than 18 years old diagnosed with severe TBI, children with either hypotension or hypoxemia had a fourfold greater risk of mortality than patients with neither hypotension nor hypoxemia [2]. Similarly, Stewart et al. examined a retrospective cohort of 180 children with severe TBI and found that 63% of patients had a severe concomitant injury, most often to the lung [3]. Other injuries frequently seen included blunt abdominal trauma to the liver and/or spleen as well as femur and pelvic fractures, all of which are risk factors for developing hypotension.

By design, the commonly used animal models of TBI, controlled cortical impact (CCI), fluid percussion injury (FPI), penetrating ballistic-like brain injury (PBBI), and diffuse injury do not incorporate secondary insults into the model. This has permitted researchers to gain insight into the pathophysiology of head trauma without the confounding influence of factors such as hypotension, hypoxia, or polytrauma. However, when modeling treatment strategies and in efforts to map basic research to a heterogeneous patient population, expanding these models beyond isolated head injury to include secondary insults such as hypoxemia, hypotension, long-bone injury, and radiation exposure is necessary. Here, we review these models and their contribution to the literature on TBI, and then move on to describe a murine model of TBI plus hemorrhage shock (HS) and resuscitation in detail [4, 5].

1.1 TBI + Hypoxemia

Hypoxia is common in patients presenting with severe TBI. Overall, 38% of adults and 8.7% of children with severe TBI will present to the hospital with hypoxia or hypoxemia ($SaO_2 < 92\%$ or $PaO_2 < 60$ mmHg respectively) [6, 7]. Additionally, in the setting of military and civilian air transport, patients may be inadvertently subject to hypoxemia for prolonged periods of time while en route to higher levels of medical care. Clinical studies have found a weak association between systemic hypoxemia and morbidity/mortality, particularly when compared to hypotension. However, PaO_2 has been shown to be an important determinant of brain tissue oxygenation (PbO_2), which has recently become a therapeutic target in patients with invasive monitoring and evidence that PbO_2 levels (<10 mmHg) are associated with a worse outcome post-TBI [8].

Many of the combined TBI + hypoxia experiments have been performed in rodent models, with early results published by Ishige et al. in 1987 [9–11]. These studies subjected adult male Sprague-Dawley rats to parasagittal FPI with an impact pressure of 4.9 ± 0.3 atm, and randomized to post-injury recovery in a control air room versus a hypoxia chamber containing 13% O_2 for 30 min (PaO_2 ~35–40 mmHg). The animals who received FPI + hypoxia had worse motor function, with 71% demonstrating motor weakness or no spontaneous movement at 24 h after injury in comparison to 29% with motor weakness in the normoxic group. In addition,

T2-weighted MRI demonstrated an increase in brain edema at 24 h extending from the cortical impact area in normoxic rats to include the entire ipsilateral cortex in hypoxemic animals. Regional cerebral blood flow (CBF), measured with [^{14}C] neuroimaging, demonstrated increased hypoperfusion in the abnormal cortical regions seen on the MRI [9]. In studies from our laboratory, adult male Sprague-Dawley rats were subjected to moderate-severe CCI, then randomized to receive either mild or moderate hypoxemia (FiO_2 0.13 or 0.11, respectively), or mild hyperoxemia ($FiO_2 = 0.33$) for 30 min. In addition to exacerbating motor deficits, moderate hypoxemia after CCI augmented hippocampal CA3 neuronal death concurrent with increased terminal deoxynucleotidyl transferase-mediated biotin-dUTP nick end labeling (TUNEL) in the hippocampus, consistent with increased apoptotic cell death [12]. It is important to note that moderate hypoxemia produces hypotension in this model, likely producing a scenario of ischemia, rather than simply hypoxia. In 1989, Jenkins et al. demonstrated that combined mechanical injury and forebrain ischemia, at levels where each alone caused minimal or no neuronal loss, could lead to extensive hippocampal neuronal loss [13]. Bramlett et al. subjected Sprague-Dawley rats to parasagittal FPI followed by randomization to post-injury hypoxemia (FiO_2 0.11 for 30 min) versus normoxemia [14]. At 72 h, increased lesion volume and CA1 and CA3 neuronal injury was seen in rats made hypoxemic.

The mechanism(s) of secondary hypoxemia-induced apoptotic cell death following TBI was evaluated in a study by Mikrogianakis et al. [15] TBI was caused in adult male C57BL/6 mice using the Marmarou diffuse injury model—a Teflon impounder was placed midline between the bregma and interaural line and a 50 g weight was released from a height of 20 cm onto the impounder. Interestingly, as originally described [16], the Marmarou diffuse injury model does not cause neuronal death or brain edema unless a secondary insult is applied. Animals were randomized to sham, TBI alone, hypoxemia alone ($FiO_2 = 0.09$ for 30 min), or TBI + hypoxemia ($FiO_2 = 0.09$ for 10, 20, or 30 min). Increased caspase 3 enzyme activity (as measured with fluorogenic caspase substrate Ac-Asp-Glu-Val-Asp-AFC) was seen in the TBI + hypoxemia group compared with TBI group alone. Immunoblotting for neuronal apoptosis inhibitor protein 1 (NAIP-1), which has been shown to inhibit a variety of apoptotic triggers, showed that peak levels of NAIP-1 were delayed (1–7 days) in the hypoxemia group. There was no effect of hypoxemia without TBI on the expression of NAIPs.

More recently, studies evaluating the neuro-inflammatory response to TBI + hypoxemia have been published [17–19]. Hellewell et al. evaluated the cellular inflammatory response using a model of diffuse brain injury. Adult Sprague-Dawley rats were anesthetized, a steel disc surgically adhered to the skull, and placed on a foam mattress to partially absorb and distribute the force of the impact. A 450 g weight was then dropped from a height of 2 m to induce injury. Rats randomized to hypoxemia were exposed to $FiO_2 = 0.12$ for 30 min, previously shown to produce an SaO_2 of 47 ± 4.3% and a PaO_2 of 48 ± 3.8 mmHg. Histologically, rats suffered diffuse axonal injury as demonstrated by immunohistochemical staining for beta-amyloid precursor protein. In areas of concentrated axonal injury, such as the corpus callosum and brainstem, microglia activation and astrogliosis was also seen. Subsequently, Yan et al. [18] evaluated the cytokine response in this model of TBI + hypoxemia. In addition to elevated levels of IL-6 in the brains of rats exposed to TBI + hypoxemia at 24 h (vs. TBI + normoxemia), there was a greater increase in the TNFα and IL1-β levels at 2 h. This coincided with impaired recovery of sensorimotor function on Rotarod testing in the 2 weeks post-injury.

In a model of hypobaric hypoxemia, adult C57BL/6 mice were used to simulate air versus ground evacuation after head injury, which is clinically relevant to military aeromedical evacuation [19]. Following blunt trauma induced by a weight drop model using a 415 g weight with a diameter of 1 cm from a height of 1.5 cm, mice randomized to hypoxemia were brought to a hypobaric chamber pressurized to a cabin altitude of 8800 ft for 5 h; this corresponded to an SaO_2 of ~85–90%. Unlike other blunt trauma models, apnea and posttraumatic seizures were not seen in this model. There was; however, 11% mortality in the animals after head injury. Results of this study showed increased concentration of brain IL-6 and MIP-1α at 24 h in animals after trauma that were exposed to hypobaric conditions at 3 h after trauma but not 24 h after trauma. There was also an increase in serum NSE levels with exposure to hypobaric hypoxemia at 3 h post-TBI but not 24 h post-TBI relative to normoxemic animals [19]. Certainly there are risks to delaying transport and more definitive care of critically ill head injured patients that are not accounted for in this model. However, this study demonstrates how combined injury models can be used to more accurately model a clinical situation and thereby devise treatment strategies to optimize outcomes.

Posttraumatic hypothermia, which has been shown to be neuroprotective in several animal models has not translated into pediatric or adult outcomes, and was tested as a treatment for TBI + hypoxemia by Matsushita et al. in the lateral FPI model previously described by Bramlett et al. [14, 20]. Following a 30-min period of hypoxemia, animals were either immediately cooled to 33 °C or maintained at 37 °C (based on temporalis temperature probes) for 4 h post-injury. Hypothermia was induced by surface cooling using cold air. In the hypothermia group, rats were either rewarmed quickly (15 min) using a heating lamp, or slowly brought back to 37 °C by being placed in a temperature-controlled chamber. After TBI + hypoxemia, a greater than 50% decrease in lesion volume in rats managed with hypothermia and slow rewarming (vs. other temperature groups) was observed.

1.2 TBI + Radiation

Combined injury models of TBI and radiation exposure are being developed to simulate radiologic terrorism, battlefield injury, or nuclear accidents. Children appear to be particularly susceptible to neurocognitive insults of cranial irradiation, possibly due to vascular injury leading to hypoxic-ischemic injury to the hippocampus and deep cortex [21]. Allen et al. have developed a combined model of TBI + whole body irradiation in juvenile (3 weeks old) and adult C57BL6J mice [22, 23]. Mice randomized to the combined injury received 4 Gray of radiation exposure and 4 weeks later, CCI was performed using a 3 mm tip, 4.5 m/s velocity, 1.5 mm depth and 150 ms dwell time. Juvenile mice randomized to isolated head injury or radiation showed no neurocognitive impairments. In contrast, juvenile mice that received combined injury showed impaired performance on Morris water-maze probe trials on days 3 and 4, suggesting impaired hippocampal-dependent spatial memory acquisition, as well as an increase in activated and $BrdU^+$ microglia. Adult mice receiving isolated radiation, trauma, or combined injury did not show any neurocognitive impairments or increased microglial activation.

1.3 TBI + Polytrauma

Several models of TBI + polytrauma have been developed, often incorporating long-bone injury, abdominal injury, and/or shock in addition to TBI [24–26]. The model developed by Probst et al. [25] and Mirzayan et al. [26] performed in 8–10-week-old C57BL6J mice incorporated a closed skull weight-drop model of TBI, and femur fracture by blunt force trauma thereby also causing soft tissue injury followed by closed reduction, splinting and volume controlled HS with 60% hemorrhage. Mortality was 25–47% in the model. Animals in the combined injury group had significantly higher levels of TNFα, CCL-2, and IL-6 at 96 h relative to animals subjected to either TBI alone or femur fracture/HS alone.

1.4 TBI + Hypotension

As previously stated, hypotension is a common comorbid factor in adult and pediatric patients with traumatic brain injury. Subsequently, multiple clinical studies of TBI + hypotension have shown increased morbidity and mortality relative to TBI alone. Manley et al. in a single-center retrospective observational study evaluated vital signs recorded in the emergency room of 107 patients with TBI who had GCS ≤ 12 and abnormal initial CT scan. Overall mortality in this cohort was 43%; however patients with >2 episodes of hypotension (SBP ≤ 90 mmHg) were eight times more likely to die than the remainder of the cohort [6]. In children, an analysis of data from the National Pediatric Trauma Registry found that hypotension significantly increased mortality

($P < 0.0001$) and was not significantly affected by whether or not patients had also suffered hypoxemia [2]. As a result of the potential effects of hypotension on outcomes from TBI and trauma in general, shock is aggressively managed in trauma patients. Typically rapid infusions of crystalloid fluids are administered initially and if necessary, followed by transfusion of blood products to restore circulating blood volume and oxygen carrying capacity. Several investigations have incorporated this resuscitation strategy into TBI + hypotension models.

The most common etiology of hypotension in trauma patients is hypovolemia due to internal or external hemorrhage (i.e., HS), and this is the etiology used in the majority of combined models of TBI + hypotension. The differential diagnosis of low blood pressure in trauma patients also includes myocardial trauma, pericardial tamponade, tension pneumothorax, fat embolism, ingestion, and spinal cord injury. However, although these injuries may cause differing patterns of neurologic injury, they are not included in current combined injury models.

1.5 TBI + HS

The models of combined TBI + HS can be divided into three categories: pressure-controlled, volume-controlled and uncontrolled hemorrhage. Uncontrolled hemorrhage, probably the most clinically applicable combined injury model, is difficult to reproduce. The pressure-controlled hemorrhage model has clinical merit since the quantification of blood loss is often not possible in trauma patients and the degree of blood loss does not have good correlation with other clinically monitored parameters. Blood pressure monitoring, often continuous via an arterial catheter, is typically performed in critically injured patients with TBI to calculate cerebral perfusion pressure, a critical determinant of CBF. Pressure control models therefore allow researchers to develop experiments that will more closely mimic clinical scenarios. However, in pressure control models, animals are typically bled to a predetermined pressure and then maintained at that pressure (often with fluid boluses) for a set period of time—which also does not mimic the clinical scenario for patients. In contrast, volume-controlled models require less monitoring and intervention to maintain the model during the experiment. Overall, it is important to find a balance between reproducibility of the model and clinical applicability of the experiments when choosing amongst the available models of TBI + hypotension.

In addition, combined TBI + HS models allow researchers to more accurately assess hemodynamic changes and intrinsic compensatory mechanisms that are affected by trauma. TBI has been shown in several studies to sensitize the cardiovascular system to the hemodynamic effects of hemorrhage [27, 28]. In general, following TBI, animals will require less hemorrhage to cause hypotension and significantly more fluid to normalize blood pressure during resuscitation. The study by Mahoney et al. [28] confirmed and expounded on these findings in a study of hypotensive adult trauma patients, citing possible brainstem involvement and altered autonomic tone or massive catecholamine surge, with ensuing neurotransmitter depletion, receptor saturation, myocardial depression and cardiovascular collapse. The mechanism of TBI is likely to affect the cardiovascular response. For example, blast-injured patients appear particularly sensitive to the effects of sustained hypotension, as a study of soldiers injured in Operation Iraqi Freedom demonstrates [29, 30].

1.6 TBI + Uncontrolled HS

Several investigators have developed models of the scenario of TBI + HS caused by uncontrolled hemorrhage. Stern et al. used the porcine FPI model (3 atm) with the addition of a 4 mm aortic tear to hemorrhage animals to a mean arterial pressure (MAP) of 30 mmHg [31]. To study the effects of initial limited resuscitation, animals were randomized to initial resuscitation to a MAP of either 60 mmHg, 80 mmHg, or no initial resuscitation. After 60 min of HS, the aortic hemorrhage was controlled and all animals were resuscitated to their baseline hemodynamic state. Mortality rates were 11 % in limited resuscitation, 50 % in aggressive resuscitation and 100 % in animals without initial resuscitation. Clearly from these results, uncontrolled hemorrhage without resuscitation had the worst mortality.

However, the animals with limited resuscitation had a survival advantage over animals resuscitated aggressively, likely due to a decrease in total hemorrhage (41 ± 18 mL/kg vs. 69 ± 32 mL/kg). In addition, there were no significant differences between limited and aggressively resuscitated animals in terms of their CBF, central venous saturation ($ScvO_2$), or $CMRO_2$. Given these results and using the same model of FPI followed by hemorrhage to a MAP of 30 mmHg and 4 mm aortic tear, White et al. performed a study of resuscitation of TBI + uncontrolled HS with hemoglobin-based oxygen carriers (HBOCs), specifically HBOC-201, which, as stated previously in regard to polynitroxylated-pegylated hemoglobin (PNPH), can be given in low volumes to stabilize hemodynamics and may be particularly useful in the prehospital or battlefield setting [32]. After a 15-min shock period, animals were randomly assigned to resuscitation with either LR, HBOC-201, or HBOC-201 in addition to an infusion of sodium nitroglycerin to act as a nitric oxide donor. In this model, there was no significant benefit of the HBOC-201 for resuscitation over LR. In contrast, Stern et al. also performed a study of HBOC-201 and showed improved outcomes relative to LR. However in this case, uncontrolled hemorrhage was caused by liver injury. After FPI (2.6 ± 0.6 atm), animals were placed in the dorsal recumbent position and the liver was exposed via laparotomy. A laceration was surgically created, resulting in approximately 25% lobectomy and consistent with a grade 3 liver laceration.

Resuscitation was divided into three phases: prehospital, emergency department, and operative/postoperative phase. Animals were randomized in the prehospital phase to receive either HBOC-201, LR, or no treatment. Despite no difference in hemorrhage volume, 6-h survival was significantly greater in animals that received HBOC-201 verses LR or no treatment (62% vs. 9% vs. 3%, respectively; $P<0.02$). Treatment with HBOC-201 significantly improved systemic hemodynamics, CPP, and brain tissue oxygenation ($PbtO_2$) compared to animals treated with LR. The different outcomes observed in these two studies may be due to the different injury pattern: either arterial high-pressure bleeding in the study by White et al. compared to splanchnic low-pressure injury in the study by Stern et al. or other less apparent differences between the two models. Clearly however, these results demonstrate the difficulty in developing effective therapeutic strategies for a heterogeneous disease process such as TBI.

1.7 TBI + Pressure-Controlled HS

In 1993, Hariri et al. published a large study about an animal model of TBI + HS [33]. Taking miniature swine and using a combination model of FPI and pressure-control hemorrhage the investigators studied the effects of volume resuscitation to high versus low central venous pressure (CVP) on intracranial hypertension and compliance. FPI was induced with a 2.5 atm barotrauma delivered to the frontal lobe over 20–25 ms; a severe injury that in the researchers' experience results in severe histopathological changes as well as development of intracranial hypertension. HS was then accomplished by withdrawal of arterial blood over 10 min to a MAP of 50–55 mmHg, a level within the lower limit of cerebral autoregulation and maintained at that level for 60 min. Pigs were then resuscitated with lactated Ringer's (LR) solution to baseline MAP with fluid rate adjusted to maintain a low or high CVP (<6 mmHg or ≥6 mmHg respectively). Catheterization of the cerebral ventricle to monitor intracranial pressure demonstrated that resuscitation of hemorrhagic shock to a high CVP in the setting of severe TBI significantly worsened intracranial hypertension (33 ± 2 mmHg vs. 24 ± 2 mmHg, $P<0.05$).

Later studies by Matsushita et al. [34] and Schültz et al. [35] applied the pressure-control hemorrhage + TBI model to rodents and obtained conflicting results, particularly in regard to histopathological outcomes. Whereas Matsushita demonstrated a doubling of lesion volume with mild hypotension (MAP 50–60 mmHg for 30 min after FPI), Schültz found no increase in lesion volume but significantly worse performance on cognitive testing such as Morris Water Maze.

A rat model of PBBI + HS was developed by Leung et al. [36]. Briefly, in the PBBI model an expandable probe is inserted 12 mm into the right frontal cortex and then inflated to 10% of the total rat brain volume. This results in severe brain injury with large lesion volume, extensive hemorrhage, and necrotic cell death. The tail artery can be cannulated to provide the addition of controlled HS. PBBI followed by 30-min pressure-controlled HS to 40–45 mmHg followed by resuscitation with LR solution exacerbated the reduction in both CBF and $PbtO_2$ and resulted in a downward shift in the lower limit of cerebral autoregulation compared with PBBI alone. In 2013, Hemerka et al. reported a pressure-control HS + CCI model in mice with a clinically relevant protocol that incorporates a shock phase (MAP 25–27 mmHg for 35 min), a 90 min prehospital phase of resuscitation with LR solution and a hospital phase where shed blood is re-infused [5]. In this model, CCI + HS animals had worse performance on cognitive testing, increased lesion volume, and exacerbated CA1 and CA3 histopathology relative to CCI alone. Although the model is quite labor intensive and technically challenging, it provides a platform for researchers to test interventions at various stages in a typical resuscitation scenario. For example, Shellington et al. [37] and Brockman et al. [38] used this model to test the application of PNPH in the prehospital setting and found that a 20 mL/kg bolus of 4% PNPH reversed shock for the duration of the 90 min prehospital phase, whereas animals resuscitated with LR required up to 180 mL/kg of volume and remained hypotensive (MAP <70 mmHg) for a significant period of time prior to re-infusion of shed blood. In addition, consistent with clinical data, Exo et al found no benefit in this model involving resuscitating with albumin rather than LR, which can possibly be related to rebound reverse osmolar effects as suggested in clinical studies of severe TBI in adults [39].

Blaisole et al. studied the effect of hyperoxic resuscitation (100% oxygen via nosecone vs. room air) in the prehospital phase and found reduced fluid requirements for resuscitation of shock and a 40% increase in CA3 neuronal survival despite a loss of antioxidant reserves [40]. Additionally, this and similar models have evaluated the pathophysiology of HS in the setting of TBI. Shein et al. evaluated the effect of HS on the inflammatory state after TBI and found a significant increase in the anti-inflammatory cytokine IL-10 [41]. Foley et al. adapted this complex model to be performed during magnetic resonance imaging (MRI) and used arterial spin labeling to show that HS exacerbated reduction in CBF in the ipsilateral and contralateral hemispheres [42].

1.8 TBI + Volume-Controlled HS

Protocols for experimental models of volume-controlled HS typically incorporate a standard incremental hemorrhage, often 30–40%, of circulating blood volume. An early report of the volume-control combined injury model was by Glass et al. who employed a porcine FPI model followed by a 40% hemorrhage over

30 min and a 60-min shock phase (MAP ~40–50 mmHg, CPP ~35–40 mmHg) [43]. Pigs were then resuscitated with shed-blood and LR to baseline cardiovascular parameters. The experiment was designed to test the hypothesis that hypercarbia would improve matching of CBF and metabolism. As expected, the researchers found that HS increased cerebral extraction of oxygen and glucose. However, in hypercarbic animals the extraction fraction was reduced and neuropathology and behavioral outcome scores were also improved in comparison to normocarbic animals. Additional studies in porcine models of FPI + volume-controlled HS were performed by Fritz et al. at two injury thresholds: moderate TBI (3.2 ± 0.6 atm) or severe TBI (4.1 ± 1.0 atm) to model the intracranial hypertension seen in the majority of cases with severe TBI [44].

Blood was withdrawn to produce a 30 % hemorrhage from 3 to 18 min after FPI (MAP ~65–70), followed by a 14-min shock phase after which the shed blood was re-infused. Animals exposed to severe FPI + HS had an immediate rise in their ICP (47 ± 14 mmHg) followed by a second peak at 5 ± 1.5 h post-injury. This was associated with a 50 % decrease in CBF and cerebral metabolic rate ($CMRO_2$) up to 24 h post-injury. In this model, there is significant axonal injury the ipsilateral hemisphere, contralateral hemisphere, and brainstem, as determined by β-amyloid precursor protein staining.

Resuscitation strategies using volume expansion by crystalloid versus colloid intravenous fluids were investigated in modified versions of the porcine model of TBI + volume-control HS by Gibson et al. [32] and Jin et al. [45] In the study by Gibson, FPI (6–8 atm) was followed by a 30 % hemorrhage via an arterial catheter over a period of 50 min (MAP ~40–60 mmHg). Two hours after FPI, animals were randomized to receive either shed blood, an equivalent volume of fresh frozen plasma, or saline bolus at 3× the volume of shed blood. Results showed that animals in the saline group required significantly more fluid resuscitation than animals that received plasma or blood ($P < 0.001$). In addition, animals resuscitated with shed blood had lower ICP, higher CPP, and a trend towards decreased mortality (0/7 vs. 5/13; $P = 0.058$). Jin et al. developed a CCI model with a 15 mm tip, 4 m/s velocity, 100 ms dwell, and 12 mm depth. Animals then had 40 % controlled hemorrhage (MAP 30–35 mmHg) over 30 min followed by a 2-h period of shock and were then randomized to resuscitation with saline at 3× volume of shed blood, 6 % hetastarch, or FFP at 1× volume shed blood. Resuscitation with colloid (hetastarch or FFP) decreased total fluid requirements and reduced brain edema and in the case of FFP, also reduced lesion volume.

Dennis et al. developed a CCI + volume-control HS model in adult C57BL6J male mice to evaluate the effect of TBI + HS on selectively vulnerable regions of the hippocampus—CA1 region typically sensitive to hypoxia or ischemia and CA3 typically vulnerable in TBI as shown in Fig. 3 [4]. In this model the severity of CCI and hemorrhage were selected based on prior and preliminary studies showing no significant neurologic injury when applied independent of each other. The CCI was performed over the left parietal cortex with a 3 mm tip impacting at 5 m/s and 1 mm of depth. Afterwards, controlled hemorrhage of 27 mL/kg in HS alone and 20 mL/kg in CCI + HS (MAP ~35–40 mmHg) caused shock for 60 or 90 min. This was followed by a 30-min prehospital phase with 6 % hetastarch resuscitation to a MAP ≥50 mmHg and then by return of shed-blood, 100 % oxygen and 6 % hetastarch targeting MAP of ≥60 mmHg.

Results of Fluoro-Jade C staining at 24 h and neuronal counts at 7 days demonstrated an increase in CA1, but not CA3, hippocampal neuronal death in the HS + CCI group. Exo et al. modified this model to assess the role of colloids in the resuscitation of trauma patients with TBI [39]. Mice were randomly assigned to either treatment with polynitroxylated albumin, 6 % hetastarch, LR, or 3 % hypertonic saline during the prehospital phase or resuscitation. Mice receiving colloids (hetastarch or polynitroxylated albumin) had higher prehospital MAP and lower fluid requirements. Hippocampal neuronal death assessed 7 days post-injury did not differ between treatment groups.

2 Materials

2.1 Material Used for TBI + HS in Mice

1. Adult C57BL/6 Mice, 12–15 weeks old, 25–30 g (Jackson Laboratories, Bar Harbor, ME).
2. Stereotaxic frame (David Kopf Instruments, Tujunga, CA) (*see* Fig. 1).
3. Controlled cortical impact device with a 3 mm flat tip impounder (Bimba, Monee, IL).
4. Operating microscope (Carl's Zeiss, Dublin, CA).
5. Anesthesia system and vaporizer.
6. Compressed air dental drill.
7. Infusion pump (Harvard Apparatus, Holliston, MA).
8. Analog volt meter.
9. Electric heating blanket and heating lamp.
10. Data acquisition software (Polyview 16; Grass Technologies, Middleton, WI).
11. Blood gas analyzer (Stat Profile; Nova Biomedical, Waltham, MA).
12. Rectal temperature probe (Physitemp, Clifton, NJ).
13. Small sterile drape.

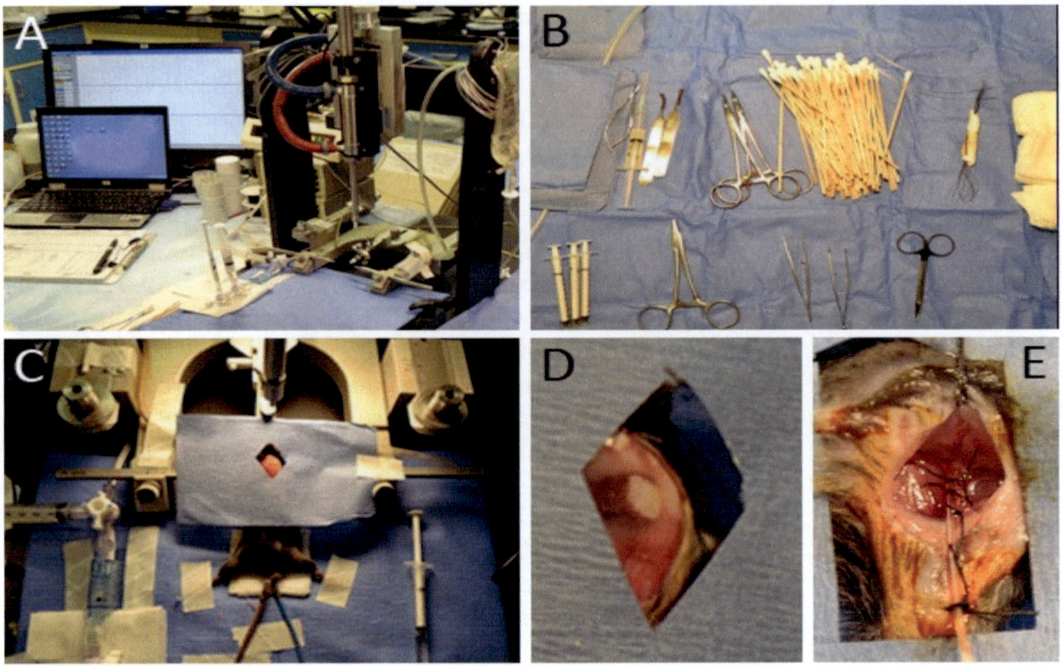

Fig. 1 TBI + HS model in mice. (**a**) Surgical station utilizing a CCI device to produce TBI. (**b**) Aseptic surgical instruments and supplies. (**c**) Stereotactic frame and surgical field for 5 mm craniotomy performed over left parietal region. (**d**) Skull immediately after CCI with bone flap replaced and secured with dental cement. (**e**) Vascular access required to produce HS. Skin is retracted medially by suture. Arterial cannula (*right*) and venous cannula (*left*) are secured with 6-0 sutures. Use of photographs approved by the institutional animal care and use committee

14. Nose cone and scavenger system.
15. Surgical kit (World Precision Instruments or Roboz Surgical Instrument Co.) (*see* Fig. 1b): Microvascular scissors, Dumont #5/45 forceps, Hemostats, Needle driver.
16. Intramedic polyethylene (PE)-50/PE-10 catheter (BD Medical, Frankline Lake, NJ) (*see* **Note 1**).
17. Cotton tipped swabs, Gauze sponges, Betadine, 6-0 silk suture, 4-0 silk suture.
18. Heparinized saline (1 unit/mL), Sodium citrate (2.2 g/100 mL).
19. Lactated Ringer's solution, 6 % hetastarch (Hextend; Hospira, Lake Forest, IL).
20. Dental cement.
21. Isoflurane, nitrous oxide, oxygen.
22. 1 % Lidocaine, 0.25 % sensorcaine.
23. (Optional) 1 F intracranial pressure monitor (Millar Instruments, Houston, TX).

24. (Optional) Brain tissue oxygen monitor.
25. (Optional) Brain temperature probe (Physitemp, Clifton, NJ).

3 Methods

3.1 Methods for TBI + Pressure-Controlled HS in Mice

1. Anesthetize the mouse in the anesthetic chamber with 2:1 N_2O:O_2 mix plus 4% isoflurane.
2. Weigh animal. Then, place supine with legs abducted at 45° with left leg towards surgeon. Reduce isoflurane to 2% via nose cone.
3. Shave left inguinal area and prep with Betadine. Cover with small drape. Under aseptic conditions perform a cut down followed by vessel dissection, separation, and vascular access. Arterial and venous catheters are placed and secured with 6-0 suture (see **Note 2**) (*see* Fig. 1).
4. Place mouse in stereotaxic frame in preparation for CCI (*see* **Note 3**). Connect arterial catheter to pressure transducer to record blood pressure and heart rate. Place rectal temperature probe and secure to table. Maintain core temperature at 38.0° ± 0.5 °C. Obtain baseline arterial blood gas and vital signs. MAP, heart rate, core temperature, brain temperature (if used), intracranial pressure (if used), and brain tissue oxygen tension (if used) are recorded every 5 min throughout the study.
5. Shave head and prep with Betadine. Under aseptic conditions, make midline incision to expose the skull. Using operating microscope and dental drill, perform a 5 mm craniotomy over left parietal cortex by following suture lines. In some studies a 1 mm burr hole is made over the right frontal lobe and a pressure monitor, temperature probe, or oxygen sensor is inserted.
6. Using a voltmeter, zero the piston tip to the dura of the exposed brain. Insert a needle electrode into the temporalis muscle ipsilateral to injury and maintain at 37° ± 0.5°. Switch to 100% medical grade room air with ~1–2% isoflurane and allow 10 min equilibration.
7. *CCI Injury*: Trigger CCI device. Typical injury level is: 5 m/s, 1.0 mm depth of deformation (*see* **Note 4**). Replace bone flap and seal with dental cement (*Refer to Fig. 2 for protocol phases*).
8. *HS Phase* (*35 min*): Maintain room air + 1% isoflurane inhalation. 5 min after CCI, withdraw an initial 2.3 mL per 100 g body weight of blood via femoral vein catheter into 1 mL tuberculin syringe containing sodium citrate (8:1 ratio of blood to citrate) over 15 min. Then, rigorously pressure control to 25–27 mmHg by removal or re-infusion of small

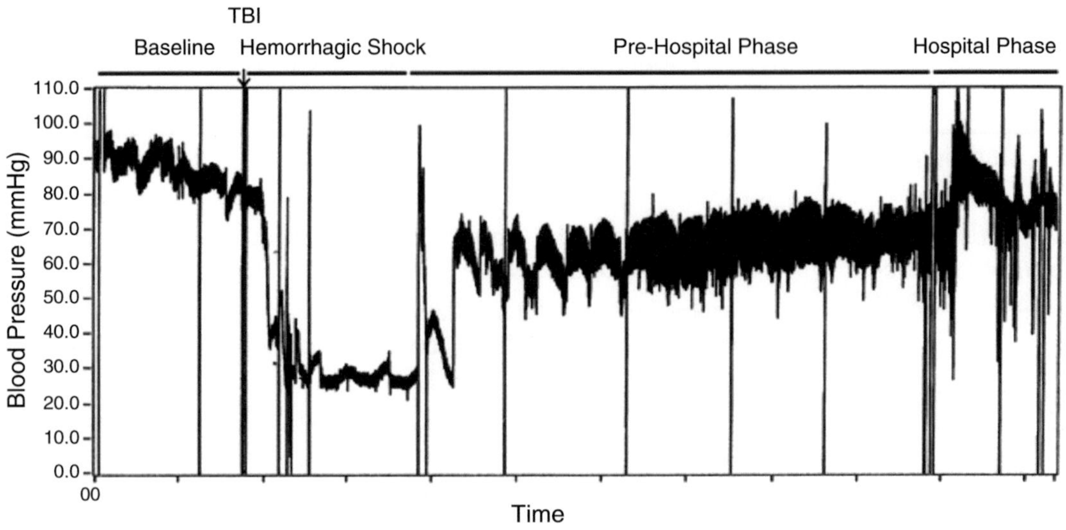

Fig. 2 Representative MAP scalar for CCI + pressure-controlled HS. The three-phase model includes a 35-min shock phase, a 90-min prehospital phase, and a 15-min hospital phase

aliquots of autologous citrated blood through the venous cannula for the remaining 20 min. A blood gas sample is taken at 30 min into the shock phase.

9. *Prehospital resuscitation phase* (*90 min*): Continue room air + 1 % isoflurane inhalation via nose cone (*see* **Note 5**). Bolus with 20 mL/kg of lactated Ringer's solution. Subsequently, every 5 min if mean arterial pressure is below the target goal of 70 mmHg, give an additional 10 mL/kg bolus (see **Note 6**). Typically, animals require a bolus every 5 min during this phase. At 85 min a blood gas sample is taken and recorded. (*see* **Note** 7).
10. *Hospital phase* (*15 min*): Switch room air to 100 % O_2 and continue isoflurane at 1–2 %. Re-infuse shed blood over 5 min via the femoral vein catheter. Obtain blood gas at 15 min.
11. Suture scalp incision with 4-0 silk suture and apply 0.25 % Sensorcaine to the wound. Turn the animal supine and remove femoral catheters. Close skin incision with 6-0 silk suture and apply 0.25 % Sensorcaine to wound site.
12. Place animal in temperature-controlled recovery chamber with 100 % O_2 for 30 min. Finally, return animal to its original housing cage.

3.2 Methods for TBI + Volume-Controlled HS in Mice

1. Please refer to Subheading 3.1, protocol **steps 1**–7, detailing urgical preparation and CCI.
2. *HS Phase* (*90 min*): Maintain room air + 1 % isoflurane inhalation 5 min after CCI, withdraw an initial 2 mL per 100 g body weight of blood via femoral vein catheter into 1 mL tuberculin

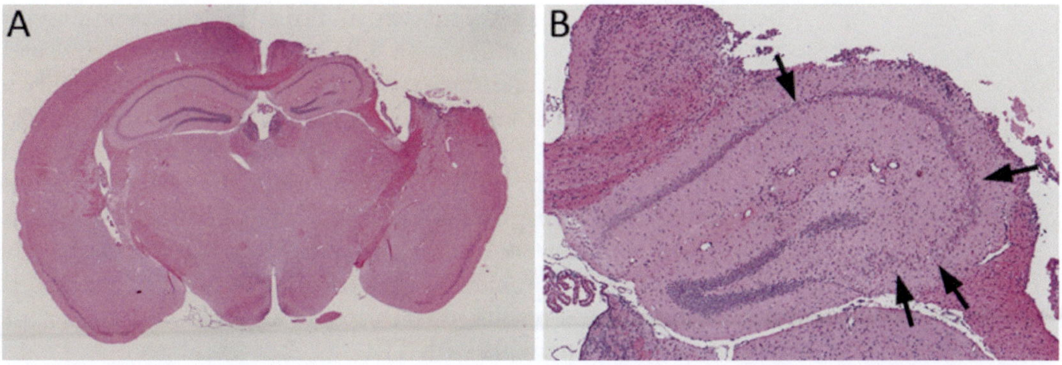

Fig. 3 Representative images from a coronal section through the dorsal hippocampus 7 days after CCI + volume controlled HS. Sections were stained with hematoxylin and eosin (H&E). (**a**) 1× view demonstrating characteristic hemorrhage and focal, full-thickness loss of cortex overlying the dorsal hippocampus. (**b**) 10× view of the ipsilateral hippocampus demonstrating eosinophilic neurodegeneration in CA1, CA2, and CA3 regions (*arrows*)

syringe containing sodium citrate (8:1 ratio of blood to citrate) over 15 min (*see* **Note 8**). Half of the blood is removed in first 5 min, with the remaining 50 % removed over the next 10 min. Continue shock phase for 90 min (*see* **Note 8**).

3. *Prehospital resuscitation phase* (*30 min*): Continue room air + 1 % isoflurane inhalation via nose cone. Bolus with 0.1 mL aliquots of 6 % hetastarch to rapidly raise MAP ≥50 mmHg.
4. *Hospital phase* (*15 min*): Switch room air to 100 % O_2 and continue isoflurane at 1 %. Re-infuse shed blood over 5 min via the femoral vein catheter. Maintain MAP ≥60 mmHg with additional 0.1 mL boluses of 6 % hetastarch. Obtain blood gas at 15 min.
5. *See* Subheading 3.1, protocol **steps 11** and **12**, for surgical recovery (Fig. 3).

3.3 Conclusions

Patients with TBI commonly present with secondary injuries that can increase mortality and worsen neurologic outcomes in survivors. Over the past 25 years, the development of large and small animal models combining TBI and secondary injury has demonstrated that the underlying pathophysiology (changes in CBF, CA1 neuronal death in addition to CA3 neuronal death, shift to anti-inflammatory cytokine profile) and response to therapy (aggressive fluid resuscitation, hyperoxia) may differ from isolated head trauma. Although these models require additional surgical specialization to perform, the investment in time and resources appears to be paying off in terms of knowledge gained. Therapies used in these models are expected to be tested in clinical trials in the near future. However, recent experience may give investigators and clinicians pause. Although resuscitation with colloids has shown beneficial effects in animal models of TBI + HS, a post-hoc review by Myburgh et al. of adults with TBI and shock resuscitated with saline or albumin in intensive care units found a higher mortality in the group treated with albumin (33.2 % vs. 20.4 % mortality; relative risk, 1.88; 95 % CI 1.31–2.70; $P < 0.001$) [46]. This difference may reflect limitations of the animal models, the post hoc analysis performed by Myburgh et al., or both [46]. Clearly, it is incumbent upon us to develop models that are reproducible and at the same time mimic the clinical condition as closely as possible to allow experimental results to translate, as hoped, to improvements in clinical care of patients with TBI.

4 Notes

1. The femoral arterial and venous catheters are made by attaching 5 cm of PE-10 tubing to 25 cm of PE-50 tubing. Using only PE-10 tubing will lead to clotting of the venous cannula. Bevel the insertion tip. Curve the venous cannula to prevent tension from developing in the line after insertion.
2. Arterial cannula should be inserted to a depth 3 mm beyond the visualized abdominal musculature. Venous cannula should be inserted to a depth 1 mm beyond abdominal musculature. This will allow consistent blood draw and blood pressure measurements. If unable to withdraw blood from the venous catheter, the arterial cannula can be used. Avoid all air bubbles in catheters and stopcocks, as they can lead to abrupt hypotension and death of the animal.
3. Mouse-appropriate ear bars should be used. Inappropriately sized ear bars may cause injury or distress to the animal.
4. In our experience, without addition of hemorrhagic shock, this level of CCI injury produces cortical injury without loss of hippocampal neurons [47, 48].
5. Room air is used during the pre-hospital resuscitation to model field resuscitation in combat care. During hospital resuscitation phase, oxygen is administered to model care received in a combat support hospital or emergency room.
6. Normal MAP in isoflurane-anesthetized adult mice is ~85 mmHg. The target value ≥70 mmHg was chosen to maintain CPP ≥ 60 mmHg given the ICP range we typically observe during HS phase in this model of 2–10 mmHg.
7. Drug therapies are usually given during the resuscitation phase either via bolus or slow infusion with syringe pump.
8. Removal of 2 mL/100 g blood over 15 min reduces MAP to 30–40 mmHg after CCI. In animals receiving HS only,

2.7 mL/100 g blood should be removed to achieve a similar degree of hypotension. Prior studies in a rodent model demonstrated absence of brain injury with this level of hemorrhagic shock [49]. After the 15-min blood withdrawal, MAP typically rises to ~45–55 mmHg transiently before settling to 30–40 mmHg for the remainder of the shock phase. A shock phase of 60 min demonstrated only rare neurodegeneration in the hippocampus at 24 h post-injury, therefore a 90 min shock phase is recommended.

Acknowledgements

NICHD T32 HD40686; U.S. Army grants W81XWH-10-1-0623 and WH81XWH-14-2-0018; NIH grant NS087978.

References

1. Chesnut RM, Marshall LF, Klauber MR, Blunt BA, Baldwin N, Eisenberg HM, Jane JA, Marmarou A, Foulkes MA (1993) The role of secondary brain injury in determining outcome from severe head injury. J Trauma 34:216–222
2. Pigula FA, Wald SL, Shackford SR, Vane DW (1993) The effect of hypotension and hypoxia on children with severe head injuries. J Pediatr Surg 28:310–314, discussion 315–316
3. Stewart TC, Alharfi IM, Fraser DD (2013) The role of serious concomitant injuries in the treatment and outcome of pediatric severe traumatic brain injury. J Trauma Acute Care Surg 75:836–842
4. Dennis AM, Haselkorn ML, Vagni VA, Garman RH, Janesko-Feldman K, Bayir H, Clark RS, Jenkins LW, Dixon CE, Kochanek PM (2009) Hemorrhagic shock after experimental traumatic brain injury in mice: effect on neuronal death. J Neurotrauma 26:889–899
5. Hemerka JN, Wu X, Dixon CE, Garman RH, Exo JL, Shellington DK, Blasiole B, Vagni VA, Janesko-Feldman K, Xu M, Wisniewski SR, Bayir H, Jenkins LW, Clark RS, Tisherman SA, Kochanek PM (2012) Severe brief pressure-controlled hemorrhagic shock after traumatic brain injury exacerbates functional deficits and long-term neuropathological damage in mice. J Neurotrauma 29:2192–2208
6. Saatman KE, Duhaime AC, Bullock R, Maas AI, Valadka A, Manley GT (2008) Classification of traumatic brain injury for targeted therapies. J Neurotrauma (Chicago, Ill: 1960) 25:719–738
7. Ramaiah VK, Sharma D, Ma L, Prathep S, Hoffman NG, Vavilala MS (2013) Admission oxygenation and ventilation parameters associated with discharge survival in severe pediatric traumatic brain injury. Child's Nerv Syst 29:629–634
8. Aji AA, Zwane E, Thompson C, Fieggen AG, Argent AC, Le Roux PD, Peter JC (2009) Brain tissue oxygen tension monitoring in pediatric severe traumatic brain injury. Part 1: relationship with outcome. Child's Nerv Syst 25:1325–1333
9. Ishige N, Pitts LH, Berry I, Carlson SG, Nishimura MC, Moseley ME, Weinstein PR (1987) The effect of hypoxia on traumatic head injury in rats: alterations in neurologic function, brain edema, and cerebral blood flow. J Cereb Blood Flow Metab 7:759–767
10. Ishige N, Pitts LH, Pogliani L, Hashimoto T, Nishimura MC, Bartkowski HM, James TL (1987) Effect of hypoxia on traumatic brain injury in rats: part 2. Changes in high energy phosphate metabolism. Neurosurgery 20:854–858
11. Ishige N, Pitts LH, Hashimoto T, Nishimura MC, Bartkowski HM (1987) Effect of hypoxia on traumatic brain injury in rats: part 1. Changes in neurological function, electroencephalograms, and histopathology. Neurosurgery 20:848–853
12. Clark RS, Kochanek PM, Dixon CE, Chen M, Marion DW, Heineman S, DeKosky ST, Graham SH (1997) Early neuropathologic effects of mild or moderate hypoxemia after controlled cortical impact injury in rats. J Neurotrauma 14:179–189
13. Jenkins LW, Moszynski K, Lyeth BG, Lewelt W, DeWitt DS, Allen A, Dixon CE, Povlishock

JT, Majewski TJ, Clifton GL et al (1989) Increased vulnerability of the mildly traumatized rat brain to cerebral ischemia: the use of controlled secondary ischemia as a research tool to identify common or different mechanisms contributing to mechanical and ischemic brain injury. Brain Res 477:211–224

14. Bramlett HM, Green EJ, Dietrich WD (1999) Exacerbation of cortical and hippocampal CA1 damage due to posttraumatic hypoxia following moderate fluid-percussion brain injury in rats. J Neurosurg 91:653–659
15. Mikrogianakis A, Shaye RE, Griffin P, Kawesa S, Lockwood J, Gendron NH, Gaboury I, Merali Z, Mackenzie AE, Hutchison JS (2007) Hypoxia alters the expression of inhibitor of apoptosis proteins after brain trauma in the mouse. J Neurotrauma 24:338–353
16. Marmarou A, Foda MA, van den Brink W, Campbell J, Kita H, Demetriadou K (1994) A new model of diffuse brain injury in rats. Part I: pathophysiology And Biomechanics. J Neurosurg 80:291–300
17. Hellewell SC, Yan EB, Agyapomaa DA, Bye N, Morganti-Kossmann MC (2010) Post-traumatic hypoxia exacerbates brain tissue damage: analysis of axonal injury and glial responses. J Neurotrauma 27:1997–2010
18. Yan EB, Hellewell SC, Bellander BM, Agyapomaa DA, Morganti-Kossmann MC (2011) Post-traumatic hypoxia exacerbates neurological deficit, neuroinflammation and cerebral metabolism in rats with diffuse traumatic brain injury. J Neuroinflammation 8:147
19. Goodman MD, Makley AT, Huber NL, Clarke CN, Friend LA, Schuster RM, Bailey SR, Barnes SL, Dorlac WC, Johannigman JA, Lentsch AB, Pritts TA (2011) Hypobaric hypoxia exacerbates the neuroinflammatory response to traumatic brain injury. J Surg Res 165:30–37
20. Matsushita Y, Bramlett HM, Alonso O, Dietrich WD (2001) Posttraumatic hypothermia is neuroprotective in a model of traumatic brain injury complicated by a secondary hypoxic insult. Crit Care Med 29:2060–2066
21. Abayomi OK (1996) Pathogenesis of irradiation-induced cognitive dysfunction. Acta Oncol (Stockholm, Sweden) 35:659–663
22. Allen AR, Eilertson K, Chakraborti A, Sharma S, Baure J, Habdank-Kolaczkowski J, Allen B, Rosi S, Raber J, Fike JR (2014) Radiation exposure prior to traumatic brain injury induces responses that differ as a function of animal age. Int J Radiat Biol 90:214–223
23. Allen AR, Eilertson K, Sharma S, Schneider D, Baure J, Allen B, Rosi S, Raber J, Fike JR (2013) Effects of radiation combined injury on hippocampal function are modulated in mice deficient in chemokine receptor 2 (CCR2). Radiat Res 180:78–88
24. Hwabejire JO, Imam AM, Jin G, Liu B, Li Y, Sillesen M, Jepsen CH, Lu J, deMoya MA, Alam HB (2013) Differential effects of fresh frozen plasma and normal saline on secondary brain damage in a large animal model of polytrauma, hemorrhage and traumatic brain injury. J Trauma Acute Care Surg 75:968–974, discussion 974–965
25. Probst C, Mirzayan MJ, Mommsen P, Zeckey C, Tegeder T, Geerken L, Maegele M, Samii A, van Griensven M (2012) Systemic inflammatory effects of traumatic brain injury, femur fracture, and shock: an experimental murine polytrauma model. Mediators Inflamm 2012:136020
26. Mirzayan MJ, Probst C, Samii M, Krettek C, Gharabaghi A, Pape HC, van Griensven M, Samii A (2012) Histopathological features of the brain, liver, kidney and spleen following an innovative polytrauma model of the mouse. Exp Toxicol Pathol 64:133–139
27. Yuan XQ, Wade CE, Clifford CB (1991) Suppression by traumatic brain injury of spontaneous hemodynamic recovery from hemorrhagic shock in rats. J Neurosurg 75:408–414
28. Mahoney EJ, Biffl WL, Harrington DT, Cioffi WG (2003) Isolated brain injury as a cause of hypotension in the blunt trauma patient. J Trauma 55:1065–1069
29. Nelson TJ, Wall DB, Stedje-Larsen ET, Clark RT, Chambers LW, Bohman HR (2006) Predictors of mortality in close proximity blast injuries during Operation Iraqi Freedom. J Am Coll Surg 202:418–422
30. Gilles EE, Nelson MD Jr (1998) Cerebral complications of nonaccidental head injury in childhood. Pediatr Neurol 19:119–128
31. Stern SA, Zink BJ, Mertz M, Wang X, Dronen SC (2000) Effect of initially limited resuscitation in a combined model of fluid-percussion brain injury and severe uncontrolled hemorrhagic shock. J Neurosurg 93:305–314
32. White NJ, Wang X, Bradbury N, Moon-Massat PF, Freilich D, Auker C, McCarron R, Scultetus A, Stern SA (2013) Fluid resuscitation of uncontrolled hemorrhage using a hemoglobin-based oxygen carrier: effect of traumatic brain injury. Shock (Augusta, Ga) 39:210–219
33. Hariri RJ, Firlick AD, Shepard SR, Cohen DS, Barie PS, Emery JM III, Ghajar JB (1993) Traumatic brain injury, hemorrhagic shock, and fluid resuscitation: effects on intracranial pressure and brain compliance. J Neurosurg 79:421–427
34. Matsushita Y, Bramlett HM, Kuluz JW, Alonso O, Dietrich WD (2001) Delayed hemorrhagic hypotension exacerbates the hemodynamic and

histopathologic consequences of traumatic brain injury in rats. J Cereb Blood Flow Metab 21:847–856
35. Schutz C, Stover JF, Thompson HJ, Hoover RC, Morales DM, Schouten JW, McMillan A, Soltesz K, Motta M, Spangler Z, Neugebauer E, McIntosh TK (2006) Acute, transient hemorrhagic hypotension does not aggravate structural damage or neurologic motor deficits but delays the long-term cognitive recovery following mild to moderate traumatic brain injury. Crit Care Med 34:492–501
36. Leung LY, Wei G, Shear DA, Tortella FC (2013) The acute effects of hemorrhagic shock on cerebral blood flow, brain tissue oxygen tension, and spreading depolarization following penetrating ballistic-like brain injury. J Neurotrauma 30:1288–1298
37. Shellington DK, Du L, Wu X, Exo J, Vagni V, Ma L, Janesko-Feldman K, Clark RS, Bayir H, Dixon CE, Jenkins LW, Hsia CJ, Kochanek PM (2011) Polynitroxylated pegylated hemoglobin: a novel neuroprotective hemoglobin for acute volume-limited fluid resuscitation after combined traumatic brain injury and hemorrhagic hypotension in mice. Crit Care Med 39:494–505
38. Brockman EC, Bayir H, Blasiole B, Shein SL, Fink EL, Dixon C, Clark RS, Vagni VA, Ma L, Hsia CJ, Tisherman SA, Kochanek PM (2013) Polynitroxylated-pegylated hemoglobin attenuates fluid requirements and brain edema in combined traumatic brain injury plus hemorrhagic shock in mice. J Cereb Blood Flow Metab 33:1457–1464
39. Exo JL, Shellington DK, Bayir H, Vagni VA, Janesco-Feldman K, Ma L, Hsia CJ, Clark RS, Jenkins LW, Dixon CE, Kochanek PM (2009) Resuscitation of traumatic brain injury and hemorrhagic shock with polynitroxylated albumin, hextend, hypertonic saline, and lactated Ringer's: effects on acute hemodynamics, survival, and neuronal death in mice. J Neurotrauma 26:2403–2408
40. Blasiole B, Bayr H, Vagni VA, Janesko-Feldman K, Cheikhi A, Wisniewski SR, Long JB, Atkins J, Kagan V, Kochanek PM (2013) Effect of hyperoxia on resuscitation of experimental combined traumatic brain injury and hemorrhagic shock in mice. Anesthesiology 118:649–663
41. Shein S, Shellington DK, Exo J, Jackson TC, Wisniewski SR, Jackson E, Vagni VA, Bayir H, Clark R, Dixon CE, Janesko KL, Kochanek PM (2014) Hemorrhagic shock shifts the serum cytokine profile from pro-to anti-inflammatory after experimental traumatic brain injury in mice. J Neurotrauma 31:1386–1395
42. Foley LM, Iqbal O'Meara AM, Wisniewski SR, Hitchens TK, Melick JA, Ho C, Jenkins LW, Kochanek PM (2013) MRI assessment of cerebral blood flow after experimental traumatic brain injury combined with hemorrhagic shock in mice. J Cereb Blood Flow Metab 33:129–136
43. Glass TF, Fabian MJ, Schweitzer JB, Weinberg JA, Proctor KG (2001) The impact of hypercarbia on the evolution of brain injury in a porcine model of traumatic brain injury and systemic hemorrhage. J Neurotrauma 18:57–71
44. Fritz HG, Walter B, Holzmayr M, Brodhun M, Patt S, Bauer R (2005) A pig model with secondary increase of intracranial pressure after severe traumatic brain injury and temporary blood loss. J Neurotrauma 22:807–821
45. Jin G, DeMoya MA, Duggan M, Knightly T, Mejaddam AY, Hwabejire J, Lu J, Smith WM, Kasotakis G, Velmahos GC, Socrate S, Alam HB (2012) Traumatic brain injury and hemorrhagic shock: evaluation of different resuscitation strategies in a large animal model of combined insults. Shock (Augusta, Ga) 38:49–56
46. Myburgh J, Cooper DJ, Finfer S, Bellomo R, Norton R, Bishop N, Kai Lo S, Vallance S (2007) Saline or albumin for fluid resuscitation in patients with traumatic brain injury. N Engl J Med 357:874–884
47. Kochanek PM, Vagni VA, Janesko KL, Washington CB, Crumrine PK, Garman RH, Jenkins LW, Clark RS, Homanics GE, Dixon CE, Schnermann J, Jackson EK (2006) Adenosine A1 receptor knockout mice develop lethal status epilepticus after experimental traumatic brain injury. J Cereb Blood Flow Metab 26:565–575
48. Foley LM, Hitchens TK, Ho C, Janesko-Feldman KL, Melick JA, Bayir H, Kochanek PM (2009) Magnetic resonance imaging assessment of macrophage accumulation in mouse brain after experimental traumatic brain injury. J Neurotrauma 26:1509–1519
49. Carrillo P, Takasu A, Safar P, Tisherman S, Stezoski SW, Stolz G, Dixon CE, Radovsky A (1998) Prolonged severe hemorrhagic shock and resuscitation in rats does not cause subtle brain damage. J Trauma 45:239–248, discussion 248–249

Chapter 23

Microdialysis as Clinical Evaluation of Therapeutic Hypothermia in Rat Subdural Hematoma Model

Shoji Yokobori, Markus S. Spurlock, Stephanie W. Lee, Shyam Gajavelli, and Ross M. Bullock

Abstract

Cerebral microdialysis (MD) is a fine laboratory technique which has been established for studying physiological, pharmacological, and pathological changes in the experimental studies of traumatic brain injury (TBI). This technique has also been well translated and widely applied to clinical bedside monitoring to provide pathophysiological analysis in severe TBI patients. The MD technique is thus well suited for straightforward translation from basic science to clinical application.

In this chapter, we describe our evaluation of MD method in acute subdural hematoma (ASDH) rat model. With 100 kDa cut-off microdialysis membrane, we could measure several biomarkers such as ubiquitin carboxy hydrolase L1 (UCH-L1), a neuronal marker and glial fibrillary acidic protein (GFAP), and a glial marker in extracellular fluid. In this experiment, we could detect that the peak of extracellular UCH-L1 in the early hypothermia group was significantly lower than in the normothermia group. Also, in the late phase of reperfusion (>2.5 h after decompression), extracellular GFAP in the early hypothermia group was lower than in the normothermia. These data thus suggested that early, preoperatively induced hypothermia could mediate the reduction of neuronal and glial damage in the reperfusion phase of ischemia/reperfusion brain injury.

Microdialysis allows for the direct measurement of extracellular molecules in an attempt to characterize metabolic derangements before they become clinically relevant. Advancements in technology have allowed for the bedside assay of multiple markers of ischemia and metabolic dysfunction, and the applications for traumatic brain injury have been well established. As clinicians become more comfortable with these tools their widespread use and potential for clinical impact with continue to rise.

Key words Microdialysis, Traumatic brain injury, Pathophysiology, Experimental model, Therapeutic hypothermia, Subdural hematoma

1 Introduction

1.1 Microdialysis Technique

Microdialysis (MD) is a technique to measure extracellular substances including amino acids and proteins in living tissue. Application of MD enables us to measure eWndogenous substances, primarily the neurotransmitters, and the infusion of drugs through the microdialysis cannula.

Firas Kobeissy et al. (eds.), *Injury Models of the Central Nervous System: Methods and Protocols*, Methods in Molecular Biology, vol. 1462, DOI 10.1007/978-1-4939-3816-2_23, © Springer Science+Business Media New York 2016

The use of membranous implants in the brain was first described in 1966 by Bito et al. [1]. Tossman and Ungerstedt also published the first data on neurotransmitter concentrations in the rat brain in 1986 [2]. Thereafter, the use of cerebral MD gained popularity in the early 1990s when commercially produced MD catheters and dialysate analyzer (CMA Microdialysis, Solna, Sweden) became available. The use of microdialysis in the neurological clinic also has become more common as MD is approved by US FDA for neuromonitoring. There are also over 250 publications attesting to its safety and utility in conditions producing acute brain damage such as severe traumatic brain injury, subarachnoid hemorrhage, and stroke [3]. With the progress of this biochemical technique in basic research and clinical application, microdialysis is now becoming a very powerful tool for translational TBI study.

1.2 Principle of Cerebral Microdialysis

MD is a technique used to monitor the chemistry of the extracellular space in living tissue, giving us a preview of what goes on in tissues, before chemical events can be reflected as changes in systemic blood levels. The MD probe is designed to mimic a blood capillary (Fig. 1). Following implantation into a tissue, a physiological salt solution (or artificial cerebrospinal fluid) is slowly pumped through the MD probe (*see* **Note 1**). In the surrounding area of the membrane, this dialysate equilibrates with the surrounding extracellular fluid. The microdialysate solution retrieved will contain a representative proportion of the tissue fluids molecules. It can then be analyzed for compounds that may have been present in this tissue compartment. It is important to recognize that there is an exchange of molecules in both directions

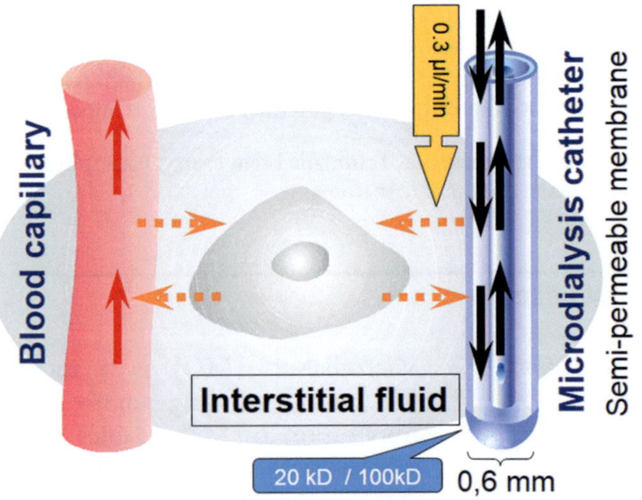

Fig. 1 Schema of principle of microdialysis (MD). A microdialysis catheter functions the same way as the capillary. With the flow of microdialysate, we can also collect some substances from interstitial fluid. The tip of microdialysis catheter has the semipermeable membrane and now two different pore sizes (cutoff size) as 20 and 100 kDa are now available

and that the difference in concentration through the membrane governs the direction of the gradient.

The absolute recovery (mol/unit time) of a substance from the tissue depends on several factors (Table 1). Other factors such as pH of the medium and degradation of the substance may also affect the recovery.

Currently, several probes with different cutoff size and length of membrane are commercially available for experimental study (*see* **Note 2**). Advances in high-performance liquid chromatography (HPLC) techniques have also made it possible to quantify changes in neurochemistry as a function of time [4], although it is still the sensitivity of the analytical method that limits the precision of microdialysis. In principle, it is possible to analyze almost every known small neurotransmitter and metabolite including the amino acids, catecholamines, acetylcholine (ACh), and neuropeptides using HPLC with electrochemical, fluorescence, ultraviolet, and conductance detectors, sometimes in combination with enzyme reactors [5–7].

When considering monitoring the endogenous substances within the brain, an important factor is mechanical damage. For example, lowered oxygenation around probe sites may cause changes in the surrounding parenchyma. Benveniste and Diemer [8] revealed that 2 days after implantation of several microdialysis probes into the rat hippocampus, the tissue immediately adjacent to the probe (within a 50 μm zone) exhibited edema, minor hemorrhages and accumulation of polymorphonuclear leukocytes. These tissue reactions are classical and unavoidable and it is imperative that the researcher is aware of these limitations, especially in chronic microdialysis studies.

1.3 Microdialysis Technique: Step by Step

The following points should be considered before starting the MD experiment. In general, it is advisable to plan a pilot study first, which allows optimizing the procedure and validation of the sensitivity of the sample analysis.

Table 1
Factors affecting recovery of substances in vitro

Length and diameter of the membrane (larger area–higher recovery)
Flow rate
Temperature
Molecular weight of the substance
Molecular shape of the substance
Molecular charge
Binding to the membrane and tubing

1. *Estimate the Properties of the Probe Membrane*

   Before measuring endogenous substances, it is always advisable to perform an in vitro experiment, to establish the dialyzing properties of the microdialysis probe for the particular substance of interest. A membrane with a low molecular weight cutoff purifies your sample by excluding large molecules, while a high molecular cutoff recovers some larger substances, such as peptides or proteins.

2. *Length of the Membrane*

   A longer membrane yields a better recovery of the substances of interest; however, the choice may be limited by the size of the structure you want to study.

3. *Perfusion Flow*

   Use a high flow if you want to remove or introduce as many molecules as possible per unit time or a low flow if you want to obtain a more concentrated dialysate. It is worth considering that a high flow is liable to disturb the physiology simply because more substances are removed.

4. *Composition of the Perfusion Fluid*

   Ideally, it should be as close as possible to the composition of the extracellular fluid (*see* **Note 3**). However, you may want to change the concentration of sodium, potassium, or calcium in order to influence the cell membrane function in the region you are studying.

5. *Type of Probe*

   A stiff probe is suitable for a stereotaxic experiment on the brain while a flexible probe may be better suited for dialysis in a peripheral organ such as adipose tissue, muscle, liver, or kidney. A brain probe may require a pre-implanted guide cannula while a subcutaneous probe may be implanted an hour or so before the start of the experiment (*see* **Note 4**).

6. *Time Needed To Obtain Steady State Conditions*

   The introduction of a probe into the tissue will always cause damage and the recovery of function will take a certain period of time. An hour or two is often used to reach "baseline conditions."

7. *Animal Awakes or Under Anesthesia Conditions?*

   Using awake animals does not necessarily mean that the conditions are more "normal." An awake animal is subject to pain and stress that may influence the results as much as the anaesthesia.

8. *Design of a Control Experiment*

   This is certainly one of the most important parts of any experimental design. One may have difficulty in determining the influence of a great number of known or unknown variables in your experiment; however, a well-designed control experiment will take care of many of these problems.

9. *Dose–Response Experiments*

Microdialysis is a wonderful technique for studying drug actions. The ease by which one can follow the time course of local drug concentrations in tissue and drug effects on local physiology is one of the really strong points of the technique.

However, it is surprising how few publications include a dose–response study, especially as we know that the qualitative action of a drug often changes as the dose changes.

10. *Sample Volume Required for Analysis*

The outlet dead volume (swivel and outlet tubing) defines the time gap between the actual event (drug injection, behavioral test, environment changes) and the moment of fraction collection. This time should be compensated with a respective delay in sample collection. An autosampler is optional but desirable. It minimizes disturbances of the animals by the researcher during sample collection. Once harvested, samples should be stored at −80 °C and analyzed as soon as possible (*see* **Note 5**).

Is a small sample volume and a high concentration (e.g., HPLC) or a large sample volume and a high amount of the particular compound (e.g., RIA—radio immuno assay) required? You may want to choose a low or a high perfusion flow, respectively.

11. *Temporal Resolution Needed the Experiment*

Frequent sampling usually means higher perfusion flow in order to get enough sample volume for the analysis.

12. *Instrument Setup*

For example, do you need to change the perfusion fluid during the experiment in order to introduce a drug or change the ionic composition of the fluid? In that case you may need a liquid switch or a pump with syringes that can be individually controlled.

1.4 Microdialysis in Acute Subdural Hematoma Animal Model for Clinical Translation of Hypothermia Therapy

One therapeutic method, posttraumatic hypothermia, has been shown in previous studies to improve histopathological and behavioral consequences of TBI using many experimental models [9–15], and different delivery paradigms. Unfortunately however, therapeutic hypothermia has not shown overall efficacy in multicenter trials probably due, at least in part, to the heterogeneous nature of the brain damage mechanisms in TBI patients [11, 16–21]. In a recent large severe TBI clinical trial [22], post hoc subgroup analysis demonstrated the possibility that early induced hypothermia might have a specific beneficial effect in the patients with acute subdural hematomas (ASDH) undergoing decompressive craniotomy. This patient subgroup with ASDH demonstrate the worst outcome of any category of severe TBI, and 60 % will die or remain severity disabled [23–25].

Based upon this data from the Clifton study, we hypothesized that hypothermia may improve outcome in acute intracranial hematomas via blunting the effects of ischemic/reperfusional (I/R) injury associated with decompressive craniotomy. To date, no well-controlled preclinical studies have been undertaken to test the efficacy of early cooling in an acute hematoma model of severe TBI. We therefore used the well-characterized rat ASDH model, to evaluate the efficacy of temperature management in reducing brain damage after ASDH.

We measured the effect of early and delayed moderate hypothermia (33 °C) with the extracellular measurement of a new neuronal biomarker: ubiquitin carboxyl-terminal hydrolase-L1 (UCH-L1) and a glial marker: glial fibrillary acidic protein (GFAP) to determine the amount of cell death, and the effect of the hypothermic therapy, in this model. We measured these two recently discovered biomarkers using 100 kDa cut-off MD probes in the rat ASDH model.

UCH-L1 has been proposed to be a sensitive and specific neuronal biomarker, able to predict injury severity and mortality after severe TBI in humans [26–29]. UCH-L1 is a compact cytosolic protein with a low molecular weight (~24 kDa), allowing measurement of UCH-L1 concentrations in extracellular fluid, relatively easily with 100 kDa MD techniques. In the past there have been no studies, laboratory or clinical that correlate MD levels of UCH-L1 and GFAP with their predictive ability for outcome.

2 Materials

Following materials and devices are required for this experiment. Also, for successful performance of the surgical procedure, the following surgical instruments, devices, and tools are required, with sterilization as applicable:

1. Adult male Sprague-Dawley rat (300–350 g).
2. Isoflurane, 0.7 l/min N_2O and 0.3 l/min O_2 for anesthesia (isoflurane level 3).
3. Polyethylene catheter (P-50, Polyethylene Tubing, Intramedic, Reorder# 427411).
4. Shielded i.v. catheter (BD Insyte Autoguard, 14GA 175IN, 2.1 × 45 mm).
5. Blanket cooling/heating system (Gaymar Medi-Therm III, Gaymar Industries, Inc., NY, USA).

6. TCAT-2 Temperature Controller with RET-3 Temperature probe and HL-1, Heat Lamp (Physitemp Instruments, Inc., Clifton, NJ).
7. Surgical scissors for skin incisions, and microsurgical scissors to cut through small vessels (Roboz Surgical Instrument, MD).
8. Operating microscope (Zeiss) and dental drill (CellPoint).
9. No. 23 gauge blunt needle (Intramedic, Becton Dickinson, Reorder #427565).
10. Cyanoacrylate glue (Loctite 404 Instant adhesive).
11. Infusion pump with Hamilton syringe (Harvard Apparatus, MA).
12. CMA 12 MD probe (CMA Microdialysis, Solna, Sweden, 100 kDa cutoff, 4 mm membrane).
13. 4 % Bovine serum albumin (Sigma).
14. A computer system for continuous blood pressure monitoring (ADinstruments, LabChart version 4.1.2, www.adinstruments.com).

3 Methods

3.1 Animal Groups

Adult male Sprague-Dawley rats were randomly divided into four treatment groups.

Animals underwent subdural hematoma induction, brain microdialysis catheter placement, tail artery cannulation, and temperature manipulation as described below (Fig. 2).

1. In the normothermia group, head temperature was maintained at normothermic levels (37 °C) during the course of the experiment.
2. The early hypothermia group underwent hypothermia induction (33 °C) 30 min prior to decompressive craniotomy and removal of the hematoma to mimic a clinical situation in which hypothermia induction could be started as soon as ASDH is diagnosed by CT scan and while the operating room is prepared. Hypothermic treatment was continued 3 h after decompression.
3. The late hypothermia group received hypothermia induction (33 °C) 30 min after decompression surgery and it was maintained for 3 h.
4. The sham group did not receive induced subdural hematoma but underwent craniotomy. Their head temperature was maintained at normothermic levels (37 °C) during the course of the experiment.

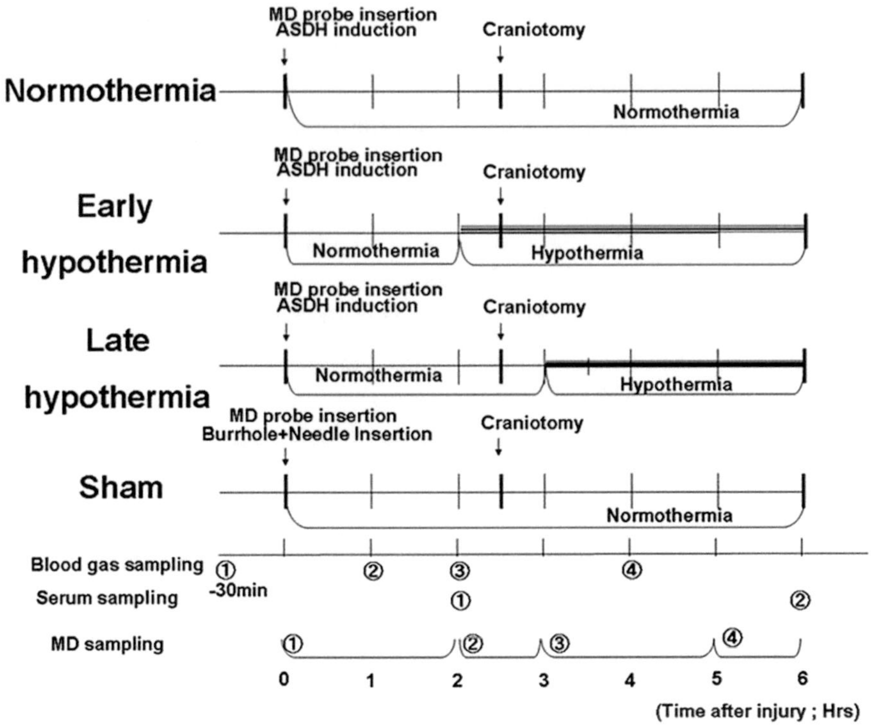

Fig. 2 Definition of each treatment group and their time course. Experimental rats were randomly allocated into four groups. In normothermia group and sham groups, the head temperature was controlled at 37 °C during the course of experiment. In the early hypothermia group, mild hypothermia (33 °C) was induced 30 min before craniotomy. In the late hypothermia group, the same level of hypothermia (33 °C) was induced 30 min after craniotomy. Microdialysate were collected in the ischemic phase (0–2 h after hematoma induction), the craniotomy phase (30 min before and after craniotomy), early reperfusion phase (0.5–2.5 h after craniotomy), and late reperfusion phase (2.5–3.5 h). Abbreviations in this figure: *ASDH* acute subdural hematoma, *MD* microdialysis

3.2 Surgical Procedure

The animals were maintained on a 12-h/12-h light/dark cycle and given food ad libitum. All animal procedures followed guidelines established by the National Institute of Health (NIH) Guide for the Care and Use of Laboratory Animals and were approved by the University of Miami's Institutional Animal Care and Use Committee.

All animals were anesthetized initially with 3% isoflurane, 70% N_2O, and a balance of 30% O_2 delivered in a Perspex chamber, with subsequent endotracheal intubation, and mechanical respiration aided with paralytic Rocuronium (0.35 mg/kg every ½ h) to control proper ventilation as previously described [30, 31]. The tail artery was cannulated with a polyethylene catheter for blood pressure monitoring (Lab Chart), blood sampling and obtaining the autologous blood needed for ASDH induction (*see* **Note 6**). Blood gas analysis was performed four times throughout the procedure (Fig. 2), to control ventilation of the animal. A PaO_2 of around 100–150 mmHg and a $PaCO_2$ of 30–40 mmHg were used to mimic clinical conditions.

3.2.1 Temperature Manipulations

The head and rectal temperatures were maintained at 33 °C in the early- and late-hypothermia groups by a combination of a cooling/heating system and local cooling fan/heating lamp (*see* **Note 7**). In the normothermia and sham rat groups, head and rectal temperatures were maintained at 37 °C during the course of the experiment. Head temperature was measured by a thermistor probe placed in the right temporalis, and estimated as brain temperature [32].

3.2.2 Subdural Hematoma Induction

Details of the method used to produce subdural hematoma are described in previous reports by our group [31, 33, 34]. A midline scalp incision was made and a 3 mm diameter burrhole was drilled 2 mm to the left of the sagittal suture and 3 mm behind the coronal suture (Fig. 3). With the aid of an operating microscope, the dura was incised and a blunt-tipped, J-shaped, No. 23 gauge needle inserted into the subdural space. Quick-setting cyanoacrylate glue was used to set the needle and seal the burrhole. The hematoma was then induced by injecting 350 μl non-heparinized autologous blood into the subdural space over a period of 7 min, allowing it to clot in situ. After injection, the induction needle was cut off and sealed. In the sham-treated group, the needle was set in place but no blood was injected.

Two and a half hours after induction of the subdural hematoma, a craniotomy measuring 15 × 6 mm was made using a saline-cooled dental drill (Fig. 3). The hematoma was then removed using saline irrigation and forceps after widely opening the dura. Hemostasis of superficial blood vessels was achieved using bipolar diathermy if needed. The scalp was closed over the craniotomy without replacing the bone to mimic clinical practice of decompressive craniotomy.

3.2.3 Extracellular Biomarker Measurement with Microdialysis

We used a CMA 12 MD probe (CMA Microdialysis, Solna, Sweden), which had an active membrane length of 4 mm and a molecular weight cutoff at 100 kDa. One hour before ASDH induction, a second burrhole was drilled 2 mm to the left of the sagittal suture and 2 mm behind the lambdoid suture for microdialysis probe insertion (Fig. 3). The probe was inserted into this burrhole at 10° from the horizontal plane and at a depth of 6 mm, as previously described [35]. This insertion technique made for optimal placement of the dialyzing membrane within the cerebral subcortical "penumbra" area [35]. Probes were precalibrated in vitro to ensure that interprobe variation was minimal. The dialysis probes were continuously perfused with physiological saline with 4 % Bovine serum albumin added at 0.3 μl/min. Microdialysis (MD) sampling was delayed by 1 h after insertion to allow the brain to adapt to the presence of the probe. Dialysate sampling began 2.5 h before craniotomy and samples were taken at four time points, i.e., ischemic phase (0–2 h after ASDH induction), craniotomy phase (0.5 h before and after craniotomy), early

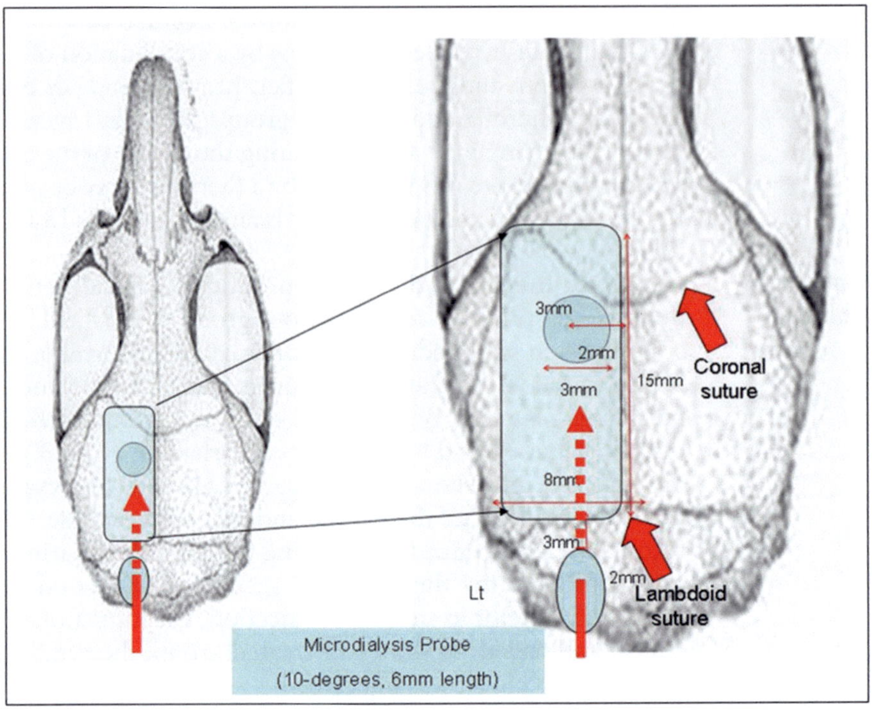

Fig. 3 Schematic representation of craniotomy, burrhole, and microdialysis catheter placement. The area of craniotomy extended from the lambdoid suture to 15 mm ahead of the lambdoid. The medial and lateral borders were the sagittal suture and superior temporal line, respectively. This made the width approximately 8 mm. The first burrhole, for hematoma induction, was 3 mm in diameter and placed 2 mm to the left of the sagittal suture and 3 mm behind the coronal suture. The second burrhole, which allowed for insertion of the microdialysis probe, was 3 mm in diameter was placed 2 mm to the left of the sagittal suture and 2 mm behind the lambdoid suture. The probe was inserted toward the front of the head, at a 10° angle from the horizontal and at a 6 mm depth from the brain surface

reperfusion phase (0.5–2.5 after craniotomy), and late reperfusion phase (2.5–3.5 h after craniotomy) (Fig. 2). Microdialysate vials were frozen and later analyzed for biomarkers.

Quantitative detection of UCH-L1 in microdialysate was performed using proprietary SW enzyme-linked immunosorbent assay (ELISA) (Banyan Biomarkers, Inc. FL, USA) and recombinant UCH-L1 as standard 68. For quantification of GFAP in microdialysate, a novel rat ELISA assay (Banyan Biomarkers, Inc. FL, USA) was used.

3.3 Results: Concentrations of Biomarkers in Cerebral Microdialysate

UCH-L1 concentration in the early induced hypothermia group was lower than in the normothermia and late hypothermia groups on the early phase of reperfusion (30 min to 2.5 h after decompression; Early; 4.9 ± 1.0 ng/dl, Late; 35.2 ± 12.1 ng/dl, Normo; 50.20 ± 28.3 ng/dl, Sham; 3.1 ± 1.3 ng/dl, Early vs. Normo; $p < 0.01$, Sham vs. Normo; $p < 0.01$: Fig. 4a). Also, on the late phase of reperfusion (>2.5 h after decompression), extracellular GFAP in the early hypothermia group was lower than in the normothermia and late hypothermia groups (Early; 5.5 ± 2.9 ng/dl, Late; 7.4 ± 3.4 ng/dl, Normo; 15.3 ± 8.4 ng/dl, Sham; 3.3 ± 1.0 ng/dl, Normo vs. Sham; $p < 0.01$; Fig. 4b).

3.4 Translation of Results

Our experiment with MD technique upheld the hypothesis that early-induced hypothermia attenuates brain damage with both neuronal and glial cell damage reduction in the reperfusional phase of I/R brain injury.

We observed that concentrations of UCH-L1 and GFAP in microdialysate were low only in the early-induced hypothermia group. As far as we know, this is the first study that has demonstrated the utility of UCH-L1 and GFAP as measured by MD as a "BIOMARKER of ASDH induced brain damage," and as a measure of moderate therapeutic hypothermia efficacy.

As shown in Fig. 4, the peak of UCH-L1 extracellular concentration was highest in the normothermia treatment group, as compared to the sham group. This indicates that the peak of neuronal injury occurs in the early reperfusion phase of ASDH (Fig. 4a). Also, the peak of the extracellular GFAP concentration in the normothermia group was significantly higher than in the sham group and occurs in the late phase of reperfusion (Normothermia; 15.3 ± 8.4 ng/dl, Sham; 3.3 ± 1.0 ng/dl, Normothermia vs. Sham; $p < 0.01$, Fig. 4b).

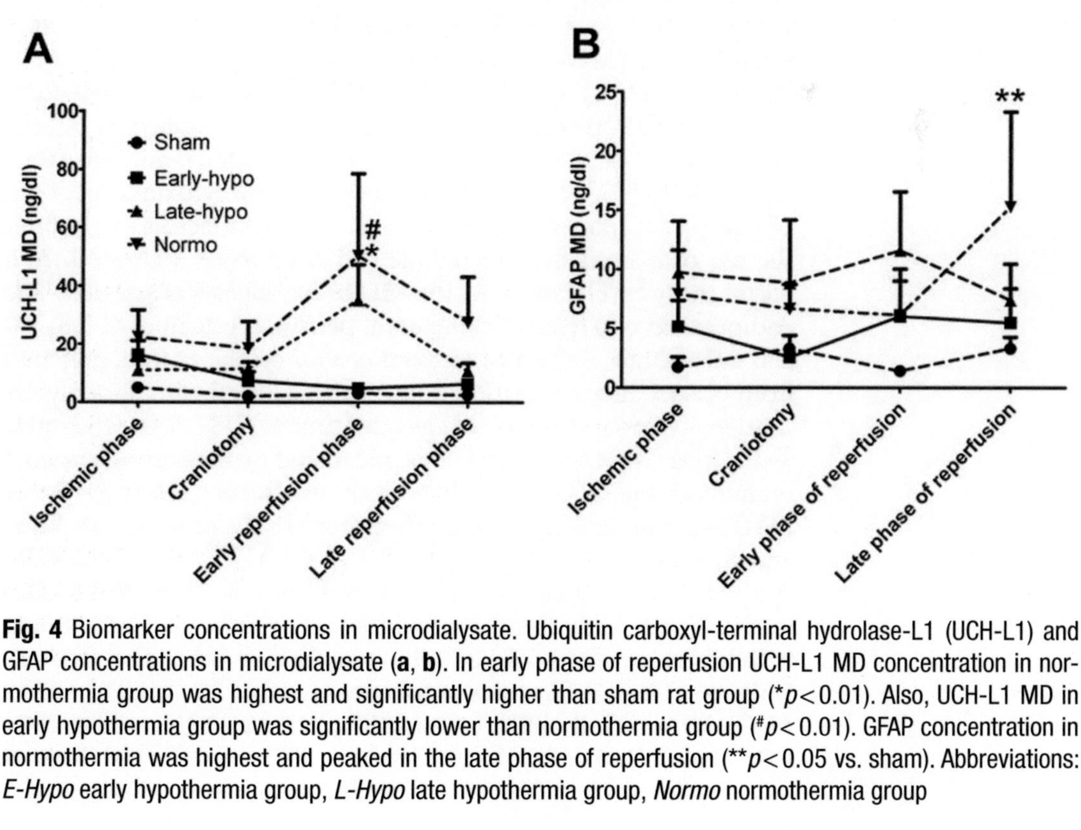

Fig. 4 Biomarker concentrations in microdialysate. Ubiquitin carboxyl-terminal hydrolase-L1 (UCH-L1) and GFAP concentrations in microdialysate (**a**, **b**). In early phase of reperfusion UCH-L1 MD concentration in normothermia group was highest and significantly higher than sham rat group (*$p < 0.01$). Also, UCH-L1 MD in early hypothermia group was significantly lower than normothermia group ($^{\#}p < 0.01$). GFAP concentration in normothermia was highest and peaked in the late phase of reperfusion (**$p < 0.05$ vs. sham). Abbreviations: *E-Hypo* early hypothermia group, *L-Hypo* late hypothermia group, *Normo* normothermia group

The peak of the extracellular concentration of UCH-L1, and subsequent peak of GFAP seemed to be lowered by early hypothermia induction (Fig. 4a, b). However, late-induced hypothermia could not to attenuate the neuronal damage in the early phase of reperfusion but reduced only the subsequent glial damage as determined by GFAP in the late phase of reperfusion. Taken together, these data suggest that early-induced hypothermia could reduce neuronal and subsequent glial injury in the delayed reperfusion phase of I/R pathophysiology. Based on the microdialysis data, the peak of glial cell damage might be occurring later than the neuronal damage, consistent with different vulnerabilities between neurons and astrocytes in I/R injury, or an astrocytic response to neuronal death. In hypoxic, ischemic brain damage models, neurons have been shown to be much more sensitive and vulnerable than astrocytes [36–39]. Astrocytes also might be more tolerant than neurons to I/R neurotoxicity, as seen in this ASDH rat model.

4 Factors Limiting Application of Microdialysis Method to Detect Cytokines Biomarkers

With recognition of the increasing role of cytokines in the unfolding of TBI pathology, the need to monitor such molecules became apparent [40, 41]. A new MD probe with five times greater molecular weight cutoff (MWCO) was introduced increasing the range from 20 kDa (CMA70) to 100 kDa (CMA71). In vitro and in vivo comparison studies revealed equivalent recovery for glucose, lactate, pyruvate, and glutamate (range 94–97% for CMA70 and 88–103% for CMA71) [42]. The recovery rates for several cytokines/peptides of interest are more complex and variable. These factors might limit the use of microdialysis technique and need to be addressed. In a recent study by Helmy et al., the MD of cytokines was assessed. The authors used two types of catheters at perfusion velocities of 0.3, 1.0 and 5.0 μl/min. Reference preparations for cytokines were prepared from plasma after incubating human whole blood with lipopolysaccharide, while for complement anaphylatoxins (C3a, C4a, C5a; m.w. 9–11 kDa) human plasma was incubated with heat-aggregated immunoglobulin G. Cytokines such as [tumor necrosis factor (TNF)-α, interleukin (IL)-1β, IL-6 and IL-10; m.w. 17–28 kDa] and chemokines (IL-8, MCP-1, IP-10, and MIG; m.w. 7–11 kDa) and complement anaphylatoxins (C3a, C4a, C5a; m.w. 9–11 kDa) were assayed. Only four chemokines passed through CMA70 and with low recovery (3–7%) at 1.0 microl/min velocity. The recovery with the 100 kDa filter was as follows: IL-1β = 75%, MCP-1 = 55%, MIG = 50%, IL-8 = 38%, C4a = 28%, IP-10 = 22%, C5a = 20%, C3a = 16%, IL-6 = 11, IL-10 = 8% and TNF-α = 4%. The highest recovery for all chemokines and anaphylatoxins were consistently at velocity 1.0 μl/min, whereas IL-1β and IL-10 recovered most efficiently at 0.3 μl/min [43]. According to the manufacturer the nominal cutoff (MWCO) is defined by a sieving coefficient threshold of 0.1. In other words for a 100 kDa molecule the recovery rate with CMA71 would be only 10%. In addition the molecules that are lower than 100 kDa may also have lower recovery rates due to variables such as shape, charge, hydrophobicity/hydrophilicity, hydrodynamic radius and interactions with other molecules (in case of multimerization) or the location of the molecule. They can be free versus inside a microparticle/exosome. Another study in TBI patients confirmed that with MD it is possible to track the proinflammatory cytokines. In this study, the authors determined the concentrations of IL-1β and IL-6 by fluorescence multiplex bead technology, and IL-10 was determined by enzyme-linked immunosorbent assay. Although different in magnitude by a factor 100, the response of IL-1β and IL-6 were similar. Both substances seemed to increase significantly after both SAH and TBI and thereafter gradually decreased. The response of both substances was stronger and more sustained in SAH patients than in TBI patients [44].

To assess the role of physicochemical parameters that influence the recovery rate of large molecules, an in vitro microdialysis recovery and delivery investigations were performed utilizing a standardized system. Analyte adsorption, pH effect, the influence of cytokine concentration and temperature of the catheter surrounding medium were assessed. A Ringer's/human albumin solution was used as microperfusate and catheter surrounding medium; interleukin 6, 8 and 10 (IL-6, IL-8, IL-10) and tumor necrosis factor alpha (TNF-α) served as model cytokines. Microdialysate was sampled ($n=3$) at flow rates of 0.3–5.0 μl/min using three linear probes. All samples were measured using a validated flow cytometry method adapted to microdialysate. Relative recoveries of the individual cytokines decreased exponentially with increasing flow rates and were not influenced by the catheter surrounding medium concentration but recovery of IL-6, IL-10, and TNF-α by the pH value. Relative recovery and relative delivery of IL-8 were of comparable extent and increased with higher temperatures. For the other cytokines, however, negative values occurred for relative delivery probably due to ultrafiltration. Clinical application of microdialysis of cytokines is principally feasible if the many influencing factors are controlled. As future perspective, in vivo microdialysis feasibility should be demonstrated.

5 Application of Microdialysis Method to Detect Therapeutic Windows

Biomarkers are a set of molecules that inform. A biomarker is an objective measure of a specific biological or pathological process that may be measured in body fluids such as cerebral spinal fluid (CSF) and blood [45]. Biochemical markers used as diagnostic tools for organ specific injuries, such as troponin for myocardial infarction, creatinine for renal failure, and pancreas amylase and lipase for acute pancreatitis. These blood tests are also used to monitor patients in intensive care settings. The challenges of accurate diagnosis and monitoring of TBI have created a need for biomarkers that reflect core elements of the TBI disease process [46]. Consistent with preclinical studies we found level of biochemicals such as glucose (G), lactate (L), and LG ratio (LGR) were high in TBI patients with GCS 3-6 ($p<0.0001$). Pyruvate level was lower in patients with GCS 7-9 ($p<0.001$). LPR was higher in patients with GCS 3-6 ($p<0.05$). High glucose, lactate level ($p<0.001$), and LPR ($p<0.01$) was observed in patients with GOS 1-3. Pyruvate level was low in patients with GOS 1-3 ($p<0.001$). LGR was higher in patient with better outcome (GOS 4-5). After craniotomy extracellular glucose and lactate were good "biomarkers" of cerebral damage in TBI patients. We consider that high extracellular lactate and low glucose is an indicator of severe neurological damage and poor outcome, because of impaired brain metabolism [47]. However the biochemical profile alone is not sufficient to monitor therapeutic efficacy and unequivocal prediction of outcomes in less severe injuries. With the advent of the novel technologies a more radical bottom-up approach (in which the body fluids of injured patients are screened for molecules and their role/presence in CNS is ascertained later) has recently been developed [48, 49]. In one such study with pooled naive and injured cortical samples (48 h post injury; rat controlled cortical impact model) that were analyzed using a differential neuroproteomics platform, the novel biomarkers ubiquitin C-terminal hydrolase-L1 (UCH-L1) [48] was discovered. UCH-L1 was previously used as a histologic marker for neurons because of its high abundance and specific expression in neurons. It is present in almost all neurons and averages 1–5 % of total soluble brain protein. UCH-L1 levels in CSF were assessed using an ELISA in a prospective case control study conducted with 66 patients. Forty-one patients with severe TBI, defined by a Glasgow coma scale (GCS) score of <8, who underwent intraventricular intracranial pressure monitoring were compared to 25 controls without TBI requiring CSF drainage for other medical reasons. Ventricular CSF was sampled from each patient at 6, 12, 24, 48, 72, 96, 120, 144, and 168 h following TBI. Injury severity was assessed by the GCS score, Marshall Classification on computed tomography, and a complicated post-injury course. Mortality was assessed at 6 weeks and long-term outcome was assessed using the Glasgow outcome score 6 months after injury. TBI patients had significantly elevated CSF levels of UCH-L1 at each time point after injury compared to uninjured controls. Overall mean levels of UCH-L1 in TBI patients was 44.2 ng/ml (±7.9) compared with 2.7 ng/ml (±0.7) in controls ($p<0.001$). There were significantly higher levels of UCH-L1 in patients with lower GCS scores at 24 h, in those with post injury complications, in those with 6-week mortality, and in those with a poor 6-month dichotomized Glasgow outcome score. These data suggest that this novel biomarker has the potential to determine injury severity in TBI patients [29, 50]. In a rat study we could demonstrate that the injured degenerate neurons were the source of the biomarker [51]. UCH-L1 complements MRI even in mild traumatic brain injury detection [52]. In this chapter we present data relating to such a biomarker (UCHLI) and how it can be used to monitor injury progression (ASDH) and the effect of therapeutic intervention (hypothermia).

6 Novel Application of Biomarker Microdialysis Method to Detect Engraftment

Recent progress in cell transplantation biology has opened yet another avenue for use of microdialysis. Cell transplant is often followed by inflammatory response that needs to address to facilitate engraftment. Such studies have not yet been performed in CNS as yet, the technique has been successfully utilized in case of liver transplants. In vivo microdialysis of inflammatory markers has been applied clinically in a liver transplant rejection and ischemia study [53]. The 100 kDa pore probe was inserted into 73 patients post-reperfusion and the inflammatory mediators complement activation product 5a (C5a), C-X-C motif chemokine 8 (CXCL8), CXCL10, interleukin-1 (IL-1) receptor antagonist, IL-6, IL-10, and macrophage inflammatory protein 1β (MIP1β) were measured. Microdialysate was collected at 0, 4, 8, 12, and 24 h after reperfusion and showed an initial increase of all inflammatory mediators that declined to low/stable values by 24 h in the non-ischemic liver grafts. The study was also able to detect a rise in CXCL10 2–5 days before clinically relevant rises alanine aminotransferases and bilirubin levels in patients suffering from graft rejection [54]. Moving forward, similar microdialysis techniques could be applied to pro-inflammatory cytokine response in severe traumatic brain injury or a penetrating ballistic injury [55].

7 Limitations of Microdialysis Technique

Several limitations of MD technique must be considered. Microdialysis is only a regional technique and cannot assess global brain damage with certainty. MD technique may be good for time-dependent analysis but weak for spatial injury analysis. If microdialysis technique is used as a monitor in clinical neurointensive care, additional data which represents global condition (e.g., intracranial pressure monitoring, frequent CT examination, xenon CT CBF mapping, PET, or MRI) will be needed.

8 Conclusions

Microdialysis has yielded important results in the field of neuromonitoring for neurologically compromised patients. It allows for the direct measurement of extracellular molecules in an attempt to characterize metabolic derangements before they become clinically relevant. Advancements in technology have allowed for the bedside assay of multiple markers of ischemia and metabolic dysfunction, and the applications for traumatic brain injury and aneurismal subarachnoid hemorrhage have been well established. As clinicians become more comfortable with these tools their widespread use and potential for clinical impact with continue to rise.

9 Notes

1. Perfusion rate of microdialysate is usually set as 0.3–2 μl/min. In clinic, injecting pump is commercially available but perfusion speed is fixed as 0.3 μl/min.
2. Cutoff molecular size of membrane is usually 20 or 100 kDa in animal experiment. Only 20 kDa cutoff probe is available for human use.
3. Artificial cerebrospinal fluid (CSF) is commercially available but, alternatively, normal saline is also available for perfusion fluid.
4. Catheter stabilization would require at least 1 h of waiting.
5. Avoid frequent defrost before batch analysis.
6. As the root of blood sampling and arterial pressure measurement, femoral artery is also available. However, when you have intent to perform behavior testing, femoral artery cannulation should be avoided because of femoral nerve injury and paralysis.
7. Dry ice is used for cooling of the wind.

Acknowledgement

The part of this work is supported by funds from NINDS R01 NS 042133 and the Miami Project to Cure Paralysis.

References

1. Bito L, Davson H, Levin E, Murray M, Snider N (1966) The concentrations of free amino acids and other electrolytes in cerebrospinal fluid, in vivo dialysate of brain, and blood plasma of the dog. J Neurochem 13:1057–1067
2. Tossman U, Ungerstedt U (1986) Microdialysis in the study of extracellular levels of amino acids in the rat brain. Acta Physiol Scand 128:9–14
3. Dunn IF, Ellegala DB, Kim DH, Litvack ZN (2006) Neuromonitoring in neurological critical care. Neurocrit Care 4:83–92
4. Kissinger PT, Shoup RE (1990) Optimization of LC apparatus for determinations in neurochemistry with an emphasis on microdialysis samples. J Neurosci Methods 34:3–10

5. Allison LA, Mayer GS, Shoup RE (1984) o-Phthalaldehyde derivatives of amines for high-speed liquid chromatography/electrochemistry. Anal Chem 56:1089–1096
6. Damsma G, Westerink BH, de Vries JB, Van den Berg CJ, Horn AS (1987) Measurement of acetylcholine release in freely moving rats by means of automated intracerebral dialysis. J Neurochem 48:1523–1528
7. Kontur P, Dawson R, Monjan A (1984) Manipulation of mobile phase parameters for the HPLC separation of endogenous monoamines in rat brain tissue. J Neurosci Methods 11:5–18
8. Benveniste H, Diemer NH (1988) Early postischemic 45Ca accumulation in rat dentate hilus. J Cereb Blood Flow Metab 8:713–719
9. Bramlett HM, Green EJ, Dietrich WD, Busto R, Globus MY, Ginsberg MD (1995) Posttraumatic brain hypothermia provides protection from sensorimotor and cognitive behavioral deficits. J Neurotrauma 12:289–298
10. Dietrich WD, Alonso O, Busto R, Globus MY, Ginsberg MD (1994) Post-traumatic brain hypothermia reduces histopathological damage following concussive brain injury in the rat. Acta Neuropathol 87:250–258
11. Dietrich WD, Bramlett HM (2010) The evidence for hypothermia as a neuroprotectant in traumatic brain injury. Neurotherapeutics 7:43–50
12. Jia F, Mao Q, Liang YM, Jiang JY (2009) Effect of post-traumatic mild hypothermia on hippocampal cell death after traumatic brain injury in rats. J Neurotrauma 26:243–252
13. Okauchi M, Kawai N, Nakamura T, Kawanishi M, Nagao S (2002) Effects of mild hypothermia and alkalizing agents on brain injuries in rats with acute subdural hematomas. J Neurotrauma 19:741–751
14. Karibe H, Zarow GJ, Graham SH, Weinstein PR (1994) Mild intraischemic hypothermia reduces postischemic hyperperfusion, delayed postischemic hypoperfusion, blood-brain barrier disruption, brain edema, and neuronal damage volume after temporary focal cerebral ischemia in rats. J Cereb Blood Flow Metab 14:620–627
15. Kawai N, Nakamura T, Okauchi M, Nagao S (2000) Effects of hypothermia on intracranial pressure and brain edema formation: studies in a rat acute subdural hematoma model. J Neurotrauma 17:193–202
16. Clifton GL, Miller ER, Choi SC, Levin HS, McCauley S, Smith KR Jr, Muizelaar JP, Wagner FC Jr, Marion DW, Luerssen TG, Chesnut RM, Schwartz M (2001) Lack of effect of induction of hypothermia after acute brain injury. N Engl J Med 344:556–563
17. Jiang JY (2009) Clinical study of mild hypothermia treatment for severe traumatic brain injury. J Neurotrauma 26:399–406
18. Marion D, Bullock MR (2009) Current and future role of therapeutic hypothermia. J Neurotrauma 26:455–467
19. McIntyre LA, Fergusson DA, Hebert PC, Moher D, Hutchison JS (2003) Prolonged therapeutic hypothermia after traumatic brain injury in adults: a systematic review. JAMA 289:2992–2999
20. Shann F (2003) Hypothermia for traumatic brain injury: how soon, how cold, and how long? Lancet 362:1950–1951
21. Timmons SD (2010) Current trends in neurotrauma care. Crit Care Med 38:S431–S444
22. Clifton GL, Valadka A, Zygun D, Coffey CS, Drever P, Fourwinds S, Janis LS, Wilde E, Taylor P, Harshman K, Conley A, Puccio A, Levin HS, McCauley SR, Bucholz RD, Smith KR, Schmidt JH, Scott JN, Yonas H, Okonkwo DO (2011) Very early hypothermia induction in patients with severe brain injury (the National Acute Brain Injury Study: Hypothermia II): a randomised trial. Lancet Neurol 10:131–139
23. Wilberger JE Jr, Harris M, Diamond DL (1991) Acute subdural hematoma: morbidity, mortality, and operative timing. J Neurosurg 74:212–218
24. Massaro F, Lanotte M, Faccani G, Triolo C (1996) One hundred and twenty-seven cases of acute subdural haematoma operated on. Correlation between CT scan findings and outcome. Acta Neurochir 138:185–191
25. Wilberger JE Jr, Harris M, Diamond DL (1990) Acute subdural hematoma: morbidity and mortality related to timing of operative intervention. J Trauma 30:733–736
26. Liu MC, Akinyi L, Scharf D, Mo J, Larner SF, Muller U, Oli MW, Zheng W, Kobeissy F, Papa L, Lu XC, Dave JR, Tortella FC, Hayes RL, Wang KK (2010) Ubiquitin C-terminal hydrolase-L1 as a biomarker for ischemic and traumatic brain injury in rats. Eur J Neurosci 31:722–732
27. Mondello S, Jeromin A, Buki A, Bullock R, Czeiter E, Kovacs N, Barzo P, Schmid K, Tortella FC, Wang KK, Hayes RL (2011) Glial neuronal ratio (GNR): a novel index for differentiating injury type in patients with severe traumatic brain injury. J Neurotrauma 29:1096–1104
28. Mondello S, Linnet A, Buki A, Robicsek S, Gabrielli A, Tepas J, Papa L, Brophy GM, Tortella F, Hayes RL, Wang KK (2012) Clinical utility of serum levels of ubiquitin C-terminal hydrolase as a biomarker for severe traumatic brain injury. Neurosurgery 70:666–675

29. Papa L, Akinyi L, Liu MC, Pineda JA, Tepas JJ 3rd, Oli MW, Zheng W, Robinson G, Robicsek SA, Gabrielli A, Heaton SC, Hannay HJ, Demery JA, Brophy GM, Layon J, Robertson CS, Hayes RL, Wang KK (2010) Ubiquitin C-terminal hydrolase is a novel biomarker in humans for severe traumatic brain injury. Crit Care Med 38:138–144
30. Kawai N, Nakamura T, Okauchi M, Nagao S (2000) Effects of hypothermia on intracranial hemodynamics and ischemic brain damage-studies in the rat acute subdural hematoma model. Acta Neurochir Suppl 76:529–533
31. Kuroda Y, Bullock R (1992) Local cerebral blood flow mapping before and after removal of acute subdural hematoma in the rat. Neurosurgery 30:687–691
32. Jiang JY, Lyeth BG, Clifton GL, Jenkins LW, Hamm RJ, Hayes RL (1991) Relationship between body and brain temperature in traumatically brain-injured rodents. J Neurosurg 74:492–496
33. Di X, Bullock R (1996) Effect of the novel high-affinity glycine-site N-methyl-D-aspartate antagonist ACEA-1021 on 125I-MK-801 binding after subdural hematoma in the rat: an in vivo autoradiographic study. J Neurosurg 85:655–661
34. Kwon TH, Chao DL, Malloy K, Sun D, Alessandri B, Bullock MR (2003) Tempol, a novel stable nitroxide, reduces brain damage and free radical production, after acute subdural hematoma in the rat. J Neurotrauma 20:337–345
35. Kwon TH, Sun D, Daugherty WP, Spiess BD, Bullock MR (2005) Effect of perfluorocarbons on brain oxygenation and ischemic damage in an acute subdural hematoma model in rats. J Neurosurg 103:724–730
36. Voloboueva LA, Suh SW, Swanson RA, Giffard RG (2007) Inhibition of mitochondrial function in astrocytes: implications for neuroprotection. J Neurochem 102:1383–1394
37. Xu L, Sapolsky RM, Giffard RG (2001) Differential sensitivity of murine astrocytes and neurons from different brain regions to injury. Exp Neurol 169:416–424
38. Kirino T (1982) Delayed neuronal death in the gerbil hippocampus following ischemia. Brain Res 239:57–69
39. Pulsinelli WA, Brierley JB, Plum F (1982) Temporal profile of neuronal damage in a model of transient forebrain ischemia. Ann Neurol 11:491–498
40. Rothwell NJ, Luheshi GN (2000) Interleukin 1 in the brain: biology, pathology and therapeutic target. Trends Neurosci 23:618–625
41. Eng LF, Ghirnikar RS, Lee YL (1996) Inflammation in EAE: role of chemokine/cytokine expression by resident and infiltrating cells. Neurochem Res 21:511–525
42. Hutchinson PJ, O'Connell MT, Nortje J, Smith P, Al-Rawi PG, Gupta AK, Menon DK, Pickard JD (2005) Cerebral microdialysis methodology—evaluation of 20 kDa and 100 kDa catheters. Physiol Meas 26:423–428
43. Waelgaard L, Pharo A, Tonnessen TI, Mollnes TE (2006) Microdialysis for monitoring inflammation: efficient recovery of cytokines and anaphylotoxins provided optimal catheter pore size and fluid velocity conditions. Scand J Immunol 64:345–352
44. Mellergard P, Aneman O, Sjogren F, Saberg C, Hillman J (2011) Differences in cerebral extracellular response of interleukin-1beta, interleukin-6, and interleukin-10 after subarachnoid hemorrhage or severe head trauma in humans. Neurosurgery 68:12–19, discussion 19
45. Biomarkers Definitions Working G (2001) Biomarkers and surrogate endpoints: preferred definitions and conceptual framework. Clin Pharmacol Ther 69:89–95
46. Mondello S, Muller U, Jeromin A, Streeter J, Hayes RL, Wang KK (2011) Blood-based diagnostics of traumatic brain injuries. Expert Rev Mol Diagn 11:65–78
47. Sanchez JJ, Bidot CJ, O'Phelan K, Gajavelli S, Yokobori S, Olvey S, Jagid J, Garcia JA, Nemeth Z, Bullock R (2013) Neuromonitoring with microdialysis in severe traumatic brain injury patients. Acta Neurochir Suppl 118:223–227
48. Ottens AK, Kobeissy FH, Golden EC, Zhang Z, Haskins WE, Chen SS, Hayes RL, Wang KK, Denslow ND (2006) Neuroproteomics in neurotrauma. Mass Spectrom Rev 25:380–408
49. Mondello S, Schmid K, Berger RP, Kobeissy F, Italiano D, Jeromin A, Hayes RL, Tortella FC, Buki A (2013) The challenge of mild traumatic brain injury: role of biochemical markers in diagnosis of brain damage. Med Res Rev 34:503–531
50. Kobeissy FH, Sadasivan S, Oli MW, Robinson G, Larner SF, Zhang Z, Hayes RL, Wang KK (2008) Neuroproteomics and systems biology-based discovery of protein biomarkers for traumatic brain injury and clinical validation. Proteomics Clin Appl 2:1467–1483
51. Shyam Gajavelli AB, Spurlock M, Diaz D, Burks S, Bomberger C, Bidot CJ, Yokobori S, Diaz J, Sanchez-Chavez J, Bullock R (2011) Immunohistochemical correlation of novel bio-

markers with neurodegeneration in rat models of brain injury. Immunocytochemistry. InTech - Open Access Publisher, Rijeka, Croatia

52. Kou Z, Gattu R, Kobeissy F, Welch RD, O'Neil BJ, Woodard JL, Ayaz SI, Kulek A, Kas-Shamoun R, Mika V, Zuk C, Tomasello F, Mondello S (2013) Combining biochemical and imaging markers to improve diagnosis and characterization of mild traumatic brain injury in the acute setting: results from a pilot study. PLoS One 8:e80296
53. Haugaa H, Thorgersen EB, Pharo A, Boberg KM, Foss A, Line PD, Sanengen T, Almaas R, Grindheim G, Waelgaard L, Pischke SE, Mollnes TE, Inge Tonnessen T (2012) Inflammatory markers sampled by microdialysis catheters distinguish rejection from ischemia in liver grafts. Liver Transpl 18:1421–1429
54. Haugaa H, Almaas R, Thorgersen EB, Foss A, Line PD, Sanengen T, Bergmann GB, Ohlin P, Waelgaard L, Grindheim G, Pischke SE, Mollnes TE, Tonnessen TI (2013) Clinical experience with microdialysis catheters in pediatric liver transplants. Liver Transpl 19:305–314
55. de Rivero Vaccari JP, Dietrich WD, Keane RW (2014) Activation and regulation of cellular inflammasomes: gaps in our knowledge for central nervous system injury. J Cereb Blood Flow Metab 34:369–375

Chapter 24

Repetitive Transcranial Magnetic Stimulation as a Novel Therapy in Animal Models of Traumatic Brain Injury

Thangavelu Soundara Rajan, Salvatore Cuzzocrea, Daniele Bruschetta, and Angelo Quartarone

Abstract

Traumatic brain injury (TBI) in humans causes a broad range of structural damage and functional deficits due to both primary and secondary injury mechanisms. Over the past three decades, animal models have been established to replicate the diverse changes of human TBI, to study the underlying pathophysiology and to develop new therapeutic strategies. However, drugs that were identified as neuroprotective in animal brain injury models were not successful in clinical trials phase II or phase III. Repetitive transcranial magnetic stimulation (rTMS) is a powerful noninvasive approach to excite cortical neurons in humans and animals, widely applied for therapeutic purpose in patients with brain diseases. In addition, recent animal studies showed rTMS as a strong neuroprotective tool. In this chapter, we discuss the rationale and mechanisms related to rTMS as well as therapeutic applications and putative molecular mechanisms. Furthermore, relevant biochemical studies and neuroprotective effect in animal models and possible application of rTMS as a novel treatment for rodent brain injury models are discussed.

Key words rTMS, TBI, Neuroprotection, Rodent brain injury models

1 Introduction

1.1 Transcranial Magnetic Stimulation: Principle and Mechanism

Transcranial magnetic stimulation (TMS) is a painless method of stimulating the brain in a non-invasive way. The stimulator produces a magnetic field comparable to that produced by an MRI scanner, but that lasts for less than a millisecond. The magnetic field is able to penetrate the scalp and skull, producing electrical currents in the region of the brain underlying the coil. Since the magnetic field falls off rapidly with increasing distance from the coil TMS seems to activate principally neural elements in the cortex or subcortical white matter (Fig. 1). The induced electrical stimulus activates a mixture of excitatory and inhibitory neurons; some are located in the area of cortex under the coil, others project axons to or from the site of stimulation [1].

Firas Kobeissy et al. (eds.), *Injury Models of the Central Nervous System: Methods and Protocols*, Methods in Molecular Biology, vol. 1462, DOI 10.1007/978-1-4939-3816-2_24,

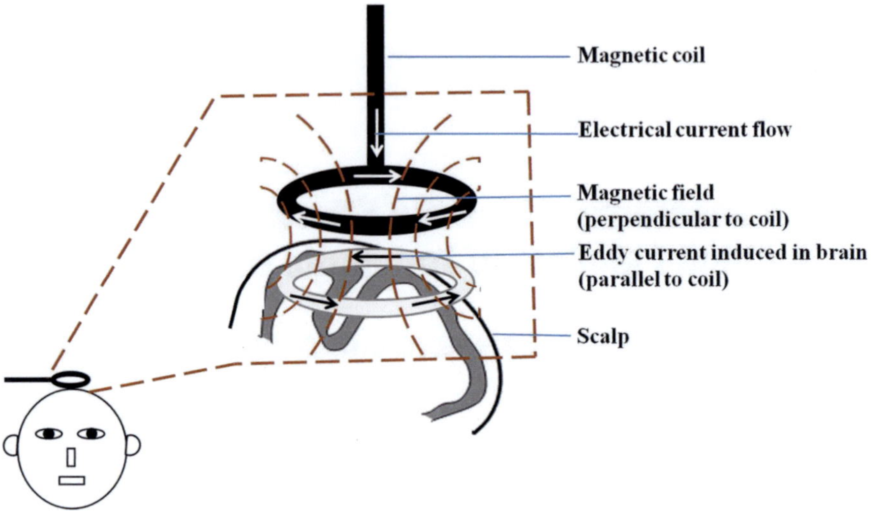

Fig. 1 Schematic illustration of the principle of transcranial magnetic stimulation (TMS)

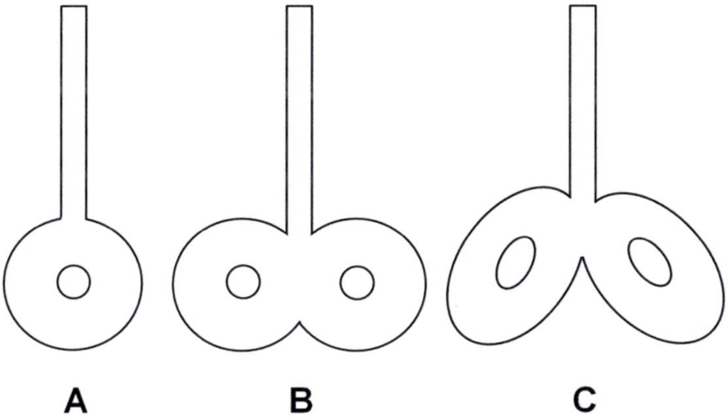

Fig. 2 Types of magnetic coils used in rTMS for humans. (**a**) Round coil, (**b**) figure-eight-shaped coil, and (**c**) double-cone coil

TMS is delivered using different types of dedicated coils (Fig. 2). The "standard" TMS circular coil consists of several circular turns of copper wire with a diameter of approximately 7 cm. This induces electric current in an annulus under the coil (without any current within the center of the coil), which covers a wide area of the brain. One way to make the stimulus more focal is to arrange the copper wire so that it forms two overlapping loops of wire in a "figure of eight" shape. Currents are then induced beneath each of the two circular loops and at the center, where they meet, the currents sum, so that stimulation under the mid-point of the figure of eight is twice that beneath the edges of the two loops. In this way, stimulation can be limited to a much smaller area in the order of 1–2 cm^2 (*see* Fig. 2) [2].

1.2 Single-Pulse Stimulation

Most of our knowledge of the action of TMS comes from studies evaluating the primary motor cortex recording electromyographic (EMG) responses in muscles on the opposite side of the body. By contrast, stimulation of most other parts of the cerebral cortex, at least with single pulses, has no obvious effects. However, in most individuals stimulation of the visual cortex generate phosphenes (bright spots of light in the visual field) that can be reported by subjects, but cannot be quantified by observers. The size of the EMG response depends on the level of activity in the cortex and the orientation of the TMS coil on the head. It is well established that the stimulus-triggered muscle response is larger when subjects are actively contracting the target muscle than if they are relaxed [3]. This might be due to the fact that in the brain and spinal cord, synaptic activation occurs more readily if the postsynaptic neurons are closer to the firing threshold than they are at normal resting potential. The orientation of the TMS coil has also an effect since it is well known that neurons are activated greatly by voltage gradients that run parallel to the axon [4].

1.3 Repetitive Transcranial Magnetic Stim-ulation (rTMS)

The initial studies on repetitive transcranial magnetic stimulation (rTMS) were prompted by the knowledge that repetitive stimulation of nerve pathways in animal models could induce long-term changes in synaptic effectiveness (long-term depression/potentiation; LTD/LTP). The prolonged after effects of rTMS depend on several factors such as the number of pulses applied, the rate of application and the intensity of each stimulus. For example, stimulation at frequencies higher than 1 Hz tends to increase rather than decrease cortical excitability. After effects depend also on the pattern of the pulses applied. For instance, Huang and colleagues used a pattern of stimulation, termed theta-burst stimulation (TBS), in which three 50 Hz pulses were applied regularly five times per second for 20–40 s. At low intensities this protocol produced suppression of motor cortex excitability, while if the burst was applied only for 2 s followed by a short pause of 8 s and then repeated, the effect was facilitatory [5]. Given that repetitive stimulation of the cortex can be epileptogenic, safety limits have been published on the parameters of rTMS that can be used in humans [6].

1.4 Therapeutic Applications of rTMS

In recent years, rTMS has been used to investigate potential therapeutic opportunities in a bewildering range of conditions based on the rationale that rTMS application may improve function in parts of the brain that are functioning suboptimally after an injury or chronic CNS diseases. In addition, as the after effects of rTMS can be excitatory or inhibitory it is possible to reduce function in parts of the brain that interfere with recovery by being overactive.

Despite rTMS has been reported to have a beneficial effect on an remarkably diverse variety of conditions such as stroke, Parkinson's disease, dystonia, writer's cramp, tinnitus, stroke, posttraumatic disorder, and aphasia [3, 7–10], many of these results are obtained from small single-center studies and are difficult to evaluate. Psychotic depression is the only condition where rTMS has proved to be beneficial in a large cohort of patients. Indeed, rTMS received US Food and Drug Administration (FDA) approval for the treatment of major depressive disorder. It is likely that a better understanding of both the effects of rTMS and the pathological processes underlying the targeted disease conditions will help to define the therapeutic potential of rTMS and the conditions that can benefit from this intervention.

1.5 Putative Molecular Mechanisms of rTMS After Effects

The mechanisms by which rTMS stimulate neurons and interferes with neural functions at the cellular and molecular level are far from clear. Nevertheless, recent studies in humans and animals have investigated some putative mechanisms associated with rTMS at the cellular and molecular level. The short term after effects of rTMS could be related to changes in neural excitability caused by shifts in ionic balance around populations of active neurons [11]. In addition, suprathreshold rTMS of the motor cortex causes muscle twitches that feed sensory information back to the motor cortex ultimately modifying its response to stimulation. These effects disappear quickly after termination of rTMS.

On the other hand, longer lasting effects are presumably caused by other mechanisms involving changes in the effectiveness of synapses between cortical neurons (long-term depression (LTD) and long-term potentiation (LTP) of synaptic connections) [12]. Evidences in human studies suggest that the after effects of rTMS depend on the glutamatergic *N*-methyl-d-aspartate (NMDA) receptor since these effects are blocked by a single dose of the NMDA-receptor antagonist dextromethorphan [13]. It is likely that BDNF, which is an up regulator of the NMDA transmission, can contribute to the rTMS long term effects. In line with this hypothesis, we have recently shown that daily 5 Hz rTMS for 5 days improves BDNF-TrkB signaling in rats by increasing the affinity of BDNF for TrkB. This results in higher tyrosine-phosphorylated TrkB, affecting the NMDA machinery in rat prefrontal cortex and in rat and human peripheral lymphocytes [10].

1.6 Animal Studies

Research during the last 15 years in animal models significantly helped to understand some of the important biochemical mechanisms that underlie rTMS. BDNF expression has been shown to be increased in the rat hippocampal areas CA3 and CA3c, the granule cell layer, and also in the parietal and the piriform cortex after long-term high-frequency rTMS [14]. Meanwhile, enhanced BDNF was noticed in rat cerebral infarction model with low-frequency rTMS [15]. In awakened rats, BDNF and GluR1 subunit of AMPA receptor were found to be up-regulated after high-frequency repetitive stimulation while in anesthetized rats, the expression was down-regulated [16].

Plasmatic BDNF, but not CSF BDNF, was increased significantly after consecutive high-frequency rTMS in rat and human. In addition, increased BDNF-TrkB binding affinity was observed in rat prefrontal cortex and in rat and human lymphocytes [10].

Chronic high-frequency rTMS increased the hippocampal neurogenesis in rats that might be involved in the anti-depressant effects of chronic TMS [17]. Another study reported that, in a chronic unpredicted mild stress rat model of depression, long-term chronic rTMS increased hippocampus cell proliferation, BDNF level, phosphorylated ERK1/2 and significantly reversed anhedonic-like behavior. The same study also suggested that high-frequency rTMS may induce long-lasting effects which could modulate neuroplasticity [18]. Single rTMS increased the NMDA binding sites of glutamate and serotonin 5-HT 1A in some rat brain areas [19]. Neurotransmitters glutamate and GABA were increased after chronic rTMS in rat hippocampus and striatum [20]. In rat nucleus accumbens shell region, dopamine concentration was induced after acute rTMS [21]. A recent study has shown enhanced neural excitability in rat hippocampal CA1 pyramidal neurons after low frequency rTMS [22]. Altogether these observations suggest that the impact of rTMS, being short-term or long-term, greatly depend on the area stimulated, the intensity and frequency of stimulation, and the duration of stimulation session.

1.7 Neuroprotective Effect

Few studies have reported neuroprotective effect associated with the use of rTMS in rodent models of cerebral ischemic. In rats with transient middle cerebral artery occlusion, chronic rTMS treatment significantly reduced cortical and striatal infarct volumes, lowered caspase-3 positive cells, and increased Bcl-2/Bax ratio [23]. Stronger Bcl-2 and weaker Bax expression after chronic rTMS treatment were also observed in subacute cerebral ischemic rat model [24]. In another study on gerbils, rTMS preconditioning followed by transient common carotid artery occlusion significantly increased the neuron density in the CA1 region which may prevent ischemic neuronal damage [25]. In rat model of vascular dementia, low and high-frequency rTMS increased the mRNA and protein expression of BDNF, NMDAR1, and synaptophysin and played a beneficial role in the restoration treatment of vascular dementia [26]. Another seminal study found that long-term high-frequency rTMS did not induce mRNA expression of inflammatory mediators in the rat central nervous system [27]. Finally, a recent study has reported that rTMS attenuated spinal cord injury-induced astroglial and microglial activation in rats [28].

1.8 rTMS as a Possible Treatment in Animal Models of Brain Injury

A growing number of data showing neuroprotective effects of rTMS in animal models such as ischemia and depression support the use of rTMS as a novel treatment for brain trauma in humans [29–32]. In addition, the application of rTMS animal models of traumatic brain injury (TBI) may help to further understand both the effects of rTMS as well as the pathophysiological mechanisms triggered by the injury. Controlled cortical impact (CCI), weight drop, vacuum deformation, and fluid percussion are the conventional methods used to create animal TBI models. More information about these techniques are described in detail elsewhere in this book (*see* Chapters 11–15). In the following section, we will discuss possible applications of rTMS in animal models of TBI. We will also provide the fundamental information necessary to perform rTMS included in the protocol that we use in our lab. All the parameters of rTMS (frequency, type of coil, mode of treatment, etc.) will be determined by the performer.

2 Materials

1. Rats (Taconic Farm, USA).
2. Mice (The Jackson Laboratory, USA).
3. Stereotaxic frame (Stoelting Wood Dale, USA).
4. Plastic restraint cones (Harvard Apparatus, USA).
5. Monopolar uninsulated 28 G stainless steel needle electrodes (Chalgren Enterprises Inc., USA) to record muscle Motor Evoked Potentials [MEP] elicited by surface electromyography (EMG).
6. Low-noise AC differential amplifier Model 1700 (AM Systems, USA).
7. Rapid rate stimulator (Magstim, UK).
8. Custom made round shaped coil (Magstim, UK; outer diameter 55 mm, inner diameter 6 mm).
9. All chemicals and reagents (Sigma-Aldrich, USA).

3 Methods

3.1 Animal Setup

1. All the animals should be housed in a 12-h light/dark cycle with food and water ad libitum. Average body weight should be around 30 g for mice and 200 g for rats.
2. After adaptation, animals need be observed and videotaped for 2 weeks to verify their activities including spontaneous behaviors, locomotive activity, and sleep/awake period.
3. Animals should be handled with the plastic cone holder for a week with a rodent coil that must be placed over their heads to simulate rTMS.
4. It is advised to keep animals in awakened state during the stimulation as rTMS may act differently in awakening and anesthetized condition [16].

5. It is recommended to cover the damaged area of the brain after TBI induction with sutures. Any other methods such as any metal tool implanted within the brain which may be excited during magnetic induction is not advisable.
6. For real stimulation, the coil should be placed directly on the skull using bi-phasic stimulation (Fig. 3) (*see* **Note 1**), and for sham stimulation, stimuli should be delivered with the coil rotated 90° about the axis of the handle and separated from the head using a 2 cm plastic spacer cube to ensure that the animal will feel the vibrations produced by the click of the TMS coil without brain stimulation [33] (*see* **Note 2**).
7. Plastic restraint cones should be used to hold the animals during the experiment.
8. All animals should be carefully monitored throughout the experiment to identify any behavioral changes occurred due to rTMS treatment.
9. Animals may be sacrificed after 1–3 days of final stimulation to avoid any acute, transient effects ensuing the last stimulation.
10. Brain tissues should be collected and stored at –80 °C for biochemical studies or in 4% phosphate-buffered formaldehyde (pH 7.4) for immunohistochemical analyses.

3.2 Shape and Size of the Coils

Custom made round coil and figure-eight shaped coils are widely used according to the size of the rodent's brain. Round rodent coils are about 0.5–2.5 cm inner diameter and 5–8 cm outer diameter in size [16, 27, 33–35]. For figure-eight shaped coils, the inner diameter is about 2–5 cm and the outer diameter is about 5–7 cm [18, 20, 36, 37].

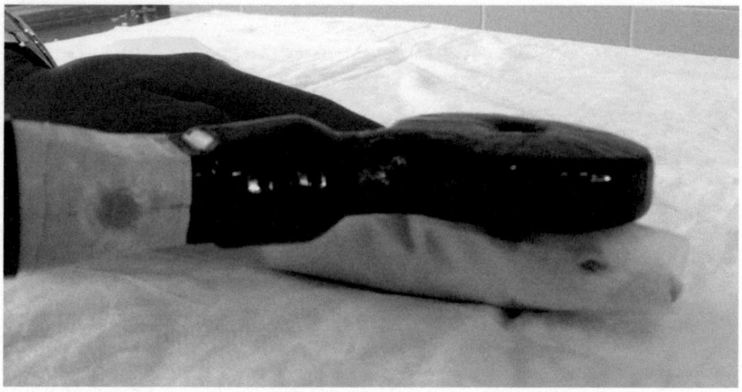

Fig. 3 Rodent coil placed directly touching the skull of a rat during rTMS treatment

3.3 Optimization of rTMS Parameters: Starting Point, Frequency and Duration, and Stimulator Intensity

Despite being a non-invasive technique, it is advised that rTMS treatment may be initiated after 24–72 h from the time of brain injury since the damaged brain is greatly vulnerable to excitation during the hyperacute stage and, therefore, rTMS application at this stage is likely to be impractical clinically [24].

3.3.1 Starting Point

3.3.2 Frequency and Duration

1. Frequency of rTMS, the number of trains of pulses per day and the duration of the treatment play a major role in activating the neural circuits differently.
2. Low frequency (≤1 Hz) rTMS may be given for longer periods (for example, 4–6 weeks) and high frequency (>1–20 Hz) rTMS may be given for shorter periods (for example, 14–21 days). High-frequency stimulation may last for very short periods also (for example, 2–10 days). It is advised to take appropriate precautions before starting the stimulation (*see* **Note 3**).

3.3.3 Stimulator Intensity

1. Number of trains of pulses for a session (or block), number of sessions per day and total number of sessions may differ depending upon the intensity of stimulator machine output (usually 20–80% of the maximal output of the stimulator), and as mentioned earlier, these parameters should be optimized by the performer.
2. The stimulator machine output shall be determined by measuring MEP of the biceps femoris (or postlateral tibial) muscle of the weak hind limb (like in the ischemia model) elicited by surface EMG. The resting motor threshold (RMT) shall be defined as the lowest stimulator output when the peak-to-peak amplitude of MEP is greater than 5% of its maximal amplitude in at least half of 5–10 trials [23, 24]. Detectable forelimb movement may also be considered for resting motor threshold measurement [38].
3. It is recommended to have an inter-block interval (for example, 1–10 min) between every successive session during repetitive stimulation to avoid heating of the coil. This inter-block interval time should be standardized before the experiment (*see* **Note 4**).

4 Notes

1. rTMS is a noninvasive tool to activate neurons in the brain. However, precautions should be taken when animal models such as rodents are subjected to magnetic stimulation due to their relatively small size of the brain. Sometimes during the stimulation, blood may leak out from the nose, the eyes, the ears, or the damaged tissue itself, owing to the strong stimulator machine output or excessive stimulation. In this condition,

stimulation should be terminated immediately. Optimizing the machine output and providing enough time interval between each train may help to stop bleeding.

2. During stimulation and/or simulation, animals should be kept calm and should not be excited or nervous since these may affect the overall effect of rTMS (unpublished observations). Animals should be held gently while performing the stimulation because holding them tightly may cause breathing problem with the consequence of death.
3. There are no recommended safety guidelines for the use of rTMS in animals. The rapid rate stimulator, its coil, and cable generate strong magnetic pulses. Therefore, a thorough study of the operating manual and appropriate precautions are needed before using the stimulator. The strong magnetic pulses produced from the coils induce eddy currents in any conductive medium such as the human body or electronic devices and metallic objects nearby. rTMS should not be performed in the vicinity of subjects wearing pacemakers, electronic or metal implants. It must not be used in an explosive ambiance or in the presence of combustible anesthetics.
4. rTMS requires a great strength of power which results in rapid heating of the coil with elevated temperature. Most of the coils have a temperature window for stimulation (for example, 4–38 °C) after which stimulator cannot produce magnetic induction until to reach optimal working temperature condition. This problem could be resolved by refrigerating the coil with a plastic bag covered in ice during the inter-block interval time. Precooling the coil may effectively help from quick coil heating. However, it is advised by the companies that cooling must be performed by using a flow of cool air from a fan or air conditioning unit since cooling with ice or via cold water immersion may cause condensation inside the coil and degrade its electrical contacts.

Disclaimer

The authors declare that there are no conflicts of interest in this protocol.

References

1. Post A, Keck ME (2001) Transcranial magnetic stimulation as a therapeutic tool in psychiatry: what do we know about the neurobiological mechanisms? J Psychiatr Res 35:193–215
2. Thielscher A, Kammer T (2004) Electric field properties of two commercial figure-8 coils in TMS: calculation of focality and efficiency. Clin Neurophysiol 115:1697–1708
3. Hallett M (2000) Transcranial magnetic stimulation and the human brain. Nature 406:147–150
4. Kammer T, Beck S, Thielscher A, Laubis-Hermann U, Topka H (2001) Motor thresholds in humans: a transcranial magnetic stimulation study comparing different pulse waveforms, current directions and stimulator types. Clin Neurophysiol 112:250–258

5. Huang YZ, Edwards MJ, Rounis E, Bhatia KP, Rothwell JC (2005) Theta burst stimulation of the human motor cortex. Neuron 45:201–206
6. Rossi S, Hallett M, Rossini PM, Pacual-Leone A, Safety of TMS Consensus Group (2009) Safety, ethical considerations, and application guidelines for the use of transcranial magnetic stimulation in clinical practice and research. Clin Neurophysiol 120:2008–2039
7. Arias-Carrión O (2008) Basic mechanisms of rTMS: Implications in Parkinson's disease. Int Arch Med 1:2. doi:10.1186/1755-7682-1-2
8. Kobayashi M, Pascual-Leone A (2003) Transcranial magnetic stimulation in neurology. Lancet Neurol 2:145–156
9. Edwards MJ, Talelli P, Rothwell JC (2008) Clinical applications of transcranial magnetic stimulation in patients with movement disorders. Lancet Neurol 7:827–840
10. Wang HY, Crupi D, Liu J et al (2011) Repetitive transcranial magnetic stimulation enhances BDNF-TrkB signaling in both brain and lymphocyte. J Neurosci 31:11044–11054
11. Kuwabara S, Cappelen-Smith C, Lin CS, Mogyoros I, Burke D (2002) Effects of voluntary activity on the excitability of motor axons in the peroneal nerve. Muscle Nerve 25:176–184
12. Cooke SF, Bliss TV (2006) Plasticity in the human central nervous system. Brain 129:1659–1673
13. Stefan K, Kunesch E, Benecke R, Cohen LG, Classen J (2002) Mechanisms of enhancement of human motor cortex excitability induced by interventional paired associative stimulation. J Physiol 543:699–708
14. Müller MB, Toschi N, Kreese AE, Post A, Keck ME (2000) Long-term repetitive transcranial magnetic stimulation increases the expression of brain-derived neurotrophic factor and cholecystokinin mRNA, but not neuropeptide tyrosine mRNA in specific areas of rat brain. Neuropsychopharmacology 23:205–215
15. Zhang X, Mei Y, Liu C et al (2007) Effect of transcranial magnetic stimulation on the expression of c-Fos and brain-derived neurotrophic factor of the cerebral cortex in rats with cerebral infarct. J Huazhong Univ Sci Technolog Med Sci 27:415–418
16. Gersner R, Kravetz E, Feil J et al (2011) Long-term effects of repetitive transcranial magnetic stimulation on markers for neuroplasticity : differential outcomes in anesthetized and awake animals. J Neurosci 31:7521–7526
17. Ueyama E, Ukai S, Ogawa A et al (2011) Chronic repetitive transcranial magnetic stimulation increases hippocampal neurogenesis in rats. Psychiatry Clin Neurosci 65:77–81
18. Feng SF, Shi TY, Fan-Yang et al (2012) Long-lasting effects of chronic rTMS to treat chronic rodent model of depression. Behav Brain Res 232:245–251
19. Kole MH, Fuchs E, Ziemann U et al (1999) Changes in 5-HT1A and NMDA binding sites by a single rapid transcranial magnetic stimulation procedure in rats. Brain Res 826:309–312
20. Yue L, Xiao-Lin H, Tao S (2009) The effects of chronic repetitive transcranial magnetic stimulation on glutamate and gamma-aminobutyric acid in rat brain. Brain Res 1260:94–99
21. Erhardt A, Silaber I, Welt T et al (2004) Repetitive transcranial magnetic stimulation increases the release of dopamine in the nucleus accumbens shell of morphine-sensitized rats during abstinence. Neuropsychopharmacology 29:2074–2080
22. Tan T, Xie J, Tong Z et al (2013) Repetitive transcranial magnetic stimulation increases excitability of hippocampal CA1 pyramidal neurons. Brain Res 1520:23–35
23. Gao F, Wang S, Guo Y et al (2010) Protective effects of repetitive transcranial magnetic stimulation in a rat model of transient cerebral ischaemia: a microPET study. Eur J Nucl Med Mol Imaging 37:954–961
24. Yoon KJ, Lee YT, Han TR (2011) Mechanism of functional recovery after repetitive transcranial magnetic stimulation (rTMS) in the subacute cerebral ischemic rat model: neural plasticity or anti-apoptosis? Exp Brain Res 214:549–556
25. Fujiki M, Kobayashi H, Abe T et al (2003) Repetitive transcranial magnetic stimulation for protection against delayed neuronal death induced by transient ischemia. J Neurosurg 99:1063–1069
26. Wang F, Geng X, Tao HY et al (2010) The restoration after repetitive transcranial magnetic stimulation treatment on cognitive ability of vasculardementia rats and its impacts on synaptic plasticity in hippocampal CA1 area. J Mol Neurosci 41:145–155
27. Okada K, Matsunaga K, Yuhi T et al (2002) The long-term high-frequency repetitive transcranial magnetic stimulation does not induce mRNA expression of inflammatory mediators in the rat central nervous system. Brain Res 957:37–41
28. Kim JY, Choi GS, Cho YW et al (2013) Attenuation of spinal cord injury-induced astroglial and microglial activation by repetitive transcranial magnetic stimulation in rats. J Korean Med Sci 28:295–299
29. Pape TL, Rosenow J, Lewis G (2006) Transcranial magnetic stimulation: a possible treatment for TBI. J Head Trauma Rehabil 21:437–451
30. Flanagan SR, Cantor JB, Ashman TA (2008) Traumatic brain injury: future assessment tools and treatment prospects. Neuropsychiatr Dis Treat 4:877–892
31. Demirtas-Tatlidede A, Vahabzadeh-Hagh AM, Bernabeu M et al (2012) Noninvasive brain

stimulation in traumatic brain injury. J Head Trauma Rehabil 27:274–292

32. Villamar MF, Santos Portilla A, Fregni F et al (2012) Noninvasive brain stimulation to modulate neuroplasticity in traumatic brain injury. Neuromodulation 15:326–338
33. Esser SK, Huber R, Massimini M et al (2006) A direct demonstration of cortical LTP in humans: a combined TMS/EEG study. Brain Res Bull 69:86–94
34. Keck ME, Engelmann M, Müller MB et al (2000) Repetitive transcranial magnetic stimulation induces active coping strategies and attenuates the neuroendocrine stress response in rats. J Psychiatr Res 34:265–276
35. Ikeda T, Kurosawa M, Morimito C et al (2013) Multiple effects of repetitive transcranial magnetic stimulation on neuropsychiatric disorders. Biochem Biophys Res Commun 436:121–127
36. Hausmann A, Weis C, Marksteiner J et al (2000) Chronic repetitive transcranial magnetic stimulation enhances c-fos in the parietal cortex and hippocampus. Brain Res Mol Brain Res 76:355–362
37. Aydin-Abidin S, Trippe J, Funke K et al (2008) High- and low-frequency repetitive transcranial magnetic stimulation differentially activates c-Fos and zif268 protein expression in rat brain. Exp Brain Res 188:249–261
38. Baek K, Chae JH, Jeong J (2012) The effect of repetitive transcranial magnetic stimulation on fear extinction in rats. Neuroscience 200:159–165

Chapter 25

Experimental Models Combining TBI, Hemorrhagic Shock, and Hypoxemia

Lai Yee Leung, Ying Deng-Bryant, Deborah Shear, and Frank Tortella

Abstract

Animal models of traumatic brain injury (TBI) provide important tools for studying the pathobiology of brain trauma and for evaluating therapeutic or diagnostic targets. Incorporation of additional insults such as hemorrhagic shock (HS) and/or hypoxemia (HX) into these models more closely recreates clinical scenarios as TBI often occurs in conjunction with these systemic insults (i.e., polytrauma). We have developed a rat model of polytrauma that combines penetrating TBI, HS and HX. Following brain trauma, HX was induced by reducing the inspired oxygen while HS was induced by withdrawing blood to lower the mean arterial pressure. The physiological, histological, and behavioral aspects of this animal model have been characterized and have demonstrated exacerbating effects of systemic insults on penetrating TBI. As such, this model may facilitate the use of simultaneous assessments of multiple mechanisms and provide a platform for testing novel diagnostic and therapeutic targets.

Key words Polytrauma, Combined neurotrauma, Traumatic brain injury, Additional insults, Blood loss, Hemorrhagic shock, Respiratory distress, Hypoxemia

1 Introduction

The leading cause of non-preventable mortality in recent conflicts in the Middle East has been traumatic brain injury (TBI), while the leading cause of potentially-survivable combat-related traumas is hemorrhage (83%), followed by airway obstruction (10%) [1]. Excessive blood loss leads to hemorrhagic shock (HS) whereas airway obstruction can cause systemic hypoxemia (HX), and the primary cause of all severe traumas in combat relates directly to hemorrhage/hypoxemia. In general, patients are faced with worse outcomes and increased morbidity when TBI occurs in conjunction with these polytrauma insults. Local tissue damage and systemic inflammatory responses further contribute to the complex injury cascades in polytrauma [2]. TBI itself causes hypoperfusion, decreased oxygen delivery, excitotoxicity, profound inflammation and metabolic dysfunction, and perturbed autoregulation of cerebral

Firas Kobeissy et al. (eds.), *Injury Models of the Central Nervous System: Methods and Protocols*, Methods in Molecular Biology, vol. 1462, DOI 10.1007/978-1-4939-3816-2_25,

blood flow, all of which account for the increased vulnerability of the injured brain to hypotensive and/or hypoxic insults.

Posttraumatic insults have been incorporated into existing TBI preclinical models to mimic the complex pathophysiology of polytrauma. Table 1 summarizes the methods and outcomes of these models (only studies using rats are listed). Notably, the

Table 1
Rat models of TBI complicated by posttraumatic insults

TBI	Posttraumatic insults	Outcomes	References
CCI (moderate)	HX (1 min post-TBI; 0.1 F_iO_2 for 30 min; resuscitated with 100% O_2 for 30 min)	Apoptosis-suppressor gene bcl-2 and its translated protein in surviving neurons; DNA damage peaks at 24 h post-injury; stress response is induced	[10–12]
FPI (moderate)	HX (pO_2 = 30–40 mmHg for 30 min)	Increase in sensorimotor and cognitive deficits; exacerbate contusion and neuronal pathological conditions	[13, 14]
FPI (moderate)	HX (0 min post-TBI; 0.1 F_iO_2 for 20 min; resuscitated with 30% O_2)	Increased contusion volume; prolonged efflux of glutamate during hypoxia; increased number of caspase-3-positive cells and TUNEL-positive cells	[15]
CCI (moderate-severe)	HX (0 min post-TBI; 0.12 F_iO_2 for 30 min)	Worsens brain edema, disrupts ionic homeostasis and blunts the normal upregulation of AQP4 after TBI	[16]
FPI (moderate)	HS (0 min post-TBI; removed 25% of total blood volume; resuscitated with LRS three times the blood loss)	TBI suppressed spontaneous hemodynamic recovery from hemorrhage and attenuates resuscitation with LRS	[17]
FPI (moderate)	HS (60 mmHg for 30 min; resuscitated with shed blood)	Worsens local histopathologic outcomes possibly through vascular mechanisms	[18]
FPI	HS (40 mmHg for 45 min; resuscitated with shed blood)	Downregulation of neuroprotective genes in both injured and uninjured neurons	[19]
FPI (moderate)	HS (5 min post-TBI; 50–60 mmHg; 39 min; blood volume removed = 6–7 ml; resuscitated with LRS three times the shed blood volume)	No potentiation of structural damage or motor deficits; delayed the speed of recovery of cognitive function	[20]

(continued)

Table 1 (continued)

TBI	Posttraumatic insults	Outcomes	References
FPI (moderate)	HS (15 min post-TBI; blood volume removed = 40 % of total blood volume)	Cardiovascular homeostatic response to hemorrhage was disrupted by TBI	[21]
CCI (mild to moderate)	HS (40 mmHg for 40 or 50 or 60 min; blood removed = 2 ml/100 g; resuscitated with LRS and shed blood with 100 % O_2); 60 min-HS starts at 1 h, 24 h or 7 days post-TBI	Therapeutic efficacy of pHBSP; detrimental effect of HS on outcome were greatest when the onset was very soon after injury	[6, 22]
FPI (moderate)	HS (5 min post-TBI; 40 mmHg for 45 min; resuscitated with shed blood)	Reduced CBF and impaired autoregulation in aged rats but not young rats after TBI + HS	[23]
PBBI (10 %)	HS (5 min post-TBI; 40 mmHg for 30 min; resuscitated with LRS three times the blood loss)	Increased occurrence and duration of cortical spreading depolarizations	[9]
FPI (2.1 atm)	Tibia fracture	Increase plasma IL-6 levels; increased callus formation	[24]
>1 Posttraumatic insults			
WD (2 m)	12 % O_2 (HX) + anesthesia-induced hypotension (30 mmHg for 10 or 30 min)	Increased neuronal damage in supraventricular cortical regions	[25]
WD (2 m)	HS (30 min post-TBI; 60–70 mmHg for 240 min) + pyrexic insult (core Temp 38.5–39.5 °C)	Astrocyte activation; no convincing potentiation of brain damage	[26]
WD (1.8 m)	Simultaneous HX (0.1 F_iO_2) + HS (40 mmHg for 15 min; resuscitated with shed blood and restore normoxia); start at 45 min or 225 min post-TBI	Early HX and HS following TBI caused more prominent metabolic disturbance	[27]
CCI (?)	HS (2 h post-TBI; 50 mmHg for 30 min; resuscitated with shed blood) + reduced intracranial compliance	Increase in lactate and lactate/pyruvate ratio; reduced intracranial compliance did not exacerbate the injury	[28]

CCI controlled cortical impact, *FPI* fluid percussion injury, *WD* weight drop, *PBBI* penetrating ballistic-like brain injury, *LRS* lactated Ringer's solution, *HX* hypoxemia, *HS* hemorrhagic shock, F_iO_2 fraction of inspired oxygen

majority of these studies combined TBI with either HS or HX while only a few studies included more than one insult. However, it is not uncommon for head trauma patients to present with multiple systemic complications including both hypotension and hypoxia [3]. Thus, an animal model of TBI that incorporates multiple systemic insults may provide greater clinical relevance for evaluating therapeutic treatments. To design a preclinical model with multiple insults, the following questions must be answered: what TBI model and what injury severity should we use? When should the posttraumatic insult be induced following TBI? What should be the duration and extent of each insult? Which insult should come first, or does it make a difference?

Our primary goal was to develop a rat model of polytrauma that would recreate clinically relevant components of a person sustaining a head injury from gunshot and/or flying shrapnel in conjunction with other extracranial traumas collectively resulting in excessive blood loss and breathing difficulty. We have previously established a rodent model of penetrating ballistic-like brain injury (PBBI) that captures several key components of a ballistic wounding event to the brain including the permanent injury tract created by the path of a bullet, the intracerebral hemorrhage caused by the ballistic trajectory, and most importantly the large temporary cavity generated when ballistic energy dissipates from the penetrating object [4]. The present protocol describes the details and procedures of our polytrauma-PBBI model. In this polytrauma model, the temporary cavity produced is approximately 10% of the total brain volume of an adult Sprague-Dawley rat and is located on the right hemisphere near the striatum. This particular PBBI severity is 100% survivable and is capable of producing substantial histopathological and behavioral changes [4, 5]. The onset of our posttraumatic insults is 5 min post-PBBI in order to more closely correspond to the superimposition of multiple insults characteristic of the polytrauma subject. Support for this approach comes from previous preclinical research demonstrating the detrimental effects of secondary insults are more prominent the sooner they are initiated following TBI thereby producing an injury model more analogous to the pathophysiology of clinical polytrauma [6]. The extent and duration of the hypotensive and hypoxemic states were chosen based on relevant published studies and our own pilot studies which demonstrated that a 30-min episode of HS at 40 mmHg combined with a 30-min episode of HX at reduced inspired oxygen (10%) does not produce any overt pathological changes in the absence of TBI (unpublished data). However, when combined with TBI, the trio of insults resulted in increased mortality and neurofunctional deficits that varied depending upon the order in which the posttraumatic insults occurred. For example, HS induced prior to HX immediately following PBBI (PBBI+HS+HX) resulted in remarkable physiological changes and significant increases in mortality rates (41%). In contrast, reversing the sequence (HX induced prior to HS; PBBI+HX+HS) resulted in significantly lower mortality rates (23%) whereas simultaneous induction of HX and HS (PBBI+HX/HS) reduced mortality rates to 0%. Critically, regardless of the order in which they occurred, the "sequential" insults (i.e., PBBI+HS+HX and PBBI+HX+HS) produced pathology that is readily distinguishable from that produced by PBBI alone [7].

The methods applied to the PBBI model can be implemented in other rat strains or rodent species. The PBBI injury severity (size of the temporary cavity in PBBI) can be adjusted from more moderate to more severe with predictable changes in outcomes measured [5]. The onset, duration, and sequence of additional pathophysiological insults may also be adjusted according to the study objectives and experimental design. In general, the polytrauma model described here can be reproduced by following these steps: (1) induction of PBBI (right, unilateral, 10 % brain volume), (2) induction of hypoxemia by reducing inspired oxygen to 10 %, (3) induction of hemorrhagic shock by withdrawing blood until the mean arterial pressure drops to 40 mmHg, and (4) fluid resuscitation. Depending on the study objectives, the outcome metrics measured in this model may include physiological changes (cerebral blood flow, brain oxygen tension, blood gas, cortical electrical activity, etc.), neurofunctional, histological, and genomic/proteomic changes as well as systemic responses to the multiple insults.

2 Material

2.1 Animals

Male adult Sprague-Dawley rats (280–320 g; Charles River Labs, Raleigh, VA, USA) were used in these experiments. Animals were housed individually under a 12-h light/dark cycle in a facility accredited by the Association for Assessment and Accreditation of Laboratory Animal Care International.

2.2 Anesthesia

1. Induction: 4.0 % Isoflurane.
2. Maintenance: 1.5–2 % Delivered in breathing air/oxygen mixture (fraction of inspired oxygen $F_iO_2 = 0.26$) (*see* **Note 1**).

2.3 Materials for Surgery

1. Homeothermic heating system (Harvard Apparatus, MA, USA).
2. Stereotaxic frame with dual manipulators (David Kopf Instruments, Model 902).
3. Isoflurane vaporizers (Matrix™ VIP 3000) and isoflurane (Fluriso™ 502017).
4. Rodent clippers (Oster Golden A5).
5. Electric drill with drill bits (Foredom K.1070 High Speed Rotary Micromotor).
6. Cotton-tipped applicator and gauze (Covidien Curity, Catalog no. 8884540500).
7. Cauterizer (Gemini Cautery Kit, Catalog no. 726067).
8. Heparin saline (20 USP units/ml—sterile) (heparin: Sagent Pharmaceuticals, NDC#25021-400-30; 0.85 % Saline: RICCA Chemical Company, Catalog no. 7200-1).
9. Lactated Ringer's solution (sterile) (Hospira, Inc., NDC#0409-7953-09).

10. Local anesthesia (1% lidocaine) (MWI/VetOne, Catalog no. 510213).
11. Syringes (3, 5, 20 ml) (Becton, Dickinson and Company, Catalog no. 309657, 309646, 302830).
12. Needle tubing adapters (24 Gauge) (Becton, Dickinson and Company, Catalog no. 8214).
13. Catheters for cannulation (Instech, BTPU-027).
14. 4-0 silk suture (Henry Schein, Becton, Dickinson and Company, Catalog no. 100-5597) and staples (MikRon Precision, Inc., Becton, Dickinson and Company, Catalog no. 205016).
15. Sterile bone wax (Ethicon, W31G).
16. Sterile surgical instruments (scalpel, blunt-tip scissors, fine-tip forceps, microclips, etc.).
17. Standard Infuse/Withdraw Pump (Pump 11 Elite; Harvard Apparatus, MA, USA).
18. PBBI device and probe (4B080; Mitre Corporation, MA or HPD-1700 Dragonfly, Inc., WV) (*see* **Note 2**). The PBBI probe (Popper & Sons Inc., Hyde Park, NY) is a 20G stainless steel tube that has fixed perforations along one end that are sealed by airtight elastic tubing.

2.4 Instruments for Physiological Recordings

1. Blood gas analyzer (ABL5, Radiometer America Inc., Westlake, OH, USA).
2. Data acquisition system (PowerLab 16/30, AD Instruments Inc., Colorado Springs, CO, USA).
3. Respiratory gas analyzer (AD Instruments Inc., Colorado Springs, CO, USA).
4. Blood pressure transducer (Harvard Apparatus, MA, USA).
5. Respiratory belt transducer (AD Instruments Inc., Colorado Springs, CO, USA).
6. A computer with LabChart software (AD Instruments Inc., Colorado Springs, CO, USA). The computer is connected to the PowerLab data acquisition system.

3 Methods

3.1 Preoperative Procedures

Preoperative care of animals in accordance with veterinary medical practices is required. Upon arrival at the institution, the animals are housed under a normal 12-h light/dark cycle for at least 1 week for acclimation to the new environment before use.

1. Sterilize all surgical materials autoclave or vaporized hydrogen peroxide.

2. Fill up two 3 ml syringes with heparin saline, attach syringes to adaptors with PE50 catheters.
3. Fill up a 5 ml syringe with lactated Ringer's solution, attach syringe to adaptor with PE50 catheter; fill a 20 ml syringe with 5 ml heparin saline.
4. Calibrate the blood pressure transducer.
5. Set up the PBBI device (for Mitre device; *see* **Note 2** for Dragonfly device): connect the PBBI probe to a piston cylinder of the PBBI device via a connecting tube. Fill the cylinder, the tube and the probe with distilled water. Bleed air using a syringe until there is no air bubbles in the system. To pressurize the piston cylinder, turn on the air compressor until the pressure reaches 60–80 psi. A computer operated air-compressed trigger is used to control the ON/OFF of the pressurized piston cylinder. When the trigger is on, a pressure pulse will be transmitted to the PBBI probe to inflate the elastic tubing at one end of the PBBI probe (Fig. 1).

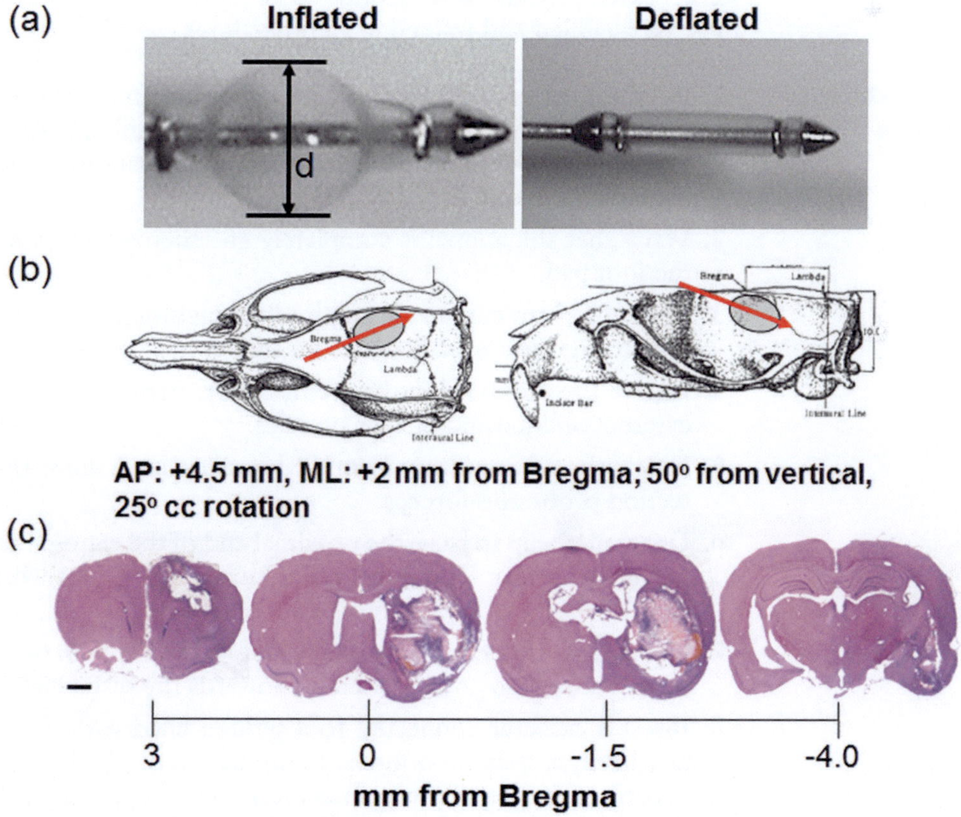

Fig. 1 (**a**) PBBI probe with inflated and deflated balloon; the diameter (denoted as 'd') or volume of the inflated balloon represents the diameter/volume of the temporary cavity that determines the injury severity (expressed in percentage of the total brain volume); (**b**) location of the probe and inflated balloon (not in scale); (**c**) hematoxylin and eosin (H&E)-stained brain sections showing the spatial progression of the lesion (7 days following PBBI; scale bar: 200 μm)

6. Calibrate the PBBI probe such that the elastic tubing on the probe is inflated to a diameter of 0.633 mm, representing 10% of the total brain volume (*see* **Note 3**).
7. Check the gas tanks (compressed air, oxygen and nitrogen) to ensure sufficient pressures.
8. Place the rat in an isoflurane chamber connected to the vaporizer. Turn on the oxygen/air and isoflurane level to 4% for about 4 min.
9. Shave hair over the dorsal head skin, as well as the right inguinal region.
10. Prior to incision, clean the incision sites with swabs of Betadine or alcohol.
11. Inject local anesthetics (1% lidocaine) subcutaneously at the incision sites.

3.2 Surgery

All surgery must be performed with aseptic procedures, including sterile gloves, masks, and aseptic techniques. Research personnel must be qualified and trained in all procedures (*see* **Note 4**).

3.2.1 Femoral Artery/Vein Cannulation

1. Place the rat on a heating pad in a dorsal recumbent position with the tail towards the operator. Insert the animal head into a nose cone system connect to the isoflurane vaporizer. Adjust the isoflurane to 2%.
2. Verify that the animal is completely anesthetized by pinching the foot pad.
3. Make a 1.5 cm cutaneous longitudinal incision at the inguinal region. Remove subcutaneous fat and connective tissues.
4. Locate the femoral bundle (vein, artery, nerve) between the external oblique and gracilis muscles.
5. Isolate the vein or artery from the bundle using a blunt-tip dissection probe and forceps.
6. Use a microclip to close the proximal end of the exposed femoral vein or artery. The distal end is closed by ligation with silk suture.
7. Use microscissors to make a small incision on the vein or artery between the two ends (preferably towards the distal end).
8. Insert a catheter connected to a syringe filled with 5 ml lactated Ringer's solution for vein cannulation or filled with 3 ml heparin saline for arterial cannulation. Remove the microclip at the proximal end and carefully advance the catheter further into the femoral vein or artery.
9. Hold the catheter in position using two pieces of silk suture—one proximal to the opening and one distal to the opening.
10. Cover the wound with saline-moistened sterile gauze.

11. Femoral artery catheterization is for blood pressure monitoring while femoral vein is for infusion of fluid for resuscitation.

3.2.2 Tail Ventral Artery Cannulation

1. Following the femoral artery/vein cannulation, keep the animal under anesthesia. Make a 1 cm mid-line incision on the ventral surface of the tail, starting at approximately 3 cm from the base of the tail.
2. Expose the artery by cutting the fascia using a blunt-tip scissors.
3. Ligate the artery distally with a silk suture and close the proximal end with a microclip.
4. Make a small incision on the artery using microscissors between the two ends (preferably towards the distal end).
5. Insert a catheter connected to a syringe filled with 3 ml heparin saline for arterial cannulation. Remove the microclip at the proximal end and carefully advance the catheter further into the artery.
6. Hold the catheter in position using two pieces of silk suture—one proximal to the opening and one distal to the opening.
7. Tail artery catheterization is for blood gas monitoring or for hemorrhagic shock procedures. Flush the line with heparin saline regularly to maintain the catheter patency.

3.2.3 Penetrating Ballistic-Like Brain Injury (PBBI)

1. Following the cannulations, transfer the anesthetized rat to a stereotaxic frame and secure its head with the ear bars.
2. Connect the femoral arterial line to the blood pressure transducer for arterial blood pressure monitoring.
3. Insert the rectal probe for monitoring core body temperature. The body temperature is regulated automatically using the homeothermic heating blanket.
4. Make an incision in the midline of the scalp (rostral-caudal) to expose the skull.
5. Clean the skull surface with cotton-tipped applicators and cauterize the bleeding spots.
6. Mark the position of the cranial window (+4.5 mm AP, +2 mm ML from Bregma) (*see* **Note 5**).
7. Adjust the manipulator arms such that the tip of the PBBI probe is aiming at the marked position, at an angle of 50° from vertical axis and 25° counter-clockwise from anterior-posterior axis. Retract the probe only in the dorsal-ventral (DV) direction.
8. Perform craniotomy at the marked position with an electric drill. The diameter of the cranial window is approximately 0.5 cm.
9. Insert the probe through the cranial window to a distance of 12 mm from dura.

10. Inflate the elastic tubing on the PBBI probe by turning on the trigger using the computer program (refer to **step 5** in Subheading 3.1; trigger manually when using Dragonfly device). The elastic tubing should inflate into an elliptically shaped balloon for less than 40 ms. The balloon will then deflate by itself.
11. Remove the probe by slowly retracting the manipulator arm in the DV direction.
12. Seal the cranial window with sterile bone wax.
13. Close the incision with silk suture or sterile wound clips.

3.3 Hypoxemia (HX)

1. Replace air with nitrogen in the inhalation gas mixture until the F_iO_2 reaches 0.1 (F_iO_2 is monitored by the respiratory gas analyzer).
2. Maintain this low oxygen concentration for 30 min (*see* **Notes 6** and **7**).
3. Withdraw 0.15 ml of blood from the tail artery. Inject the blood sample into the blood gas analyzer. The partial pressure of oxygen in the blood (P_aO_2) should be around 30–40 mmHg.
4. Switch the nitrogen to breathing air to restore normoxia ($F_iO_2 = 0.26$).

3.4 Hemorrhagic Shock (HS)

1. Disconnect the tail artery catheter from the 3 ml syringe. Connect it to a 20 ml syringe filled with 5 ml heparin saline.
2. Secure the syringe on an infuse/withdraw pump. Set the syringe diameter and withdrawal rate at 0.25 ml/100 g/min.
3. Turn on the pump (withdrawal mode) and withdraw blood from the tail artery until mean arterial pressure (MAP) reaches 40 mmHg. MAP is monitored using the PowerLab system with a blood pressure transducer.
4. Maintain the hypotensive state for 30 min. Withdraw additional blood if MAP rises above 45 mmHg (*see* **Note 6**).
5. Record the total volume of blood withdrawn. Fill a 20 ml syringe with lactated Ringer's solution for fluid resuscitation. The volume for resuscitation is three times of the blood volume withdrawn [8].
6. Connect the femoral vein catheter to the syringe filled with lactated Ringer's solution for resuscitation.
7. Turn on the pump (infusion mode) and infuse lactated ringer's solution via femoral vein at a rate of 0.3 ml/100 g/min.

3.5 Combining PBBI, HS and HX

1. Initiate HS or HX 5 min following PBBI. Allow 5-min recovery period between the insults following normoxia (if HX is done before HS) or fluid resuscitation (if HS is done before HX) (*see* **Note 8**).

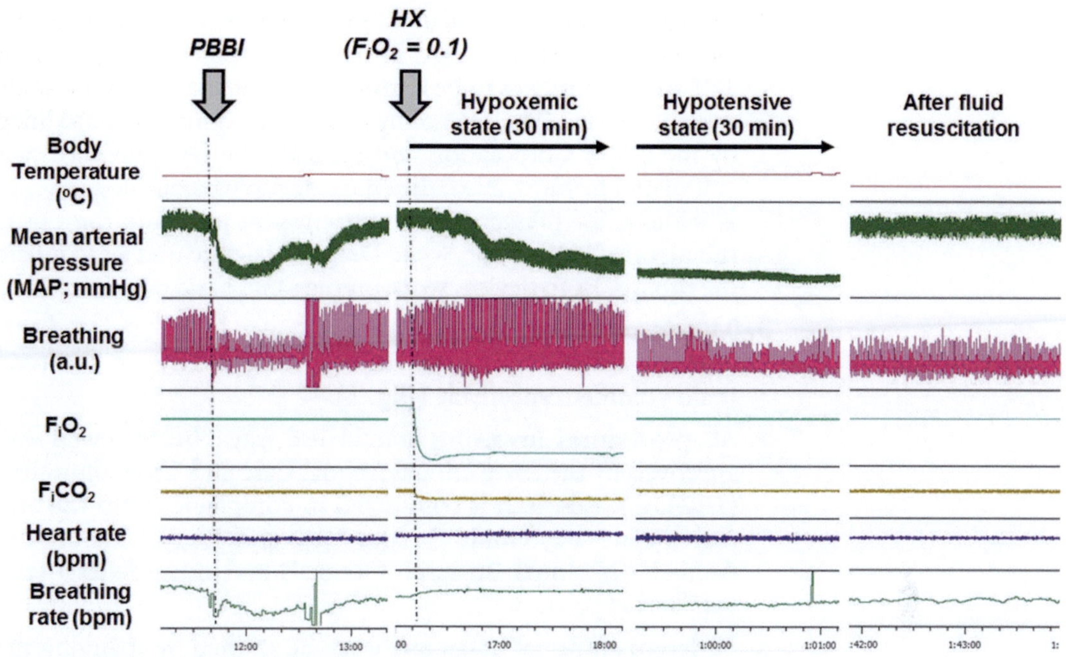

Fig. 2 Physiological monitoring in polytrauma model. Breathing rate is derived from the breathing measured from the respiratory belt. Heart rate is derived from the pulses obtained from the mean arterial pressure recording

2. At the end of the injuries, release the rat from the stereotaxic frame.
3. Remove all the catheters and ligate the vein/artery with silk suture.
4. Close the incision(s) with silk suture.
5. Apply topical antibiotics on the wounds.
6. Place the rat in a clean cage and monitor for recovery.

3.6 Physiological Monitoring

Monitoring of vital signs such as MAP, heart rate, breathing rate, and core body temperature is necessary prior to, during and following injuries in this polytrauma model as shown in Fig. 2. Multimodal neuromonitoring may also be useful in this model (*see* **Note 9**).

4 Notes

1. F_iO_2 can be adjusted according to the laboratory's routine procedure, as long as the arterial blood gas and other physiological variables are maintained within normal ranges. Pure oxygen ($F_iO_2 = 1$) is not recommended due to potential oxygen toxicity primarily affecting the central nervous system and the cardiac/pulmonary system.

2. The original PBBI device used by WRAIR was manufactured by Dragonfly, Inc. It was used in our early studies on penetrating TBI and it continues to be a strongly validated and commercially available device. The Dragonfly device was significantly modified by the Mitre Corporation and includes the computerized and automated features. Also, the Mitre device uses distilled water as a medium for pressure pulse transmission to inflate the elastic tubing on a PBBI probe, while Dragonfly device uses air to inflate the tubing. Both devices work on the same basic principle.
3. The elastic tubing on the PPBI probe may be inflated to diameters of 0.448–0.776 mm, representing 5–15 % of the total brain volume respectively (Fig. 1).
4. All procedures involving animal use must be reviewed and approved by the Institutional Animal Care and Use Committee (IACUC). Research is conducted in compliance with the animal welfare act, Guide for the Care and Use of Laboratory Animals (National Research Council) and other federal statutes and regulations.
5. Different angles of bullet entry can be studied by changing the angle and insertion location of the probe [5]. For example, to mimic a bullet entry to the side of the head, the probe would be angled at 90° and the craniotomy would be performed on the right temporal bone of the skull (+1 mm AP and +4 mm DV from Bregma) after retraction of the temporalis muscle.
6. Duration and level of hypoxemic or hypotensive state varies depending on the study design. In general, the duration ranges between 30 and 60 min. Mortality rate may be too high if the duration is greater than 60 min.
7. Hypoxemia may cause transient hypotension, tachycardia and hyperventilation. These are physiological compensatory responses to maintain oxygen homeostasis.
8. Injury profiles and mortality rates depend on the sequence of the insults. Based on our observation, HS prior to HX in animals subjected to PBBI resulted in the highest mortality rate and neurological deficits. On the other hand, simultaneous HS and HX following PBBI caused zero mortality (*see* Fig. 3 for validated data).
9. Multimodal monitoring provides a comprehensive approach to study the acute physiological changes following polytrauma. We have previously reported the overt changes in mean arterial pressure, cerebral blood flow (using laser Doppler flowmetry) and brain tissue oxygen tension (using Licox® probe) induced by hemorrhagic shock in the PBBI model. In addition, increase in cortical spreading depolarization (detected using epidural silver/silver-chloride ball electrodes) was observed that suggests substantial disruption of ionic homeostasis caused by the combined injury [9].

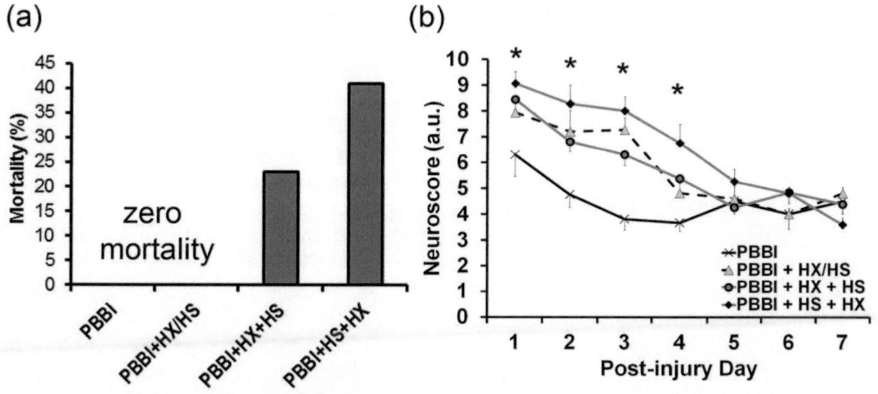

Fig. 3 (**a**) Mortality rate and (**b**) neurological deficits of polytrauma models with different sequences of additional insults. **p-value* < 0.05 compared to PBBI group

Disclaimers

The views of the authors do not purport or reflect the position of the Department of the Army or the Department of Defense (para 4-3, AR 360-5). The authors declare that there are no conflicts of interest in this protocol. This research is funded by Combat Casualty Care Research Program.

References

1. Kelly JF, Ritenour AE, McLaughlin DF et al (2008) Injury severity and causes of death from Operation Iraqi Freedom and Operation Enduring Freedom: 2003-2004 versus 2006. J Trauma 64(2 Suppl):S21–S26, discussion S26–S27
2. Keel M, Trentz O (2005) Pathophysiology of polytrauma. Injury 36(6):691–709
3. Manley G, Knudson MM, Morabito D et al (2001) Hypotension, hypoxia, and head injury: frequency, duration, and consequences. Arch Surg 136(10):1118–1123
4. Williams AJ, Hartings JA, Lu XC et al (2005) Characterization of a new rat model of penetrating ballistic brain injury. J Neurotrauma 22(2):313–331
5. Williams AJ, Ling GS, Tortella FC (2006) Severity level and injury track determine outcome following a penetrating ballistic-like brain injury in the rat. Neurosci Lett 408(3):183–188
6. Navarro JC, Pillai S, Cherian L et al (2012) Histopathological and behavioral effects of immediate and delayed hemorrhagic shock after mild traumatic brain injury in rats. J Neurotrauma 29(2):322–334
7. Leung LY, Wei G, Shear DA et al (2013) Polytrauma and penetrating ballistic-like brain injury (PBBI): effects of multiple secondary insults and their sequence on acute physiology, neurological deficits and mortality. Int J Neurotrauma 15:A95–A95, Mary Ann Liebert, INC
8. Salomone JP, Pons PT (2007) PHTLS: prehospital trauma life support. Elsevier Mosby, St. Louis
9. Leung LY, Wei G, Shear DA et al (2013) The acute effects of hemorrhagic shock on cerebral blood flow, brain tissue oxygen tension, and spreading depolarization following penetrating ballistic-like brain injury. J Neurotrauma 30(14):1288–1298

10. Chen M, Clark RS, Kochanek PM et al (1998) 72-kDa heat shock protein and mRNA expression after controlled cortical impact injury with hypoxemia in rats. J Neurotrauma 15(3): 171–181
11. Clark RS, Chen J, Watkins SC et al (1997) Apoptosis-suppressor gene bcl-2 expression after traumatic brain injury in rats. J Neurosci 17(23):9172–9182
12. Clark RSB, Chen M, Kochanek PM et al (2001) Detection of single- and double-strand DNA breaks after traumatic brain injury in rats: comparison of in situ labeling techniques using DNA polymerase I, the Klenow fragment of DNA polymerase I, and terminal deoxynucleotidyl transferase. J Neurotrauma 18(7):675–689
13. Bramlett HM, Dietrich WD, Green EJ (1999) Secondary hypoxia following moderate fluid percussion brain injury in rats exacerbates sensorimotor and cognitive deficits. J Neurotrauma 16(11):1035–1047
14. Bramlett HM, Green EJ, Dietrich WD (1999) Exacerbation of cortical and hippocampal CA1 damage due to posttraumatic hypoxia following moderate fluid-percussion brain injury in rats. J Neurosurg 91(4):653–659
15. Matsushita Y, Shima K, Nawashiro H et al (2000) Real time monitoring of glutamate following fluid percussion brain injury with hypoxia in the rat. Acta Neurochir Suppl 76:207–212
16. Taya K, Marmarou CR, Okuno K et al (2010) Effect of secondary insults upon aquaporin-4 water channels following experimental cortical contusion in rats. J Neurotrauma 27(1): 229–239
17. Yuan XQ, Wade CE (1992) Traumatic brain injury attenuates the effectiveness of lactated Ringer's solution resuscitation of hemorrhagic shock in rats. Surg Gynecol Obstet 174(4):305–312
18. Matsushita Y, Bramlett HM, Kuluz JW et al (2001) Delayed hemorrhagic hypotension exacerbates the hemodynamic and histopathologic consequences of traumatic brain injury in rats. J Cereb Blood Flow Metab 21(7): 847–856
19. Hellmich HL, Garcia JM, Shimamura M et al (2005) Traumatic brain injury and hemorrhagic hypotension suppress neuroprotective gene expression in injured hippocampal neurons. Anesthesiology 102(4):806–814
20. Schutz C, Stover JF, Thompson HJ et al (2006) Acute, transient hemorrhagic hypotension does not aggravate structural damage or neurologic motor deficits but delays the long-term cognitive recovery following mild to moderate traumatic brain injury. Crit Care Med 34(2):492–501
21. McMahon CG, Kenny R, Bennett K et al (2008) Modification of acute cardiovascular homeostatic responses to hemorrhage following mild to moderate traumatic brain injury. Crit Care Med 36(1):216–224
22. Robertson CS, Cherian L, Shah M et al (2012) Neuroprotection with an erythropoietin mimetic peptide (pHBSP) in a model of mild traumatic brain injury complicated by hemorrhagic shock. J Neurotrauma 29(6): 1156–1166
23. Hawkins BE, Cowart JC, Parsley MA et al (2013) Effects of trauma, hemorrhage and resuscitation in aged rats. Brain Res 1496:28–35
24. Maegele M, Riess P, Sauerland S et al (2005) Characterization of a new rat model of experimental combined neurotrauma. Shock 23(5):476–481
25. Yamamoto M, Marmarou CR, Stiefel MF et al (1999) Neuroprotective effect of hypothermia on neuronal injury in diffuse traumatic brain injury coupled with hypoxia and hypotension. J Neurotrauma 16(6):487–500
26. Lammie GA, Piper IR, Thomson D et al (1999) Neuropathologic characterization of a rodent model of closed head injury--addition of clinically relevant secondary insults does not significantly potentiate brain damage. J Neurotrauma 16(7):603–615
27. Geeraerts T, Friggeri A, Mazoit JX et al (2008) Posttraumatic brain vulnerability to hypoxia-hypotension: the importance of the delay between brain trauma and secondary insult. Intensive Care Med 34(3):551–560
28. Salci K, Enblad P, Goiny M et al (2010) Metabolic effects of a late hypotensive insult combined with reduced intracranial compliance following traumatic brain injury in the rat. Ups J Med Sci 115(4):221–231

Chapter 26

Experimental Models Combining Traumatic Brain Injury and Hypoxia

Eric P. Thelin

Abstract

Traumatic brain injury (TBI) is one of the most common causes of death and disability, and cerebral hypoxia is a frequently occurring harmful secondary event in TBI patients. The hypoxic conditions that occur on the scene of accident, where the airways are often obstructed or breathing is in other ways impaired, could be reproduced using animal TBI models where oxygen delivery is strictly controlled throughout the entire experimental procedure. Monitoring physiological parameters of the animal is of utmost importance in order to maintain an adequate quality of the experiment. Peripheral oxygen saturation, O_2 pressure (pO_2) in the blood, or fraction of inhaled O_2 (FiO_2) could be used as goals to validate the hypoxic conditions. Different models of traumatic brain injury could be used to inflict desired injury type, whereas effects then could be studied using radiological, physiological and functional tests. In order to confirm that the brain has been affected by a hypoxic injury, appropriate substances in the affected cerebral tissue, cerebrospinal fluid, or serum should be analyzed.

Key words Hypoxia, Hypoxic, Traumatic brain injury, Controlled cortical impact, CCI, Fluid percussion, Ventilation, Animal model, Monitoring, Oxygen saturation, Posttraumatic hypoxia

1 Introduction

Up to 45 % of severe TBI patients suffer from hypoxia [1], a condition which subsequently has been shown to correlate with worse outcome [2–4]. Cerebral hypoxia may lead to cerebral ischemia, an irreversible secondary brain injury frequently seen in autopsy materials of TBI patients [5]. In experimental conditions, hypoxic TBI seems to lead to an exacerbated cerebral inflammation [6, 7], an aggravated neuronal death and lesion size [7–10], a detrimental effect on the blood-brain barrier (BBB) and edema formation [6, 11, 12], as well as a worse functional outcome [6, 7, 12–15]. These models usually involve a standard traumatic brain injury method, either diffuse or focal, with a following period of hypoxia and mutually involve rigorous monitoring of physiological parameters of the animal in order to validate correct hypoxic conditions [6, 7,

Firas Kobeissy et al. (eds.), *Injury Models of the Central Nervous System: Methods and Protocols*, Methods in Molecular Biology, vol. 1462, DOI 10.1007/978-1-4939-3816-2_26,

10–14, 16]. Detrimental effects of hypoxia in TBI provide an interesting platform to study potentially beneficial therapies, such as hypothermia or erythropoietin.

The hypoxic TBI models described in the literature are all similar in their setup, as is shown in Table 1.

This chapter serves as a guide through a standard procedure of a hypoxic TBI, a controlled cortical impact (CCI) injury on intubated, and mechanically ventilated, Sprague-Dawley (SD) rats.

2 Materials

2.1 Medications

1. Inhalation anesthetics, isoflurane (5% induction, 1–1.5% continuous).
2. Local anesthetics (LA), Bupivacaine, Marcain® 0.25%.
3. Opioids, buprenorphine, Temgesic® 0.05 mg/kg.
4. Euthanasia agent, sodium pentobarbital 100 mg/kg.

2.2 Ventilation/Respiration

1. Ventilator (Rodent ventilator, Ugo Basile, Gemonio, Italy).
2. Capnograph (Datex Oscar Oxy, GE Healthcare, Little Chalfont, UK).
3. Gas Mixer with vaporizer (re-built pediatric gas mixer, Dameca A/S, currently Philip Electronics, Amsterdam, The Netherlands) (Vaporizer PPVE, Penlon, Oxford, UK).
4. Pulse Oxymeter (MouseSTAT™, Kent Scientific, Torrington, CT, USA).
5. Blood gas analyzer (ABL800 FLEX analyzer, Radiometer Medical, Brønshøj, Denmark).
6. Reservoir (medium-sized plastic glove in plastic, airtight box).
7. Portable oxygen analyzer (TED 60-T, Teledyne Electronic Devices, Thousand Oaks, CA, USA).
8. Y connector (Part # YJ-38-HDPE, Industrial Specialties Mfg, CO, USA).
9. Intubation tube—White angiocatheter (17 Gauge).
10. Laryngoscope blade—Formatted spatula with fiber-optic tubing.
11. Tubing (oxygen bubble tubing).

2.3 Traumatic Brain Injury Device/Procedure

1. *Head Impactor* (TBI 0310, Precision Systems and Instrumentation LLC, Lexington, KY, USA).
2. *Stereotaxic frame* (Model 900, Agnthos, Stockholm, Sweden).
3. *Heating pad* (Temperature Control Unit HB 101/2, Panlab, Harvard Apparatus, Barcelona, Spain).

Table 1
Illustrating published hypoxic TBI models

Name	Injury device	Injury location	Intubation	Inhaled gas	Oxygen level during hypoxia	Hypoxia target	Treatment	Duration of hypoxia	Euthanasia	Survival time	Outcome	Conclusion
Ishige et al. 1987 [11, 38]	Fluid percussion (lateral)	Right parietal cortex	No, box	NA	13 % O_2	4.7–5.3 kPa pO_2	None	30 min	NA	24 h?	Neurologic examination, MRI, CBF, tissue gravity, EEG, histopathology	Deleterious effects of TBI + hypoxia
Tanno et al. 1992 [11]	Fluid percussion (lateral)	Right parietal cortex	No, box	NA	10 % O_2	4.0–4.6 kPa pO_2	None	45 min	Anes thetized	1 h, 3 h, 6 h, 24 h	Immuno cytochemistry (IgG och HRP)	Hypoxia exacerberates BBB permeability
Clark et al. 1997 [39]	Controlled cortical impact	Left parietal cortex	Yes	Isoflurane	6 % O_2	11 % FiO_2 (6 kPa pO_2)	None	30 min	NA	6 h, 1 d, 3 d	BCL-2 activity and expression, TUNEL	BCL-2 and TUNEL increased after hypoxic TBI
Clark et al. 1997 [14]	Controlled cortical impact	Left parietal cortex	Yes	Isoflurane	6 % and 8 % O_2	11 % and 13 % FiO_2	None	30 min	Anes thetized	6 h, 24 h, 72 h, 7 d	Beam balance, contusion size, neuronal death (TUNEL)	No increase of contusion volume, worse functional outcome, more neuronal cell death in hypoxic TBI
Dave et al. 1997 [40]	Fluid percussion (parasaggital)	Right parietal cortex	No, mask	Halothane	13 % O_2	5.1–5.3 kPa pO_2	None	30 min	Decap itation	30 min	c-fos mRNA (Northern Blot)	c-fos higher in hypoxic animals
Chen et al. 1998 [35]	Controlled Cortical Impact	Left parietal cortex	Yes	Isoflurane	6 % O_2	11 % FiO_2 (6 kPa pO_2)	None	30 min	Anes thetized	6 h, 8 h, 24 h, 72 h, 168 h	Hsp72 activity and expression (western blot, in situ hybridization, immuno histochemistry)	hsp72 protein and mRNA is increased
Yamamoto et al. 1999 [8]	Traumatic axonal injury, diffuse	Midline	Yes	Halothane	10 % O_2	12 % FiO_2	Hypo thermia (30 °C)	10 or 30 min	NA	4 h, 24 h	Neuronal quantification	Positive effect of hypothermia
Bramlett et al. 1999 [41]	Fluid percussion	Right parietal cortex	Yes	Halothane	11 % O_2	4.0–5.3 kPa pO_2	None	30 min	NA	72 h	Contusion size, neuronal death	TBI + hypoxia increased contusion volume and neuronal death

(continued)

Table 1 (continued)

Name	Injury device	Injury location	Intubation	Inhaled gas	Oxygen level during hypoxia	Hypoxia target	Treatment	Duration of hypoxia	Euthanasia	Survival time	Outcome	Conclusion
Bramlett et al. 1999 [13]	Fluid Percussion	Right parietal cortex	Yes	Halothane	11 % O_2	4.0–5.3 kPa pO_2	None	30 min	NA	30 d	Limb placing, beam-walking, rotarod, spatial navigation.	Deleterious effects of TBI + hypoxia
Robertson et al. 2000 [42]	Controlled cortical impact	Left parietal cortex	Yes	Isoflurane	6 % O_2	11 % FiO_2	Hypo thermia (32 °C)	30 min	NA	21 d	Beam balance, beam walking, Morris water maze, lesion size.	No significant effect of hypothermia
Matsushita et al. 2001 [10]	Fluid Percussion	Right parietal cortex	Yes	Halothane	11 % O_2	4.0–5.3 kPa pCO_2	Hypo thermia (33 °C)	30 min	Anes thetized	3 d	Contusion size, neuronal death	Mild hypothermia may be neuro-protective
Hallaman et al. 2004 [15]	Drop device and lateral fluid percussion	Midline and right parietal cortex	Yes	Isoflurane	11 % O_2	11 % Fi O_2	None	10 min	Sodium Pento barbital	24 h, 48 h, 7 d (42 d)	Radial arm maze, beam walk, inclined plane, Morris water maze, cell death (FJ)	Fluid percussion worse than drop device + hypoxia
Gao et al. 2010 [9]	Traumatic axonal injury, diffuse	Midline	Yes	NA	10 % O_2	?	Hypo thermia (33 °C)	10 × 3 min	NA	6 h	Axonal damage (IHC APP), arteriolar reactivity	Hypothermia protected axonal and vascular injury
Hellewell et al. 2010 [7]	Traumatic axonal injury, diffuse	Midline	Yes	Isoflurane	12 % O_2	50 % sat	None	30 min	NA	1 d, 7 d, 14 d	Axonal damage (Beta-APP, NF-H), inflammation (CD68, microglia), astrocyte activation	Hypoxic TBI exacerbates neuro-inflammation and axonal injury

Yan et al. 2011 [6]	Traumatic axonal injury, diffuse	Midline	Yes	Isoflurane	12 % O_2	50 % sat	None	30 min	NA	2 h, 1 d, 2 d, 3 d, 6 d, 14 d	Microdialysis, brain edema, sensorimotor functions, open field test, cytokines	Increased cerebral inflammation in hypoxia + TBI
Hellewell et al. 2013 [43]	Traumatic axonal injury, diffuse	Midline	Yes	Isoflurane	12 % O_2	50 % sat	Eryth ropoetin	30 min	NA	1 d, 7 d, 14 d	Rotarod, open field test, axonal pathology (NF-200), lesion size, MAP2, EPO-receptor, inflammation	EPO is beneficial after hypoxia + TBI

Male SD is used in all models. Fluid percussion, CCI and diffuse injury models are all used. Isoflurane is more common in later models of hypoxic TBI. Different targets to determine hypoxia are used, as well as hypoxic gas provided. *CBF* cerebral blood flow, *IgG* immunoglobulin G, *HRP* Horseradish peroxidase, *EEG* electroencephalography, *FJ* Fluoro-Jade, *IHC* immunohistochemistry, *APP* amyloid precursor protein, *NF-H* neurofilament-heavy chain, *EPO* erythropoetin, *MAP-2* microtubule-associated protein 2, *NF-200* hypophosphorylated neurofilament heavy chain

4. *Surgical drill*—0.5 mm in diameter, diamond tip (Microspeed 317 IN; Silfradent, Forli, Italy).
5. *Surgical microscope* (Wild Heerbrugg M3C Stereozoom Microscope, Leica, Wetzlar, Germany).
6. A scalpel (replaceable blades).
7. Electronic shaver.
8. Electronic scale.
9. Suturing material (needle holder, Vicryl® 4-0).
10. Retractors.
11. Tweezers.
12. Syringes (1–5 mL for medical injections, blood gas sampling, etc.).
13. Cotton swabs.
14. Saline for wound cleaning.

2.4 Animals and Handling

The outlined experiment focuses on rats, but could also be applied for other small rodents. For larger animals, special care, methods, handling, and drugs need to be provided (*see* **Notes 1–4**). As seen in Table 1, Sprague-Dawley (SD) rats are the most frequently used in hypoxic TBI experiments [6–9]. If different species are used (like Dark Agouti (DA) or Piebald Virol Glaxo (PVG)) one should take into account their different metabolic conditions [17], and immunological response following traumatic brain injury when interpreting the results [18–20].

2.5 Medications

There are different providers of medications; this provides a list including generic names for the most frequently used and necessary medications for this type of animal experimentation. Except for the medications listed here, make sure to use physiological saline and eye gel in every experiment.

1. Inhalation anesthetics, isoflurane (5 % induction, 1–1.5 % continuous)

   Isoflurane, a halogenated ethyl-methyl ether, is an inhalation anesthetic with short induction time and a good anesthetic effect. Halothane (Fluothane®), as inhalation agent, has been discontinued due to its detrimental side effects (hepatotoxicity).

2. Local anesthetics (LA), Bupivacaine, Marcain® 0.25 %

   In order to improve nociceptive operational conditions, and also to improve control over stimuli that affects the animal during surgery, it is important to provide the animal with adequate local anesthesia (LA). Lidocaine, bupivacaine, ropivacaine, or other equivalents in suitable concentrations and volumes are recommended. Bupivacaine lasts longer and is thus preferable in hypoxic TBI. Inject subcutaneously (sc) in the area where the skin incision is to be performed. Avoid intravenous injection by needle aspiration.

3. Opioids, buprenorphine, Temgesic® 0.05 mg/kg

   LA is usually not sufficient to fully minimize painful sensations in the animal. Just like in humans, more adequate medication is necessary. Opioids are an important step in the preoperative anesthesia, especially in hypoxic TBI models, since opioids are potent respiratory inhibitors, as well as good anesthesia and analgesia agents [21]. The ability to limit spontaneous respiration leads to an increased control of the animal's ventilation, which is a necessity in many hypoxic TBI models. Buprenorphine, fentanyl, or equivalent are recommended.

4. Euthanasia agent, sodium pentobarbital 100 mg/kg

   Pentobarbital, a potent barbiturate, will lead to respiratory arrest in the animal within minutes if 100 mg/kg is injected intraperitoneally. Sodium pentobarbital is not readily available worldwide, equivalent dose of any barbiturate is then adequate for euthanasia.

2.6 Ventilator and Monitoring Setup

Many of the materials described in the ventilator setup are common elements in a clinical environment. It is recommended to establish a contact with a hospital, or a technical facility with adequate knowledge of hospital equipment, that may assist in constructing some of the parts needed. Figure 1 systematically displays the ventilator setup.

Important: Human and animal samples should not be mixed. However, discarded, yet functional, equipment designated for use on humans could sometimes be used to analyze animal samples.

1. *Ventilator* (rodent ventilator, Ugo Basile, Gemonio, Italy)

   There are several commercially available rodent ventilators that are suitable for hypoxic TBI models. The ventilator will let you adjust the tidal volume and respiratory rate. There are charts to follow which will let you determine the stroke volume depending on the size of your animal. The normal respiratory rate of rats is 70–110 breaths per minute, but could be adjusted if hyperventilation is deemed necessary for the hypoxic TBI model. You will also need an induction chamber before the rat is hooked up to the ventilator.

2. *Capnograph* (Datex Oscar Oxy, GE Healthcare, Little Chalfont, UK)

   As the exhaled gas is brought back to the ventilator, it is then possible to determine the FiO_2 and the end tidal pCO_2 with a capnograph connected to the ventilator. This will make it possible to confirm gas delivered to the rat, and that the rat has an adequate metabolism.

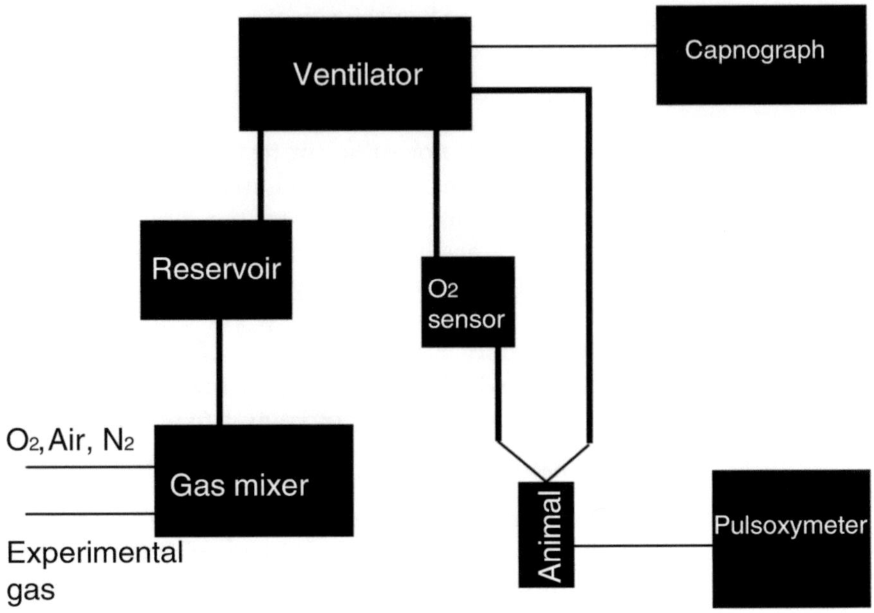

Fig. 1 Illustrating a simplified model of the ventilator setup. The gas inflow reaches the gas mixer. The gas is hooked up to the reservoir which will fill when gas is being transferred. The ventilator will rhythmically extract gas from the reservoir into the animal. The percentage of O_2 in the inhaled gas is measured with an O_2-sensor. The animal is monitored using a pulse oximeter. The exhaled gas flows through a Capnograph in order to analyze the fraction of inspired oxygen (FiO_2) and the end tidal pCO_2 in real time

3. *Gas mixer with vaporizer* (re-built pediatric gas mixer, Dameca A/S, currently Philip Electronics, Amsterdam, The Netherlands) (Vaporizer PPVE, Penlon, Oxford, UK). Gas mixers are frequently used in hospital environments, but may be purchased commercially. They usually mix air with oxygen (O_2), but are often able to add other gases to the mixture, in the clinic commonly nitrous oxide (N_2O, "laughing gas"). These devices could be adapted to work for animal research, preferably gas mixers used in pediatrics and neonatal care since the volumes are more compatible with rodents and smaller mammals. In hypoxic TBI models, hypoxic gases (O_2 6–13%) are often connected to the gas mixer in order to stimulate hypoxic conditions in the animal [6, 14]. It is also a possibility to directly connect nitrogen gas and to thereby make your own hypoxic gas mixture.
4. *Pulse Oxymeter* (MouseSTAT™, Kent Scientific, Torrington, CT, USA)

Even if human equipment is possible to use, there are today several good commercial pulse oxymeters optimized for animal use. This will let you monitor pulse, saturation, and blood pressure (with appropriate extensions).

5. *Blood gas analyzer* (ABL800 FLEX analyzer, Radiometer Medical, Brønshøj, Denmark)

   Human blood gas analyzers are possible to use. Such analyzer can provide detailed information of the animals' physiological parameters, including pO_2, pCO_2, glucose, lactate, and electrolytes.

6. *Reservoir* (medium-sized plastic glove in plastic, airtight box)

   A gas reservoir is a storage volume where gas from the gas mixer is stored and mixed before getting regularly pumped into the ventilator. All ventilators need a reservoir of different size depending on the volumes of gases used. The gas reservoir should be flexible to be allowed to expand and hold several stroke volumes for the rat, yet airtight in order to minimize inhalation of sedative gas for the operator. If not a premanufactured gas reservoir is available, a plastic glove with a small cut in it will suffice Fig. 2.

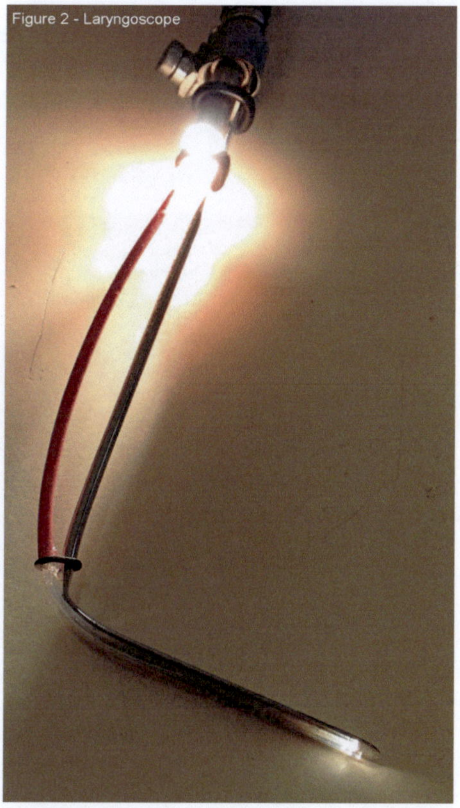

Fig. 2 Example of an air reservoir. The air reservoir made of a standard plastic glove (small cut in one finger) connected to the tube. It is being placed in an airtight chamber with ventilation in order to remove excess Isoflurane

7. *Portable oxygen analyzer* (TED 60-T, Teledyne Electronic Devices, Thousand Oaks, CA, USA)

   In order to certify that a correct amount of oxygen is inhaled, it is useful to have a portable oxygen analyzer connected to the Y-connector. This will reveal the exact concentration when changing to hypoxic gas mixtures and back.

8. *Y connector* (Part # YJ-38-HDPE, Industrial Specialties Mfg, CO, USA)

   The Y connector receives gas from the ventilator, pushing it into the animal, before leading it back again to the ventilator. Make sure to get a Y connector that fits with the angiocatheter, or whatever tube you use for intubation.

9. *Intubation tube—White angiocatheter (17 Gauge)*

   Intubation tubes are used to intubate the animal, which in this case means the hollow plastic tube. The needle may be adapted to work as a stylet to facilitate intubation (*see* Subheading 3).

10. *Laryngoscope blade—formatted spatula with fiber-optic tubing*

    Rodent laryngoscopes are commercially available, yet it is possible to construct your own using a bent spatula with fiber-optic tubing as light source (Fig. 3).

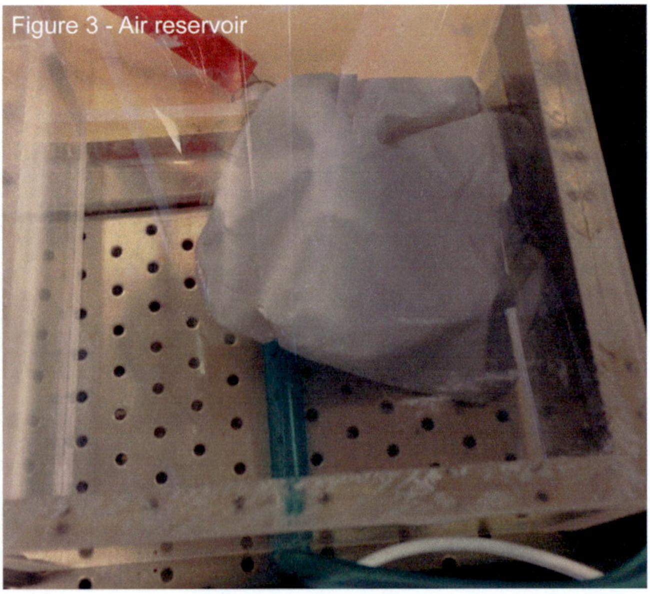

Fig. 3 Example of a rat laryngoscope. Laryngoscope made of a small spatula connected to a fiber-optic light in order to provide adequate lighting conditions for intubation

11. *Tubing* (oxygen bubble tubing)

    When choosing tubes for your model, it is important to use tubes that are built to transport oxygen and reduce leakage from the tube and connections in the system. Bubble tubing is preferable to certify tight connections, and a color (green or pink) to increase visibility.

2.7 Injury Device Setup

The equipment needed is dependent on the type of injury inflicted to the animal. Nevertheless, the following instruments are always necessary:

1. A scalpel (replaceable blades).
2. Electronic shaver.
3. Electronic scale.
4. Suturing material (needle holder, Vicryl® 4-0).
5. Retractors.
6. Tweezers.
7. Syringes (1–5 mL for medical injections, blood gas sampling, etc.).
8. Cotton swabs.
9. Saline for wound cleaning.

As mentioned, if your lab is not close to a hospital, I recommend contacting a clinical department that might be able to provide you with discarded or used equipment. This chapter describes a setup, but several devices from different manufacturers are of course possible.

2.8 Head Impactor (TBI 0310, Precision Systems and Instrumentation LLC, Lexington, KY, USA)

The injury device is a commercially available CCI device and has been used in previous studies [22, 23]. It allows modification of dwell time, impact speed, and impact depth making it possible to produce different injury severity in the same experiment.

2.9 Stereotaxic Frame (Model 900, Agnthos, Stockholm, Sweden)

Any commercially available stereotaxic frame modified for use on rats, that allow ample room around the head, is preferable. Be sure to remove anything that might affect the injury device. The impactor (TBI 0310) allows for quite high, and wide, frames.

2.10 Heating Pad (Temperature Control Unit HB 101/2, Panlab, Harvard Apparatus, Barcelona, Spain)

A commercially available heating pad, using a rectal probe to monitor body temperature.

2.11 Surgical Drill (Microspeed 317 IN; Silfradent, Forli, Italy)

This is originally a drill for dental use, but is perfect to perform a small craniectomy. Diamond tip, 0.5 diameter millimeter drill or equivalent should be used.

2.12 Surgical Microscope (Wild Heerbrugg M3C Stereozoom Microscope, Leica, Wetzlar, Germany)

A surgical microscope with the capacity to magnify up to 40× will be necessary to perform the craniectomy.

3 Methods

This section will guide you through a standard hypoxic TBI model using a CCI device.

3.1 Preoperative Conditions

3.1.1 Procedure

1. Put the animal in the airtight induction chamber (*see* **Note 5**).
2. Distribute isoflurane 5% (with 100% O_2 as carrier) induction until the animal is fully unconscious (*see* **Note 6**).
3. Continue swiftly with intubation procedures.

3.2 Intubation

In many hypoxic TBI models, it is necessary to control the animal's respiration. Even if it is possible to do this with a mask in the case of gas anesthesia, it is recommended to intubate the animal and subsequently connect it to a respirator. Not only will the respirator give you full control over the respiration of the animal, it will also allow adding different gasses to the air mixture inhaled by the rat, which is usually the most important aspect of many hypoxic models of TBI. The following will describe the method for smaller animals.

The intubation process is of course not always identical, depending on factors such as the animal's size, tools provided etc. It is recommended to use a sliding plane (45°) to gain access to the trachea of the animal. In all cases of intubation, it is also important to have adequate lighting conditions. The ideal situation is to intubate in a dark environment with light only emitted by the intubation device (lamp on tip of the laryngoscope blade) used to visualize the trachea. Transillumination through the pharyngoepiglottic region (through the skin) is also possible to increase visualization [24]. There are commercially available intubation equipment. For animals larger than rodents, pre-oxygenation with mask is often necessary.

The size of the intubation tube should depend on which animal is used. There are commercially available tubes to be used on different animals/sizes. The tube should not be too large, since it will increase the risk of damaging the larynx and trachea of the animal. If the tube is too small, it might not be adequate to sedate the animal or it could allow too much gas to pass next to the tube hence decrease the control of the animal's respiration. The tube should not be entirely rigid as it requires some flexibility throughout the experiment. A 16-gauge angiocatheter (without the needle) is optimal to use, and the including needle could be reformatted to fit as a stylet to facilitate the procedure [24]. The tube is often stitched in the cheek of the animal in order to stay in place. This is however not necessary if your stereotactic frame allows you to connect the tube to it, which means less unnecessary pain for the animal.

3.2.1 Procedure

1. Place the animal on its back, preferably with its front teeth hooked to a rubber band or equivalent, lying in a 45° angle, which will give you increased accessibility to the larynx and trachea.
2. A laryngoscope should be used to carefully elevate the larynx to visualize vocal cords.
3. It is important to time the intubation when the rat inhales and retracts the vocal cords. When inserted, it is not entirely easy to visualize if the tube is in the correct place. Though it is possible to feel the texture of cartilage in the trachea which means that the tube is correctly located (*see* **Notes 7–9**).
4. Get a firm grip of the animal and the tube, and swiftly connect it to the ventilator/stereotaxic frame.

3.3 Preoperative Local Anesthesia and Analgesia

Attach the rat to the stereotaxic frame, with the ear bars firmly fixated, and the animal hooked up to the ventilator. If your animal weighs 300 g, and your respiratory rate is 90 breaths per minute, set the tidal volume to 1.5 ml. See chart for more specific information (CWE Incorporated, SAR-830 series Small Animal Ventilator, Instruction Manual, page 14) (*see* **Note 11**).

3.3.1 Procedure

1. Shave the head of the rat, preferably with an electronic shaver.
2. Dilute bupivacaine (Marcain®) to 0.25 %; inject about 0.2 mL in the skin in the midline of the skull.
3. Inject 0.05 mg/kg buprenorphine (Temgesic®) subcutaneously (*see* **Note 10**).

3.4 Surgery: Induction of TBI

When all preoperative medication has been administered, it is time to prepare for the surgical event that will lead to the traumatic injury. This method will now describe a common trauma procedure: the controlled cortical impact (CCI).

3.4.1 Procedure

1. A major area of the right parietal bone is removed (*see* **Note 12**) using a surgical drill (Microspeed 317, Ø0.5 mm).
2. After craniectomy, the stereotaxic frame is moved to the CCI-device.
3. The impact center is adjusted,
   (a) mm right of the central suture
   (b) mm posterior to the lambda.

4. The exact location of the impact is adjusted using the built in stereotaxic levers.
5. A piston, 3 mm in diameter is used to impact a 3 mm deep lesion in the right parietal lobe (an injury described as a severe TBI [25]) (*see* **Note 13**).
6. Following the impact, the animal is removed from the CCI device, and is set to inhale either:
7. A hypoxic- (11 % O_2/88 % N_2) gas mixture—saturation should be aimed at 50 % (FiO_2 11 %).
8. Or normoxic (22 % O_2/78 % N_2) gas mixture—normoxic group >90 % (FiO_2 21 %).
   (a) With an appropriate amount of isoflurane to maintain sedation (1–1.5 %) for 30 min.
9. The head incision is stitched using Vicryl® 4-0.
10. When 30 min has passed, a blood gas sample is extracted from the tail artery, a wound that should be later stitched with Vicryl® 4-0.
11. Remember to mark the ear of the rat for further identification.

3.5 Per-operative Monitoring

Usually when performing animal experiments, pulse oximetry, or visual control that the rat is breathing and having a pulse, might be enough to monitor a rat (or another small rodent). In hypoxic TBI models however, monitoring the rat's physiological parameters is extremely important since the goal of the model is based on certain hypoxic thresholds (*see* **Note 15**).

3.5.1 Pulse

The pulse rate is easily detectable using a pulse oximetry device on the tail or paw of the animal. The pulse rate varies a lot during surgery, the normal heart rate for a rat is around 330–365 beats per minute (bpm) [26]. Passing the lower range usually means that an excessive amount of sedation has been used and/or the rat is dying, both should prompt an immediate response. The pulse usually follows blood pressure, which could decrease following a major bleeding or a hypoxic state. If the upper range were passed, it could mean that the rat is starting to wake up, or is sensitive to per-operative nociception. Therefore, the aim of the model is to keep the pulse rate within a certain interval.

3.5.2 Blood Pressure

Blood pressure (BP) in rats can be measured using either tail cuff plethysmography or radio-telemetry, which are both noninvasive (commonly used) or by intra-arterial catheters, which is invasive procedure. Both rats and mice have almost the same blood pressure as humans of 120 mmHg systolic and 80 mmHg diastolic pressure. An invasive intra-arterial catheter is, compared to the noninvasive monitoring, more accurate for BP monitoring [27]. BP usually decreases in hypoxic models [6, 13, 14] providing an additional systemic insult (hypotension) in systemic hypoxia models.

3.5.3 Peripheral Oxygen Saturation

The definition of hypoxia in TBI is usually defined as a peripheral oxygen saturation <90% [28]. That is why peripheral oxygen saturation is a common target in several models of hypoxic TBI [6, 7]. Another reason is that it is a non-invasive and readily available provider of real-time monitoring which allows for swift changes in ventilator settings in order to optimize the physiological conditions. Peripheral saturation is, however, not perfect because the peripheral perfusion of the animal decreases during hypoxic conditions and so does the quality of the measurements (*see* **Notes 16** and **18**).

3.5.4 End Tidal pCO_2

The exhaled pCO_2 from the animal could be measured using a capnograph. This will give you an indirect measurement of the pCO_2 in the blood which could guide the treatment. A low pCO_2 (less than 4.0) could indicate hyperventilation, but also that the metabolism is too low and that the amount of isoflurane should be decreased (*see* **Note 17**).

3.5.5 Fraction of Inspired Oxygen (FiO_2)

If more advanced monitoring equipment is available, the fraction of inspired oxygen (FiO_2) should be monitored. The FiO_2 is the assumed percentage of oxygen concentration participating in gas exchange in the lung parenchyma [29]. FiO_2 could be used instead of peripheral saturation since it provides better non-invasive real time monitoring [14]. Furthermore, the correlation between FiO_2 and pO_2 has been shown to be $r = 0.995$ in rats [30].

3.5.6 Blood Gases

While peripheral oxygen saturation and FiO_2 provide an indirect measurement of the actual inhaled fraction of oxygen, the blood gas will be the proof that the rat has actually been affected by hypoxia. Several physiological parameters may be analyzed using standard blood gas equipment. At least one blood gas sample should be acquired, preferably after the hypoxic gas has been inhaled for a designated time. Obtaining of additional blood gas samples is recommended.

Procedure

1. Make an incision in the midline, ventral side, of the root of the rat's tail (optional)—it is possible to do it through the skin. Be sure to stitch the incision if you make one!
2. The artery is located a few millimeters below the skin.
3. Insert a syringe horizontally with a thin hypodermic needle (25–26 gauge) into the dissected artery.
4. You will need about 1.5 mL to analyze in the blood gas analyzer (*see* **Notes 19** and **20**).

Blood Gas Parameters (See Note 21)

1. *pO2*

   pO_2 is the partial pressure of oxygen, amount of O_2, in the blood. During normoxic conditions it is around 16–18 kPa in rats, but may decrease to 4–6 kPa during hypoxic conditions [13].

2. *pCO2*

   pCO_2 is the partial pressure of carbon dioxide, amount of CO_2, in the blood. The cerebral blood flow is correlated to the pCO_2 levels [31], and hyperventilation ($pCO_2 > 4.5$) is sometimes used as a treatment regime for high intracranial pressure. Normal levels should be around 4.5–5.3 kPa, and should not differ between normoxic and hypoxic animals [13].

3. pH

   pH is a measurement of the acidity/basicity of the blood. Prolonged periods of hypoxia will lower the pH to 7.26–7.32, while normal levels are around 7.4 in SD rats [13].

4. *Lactate*

   Lactate is produced during anaerobic metabolism, a condition common during hypoxic conditions. Normal reference levels are 0.4–2.2 mmol/L, but levels around 5.0 mmol/L are present during hypoxic conditions [7].

5. *Temperature*

   The normal body temperature of a rat is 35.9–37.5 °C and should be maintained throughout the experiment. Preferably by placing the animal on a heating pad after intubation, inside the stereotactic frame, and never removing the animal during the surgery. Since many experiments are often long in duration, it is important to ensure an even core temperature throughout the experiment.

6. *More advanced monitoring*

   Other monitoring techniques in hypoxic TBI are available, such as intracranial pressure monitoring [32], brain tissue microdialysis monitoring (cerebral metabolism, including lactate, pyruvate, glucose, and glycerol) [6] and magnet resonance imaging (MRI) [12]. These are however more advanced techniques that are not necessary for basic hypoxic TBI models.

3.6 Postoperative Examination

3.6.1 Procedure

1. When the monitoring time after injury has passed (turn off the isoflurane and shift inhalation to 100% O_2).
2. Let the animal increase pCO_2 to at least 4.0 kPa prior to extubation to ensure spontaneous breathing.

3. Let the animal stay in the induction chamber filled with 100% O_2 until conscious, and then return it to its cage.
4. Be sure to monitor the rat frequently the first hours after trauma, then at least once per day and be sure to note any changes and deteriorations in the rat (*see* **Note 14**).

3.7 Functional Tests

To determine the severity of the lesion inflicted, neurofunctional tests are often used. Since many of the focal injuries inflicted are in the fronto-parietal region, sensoric and motoric tests are the most commonly used. Neurologic assessment [12], rotarod, limb placing (walking analysis), and beam walking are commonly used [6, 13, 14]. Please see more detailed descriptions of these methods.

3.8 Euthanasia

3.8.1 Procedure

1. Euthanasia is performed by injecting of sodium pentobarbital, 100 mg/kg, intraperitoneally.
2. As the animal loses consciousness,
   (a) Blood should be extracted through puncture of the (still pulsating) heart with a hypodermic needle (preferably a 19 gauge or bigger). (A total of 4 mL is usually easy to collect in a 250 g SD rat.)
   (b) Cerebrospinal fluid could be collected through the cisterna magna; please see reference for more details [33].
3. The rat is then perfused using normal saline, followed by formaldehyde (depending if you want fresh frozen tissue or not). The brain should be further collected for immunohistochemistry, western blot, flow cytometry, or other desired analyses.

3.9 Analysis

In order to ensure exhibited hypoxia in the cerebral tissue, using specific antibodies against proteins released, or expressed, during hypoxic conditions is important. Hypoxia-inducible factor-1α (HIF-1α) [34], heat-shock protein (HSP) 70 [34], HSP 72 [35], or annexin V (A5) [34] are different substances that may be analyzed since they have shown to increase in the damaged brain after hypoxic/ischemic injury.

4 Notes

1. Animals should be handled with special caution since when affected by hypoxia; they may need more attention and treatment. If it looks like an animal is suffering in any way [36], euthanasia should be considered in order to prevent unnecessary discomfort.
2. Spacious cages, where the animal can move around freely, are recommended to provide adequate stimulation between experiments and to improve functional tests after TBI. Normal, or

enhanced, ventilation where the animal is being kept is important for recovery after hypoxic TBI.

3. While per-operative hypothermia have been shown to improve brain functions after hypoxic TBI [8, 9], prolonged hypothermia in conscious animals is probably detrimental and would only increase stress in the injured animal. However, in ischemic stroke models, prolonged mild hypothermia has been shown to improve functional outcome and lesion volume [37]. To prevent temperature bias in your experimental model, normal room temperature is recommended.
4. Food and water should be kept ad libitum before and after experimental procedures in order to optimize the nutritional conditions in the animal and therefore be able to remove such factor as potential bias. Since hypoxia induces metabolic stress to the animals, the animal should be as nutritionally well prepared for the experiment as possible, provided that the experimental model allows for it.
5. Something that you might forget is to weigh the animal; therefore a tip is to place the scale on the induction chamber and to take note while the animal is sedated.
6. A good way to know when the animal is sufficiently sedated is when it has about 3 s between each breath (respiratory rate 20/min, SD rat). This will give a time frame of about 100 s until the animal wakes up to perform the intubation procedure, and to further connect the animal to the ventilator for continuous sedative administration. If the animal wakes up, be sure to put it back in the induction box and wait until full sedation.
7. For SD rat intubation, a white angiocatheter (17 gauge) is possible to use. For Dark Agouti (DA) rats of approximately the same size, the white angiocatheter is usually too large, a green angiocatheter (18 gauge) works better. One is able to fit the angiocatheter to the stereotaxic frame, without having to stitch it in the cheek of the rat, *see* Fig. 4.
8. Hints that will indicate correct tube placement is rhythmic breathing when connected to respirator, rhythmic fogging of the tube when the animal exhales and, if end-tidal pCO_2 is measured, there will be a pulsating curve on the capnography. If the intubation is esophageal, the rat will wake up after about 30–60 s.
9. If you fail with the intubation more than 2–3 times, you might cause some hemorrhage in the larynx of the rat. Do not continue after this. Keep the rat in 100% O_2 for a while until it gets fully conscious and determine that you do not hear any wheezing sounds while the rat is breathing. If this wheezing continuous, and the rat behavior is altered in an unnatural way, consider euthanasia.
10. We have noticed that the use of opioids leads to a lower usage of per-operative isoflurane. This is favorable in many ways since

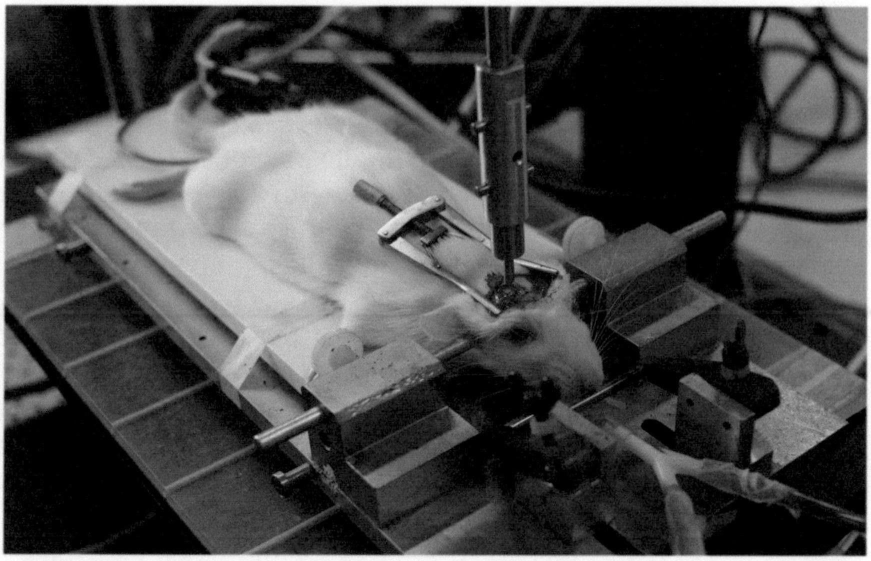

Fig. 4 Surgical setup. The figure shows a SD rat in stereotaxic frame, connected to the ventilator through a Y connector and angiocatheter. The animal is located on a heating pad, rectal probe inserted. Hemicraniectomy has been performed over the right parietal cortex. Pulse oximeter clip is located on the left back paw of the animal

a too high use of isoflurane was the most common risk of mortality (since it probably led to cardiac arrest) in our experimental setting.

11. Check that your ventilator settings are actually delivering the amount of volumes it claims. This is easily done by turning a full beaker upside down in water and inserting the y-connector into the beaker and turning on the ventilator. The volume of the bubbles that emerge will remove the same volume of water; hence, after five "respirations," it is possible to determine the amount of gas the rat will actually inhale. This is very important, even with a modern ventilator, since gas leakage could be present anywhere in your model.
12. Ensure to shave the head of the rat prior to injection of local anesthesia in order to better visualize the surgical area.
13. The craniectomy in my model was about 4 (wide) × 6 (long) mm, to ensure that the impactor does not touch the bone. Make sure to stay about 1 mm away from the midline suture to avoid damage to the superior sagittal sinus. Use a lot of saline to avoid injury because of drying of the area, and note that the microscope lamp usually makes the surgery area warmer, which further increases the need for saline.
14. Before returning the animal to their cage, be sure to wash away any blood since it will attract other rats in the cage to chew on the stitches. If any suture should open up, be sure to re-stitch.

15. It takes about 10 s for a change on the ventilator to give an effect on the peripheral saturation. This of course depends on the ventilator setting, size of reservoir, and the size of the animal.
16. Let the animal have a spontaneous breathing, around one breath every 3–4 s. It will be your best insurance that the rat is still alive since pulse and saturation are two parameters that may quickly disappear if the perfusion drops, and the process is irreversible.
17. Make sure that the end tidal pCO_2 does not go below 4.0 kPa since it usually means that the rat is about to die from too much isoflurane; immediately decrease the flow of isoflurane.
18. Be sure to measure the saturation at the same location in each animal. Usually, the pulse oximeter will show saturation a couple of units lower than the actual level of saturation as measured by the arterial blood gas.
19. Use designated syringes for the blood gas; these are often coated with heparin to avoid coagulation.
20. Coat the syringes with heparin if pre-coated syringes are unavailable.
21. pO_2 in the blood gas, as well as peripheral oxygenation, may provide an inadequate picture of the hypoxic conditions in your animals depending on when in the experiment you sample it. However, lactate is more stable and high lactate concentrations in the blood indicate that anaerobic conditions have been present.

References

1. Jeremitsky E, Omert L, Dunham CM, Protetch J, Rodriguez A (2003) J Trauma 54:312–319
2. Chi JH, Knudson MM, Vassar MJ, McCarthy MC, Shapiro MB, Mallet S, Holcroft JJ, Moncrief H, Noble J, Wisner D, Kaups KL, Bennick LD, Manley GT (2006) J Trauma 61:1134–1141
3. McHugh GS, Engel DC, Butcher I, Steyerberg EW, Lu J, Mushkudiani N, Hernandez AV, Marmarou A, Maas AI, Murray GD (2007) J Neurotrauma 24:287–293
4. Jones PA, Andrews PJ, Midgley S, Anderson SI, Piper IR, Tocher JL, Housley AM, Corrie JA, Slattery J, Dearden NM et al (1994) J Neurosurg Anesthesiol 6:4–14
5. Graham DI, Ford I, Adams JH, Doyle D, Teasdale GM, Lawrence AE, McLellan DR (1989) J Neurol Neurosurg Psychiatry 52:346–350
6. Yan EB, Hellewell SC, Bellander BM, Agyapomaa DA, Morganti-Kossmann MC (2011) J Neuroinflammation 8:147
7. Hellewell SC, Yan EB, Agyapomaa DA, Bye N, Morganti-Kossmann MC (2010) J Neurotrauma 27:1997–2010
8. Yamamoto M, Marmarou CR, Stiefel MF, Beaumont A, Marmarou A (1999) J Neurotrauma 16:487–500
9. Gao G, Oda Y, Wei EP, Povlishock JT (2010) J Cereb Blood Flow Metab 30:628–637
10. Matsushita Y, Bramlett HM, Alonso O, Dietrich WD (2001) Crit Care Med 29:2060–2066
11. Tanno H, Nockels RP, Pitts LH, Noble LJ (1992) J Neurotrauma 9:335–347
12. Ishige N, Pitts LH, Berry I, Carlson SG, Nishimura MC, Moseley ME, Weinstein PR (1987) J Cereb Blood Flow Metab 7:759–767
13. Bramlett HM, Dietrich WD, Green EJ (1999) J Neurotrauma 16:1035–1047
14. Clark RS, Kochanek PM, Dixon CE, Chen M, Marion DW, Heineman S, DeKosky ST, Graham SH (1997) J Neurotrauma 14:179–189

15. Hallam TM, Floyd CL, Folkerts MM, Lee LL, Gong QZ, Lyeth BG, Muizelaar JP, Berman RF (2004) J Neurotrauma 21:521–539
16. Robertson CS, Valadka AB, Hannay HJ, Contant CF, Gopinath SP, Cormio M, Uzura M, Grossman RG (1999) Crit Care Med 27:2086–2095
17. Strohl KP, Thomas AJ, St Jean P, Schlenker EH, Koletsky RJ, Schork NJ (1997) J Appl Physiol 82:317–323
18. Bellander BM, Lidman O, Ohlsson M, Meijer B, Piehl F, Svensson M (2010) Exp Brain Res 205:103–114
19. Al Nimer F, Lindblom R, Strom M, Guerreiro-Cacais AO, Parsa R, Aeinehband S, Mathiesen T, Lidman O, Piehl F (2013) Brain Behav Immun 27:109–122
20. Gunther M, Al Nimer F, Gahm C, Piehl F, Mathiesen T (2012) Acta Neurochir (Wien) 154:689–697
21. Matthes HW, Maldonado R, Simonin F, Valverde O, Slowe S, Kitchen I, Befort K, Dierich A, Le Meur M, Dolle P, Tzavara E, Hanoune J, Roques BP, Kieffer BL (1996) Nature 383:819–823
22. Anderson KJ, Scheff SW, Miller KM, Roberts KN, Gilmer LK, Yang C, Shaw G (2008) J Neurotrauma 25:1079–1085
23. Norris CM, Scheff SW (2009) J Neurotrauma 26:2269–2278
24. Rivard AL, Simura KJ, Mohammed S, Magembe AJ, Pearson HM, Hallman MR, Barnett SJ, Gatlin DL, Gallegos RP, Bianco RW (2006) J Invest Surg 19:267–274
25. Clark RS, Schiding JK, Kaczorowski SL, Marion DW, Kochanek PM (1994) J Neurotrauma 11:499–506
26. Coleman TG (1980) Am J Physiol 238:H515–H520
27. Plehm R, Barbosa ME, Bader M (2006) Methods Mol Med 129:115–126
28. Bratton SL, Chestnut RM, Ghajar J, McConnell Hammond FF, Harris OA, Hartl R, Manley GT, Nemecek A, Newell DW, Rosenthal G, Schouten J, Shutter L, Timmons SD, Ullman JS, Videtta W, Wilberger JE, Wright DW (2007) J Neurotrauma 24(Suppl 1):S7–S13
29. Allardet-Servent J, Forel JM, Roch A, Guervilly C, Chiche L, Castanier M, Embriaco N, Gainnier M, Papazian L (2009) Crit Care Med 37:202–207, e204–e206
30. Rolett EL, Azzawi A, Liu KJ, Yongbi MN, Swartz HM, Dunn JF (2000) Am J Physiol Regul Integr Comp Physiol 279:R9–R16
31. Reivich M (1964) Am J Physiol 206:25–35
32. Kusaka G, Calvert JW, Smelley C, Nanda A, Zhang JH (2004) J Neurosci Methods 135:121–127
33. Pegg CC, He C, Stroink AR, Kattner KA, Wang CX (2010) J Neurosci Methods 187:8–12
34. Zhang X, Deguchi K, Yamashita T, Ohta Y, Shang J, Tian F, Liu N, Panin VL, Ikeda Y, Matsuura T, Abe K (2010) Brain Res 1343:143–152
35. Chen M, Clark RS, Kochanek PM, Chen J, Schiding JK, Stetler RA, Simon RP, Graham SH (1998) J Neurotrauma 15:171–181
36. Mayer J (2007) Lab Anim 36:43–48
37. Yanamoto H, Nagata I, Niitsu Y, Zhang Z, Xue JH, Sakai N, Kikuchi H (2001) Stroke 32:232–239
38. Ishige N, Pitts LH, Hashimoto T, Nishimura MC, Bartkowski HM (1987) Neurosurgery 20:848–853
39. Clark RS, Chen J, Watkins SC, Kochanek PM, Chen M, Stetler RA, Loeffert JE, Graham SH (1997) J Neurosci 17:9172–9182
40. Dave JR, Bauman RA, Long JB (1997) Neuroreport 8:395–398
41. Bramlett HM, Green EJ, Dietrich WD (1999) J Neurosurg 91:653–659
42. Robertson CL, Clark RS, Dixon CE, Alexander HL, Graham SH, Wisniewski SR, Marion DW, Safar PJ, Kochanek PM (2000) Crit Care Med 28:3218–3223
43. Hellewell SC, Yan EB, Alwis DS, Bye N, Morganti-Kossmann MC (2013) J Neuroinflammation 10:156

Chapter 27

Animal Models of Posttraumatic Seizures and Epilepsy

Alexander V. Glushakov, Olena Y. Glushakova, Sylvain Doré, Paul R. Carney, and Ronald L. Hayes

Abstract

Posttraumatic epilepsy (PTE) is one of the most common and devastating complications of traumatic brain injury (TBI). Currently, the etiopathology and mechanisms of PTE are poorly understood and as a result, there is no effective treatment or means to prevent it. Antiepileptic drugs remain common preventive strategies in the management of TBI to control acute posttraumatic seizures and to prevent the development of PTE, although their efficacy in the latter case is disputed. Different strategies of PTE prophylaxis have been showing promise in preclinical models, but their translation to the clinic still remains elusive due in part to the variability of these models and the fact they do not recapitulate all complex pathologies associated with human TBI. TBI is a multifaceted disorder reflected in several potentially epileptogenic alterations in the brain, including mechanical neuronal and vascular damage, parenchymal and subarachnoid hemorrhage, subsequent toxicity caused by iron-rich hemoglobin breakdown products, and energy disruption resulting in secondary injuries, including excitotoxicity, gliosis, and neuroinflammation, often coexisting to a different degree. Several in vivo models have been developed to reproduce the acute TBI cascade of events, to reflect its anatomical pathologies, and to replicate neurological deficits. Although acute and chronic recurrent posttraumatic seizures are well-recognized phenomena in these models, there is only a limited number of studies focused on PTE. The most used mechanical TBI models with documented electroencephalographic and behavioral seizures with remote epileptogenesis include fluid percussion, controlled cortical impact, and weight-drop. This chapter describes the most popular models of PTE-induced TBI models, focusing on the controlled cortical impact and the fluid percussion injury models, the methods of behavioral and electroencephalogram seizure assessments, and other approaches to detect epileptogenic properties, and discusses their potential application for translational research.

Key words Posttraumatic epilepsy, Posttraumatic seizures, Epileptogenesis, Animal model, Rodents, CCI, TBI, Fluid percussion injury, Behavioral seizures, Electroencephalography (EEG)

1 Introduction

1.1 Development of Posttraumatic Epilepsy (PTE) Following Brain Injuries

Traumatic brain injury (TBI) afflicts over ten million people worldwide and around 1.7 million in the United States alone annually, and is also a hallmark injury of the wars in Iraq and Afghanistan [1]. TBI often results in the development of devastating chronic consequences that significantly affect the quality of

Firas Kobeissy et al. (eds.), *Injury Models of the Central Nervous System: Methods and Protocols*, Methods in Molecular Biology, vol. 1462, DOI 10.1007/978-1-4939-3816-2_27, © Springer Science+Business Media New York 2016

life of survivors. PTE is one such common consequence of brain injury [2–5] and TBI is among the major causes of acquired epilepsy [6], accounting for about 5 % of the over three million all of epilepsy cases in the US. One remarkable feature of PTE is the variable, often very prolonged, latency from injury to epilepsy, which can range from weeks to years [2, 3, 7]. In general, according to the International League Against Epilepsy (ILAE), epilepsy is defined as a brain disorder that is characterized by either of the following conditions: (a) Operational (practical) clinical definition of epilepsy at least two unprovoked (or reflex) seizures occurring >24 h apart; (b) One unprovoked (or reflex) seizure and a probability of further seizures similar to the general recurrence risk (at least 60 %) after two unprovoked seizures occurring within the next 10 years, or (c) Diagnosis of an epilepsy syndrome [8]. The development of acute posttraumatic seizure (PTS) occurring within weeks after TBI is a common complication of TBI and is often thought to be associated with the progression of PTE further along in life. The reoccurrence of epileptic seizures within the next 2 years in TBI patients with a single acute PTS might be as high as 86 % and the remission rates with longer terms are 25–40 % [9]. The PTS resulting from brain injury or surgery often are categorized as "early" or "late" PTS, occurring within a week or after a week following injury, respectively. "Late" PTS may even occur months or years after initial injury; PTE is a disorder that is characterized by reoccurring "late" seizures.

The etiology of PTE, as well as other types of epilepsy in general, is currently poorly understood and there is little known about the pathophysiological processes involved in the generation of individual seizures or the progression of PTE. The need for better understanding of the detailed molecular mechanisms underlying these processes is among the major translational challenges for those who are developing treatment and prophylactic strategies [1, 10]. Treatment with antiepileptic drugs (AEDs) aimed at controlling seizures remains the principal option for most epilepsy patients, although almost 30 % of those who suffer from epilepsy are resistant to such medications [11–13].

1.1.1 Risk Factors for the Devel-opment of PTE

Understanding of the risk factors for the development of PTE after TBI is important so that preclinical research can allow for reliable modeling of the clinical features as well as better control of the conditions to assure the reproducibility of the results and decrease the variability of the outcome measures in animal models. Intensive studies performed over the past century documented that penetrating TBI is a well-established risk factor for PTE with significantly higher incidence compared to closed head injuries, including contusions and concussions, although these studies were focused mostly on severe brain injuries from wounds garnered in war, with up to 53 % of patients having penetrating wounds [3, 7], reviewed in [9]. Historically, anatomical brain injuries from a blast have not been commonly detected. However, in recent conflicts, blast exposure has been the most common mechanism of injury, accounting for over 80 % of TBI cases in veterans of the Iraq and Afghanistan wars. The risk factors and prevalence of PTE in blast-induced TBI is currently poorly understood, in part because the occurrence of PTE is normally observed over several years after initial injury and the latency may be even more prolonged with mild injuries. Blast-induced TBI is often associated with persistent post-concussive syndrome, posttraumatic stress disorder (PTSD), and chronic pain, and may pose a significant risk of PTE [14]. Limited clinical data suggest that the prevalence of psychogenic nonepileptic seizures is more characteristic in blast-induced TBI [15]. A small study including veterans who were assigned an outpatient diagnosis of both epilepsy and TBI has shown that the diagnosis of PTE was clinically confirmed in 18 % of the sample, whereas a diagnosis of nonepileptic, or psychogenic, seizures was suspected in 44 % of the patients [15]. Similar tendencies are observed in TBI patients from the civilian population [16]. In children, PTS following TBI affects 12–35 % of children after TBI and the occurrence of PTS is associated with worse cognitive and functional outcome after adjustment for injury severity [17].

Significant risk factors for the development of both "early" (within 1 week following TBI) and "late" (after 1 week following TBI) PTS include acute intracerebral and subdural hematomas, surgery for an intracerebral hematoma, increased injury severity (Glasgow Coma Scale in the severe range of 3–8), depressed skull fractures, and dural penetration [9, 16]. Younger age is a risk factor for the development of "early" PTS, whereas older age at the time of injury (over 65 years) is a risk factor for "late" PTS [9]. In addition, the risk factors include preexisting conditions such as chronic alcoholism. The age-related occurrence of the seizures and acquired epilepsies are also observed in other brain injuries. For example, post-stroke seizures and post-stroke epilepsy are widely recognized neurological consequences of stroke, especially in the elderly population [18].

1.1.2 Clinical Management of TBI and PTE

Despite extensive research and limited success with preclinical trials with numerous neuroprotective strategies, the translational viability of such approaches is still elusive [19]. To date, the reduction in elevated intracranial pressure using decompressive craniectomy remains one of the common symptomatic treatments with proven efficacy for reducing mortality following TBI. However, such life-saving treatment is associated with acute post-craniectomy seizures with a high risk (76.2 %) of development of epilepsy [20]. Due to the increased risk of seizures after TBI and the possible association of acute seizures with the development of PTE, several common antiepileptic drugs (AEDs) are routinely used in clinical practice, although their efficacy in the prophylaxis of PTE is disputed [21–28]. Phenytoin and, most recently, levetiracetam are commonly used in the clinic to control PTS and as prophylaxis for PTE [21–23]. These AEDs have different mechanisms of action and, accordingly, differentially affect seizure outcomes in preclinical models. For example, a recent study showed that none of the drugs studied (phenytoin, phenobarbital, carbamazepine, valproate, or some combination therein) gave reliable evidence that it prevented, or even suppressed, epileptic seizures after TBI [25].

Phenytoin is considered a classic anticonvulsant, exhibiting an immediate effect on electrically or chemically induced seizures, whereas levetiracetam is not effective in such acute models but demonstrates antiepileptic properties in chronic seizure models. In addition, limited preclinical data suggest that chronic treatment with levetiracetam may also have neuroprotective properties [29], whereas chronic treatment with phenytoin is deleterious [30], although these studies neither assessed seizures nor other PTE outcomes. Interestingly, certain AEDs are commonly used in the clinic for treatment and prophylaxis of chronic headaches and migraine, suggesting the presence of common mechanisms involved in the progression of these disorders. In general, a relationship between epilepsy and migraine has long been recognized, but the nature of this interaction is still debated [31]. Chronic posttraumatic headaches or migraine occurs in up to 90 % of TBI patients [32], whereas PTE occurs in approximately 10 % of TBI patients with closed head trauma and in over 50 % in patients with penetrating TBI [2–5].

In addition, a significant proportion of TBI patients suffer "silent" non-convulsive seizures, including many without evident behavioral manifestations that can only be conclusively diagnosed by electroencephalogram (EEG) [33, 34]. The incidence of non-convulsive seizures has been reported to be from 18 to 38 % in neurointensive care patients [35–40], with the incidence in patients with brain lesions even higher (up to 55 %) [41].

1.2 Putative Mechanisms Underlying the Development of PTE

TBI causes potentially epileptogenic brain damage through several coexisting mechanisms, including mechanical damage, and primary and secondary excitotoxicity associated with ischemia, intracerebral and subdural hemorrhage, and neuroinflammation [12]. In addition, it is widely recognized that different acute brain injuries involving many common injury mechanisms such as ischemic and hemorrhage strokes are associated with increased prevalence of acute seizures and acquired epilepsy [18]. The anatomical brain pathologies resulting from these brain injuries often resemble brain pathologies observed in epilepsy patients, suggesting the existence of common mechanisms in these neurological disorders.

The clinical and pathophysiological features of PTE closely resemble many features of temporal lobe epilepsy. Well-recognized hallmarks of PTE that could be reproduced in animal TBI/PTE models include: (a) chronic seizure susceptibility; (b) persistent hyperexcitability in the dentate gyrus; (c) hippocampal sclerosis selective cell loss in the hilus of the dentate gyrus and CA3 area of the hippocampus, and (d) synaptic reorganization of mossy fiber.

1.2.1 Cortical and Hippocampal Damage After TBI Associated with Epileptogenesis

It is believed that one of the most important factors associated with acute seizures and the development of PTE is the level of mechanical brain damage and brain lesions resulting from secondary neurodegenerative processes. It is well recognized that seizure susceptibility is associated with the severity of TBI and that total brain volume loss predicts seizure frequency [2]. In addition, experimental and clinical data suggest that hippocampal damage resulting from TBI is also associated with recurrent seizure activity and might be involved in the etiology of PTE [42, 43]. In human TBI patients, reduction of hippocampal volume and alterations in cortico-subcortical connectivity are well-recognized phenomena following TBI and the levels of anatomical pathology associated with cognitive and neurological outcomes [44–47]. Importantly, hippocampal atrophy after TBI reported in humans resembles the hippocampal sclerosis often associated with temporal lobe epilepsy [44], and the presence of such hippocampal sclerosis is associated with poor prognosis and a significantly higher rate of AED resistance [48]. Selective hippocampal cell death has been reported in several common preclinical TBI models, including fluid percussion injury (FPI) controlled cortical impact (CCI; mouse) [49], and weight-drop impact acceleration injury [43]. Further, the data suggest that an early and selective cell loss in the hilus of the dentate gyrus and CA3 areas of the hippocampus is associated with enhanced susceptibility to pentylenetetrazole (PTZ)-induced seizures up to 15 weeks following experimental TBI [43].

1.2.2 Ischemia and Hypoxia Following TBI

Brain ischemia and hypoxia associated with primary vascular damage and secondary injuries are common features of TBI, affecting overall neurological outcomes and mortality [50]. It is well recognized that brain ischemia is associated with glutamate-induced excitotoxicity, and thus, the lowering of the seizure threshold resulting from depolarization is believed to be one of the major contributory factors to seizure. Although ischemia following TBI is commonly associated with severe head injury, there is also evidence of its association with moderate or mild head injuries [51]. Clinical and preclinical studies suggest that the persistent microvascular damage that occurs at chronic stages following TBI resembles microvascular pathologies associated with Alzheimer disease and might be involved in chronic neurodegenerative processes associated with these disorders [52]. On the other hand, the evidence indicates that different ischemic brain injuries are associated with the increased risk of the development of acute seizures and epilepsies. Acute seizures and the increased prevalence of epilepsy are widely recognized neurological consequences of cerebral ischemia. The stroke is also the primary cause of symptomatic epilepsy in the elderly population [18]. In hospital settings, up to 40 % of stroke patients experience seizures manifested as clinical motor convulsions [53, 54]. However, similar to TBI patients, a significant proportion of stroke patients suffer seizures that may or may not have evident behavioral manifestations and that could be conclusively diagnosed only by EEG [33, 34, 53, 54]. Convulsive and non-convulsive seizures are associated with poor outcome and increased mortality [33, 55]. In addition, non-convulsive seizures are also connected with a decline in patient outcome after ischemic brain injury, and are significantly more refractory to AED therapy as compared to convulsive seizures [28, 37, 56].

1.2.3 Hemorrhage and Hemoglobin Breakdown Products

In the hospital setting, the proportion of stroke patients experiencing seizures varies from 1 to 40 %, which may be due to differences in study design and diagnostic criteria [53, 54]. ICH and subarachnoid hemorrhage (SAH) are associated with increased risk for early offset seizures, as defined as 1–15 days after the hemorrhagic stroke [53, 55, 57]. Most seizures associated with ICH occur at onset or within the first 24 h [57, 58]. In about 18 % of patients with SAH, seizures occur before admission and in 4–10 % of patients after admission to hospital; in aneurysmal SAH, the percentage is higher (10–26 %) [59, 60]. The hemorrhagic component associated with acute stroke is associated with increased risk of development of epilepsy and poor quality of life [53, 54, 61, 62].

Although the detailed mechanisms involving in seizure initiation and epileptogenesis associated with ICH are not currently established, the data obtained in animal models suggest that hemolysis, the formation of reactive hemoglobin breakdown products, and accumulation of free iron and iron-rich compounds such as hemosiderin may play significant role [63–65]. Interestingly, recent preclinical data demonstrate chronic progressive punctate deposition of hemosiderin in the traumatic brain [66]. The experimental data suggest that in an iron-induced epilepsy model, which is often referred to as a PTE model, the main mechanisms underlying seizure activity involve *N*-methyl-d-aspartate (NMDA)-receptor-mediated glutamate excitotoxicity, formation of reactive oxygen species, and subsequent membrane lipid peroxidation, and neuronal cell death [64, 65, 67, 68].

1.2.4 Blood–Brain Barrier (BBB) Disruption in PTS and PTE

The role of BBB dysfunction has long been suggested as a factor of epileptogenesis [69, 70]. A concern for the possible role of the BBB in the various epilepsies has been based on ultrastructural studies that demonstrated increased micropinocytosis in cerebral capillaries during seizures. Preclinical results indicate that evolving white matter degeneration following experimental TBI is associated with significantly delayed microvascular damage and focal microbleeds that are temporally and regionally associated with the development of punctate BBB breakdown and progressive inflammatory responses [66]. Experimental and clinical evidence suggest that BBB pathologies are associated with several potentially epileptogenic abnormalities connected to brain injuries such as infiltration of erythrocytes and other blood cells and blood components resulting in hemoglobin-related products which collectively are associated with inflammatory mechanisms with augmented neuronal activity [70, 71]. Interestingly, acute and chronic BBB pathologies and neurodegeneration in TBI models are associated with biomarker responses similar to those observed in one of a most common seizure model induced by kainic acid neurotoxicity [66, 72, 73].

1.2.5 Neuroinflammation

Neuroinflammation is characterized by the activation of several neuronal and glial pathways in response to insult or injury, primarily involving microglia that are considered to be the resident innate immune cells in the brain, resulting in the release of proinflammatory cytokines and other mediators. Infiltrating macrophages could also trigger a parallel inflammatory cascade. A growing body of experimental and clinical evidence suggests that neuroinflammation involving the increased production of various proinfammatory molecules. Among them, the so-called proinflammatory prostaglandins results from the upregulation of cyclooxygenase (COX) and prostaglandin E synthase (PGES) enzymes in the head and the spinal cord after trauma [30, 32, 74–76] and these prostaglandins consequently affect related neurological disorders such as stroke, seizure disorders, and epilepsy [77, 78]. Several preclinical studies using different animal models demonstrated COX-2 upregulation during the seizures, mainly in hippocampal neurons and glia [79, 80], and this upregulation was associated with increased neuronal activity. There are two major isotypes of COX enzymes expressed in the brain: COX-1 and COX-2, often called constitutive and inducible, respectively, that are encoded by different genes and differently respond to traumatic insults or toxic stimuli. Although both COX-1 and COX-2 are constitutively expressed in the brain, the levels of their expression may change during pathological conditions [81]. Data suggest that the COX-2 expression is primarily increased in the brain in response to a variety of brain injuries and diseases involving excitotoxic brain damage, ischemic and hypoxic insults, and ICH, and increased COX-2 immunoreactivity has been demonstrated in both microglia and neurons [82–84]. A number of studies in rats and mice subjected to TBI demonstrated that COX-2 expression is increased mainly in ipsilateral cortex and hippocampus [74–76, 85, 86]. However, the level of COX-2 expression is more profound in the hippocampus, with significant increases starting within hours after injury and persisting up to 2 weeks [30, 32, 87]. The increased COX-2 mRNA immunoreactivity was localized through neuronal and astrocytic markers: MAP2 and GFAP, respectively [87]. Additionally, some preclinical data suggest that both the selective and non-selective COX-2 inhibitors currently in use clinically might also be used for the treatment of neurological disorders, notably ischemic stroke [88–92] and TBI [30, 32, 74, 75], though studies have also shown that COX-2 induction following TBI may play a protective role [75, 76].

However, one has to be consistently careful of well-documented adverse side effects of COX-2 inhibitors [93]. Similarly, targeting downstream COX-2 pathways such as selective prostaglandin receptors PGE_2 EP1 and $PGF_{2\alpha}$ FP receptors have been suggested as an alternative to COX-2 inhibitors affecting total prostaglandin production. Pharmacological blockade or genetic deletion of FP receptor improves some outcomes in mouse TBI model [94]. However, the role of the EP1 receptor is complex and its blockade differentially affects outcomes in different neurological conditions such as ischemia [88–92], hemorrhagic stroke [85, 95], excitotoxicity [88, 89, 91], epilepsy [96, 97], surgical brain injury [98], and TBI [49, 99]. Interestingly, blockade of the EP1 receptor was tested as a strategy in the treatment of epilepsy [96, 97], but to reach a desired effect, it required up to a 1000-fold higher dose of one of the commonly used antagonists (i.e., SC-51089) [97] than that used in models of stroke and NMDA-induced excitotoxicity [100].

Increased prostaglandin levels has been described in a wide range of animal models of epilepsy, including with both chemically (e.g., kainic acid [80, 101–103], NMDA [104], PTZ [105–110], picrotoxin [106], and pilocarpine [111]) or electrically induced seizures, as well as during seizures in epileptic and convulsion-susceptible strains [112, 113]. One difference between these models is $PGF_{2\alpha}$, one of the major prostanoids increased during seizures, and the data indicate that this increase is due to augmented neuronal activity [106, 108, 113]. On the other hand, there is experimental evidence that in rodent PTZ and kainic acid mouse models, the increased levels of $PGF_{2\alpha}$ preceded epileptic activity [114], which might be considered a factor modulating seizure susceptibility. Although $PGF_{2\alpha}$ is elevated during seizures, its role has long been disputed. Interestingly, recent data suggested the differential roles of the $PGF_{2\alpha}$ cognate FP receptor in TBI and seizure models. The pharmacological blockade of the FP receptor after TBI with a selective antagonist (i.e., AL-8010) improves neurological outcomes and hippocampal swelling [94]. In contrast, the data obtained in immature and adult models of kainic acid-induced seizures demonstrated that the decrease in $PGF_{2\alpha}$ levels or blockade of FP receptors worsened seizure outcomes [115, 116].

1.3 Animal Models of PTE

Several animal models have been developed to reproduce the mechanics of TBI, reflect its anatomical pathologies, and replicate its neurological deficits. Historically, the development of preclinical models was focused on using larger animals, which were further adapted to rodent species. Although larger animals with gyrencephalic brains are likely more closely reflect neurophysiological and biomechanical aspects of human TBI, recent rodent models have become standard in TBI and epilepsy research because of their modest cost, small size, and accelerated development compared to larger animals, in addition to some of their well-characterized outcome measures. Currently, the most widely used preclinical models in TBI research include FPI [117], CCI injury [118, 119], weight-drop impact acceleration injury [120], and blast injury [121].

Acute and chronic recurrent PTS is a well-recognized and commonly observed phenomena observed in most of these models using different species except with blast injuries. However, there is only a limited number of studies using these models that are specifically focused on PTE, probably because of its chronic nature and complicated outcome assessment. The most commonly used mechanical TBI models with documented electroencephalographic and behavioral seizures and remote epileptogenesis include FPI, CCI, and weight-drop.

Recent clinical data suggest the utility of modern neuroimaging techniques for the prognosis of long-term TBI outcomes; that would likely also be useful for revealing the mechanisms underlying PTE. Nonetheless, prospective clinical studies employing advanced brain imaging and EEG monitoring, as well as preclinical research and development of improved animal models closely reflecting all features and aspects of human disease, are still warranted to better understand the phenotypes and etiopathology of PTE [122].

Preclinical models of brain contusion and concussion are among the most widely used models of PTE, as they represent the biomechanics, brain pathology, and neurological deficits characteristic of human TBI. In contrast, even some clinical and experimental data suggest epileptogenic alterations following blast injury preclinical studies focusing on PTS or PTE are still warranted.

1.3.1 PTS in Animal Models of Blast Injury

In blast animal models, the occurrence of PTS or EEG epileptiform is not generally assessed for reported outcomes [123]. There are several models of blast exposure-induced TBI often referred to as overpressure injury (OBI) that produce different severities of TBI. A lack of seizure data in blast models might be due to fewer blast studies compared to other models; it is also possible that the impact magnitudes used in most of these studies was insufficient to produce injury with obvious seizure outcomes.

Only a few studies employed EEG in the blast models and the data suggest that blast affects electrophysiological properties associated with brain injury at acute time points similar to those observed in contusive or concussive models of TBI, but provided no evidence of epileptiform activity observed in other models of

TBI and stroke. For example, the exposure of goats to a biphasic blast waveform comprising incident and reflection peaks caused alterations in EEG waveforms including an increase in amplitude and a decrease in frequency reflecting brain depression that was associated with substantial changes in gross brain pathology assessed microscopically, including SAH and parenchymal hemorrhage, enlarged perivascular space, vascular dilatation, and congestion accompanied by elevated serum concentrations of S-100β and specific enolase at acute time points [124]. In rabbits, blast exposure caused characteristic systemic responses (e.g., mean arterial pressure, bradycardia, fast and shallow respiration) and significant increases in both frequency and amplitude of EEG activity, but there were no anatomical changes such as hemorrhage or brain lesions [125]. Acute PTSs were reported following severe blast and, based on our best knowledge, there are no published studies to date reporting the occurrence of chronic seizures or epileptogenic changes.

1.3.2 PTS and PTE in FPI Models

The FPI model is one of the most commonly used models of TBI and is one of the most commonly used models of PTS and PTE. The main advantages of this model are a relatively simple setup and a possibility for inducing pathophysiological features of human brain trauma (with the intrinsic caveats of the given species brain anatomy), including intracranial hemorrhage, brain swelling (edema), changes in intracranial pressure, and progressive grey matter lesions [126, 127]. It is also often considered as model of diffuse TBI. However, one of the disadvantages of this model is a requirement for precise surgical procedures and control of the injury parameters in the commonly used devices where even minor deviations may result in significant variability of the outcomes and relatively high mortality compared to other models. However, recently a pneumatic instrument with microprocessor-controlled pressure and dwell time has been developed to address operational concerns and reproducibility issues presented in standard FPI devices [128]. Initially, the FPI model was developed for use in cats and rabbits [129–131] and was soon adapted for use in rats [132, 133] and mice [134]. The FPI model produces a rapid controlled fluid pressure pulse applied directly to the dura mater through a craniotomy window, causing a brief deformation of brain tissue. The change in pressure is produced by moving a pendulum striking the plunger of a cylinder filled with a fluid (e.g., physiological solution, PBS, ACSF).

There are three major variations of this model based on the position of craniotomy: midline, parasagittal, and lateral. Of these three variations, the lateral FPI in rodents is currently the most commonly used for small animals [133, 134]. FPI model has been used in many different species, including cats [135], rabbits [130], dogs and sheep [136], rats [137], mice [134, 138], and pigs [137, 139].

The pressure is a variable that allows for control of injury severity in the FPI model, affecting overall neurological and anatomical outcomes. In rats, lateral FPI produces a combination of focal cortical contusion and diffuse subcortical neuronal injury extended to the hippocampus and thalamus, but the progression of delayed neuronal cell death following FPI brain injury is slower than in the CCI model and the tissue loss and the glial scarring progress within a year after initial impact [140, 141]. Similarly, cognitive dysfunction and neurological impairment persist for more than a year following severe FPI [132].

In midline FPI models, at acute tine points, hemorrhage is present in the corpus callosum, fimbria hippocampi, and thalamus, and to a lesser extent, in the brain stem [117]. The mossy fiber sprouting in the dentate gyrus revealed by Timm's sulfide silver staining can be observed at both acute and chronic time points [142]. The anatomical pathology in the FPI model is associated with neurological locomotor deficits persistent within the first week of injury, even at moderate magnitudes [117].

An investigation of acute EEG changes after FPI with different pressure magnitudes reflecting injury severity has shown significant decreases in EEG amplitude immediately after initial impact injury, and decreases in delta frequency band (1–4 Hz) activity with more profound changes at higher injury severity [137]. In addition, the changes in waveform dynamics are important for acute outcomes. A study of the effects of peak pressure on neurological outcomes performed in juvenile rats using a programmable FPI device has shown that FPI with the parameter of fast-rising pressure, which could be translated as a primary concussive injury, caused less profound acute neurological outcomes (including fewer incidences of seizure) and decreased acute neurodegeneration of hilar neurons within hours after injury compared to a standard rise in injury pressure parameters [143]. However, one week after injury, the differences in hilar cell loss and dentate hyperexcitability between fast and standard peak pressure rates were no longer significant.

Alteration of Gene Expression After FPI Associated with Epilepsy and Seizures

Excitotoxic neurodegeneration in the hippocampus is evident within the first hours following FPI neuronal hilar cell loss and neuronal cell death has progressed within the first week or weeks after FPI; the pattern of cell death resembles, to a certain extent, the pattern of neuronal death observed in some patients with temporal lobe epilepsy [143, 144]. Lateral FPI injury leads to enhanced expression of the selected immediate early gene markers associated with neuronal hyperactivation within 24 h after injury, including brain-derived neurotrophic factor (BDNF) and c-fos in the ipsilateral hippocampus and Bax in both ipsilateral and contralateral hippocampi, and these changes in the immediate early gene responses are associated with overactivation of NMDA receptors [145]. The overactivation of the glutamate receptors seems to play a major role in the hyperexcitability characteristic of FPI, resulting in excitotoxic cell death. One week after FPI injury, there is neuronal hilar cell loss assessed by silver staining and increased dentate field excitability in response to stimulation of the perforant path [142, 143]. The electrophysiological changes recorded in vitro demonstrate a long-lasting increase in the frequency of spontaneous inhibitory postsynaptic currents (IPSCs) that were associated with the augmented activity of inotropic glutamate receptors [142].

Electrophysiological recordings from slices obtained from FPI-injured rats has demonstrated that the injury caused a persistent decrease in the threshold to induction of seizure-like electrical activity in response to high-frequency tetanic stimulation in the ipsilateral hippocampus [142]. In vitro studies have shown electrophysiological changes resulting in neuronal hyperexcitability and associated with reactive gliosis such as abnormalities in potassium conductance and local field potential recordings, suppressed paired-pulse facilitation, enhanced evoked and miniature glutamatergic excitatory postsynaptic potential, and decreased in the frequency of miniature inhibitory postsynaptic currents (IPSCs) [146–149]. When in vitro electrophysiological recordings from rat slices that underwent FPI were compared to the data obtained from the animals with chronic pilocarpine-induced epilepsy, the results revealed that the changes within the limbic system resulting from TBI are qualitatively similar but quantitatively less severe than changes in rats with chronic temporal lobe epilepsy [150].

Acute PTS and the Development of Epilepsy After FPI

In a midline FPI model, tonic–clonic seizure-like convulsions were noted in early studies, although the convulsions were evident only in animals injured at higher levels of pressure, and the occurrence of PTS was associated with increased mortality [117]. The seizures were observed in only in 11–15 % of surviving rats who underwent FPI, whereas the incidence of seizures was much greater in nonsurviving rats (33–80 %); the occurrence of convulsions was not related to the magnitude of injury [117]. Another study has shown that PTSs were observed in about 50 % of rats within 15 min after FPI [151].

In rats, single severe lateral FPI in rats is sufficient to induce spontaneous chronic seizures evident on intracortical EEG recordings associated with behavioral manifestations [146, 152]. Interestingly, experimental PTE outcomes in lateral FPI strongly depend on the location of injury [153]. Epileptiform activity without behavioral seizure manifestation occurs 5 weeks after lateral FPI and this activity progresses with time; by 33 weeks after impact, nearly all rats who underwent FPI exhibited spontaneous convulsive and non-convulsive seizures. In addition, the latter study reported the occurrence of non-convulsive seizures of lesser severity and duration in some control animals and in over 50 % of rats who underwent sham procedures [154].

PTZ Challenge Test as a Measure of Epileptogenesis After FPI

The occurrence of acute and chronic spontaneous behavioral seizures is strongly associated with injury severity. The seizure could be only observed following FPI with the pressure magnitudes corresponding to severe TBI, whereas in the models with moderate injuries, behavioral seizures are not evident [43, 142, 155]. In addition, chronic spontaneous seizures generally develop within several weeks after induction of FPI.

Thus, to study epileptogenic changes in mild or moderate FPI models or at subacute time points, surrogate tests of epileptogenesis employing induction of seizures with ether bolus injection of infusion of PTZ to assess seizure severity using the Racine's scale or measure seizure threshold. The data using a bolus injection of PTZ at a dose below the seizure threshold in non-injured animals have shown that FPI causes an increase in seizure susceptibility [156]. Persistently enhanced susceptibility to PTZ-induced seizures was observed for at least 6 months after FPI, even in animals that had no spontaneous seizures [43]. However, seizure susceptibility in injured rats was comparable to that in control animals with the lowest number of neurons immunoreactive for parvalbumin, a calcium-binding albumin protein, in the reticular nucleus of the thalamus [157].

The dependence of seizure susceptibility on the magnitude of injury suggests that epileptogenesis is associated with anatomical damage and neuroprotective strategy following TBI and will be also beneficial against the development of PTE. Neuroprotective treatment with hypothermia for 4 h immediately after FPI significantly decreased PTZ-induced seizure susceptibility and attenuated mossy fiber sprouting, a hallmark of the epileptic brain compared with normothermic controls 12 weeks after recovery following experimental TBI [158]. In addition, increased seizure susceptibility in FPI-injured rats determined using the PTZ challenge was associated with enhanced neuronal activity detected on BOLD-fMRI, but not EEG abnormalities assessed 1 week before the PTZ test [159]. On the other hand, induction of seizures with PTZ in rats who underwent FPI resulted in worsened anatomical outcomes such as lesion size and neuronal cell death in the CA3 region of the hippocampus and mossy fiber sprouting [156].

Pharmacological Interventions Affecting TBI and PTE Outcomes in FPI Model

A limited number of studies have shown that some pharmacological interventions may ameliorate PTS and attempts were made to test whether AEDs used clinically could protect from experimental TBI. Interestingly, rats administered tacrolimus, a calcineurin inhibitor, acutely after TBI showed significantly fewer non-convulsive seizures than untreated rats months after FPI, but a similar degree of cortical atrophy [154]. In addition, attempts were made to use common AEDs for neuroprotection following FPI. A study performed with the slices from rats receiving treatment with felbamate after FPI exhibited long-term potentiation in the CA1 region of the hippocampus, which is suppressed in untreated animals, suggesting that the neuroprotective property of felbamate is neuroprotective against traumatic neuronal injury [160]. Another study has shown no significant effects of lacosamide on anatomical damage or functional recovery [161].

1.3.3 PTS and PTE in CCI Injury Model

CCI is one of the most commonly used TBI models, primarily due to the tight control of injury parameters and the resemblance of several anatomical and neurological outcomes with those observed in man following head trauma [119, 162–165]. Applying different CCI parameters allows for the production of experimental TBI with different magnitudes affecting a pattern of excitotoxic, ischemic, and hemorrhagic injury components that can differently affect epileptogenic cascades in the brain following TBI.

Thus, the CCI model allows for the reproduction, in an experimental animal, of several of the characteristics of human TBI, and anatomical and neurological deficits, to a certain degree in a controlled fashion [119, 162–165]. The CCI model study was initially developed as a TBI model in ferrets to reflect features of closed head trauma commonly observed in human TBI with distinctive characteristics of axonal injury caused by deformation of brain tissue that results in delayed axonal degeneration due to secondary injury that spreads over time via the white matter tract [162, 166]. Further CCI models were extended to rodent species and currently there is one of the most characterized and commonly used preclinical models for studying various secondary consequences of TBI such as gliosis and activation of neuroinflammatory cascades [119, 163–165]. To date, the CCI model has been adopted for many different species, including ferrets [118], rats [119, 167], mice [164], pigs [168, 169], and monkeys [170].

In the CCI model, the injury to the brain is induced by compression of brain tissues with a rigid impactor directly positioned to the cortical surface with intact dura through the craniectomy window using a specific pneumatically or electromagnetically driven impact device, allowing for the tight control of the velocity of impact, the dwelling distance, and the time of compression [119, 171]. However, there are studies suggesting some advantages of the electromagnetic versus pneumatic devices [171].

Furthermore, using impactor tips with different diameters allows researchers to add another variable to change the injury severity as well as to scale the magnitudes to animal size. Both types of device are widely used in TBI research and are commercially available. The shape of the impactor (flat of rounded) may also affect biomechanical features as well as outcomes of experimental TBI. The craniotomy (or craniectomy) procedure is crucial for the reproducibility of the injury as even minor damage to the dura mater or bleeding during the surgery may dramatically affect anatomical behavioral outcomes and increase the variability of quantitative measures. Two types of craniotomy surgeries are commonly performed in the CCI model: without closing the craniotomy window or covering the opening with the bone flap secured with bone wax. However, replacing the bone flap after craniotomy surgery even without induction of CCI can produce considerable brain lesions [172]. In contrast, the craniotomy in sham animals without bone flap replacement causes no detectable brain lesions [69, 168]. Furthermore, based on our unpublished observations in both rats and mice, a craniotomy site without bone flap replacement in sham animals starts to heel within 48 h after surgery and is completely healed with longer time points; the bone looks normal with no macroscopic changes within 1–3 months, whereas in animals who underwent CCI, the craniotomy site could be easily detected by lesser bone thickness.

The levels of anatomical alterations in the brain after CCI depends on the injury magnitude from no obvious or significant macroscopic or microscopic changes at low levels to significant cortical contusion and subdural and intraparenchymal hemorrhages that progress with time and complete loss of brain tissue within the injury site, often referred as to cavitation, due to delayed necrosis at higher injury levels [119, 164].

Distinctive anatomical hallmarks that mimic human TBI observed in rodents include concussion [119], diffuse axonal injury, acute subdural and intraparenchymal hematomas [119], loss of cortical tissue [164], delayed hippocampal atrophy, neuronal degeneration in the cortex, hippocampus, and thalamus [164, 173], vascular damage [174], disruption of the BBB and extravasation of blood content [164], and gliosis [66, 164]. However, diffuse axonal injury in rodent CCI is less distinctive than that described in ferrets [162]. The severity of histopathological alterations show correlation with incising in velocity and depth of cortical deformation [171, 172, 175].

The functional changes observed in CCI injury in rodents include both cognitive and motor deficits [171, 176–179]. The motor deficits are transient and most resolve within months after CCI induction, whereas cognitive deficits are persistent and could be observed even at 1 year after injury [176–179]. Cognitive deficits are commonly observed at chronic to subchronic time points and are commonly associated with anatomical changes in the brain such as cortical and hippocampal atrophy, progressive decrease in cerebral circulation [180], Aβ and Tau pathologies, and the level of impairment strictly dependents on the magnitude of the adjustable CCI parameters. In addition, recent behavioral studies suggest changes associated with emotional deficits [178, 180], which are also characteristic for human TBI; although changes in emotional behavior are less dependent on the CCI magnitude compared with cognitive outcomes [178].

Strain- and Gender-Dependent Differences in Outcomes in CCI Models

Strain-dependent differences in the rodent species have been extensively reported in many animal models of brain injuries, including models of TBI and seizures/epilepsy [176]. The mouse strain should be considered especially carefully as the mice from different strains have different sensitivities to excitotoxic injuries, reflecting the differences in seizure susceptibility in both chemically induced and reflex seizures models. However, in a mouse CCI model, the animals from different strains subjected to either to sham or experimental injury show significant differences in functional rather than anatomical outcomes. The differences are most evident in behavioral responses for performing cognitive tasks (e.g., learning in the Morris water maze and Barnes circular maze) [175, 181], whereas the differences in the performance of motor tasks are generally not significant among strains [175].

A study performed in three mouse strains commonly used as background in genetically altered mice (i.e., C57BL/6, FVB/N, and 129/SvEMS) showed that there was no significant difference in body weight between mice subjected to sham surgery or CCI injury or lesion volume induced by CCI within any of the three different strains [175], and these data are consistent with the results of several other studies. Importantly, the posttraumatic mossy fiber sprouting, an outcome implicated in epileptogenic alterations in the brain, was qualitatively similar among CD-1, C57BL/6, and FVB/N mouse strains [181].

Several studies performed using the CCI model suggest that gender does not critically affect several important CCI outcomes such as lesion severity, which is quantified using cresyl violet staining [168] and silver staining volume [168, 171, 177]. However, some gender differences were observed in temporal changes with silver staining following CCI, while the rate of resolution of staining between 48 h and 4 weeks was similar [177]. Interestingly, some outcomes following CCI showing no gender difference are in contrast to that seen in the weight-drop injury paradigm [177]. In addition, some gender-specific differences were observed in PTS activity after CCI where 33 % of male and 25 % of female wild type mice exhibited seizure activity within 2 h after CCI; also, some difference in seizure incidence was observed in A1 adenosine receptor knockout male and female mice [182].

Acute and Chronic PTS Assessment After CCI

The studies performed in rodent CCI models revealed the manifestation of acute and chronic behavioral seizures and EEG abnormalities associated with epileptiform activity and hyperexcitability in the brain [173]. The occurrence of acute and chronic PTS after CCI is an important outcome that represents clinical features of human TBI. In addition to behavioral seizures and EEG epileptiform discharges, CCI-induced injury results in several chronic alterations in the brain such as mossy fiber sprouting and delayed hippocampal lesions that are often considered hallmarks of temporal lobe epilepsy.

In wild type control mice following CCI injury, seizures are generally observed within the first hours, and these seizures are characterized by mild-to-moderate severity and short duration (generally within a few seconds), whereas in genetically modified mice lacking the A1 adenosine receptor, the severity of seizures was significantly higher, including the occurrence of tonic–clonic seizures and the development of lethal status epilepsy [182]. Interestingly, the study also showed that A1 adenosine receptor knockout mice exhibited enhanced microglial responses compared to wild type mice, especially in the ipsilateral and contralateral cortices and thalami, and in the ipsilateral CA3 and contralateral CA1 regions of the hippocampus, whereas lesion and cortical volumes were not different between knockout and wild-type mice and Fluoro-Jade staining revealed no neuronal death in CA1 or CA3 regions, suggesting a critical role for microglia in epileptogenic changes after TBI [183]. A study performed in rats has shown that following CCI, 67 % exhibit transient epileptiform discharges that were significantly altered by treatment with lisuride, an agonist of dopamine and a partial agonist of serotonin receptors, including total duration of epileptiform discharges, spectral characteristics such as reduction in delta power; it also resulted in an increase in the mean EEG frequency representing anticonvulsive effects [184]. Although lisuride has significant anticonvulsive effects, it did not have significant effects on secondary brain damage [184].

In the mouse CCI model, spontaneous behavioral seizures occur after several weeks following induction of the injury and the seizure incidence depends on the severity of CCI parameters such as dwell distance [185]. When using 0.5 and 1 mm injury depths, seizures were observed in 20 and 36 % of mice, respectively, and both parameters induced mossy fiber sprouting [185]. A later chronic study performed in mice employing video-EEG monitoring for 16 weeks following CCI has shown that most PTS occur after 10 weeks following TBI [186]. This study was focused on the mammalian target of rapamycin complex 1 (mTORC1) pathway in posttraumatic epileptogenesis and the results showed that inhibition of this pathway with the specific inhibitor (i.e., rapamycin) for 30 days after CCI decreased neuronal degeneration and mossy fiber sprouting, and decreased the frequency of seizures and the rate of PTE [186]. Another study demonstrated that although rats did not develop spontaneous behavioral seizures for up to 8 weeks after CCI, they have increased susceptibility to bicuculline-induced seizures 1 week after injury, suggesting an increased hyperexcitability [187]. A chronic 9-month study performed in C57BL/6S adult male mice that included three 2-week continuous video-EEG monitoring showed that after CCI, 9 % of mice exhibited electrographic seizures and 82 % of mice exhibited epileptiform spiking on EEG and increased susceptibility to the PTZ-induced seizure threshold test [188]. These changes developed within 6 months after CCI and did not further progress for up to 9 months [188].

Of interest is a PTE model induced by CCI in the immature rat, which is often called a pediatric PTE model. In this model, CCI is applied to 2- to 3-week-old-rats and epileptogenic changes occurred during maturation [17, 189]. CCI performed in 16- to 18-week-old rats resulted in a sustained reduction of the minimal clonic seizure threshold at 2 months of age [17]. A study using chronic video-EEG monitoring within 4–11 months after CCI showed the occurrence of EEG spiking in 87.5 % of rats who underwent CC and in rare cases even spontaneous recurrent seizures [189].

Electrophysiological In Vitro Assessment of Epileptogenic Changes Induced by CCI

These investigations usually employ patch-clamp, intracellular, or extracellular recording techniques, allowing registration of evoked and spontaneous events at different time points following experimental TBI. The progressive development of neocortical hyperexcitability underlying spontaneous epileptiform firing could be detected 2 weeks following CCI [173]. The mechanisms underlying this hyperexcitability are complex and poorly understood. For example, CCI produces a transient decrease in the expression of the Kv4.2 subunit of the $I_{K(A)}$ channel and a reduction in $I_{K(A)}$ currents in the ipsilateral CA1 pyramidal neurons 1 week after CCI but these changes were no longer significant 8 weeks after injury [187].

Intracellular and extracellular recordings obtained from layer V in neocortical slices from rats who underwent CCI revealed that characteristically evoked epileptiform discharges are evident even in the first week after injury, whereas spontaneous epileptiform discharges are evident after 2 weeks [173]. The data suggest that the hyperexcitability observed following CCI is associated with augmentation of excitatory reduction of inhibitory synaptic activities. Whole-cell patch-clamp recordings in granule cells revealed a reduction in spontaneous and miniature IPSC frequency in GABAergic hilar interneurons after 8–13 weeks of CCI injury, whereas the action potential and EPSC frequencies were increased [190]. The abnormalities in GABAergic transmission in hippocampal dentate granule cells resulted from CCI injury includes differential alterations in the function of synaptic and extrasynaptic $GABA_A$ receptors [126]. In addition, CCI results in the altered expression of both $GABA_A$ and glutamate receptor subunits, and, remarkably, an increase in NR2B protein and decrease in GluR1 protein expression [191].

2 Materials

2.1 Equipment and Instruments for General Surgical Procedures in Rodents

1. Stereotaxic apparatus (e.g., David Kopf Instruments, Tujunga, CA, USA) equipped with anesthesia nose cone (if gas anesthesia is used).
2. Anesthesia machine for gas anesthesia (e.g., tabletop nonrebreathing anesthesia machine, Harvard Apparatus, Holliston, MA, USA or similar).

3. High speed dental drill or surgical micro drill (e.g., Complete Bone Micro Drill System, Harvard Apparatus; OmniDrill25, World Precision Instruments, Sarasota, FL USA) and 0.5 mm diameter bit for craniotomy procedure (World Precision Instruments).
4. Closed loop temperature control system for small rodents Homeothermic Monitoring System, Harvard Apparatus). Alternatively, animal's body temperature could be monitored during surgery using a low-cost electronic thermometer (MicroTherma 2T Hand Held Thermometer, Braintree Scientific Inc., Braintree, MA, USA) equipped with a rodent rectal probe (Braintree Scientific Inc.) and temperature-controlled heating pad (e.g., Fine Science Tools, Vancouver, BC, Canada).
5. Recovery temperature/humidity-controlled chamber (e.g., Small Animal Recovery Chamber/Warming Cabinet, Harvard Apparatus).
6. Surgical instruments: forceps, scissors, scalpel.
7. Surgical retractor (optional) (e.g., small animal retractor system Harvard Apparatus).
8. Reflex 7 or Reflex 9 wound closure system (CellPoint Scientific, Inc., Gaithersburg, MD, USA) for mice or rats, respectively. Alternatively, surgical sutures can be used.
9. Isoflurane (or other desired anesthesia gas).
10. Supplies for aseptic surgical preparation. Although the aseptic technique is critical for the animal surgical studies, specific aseptic techniques might be selected based on the internal guidelines or recommendation of institutional committee (e.g., Betadine, 2 % chlorhexidine, 70 % isopropyl alcohol).
11. Sterile gauze sponges, cotton tips, sterile surgical spears (e.g., BIO CEL™ and BIO SPEARS™, Biotech Visioncare, India).
12. Eye ointment.
13. Triple antibiotic ointment (optional).
14. Dental cement and/or veterinary acrylic adhesive (e.g., LiquiVet).

2.2 Equipment for Induction of Experimental TBI (Depending on the Experimental Model Used in the Study)

1. Cortical impactor device [e.g., electromagnetically driven impactor devices: PCI3000 PinPoint Precision Cortical Impactor (Hatteras Instruments, Cary, NC, USA) or Leica Impact One™ Stereotaxic Impactor (Leica), pneumatically driven impactor devices: Pneumatic (Cortical) Impact Device AMS 201 (AmScien Instruments LLC,, Richmond, VA, USA), TBI-0310 Impactor (Precision Systems and Instrumentation, LLC, Fairfax Station, VA].

2. Fluid Percussion Injury device (Custom Design & Fabrication, Virginia Commonwealth University Medical Center, Richmond, Virginia, USA). FPI assembly includes the base, pendulum mast, graduated protractor, ball bearing-enhanced impact hammer assembly, acrylic fluid cylinder and bracketry, transducer housing, transfer tube, and a digital transducer. The device is completed with a self-calibrating Trauma Inducer Pressure Inducer Amplifier.

2.3 Materials for EEG Monitoring and Seizure/Epilepsy Assessment

2.3.1 Setup for EEG/Video Seizure Monitoring and Analysis

1. Biopotential amplifier (type and number of channels depend on needs) (e.g., Bio Potential Amplifier or Electroencephalogram Amplifier Modules, Harvard Apparatus).
2. Analog-to-digital converter for digitizing of EEG (e.g., Model#: PCI-6221, 16-Bit, 250 kS/s, 16 Analog Inputs or Model#: USB-6008, 12-Bit, 10 kS/s Low-Cost Multifunction DAQ with required accessories, National Instruments Corp., Austin, TX, USA).
3. Tether and commutator to connect the biopotential amplifier to the digitizer (e.g., Cat#: 8204 3-channel mouse commutator/swivel, Pinnacle Technology Inc.).
4. Preamplifier Mouse Preamplifier for 3-Channel System (optional, depending on the equipment used, e.g., 8202, Pinnacle Technology Inc.).
5. 8-Pin Surface Mount Connector for Mice (8415-SM, Pinnacle Technology Inc.).
6. EEG headmounts (#8201-EEG headmounts, Pinnacle Technology Inc., Lawrence, KS).
7. Two-component epoxy for ensuring solid connections at the electrodes (e.g., 8226 Silver Epoxy for use with mouse #8201 or #8201-EEG headmounts).
8. Mouse or rat stainless EEG screws (e.g., 8209: 0.10″ or 8212: 0.12″, Pinnacle Technology Inc.).
9. EEG signal acquisition software (e.g., AcqKnowledge BioPac Systems Inc., Santa Barbara, California). Alternatively, a custom specific for the study acquisition/analysis program can be written using Matlab software (Mathworks Co., Natick, MA).
10. Digital video surveillance system with infrared camera to enable night recording (e.g., Automated Video Systems).

2.3.2 Materials and Equipment: PTZ Challenge Seizure Tests

1. Pentylenetetrazole (CAS Number 54-95-5) (e.g., Cat#: P6500, Sigma-Aldrich, St. Louis, MO, USA).
2. ~10″ diameter circular mouse cage for performing PTZ challenge test (e.g., Pinnacle Technology Inc., Lawrence, KA, USA) (*see* **Note 1**).

3. Syringe infusion pump for PTZ-induced seizure threshold test (e.g., Standard Infusion Only PHD Ultra Syringe Pump, Harvard Apparatus) (*see* **Note 2**).
4. Syringe (depend on the syringe pump used).
5. 30 G Needles.
6. Polyethylene tubing (e.g., Cat#: PE10, Braintree Scientific Inc.; Cat#: 51150, Stoelting Co., Wood Dale, IL 60191 USA).
7. Medical tape to fix a needle to animal's tail (optional) (e.g., 3M Transpore Transparent Tape).

3 Methods

This section describes common protocols for induction and detection of acute PTS and/or chronic PTE based on the common TBI models: CCI and FPI. The main differences between these experimental models are reflected in the method of TBI induction, whereas most of the general surgical procedures such as craniotomy and electrode implantation, and EEG and behavioral seizure/epilepsy assessments are the same for both these models.

3.1 Animal Selection

Prior experiments, animals should be examined for their general appearance and health condition. The animals showing any unexpected diseases or pathology such as skin lesions, wound from fighting, ophthalmic pathologies should not be used. In addition, the animals should be excluded from the experiments if any phenotypic alternations from group characteristic are present including size and weight, fur condition and/or color, skull and body shape, and behavioral baseline tests (*see* **Note 3**). Each animal should be assigned a unique identification and the animals should be randomized into the study groups. It is the best approach to use littermates if available.

3.2 Surgical Procedures

3.2.1 Surgery Preparation

1. Prior to all procedures, anesthesia has been induced with 4 % isoflurane in a 25 % oxygen-in-air mixture in the anesthesia induction chamber. Exposure to isoflurane for about 1 min is normally enough induce deep anesthesia that allows for removing animal's hair within surgery site and transferring to the stereotactic apparatus equipped with nose cone for maintaining anesthesia and heating pad.
2. Animal's hair is removed using any of the three methods that are currently commonly used: clipping with an electrical rodent hair clipper, using a chemical depilatory agent, or shaving with a razor.
3. Animal is transferred to the stereotaxic apparatus and connected to nose cone. During all surgical procedures, mice are maintained on 2 % isoflurane anesthesia via nose cone.

4. Eye ointment is applied on animal's eyes to prevent drying of cornea.
5. A lubricated rectal probe is placed in the rectum to monitor animal's body temperature. The body temperature is maintained with a controlled heating pad.
6. Surgery site is prepared by appropriate aseptic techniques, e.g., washing with 2 % chlorhexidine and sterile saline for three times, starting from center and moving to edge.

3.3 Electrode Implantation for PTS Detection

For acute recording of acute PTS the electrodes are implanted before TBI induction, whereas for recording of spontaneous seizures for characterization of the experimental PTE, the electrodes are implanted weeks or months after experimental TBI.

3.3.1 General Consideration for Selection of Method of EEG Monitoring and Type of Electrodes

Human PTE is heterogeneous in terms of seizure durations and semiologies and may represent simple or complex partial seizures and secondarily generalized seizures. Generally, PTE may represent simple or complex partial seizures and secondarily generalized seizures. Although initial seizures in PTE most commonly have a focal neocortical origin, both partial neocortical epilepsy and TLE might be developed [2]. For selection of the method for EEG monitoring it is important to take into consideration the characteristic brain pathologies characterized by significant loss of brain tissue and hippocampal degeneration especially at chronic stages (Fig. 1). Both single-channel and multiple-channel array EEG recording systems have been described for the assessment spontaneous seizures following TBI. The methods of EEG acquisition include unipolar, bipolar and average reference, and, in general, the same basic information could be obtained by any of these methods [192]. The choice of a method of EEG acquisition depends to a large extent on the specific problem addressed in the study and on the personal preference of the investigator. The epidural electrodes (e.g., stainless steel, Ag/AgCl) are used for both bipolar and unipolar recordings. Commercially available stainless steel screw electrodes (e.g., Plastics One, Pinnacle Technology) are an example of electrodes suitable for chronic EEG monitoring. However, using unipolar recording techniques allows to determine only overall changes in EEG activity. The EEG waveform parameters strongly depend on the location of the electrodes. Reference electrodes could be pleased on the distal part of skull or other animal's body part (e.g., tail, cheek). The placing of electrodes at the margins of TBI injury on the ipsilateral side and the corresponding electrodes in the contralateral side allow to measure changes in EEG activity measured as a difference between the two hemispheres versus one reference electrode [193].

3.3.2 Selection of Position of EEG Electrode Placement

EEG electrode positions depend on the method of acquisition and available equipment:

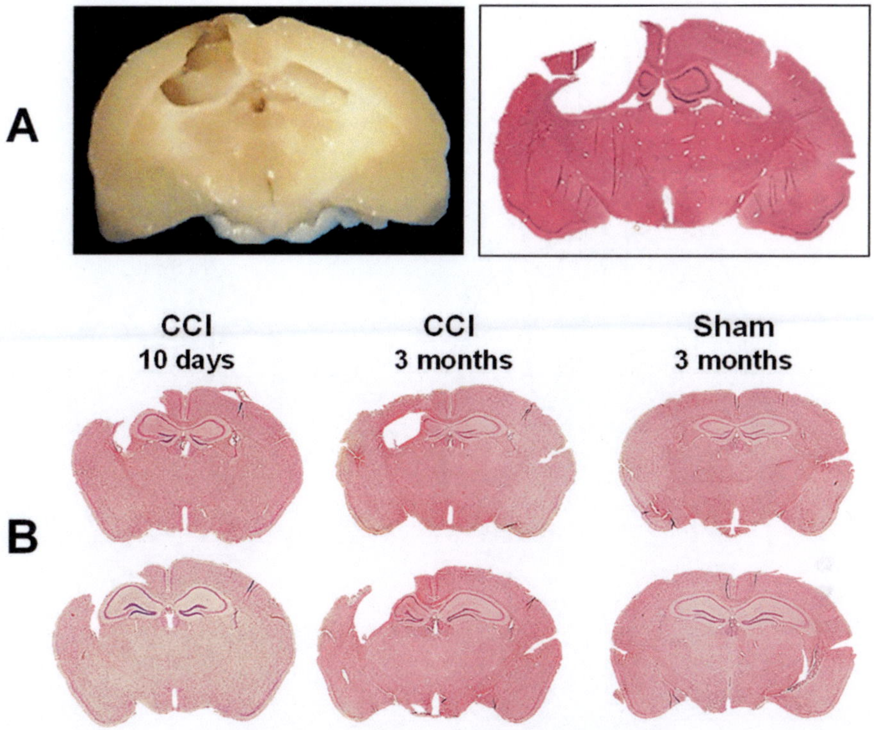

Fig. 1 Chronic brain pathology following CCI. Representative photograph of paraformaldehyde-perfused rat brain spacemen collected at 3 months after induction of CCI and corresponding hematoxylin–eosin (H&E)-stained coronal 8-μm paraffin section. The photographs demonstrate marked brain degradation at chronic stages resulted in loss of cortical tissue (cavitation) and significant reduction in hippocampal size ipsilateral to the injury at chronic time points (**a**). Representative photographs of H&E-stained coronal 8-μm paraffin mouse brain sections demonstrating progression of hippocampal pathology at subacute and chronic time points after CCI. At subacute time point (10 days after CCI) the ipsilateral hippocampus is enlarged due to focal edema that is still evident at this time point, whereas at chronic time point after CCI, the ipsilateral hippocampus is reduced in size due to neurodegenerative processes resulting in neuronal death and loss of brain tissue. Cortical cavitations in CCI-injured mice are presented at both time points. In contrast, no obvious brain pathology is observed in animals at 3 months after sham injury (**b**)

1. Position of the single-channel epidural EEG electrode implantation is at the caudal edge of the TBI lesion.
2. For the bipolar recording, two stainless steel screws are positioned fronto-lateral and occipito-parasagittal to injury site, respectively.
3. For multichannel recording, the electrodes are positioned ipsilateral and contralateral to injury hemispheres [194]. In the ipsilateral hemisphere, one frontal and one parietal electrode are positioned rostral and caudal to craniotomy, respectively. In the ipsilateral hemisphere, one frontal and one parietal electrode are inserted in the corresponding positions. An additional ground electrode is positioned into the occipital bone caudal to lambda. The schematic of default EEG multiple electrode placements in rodents is shown in Fig. 2.

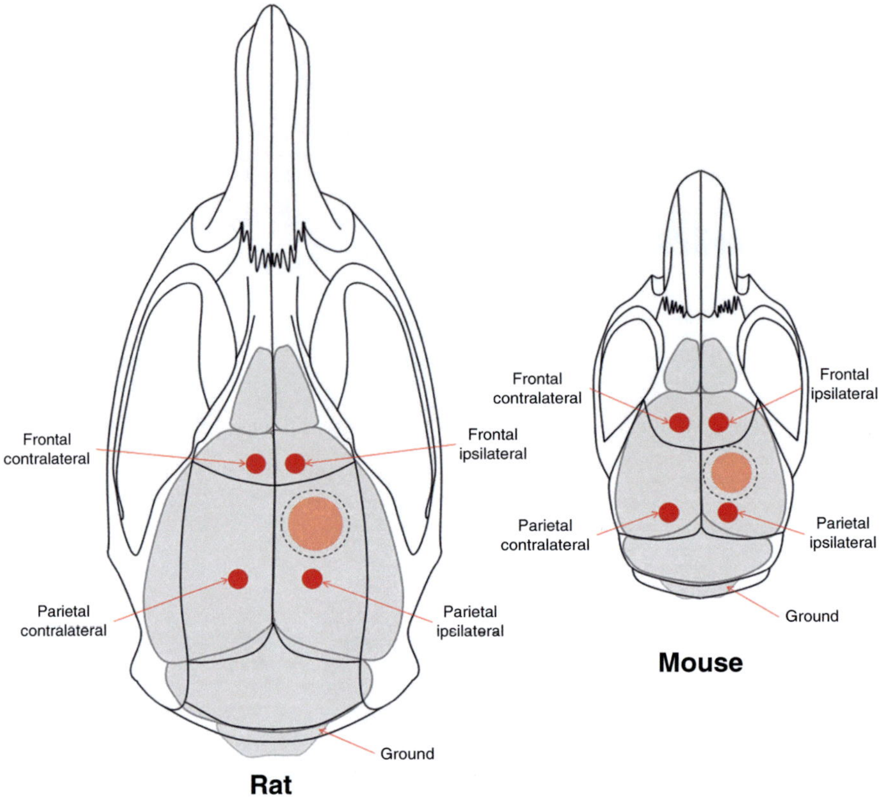

Fig. 2 Schematic of the EEG electrode placement in rodents. Position of the EEG leads in rat and mouse TBI models for detection of the PTS and PTE. *Scattered line* represents positions of craniotomy and *red filled areas* represent the sites of injury. Ground electrode if required could be positioned over cerebellum

3.3.3 Procedures of EEG Electrode Implantation

1. Epidural stainless still screw electrodes are inserted prior to CCI induction proximally from the craniotomy boundary and at the same position contralaterally. Alternatively, scalp electrodes could be used [182].
2. Metal parts are insulated with dental cement.
3. Baseline EEG recordings are performed for at least 10–30 min under anesthesia downtitrated to about 1 % of isoflurane when animal still has negative pinch toe reflex.
4. The EEG recording to detect acute PTS will be resumed after induction of experimental TBI CCI (see below).

3.3.4 Craniotomy Procedures

1. Under 2 % isoflurane anesthesia, central skin incision is made and connecting tissues covering skull are removed with cotton tips. Normally, spreading skin to sides allows for performing all further surgical procedures and other manipulations when gently holding the skin with a cotton tip. Using a surgical restrainer might be helpful.
2. Craniotomy site according to desired coordinates is marked using appropriate marking tool. Generally in PTE models, cir-

cular craniotomy with appropriate diameter to allow freely access for rigid impactor for CCI model or attach connector to FPI is positioned in the center of temporal bone.

3. After the skull is exposed with a central skin incision and soft tissue is removed with a cotton tip, circular craniotomy, ~4 mm in diameter, is made in the middle of the right parietal bone, about 0.5 mm from sagittal, coronal, and lambdoid sutures.
4. Circular craniotomy is made using an electric high-speed drill with 0.5 mm diameter bit periodically applying sterile saline to prevent drying of skull bone and overheating. Visual control using a low-power microscope during the whole procedure is highly preferred. It is critical to continue drilling removing a small amount of bone tissue around the edge until the bone cracks. Generally, the craniotomy procedure using electric drill takes less than 5 min. Alternatively, trephine device (preferably manual) might be used. However, when using trephine, it is more difficult to control depth to prevent damage of dura mater.
5. Bone flap is gently removed with forceps. If craniotomy is performed properly, the bone flap moves freely. If bone flap is still attached to the skull bone at some points, there is a high risk for damaging dura mater (*see* **Note 4**).
6. The exposed brain is washed with warm sterile saline to remove bone dust. Brain tissue should be moist all the time.

3.3.5 CCI Procedures

1. Excess of saline should be gently removed with cotton tip before position of impactor.
2. The impactor is positioned on the site of injury under visual control and or using a sound sensor if equipped (*see* **Note 5**). Visual control under magnification allows establishing a starting position more precisely.
3. The impact with desired parameters is executed and the impactor is retracted.
4. Occasional minor bleeding is stopped with cotton tips from edge of craniotomy not touching brain surface and washed with warm saline.

The example CCI parameters for mice are as follows: impact tip diameter 3–4 mm, velocity 3 m/s, compression time 100–200 ms, and a compression distance of 1–1.5 mm; using these parameters nearly 100 % of WT C57BL/6 mice exhibit behavioral PTS within 1 h after CCI. Characteristic subacute and chronic brain pathologies following CCI with aforementioned parameters are shown in Fig. 1b.

3.3.6 Lateral FPI Procedures

The general craniotomy procedures are the same as those described above.

1. After performing craniotomy, Luer lock hub fitting is attached to animal's scull with acrylic adhesive to provide tight seal and

filled with sterile saline. Brain tissue should be kept moist with sterile saline.

2. Animal is removed from the stereotaxic apparatus and allowed to recover for ~1 h in a temperature controlled recovery chamber.
3. Before induction of injury, the high-pressure tubing of FPI device is connected and filled with sterile water and ensure that it is free of air bubbles and FPI device is settled up to deliver injury pulse free of noise as monitored using oscilloscope.
4. Prior to TBI procedures, deep anesthesia is induced with 4 % isoflurane in a 25 % oxygen-in-air mixture in the anesthesia induction chamber.
5. Animal is placed on the platform on its side close to FPI device and connected to the high-pressure tubing of the FPI device.
6. Experimental TBI is delivered by releasing pendulum to produce a single inquiry pulse. The exact pressure is measured and recorded (the transducer amplifier is calibrated at 10 mV per 1 PSI pressure).
7. After completion of the experimental TBI, animal is returned to surgery setup and maintained under isoflurane anesthesia to complete surgery.
8. Luer lock fitting used for connection of the animal to the FPI device is removed from the skull and the incision is closed.

3.3.7 Completion of Surgery

1. For animals intended for chronic PTS/PTE assessments after surgical procedures, the incision is closed using wound clips (or sutures).
2. Each animal receives an intraperitoneal injection of warm saline to prevent dehydration and transferred to a temperature-controlled recovery chamber for at least 1 h.

3.3.8 Procedures in Sham Animals

The animals assigned into the sham groups undergo craniotomy procedures only. In animal TBI models, craniotomies with replacing bone flap result in considerable morphological, inflammatory, and behavioral alternations [172], and may cause an increase in allodynia [195]. In the latter case, an increase in allodynia following above-described sham procedures was observed in mice but not in rats [195].

3.4 Acute and Chronic Spontaneous PTS Assessment

3.4.1 General Consideration for EEG/Video Monitoring

The EEG recording to detect acute PTS will be resumed immediately after induction of CCI when animal still under anesthesia of or after waking up after anesthesia using an EEG/video recording setup for recordings in freely moving animals.

For chronic EEG/behavioral seizure activity the EEG and video recordings are continuously performed to assess remote spontaneous seizure activity, which is expected to occur within 2 months after CCI. The approaches for EEG and video monitoring of PTE development require continuous recordings in epoch with duration from weeks-to-months and computerized analysis for EEG episode detection [196–198]. Alternatively, the animals could be monitored for the spontaneous PTS using random 1–2 h intervals within the same time frame [185]. The EEG episodes are confirmed with synchronized video recording for behavioral seizure manifestation which are scored using the Racine's scale.

3.4.2 EEG Data Analysis

Computer-assisted EEG spectral analysis is performed on selected epochs of fixed duration (e.g., 60 s) of cortical to evaluate changes in EEG power [199]. Compressed spectral arrays, spectral trend graphs, and spectrographic analyses are also used to evaluate EEG changes. The respective power scores across each frequency band are valuated at fixed time points following and compared with baseline and the corresponding controls (e.g., sham-operated animals or vehicle injection). Changes in power are assessed across four following frequency bands:

1. Delta (0–4 Hz).
2. Theta (4–8 Hz).
3. Alpha (8–12 Hz).
4. Beta (12–30 Hz).

An EEG seizure episode is defined as repetitive spikes or spike-and-wave discharges recurring at frequencies of >1 Hz with amplitude greater than spontaneous activity and duration greater than 2–10 s. The requiring EEG seizures with a short interval between two episodes (e.g., less than 10 s) could be considered as a single event.

Chronic epilepsy is defined as appearance of recurring paroxysmal events detected by EEG and/or behavioral manifestations. Based on these criteria, the following behavioral/EEG seizure parameters are included in the statistical analyses:

1. Incidence of the chronic PTS is determined as a percentage of animals that have at least one unprovoked seizure.
2. Incidence of the PTE is determined as a percentage of animals that have at least two unprovoked seizure.
3. Onset latency of the PTS/PTE is determined as the time between the induction of experimental TBI and the first episode or recurrent EEG seizures.
4. PTS frequency is determined as the number of EEG episodes observed in individual animals for a fixed period time.

5. PTS episode duration is determined as an average duration between the onset and offset of each EEG seizure episode.
6. Total PTS duration is determined as the sum of episode durations.
7. Time course of PTS frequency, defined as the hourly EEG seizure frequency up to 2 months.

3.4.3 Assessment of Behavioral Seizure Severity

Behavioral acute or chronic seizures are confirmed and scored based on the revised Racine's seizures severity scale. This scaling system has been adapted to different seizure models including, electrically and chemically induced seizure models [200–203]. The same scaling system is also used for seizure severity assessment for PTZ challenge test. The system is based on the behavioral manifestation with the following hallmarks reflecting the discrete seizure scores: 0, normal response; 1, ear and facial twitching; 2, myoclonic twitches or waves of the whole body; 3, myoclonic jerks and/or rearing; 4, clonic–tonic seizures; 5, generalized clonic–tonic seizures, loss of postural control; 6, death (*see* **Note 6**).

In addition, the behavioral seizure parameters such as incidence frequency, seizure episode duration, total duration, onset latency, and time course will be defined and assessed similarly to corresponding parameters of EEG seizures.

3.4.4 PTZ Challenge Seizure Tests

There are two types of PTZ seizures induced by bolus injection or determination of seizure threshold using slow PTZ infusion.

PTZ Challenge with Bolus Injection

1. Animal is allowed to acclimatize in 10″ d circular mouse cage (Pinnacle Technology Inc., Lawrence, KA) and 10–30 min baseline EEG is recorded.
2. Animal are removed from the chamber injected intraperitoneally with PTZ (50 mg/kg for rat or 80 mg/kg for mouse).
3. The seizures are graded using Racine's scale during 30–60 min with a constant bin intervals selected from 30 s for up to 5 min.

PTZ-Induced Seizure Threshold Test

A PTZ-induced clonic seizure threshold test is widely used as surrogate test for seizure susceptibility [204]. The PTZ-threshold test is performed in all animal groups before terminal time point as a quantitative test of epileptogenesis in PTE [43, 188]. The value is expressed as an amount of PTZ infused via tail vein required to induce clonic seizure in a freely moving animal. However, because of the possible day-to-day variability in the PTZ threshold, appropriate control groups should be included for each testing.

PTZ solution (0.5 % in saline) will be infused into the lateral tail vein at a constant rate of 0.5 mL/min using a and an infusion pump, which is connected by polyethylene tubing to the 30-G dental needle secured to the tail with a narrow piece of adhesive tape (optional).

Infusion is halted when forelimb clonus followed by full clonus of the body is observed. The minimum dose of PTZ (mg/kg) needed to induce a clonic seizure will be measured as an index of clonic seizure threshold.

In addition, the PTZ test allows analysis of different convulsive end points including the following observed seizure phases: general excitation with myoclonic twitches of the whole body, generalized clonic seizures with loss of righting reflex and tonic hind limb extension.

3.4.5 Exclusion Criteria in TBI/PTE Animal Models

The experimental results could be influences by several factors that are not related to the experimental model or treatment such as those results from accidents, human error (e.g., incorrect procedures or treatment) or animal's health condition resulted from sickness (e.g., infection diseases or wound resulted from fighting). Importantly, because the number of animal experimental groups based on power analysis for each particular study, the presence even a single animal with erroneous outcome measure may lead to marked increase of the standard deviation and loss of statistical significance resulting in wrong experiment interpretation and conclusions. Thus, recognizing and exclusion of the animals with deficiencies or alterations that are not related to the study is critical to ensure quality of the experimental results and reproducibility (*see* also Subheading 3.1 and **Note 7**).

Exclusion Criteria Related to Surgical Procedures

1. Damaged dura mater, meninges, and/or brain tissue during craniotomy.
2. Subdural hemorrhage after craniotomy
3. Excessive bleeding during or after craniotomy
4. Accidental incorrect position of impactor causing brain deformation (e.g., exceeding ~50 μm).
5. Prolonged surgery or anesthetic exposure (e.g., more than 30 min due to any reasons).
6. The following applies for wild type animals, whereas for genetically modified animals they should not be considered as outcomes unless the study confirms that any of these reflects outlier characteristics:
   (a) Excessive bleeding after CCI impact.
   (b) Any alteration in anesthesia response reflected in animal behavior (e.g., waking-up, abnormal breathing pattern) or response to test (e.g., toe pinch reflex) when animal maintained at 2 % isoflurane. In this case, anesthesia should be adjusted to the proper level and the same anesthesia settings should be maintained in all experimental groups.

Exclusion Criteria Post-surgery

1. Humane end point criteria defined by the animal use protocol or institutional committee.
2. Removing clips (extremely rare event with proper wound closure technique).
3. Removing EEG electrode headset.
4. Any unexpected diseases or pathology in WT mice w/o treatment.
5. Fighting, cage flooding, mishandling (e.g., accidental falling), etc.

Exclusion Criteria Based on Anatomical Assessment of Brain Pathology

1. Broken skull or bone pieces (or foreign articles) in brain tissue.
2. Incorrect implantation of EEG electrodes.
3. Brain pathology that is not associated with experimental TBI, excessive hematoma and/or cavitation statistically distinct from the same group and injury parameters, excessive brain pathology in sham animals.

4 Notes

1. Circular chamber is used for grading of the seizures induced with bolus PTZ injection to prevent potential injuries of animal during progression of violent seizures. In contrast, PTZ threshold test is normally performed on the table top.
2. Low-cost infusion pumps are generally not suitable for the PTZ threshold test because of low precision of the delivery of small volumes of solution.
3. In genetically modified animals, phenotypic or baseline behavioral characteristics might be different from those observed in their wild type controls.
4. Damaged dura mater, excessive bleeding, or subdural hemorrhage after craniotomy procedure should be considered as exclusion criteria.
5. If using sound sensor it is critical to verify if impactor tip is positioned properly and touches the brain surface. Excessive amount of saline may result in a false sound signal.
6. Generally, seizures with a Racine score of 2 or lower are considered as "non-convulsive" and those with a Racine score of 3 and above are identified as "convulsive."
7. The number of animals that are excluded from analyses and condition needs to be reported to determine success rate and reproducibility of the technique used in the study.

References

1. Maas AI, Stocchetti N, Bullock R (2008) Moderate and severe traumatic brain injury in adults. Lancet Neurol 7:728–741
2. Raymont V, Salazar AM, Lipsky R, Goldman D, Tasick G, Grafman J (2010) Correlates of posttraumatic epilepsy 35 years following combat brain injury. Neurology 75:224–229
3. Salazar AM, Jabbari B, Vance SC, Grafman J, Amin D, Dillon JD (1985) Epilepsy after penetrating head injury. I. Clinical correlates: a report of the Vietnam Head Injury Study. Neurology 35:1406–1414
4. Scher AI, Wu H, Tsao JW, Blom HJ, Feit P, Nevin RL, Schwab KA (2011) MTHFR C677T genotype as a risk factor for epilepsy including post-traumatic epilepsy in a representative military cohort. J Neurotrauma 28:1739–1745
5. Kazemi H, Hashemi-Fesharaki S, Razaghi S, Najafi M, Kolivand PH, Kovac S, Gorji A (2012) Intractable epilepsy and craniocerebral trauma: analysis of 163 patients with blunt and penetrating head injuries sustained in war. Injury 43:2132–2135
6. Bruns J Jr, Hauser WA (2003) The epidemiology of traumatic brain injury: a review. Epilepsia 44(Suppl 10):2–10
7. Annegers JF, Hauser WA, Coan SP, Rocca WA (1998) A population-based study of seizures after traumatic brain injuries. N Engl J Med 338:20–24
8. Fisher RS, Acevedo C, Arzimanoglou A, Bogacz A, Cross JH, Elger CE, Engel J Jr, Forsgren L, French JA, Glynn M, Hesdorffer DC, Lee BI, Mathern GW, Moshe SL, Perucca E, Scheffer IE, Tomson T, Watanabe M, Wiebe S (2014) ILAE official report: a practical clinical definition of epilepsy. Epilepsia 55:475–482
9. Frey LC (2003) Epidemiology of posttraumatic epilepsy: a critical review. Epilepsia 44(Suppl 10):11–17
10. Cabral RJ, King TT, Scott DF (1976) Epilepsy after two different neurosurgical approaches to the treatment of ruptured intracranial aneurysm. J Neurol Neurosurg Psychiatry 39:1052–1056
11. Loscher W, Potschka H (2005) Drug resistance in brain diseases and the role of drug efflux transporters. Nat Rev Neurosci 6:591–602
12. Diaz-Arrastia R, Agostini MA, Frol AB, Mickey B, Fleckenstein J, Bigio E, Van Ness PC (2000) Neurophysiologic and neuroradiologic features of intractable epilepsy after traumatic brain injury in adults. Arch Neurol 57: 1611–1616
13. Kwan P, Brodie MJ (2000) Early identification of refractory epilepsy. N Engl J Med 342:314–319
14. Risdall JE, Menon DK (2011) Traumatic brain injury. Philos Trans R Soc Lond B Biol Sci 366:241–250
15. Chen LL, Baca CB, Choe J, Chen JW, Ayad ME, Cheng EM (2014) Posttraumatic epilepsy in operation enduring freedom/operation Iraqi freedom veterans. Mil Med 179:492–496
16. Temkin NR (2003) Risk factors for posttraumatic seizures in adults. Epilepsia 44(Suppl 10):18–20
17. Statler KD, Swank S, Abildskov T, Bigler ED, White HS (2008) Traumatic brain injury during development reduces minimal clonic seizure thresholds at maturity. Epilepsy Res 80:163–170
18. Kramer G (2001) Epilepsy in the elderly: some clinical and pharmacotherapeutic aspects. Epilepsia 42(Suppl 3):55–59
19. Roberts I, Schierhout G, Alderson P (1998) Absence of evidence for the effectiveness of five interventions routinely used in the intensive care management of severe head injury: a systematic review. J Neurol Neurosurg Psychiatry 65:729–733
20. Huang YH, Liao CC, Chen WF, Ou CY (2015) Characterization of acute post-craniectomy seizures in traumatically brain-injured patients. Seizure 25:150–154
21. Torbic H, Forni AA, Anger KE, Degrado JR, Greenwood BC (2013) Use of antiepileptics for seizure prophylaxis after traumatic brain injury. Am J Health Syst Pharm 70:759–766
22. Beghi E (2003) Overview of studies to prevent posttraumatic epilepsy. Epilepsia 44(Suppl 10):21–26
23. Jones KE, Puccio AM, Harshman KJ, Falcione B, Benedict N, Jankowitz BT, Stippler M, Fischer M, Sauber-Schatz EK, Fabio A, Darby JM, Okonkwo DO (2008) Levetiracetam versus phenytoin for seizure prophylaxis in severe traumatic brain injury. Neurosurg Focus 25, E3
24. Pagni CA, Zenga F (2005) Posttraumatic epilepsy with special emphasis on prophylaxis and prevention. Acta Neurochir Suppl 93:27–34
25. Temkin NR (2009) Preventing and treating posttraumatic seizures: the human experience. Epilepsia 50(Suppl 2):10–13

26. Kinirons P, McCarthy M, Doherty CP, Delanty N (2006) Predicting drug-resistant patients who respond to add-on therapy with levetiracetam. Seizure 15:387–392
27. Michelucci R (2006) Optimizing therapy of seizures in neurosurgery. Neurology 67: S14–S18
28. Treiman DM, Meyers PD, Walton NY, Collins JF, Colling C, Rowan AJ, Handforth A, Faught E, Calabrese VP, Uthman BM, Ramsay RE, Mamdani MB (1998) A comparison of four treatments for generalized convulsive status epilepticus. Veterans Affairs Status Epilepticus Cooperative Study Group. N Engl J Med 339:792–798
29. Zou H, Brayer SW, Hurwitz M, Niyonkuru C, Fowler LE, Wagner AK (2013) Neuroprotective, neuroplastic, and neurobehavioral effects of daily treatment with levetiracetam in experimental traumatic brain injury. Neurorehabil Neural Repair 27:878–888
30. Cernak I, O'Connor C, Vink R (2002) Inhibition of cyclooxygenase 2 by nimesulide improves cognitive outcome more than motor outcome following diffuse traumatic brain injury in rats. Exp Brain Res 147:193–199
31. Velioglu SK, Ozmenoglu M (1999) Migraine-related seizures in an epileptic population. Cephalalgia 19:797–801, discussion 766
32. Cernak I, O'Connor C, Vink R (2001) Activation of cyclo-oxygenase-2 contributes to motor and cognitive dysfunction following diffuse traumatic brain injury in rats. Clin Exp Pharmacol Physiol 28:922–925
33. Jordan KG (2004) Emergency EEG and continuous EEG monitoring in acute ischemic stroke. J Clin Neurophysiol 21:341–352
34. Menon B, Shorvon SD (2009) Ischaemic stroke in adults and epilepsy. Epilepsy Res 87:1–11
35. Jordan KG (1993) Continuous EEG and evoked potential monitoring in the neuroscience intensive care unit. J Clin Neurophysiol 10:445–475
36. Privitera MD, Strawsburg RH (1994) Electroencephalographic monitoring in the emergency department. Emerg Med Clin North Am 12:1089–1100
37. Jordan KG (1995) Neurophysiologic monitoring in the neuroscience intensive care unit. Neurol Clin 13:579–626
38. Bergsneider M, Hovda DA, Shalmon E, Kelly DF, Vespa PM, Martin NA, Phelps ME, McArthur DL, Caron MJ, Kraus JF, Becker DP (1997) Cerebral hyperglycolysis following severe traumatic brain injury in humans: a positron emission tomography study. J Neurosurg 86:241–251
39. Claassen J, Mayer SA, Kowalski RG, Emerson RG, Hirsch LJ (2004) Detection of electrographic seizures with continuous EEG monitoring in critically ill patients. Neurology 62:1743–1748
40. Vespa PM, O'Phelan K, Shah M, Mirabelli J, Starkman S, Kidwell C, Saver J, Nuwer MR, Frazee JG, McArthur DA, Martin NA (2003) Acute seizures after intracerebral hemorrhage: a factor in progressive midline shift and outcome. Neurology 60:1441–1446
41. Grand'Maison F, Reiher J, Leduc CP (1991) Retrospective inventory of EEG abnormalities in partial status epilepticus. Electroencephalogr Clin Neurophysiol 79:264–270
42. Cavazos JE, Jones SM, Cross DJ (2004) Sprouting and synaptic reorganization in the subiculum and CA1 region of the hippocampus in acute and chronic models of partial-onset epilepsy. Neuroscience 126:677–688
43. Golarai G, Greenwood AC, Feeney DM, Connor JA (2001) Physiological and structural evidence for hippocampal involvement in persistent seizure susceptibility after traumatic brain injury. J Neurosci 21:8523–8537
44. Ariza M, Serra-Grabulosa JM, Junque C, Ramirez B, Mataro M, Poca A, Bargallo N, Sahuquillo J (2006) Hippocampal head atrophy after traumatic brain injury. Neuropsychologia 44:1956–1961
45. Palacios EM, Sala-Llonch R, Junque C, Fernandez-Espejo D, Roig T, Tormos JM, Bargallo N, Vendrell P (2013) Long-term declarative memory deficits in diffuse TBI: correlations with cortical thickness, white matter integrity and hippocampal volume. Cortex 49:646–657
46. Hickey RW, Adelson PD, Johnnides MJ, Davis DS, Yu Z, Rose ME, Chang YF, Graham SH (2007) Cyclooxygenase-2 activity following traumatic brain injury in the developing rat. Pediatr Res 62:271–276
47. Miller G (2011) Neuropathology. A battle no soldier wants to fight. Science 333:517–519
48. Menzler K, Thiel P, Hermsen A, Chen X, Benes L, Miller D, Sure U, Knake S, Rosenow F (2011) The role of underlying structural cause for epilepsy classification: clinical features and prognosis in mesial temporal lobe epilepsy caused by hippocampal sclerosis versus cavernoma. Epilepsia 52:707–711
49. Glushakov AV, Galvis JM, Solaski SL, Doré S (2015) Hippocampal degeneration after traumatic brain injury: the roles of the PGE2 EP1 receptor. J Trauma Care 1:1007
50. Hellewell SC, Yan EB, Agyapomaa DA, Bye N, Morganti-Kossmann MC (2010) Post-traumatic hypoxia exacerbates brain tissue

damage: analysis of axonal injury and glial responses. J Neurotrauma 27:1997–2010

51. Teasdale GM, Graham DI (1998) Craniocerebral trauma: protection and retrieval of the neuronal population after injury. Neurosurgery 43:723–737, discussion 737-728
52. Franzblau M, Gonzales-Portillo C, Gonzales-Portillo GS, Diamandis T, Borlongan MC, Tajiri N, Borlongan CV (2013) Vascular damage: a persisting pathology common to Alzheimer's disease and traumatic brain injury. Med Hypotheses 81:842–845
53. Ferro JM, Pinto F (2004) Poststroke epilepsy: epidemiology, pathophysiology and management. Drugs Aging 21:639–653
54. Berges S, Moulin T, Berger E, Tatu L, Sablot D, Challier B, Rumbach L (2000) Seizures and epilepsy following strokes: recurrence factors. Eur Neurol 43:3–8
55. Bladin CF, Alexandrov AV, Bellavance A, Bornstein N, Chambers B, Cote R, Lebrun L, Pirisi A, Norris JW (2000) Seizures after stroke: a prospective multicenter study. Arch Neurol 57:1617–1622
56. Maganti R, Gerber P, Drees C, Chung S (2008) Nonconvulsive status epilepticus. Epilepsy Behav 12:572–586
57. Silverman IE, Restrepo L, Mathews GC (2002) Poststroke seizures. Arch Neurol 59:195–201
58. Berger AR, Lipton RB, Lesser ML, Lantos G, Portenoy RK (1988) Early seizures following intracerebral hemorrhage: implications for therapy. Neurology 38:1363–1365
59. Rhoney DH, Tipps LB, Murry KR, Basham MC, Michael DB, Coplin WM (2000) Anticonvulsant prophylaxis and timing of seizures after aneurysmal subarachnoid hemorrhage. Neurology 55:258–265
60. Pinto AN, Canhao P, Ferro JM (1996) Seizures at the onset of subarachnoid haemorrhage. J Neurol 243:161–164
61. Claassen J, Peery S, Kreiter KT, Hirsch LJ, Du EY, Connolly ES, Mayer SA (2003) Predictors and clinical impact of epilepsy after subarachnoid hemorrhage. Neurology 60:208–214
62. Burn J, Dennis M, Bamford J, Sandercock P, Wade D, Warlow C (1997) Epileptic seizures after a first stroke: the Oxfordshire Community Stroke Project. BMJ 315:1582–1587
63. Willmore LJ, Sypert GW, Munson JV, Hurd RW (1978) Chronic focal epileptiform discharges induced by injection of iron into rat and cat cortex. Science 200:1501–1503
64. Willmore LJ, Triggs WJ (1991) Iron-induced lipid peroxidation and brain injury responses. Int J Dev Neurosci 9:175–180
65. Kucukkaya B, Aker R, Yuksel M, Onat F, Yalcin AS (1998) Low dose MK-801 protects against iron-induced oxidative changes in a rat model of focal epilepsy. Brain Res 788:133–136
66. Glushakova OY, Johnson D, Hayes RL (2014) Delayed increases in microvascular pathology after experimental traumatic brain injury are associated with prolonged inflammation, blood-brain barrier disruption, and progressive white matter damage. J Neurotrauma 31:1180–1193
67. Willmore LJ, Rubin JJ (1981) Antiperoxidant pretreatment and iron-induced epileptiform discharges in the rat: EEG and histopathologic studies. Neurology 31:63–69
68. Willmore LJ, Hiramatsu M, Kochi H, Mori A (1983) Formation of superoxide radicals after FeCl3 injection into rat isocortex. Brain Res 277:393–396
69. Glushakov AV, Robbins SW, Bracy CL, Narumiya S, Dore S (2013) Prostaglandin F2 alpha FP receptor antagonist improves outcomes after experimental traumatic brain injury. J Neurotrauma 30:A163
70. Oby E, Janigro D (2006) The blood-brain barrier and epilepsy. Epilepsia 47:1761–1774
71. Tomkins O, Shelef I, Kaizerman I, Eliushin A, Afawi Z, Misk A, Gidon M, Cohen A, Zumsteg D, Friedman A (2008) Blood-brain barrier disruption in post-traumatic epilepsy. J Neurol Neurosurg Psychiatry 79:774–777
72. Huang XJ, Glushakova O, Mondello S, Van K, Hayes RL, Lyeth BG (2015) Acute temporal profiles of serum levels of UCH-L1 and GFAP and relationships to neuronal and astroglial pathology following traumatic brain injury in rats. J Neurotrauma 32:1179–1189
73. Glushakova OY, Jeromin A, Martinez J, Johnson D, Denslow N, Streeter J, Hayes RL, Mondello S (2012) Cerebrospinal fluid protein biomarker panel for assessment of neurotoxicity induced by kainic acid in rats. Toxicol Sci 130:158–167
74. Gopez JJ, Yue H, Vasudevan R, Malik AS, Fogelsanger LN, Lewis S, Panikashvili D, Shohami E, Jansen SA, Narayan RK, Strauss KI (2005) Cyclooxygenase-2-specific inhibitor improves functional outcomes, provides neuroprotection, and reduces inflammation in a rat model of traumatic brain injury. Neurosurgery 56:590–604
75. Strauss KI, Barbe MF, Marshall RM, Raghupathi R, Mehta S, Narayan RK (2000) Prolonged cyclooxygenase-2 induction in neurons and glia following traumatic brain injury in the rat. J Neurotrauma 17:695–711
76. Dash PK, Mach SA, Moore AN (2000) Regional expression and role of cyclooxygen-

ase-2 following experimental traumatic brain injury. J Neurotrauma 17:69–81
77. Yang T, Zhou D, Stefan H (2010) Why mesial temporal lobe epilepsy with hippocampal sclerosis is progressive: uncontrolled inflammation drives disease progression? J Neurol Sci 296:1–6
78. Desjardins P, Sauvageau A, Bouthillier A, Navarro D, Hazell AS, Rose C, Butterworth RF (2003) Induction of astrocytic cyclooxygenase-2 in epileptic patients with hippocampal sclerosis. Neurochem Int 42:299–303
79. Turrin NP, Rivest S (2004) Innate immune reaction in response to seizures: implications for the neuropathology associated with epilepsy. Neurobiol Dis 16:321–334
80. Takemiya T, Maehara M, Matsumura K, Yasuda S, Sugiura H, Yamagata K (2006) Prostaglandin E2 produced by late induced COX-2 stimulates hippocampal neuron loss after seizure in the CA3 region. Neurosci Res 56:103–110
81. Yasojima K, Schwab C, McGeer EG, McGeer PL (1999) Distribution of cyclooxygenase 1 and cyclooxygenase-2 mRNAs and proteins in human brain and peripheral organs. Brain Res 830:226–236
82. Toti P, DE Felice C, Schürfeld K, Stumpo M, Bartolommei S, Lombardi A, Petraglia E, Buonocore G (2001) Cyclooxygenase-2 immunoreactivity in the ischemic neonatal human brain. An autopsy study. J Submicrosc Cytol Pathol 33:245–249
83. Tomimoto H, Akiguchi I, Wakita H, Lin JX, Budka H (2000) Cyclooxygenase-2 is induced in microglia during chronic cerebral ischemia in humans. Acta Neuropathol (Berl) 99:26–30
84. Tomimoto H, Shibata M, Ihara M, Akiguchi I, Ohtani R, Budka H (2002) A comparative study on the expression of cyclooxygenase and 5-lipoxygenase during cerebral ischemia in humans. Acta Neuropathol (Berl) 104:601–607
85. Kunz T, Marklund N, Hillered L, Oliw EH (2002) Cyclooxygenase-2, prostaglandin synthases, and prostaglandin H2 metabolism in traumatic brain injury in the rat. J Neurotrauma 19:1051–1064
86. Ahmad M, Rose ME, Vagni V, Griffith RP, Dixon CE, Kochanek PM, Hickey RW, Graham SH (2008) Genetic disruption of cyclooxygenase-2 does not improve histological or behavioral outcome after traumatic brain injury in mice. J Neurosci Res 86:3605–3612
87. Strauss KI (2008) Antiinflammatory and neuroprotective actions of COX2 inhibitors in the injured brain. Brain Behav Immun 22:285–298
88. Kawano T, Anrather J, Zhou P, Park L, Wang G, Frys KA, Kunz A, Cho S, Orio M, Iadecola C (2006) Prostaglandin E_2 EP1 receptors: downstream effectors of COX-2 neurotoxicity. Nat Med 12:225–229
89. Ahmad AS, Saleem S, Ahmad M, Doré S (2006) Prostaglandin EP1 receptor contributes to excitotoxicity and focal ischemic brain damage. Toxicol Sci 89:265–270
90. Abe T, Kunz A, Shimamura M, Zhou P, Anrather J, Iadecola C (2009) The neuroprotective effect of prostaglandin E2 EP1 receptor inhibition has a wide therapeutic window, is sustained in time and is not sexually dimorphic. J Cereb Blood Flow Metab 29:66–72
91. Ahmad AS, Kim YT, Ahmad M, Maruyama T, Doré S (2008) Selective blockade of PGE_2 EP1 receptor protects brain against experimental ischemia and excitotoxicity, and hippocampal slice cultures against oxygen-glucose deprivation. Neurotox Res 14:343–351
92. Zhen G, Kim YT, Li RC, Yocum J, Kapoor N, Langer J, Dobrowolski P, Maruyama T, Narumiya S, Doré S (2011) PGE(2) EP1 receptor exacerbated neurotoxicity in a mouse model of cerebral ischemia and Alzheimer's disease. Neurobiol Aging 33:2215–2219
93. Topol EJ (2004) Failing the public health—rofecoxib, Merck, and the FDA. N Engl J Med 351:1707–1709
94. Glushakov AV, Robbins SW, Bracy CL, Narumiya S, Doré S (2013) Prostaglandin F2alpha FP receptor antagonist improves outcomes after experimental traumatic brain injury. J Neuroinflammation 10:132
95. Willmore LJ, Ueda Y (2009) Posttraumatic epilepsy: hemorrhage, free radicals and the molecular regulation of glutamate. Neurochem Res 34:688–697
96. Oliveira MS, Furian AF, Rambo LM, Ribeiro LR, Royes LF, Ferreira J, Calixto JB, Mello CF (2008) Modulation of pentylenetetrazol-induced seizures by prostaglandin E2 receptors. Neuroscience 152:1110–1118
97. Fischborn SV, Soerensen J, Potschka H (2010) Targeting the prostaglandin E2 EP1 receptor and cyclooxygenase-2 in the amygdala kindling model in mice. Epilepsy Res 91:57–65
98. Khatibi NH, Jadhav V, Matus B, Fathali N, Martin R, Applegate R, Tang J, Zhang JH (2011) Prostaglandin E2 EP1 receptor inhibition fails to provide neuroprotection in surgically induced brain-injured mice. Acta Neurochir Suppl 111:277–281
99. Glushakov AV, Fazal JA, Narumiya S, Dore S (2014) Role of the prostaglandin E2 EP1

receptor in traumatic brain injury. PLoS One 9, e113689
100. Shimamura M, Zhou P, Casolla B, Qian L, Capone C, Kurinami H, Iadecola C, Anrather J (2013) Prostaglandin E2 type 1 receptors contribute to neuronal apoptosis after transient forebrain ischemia. J Cereb Blood Flow Metab 33:1207–1214
101. Baik EJ, Kim EJ, Lee SH, Moon C (1999) Cyclooxygenase-2 selective inhibitors aggravate kainic acid induced seizure and neuronal cell death in the hippocampus. Brain Res 843:118–129
102. Kawaguchi K, Hickey RW, Rose ME, Zhu L, Chen J, Graham SH (2005) Cyclooxygenase-2 expression is induced in rat brain after kainate-induced seizures and promotes neuronal death in CA3 hippocampus. Brain Res 1050:130–137
103. Yoshikawa K, Kita Y, Kishimoto K, Shimizu T (2006) Profiling of eicosanoid production in the rat hippocampus during kainate-induced seizure: dual-phase regulation and differential involvement of COX-1 and COX-2. J Biol Chem 281:14663–14669
104. Toscano CD, Kingsley PJ, Marnett LJ, Bosetti F (2008) NMDA-induced seizure intensity is enhanced in COX-2 deficient mice. Neurotoxicology 29:1114–1120
105. Berchtold-Kanz E, Anhut H, Heldt R, Neufang B, Hertting G (1981) Regional distribution of arachidonic acid metabolites in rat brain following convulsive stimuli. Prostaglandins 22:65–79
106. Steinhauer HB, Anhut H, Hertting G (1979) The synthesis of prostaglandins and thromboxane in the mouse brain in vivo. Influence of drug induced convulsions, hypoxia and the anticonvulsants trimethadione and diazepam. Naunyn Schmiedebergs Arch Pharmacol 310:53–58
107. Steinhauer HB, Hertting G (1981) Lowering of the convulsive threshold by non-steroidal anti-inflammatory drugs. Eur J Pharmacol 69:199–203
108. Forstermann U, Heldt R, Hertting G (1983) Increase in brain prostaglandins during convulsions is due to increased neuronal activity and not to hypoxia. Arch Int Pharmacodyn Ther 263:180–188
109. Forstermann U, Heldt R, Hertting G (1983) Effects of intracerebroventricular administration of prostaglandin D2 on behaviour, blood pressure and body temperature as compared to prostaglandins E2 and F2 alpha. Psychopharmacology (Berl) 80:365–370
110. Forstermann U, Heldt R, Knappen F, Hertting G (1982) Potential anticonvulsive properties of endogenous prostaglandins formed in mouse brain. Brain Res 240:303–310
111. Naffah-Mazzacoratti MG, Bellissimo MI, Cavalheiro EA (1995) Profile of prostaglandin levels in the rat hippocampus in pilocarpine model of epilepsy. Neurochem Int 27:461–466
112. Okada K, Yuhi T, Tsuji S, Yamashita U (2001) Cyclooxygenase-2 expression in the hippocampus of genetically epilepsy susceptible El mice was increased after seizure. Brain Res 894:332–335
113. Forstermann U, Seregi A, Hertting G (1984) Anticonvulsive effects of endogenous prostaglandins formed in brain of spontaneously convulsing gerbils. Prostaglandins 27:913–923
114. Folco GC, Longiave D, Bosisio E (1977) Relations between prostaglandin E2, F2alpha, and cyclic nucleotides levels in rat brain and induction of convulsions. Prostaglandins 13:893–900
115. Kim HJ, Chung JI, Lee SH, Jung YS, Moon CH, Baik EJ (2008) Involvement of endogenous prostaglandin F2alpha on kainic acid-induced seizure activity through FP receptor: the mechanism of proconvulsant effects of COX-2 inhibitors. Brain Res 1193:153–161
116. Chung JI, Kim AY, Lee SH, Baik EJ (2013) Seizure susceptibility in immature brain due to lack of COX-2-induced PGF2alpha. Exp Neurol 249:95–103
117. Dixon CE, Lyeth BG, Povlishock JT, Findling RL, Hamm RJ, Marmarou A, Young HF, Hayes RL (1987) A fluid percussion model of experimental brain injury in the rat. J Neurosurg 67:110–119
118. Lighthall JW (1988) Controlled cortical impact: a new experimental brain injury model. J Neurotrauma 5:1–15
119. Dixon CE, Clifton GL, Lighthall JW, Yaghmai AA, Hayes RL (1991) A controlled cortical impact model of traumatic brain injury in the rat. J Neurosci Methods 39:253–262
120. Marmarou A, Foda MA, van den Brink W, Campbell J, Kita H, Demetriadou K (1994) A new model of diffuse brain injury in rats. Part I: pathophysiology and biomechanics. J Neurosurg 80:291–300
121. Cernak I, Savic J, Malicevic Z, Zunic G, Radosevic P, Ivanovic I, Davidovic L (1996) Involvement of the central nervous system in the general response to pulmonary blast injury. J Trauma 40:S100–S104
122. Diaz-Arrastia R, Agostini MA, Madden CJ, Van Ness PC (2009) Posttraumatic epilepsy: the endophenotypes of a human model of epileptogenesis. Epilepsia 50(Suppl 2):14–20

123. Kovacs SK, Leonessa F, Ling GS (2014) Blast TBI models, neuropathology, and implications for seizure risk. Front Neurol 5:47
124. Li BC, Li Y, Xu C, Wang J, Chen Z, Li G, Zhang J, Hu S, Wang L, Feng H (2014) Blast-induced traumatic brain injury of goats in confined space. Neurol Res 36:974–982
125. Cernak I (2010) The importance of systemic response in the pathobiology of blast-induced neurotrauma. Front Neurol 1:151
126. Graham DI, McIntosh TK, Maxwell WL, Nicoll JA (2000) Recent advances in neurotrauma. J Neuropathol Exp Neurol 59:641–651
127. Thompson HJ, Lifshitz J, Marklund N, Grady MS, Graham DI, Hovda DA, McIntosh TK (2005) Lateral fluid percussion brain injury: a 15-year review and evaluation. J Neurotrauma 22:42–75
128. Kabadi SV, Hilton GD, Stoica BA, Zapple DN, Faden AI (2010) Fluid-percussion-induced traumatic brain injury model in rats. Nat Protoc 5:1552–1563
129. Hayes RL, Stalhammar D, Povlishock JT, Allen AM, Galinat BJ, Becker DP, Stonnington HH (1987) A new model of concussive brain injury in the cat produced by extradural fluid volume loading: II. Physiological and neuropathological observations. Brain Inj 1:93–112
130. Hartl R, Medary M, Ruge M, Arfors KE, Ghajar J (1997) Blood-brain barrier breakdown occurs early after traumatic brain injury and is not related to white blood cell adherence. Acta Neurochir Suppl 70:240–242
131. Stalhammar D, Galinat BJ, Allen AM, Becker DP, Stonnington HH, Hayes RL (1987) A new model of concussive brain injury in the cat produced by extradural fluid volume loading: I. Biomechanical properties. Brain Inj 1:73–91
132. Pierce JE, Smith DH, Trojanowski JQ, McIntosh TK (1998) Enduring cognitive, neurobehavioral and histopathological changes persist for up to one year following severe experimental brain injury in rats. Neuroscience 87:359–369
133. McIntosh TK, Noble L, Andrews B, Faden AI (1987) Traumatic brain injury in the rat: characterization of a midline fluid-percussion model. Cent Nerv Syst Trauma 4:119–134
134. Carbonell WS, Maris DO, McCall T, Grady MS (1998) Adaptation of the fluid percussion injury model to the mouse. J Neurotrauma 15:217–229
135. Sullivan HG, Martinez J, Becker DP, Miller JD, Griffith R, Wist AO (1976) Fluid-percussion model of mechanical brain injury in the cat. J Neurosurg 45:521–534
136. Millen JE, Glauser FL, Fairman RP (1985) A comparison of physiological responses to percussive brain trauma in dogs and sheep. J Neurosurg 62:587–591
137. Dixon CE, Lighthall JW, Anderson TE (1988) Physiologic, histopathologic, and cineradiographic characterization of a new fluid-percussion model of experimental brain injury in the rat. J Neurotrauma 5:91–104
138. Alder J, Fujioka W, Lifshitz J, Crockett DP, Thakker-Varia S (2011) Lateral fluid percussion: model of traumatic brain injury in mice. J Vis Exp 54:pii 3063
139. Pfenninger EG, Reith A, Breitig D, Grunert A, Ahnefeld FW (1989) Early changes of intracranial pressure, perfusion pressure, and blood flow after acute head injury. Part 1: an experimental study of the underlying pathophysiology. J Neurosurg 70:774–779
140. Hicks R, Soares H, Smith D, McIntosh T (1996) Temporal and spatial characterization of neuronal injury following lateral fluid-percussion brain injury in the rat. Acta Neuropathol 91:236–246
141. Zink BJ, Walsh RF, Feustel PJ (1993) Effects of ethanol in traumatic brain injury. J Neurotrauma 10:275–286
142. Santhakumar V, Ratzliff AD, Jeng J, Toth Z, Soltesz I (2001) Long-term hyperexcitability in the hippocampus after experimental head trauma. Ann Neurol 50:708–717
143. Neuberger EJ, Abdul Wahab R, Jayakumar A, Pfister BJ, Santhakumar V (2014) Distinct effect of impact rise times on immediate and early neuropathology after brain injury in juvenile rats. J Neurosci Res 92:1350–1361
144. Lowenstein DH, Thomas MJ, Smith DH, McIntosh TK (1992) Selective vulnerability of dentate hilar neurons following traumatic brain injury: a potential mechanistic link between head trauma and disorders of the hippocampus. J Neurosci 12:4846–4853
145. Wang Y, Hameed MQ, Rakhade SN, Iglesias AH, Muller PA, Mou DL, Rotenberg A (2014) Hippocampal immediate early gene transcription in the rat fluid percussion traumatic brain injury model. Neuroreport 25:954–959
146. D'Ambrosio R, Fairbanks JP, Fender JS, Born DE, Doyle DL, Miller JW (2004) Post-traumatic epilepsy following fluid percussion injury in the rat. Brain 127:304–314
147. D'Ambrosio R, Maris DO, Grady MS, Winn HR, Janigro D (1999) Impaired K(+) homeostasis and altered electrophysiological proper-

ties of post-traumatic hippocampal glia. J Neurosci 19:8152–8162

148. Cao R, Hasuo H, Ooba S, Akasu T, Zhang X (2006) Facilitation of glutamatergic synaptic transmission in hippocampal CA1 area of rats with traumatic brain injury. Neurosci Lett 401:136–141

149. Toth Z, Hollrigel GS, Gorcs T, Soltesz I (1997) Instantaneous perturbation of dentate interneuronal networks by a pressure wave-transient delivered to the neocortex. J Neurosci 17:8106–8117

150. Coulter DA, Rafiq A, Shumate M, Gong QZ, DeLorenzo RJ, Lyeth BG (1996) Brain injury-induced enhanced limbic epileptogenesis: anatomical and physiological parallels to an animal model of temporal lobe epilepsy. Epilepsy Res 26:81–91

151. Dewitt DS, Kong DL, Lyeth BG, Jenkins LW, Hayes RL, Wooten ED, Prough DS (1988) Experimental traumatic brain injury elevates brain prostaglandin E2 and thromboxane B2 levels in rats. J Neurotrauma 5:303–313

152. D'Ambrosio R, Fender JS, Fairbanks JP, Simon EA, Born DE, Doyle DL, Miller JW (2005) Progression from frontal-parietal to mesial-temporal epilepsy after fluid percussion injury in the rat. Brain 128:174–188

153. Curia G, Levitt M, Fender JS, Miller JW, Ojemann J, D'Ambrosio R (2011) Impact of injury location and severity on posttraumatic epilepsy in the rat: role of frontal neocortex. Cereb Cortex 21:1574–1592

154. Campbell JN, Gandhi A, Singh B, Churn SB (2014) Traumatic brain injury causes a tacrolimus-sensitive increase in non-convulsive seizures in a rat model of post-traumatic epilepsy. Int J Neurol Brain Disord 1:1–11

155. Kharatishvili I, Nissinen JP, McIntosh TK, Pitkanen A (2006) A model of posttraumatic epilepsy induced by lateral fluid-percussion brain injury in rats. Neuroscience 140:685–697

156. Bao YH, Bramlett HM, Atkins CM, Truettner JS, Lotocki G, Alonso OF, Dietrich WD (2011) Post-traumatic seizures exacerbate histopathological damage after fluid-percussion brain injury. J Neurotrauma 28:35–42

157. Huusko N, Pitkanen A (2014) Parvalbumin immunoreactivity and expression of GABAA receptor subunits in the thalamus after experimental TBI. Neuroscience 267:30–45

158. Atkins CM, Truettner JS, Lotocki G, Sanchez-Molano J, Kang Y, Alonso OF, Sick TJ, Dietrich WD, Bramlett HM (2010) Post-traumatic seizure susceptibility is attenuated by hypothermia therapy. Eur J Neurosci 32: 1912–1920

159. Mishra AM, Bai X, Sanganahalli BG, Waxman SG, Shatillo O, Grohn O, Hyder F, Pitkanen A, Blumenfeld H (2014) Decreased resting functional connectivity after traumatic brain injury in the rat. PLoS One 9, e95280

160. Wallis RA, Panizzon KL (1995) Felbamate neuroprotection against CA1 traumatic neuronal injury. Eur J Pharmacol 294:475–482

161. Pitkanen A, Immonen R, Ndode-Ekane X, Grohn O, Stohr T, Nissinen J (2014) Effect of lacosamide on structural damage and functional recovery after traumatic brain injury in rats. Epilepsy Res 108:653–665

162. Lighthall JW, Goshgarian HG, Pinderski CR (1990) Characterization of axonal injury produced by controlled cortical impact. J Neurotrauma 7:65–76

163. Hamm RJ, Dixon CE, Gbadebo DM, Singha AK, Jenkins LW, Lyeth BG, Hayes RL (1992) Cognitive deficits following traumatic brain injury produced by controlled cortical impact. J Neurotrauma 9:11–20

164. Smith DH, Soares HD, Pierce JS, Perlman KG, Saatman KE, Meaney DF, Dixon CE, McIntosh TK (1995) A model of parasagittal controlled cortical impact in the mouse: cognitive and histopathologic effects. J Neurotrauma 12:169–178

165. Brody DL, Mac Donald C, Kessens CC, Yuede C, Parsadanian M, Spinner M, Kim E, Schwetye KE, Holtzman DM, Bayly PV (2007) Electromagnetic controlled cortical impact device for precise, graded experimental traumatic brain injury. J Neurotrauma 24: 657–673

166. Lighthall JW, Dixon CE, Anderson TE (1989) Experimental models of brain injury. J Neurotrauma 6:83–97

167. Sanders MJ, Dietrich WD, Green EJ (1999) Cognitive function following traumatic brain injury: effects of injury severity and recovery period in a parasagittal fluid-percussive injury model. J Neurotrauma 16:915–925

168. Alessandri B, Heimann A, Filippi R, Kopacz L, Kempski O (2003) Moderate controlled cortical contusion in pigs: effects on multiparametric neuromonitoring and clinical relevance. J Neurotrauma 20:1293–1305

169. Manley GT, Rosenthal G, Lam M, Morabito D, Yan D, Derugin N, Bollen A, Knudson MM, Panter SS (2006) Controlled cortical impact in swine: pathophysiology and biomechanics. J Neurotrauma 23:128–139

170. King C, Robinson T, Dixon CE, Rao GR, Larnard D, Nemoto CE (2010) Brain temperature profiles during epidural cooling with the ChillerPad in a monkey model of traumatic brain injury. J Neurotrauma 27:1895–1903

171. Hall ED, Sullivan PG, Gibson TR, Pavel KM, Thompson BM, Scheff SW (2005) Spatial and temporal characteristics of neurodegeneration after controlled cortical impact in mice: more than a focal brain injury. J Neurotrauma 22:252–265
172. Goodman JC, Cherian L, Bryan RM Jr, Robertson CS (1994) Lateral cortical impact injury in rats: pathologic effects of varying cortical compression and impact velocity. J Neurotrauma 11:587–597
173. Yang L, Afroz S, Michelson HB, Goodman JH, Valsamis HA, Ling DS (2010) Spontaneous epileptiform activity in rat neocortex after controlled cortical impact injury. J Neurotrauma 27:1541–1548
174. Sutton RL, Lescaudron L, Stein DG (1993) Unilateral cortical contusion injury in the rat: vascular disruption and temporal development of cortical necrosis. J Neurotrauma 10: 135–149
175. Saatman KE, Feeko KJ, Pape RL, Raghupathi R (2006) Differential behavioral and histopathological responses to graded cortical impact injury in mice. J Neurotrauma 23:1241–1253
176. Dixon CE, Kraus MF, Kline AE, Ma X, Yan HQ, Griffith RG, Wolfson BM, Marion DW (1999) Amantadine improves water maze performance without affecting motor behavior following traumatic brain injury in rats. Restor Neurol Neurosci 14:285–294
177. Dixon CE, Kochanek PM, Yan HQ, Schiding JK, Griffith RG, Baum E, Marion DW, DeKosky ST (1999) One-year study of spatial memory performance, brain morphology, and cholinergic markers after moderate controlled cortical impact in rats. J Neurotrauma 16:109–122
178. Vink R, Mullins PG, Temple MD, Bao W, Faden AI (2001) Small shifts in craniotomy position in the lateral fluid percussion injury model are associated with differential lesion development. J Neurotrauma 18:839–847
179. Fox GB, Fan L, Levasseur RA, Faden AI (1998) Sustained sensory/motor and cognitive deficits with neuronal apoptosis following controlled cortical impact brain injury in the mouse. J Neurotrauma 15:599–614
180. Kochanek PM, Hendrich KS, Dixon CE, Schiding JK, Williams DS, Ho C (2002) Cerebral blood flow at one year after controlled cortical impact in rats: assessment by magnetic resonance imaging. J Neurotrauma 19: 1029–1037
181. Hunt RF, Haselhorst LA, Schoch KM, Bach EC, Rios-Pilier J, Scheff SW, Saatman KE, Smith BN (2012) Posttraumatic mossy fiber sprouting is related to the degree of cortical damage in three mouse strains. Epilepsy Res 99:167–170
182. Kochanek PM, Vagni VA, Janesko KL, Washington CB, Crumrine PK, Garman RH, Jenkins LW, Clark RS, Homanics GE, Dixon CE, Schnermann J, Jackson EK (2006) Adenosine A1 receptor knockout mice develop lethal status epilepticus after experimental traumatic brain injury. J Cereb Blood Flow Metab 26:565–575
183. Haselkorn ML, Shellington DK, Jackson EK, Vagni VA, Janesko-Feldman K, Dubey RK, Gillespie DG, Cheng D, Bell MJ, Jenkins LW, Homanics GE, Schnermann J, Kochanek PM (2010) Adenosine A1 receptor activation as a brake on the microglial response after experimental traumatic brain injury in mice. J Neurotrauma 27:901–910
184. Zweckberger K, Simunovic F, Kiening KL, Unterberg AW, Sakowitz OW (2010) Anticonvulsive effects of the dopamine agonist lisuride maleate after experimental traumatic brain injury. Neurosci Lett 470:150–154
185. Hunt RF, Scheff SW, Smith BN (2009) Posttraumatic epilepsy after controlled cortical impact injury in mice. Exp Neurol 215: 243–252
186. Guo D, Zeng L, Brody DL, Wong M (2013) Rapamycin attenuates the development of posttraumatic epilepsy in a mouse model of traumatic brain injury. PLoS One 8, e64078
187. Lei Z, Deng P, Li J, Xu ZC (2012) Alterations of A-type potassium channels in hippocampal neurons after traumatic brain injury. J Neurotrauma 29:235–245
188. Bolkvadze T, Pitkanen A (2012) Development of post-traumatic epilepsy after controlled cortical impact and lateral fluid-percussion-induced brain injury in the mouse. J Neurotrauma 29:789–812
189. Statler KD, Scheerlinck P, Pouliot W, Hamilton M, White HS, Dudek FE (2009) A potential model of pediatric posttraumatic epilepsy. Epilepsy Res 86:221–223
190. Hunt RF, Scheff SW, Smith BN (2011) Synaptic reorganization of inhibitory hilar interneuron circuitry after traumatic brain injury in mice. J Neurosci 31:6880–6890
191. Kharlamov EA, Lepsveridze E, Meparishvili M, Solomonia RO, Lu B, Miller ER, Kelly KM, Mtchedlishvili Z (2011) Alterations of GABA(A) and glutamate receptor subunits and heat shock protein in rat hippocampus following traumatic brain injury and in posttraumatic epilepsy. Epilepsy Res 95:20–34

192. Osselton JW (1965) Acquisition of EEG data by bipolar, unipolar and average reference methods: a theoretical comparison. Electroencephalogr Clin Neurophysiol 19:527–528

193. Stover JF, Sakowitz OW, Beyer TF, Dohse NK, Kroppenstedt SN, Thomale UW, Schaser KD, Unterberg AW (2003) Effects of LY379268, a selective group II metabotropic glutamate receptor agonist on EEG activity, cortical perfusion, tissue damage, and cortical glutamate, glucose, and lactate levels in brain-injured rats. J Neurotrauma 20:315–326

194. Frey L, Lepkin A, Schickedanz A, Huber K, Brown MS, Serkova N (2014) ADC mapping and T1-weighted signal changes on post-injury MRI predict seizure susceptibility after experimental traumatic brain injury. Neurol Res 36:26–37

195. Macolino CM, Daiutolo BV, Albertson BK, Elliott MB (2014) Mechanical alloydnia induced by traumatic brain injury is independent of restraint stress. J Neurosci Methods 226:139–146

196. Sanchez JC, Alba N, Nishida T, Batich C, Carney PR (2006) Structural modifications in chronic microwire electrodes for cortical neuroprosthetics: a case study. IEEE Trans Neural Syst Rehabil Eng 14:217–221

197. Nandan M, Talathi SS, Myers S, Ditto WL, Khargonekar PP, Carney PR (2010) Support vector machines for seizure detection in an animal model of chronic epilepsy. J Neural Eng 7:036001

198. MacLennan AJ, Carney PR, Zhu WJ, Chaves AH, Garcia J, Grimes JR, Anderson KJ, Roper SN, Lee N (2001) An essential role for the H218/AGR16/Edg-5/LP(B2) sphingosine 1-phosphate receptor in neuronal excitability. Eur J Neurosci 14:203–209

199. Lu XC, Williams AJ, Tortella FC (2001) Quantitative electroencephalography spectral analysis and topographic mapping in a rat model of middle cerebral artery occlusion. Neuropathol Appl Neurobiol 27:481–495

200. Cao W, Glushakov A, Shah HP, Mecca AP, Sumners C, Shi P, Seubert CN, Martynyuk AE (2011) Halogenated aromatic amino acid 3,5-dibromo-D: -tyrosine produces beneficial effects in experimental stroke and seizures. Amino Acids 40:1151–1158

201. Cao W, Shah HP, Glushakov AV, Mecca AP, Shi P, Sumners C, Seubert CN, Martynyuk AE (2009) Efficacy of 3,5-dibromo-L-phenylalanine in rat models of stroke, seizures and sensorimotor gating deficit. Br J Pharmacol 158:2005–2013

202. Lopez PH, Ahmad AS, Mehta NR, Toner M, Rowland EA, Zhang J, Dore S, Schnaar RL (2011) Myelin-associated glycoprotein protects neurons from excitotoxicity. J Neurochem 116:900–908

203. Martynyuk AE, Seubert CN, Yarotskyy V, Glushakov AV, Gravenstein N, Sumners C, Dennis DM (2006) Halogenated derivatives of aromatic amino acids exhibit balanced antiglutamatergic actions: potential applications for the treatment of neurological and neuropsychiatric disorders. Recent Pat CNS Drug Discov 1:261–270

204. Moezi L, Shafaroodi H, Hojati A, Dehpour AR (2011) The interaction of melatonin and agmatine on pentylenetetrazole-induced seizure threshold in mice. Epilepsy Behav 22:200–206

Chapter 28

Closed-Head TBI Model of Multiple Morbidity

Floyd J. Thompson, Jiamei Hou, and Prodip K. Bose

Abstract

Successful therapy for TBI disabilities awaits refinement in the understanding of TBI neurobiology, quantitative measurement of treatment-induced incremental changes in recovery trajectories, and effective translation to human TBI using quantitative methods and protocols that were effective to monitor recovery in preclinical models. Details of the specific neurobiology that underlies these injuries and effective quantitation of treatment-induced changes are beginning to emerge utilizing a variety of preclinical and clinical models (for reviews see (Morales et al., Neuroscience 136:971–989, 2005; Fujimoto et al., Neurosci Biobehav Rev 28:365–378, 2004; Cernak, NeuroRx 2:410–422, 2005; Smith et al., J Neurotrauma 22:1485–1502, 2005; Bose et al., J Neurotrauma 30:1177–1191, 2013; Xiong et al., Nat Rev Neurosci 14:128–142, 2013; Xiong et al., Expert Opin Emerg Drugs 14:67–84, 2009; Johnson et al., Handb Clin Neurol 127:115–128, 2015; Bose et al., Brain neurotrauma: molecular, neuropsychological, and rehabilitation aspects, CRC Press/Taylor & Francis, Boca Raton, 2015)). Preclinical models of TBI, essential for the efficient study of TBI neurobiology, benefit from the setting of controlled injury and optimal opportunities for biometric quantitation of injury and treatment-induced changes in the trajectories of disability. Several preclinical models are currently used, and each offer opportunities for study of different aspects of TBI primary and secondary injuries (for review see (Morales et al., Neuroscience 136:971–989, 2005; Xiong et al., Nat Rev Neurosci 14:128–142, 2013; Xiong et al., Expert Opin Emerg Drugs 14:67–84, 2009; Johnson et al., Handb Clin Neurol 127:115–128, 2015; Dixon et al., J Neurotrauma 5:91–104, 1988)). The closed-head, impact-acceleration model of TBI designed by Marmarou et al., 1994 (J Neurosurg 80:291–300, 1994), when used to produce mild to moderate TBI, produces diffuse axonal injuries without significant additional focal injuries of the brain (Morales et al., Neuroscience 136:971–989, 2005; Foda and Marmarou, J Neurosurg 80:301–313, 1994; Kallakuri et al., Exp Brain Res 148:419–424, 2003). Accordingly, use of this preclinical model offers an opportunity for (a) gaining a greater understanding of the relationships of TBI induced diffuse axonal injuries and associated long term disabilities, and (b) to provide a platform for quantitative assessment of treatment interactions upon the trajectories of TBI-induced disabilities. Using the impact acceleration closed head TBI model to induce mild/moderate injuries in the rat, we have observed and quantitated multiple morbidities commonly observed following TBI in humans (Bose et al., J Neurotrauma 30:1177–1191, 2013). This chapter describes methods and protocols used for TBI-induced multiple morbidity involving cognitive dysfunction, balance instability, spasticity and gait, and anxiety-like disorder.

Key words Diffuse axonal injury, Cognitive impairment, Balance instability, Anxiety spasticity

Firas Kobeissy et al. (eds.), *Injury Models of the Central Nervous System: Methods and Protocols*, Methods in Molecular Biology, vol. 1462, DOI 10.1007/978-1-4939-3816-2_28, © Springer Science+Business Media New York 2016

1 Introduction

The most common type of brain injury is produced by impact acceleration TBI due to motor vehicle accident, falls [1, 2], or sports-related impacts [3]. Even moderate injuries produce enduring cognitive, anxiety, vestibulomotor (balance), and motor impairments including locomotor disorder and spasticity [4, 5]. Although these injuries include concussion, contusion, and hemorrhage, it is the diffuse axonal injuries of the brain and brainstem that produce the greatest potential for disconnection and disability [3, 6–9]. Successful therapy for TBI disabilities awaits refinement in the understanding of TBI neurobiology, quantitative measurement of treatment-induced incremental changes in recovery trajectories, and effective translation to human TBI using quantitative methods and protocols that were effective to monitor recovery in preclinical models. Details of the specific neurobiology that underlies these injuries and effective quantitation of treatment-induced changes are beginning to emerge utilizing a variety of preclinical and clinical models (for reviews see: [1, 4, 10–16]). Preclinical models of TBI, essential for the efficient study of TBI neurobiology, benefit from the setting of controlled injury and optimal opportunities for biometric quantitation of injury and treatment-induced changes in the trajectories of disability.

Several preclinical models are currently used, and each offers opportunities for study of different aspects of TBI primary and secondary injuries (for review see: [1, 10, 14, 15, 17]). The closed-head, impact-acceleration model of TBI designed by Marmarou et al., 1994 [18], when used to produce mild to moderate TBI, produces diffuse axonal injuries without significant additional focal injuries of the brain [10, 19, 20]. Accordingly, use of this preclinical model offers an opportunity for a) gaining a greater understanding of the relationships of TBI induced diffuse axonal injuries and associated long term disabilities, and b) to provide a platform for quantitative assessment of treatment interactions upon the trajectories of TBI-induced disabilities. Using the impact acceleration closed-head TBI model to induce mild/moderate injuries in the rat, we have observed and quantitated multiple morbidities commonly observed following TBI in humans [4]. This chapter describes methods and protocols used for TBI-induced multiple morbidity involving cognitive dysfunction, balance instability, spasticity and gait, and anxiety-like disorder. Although a criticism of the impact-acceleration model has been the lack of demonstration of persisting behavioral disabilities [10], we observed stable deficits and enduring immunohistochemical changes beyond 2 months post-injury.

The appearance of multiple disabilities following TBI may suggest that injury occurred in multiple regions of the brain.

On the other hand, our recent studies highlight the potential for multiple morbidity derived from injury of particular systems (such as the central noradrenergic system) that, in turn, influence the functions of multiple regions; injury at the brainstem level may have significant upstream (e.g., decrease DβH expression in the hippocampus) and downstream (significant decrease in DβH expression in lumbar spinal cord) impact [4]. A comprehensive list of behavioral and physiological analyses was used to demonstrate persisting behavioral disabilities in this model. Serial learning deficits were tested using a Morris water maze. Changes in balance were tested by assessing walking performance on a rotorod. Motor disabilities were quantitated using longitudinal measures of lower limb joint torques to quantitate longitudinal patterns of spastic hypertonia (see also, [21, 22]). Gait disabilities were quantitated using footprint analysis of locomotor patterns. Changes in long tract conduction patterns were assessed using TMS elicited motor evoked potentials. Anxiety-like behaviors were assessed using patterns of activity within an elevated plus maze. Recently, we utilized these methods to evaluate the nature and magnitude of TBI-induced spasticity and other TBI disabilities and how recovery trajectories were influenced by acute treatment using intrathecal baclofen [4]. In addition, TBI and treatment-induced changes in the immunohistochemical expression of selected signaling agents and receptors known to influence the excitability of neurons in selected neural regions critical to the tested functions were evaluated and compared. Details of the methods and protocols that were used in our studies of TBI-induced multiple morbidity will be described to enhance the opportunity for these approaches to be of value to future preclinical and clinical assessments of TBI disability and treatment.

2 Materials

1. Closed-head impact-acceleration TBI.
2. Animal: Young adult Sprague-Dawley rats (Charles River Laboratory).
3. Anesthesia: A combination of ketamine (80 mg/kg) and xylazine (5 mg/kg) and incision site infiltrated with 0.25 % bupivacaine (3 mg/kg).
4. Analgesics: Buprenex (0.05 mg/kg) and ketoprofen (5 mg/kg).
5. Surgical area preparation: Betadine scrub (7.5 %) and 70 % alcohol wipe.
6. Custom-made weight drop system: Modified Marmarou weight drop system with a laser pointer; plexiglass box with foam pad, and a stainless disk.

7. Material for behavioral/physiological experiment.
   (a) Morris water maze (MWM), a black plexiglas platform, submersible water heating coil, and a computerized video tracking system (Noldus EthoVision version XT 8.0).
   (b) A Rotarod (Economex; Columbus Instrument, Columbus, Ohio).
8. Instruments for velocity-dependent ankle torques.
   (a) An electromechanical shaker; Model 405, Ling Dynamic systems, Royston Herts, UK).
   (b) A force transducer (LVDT) (Model FT-03, Grass Instruments, Quincy, MA, USA).
   (c) EMGs recording electrodes; LabView signal acquisition hardware and software system (National Instrument, USA).
9. Custom-designed animal trunk-restraint system.
10. Local anesthesia, a xylocaine 2% jelly (Lidocaine HCl, Astra USA Inc.), and a topical antibiotic ointment (Fougera Altana Inc., USA).
11. Footprint analysis of locomotor patterns: Catwalk apparatus (Noldus Information Technology, Leesburg, VA) and EthoVision Catwalk software.
12. Motor-evoked potentials: Transcranial magnetic stimulator (Magstim, UK); LabView hardware and software (National Instruments).
13. Anxiety-like behaviors: An elevated plus maze, a Noldus EthoVision hardware and software (Noldus Information Technology, The Netherlands).
14. Histology and immunohistochemistry (IHC) experiments. Reagents for Luxol fast Blue and H&E.
    (a) *Antibodies for IHC*
       - GAD67, 1:4000; Chemicon International
       - GABAb, 1:4500; Chemicon International
       - DBH, 1:6500 Chemicon International
       - BDNF, 1:2500; Chemicon International

3 Methods

3.1 Closed-Head Impact-Acceleration TBI

The impact acceleration procedure uses a weight drop method to produce controlled levels of TBI as originally described by Marmarou and colleagues [18, 19]. For the TBI procedure, young adult Sprague-Dawley rats are anesthetized using a combination of ketamine (80 mg/kg) and xylazine (5 mg/kg). Buprenex (0.05 mg/kg) and ketoprofen (5 mg/kg) are used as analgesics during surgery and 1 day post-surgery. Using aseptic surgical techniques, the dorsal surface of the head is shaved, the incision site infiltrated with 0.25 % bupivacaine (3 mg/kg), and the surgical area cleaned by repeating cycles of Betadine scrub (7.5 %) followed by 70% alcohol wipe, and a final application of Betadine solution (5.0%). The animal's body temperature is monitored and maintained at 37–39 °C using a clean and sanitized temperature-controlled blanket. The animal and the surrounding surgery field are covered with a sterile drape. A skin incision is made to expose the dorsal midline of the cranium. Fascia and soft tissue are retracted to expose the landmarks of lambda and bregma on the cranium. A stainless disk (10 mm diameter × 2 mm height) is attached to the exposed skull with dental acrylic. The animal is then placed on a 12 × 12 × 43 cm foam pad (specifically 2500 N/m spring rate, acquired from Foam to Size Inc., Ashland, VA) contained in plexiglass box. Use of foam with these dimensions and spring rate is critical to standardize the acceleration-deceleration parameters of this model. The cortical impact trauma device is maneuvered over the animal and aligned with the stainless disk as shown in Fig. 1.

A laser pointer placed in the center of the top of the tube is used to enhance the precision of the alignment of the tube and the helmet. A 450 g impactor, with a 6 mm diameter tip, is dropped through thermoplastic (Delrin) tube from 1.0, 1.25, or 1.5 m heights, depending on the desired intensity (mild, mild/moderate, moderate, respectively) of the injury. The 450 g impactor strike

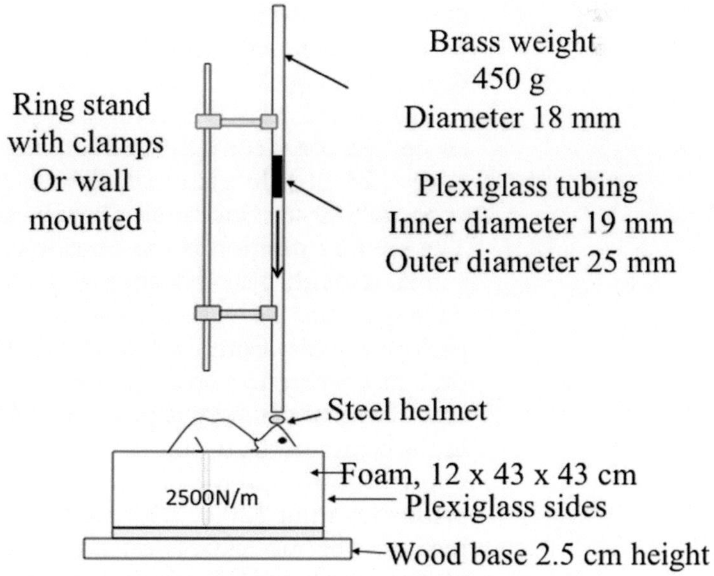

Fig. 1 Closed-head injury setup. The apparatus and protocol (after Marmarou et al. 1994) produce repeatable, graded, impact acceleration TBIs. The 2500 N/M foam is a critical element in the standardization of the injury

from 1.5 m height on the stainless disk affixed to the cranium of the animal will produce an impact acceleration TBI with a force profile of 0.2 ms, and <0.25 mm compression of the skull [18]. The stainless disk serves to diffuse the force of the impactor, and decreases the probability of skull fracture. Further, the elastic properties of the foam pad (Foam to Size, 2500 N/m) on which the head is resting, produces a controlled, low impedance path for acceleration/ deceleration of the head which minimizes skull compression and focal brain injury for this type of closed-head TBI [18–20]. High speed video of the procedure in our laboratory revealed that immediately following a single impact, the head accelerates deeply into the foam which envelopes the head while deflecting the impactor away from the strike site to prevent unintended additional “2nd hits.” Following injury, the stainless disk is removed from the cranium and the skin closed with sterile stainless wound clips. The animals are monitored closely for post-impact apnea, which is rarely observed using drop heights <1.5 m.

3.2 Behavioral/ Physiological Experimental Methodology

3.2.1 Cognitive Disability (Cog) Testing: Morris Water Maze (MWM)

The MWM is used to investigate cognitive deficits based upon measures of spatial learning. The MWM tests the rodent’s ability to spatially orient using reference cues to locate a hidden escape platform. The MWM is a 182 cm diameter and 60 cm high black plastic pool filled with water to depth of 27 cm. Water temperatures is maintained at 23–26 °C by a submersible heating coil. A black Plexiglas platform 10 cm in diameter and 25-cm high, is located 2 cm below the water surface, to serve as hidden goal platform. The maze is located in a 3.0 × 3.0 m room with numerous extra maze cues (posters, pictures, and ceiling mounted video camera) that remain constant throughout the several days of testing as depicted in Fig. 2.

The standard MWM test for serial learning consists of four trials per day for 4 consecutive days, beginning POday 11 following injury [23–25]. In each trial, rats are placed manually in the pool at one of four start locations. These locations are randomly selected. The starting position is the outside center of each quadrant, the animal is gently placed in the water with the nose toward the wall. Each quadrant is identified as south, west, north, and east. The rat performs a trial from each of the four possible start locations on each day, where the order is randomized for each animal on each day. The goal platform is positioned 45 cm from the outside wall in the center of the southeast quadrant of the maze. Rats are given up to 120 s to find the hidden platform. If the rat fails to find the platform within 120 s, it is placed on the platform by the experimenter. The rat remains on the platform for 30 s before being placed in a heated incubator (28 °C) between trials. There is a minimum of 4-min inter-trial interval. All animals are allowed to dry completely before being returned to their home vivarium. Swim speeds for each animal for each trial are calculated to insure

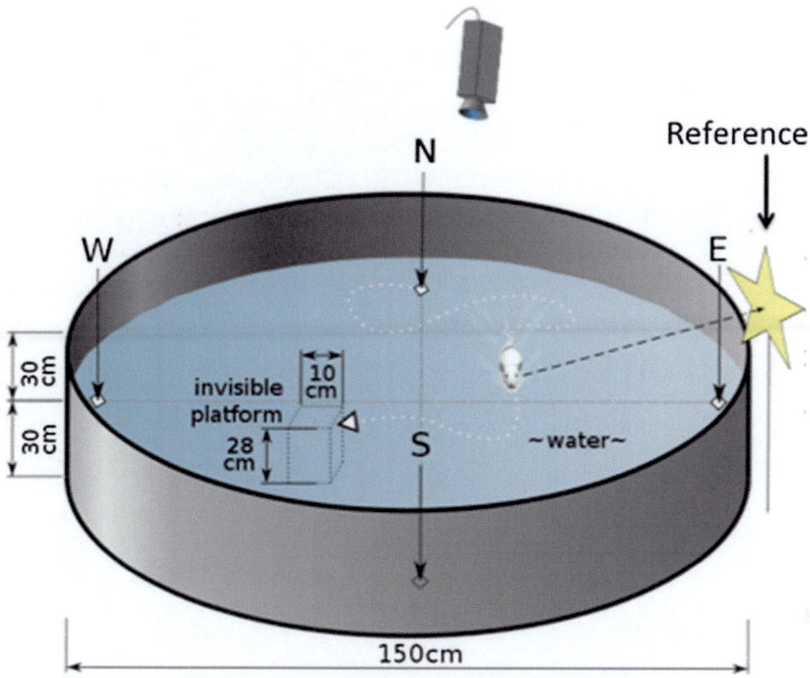

Fig. 2 Morris water maze (MWM). The MWM and serial testing protocols (e.g., Hamm et al. 1992) provide a standardized means to evaluate serial learning following TBI in the rat

that latency differences are not due to impaired motor function. The animals are video taped during the Morris water maze test. A computerized video tracking system (Noldus EthoVision version XT 8.0) is used to record each animal's latency to reach the goal platform, swim pattern, total distance swam, and cumulative swim distance. In addition to search latency, several other parameters are recorded including, total path length. Therefore, the average swim speed is determined from the path length and the search latency data. After the 4-day training session, a probe trial is performed which consists of placing the animal in the tank with no escape platform. The animal is placed in the quadrant that is opposite the previous goal quadrant for 60 s to record the subsequent fictive platform crossings, the time and path-length that the animal spends in each quadrant. The probe trial is conducted immediate after the end of 4 days trial to test acquisition of learned behavior (acquisition memory) and 24 h thereafter to test retention of memory. A discrimination index (DI) is calculated using time spent in the goal and in the opposite quadrants.

3.2.2 Vestibulomotor Testing: Rotarod

The rotarod is designed for testing coordination and impairment of locomotor ability of the rat using the well-established principle of a rotating beam on which an animal is challenged to maintain its balance. The animal's task is to walk on the beam as it rotates (Fig. 3).

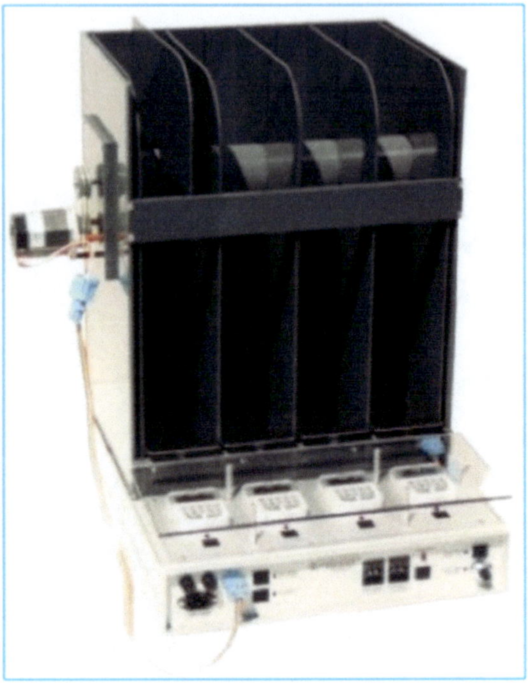

Fig. 3 Rotorod device. This device and protocols (e.g., Ham 2001) can provide standardized testing of balance disability following TBI in the rat

The length of time the animal is able to maintain itself on the rotating beam and the speed of the beam at which the animal falls are parameters quantitating an animal's balance ability. The rotarod we use features a variable speed rod rotation in the range of 0–99 RPM, four individual electronic timers with a fall sensor for each animal ("Economex" Columbus Instrument, Columbus, Ohio).

The timers are activated by the operator at the moment the animal is placed on the rod and automatically stopped when the animal drops from the rotating beam. Stainless steel compartments below the rod restrain the animal from escape once it drops from the beam. The standard protocol is one adopted from Hamm [26]. The animal is allowed to remain stationary for 10 s at 0 rpm. The rotational speed is then slowly increased to 3 rpm for 10 s and is steadily increased (automatic preset) by 3 rpm in 10 s intervals until the maximum rpm of 30 is reached. The animals remain on the device at this speed for another 20 s until the 2-min test period has elapsed. Animals are tested using three trials per day with a 10 min inter-trial interval pre-injury and typically at multiple weekly time points following injury. Typically rotorod assessments are performed pre-injury and at post-injury intervals to permit a repeated measures statistical paradigm and changes expressed as percentage of pre-injury values. Alternatively, time-matched normal controls are assessed in tandem with the pre- and post-injury animals to permit between group statistical paradigm and TBI changes expressed as percentage of rotorod performance in normal animals.

3.2.3 Spasticity (Motor Disability) Testing: Velocity-Dependent Ankle Torques and EMGs Recording

The velocity-dependent lengthening resistance of the triceps surae muscles is tested by measuring ankle torque during rotations of the foot across a broad range of rotation velocities. [4, 21, 22, 27]. Since the triceps surae muscles insert as an aponeurosis upon the calcaneous, controlled rotation of the foot produces a controlled lengthening of the triceps surae muscles. The resistive force (torque) opposing foot rotation is produced by the passive stretch and (at threshold and higher velocities for the triceps surae stretch reflex) the active contraction of the triceps muscles. Rats are immobilized in a custom designed trunk restraint, without trauma or apparent agitation. All recordings are performed in awake animals. The proximal portion of the hind limbs to the mid-shank are individually secured in a form fitted cast that immobilizes the limb while permitting normal range of ankle rotation (60–160°). The lengthening resistance of the triceps surae muscles is measured indirectly by quantitating ankle torque during 12.24-degree dorsiflexion rotations of the ankle from 95° through 83° (Fig. 4a). Contact with the foot is achieved using a form-fitted cradle aligned with the dorsal edge of the central footpad 2.6 cm distal to the ankle joint. Ankle torque (T) is assessed as the product of gram force (F, 980.7 dynes = 1 gf), the length of moment (torque) arm (l) (2.6 cm distance from ankle joint to stirrup contact point), and sin function of displacement force vector, e.g., $T = F \times l \times \sin\theta$ (*see* [21, 22, 27], for details of this measure). The angle of contact between the displacement shaft and the moment arm is 95°. The neural activity of the triceps surae muscle is measured using transcutaneous EMG electrodes. The electrode is inserted in a skin fold over the distal soleus muscle just proximal to the aponeurotic convergence of the medial and lateral gastrocnemii into the tendonocalcaneousus. The reference electrode is placed in a skin fold over the greater trochanter. A xylocaine 2 % jelly (Lidocaine HCl, Astra USA Inc.) is applied

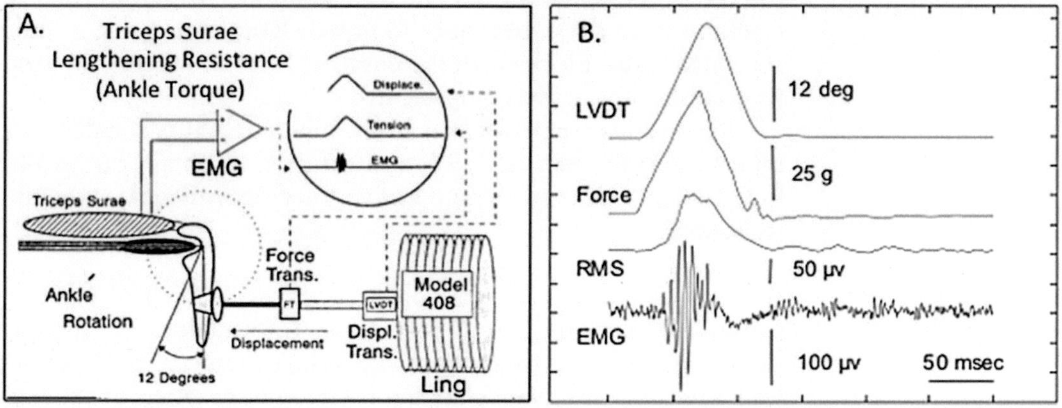

Fig. 4 (**a**) Measurement of velocity-dependent ankle torque and triceps surae EMG. This apparatus utilizes a Ling shaker, an LVDT, a Grass force transducer, and EMG recordings to quantitate velocity dependent lengthening of the triceps surae muscles. (**b**) Assessment across a wide range of velocities can provide quantitation of changes in threshold and velocity-dependent spasticity in muscles that extend the foot

over the electrode insertion points to minimize pain during recording. A topical antibiotic ointment (a combination of Bacitracin, Neomycin and Polymyxin B, Fougera Altana Inc., USA) is also applied on these areas after removing the electrodes at the end of each trial to reduce the chance of infection. Before each trial, the urinary bladder is voided to minimize visceral-somatic influence. Controlled dorsiflexion is achieved through the use of an electromechanical shaker (Model 405, Ling Dynamic systems, Royston Herts, UK). A force transducer (LVDT) (Model FT-03, Grass Instruments, Quincy, MA, USA) is mounted on an axel and configured for recording using red/black spring tension to provide a recording range of 10 mg to 200 g. The output shaft of the ling shaker is joined in series through the transducer to reach the foot cradle that contacts the central footpad. Therefore since the FT-03 transducer shaft is free to move, its 1.1 mm of beam travel is only produced by a 200 g resistance, which is typically more than two times greater than any peak force measured. Raw EMG and root mean square (RMS, i.e., a 0.96 DC equivalent of the full wave rectified AC signals to a digital RMS) of EMG bursts are recorded on additional channels of the signal acquisition system. EMG magnitude is reported as mean RMS magnitude of the EMG bursts time-locked to ankle dorsiflexion. Collectively, this arrangement allows simultaneous monitoring of triceps surae EMG, resistive force (ankle torque), and velocity of shaft displacement, Fig. 4b. Recorded data are processed using LabView signal acquisition hardware and software system (National Instrument, USA). Dorsiflexions of 12.24° are performed with a 3-s intervals at 49°, 136°, 204°, 272°, 350°, 408°, 490°, and 612° per second. In normal animals, the triceps surae EMS reveals that the threshold for the triceps surae stretch reflex is around 272°/s. Ankle torques observed at 49–272°/s foot rotations are, accordingly, attributed to passive elastic properties of the lengthening tissues. With the appearance of the EMG, the ankle torques reveal an additional magnitude that we attribute to an active resistance to muscle lengthening due to muscle contraction. Blockade of the nerve by local or systemic anesthesia abolishes the active component.

Recordings are performed at pre-injury and at particular post-injury intervals such as 7, 28, and 60 days to obtain progressive longitudinal outcome measures of velocity-dependent lengthening resistance.

3.2.4 Footprint Analysis of Locomotor Patterns

Animals are trained to walk on a 10 cm wide by 120 cm long Catwalk apparatus [28, 29] (Noldus Information Technology, Leesburg, VA) in a darkroom. The animals traverse the enclosed walkway that has a glass plate floor. Green light enters at the long edge of the glass plate striking the surface below the critical angle for transmission and is internally reflected. Accordingly, light escapes and is scattered by the contact of the rat's paws with the

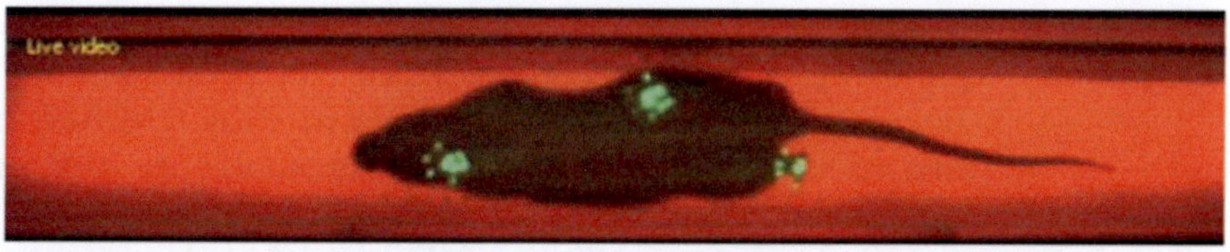

Fig. 5 The photo illustrates cat walk capture of foot falls during walking. The catwalk apparatus and Ethvision XT software provide comprehensive assessment of changes in gait following TBI in the rat

glass plate [30, 31]. The paw prints illuminated during contact (Fig. 5) are captured by a high-speed video camera positioned below the walkway. The digital images are then transferred to the computer for data analysis using a program (Catwalk) that captures, indexes, stores, and analyzes multiple parameters of gait dynamics according to temporal/spatial patterns of footfalls [28]. During the first walking pass, the bright areas (bright spots which provide high-resolution image of each footprints) are labeled and indexed. In subsequent passes, these footprints are interactively categorized (right forepaw, left forepaw, right hindpaw, left hindpaw). The temporal/spatial positions of footprint spots including quantitative measures of size and intensity along with a tag (indicating which category relative to each paw) are stored in separate files. The Catwalk program utilizes algorithms to assign tags to each paw and assesses the temporal/spatial pattern of footfalls. Three non-stop runs are acquired for each animal pre-injury, and typically at weekly intervals post injury. Analysis of normal step sequence and interlimb sequence regularity are calculated and graphical representations of paw placement are generated using EthoVision Catwalk software.

3.2.5 Motor-Evoked Potentials

Transcranial magnetic motor-evoked potentials (tcMMEPs) are used (Magstim, UK) as an index of activation status of the cortico-motor and spinal motor system outcome following TBI. tcMMEPs have become an important tool for in situ evaluation of the integrity of locomotor pathways and provides a non-invasive neurophysiologically quantitative measure of the functional status of motor system integrity following injury, disease, and therapy. Our studies use magnetic pulse stimulation applied transcranially to the animal's cortex to activate intracranial structures using an electromagnetic coil (25 mm) placed to provide maximal field strength over the center of the cranium at bregma. The lowest threshold regions for evoking hindlimb muscle activity using intracortical microstimulation are found 0–1.5 mm coronal, –2.0 to 2.5 mm sagittal [32]. This stimulation is supplied by TMS (Magstim Rapid) using a single magnetic pulse with stimuli intensities from 30 to 70% (in 10% increments) Max stimuli intensities to produce a recruitment curve of motoneuron excitability. These percentages correspond to field intensities of 1, 1.4, 1.8, 2.2, and 2.4 T, respectively. The magnetic pulses are delivered by computer Lab View programming (National Instrument hardware and software) to elicit tcMMEP responses that assess motoneuron excitability thresholds pre- and post-injury and in response to various treatments. The onset latency time and amplitude are recorded from left and right soleus muscles, tibialis anterior muscles, and muscles from forelimb flexors using intracutaneous electrodes. Rats are immobilized in a custom designed trunk restraint, without trauma or apparent agitation during this stimulation and recording. All recordings are performed on awake animals at pre- and at various post-injury intervals.

3.2.6 Anxiety-Like Behaviors

Anxiety-like behavior is assessed using an elevated plus maze that provides an opportunity for tested animals to explore open or closed portions of the maze. The activity path and time distribution between open and closed areas provide quantitative measures of these activities. Normal animals spend approximately equal time in the open and closed portions of the maze. Whereas, anxiety like behavior is expressed in proportion to time spent >50% in the closed portions of the maze. A Noldus EthoVision experimenter program is used to video and record the amount of time the animal spends in the open or closed portions of the maze (EPM) (Fig. 6). Each animal is initially placed in the center piece and

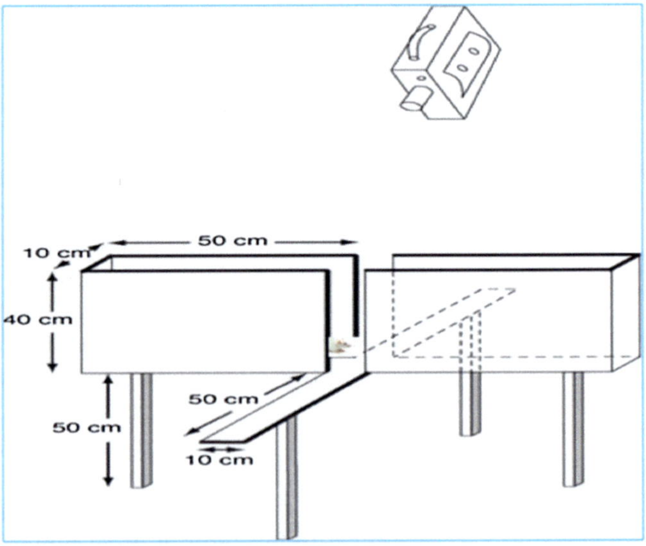

Fig. 6 Elevated plus maze (EPM). The EPM capitalizes upon the rats curiosity to explore versus its intrinsic instinct for staying hidden. Noldus EthoVision software utilizes video recording of the test session to quantitate activity, path length, time distribution, and entries on each portion of the maze

allowed 5 min to explore the maze. Each animal is tested three times during the course of the recording session. Between trials, the animal is removed, and the EPM is wiped down with 70% alcohol and allowed to dry before the testing is continued. A minimum interval of 3 weeks is used before a repeat test for each animal. Typically, testing is conducted before injury, and at 4 and 8 weeks post-injury. Relevant data include the time distribution in open vs. closed portions of the maze and the distribution of activity (path lengths) of movement. Data in TBI animals is compared with that recorded in time-matched normal control animals sequenced between the recording trials in the TBI animals during the same testing session.

3.3 Histology and Immunohistochemistry (IHC) Experiments

Standard histology (Luxol fast Blue and H&E) and fluorescent IHC is performed using selected cortical, brainstem, and lumbar spinal segments (L_3–L_5) in TBI and time-matched normal control animals that are euthanized at selected post-injury time points. The animals are deeply anesthetized using ketamine (110 mg/kg) anesthesia [33, 34], and perfused through the left ventricle with phosphate buffer saline (0.1 M PBS, 35 °C, pH 7.4) followed by 4% *para*-formaldehyde in phosphate buffer (0.1 M, pH 7.4, 4 °C). After postfixation (overnight in same fresh fixative) and cryoprotection (30% sucrose in PBS), the tissue (brain, brainstem, and lumbar spinal cord) specimens are sectioned by Cryostat (40 μm thickness) and processed for IHC. Fluorescent IHC is conducted for GAD_{67}, $GABA_b$, DBH, and BDNF using different sets of sections from same animal. The sections are washed in 0.01 M PBS, and incubated overnight (at 4 °C) with monoclonal or polyclonal appropriate primary antibodies (Chemicon International). This procedure is followed by incubation with secondary antibodies conjugated to Alexa Fluorchrome (highly cross-absorbed antibodies) (Molecular Probes). In control experiments, the primary antibody is either preabsorbed with an excess amount of the immunogenic peptide (1 μm/ml) or replaced by preimmune serum at the same dilution. The sections are washed thoroughly with PBS (0.01 M, pH 7.4) after the antibody incubation. After washing with phosphate buffer (PB), the sections are coverslipped with anti-fading mounting media Vectashield (Vector Laboratory). All slides are stored at 4 °C and fluorescent images were captured within 1 week after completion of IHC for light morphometric studies. Since spasticity measurements are typically performed for hindlimb muscles, lumbar spinal cord segments are also used for IHC. An example is shown using DBH staining to evaluate noradrenergic cell loss in nucleus locus coeruleus, Fig. 7. Histological studies of impact injuries in impact-constrained [35] vs. impact acceleration [4, 19] reveal marked differences in the spatial extent of neuronal changes.

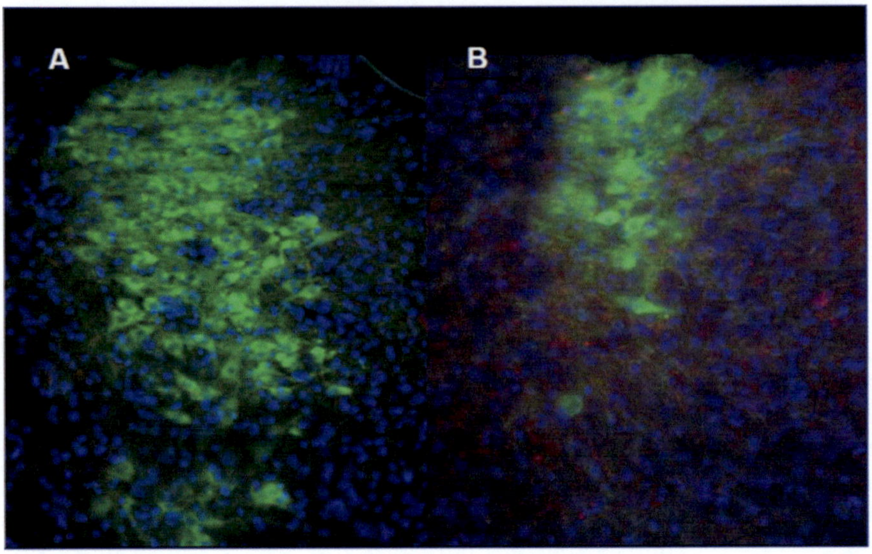

Fig. 7 Nucleus locus coeruleus, dopamine beta hydroxylase (DβH) staining. Comparison of labeled cells in normal and TBI animals provides a means to evaluate post-injury cell specific to norepinephrine synthesis

4 Summary Comments

The disappointing lack of success in human clinical trials testing experimental agents to enhance recovery following TBI has, in part, been attributed to the lack of use of sensitive quantitative measures to assess outcome of TBI disability and recovery [2, 14, 36, 37]. Accordingly, there is an urgent need for translatable methods and protocols to facilitate the expansion of quantitative measures in the experimental clinical setting [15]. The methods and protocols that we used in these preclinical studies provided quantitative measures of disability and incremental changes in recovery that are highly translatable to evaluate the impact of TBI and treatment on the trajectory of disability and recovery in human TBI. Accordingly, it is anticipated that the evidence-based incorporation of quantitative methods to assess long-term outcome in human TBI clinical trials will steadily progress to enhance the detection of potentially essential data regarding incremental changes in neurobiology that underlies therapeutic enhancement of cognitive, behavioral, and motor/sensory recovery.

References

1. Xiong Y, Mahmood A, Chopp M (2013) Animal models of traumatic brain injury. Nat Rev Neurosci 14:128–142
2. Zitnay GA, Zitnay KM, Povlishock JT, Hall ED, Marion DW, Trudel T, Zafonte RD, Zasler N, Nidiffer FD, DaVanzo J, Barth JT (2008) Traumatic brain injury research priorities: the Conemaugh International Brain Injury Symposium. J Neurotrauma 25: 1135–1152

3. Ling H, Hardy J, Zetterberg H (2015) Neurological consequences of traumatic brain injuries in sports. Mol Cell Neurosci 66: 114–122
4. Bose P, Hou J, Nelson R, Nissim N, Parmer R, Keener J, Wacnik PW, Thompson FJ (2013) Effects of acute intrathecal baclofen in an animal model of TBI-induced spasticity, cognitive, and balance disabilities. J Neurotrauma 30:1177–1191
5. Johnson VE, Stewart W, Smith DH (2013) Axonal pathology in traumatic brain injury. Exp Neurol 246:35–43
6. Adams JH, Graham DI, Gennarelli TA, Maxwell WL (1991) Diffuse axonal injury in non-missile head injury. J Neurol Neurosurg Psychiatry 54:481–483
7. Povlishock JT, Pettus EH (1996) Traumatically induced axonal damage: evidence for enduring changes in axolemmal permeability with associated cytoskeletal change. Acta Neurochir Suppl 66:81–86
8. Smith DH, Meaney DF, Shull WH (2003) Diffuse axonal injury in head trauma. J Head Trauma Rehabil 18:307–316
9. Buki A, Povlishock JT (2006) All roads lead to disconnection?—Traumatic axonal injury revisited. Acta Neurochir (Wien) 148:181–193, discussion 193-184
10. Morales DM, Marklund N, Lebold D, Thompson HJ, Pitkanen A, Maxwell WL, Longhi L, Laurer H, Maegele M, Neugebauer E, Graham DI, Stocchetti N, McIntosh TK (2005) Experimental models of traumatic brain injury: do we really need to build a better mousetrap? Neuroscience 136:971–989
11. Fujimoto ST, Longhi L, Saatman KE, Conte V, Stocchetti N, McIntosh TK (2004) Motor and cognitive function evaluation following experimental traumatic brain injury. Neurosci Biobehav Rev 28:365–378
12. Cernak I (2005) Animal models of head trauma. NeuroRx 2:410–422
13. Smith DC, Modglin AA, Roosevelt RW, Neese SL, Jensen RA, Browning RA, Clough RW (2005) Electrical stimulation of the vagus nerve enhances cognitive and motor recovery following moderate fluid percussion injury in the rat. J Neurotrauma 22:1485–1502
14. Xiong Y, Mahmood A, Chopp M (2009) Emerging treatments for traumatic brain injury. Expert Opin Emerg Drugs 14:67–84
15. Johnson VE, Meaney DF, Cullen DK, Smith DH (2015) Animal models of traumatic brain injury. Handb Clin Neurol 127:115–128
16. Bose P, Hou J, Thompson FJ (2015) Traumatic brain injury (TBI)-induced spasticity: neurobiology, treatment, and rehabilitation. In: Kobeissy FH (ed) Brain neurotrauma: molecular, neuropsychological, and rehabilitation aspects. CRC Press/Taylor & Francis, Boca Raton
17. Dixon CE, Lighthall JW, Anderson TE (1988) Physiologic, histopathologic, and cineradiographic characterization of a new fluid-percussion model of experimental brain injury in the rat. J Neurotrauma 5:91–104
18. Marmarou A, Foda MA, van den Brink W, Campbell J, Kita H, Demetriadou K (1994) A new model of diffuse brain injury in rats. Part I: pathophysiology and biomechanics. J Neurosurg 80:291–300
19. Foda MA, Marmarou A (1994) A new model of diffuse brain injury in rats. Part II: morphological characterization. J Neurosurg 80: 301–313
20. Kallakuri S, Cavanaugh JM, Ozaktay AC, Takebayashi T (2003) The effect of varying impact energy on diffuse axonal injury in the rat brain: a preliminary study. Exp Brain Res 148:419–424
21. Bose P, Parmer R, Thompson FJ (2002) Velocity-dependent ankle torque in rats after contusion injury of the midthoracic spinal cord: time course. J Neurotrauma 19: 1231–1249
22. Thompson FJ, Browd CR, Carvalho PM, Hsiao J (1996) Velocity-dependent ankle torque in the normal rat. Neuroreport 7: 2273–2276
23. Hamm RJ, Dixon CE, Gbadebo DM, Singha AK, Jenkins LW, Lyeth BG, Hayes RL (1992) Cognitive deficits following traumatic brain injury produced by controlled cortical impact. J Neurotrauma 9:11–20
24. Hamm RJ, Lyeth BG, Jenkins LW, O'Dell DM, Pike BR (1993) Selective cognitive impairment following traumatic brain injury in rats. Behav Brain Res 59:169–173
25. Hamm RJ, Temple MD, Pike BR, Ellis EF (1996) The effect of postinjury administration of polyethylene glycol-conjugated superoxide dismutase (pegorgotein, Dismutec) or lidocaine on behavioral function following fluid-percussion brain injury in rats. J Neurotrauma 13:325–332
26. Hamm RJ (2001) Neurobehavioral assessment of outcome following traumatic brain injury in rats: an evaluation of selected measures. J Neurotrauma 18:1207–1216

27. Wang DC, Bose P, Parmer R, Thompson FJ (2002) Chronic intrathecal baclofen treatment and withdrawal: I. Changes in ankle torque and hind limb posture in normal rats. J Neurotrauma 19:875–886
28. Hamers FP, Lankhorst AJ, van Laar TJ, Veldhuis WB, Gispen WH (2001) Automated quantitative gait analysis during overground locomotion in the rat: its application to spinal cord contusion and transection injuries. J Neurotrauma 18:187–201
29. Hamers FP, Koopmans GC, Joosten EA (2006) CatWalk-assisted gait analysis in the assessment of spinal cord injury. J Neurotrauma 23:537–548
30. Betts RP, Paschalis C, Jarratt JA, Jenner FA (1978) Nerve fibre refractory period in patients treated with rubidium and lithium. J Neurol Neurosurg Psychiatry 41:791–793
31. Clarke KA (1992) A technique for the study of spatiotemporal aspects of paw contact patterns, applied to rats treated with a Trh analog. Behav Res Method Instrum 24:407–411
32. Gu X, Staines WA, Fortier PA (1999) Quantitative analyses of neurons projecting to primary motor cortex zones controlling limb movements in the rat. Brain Res 835:175–187
33. Thompson FJ, Reier PJ, Lucas CC, Parmer R (1992) Altered patterns of reflex excitability subsequent to contusion injury of the rat spinal cord. J Neurophysiol 68:1473–1486
34. Thompson FJ, Reier PJ, Parmer R, Lucas CC (1993) Inhibitory control of reflex excitability following contusion injury and neural tissue transplantation. Adv Neurol 59:175–184
35. Maruichi K, Kuroda S, Chiba Y, Hokari M, Shichinohe H, Hida K, Iwasaki Y (2009) Graded model of diffuse axonal injury for studying head injury-induced cognitive dysfunction in rats. Neuropathology 29:132–139
36. Tolias CM, Bullock MR (2004) Critical appraisal of neuroprotection trials in head injury: what have we learned? NeuroRx 1:71–79
37. Stein DG (2015) Embracing failure: what the phase III progesterone studies can teach about TBI clinical trials. Brain Inj 29:1259–1272

Part V

Outcome Measures in Brain Injury Models

Chapter 29

Cognitive Evaluation Using Morris Water Maze in Neurotrauma

Ying Deng-Bryant, Lai Yee Leung, Krista Caudle, Frank Tortella, and Deborah Shear

Abstract

The Morris water maze (MWM) task is one of the most widely used and versatile tools in behavioral neuroscience for evaluating spatial learning and memory. With regard to detecting cognitive deficits following central nervous system (CNS) injuries, MWM has been commonly utilized in various animal models of neurotrauma, such as fluid percussion injury (FPI), cortical controlled impact (CCI) injury, weight-drop impact injury, and penetrating ballistic-like brain injury (PBBI). More importantly, it serves as a therapeutic index for assessing the efficacy of treatment interventions on cognitive performance following neurotrauma. Thus, it is critical to design an MWM testing paradigm that is sensitive yet discriminating for the purpose of evaluating potential therapeutic interventions. In this chapter, we discuss how multiple test manipulations, including the size of platform, numbers of trials per day, the frequency of retesting intervals, and the texture of platform surface, impact MWM's ability to detect cognitive deficits using a rat model of PBBI.

Key words Neurotrauma, Traumatic brain injury, Morris water maze, Cognitive function, Spatial learning, Memory, Cognitive deficit

1 Introduction

The MWM task utilizes the animals' inclination to escape the water to locate the standing platform as a means of motivation for learning and memorization [1]. Based on this concept, the MWM is primarily designed for measuring spatial learning and memory, and has been extensively used and well accepted as a method for evaluating the outcomes of experimental models of neurological disorders, such as aging and neurotrauma, and testing therapeutic effects of potential treatment interventions [2]. The standard MWM spatial learning test encompasses four trials per day for 5 consecutive days [1]. It is routinely used to evaluate the animal's ability to acquire spatial elements/cues relevant to the platform that are subsequently processed and consolidated, and later

Firas Kobeissy et al. (eds.), *Injury Models of the Central Nervous System: Methods and Protocols,* Methods in Molecular Biology, vol. 1462, DOI 10.1007/978-1-4939-3816-2_29, © Springer Science+Business Media New York 2016

referenced to for the purpose of locating the platform [3]. Because the "visuospatial navigation" process in rodent performance also contributes to human everyday cognitive process, the MWM task renders great importance in the study of neurodegenerative diseases where cognition is impaired [4]. More importantly, it can be utilized to assess efficacy of therapeutic treatments on cognitive deficits in experimental models that would provide critical information for clinical studies [5, 6].

The spatial learning and navigation aspect of the MWM task is considered largely hippocampus dependent and therefore lesions specifically to the hippocampus have been shown to impair the performance in the MWM spatial learning test [7, 8]. However, there are increasing numbers of studies have indicated the significant involvement of cortical and subcortical regions in spatial learning and search strategies in the MWM. For example, focal injuries to the medial thalamus have been shown to impair search strategy and swimming behavior; however, they do not affect spatial mapping and navigation [9]. Furthermore, striatum has been indicated in habit learning and plays an important role in stimulus response; therefore influencing thigmotaxic behavior (i.e., wall-hugging behavior) in the MWM following injury [10, 11]. In addition, prefrontal cortical injured rats were shown to exhibit thigmotaxic swim pattern partly due to their inability to initiate search strategy [12, 13]. It is also suggested that parietal cortex is critical in spatial information processing and lesions to this region impairs the ability to incorporate spatial cues in the MWM [14].

Of particular relevance to this chapter, MWM has been used in multiple animal models of traumatic brain injury (TBI) to determine cognitive deficits as a result of the direct trauma to the brain [12, 15–17]. Typically, animals (e.g., rodent) that sustained a TBI show increased latencies to locate the hidden platform in the spatial learning test compared to non-injured animals (e.g., sham controls). It is considered that the injured animals have difficulties acquiring the spatial elements/cues of the MWM task, therefore exhibiting a reduced capability to master the strategy for locating the platform [15]. Consequently, the missing platform test (i.e., probe trial), performed immediately after the spatial learning test, often reveals the inability of the animal searching for the platform in the target zone where the platform was located [12]. Additionally, the TBI-injured animal has been shown to swim in loops in the outer annulus of the pool (i.e., thigmotaxis) that may indicate anxiety as well as impairment in navigation strategies [11]. With regard to working memory, a uniquely designed hidden platform test (i.e., one-trial learning) can be used to evaluate the short-term storage of trial-specific memory, and was shown to be valuable in detecting working memory deficits in rodents following TBI [18, 19].

In this chapter, we primarily focus on the application of MWM task in evaluating cognitive dysfunction in a rat model of military-relevant PBBI. This model simulates the large temporary

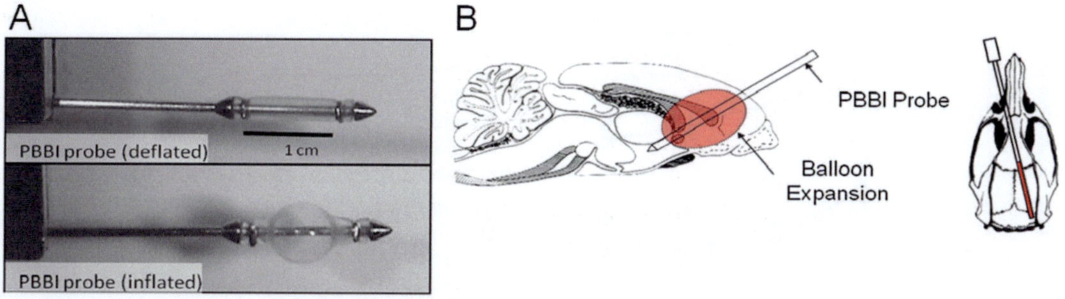

Fig. 1 The PBBI probe was constructed from a 20G stainless steel tube, with fixed perforations along one end that are sealed by airtight elastic tubing (**a**). The probe was secured on the probe holder with the un-perforated end attached to the pulse generator; angled 50° from the vertical axis and 25° counterclockwise from the midline, and stereotactically advanced through a cranial window (1 cm diameter) drilled on the dorsal surface of the skull (+4.5 mm AP; +2.0 mm ML from bregma) to a depth of 1.2 cm (from dura) (**b**)

cavity caused by energy dissipation from a penetrating bullet round (Fig. 1) [20]. As demonstrated in our previous reports on the model characterization, PBBI causes damage to multiple brain regions, including prefrontal cortex, striatum (caudate nucleus), and extend to pallidum, thalamic nucleus, and lateral amygdala [21, 22]. It should be noted that majority of the cavity formation was developed over the first week post-injury, and the size of the cavity remained relative constant out to 10 weeks post-injury as examined by hematoxylin and eosin (H&E)-stained brain sections [12, 23].

In order to identify the optimal MWM testing parameters for assessing cognitive abnormalities in the PBBI model, we designed a series of experiments assessing the effects of manipulating (1) platform size (10 cm vs. 15 cm diameter), (2) number of trials per day (TPD), (3) exposure to repeated MWM testing (i.e., weekly vs. biweekly), and (4) platform texture (scored vs. deeply grooved surface). Unless otherwise noted, MWM testing was initiated 2 weeks post-injury. In *experiment 1* (platform size), the results showed that average latencies to find the hidden platform were significantly increased by 136 ± 9 % in PBBI rats exposed to the smaller (10 cm) platform and by 52 ± 16 % in PBBI rats exposed to the larger (15 cm) platform vs. respective sham control groups. PBBI rats showed significantly higher latencies when exposed to the smaller platform than the larger platform (PBBI_10cm = 69 ± 3 s vs. PBBI_15cm = 40 ± 4 s; $p < 0.05$). However, platform size did not affect sham performance (sham_10cm = 29 ± 1 s vs. sham_15cm = 26 ± 2 s; $p > 0.05$) (Fig. 2). In *experiment 2* comparing learning curves in animals exposed to 4 TPD (×5 days) or 20 TPD, the results revealed the average latency (across 20 trials) increased by 136 ± 9 % (PBBI_4TPD) and 148 ± 17 % (PBBI_20TPD) vs. respective sham control groups ($p < 0.05$). TPD had no significant effect on sham performance or between respective injured groups (Fig. 3). In *experiment 3* assessing the

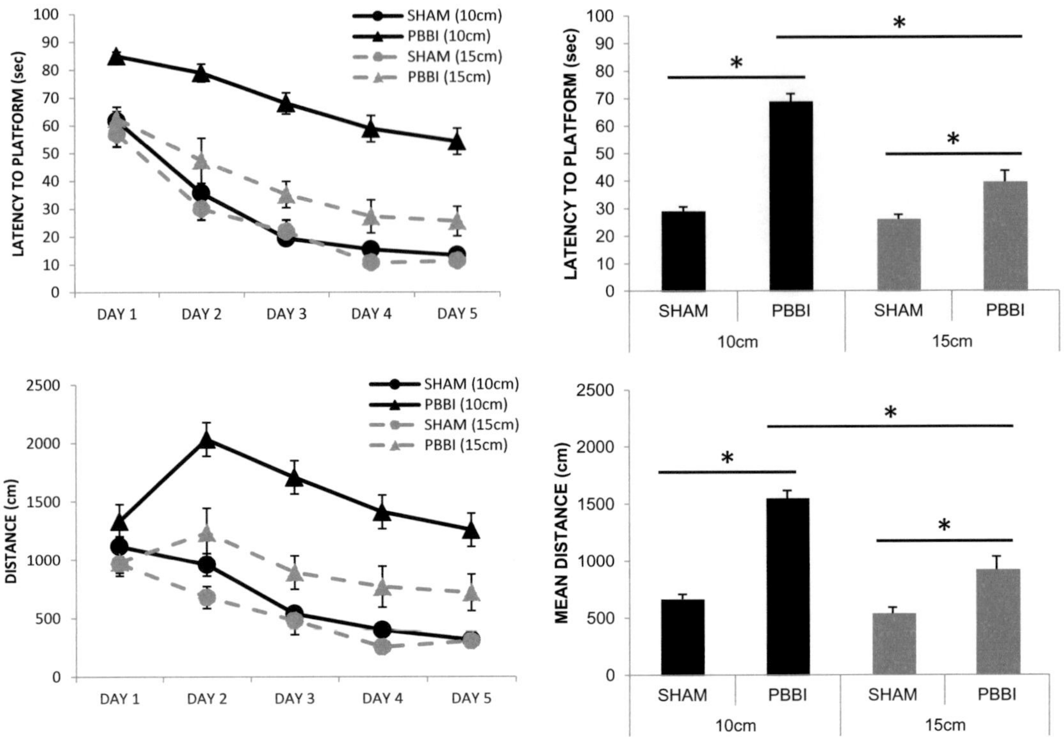

Fig. 2 Results of platform size (10 cm vs. 15 cm diameter) showed that average latencies to find the hidden platform and swim distances were significantly increased in PBBI rats exposed to the smaller (10 cm) platform and the larger (15 cm) platform vs. respective sham control groups. PBBI rats showed significantly higher latencies and swim distances when exposed to the smaller platform than the larger platform ($p<0.05$). However, platform size did not affect sham performance ($p>0.05$)

effects of repeated MWM testing using either a 1-week or a 2-week delay between retesting periods, the results showed that PBBI caused significant cognitive deficits during initial exposure to the MWM at either 1 (66 ± 8 s) or 2 weeks (68 ± 8 s) post-injury. When retested for retention 1 week later, PBBI rats displayed significant (49 %) improvement in the MWM task, albeit not to sham levels. In contrast, when re-testing was delayed by 2 weeks the PBBI animals failed to show any significant improvement (Fig. 4). In *experiment 4* (platform texture), the results demonstrated that PBBI caused significant cognitive deficits during exposure to MWM using platform using standard (scored) surface (67 ± 4 s) or textured (deeply grooved) surface (48 ± 3 s) post-injury vs. repetitive sham controls groups ($p<0.05$). PBBI rats showed significantly lower latencies (28 %) when exposed to the textured platform than the standard platform ($p<0.05$). However, platform texture did not affect sham performance ($p>0.05$) (Fig. 5). In summary, these results indicate cognitive performance following PBBI is most strongly influenced by platform size, the frequency of retesting (weekly vs. biweekly) intervals, and the texture of platform surface,

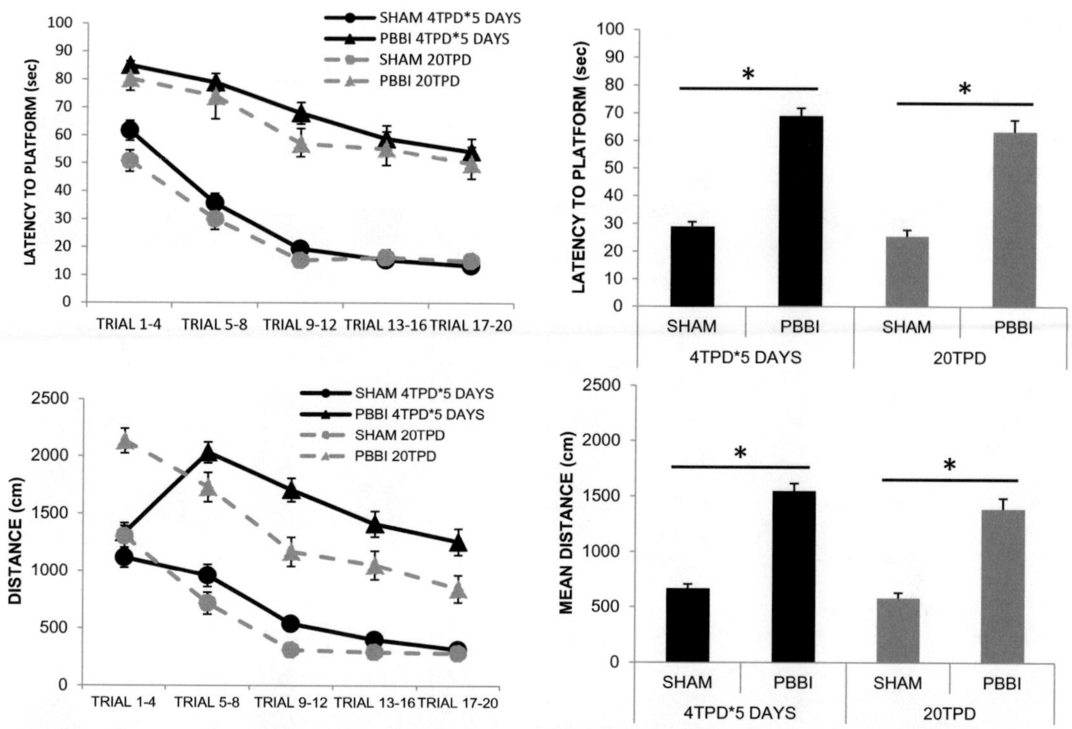

Fig. 3 Results of number of trials per day (TPD) revealed the average latencies and swim distances (across 20 trials) significantly increased in PBBI rats exposed to 4TPD for 5 consecutive days and PBBI rats exposed to 20TPD in 1 day vs. respective sham control groups ($p < 0.05$). TPD had no significant effect on sham performance or between respective injured groups

but not the numbers of trials per day. Critically, these results also provide valuable information for designing a discriminative testing paradigm for the purpose of screening preclinical therapeutics using the PBBI model.

2 Materials

2.1 Animals

1. Male adult Sprague-Dawley rats (280–320 g; Charles River Labs, Raleigh, VA, USA) were used in these experiments. Animals were housed individually under a 12-h light/dark cycle in a facility accredited by the Association for Assessment and Accreditation of Laboratory Animal Care International.

2.2 Equipment

1. A chemical-resistant polyethylene circular water maze basin in black or dark blue color for rats (diameter: 175 cm, depth: 75 cm).
2. A clear, height-adjustable, circular Plexiglas platform (MED Associate Inc. ENV-596M-X).
3. Noldus EthoVision detection system, including EthonVision XT, computer, and video camera (Leesburg, VA, USA).

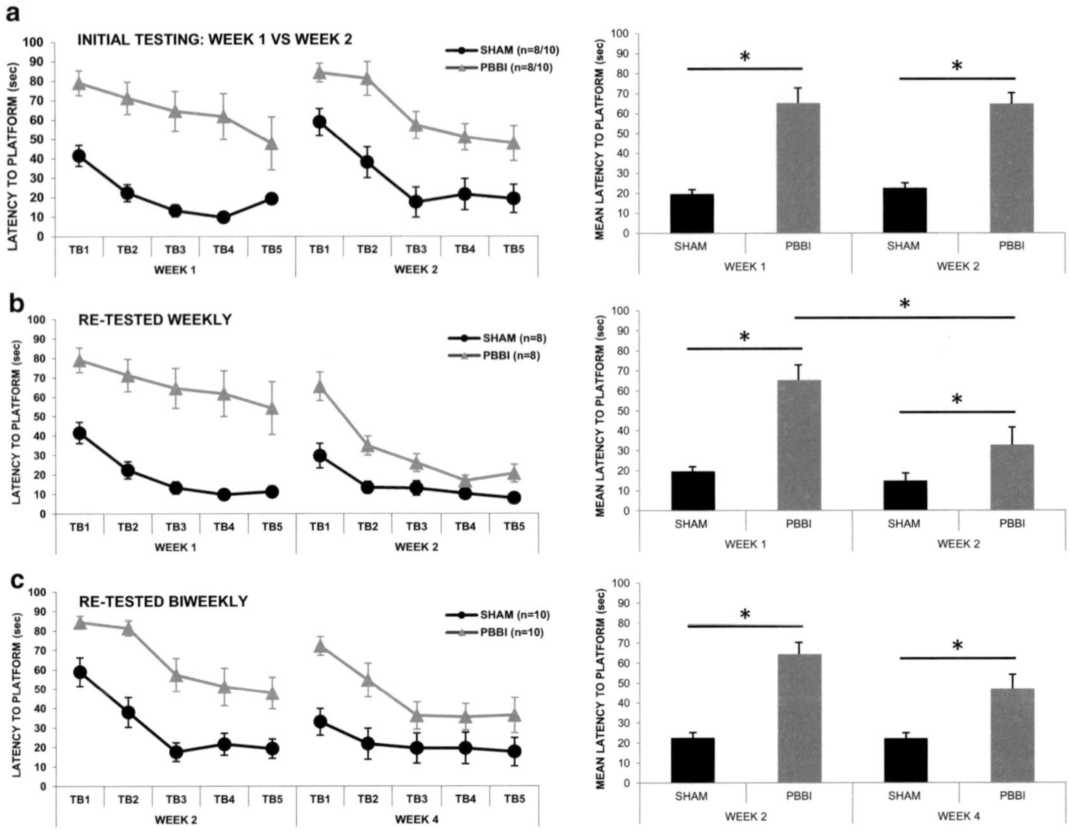

Fig. 4 Results of repeated MWM testing (i.e., weekly vs. biweekly) showed that PBBI caused significant cognitive deficits during initial exposure to the MWM at either 1 (65 ± 8 s) or 2 weeks (64 ± 8 s) post-injury. When retested for retention 1 week later, PBBI rats displayed significant (49 %) improvement in the MWM task, albeit not to sham levels ($p < 0.05$). In contrast, when retesting interval was extended to 2 weeks, the PBBI animals failed to show any significant improvement ($p > 0.05$)

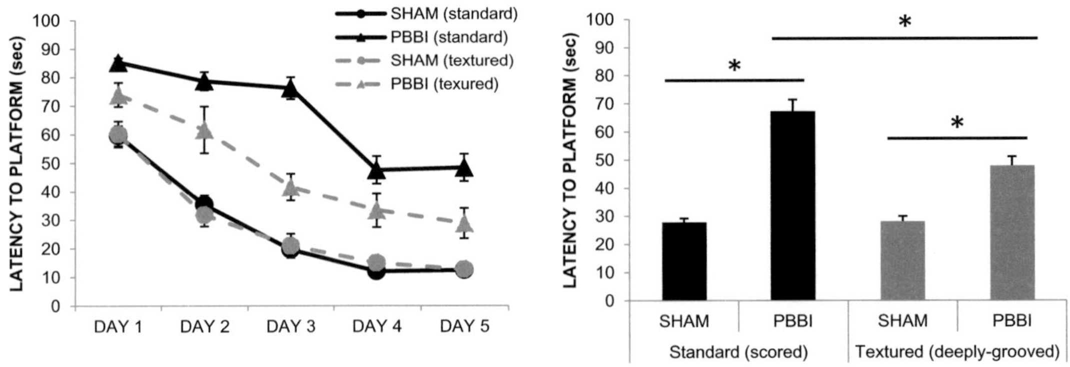

Fig. 5 Results of platform texture showed that PBBI caused significant cognitive deficits during exposure to the MWM using platform with standard (scored) surface (67 ± 4 s) or textured (deeply grooved) surface (48 ± 3 s) post-injury vs. respective sham controls groups ($p < 0.05$). PBBI rats showed significantly lower latencies (28 %) when exposed to the textured platform than the standard platform ($p < 0.05$). However, platform texture did not affect sham performance ($p > 0.05$)

4. A stopwatch (Fisher Scientific, Catalog no. 14-648-1).
5. A chemical-resistant science lab table for electronic device, such as computer and monitor.
6. Visual cues strategically placed on the wall of the testing room, e.g., highly visible geometric squares with black and white color.
7. A floor lamp (VWR, Catalog no. 500041-082) with red safe-light bulb (Fisher Scientific, Catalog no. NC0253883).
8. Flat vinyl heating blankets with warm circulating water powered by special electrical pumps (Fisher Scientific, Catalog no. NC9318897, NC9831543).
9. Cart for transporting animals (Leonard Paper Co. Catalog no. RM-4520-88).
10. Autoclavable shelf rack for holding rodent cages.
11. Portable, battery-powered swimming pool vacuum (Amazon.com Inc. B0015UCRXM).
12. Portable, submersible sump pump (Amazon.com Inc. B000X05G1A).
13. 70% Isopropyl for cleaning the apparatus (Fisher Scientific, Catalog no. 04-355-63).

3 Methods

3.1 Procedure Room Setup

1. Designate a large, quiet room for testing that will prevent interference and minimize distraction (*see* **Note 1**).
2. Place the floor lamp with red safelight bulb in a corner of the room for easy visualization.
3. Hang geometric square visual cues with black and white color on one side of the wall as a means of spatial reference for the animal in the maze.
4. Place the shelf rack along one side of the wall with multiple heating blankets on each shelf to cover the surface space.
5. Place the water maze basin in the center of the room.
6. Set up the computer-supported Noldus EthoVision system on a science lab table in a corner of the room.
7. Mount the camera on the ceiling centered above the water maze basin for videotaping object movement.

3.2 Morris Water Maze Procedure

3.2.1 Pre-training

1. To avoid stress introduced by human handling and transport, handle the animals prior to any water maze procedure. Handling should be daily for consecutive 3 days for at least 5 min per day before the start of testing.
2. Place the animals in the testing room for 30 min prior to water maze procedure to allow for habituation to the testing environment.

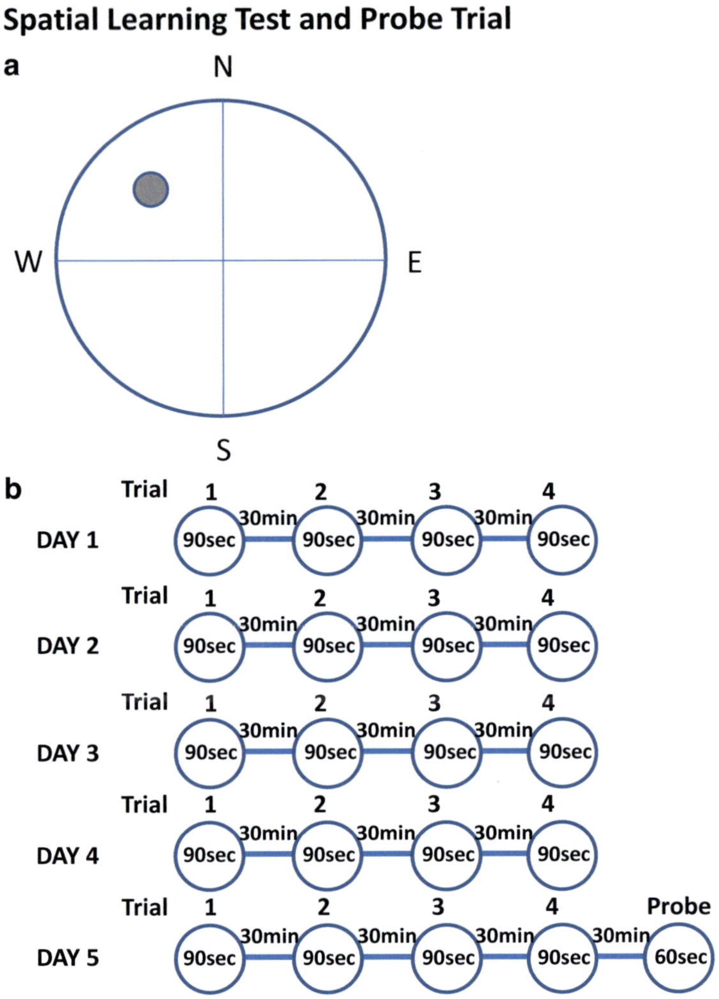

Fig. 6 Schematic illustration of spatial learning test design (**a**) and procedure (**b**)

3.2.2 Water Maze Basin Setup

1. Fill the water maze basin with clear water at ambient temperature (approximately 22 °C) to a depth of 60 cm to cover a clear circular platform.
2. Designate four equally spaced points on the circular rim of the water maze basin and define as North, South, East, and West. The equally spaced four quadrants of the water maze basin will then be defined as Northeast (NE), Northwest (NW), Southeast (SE), and Southwest (SW) quadrants (Fig. 6a).
3. Place the clear Plexiglas platform in the center of the NW quadrant, which is approximately 35 cm from the wall the basin.
4. Adjust the height of the platform to be submerged to a depth of 2.5 cm from the water surface.
5. Turn on the red safelight and turn off the regular light before testing.

3.2.3 Spatial Learning and Acquisition (Hidden Platform Test)

1. Rats will be given four trials per day with 30 min inter-trial interval (ITI) for 5 consecutive days (Fig. 6b).
2. The platform position remains constant throughout each set of trials in order to test spatial learning and acquisition.
3. Based on the platform's location in the NW quadrant, the North and West starting positions will be designated as short arms of the maze, whereas the South and East starting positions will be designated as long arms of the maze.
4. At the start of each trial, place the rat in the pool with the snout facing the basin wall at one of four starting positions: North, South, East, and West (*see* **Note 3**).
5. Pseudo-randomly determine the starting positions for each trial within a day, alternating between short and long arms.
6. Change the order of the starting position on each subsequent day of testing.
7. Allow each rat to freely swim in the pool until it finds the submerged platform or until a maximum of 90 s has elapsed.
8. When the rat successfully mounts onto the platform, it will be allowed to rest on the platform for 10 s (time it using a stopwatch) before being removed from the pool and placed back in its home cage.
9. If the rat does not find the platform in 90 s, it will be manually guided to the platform and allowed to rest on the platform for 10 s (time it using a stopwatch) before being removed from the pool and placed back in its home cage (*see* **Notes 4–6**).
10. Between trials, rats will be allowed to rest and warm up in its home cage with water-circulating heating blankets placed underneath each cage for 30 min.
11. Latency to find the hidden platform (sec), distance to the hidden platform (cm), and swim speeds (cm/s) will be recorded as primary outcome metrics for further statistical analysis (*see* **Note 2**).

3.2.4 Spatial Memory Retention (Probe Trial)

1. Rats will be given the probe trial on day 5 following the last trial of the hidden platform test (Fig. 6b).
2. In the probe trial, the platform will be removed (i.e., missing platform test) to assess rat's memory retention for the previous platform location.
3. Place the rat in the pool with the snout facing the basin wall at one of the long-arm starting positions.
4. Allow each rat to freely swim in the pool until a maximum of 60 s elapsed.
5. Remove the rat from the pool at the end of 60 s, and place it back to its home cage to rest (*see* **Notes 4–6**).

6. An annulus ring circumscribed around the previous target location is defined as the target zone in the EthoVision computer software tracking system for assessing the memory retention of the previous platform location.
7. An annulus ring circumscribed the outer rim of the maze is defined as the thigmotaxic annulus in the EthoVision computer software tracking system for assessing stress-related wall-hugging behavior.
8. The percent time spent swimming in thigmotaxic annulus, and percent time spent searching in the target zone for the missing platform during the probe trial will be recorded as primary outcome metrics for further statistical analysis.

3.2.5 Working Memory (One-Trial Learning Test)

1. Rats will be given two sets of trials for 1 day. Additional testing days can be added depending on the design of the test for different animal models. Each set of trials consists of two trials with 4-min ITI (Fig. 7b).
2. The platform position will be rotated among four different designs: (1) the center of a quadrant, (2) the outer corner of a quadrant (i.e., near the basin wall), (3) the inner corner of a quadrant (i.e., near the basin center), and (4) the midpoint between the center point of the basin and one of the starting positions (North, South, East, or West) (Fig. 7a).
3. The platform position and the starting position remains constant within the same set of trials, but will be changed between trial sets in order to test working memory (*see* **Note 3**).
4. For each additional day of testing, the platform position and the starting position will be redesigned to avoid replication.
5. Based on the platform's location in the maze, place the rat in the pool with the snout facing the basin wall at one of the long-arm starting positions.

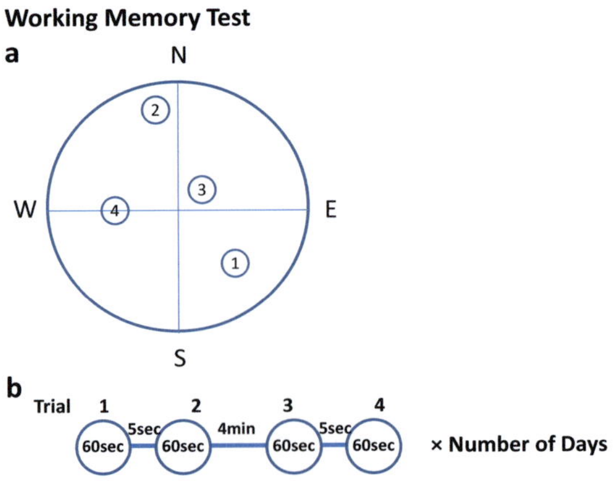

Fig. 7 Schematic illustration of working memory test design (**a**) and procedure (**b**)

6. Pseudo-randomly determine the platform location and starting positions for each trial set.
7. Allow each rat to freely swim in the pool until it finds the submerged platform or until a maximum of 60 s has elapsed.
8. When the rat successfully mounts onto the platform, it will be allowed to rest on the platform for 10 s (time it using a stopwatch) before being removed from the pool and placed back in its home cage.
9. If the rat does not find the platform in 60 s, it will be manually guided to the platform and allowed to rest on the platform for 10 s (time it using a stopwatch) before being removed from the pool and placed back in its home cage (*see* **Notes 4–6**).
10. The latency to find the hidden platform over the two trials within a trial set on a given day will be recorded as primary outcome metrics for further statistical analysis.
11. The delta difference between the two trials within a trial set as an indicator for one-trial learning ability will be calculated for further statistical analysis.

3.3 Statistical Analysis

1. Spatial learning and acquisition test: average the latency (s), distance (cm), and swim speeds (cm/s) for each experimental group on each testing day. A two-way repeated measures analysis of variance (ANOVA) followed by Fisher's LSD post hoc tests is used. Further, calculate the overall means of each experimental group across the 5 days of testing. A one-way analysis of variance (ANOVA) followed by Fisher's LSD post hoc tests is used.
2. Probe trial: Calculate the percent time in thigmotaxic annulus, and target zone for each rat (%). Average the percentage for each experimental group. A one-way analysis of variance (ANOVA) followed by Fisher's LSD post hoc tests is used.
3. Working memory test: Average the latency (s) over the two trials within a trial set on a given day for each experimental group. A two-way repeated measures analysis of variance (ANOVA) followed by Fisher's LSD post hoc tests is used. Calculate the delta difference between the two trials within a trial set. A One-way analysis of variance (ANOVA) followed by Fisher's LSD post hoc tests is used.

4 Notes

1. Diffuse lighting is utilized in the darkroom for rats to visualize the visual cues, such as dim light generated from the computer monitor and the peep window of the entry door.
2. In the hidden platform experiments using Sprague-Dawley rats, the platform is rendered invisible using a clear Plexiglas

platform in clear water in a darkroom. White coloring agents are not used due to the white fur of the Sprague-Dawley rats.

3. Do not place the rat in adjacent quadrants in consecutive trials in any task design to avoid the rat adopting a search strategy (e.g., right turns or left turns) to locate the platform.
4. Dry each rat thoroughly with paper towel or clean cloth after each trial before placing back into the home cage. This will allow the rats regain normal body temperature quickly and be ready for the next trial. Rats that experience cold temperatures may behave differently in swimming in the pool.
5. Use swimming pool vacuum to clean up animal feces submerged in the pool between trials.
6. Use cleaning agents to wipe the basin wall after experiments to eliminate residue contaminants.

Disclaimers

The views of the authors do not purport or reflect the position of the Department of the Army or the Department of Defense (para 4-3, AR 360-5). The authors declare that there are no conflicts of interest in this protocol. This research is funded by Combat Casualty Care Research Program.

References

1. Morris R (1984) Developments of a water-maze procedure for studying spatial learning in the rat. J Neurosci Methods 11:47–60
2. McNamara RK, Skelton RW (1993) The neuropharmacological and neurochemical basis of place learning in the Morris water maze. Brain Res Brain Res Rev 18:33–49
3. D'Hooge R, De Deyn PP (2001) Applications of the Morris water maze in the study of learning and memory. Brain Res Brain Res Rev 36:60–90
4. Hellawell DJ, Taylor RT, Pentland B (1999) Cognitive and psychosocial outcome following moderate or severe traumatic brain injury. Brain Inj 13:489–504
5. Shear DA, Galani R, Hoffman SW, Stein DG (2002) Progesterone protects against necrotic damage and behavioral abnormalities caused by traumatic brain injury. Exp Neurol 178:59–67
6. Shear DA, Tate MC, Archer DR, Hoffman SW, Hulce VD, Laplaca MC, Stein DG (2004) Neural progenitor cell transplants promote long-term functional recovery after traumatic brain injury. Brain Res 1026:11–22
7. Morris RG, Garrud P, Rawlins JN, O'Keefe J (1982) Place navigation impaired in rats with hippocampal lesions. Nature 297:681–683
8. Smith DH, Okiyama K, Thomas MJ, Claussen B, McIntosh TK (1991) Evaluation of memory dysfunction following experimental brain injury using the Morris water maze. J Neurotrauma 8:259–269
9. Cain DP, Boon F, Corcoran ME (2006) Thalamic and hippocampal mechanisms in spatial navigation: a dissociation between brain mechanisms for learning how versus learning where to navigate. Behav Brain Res 170:241–256
10. Packard MG (2009) Anxiety, cognition, and habit: a multiple memory systems perspective. Brain Res 1293:121–128
11. Devan BD, McDonald RJ, White NM (1999) Effects of medial and lateral caudate-putamen lesions on place- and cue-guided behaviors in the water maze: relation to thigmotaxis. Behav Brain Res 100:5–14

12. Shear DA, Lu XC, Bombard MC, Pedersen R, Chen Z, Davis A, Tortella FC (2010) Longitudinal characterization of motor and cognitive deficits in a model of penetrating ballistic-like brain injury. J Neurotrauma 27:1911–1923
13. Hoffman SW, Fulop Z, Stein DG (1994) Bilateral frontal cortical contusion in rats: behavioral and anatomic consequences. J Neurotrauma 11:417–431
14. DiMattia BD, Kesner RP (1988) Spatial cognitive maps: differential role of parietal cortex and hippocampal formation. Behav Neurosci 102:471–480
15. Hamm RJ, Dixon CE, Gbadebo DM, Singha AK, Jenkins LW, Lyeth BG, Hayes RL (1992) Cognitive deficits following traumatic brain injury produced by controlled cortical impact. J Neurotrauma 9:11–20
16. Bramlett HM, Green EJ, Dietrich WD (1997) Hippocampally dependent and independent chronic spatial navigational deficits following parasagittal fluid percussion brain injury in the rat. Brain Res 762:195–202
17. Beaumont A, Marmarou A, Czigner A, Yamamoto M, Demetriadou K, Shirotani T, Marmarou C, Dunbar J (1999) The impact-acceleration model of head injury: injury severity predicts motor and cognitive performance after trauma. Neurol Res 21:742–754
18. Fox GB, Faden AI (1998) Traumatic brain injury causes delayed motor and cognitive impairment in a mutant mouse strain known to exhibit delayed Wallerian degeneration. J Neurosci Res 53:718–727
19. Hamm RJ, Temple MD, Pike BR, O'Dell DM, Buck DL, Lyeth BG (1996) Working memory deficits following traumatic brain injury in the rat. J Neurotrauma 13:317–323
20. Williams AJ, Hartings JA, Lu XC, Rolli ML, Dave JR, Tortella FC (2005) Characterization of a new rat model of penetrating ballistic brain injury. J Neurotrauma 22:313–331
21. Williams AJ, Hartings JA, Lu XC, Rolli ML, Tortella FC (2006) Penetrating ballistic-like brain injury in the rat: differential time courses of hemorrhage, cell death, inflammation, and remote degeneration. J Neurotrauma 23:1828–1846
22. Williams AJ, Ling GS, Tortella FC (2006) Severity level and injury track determine outcome following a penetrating ballistic-like brain injury in the rat. Neurosci Lett 408:183–188
23. Chen Z, Tortella FC, Dave JR, Marshall VS, Clarke DL, Sing G, Du F, Lu XC (2009) Human amnion-derived multipotent progenitor cell treatment alleviates traumatic brain injury-induced axonal degeneration. J Neurotrauma 26:1987–1997

Chapter 30

Assessment of Cognitive Function in the Water Maze Task: Maximizing Data Collection and Analysis in Animal Models of Brain Injury

Mark D. Whiting and Olga N. Kokiko-Cochran

Abstract

Animal models play a critical role in understanding the biomechanical, pathophysiological, and behavioral consequences of traumatic brain injury (TBI). In preclinical studies, cognitive impairment induced by TBI is often assessed using the Morris water maze (MWM). Frequently described as a hippocampally dependent spatial navigation task, the MWM is a highly integrative behavioral task that requires intact functioning in numerous brain regions and involves an interdependent set of mnemonic and non-mnemonic processes. In this chapter, we review the special considerations involved in using the MWM in animal models of TBI, with an emphasis on maximizing the degree of information extracted from performance data. We include a theoretical framework for examining deficits in discrete stages of cognitive function and offer suggestions for how to make inferences regarding the specific nature of TBI-induced cognitive impairment. The ultimate goal is more precise modeling of the animal equivalents of the cognitive deficits seen in human TBI.

Key words Traumatic brain injury, Morris water maze, Cognitive impairment, Memory, Animal models

1 Introduction

Traumatic brain injury (TBI) produces a range of impairments across the motor, cognitive, and emotional domains. Deficits in anterograde and retrograde memory are common following TBI, as are deficits in speed of information processing and executive function [1–3]. Animal models have been used extensively for preclinical evaluation of therapeutic interventions following brain injury, as well as to understand the pathophysiological, biomechanical, and behavioral consequences of traumatic injury. The Morris water maze (MWM) is frequently used to assess the degree of functional impairment produced by injury as well as the potential therapeutic effects of post-injury interventions. Known primarily as a spatial memory task, inferences are often drawn between TBI and disruption of hippocampally dependent memory processes.

Firas Kobeissy et al. (eds.), *Injury Models of the Central Nervous System: Methods and Protocols*, Methods in Molecular Biology, vol. 1462, DOI 10.1007/978-1-4939-3816-2_30, © Springer Science+Business Media New York 2016

However, performance in the MWM is mediated by a complex interplay of cognitive and non-cognitive processes and is dependent on intact functioning in numerous brain regions. Careful characterization of how animal models of TBI-induced deficits in the MWM is of critical importance if findings from preclinical studies are to have maximum translational impact to human TBI.

In this chapter, we review some of the fundamental considerations involved in utilizing the MWM as a tool in preclinical TBI research, with an emphasis on carefully planned analysis of performance data that is reflective of underlying cognitive processes of interest. We examine common dependent measures used when analyzing water maze data and discuss the pros and cons of various outcomes. We also present a theoretical framework for examining deficits in specific aspects of cognitive function following experimental TBI, with the ultimate objective being reliable and valid modeling of the animal equivalents of those deficits observed in human TBI.

2 Materials

1. Round polypropylene tank (pool) with smooth interior and non-cloggable drain; 140 centimeter (cm) diameter for mice or 180 cm diameter for rats (Harvard Apparatus, MA).
2. Circular or square Plexiglas goal platform; 10 cm diameter will accommodate mice and rats; base and height recommended to be 30 cm (Harvard Apparatus, MA).
3. Small "flag" for visible platform spatial cue (remove plunger from plastic syringe [without needle]; invert the syringe and place a piece of tape around the end where the needle attaches to create a "flag").
4. 2-4 Rubber bands.
5. Water warmed to 24–27 °C.
6. Submersible and programmable water heater (Aquacave, Inc., IL).
7. Small fish net.
8. Nontoxic white paint.
9. Yard stick or broom stick for stirring the paint in the pool.
10. Opaque, ceiling-mounted curtains.
11. Four highly visible spatial cues.
12. Computer connected to a video tracking system (Ethovision XT, Noldus, VA).
13. Stopwatch.
14. Heat lamp or incubator.
15. Holding cage for animals during testing.

3 Methods

1. The day before testing begins, place the goal platform inside of the pool.
2. Fill the pool with water to a depth that is 2–4 cm above the goal platform for memory testing.
   (a) During visible platform training, fill the pool with water to a depth that is level with the top of the goal platform.
3. Add 1–2 cups of nontoxic white paint to the water to make it cloudy.
4. Place the water heater in the pool to warm the water to 24–27 °C.
5. Imagine that the pool is divided into four quadrants. A spatial cue should be positioned on the wall/curtain near the center of each quadrant.
   (a) Spatial cues should remain constant through an entire project.
6. On the day of testing, transport animals from housing room to testing room 15–30 min before testing is scheduled to begin.
7. Move animals to holding cages to be used during testing. Use of holding cages will prevent water buildup in home cages.
8. Remove the water heater from the pool.
9. Use a yard stick or broom stick to stir the pool and redistribute any paint that has gathered on the bottom of the pool.
10. The goal platform should be placed in the center of a quadrant.
    (a) The goal platform will remain in the same location throughout memory testing
    (b) The goal platform will move to a new position for each trial during visible platform testing.
11. Set up visible tracking system to acquire data during testing trials.
12. Animals complete four trials per day for 5 days during memory testing; maximum trial length is 60 s for mice and 120 s for rats.
    (a) Animals complete four trials per day for 3 days during visible platform testing; maximum trial length is 60 s for mice and 120 s for rats.
13. Ensure that the visible tracking software is prompted to begin data collection.
14. Gently place the animal in the pool facing the wall.
15. A stopwatch should be used in addition to video tracking software to document latency to goal platform.

16. If the animal reaches the goal platform, record the latency and let the animal remain on the platform for an additional 15–20 s.
17. In the animal does not reach the goal platform at the end of a trial, gently guide the animal to the platform without removing it from the water. Let the animal remain on the platform for 15–20 s.
    (a) The investigator should step behind the curtain or wall while the animal sits on the platform at the end of a trial. The investigator will likely have to spend additional time with some animals that display anxiety sitting on the platform alone. In the case of anxious animals, patience is essential. In the event that an animal will not sit on the platform alone, an investigator may hold the animal on the platform for 10–15 s; however, this is not recommended more than a few times. Extremely anxious animals that do not learn to sit on the platform should be discarded from the study.
18. Place the animal under a heat lamp or in a heated incubator between trials.
19. Use the small fish net to scoop out any feces floating in the pool.
    (a) A yard stick or broom stick can be used to stir the pool; however, do not start another trial until all water movement has subsided.
20. Repeat remaining trials with subsequent subjects.
    (a) Animals should be placed into the pool from one of four different start locations at the beginning of each trial. Start locations are often referred to by direction (e.g., north, south, east, west). Thus, an animal will begin a trial from the north position, the south position, the east position, and the west position within each testing day. The order of drop-in locations should change from day to day.
21. When all subjects have finished testing for the day, place the heater back into the pool, and save data files.
22. Allow animals to warm under the heat lamp or in the incubator for 10–15 min before placing in home cages.
23. Transport animals back to housing room.

4 Experimental Considerations

Data obtained from behavioral tasks such as the MWM represent an animal's performance, yet interpretations of performance are frequently centered on cognitive or neurobiological constructs such as memory. However, poor performance in the MWM may result from impairment to processes that are essentially non-mnemonic in nature, such as impairment in sensory processing, motor deficits, or motivation. In this way, the MWM is integrative in nature: a number of mnemonic and non-mnemonic processes working simultaneously and interdependently result in the animal's performance. Furthermore, a number of brain regions appear to participate in various stages of MWM task performance. While generally thought of as a hippocampally dependent task, deficits in various versions of the water maze task have been demonstrated following lesions to the orbitofrontal cortex [4], perirhinal cortex [5], striatum [6], and cingulate cortex [7], among many others. In experimental models of TBI, where selective damage to the hippocampus has been observed [8, 9], there has frequently been a focus on correlating spatial memory impairments with TBI-induced hippocampal damage. However, it is unclear what role damage to other brain regions plays in TBI-induced spatial memory impairments. While a complete understanding of how TBI impairs water maze performance remains to be determined, investigators employing the MWM as an outcome will benefit from keeping in mind that the integrative nature of the task requires the functional integrity in numerous brain systems [10], many of which may be damaged or rendered temporarily dysfunctional following experimental TBI.

4.1 General Protocol and Experimental Timing

A typical MWM protocol involves repeated daily trials over several days (*see* [10] for descriptions of standard and alternative MWM protocols). Our practice has been to use 20 trials across 5 days (four trials per day from four different start locations), although other investigators have used more or less training. Several reports suggest that the total number of trials, rather than the spacing of trials, is the prime determinant in overall performance in healthy animals. However, investigators should consider the possibility that animals subjected to experimental TBI may undergo behavioral fatigue more rapidly than healthy animals; thus, the use of massed trials in a single day should only be used when necessary to answer specific questions that cannot be answered with a multiple-day protocol. Every effort should be made to keep the inter-trial interval consistent among all animals (*see* **Note 1**).

The pre- and post-injury timing of MWM testing in preclinical TBI studies is an additional important consideration. To examine retrograde memory impairment, animals must be pre-trained in the MWM prior to injury; yet the optimal time to perform this pre-training must be carefully considered. Recent memories, which have not been fully consolidated, are more susceptible to disruption or injury than remote memories. Thus, investigators wishing to examine retrograde memory impairment in experimental TBI should consider which retrograde memory processes (i.e., consolidation, storage, or retrieval) are of greatest interest and design their pre-training accordingly. Injury administered soon after pre-training (e.g., within a few days) would presumably disrupt the consolidation of a recently formed spatial memory. To assess storage or retrieval processes, it may be necessary to wait several weeks between pre-training and injury, when the spatial memory is considered remote and consolidated [11]. The post-injury timing of MWM training is equally important. During the acute post-injury period, animals receiving moderate to severe TBI may display motor, sensory, and motivational deficits that prohibit successful water maze performance. At mild levels of injury, transient deficits in the MWM [12, 13] may be more reflective of acute posttraumatic amnesia and may not reflect a persistent learning and memory impairment such as that seen in more severe levels of injury. Furthermore, many post-injury interventions of interest will require multiple days of administration for effectiveness; thus, testing may need to be delayed until the intervention has been given sufficient time to produce the desired effect. Depending on the severity of injury, the water maze may not be sensitive enough to detect group differences far into the post-injury period (*see* **Note 2**). Although subtle performance deficits in the MWM have been observed as late as 1 year post injury [9], many labs report only transient deficits that persist for up to a few weeks following injury. These obstacles can be overcome, to some extent, through careful planning and analysis and the use of appropriate dependent variables (*see* Subheading 3).

4.2 Testing Room and Pool

The apparatus and stimuli utilized during MWM testing can significantly influence animal behavior and should be carefully considered. MWM testing should take place in a secluded room with limited noise and no windows. This space should be in close proximity to animal housing to reduce transportation time and associated stress (*see* **Note 3**). Within the testing room, one should have access to a small desk with a computer to control a mounted video tracking system above the pool. Importantly, a barrier (e.g., ceiling-mounted opaque curtains) should surround the pool to prevent a test subject from seeing the investigator, computer, and other subjects during test trials (*see* **Note 4**). A pool can be commercially purchased from several vendors and is typically made out of durable plastic or aluminum. Preference should be given to a plastic white pool with a consistent diameter (140 cm for mice and 180 cm for rats) along the outer wall (75 cm). Pools with angled walls reduce floor space and limit locations for a goal platform that has a large base. The white color will complement the opaque water and exclude the need to paint the inside of the pool (*see* **Note 5**); however, one must wipe the pool walls daily. Intra-maze cues that arise from repeated use, such as splash or scratch marks, can potentially influence search strategy. A goal platform made of clear Plexiglas is ideal for MWM testing. A square or round base of 30 cm and a

10 cm platform provides adequate support and stability for both mice and rats. Most vendors offer platforms at a fixed height; however, some sell adjustable platforms that are controlled manually or by a hydraulic system. Hydraulic platforms are particularly useful during probe trials where the investigator can reward an animal that recalls the position of the platform (*see* Subheading 6). A submergible water heater should be used to keep the water temperature constant (24–27 °C) across testing days (*see* **Note 6**).

4.3 Spatial Cues

Four highly visual extra-maze spatial cues should be hung around the pool. The position of these cues often correlates with the drop-in locations utilized during memory testing and should be hung at a height that is clearly visible from the water level. Spatial cues should be similar in size and are often made from laminated white and black poster board depicting clear geometric shapes (*see* **Note 7**). Animals learn very quickly where an investigator enters and exits the testing area, which can result in certain animals showing a preference for the pool quadrant closest to this area. Ideally, the investigator will enter and exit the testing area from multiple locations to avoid reinforcing a spatial preference. Investigators should be cautious to avoid sudden or frequent behaviors that result in movement of barriers, such as curtains, around the pool. Curtain waving can alert the animal to the investigators location and influence behavior.

4.4 Video Tracking System

A quality video tracking system is needed for detailed data acquisition and analysis. Complete systems include a mounted video camera over the swimming pool that is connected to a computer running a tracking software program. Software programs for tracking animal behavior are available through multiple vendors. Preference should be given to programs that allow an investigator to identify specific regions of interest including quadrants, platform location, and thigmotaxic zone. Variables measured such as location, distance, and swim speed should be linked to these regions. Automated experiments establish consistency between subjects and increase efficiency in starting and stopping trials (*see* **Note 8**). Finally, video files of each trial facilitate swim path analysis as well as identification of spatial preferences within the pool via heatmaps.

4.5 Investigator

The stimulus of the investigator should not be underestimated as there is great responsibility in consistently handling all subjects during MWM testing (*see* **Note 9**). Inexperienced investigators that are hesitant or anxious should be supervised by a trained colleague, and pilot groups of animals should be run before actual data collection begins. Ideally, one to two designated investigators will run the animals at the same time each day and remain blinded to experimental groups to decrease variability and bias.

5 Dependent Measures Used in the Morris Water Maze

5.1 Goal Latency

Of the commonly used dependent measures in the water maze, latency to find the platform may actually provide the least amount of information regarding true spatial learning. This is because non-spatial search strategies and random platform encounters influence goal latency. Although random platform encounters are few and the effects on goal latency are likely washed out across many subjects, they still contribute to the overall amount of statistical noise and make goal latency a less than ideal outcome. Further, it has been *note*d that rats with hippocampal damage, as well as aged rats, may use a circling strategy at a fixed distance from the wall of the pool to solve the water maze task [14]. While a circling strategy will reduce latency and total path length, it is not as effective over the long run as a true spatial strategy. Using goal latency only, with no analysis of the animals track through the maze, it is impossible to determine the type of search strategy the animal has adopted or make inferences regarding an animal's knowledge of the spatial location of the goal platform.

5.2 Goal Proximity

Ideally, shorter latencies to locate the goal platform will correlate with more efficient, spatially localized search strategies. As discussed above, a number of factors can lead to lower latencies in the absence of effective spatial search strategies. The solution is to use an outcome that is independent of the length of the trial or the total distance traveled. Goal proximity, or average distance to the goal platform across the entire trial, provides a more direct measure of spatial learning [14]. Unlike latency, proximity is not influenced by random encounters with the goal platform or other non-spatial searching strategies. To the extent that the water maze is often used to assess deficits in hippocampally dependent spatial learning and memory, proximity is superior to latency as a measure of the spatial nature of an animal's search pattern. Averaged across all trials during a given training session, goal proximity more accurately reflects an animal's search strategy during both acquisition and probe trials. Also known as Gallagher's measure, certain tracking programs may refer to proximity as mean proximity, cumulative proximity, or simply P.

5.3 Thigmotaxia

One issue that remains unclear regarding TBI-induced impairments in learning and memory is why injured animals display such drastic impairments in the MWM during the initial acquisition trials. Arguably, impaired performance during the first few trials of the water maze is not a truly spatial learning deficit. More likely, this represents an animal's inability to integrate the procedural requirements of the task with the spatial requirements, a type of learning that may be dependent on the striatum [15], another structure known to be vulnerable to TBI [16, 17]. After learning

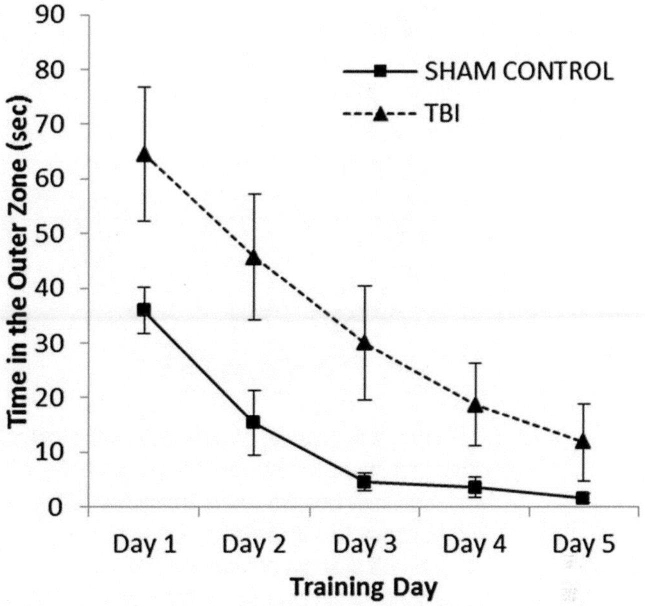

Fig. 1 Illustration of thigmotaxic behavior (perimeter swimming) in Long-Evans rats following moderate TBI or sham injury. The time spent in the outer perimeter of the water maze is shown for each of 5 days of water maze acquisition. Data points are mean ± SEM (n = 10/group). The rapid decline in thigmotaxic behavior observed in sham controls is thought to be indicative of the adoption of spatial searching strategies; animals with TBI are slow to adopt a truly spatial search strategy

the procedural requirements of the task, animals must then shift to a spatial strategy using extra-maze cues in order to solve the task efficiently. Thus, measures that provide information about the adoption of strategies may contribute to the overall picture of how TBI disrupts behavioral processes. We have found that measures of thigmotaxia, or perimeter searching, provide valuable information regarding the shift to a spatial searching strategy and that this effect is especially powerful during the first days of training. While all animals tend to exhibit thigmotaxia during the first few trials, healthy animals quickly learn that this strategy is ineffective and begin searching throughout the pool. On the other hand, animals with TBI display thigmotaxic behavior for several days of training, indicating a failure to shift from non-spatial to spatial search strategies similar to sham animals (*see* Fig. 1; *see* **Note 10**). As with proximity, this measure provides important information not reflected in latency to find the platform.

5.4 Comparison of Dependent Variables Used in the Water Maze

Figure 2 illustrates vividly the need to incorporate multiple dependent measures in order to make inferences about cognitive performance in the water maze task. Panels a and b are track plots from a TBI-injured animal and a sham control, respectively, on day 2, trial 4 of water maze training. Clearly, the animals have adopted different

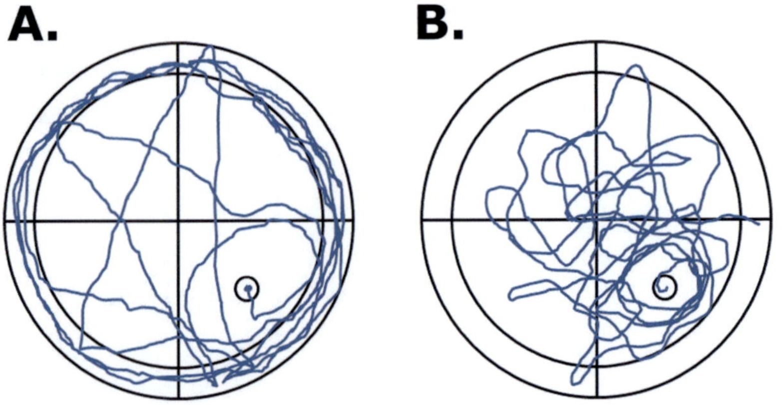

Fig. 2 Illustration of different search strategies adopted by TBI injured (panel **a**) and sham control animals (panel **b**) by trial 4, day 2 of water maze acquisition training. Despite the dramatically different search paths through the maze, the latency to the platform for these two animals was nearly identical (TBI panel **a** = 85 s, sham control panel **b** = 88 s), illustrating the importance of using additional dependent variables such as proximity and thigmotaxia

strategies at this point. The injured animal is combining thigmotaxia with a "bouncing" strategy, wherein it sporadically pushes off of the pool wall and swims across the central portion of the maze, while the sham animal is using a spatial strategy that most likely incorporates the extramaze cues. However, the latencies to find the goal platform between the two animals are nearly identical. Unlike latency, goal proximity captures this difference quantitatively, and the measure of thigmotaxia further supports the notion that only the sham animal has adopted a truly spatial search strategy.

The dependent measures described above not only provide distinct information about the behavioral processes involved in learning but also provide varying levels of statistical sensitivity. This can be seen in power calculations for each of the dependent measures. Using an existing data set from our lab (MW), we found that the observed power values for latency and proximity were 0.85 and 0.96, respectively (*see* Table 1). Thus, not only does proximity provide a better measure of true spatial learning and memory, but it also offers greater statistical power and thus allows more room for detecting treatment effects. Proximity values tend to have less inherent variability compared to latency, with standard errors nearly half that of latency. Thigmotaxia has a power of 0.80 that, while the lowest of the three measures, still provides an acceptable level of statistical power. These power values must also be interpreted in light of what information is being provided. While latency and proximity are generally interpreted in terms of ability to solve the spatial navigation task, thigmotaxia is more reflective of strategy adoption during the early days of testing and thus this measure provides qualitatively different information.

Table 1
Observed power, effect size (Eta-squared), and standard error values for three dependent variables commonly used in the Morris water maze

	Observed power	Effect size	Standard error
Latency	0.85	0.37	24.7
Proximity	0.96	0.48	13.2
Thigmotaxia	0.80	0.34	20.4

Data is from a 2 (sham vs. TBI) × 5 (day) mixed ANOVA. Proximity is the most sensitive measure, and yet each provides different information about an animal's performance in the maze

6 Probe Trials

Removing the goal platform from the MWM for a probe trial has been used frequently to make inferences regarding an animal's knowledge of the exact spatial location of the goal platform. Performance in the probe trial is often assessed with the amount of time spent in the goal quadrant, the number of platform crossings, or with goal proximity. The use of probe trials to make between-group inferences in TBI research is not a decision that should be taken lightly, as several considerations could confound interpretations of an animal's performance during a probe trial. First, it should be recognized that the non-reinforcement of a previously-reinforced behavior is an extinction trial; thus, probe trials should only be conducted at the end of acquisition training unless strongly justified by the question of interest. After a period of non-reinforcement, the most efficient strategy to adopt is to search adjacent areas of the pool for the platform. Indeed, we have found that in a single, 60-s probe trial following 5 days of MWM training, sham animals will display goal proximity scores lower (i.e., closer to the platform) than animals with TBI only during the first 30 s of the probe trial. If the probe trial is allowed to continue for a full minute, sham animals will resume a spatial searching strategy in other areas of the pool. In contrast, animals with TBI demonstrate perseverative behavior by remaining closer to the goal location across the entire probe trial. Therefore, it is worthwhile to examine probe trial data in blocks of 15 or 30 s, as differential performance may be observed as the probe trial progresses. Another potential means of overcoming the extinction problem is to provide a rewarded probe trial or a reminding procedure introduced at the end of the probe trial (*see* 11). In water mazes equipped with an Atlantis-style telescoping platform, the platform is remotely raised at the end of the probe trial or when the animal hovers over the previous platform location for a given amount of time.

Even without an Atlantis-style goal platform, the experimenter can place the platform back in the pool manually and place the animal on the platform for a rest period. This procedure has been used previously in TBI models with some success [18]. Similar to our findings that goal proximity is the most sensitive measure for assessing water maze performance across multiple days of acquisition training, Maei et al. [19] used Monte Carlo simulations to demonstrate that goal proximity consistently outperforms other dependent measures in its ability to detect group differences during a probe trial in the water maze.

7 Controlling for Non-cognitive Processes

Successful completion of a test trial in the water maze requires animals to utilize many senses and behaviors, thus investigators should appreciate and adequately control for the potential influence of non-cognitive processes. Sensorimotor and visual impairments are common non-cognitive processes that can influence water maze performance, and fortunately investigators can characterize these behaviors with minimal effort. An internal control that is easily acquired from even basic video tracking systems is swim speed, which is commonly used to identify global motor and motivational deficits. Swim speed should be similar across all experimental groups during memory testing regardless of differences in latency to reach the hidden platform. Animals that display significantly faster or slower swim speeds might have a true swimming motor deficit or an anxiety induced motor deficit. In anticipation of this potential confound, investigators are advised to run the water maze in conjunction with additional motor and sensory assessments such as a non-spatial swimming task, open-field, or elevated plus maze test.

Cued learning is another way to identify potential visual, motor, or motivational deficits and was incorporated in the original water maze study. It has since been used extensively to control for non-mnemonic processes and has recently been described as a potentially useful tool in cognitive rehabilitation following TBI [20]. Procedurally, cued learning is the same as the hidden platform version of water maze testing. A subject must visualize a spatial cue that is associated with the location of a goal platform in order to escape swimming in the water. However, during cued learning the water level is adjusted so that it is slightly below the surface of the goal platform and a spatial cue, most commonly a small flag, is attached to one side of the platform. The position of the goal platform is moved between trials so that the extramaze spatial cues are irrelevant to its location, and the only reliable marker of the goal location is the flag attached to the platform. Cued learning can be completed before or after memory testing.

We have found that administering it before in mice is particularly useful in reducing the number of platform swim-overs and jump-offs during the first few trials of memory testing. Rats tend to procedurally acquire the memory task better than mice and display fewer swim-overs and jump-offs, so cued learning can be completed after memory testing.

8 Evaluation of Specific Aspects of Memory Impairment Following TBI

Memory formation may be subdivided into several distinct processes, and each stage may be mediated by distinct cognitive and neurobiological substrates. A number of studies using electrolytic lesion and reversible pharmacological inactivation of specific receptors have provided important insight into the dissociable nature of these memory processes. For instance, *N*-methyl-D-aspartate (NMDA) receptor antagonists administered during learning of a hippocampally dependent task interfere with the formation of spatial memory; yet NMDA blockade does not appear to affect the retrieval of a previously formed memory [21, 22]. On the other hand, administration of alpha-amino-3-hydroxy-5-methyl-4-isoxazolepropionic acid (AMPA) receptor antagonists has been shown to disrupt both consolidation [23] and retrieval [24] processes. Although some uncertainties exist as to the specific role various brain structures and receptor types play in each of these processes, it is clear that dividing memory into distinct processes such as encoding, consolidation, storage, and retrieval offers a greater level of explanatory power compared to a view in which memory is seen as a single construct. This dynamic view of memory may hold especially strong explanatory power in conditions where the various stages of the memory process may be differentially affected by disease or injury, such as TBI.

Figure 3 presents one possible view of memory formation as a series of processes. The stages of processing within the shaded center area are cognitive processes that are thought to be mediated by distinct neurobiological substrates. The non-associative processes outside of the shaded region are processes that may interfere with performance in a learning and memory task, and yet are not strictly cognitive processes. There are two important points to be taken from Fig. 3. First, to make inferences about the cognitive processes in the shaded region, the non-associative processes outside of this region must be controlled for. Second, the memory processes in the shaded area are not directly observable. Because we cannot directly observe processes such as encoding and consolidation, deficits in these processes may appear superficially similar when we analyze MWM data. A standard water maze analysis provides information regarding the overall degree of learning and memory impairment, but provides little or no information regarding which

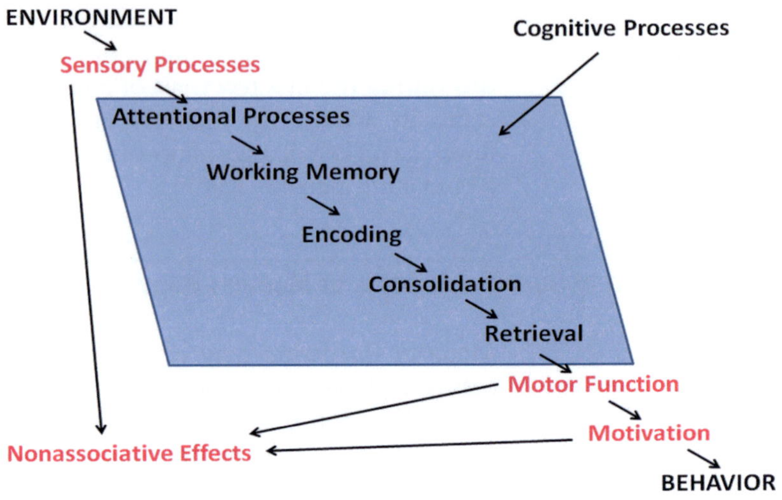

Fig. 3 Schematic representation of processes involved in performance of a learning and memory task

memory processes are responsible for the observed deficit, or whether the deficit is even mnemonic in nature.

Studies in both the human and experimental TBI literature have attempted to determine how TBI disrupts specific stages of learning and memory formation. We have previously reported that anterograde memory impairment in the MWM appears to be mediated primarily by encoding deficits, while retrograde memory processes appeared to be mediated by deficits in both consolidation and retrieval [18]. Others have suggested that consolidation deficits may play explain the impairment of a conditioned fear response following experimental TBI [25]. Findings from studies in human TBI likewise provide conflicting information regarding the role of each memory process in mediating memory impairment, with some studies suggesting encoding deficits are the primary mediator [26] and others suggesting consolidation deficits predominate [27, 28]. It seems likely that each of these memory processes would be disrupted to some extent, and factors such as the locus and severity of injury, the elapsed time between injury and memory testing, and a host of other variables may determine which memory processes are most dysfunctional in a given task. Given that such factors are relatively difficult to control in human TBI studies, experimental animal models should be used to shed light on the exact nature of TBI-induced memory deficits under a range of experimental parameters.

8.1 Learning Indices

In studies of TBI-induced memory deficits, memory impairment in injured animals becomes obscured after extensive training. In studies of moderate TBI, the most pronounced deficits occur during the initial trials, as sham and injured animals adopt different strategies. After several days of testing, these differences may be minimized, leaving little room to determine the efficacy of therapeutic interventions. However, a learning index may reveal information not seen in a standard group-by-day analysis and can provide information regarding the specific stages of learning and memory that contribute to the performance deficit. Here we present two indices of learning, an acquisition index and a savings index, that represent different aspects of MWM learning across repeated days of training in the water maze. Each may be used with any of the dependent variables commonly used in the MWM. The acquisition index is simply a measure of within-day learning, averaged across all days of training, so that in a multiple day water maze protocol the acquisition index would be

$$t_F d_1 - t_L d_1 + \cdots + t_F d_k - t_L d_{k/k}$$

where t_F and t_L are the first and last trials of the day (d), respectively, and k is the number of days of training in the water maze. This index will often reveal differences in within-day learning between sham and injured animals, as sham animals tend to show much greater within-day improvements relative to injured animals.

Another important question when analyzing behavioral data from any multiple-day learning and memory task is whether animals can transfer what is learned in 1 day of training to subsequent training sessions. A savings index is thus an indirect measure of memory consolidation, storage, and retrieval processes, although it alone cannot distinguish among the three. In a multiple-day water maze task, the savings index would be

$$t_L d_1 - t_F d_2 + \cdots + t_L d_{k-1} - t_F d_{k/k}$$

where the difference is computed between the last trial of a given day and the first trial of the subsequent day, averaged across all days. Recent studies examining the effects of brain injury induced neuroinflammation in a mouse model [29, 30] of Alzheimer's disease (AD) demonstrate the utility of these learning indices. Although latency to reach the submerged goal platform was similar between brain injured non-transgenic and R1.40 mice at 4 months post-injury, the acquisition index revealed that R1.40 mice exhibited tremendous improvement within any given testing day compared to non-transgenic mice. In contrast, the savings index showed that R1.40 mice performed significantly worse than non-transgenic mice between testing days. These results highlight a potential deficit in transferring learned information from one testing day to the next, which would not have been identified otherwise [31].

8.2 Rate of Forgetting

In addition to the acquisition and savings indices described above, another method of gaining insight into specific memory processes disrupted by TBI is rate of forgetting. We have previously used rate of forgetting in both short-term [32] and long-term [18] memory tasks to examine memory deficits following experimental TBI. Given that the amount of information recalled after a delay is dependent on the amount of information originally encoded into memory, rate of forgetting analyses require that experimental groups be matched on initial levels of acquisition. At mild-to-moderate levels of injury, this may require additional acquisition trials to bring injured animals to the performance level of controls; at more severe levels of TBI, it may be impossible to match injured and control groups. More rapid forgetting following similar levels of acquisition is generally thought to be indicative of a consolidation deficit [27], yet retrieval deficits cannot be ruled out when rapid forgetting is observed. Used in conjunction with the previously discussed reminding procedure, it may be possible to distinguish between impaired consolidation and impaired retrieval: if the reminding procedure produces no benefit, one can infer impaired consolidation processes as the primary deficit.

9 Summary and Conclusions

The MWM has proven to be particularly useful for assessing the effects of TBI in experimental animal models as well as assessing the benefits of post-injury therapeutic interventions. The task is quite flexible in that it can be altered to examine specific types of memory (e.g., working memory, place recall, discrimination), and it is sensitive enough to examine impairments in specific stages of learning and memory formation (e.g., encoding, consolidation, retrieval) when appropriate methodology and analyses are employed. Investigators using the MWM to examine deficits induced by TBI should consider the various mnemonic and non-mnemonic processes which may contribute to impaired performance and carefully design their experiments so that they most closely reflect the underlying processes of interest.

There are several arguments for utilizing animal models to more precisely define the nature of TBI-induced memory impairment. First, armed with a better understanding of the specific nature of the cognitive deficits induced by TBI, therapeutic and rehabilitative strategies can be specifically tailored to those deficits. Second, by assessing specific aspects of cognitive functioning, we develop a better correlation between the behavioral processes being studied and the anatomical locus of injury. Third, isolating specific cognitive processes often reduces the statistical noise inherent in behavioral data, thus providing increased statistical sensitivity. Lastly, it is possible that subgroups of TBI patients suffer from deficits in discrete stages of the memory process. Identification of the nature of this memory impairment in animal models can potentially pave the way for translational studies in humans and the development of targeted rehabilitative strategies.

10 Notes

1. The recommended inter-trial interval for MWM testing is 15 min.
2. It is recommended that both sham and injured subjects are included in experimental testing groups. Thus one maintains a constant control to document brain vs. sham injury induced behavioral deficits over time.
3. A habituation period of 15–30 min is recommended for animals transported from a housing room to the behavioral testing room.
4. The surrounded curtain should remain 3–4 ft away from the perimeter of the pool, which should provide ample space for the investigator to walk around the pool.
5. In an attempt to "hide" the goal platform, many investigators add white nontoxic paint to the water. This is a much safer option than household paint or powdered milk, which have both been used in the past.
6. The water temperature is very important and can significantly influence animal behavior. Cold water can induce hypothermia and motor deficits, while the comfort of warm water elicits less motivation to locate the goal platform.
7. Spatial cues should remain constant during the entire length of an experiment, particularly if multiple groups of animals are being run over time with plans of combining data for statistical analysis.
8. A stopwatch should be used in addition to any software program. This provides the investigator with multiple tools to accurately document the beginning and end of a trial. Maintaining a paper copy of latency data is also common practice in many laboratories.
9. Auditory stimulation from investigators should also be considered. During testing, investigators should refrain from talking or whisper. Music or videos should not be played during testing. Likewise, environmental noises should be kept at a minimum.
10. Animals that display thigmotaxia throughout the entire testing protocol are often excluded from the study, particularly in cases where a majority of subjects are learning the procedures of the task.

References

1. Capruso DX, Levin HS (1992) Cognitive impairment following closed head injury. Neurol Clin 10:879–893
2. Levin HS (1990) Memory deficit after closed-head injury. J Clin Exp Neuropsychol 12: 129–153
3. McDonald BC, Flashman LA, Saykin AJ et al (2002) Executive dysfunction following traumatic brain injury: neural substrates and treatment strategies. NeuroRehabilitation 17:333–344
4. Vafaei AA, Rashidy-Pour A (2004) Reversible lesion of the rat's orbitofrontal cortex interferes with hippocampus-dependent spatial memory. Behav Brain Res 149:61–68
5. Liu P, Bilkey DK (1998) Perirhinal cortex contributions to performance in the Morris water maze. Behav Neurosci 112:304–315
6. McDonald RJ, King AL, Foong N et al (2008) Neurotoxic lesions of the medial prefrontal cortex or medial striatum impair multiple-location place learning in the water task: evidence for neural structures with complementary roles in behavioral flexibility. Exp Brain Res 187:419–427
7. Sutherland RJ, Whishaw IQ, Kolb B (1988) Contributions of cingulate cortex to two forms of spatial learning and memory. J Neurosci 8:1863–1872
8. Colicos MA, Dixon CE, Dash PK (1996) Delayed, selective neuronal death following experimental cortical impact injury in rats: possible role in memory deficits. Brain Res 739:111–119
9. Dixon CE, Kochanek PM, Yan HQ (1999) One-year study of spatial memory performance, brain morphology, and cholinergic markers after moderate controlled cortical impact in rats. J Neurotrauma 16:109–122
10. Terry AV (2009) Spatial navigation (water maze) tasks. In: Buccafusco JJ (ed) Methods of behavior analysis in neuroscience, 2nd edn. CRC, Boca Raton
11. Martin SJ, de Hoz L, Morris RGM (2005) Retrograde amnesia: neither partial nor complete hippocampal lesions in rats result in preferential sparing of remote spatial memory, even after reminding. Neuropscyhologia 43:609–624
12. Hylin MJ, Orsi SA, Zhao J et al (2013) Behavioral and histopathological alterations resulting from mild fluid percussion injury. J Neurotrauma 30:702–715
13. Shultz SR, MacFabe DF, Foley KA et al (2011) A single mild fluid percusiion injury induced short-term behavioral and neuropathological changes in the Long-Evans rat: support for an animal model of concussion. Behav Brain Res 224:326–335
14. Gallagher M, Burwell R, Burchinal M (1993) Severity of spatial learning impairment in aging: development of a learning index for performance in the Morris water maze. Behav Neurosci 107:618–626
15. Devan BD, Goad EH, Petri HL (1996) Dissociation of hippocampal and striatal contributions to spatial navigation in the water maze. Neurobiol Learn Mem 66:305–323
16. Huang EY, Tsai TH, Kuo TT (2014) Remote effects on the striatal dopamine system after fluid percussion injury. Behav Brain Res 267:156–172
17. Wagner AK, Sokoloski JE, Ren D (2005) Controlled cortical impact injury affects dopaminergic transmission in the rat striatum. J Neurochem 95:457–465
18. Whiting MD, Hamm RJ (2008) Mechanisms of anterograde and retrograde memory impairment following experimental traumatic brain injury. Brain Res 1213:69–77
19. Maei HR, Zaslavsky K, Teixeira CM et al (2009) What is the most sensitive measure of water maze probe test performance? Front Integr Neurosci doi 3:4. doi:10.3389/neuro.07.004.2009
20. Wagner AK, Brayer SW, Hurwitz M et al (2013) Non-spatial pre-training in the water maze as a clinically relevant model for evaluation learning and memory in experimental TBI. Neurobiol Learn Mem 106:71–86
21. Steele RJ, Morris RG (1999) Delay-dependent impairment of a matching-to-place task with chronic and intrahippocampal infusion of the NMDA-antagonist D-AP5. Hippocampus 9:118–136
22. Day M, Langston R, Morris RG (2003) Glutamate-receptor-mediated encoding and retrieval of paired-associate learning. Nature 424:205–209
23. Riedel G, Micheau J, Lam AG et al (1999) Reversible neural inactivation reveals hippocampal participation in several memory processes. Nat Neurosci 2:898–905
24. Bast T, da Silva BM, Morris RG (2005) Distinct contributions of hippocampal NMDA and AMPA receptors to encoding and retrieval of one-trial place memory. J Neurosci 25:5845–5856
25. Lifshitz J, Witgen BM, Grady MS (2007) Acute cognitive impairment after lateral fluid percussion brain injury recovers by 1 month: evaluation by conditioned fear response. Behav Brain Res 177:347–357

26. DeLuca J, Schultheis MT, Madigan NK et al (2000) Acquisition versus retrieval deficits in traumatic brain injury: implications for memory rehabilitation. Arch Phys Med Rehabil 81: 1327–1333
27. Vanderploeg RD, Crowell TA, Curtiss G (2001) Verbal learning and memory deficits in traumatic brain injury: encoding, consolidation, and retrieval. J Clin Exp Neuropsychol 23:185–195
28. Vanderploeg RD, Donnell AJ, Belanger HG et al (2014) Consolidation deficits in traumatic brain injury: the core and residual verbal memory deficit. J Clin Exp Neuropsychol 36:58–73
29. Lamb BT, Sisodia SS, Lawler AM et al (1993) Introduction and expression of the 400 kilobase amyloid precursor protein gene in transgenic mice [corrected]. Nat Genet 5:22–30
30. Lamb BT, Call LM, Slunt HH et al (1997) Altered metabolism of familial Alzheimer's disease-linked amyloid precursor protein variants in yeast artificial chromosome transgenic mice. Hum Mol Genet 6:1535–1541
31. Kokiko-Cochran ON, Ransohoff L, Veenstra M et al (2016) Altered neuroinflammation, neurodegeneration and behavior following traumatic brain injury in a mouse model of Alzheimer's disease. J Neurotrauma 33: 625–640
32. Whiting MD, Hamm RJ (2006) Traumatic brain injury produces delay-dependent memory impairment in rats. J Neurotrauma 23: 1233–1240

Chapter 31

Detecting Behavioral Deficits Post Traumatic Brain Injury in Rats

Hibah O. Awwad

Abstract

Traumatic brain injury (TBI), ranging from mild to severe, almost always elicits an array of behavioral deficits in injured subjects. Some of these TBI-induced behavioral deficits include cognitive and vestibulomotor deficits as well as anxiety and other consequences. Rodent models of TBI have been (and still are) fundamental in establishing many of the pathophysiological mechanisms of TBI. Animal models are also utilized in screening and testing pharmacological effects of potential therapeutic agents for brain injury treatment. This chapter details validated protocols for each of these behavioral deficits post traumatic brain injury in Sprague-Dawley male rats. The elevated plus maze (EPM) protocol is described for assessing anxiety-like behavior; the Morris water maze protocol for assessing cognitive deficits in learning memory and spatial working memory and the rotarod test for assessing vestibulomotor deficits.

Key words Traumatic brain injury, Morris water maze, Rotarod, Elevated plus maze, Anxiety, Memory, Learning, Vestibulomotor

1 Introduction

Behavioral deficits following mild-to-moderate traumatic brain injury (TBI) are often presented with symptoms such as headache, dizziness (vestibular pathology), nausea/vomiting, forgetfulness, irritability and lack of concentration (cognition deficits), depression, and anxiety [1–5]. As the severity of TBI increases, patients may experience additional symptoms such as sensorimotor dysfunction (slowed speech and blurred vision), insomnia, convulsions or seizures, and consequences of prolonged loss of consciousness (>30 min) [1–5]. More than half of TBI patients are diagnosed with coexisting major depressive disorder, post-traumatic stress disorder (PTSD) or generalized anxiety disorder (GAD) [6–10]. Anxiety is often the result of a mood imbalance due to changes in the neuronal circuitry that include the amygdala and hippocampus [11].

Firas Kobeissy et al. (eds.), *Injury Models of the Central Nervous System: Methods and Protocols*, Methods in Molecular Biology, vol. 1462, DOI 10.1007/978-1-4939-3816-2_31, © Springer Science+Business Media New York 2016

The elevated plus maze (EPM) is a widely accepted method for measuring anxiety-like behavior in rats and is explained in this chapter [12, 13]. Memory deficits and difficulty in concentrating are persistent symptoms in many patients and can be detected up to 10 years after a TBI incident [10, 14]. Learning and memory assessment in rodents using a water maze was initially described by Richard Morris using the now called "Morris Water Maze" [15, 16]. The MWM has been utilized for more than two decades to assess a variety of types of learning and memory retention measures. There is a plethora of literature and multiple protocols for measuring various types of cognitive function, therefore it is important to know beforehand which type of cognitive function the user wants to assess. A discussion of the multiple protocols available is beyond the scope of this chapter, but has been very adequately addressed by Morris, as well as Vorhees and Williams [15–17]. Two protocols, learning memory (spatial memory acquisition) and matching-to-sample working memory protocols using the MWM, are described in this chapter. Vestibular pathology is a comorbid condition when coexisting with TBI, especially blast-related TBI [18, 19]. Vestibular pathology affects the daily function of TBI patients and often prevents them from operating machinery, driving vehicles or working in elevated areas. This chapter addresses the most sensitive method for testing vestibulomotor function: the rotarod apparatus [20–22].

As explained above, this chapter will focus on methods that will allow the measurement of behavioral deficits such as anxiety, cognitive and vestibulomotor deficits in rodent models of TBI. This chapter is written with details that will prepare the "novel" user to perform the experiment. The EPM, MWM and Rotarod protocols in this chapter are described for male Sprague-Dawley rats. These protocols are applicable to mice and other rodents as well. However, investigators need to be aware that variables such as the age [23], strain (Fisher 344 vs. Sprague-Dawley; albino vs. non-albino) [24], gender (female or hormone-treated) and species (rats vs. mice) [22] of the rodent being used may affect behavioral outcomes and additional considerations or adjustments to the protocol might be necessary. It is also important to note that these widely used protocols are not the only method but are one of many methods to assess the respective behavioral deficits associated with TBI.

Rodent models of TBI vary widely in their mechanism of injury such that some TBI models are open-head and others are closed-head TBI [25–29]. Open-head TBI models involve either surgical or penetrating procedures where the skull is exposed and/or the dura mater is disrupted and almost always involve a sutured wound on the rat's head. On the other hand, some TBI models are closed head injuries that involve either a mechanical impact, blast wave or drug exposure to induce the brain injury. In most TBI models, behavioral deficits are detectable post-injury; however, it is very important to pay close attention to the method of injury and the method of choice for assessing behavioral deficits related to TBI. Anesthesia, surgical procedures, recovery, and postoperational complications may complicate the interpretation of the results and it is very important to ensure that the outcomes measured are related solely to TBI. Accordingly, the investigator designing and performing the experiment is responsible for determining the necessary changes that need to be adapted for their specific TBI model.

2 Materials

2.1 Elevated Plus Maze

1. Sprague-Dawley male rats (225–250 g) (housed 2/cage).
2. Rat elevated plus maze.
   (a) SPECIFICATIONS.

   EPM is a plus "+" shaped maze that is elevated around 2–3 ft above the ground. The open and closed arms are each about 50 cm long and 10 cm wide with a center zone of 10 cm × 10 cm connecting them. The closed arms have walls that are 40 cm high surrounding the three sides of the arms, except for the entrance to each arm from the center zone. The color of the EPM must be a color that will allow enough contrast with the rat's white color for the video-tracking software to detect motion and be able to track the rat without difficulties (*see* **Notes 1** and **2**).
3. Computer with required specifications to run the video-tracking software of your choice.
4. Video camera or webcam installed on the ceiling such that it is centered directly above the apparatus with the viewing eyepiece facing the center of the EPM (specifications compliant with video-tracking requirements) (*see* **Note 3**).
5. Video-tracking software (many are available from vendors, ANY-Maze, Stoelting Inc. is a user-friendly software that the author has used).
   (a) *Details for setting criteria in the software setup:*
   - Assign three areas on the software: closed arms, open arms, and center zone as shown in Fig. 1.
   - Verify these areas using the camera view-overlay view in the software window. Due to the 3D aspects of the plus maze, it is encouraged to place a pen/ruler on the border of each zone on the plus maze and then view it on the camera to confirm that the assigned zones in the software setup match the actual zones on the apparatus.

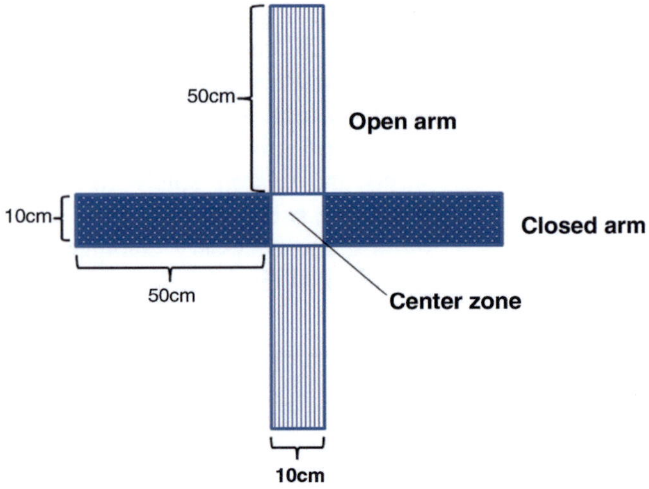

Fig. 1 A diagram representative of the dimensions of a rat elevated plus maze apparatus. The area of both open arms (*striped area*) and both closed arms (*dotted area*) are 50 cm × 10 cm each. The closed arms walls are 40 cm high (not shown). The center zone is 10 cm × 10 cm. The borders of the center zone are the entry and exit points used by the software in tracking the rat's movement on the EPM

- The center of the body of the rat is assigned as the tracking point for software calculations regarding the time spent in the arms or number of entries into the arms (*see* **Note 4**).

6. Extra animal cage.
7. 10% ethanol solution.
8. Sponge.
9. A clicker, or computer mouse, that has a long wire, to be able to start the video-tracking at the same time the rat is placed on the EPM (Optional) (*see* **Note 5**). Newer versions of video-tracking softwares now have a feature that allows the user to set up the video-tracking to start recording as soon as the animal is placed in the camera view and/or the experimenter leaves the room.

2.2 Morris Water Maze

1. Sprague-Dawley male rats (225–250 g) (housed 2/cage).
2. Water faucet nearby.
3. Water hose connected to the water faucet to fill the water tank as needed.
4. Water tank: a large circular water tank (6–7 ft (180–210 cm) in diameter and walls that are 2 ft high). The top half of the water tank should be painted black. Water is filled half-way to a water level of 30 cm and mixed with nontoxic Tempera black paint. A graded measuring stick is useful to keep track of water levels and island location over the span of the experiment (*see* **Note 6**).

5. Nontoxic Tempera paint (black) (*see* **Note 7**).
6. Platform: consists of a base connected to a pole that holds the circular island of 10 cm diameter flat at the water surface.
   (a) Platforms are available from vendors and may be custom made in-house if desired.
   (b) The platform should have a heavy base to prevent it from floating, tipping over or shifting its location between trials as the rat swims around or tries to climb onto it.
   (c) A platform with adjustable height is recommended; this will allow the platform/island to appear above the water level for the cued navigation protocol (31 cm high) or to be hidden under the water level for learning and memory protocols (28 cm high). Alternatively, if the platform is of a fixed height, the water level could be adjusted to have the platform under (2 cm) or above (1 cm) water level.
7. Four distal cues: The cues are white poster sheets ($2' \times 2'$ or $3' \times 3'$) with various filled and hollow black and white shapes having a distinguishable pattern of your choice (for example one with stripes, one with circles, one with squares, and one with triangles). These cues are hung on the wall or curtain around the water maze at the designated North, South, East, and West entry points of the water maze as shown in Fig. 2. An acceptable distance to place the cues is about 1 ft away from the side of the water tank and a height of 2 ft above the edge of the water tank. This will allow the rats to see the visual cues as they are swimming and when sitting on the platform during inter-trial periods.
8. Computer with required specifications to run the video-tracking software of your choice.
9. Video camera or webcam installed on the ceiling such that it is centered directly above the apparatus with the viewing eye-piece facing the center of the MWM (specifications compliant with video-tracking requirements) (*see* **Note 3**).
10. Video-tracking software (many are available from vendors, ANY-Maze, Stoelting Inc. is a user-friendly software that the author has used).
11. Details for setting criteria in the software setup:
    (a) The user should arbitrarily assign four points as N, E, S, and W, as shown in Fig. 2. Accordingly, the four quadrants (NE, NW, SE, SW) relative to these 4 points are also assigned as shown in Fig. 2. All areas on the water tank, island area (blue) and circle surrounding the island (grey) should be assigned by the user, via the software, and verified using the camera viewing window to confirm that the assigned zones in the software setup do match the actual zones on the apparatus (*see* **Notes 8** and **9**).

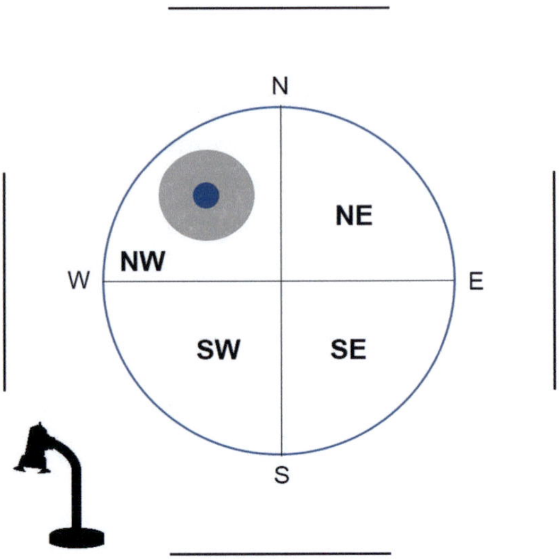

Fig. 2 Schematic diagram of the Morris Water Maze tank. The water tank is 6 ft in diameter and the height of the walls are 2 ft. The user assigns the four relative points, N, E, W, S and the four quadrants (NE, NW, SE, SW) accordingly. Visual cues are placed on the four sides around the water maze. A floor lamp light is used to keep enough contrast lighting in the room for both the user and camera to detect and follow the rat's motion. The island (*blue circle*) is assigned as the target zone and is placed in the designated quadrant per the protocol being used. The island is shown in the NW quadrant in the picture. The *grey circle* is assigned as the circle area surrounding the target zone for data analysis with regards to the proximity of the rat to the island

12. Water pump (optional): an efficient way to pump the water out of the tank after the last day of experiments if the tank does not have a drain opening at the bottom of the tank.
13. Short floor/standing lamp with dark shade (i.e., light bulb facing downward and not directly seen by the rat while the test is being done).
14. Extra cages.
15. Skimmer pool net.
16. Empty bucket with towels.
17. Heat lamp to allow the rats to dry faster and avoid hypothermia.
18. Flat block of any hard material (6″ × 10″—acrylic with holes in it or wooden flat board) to help mix the water and the tempera paint on a regular basis.
19. Handheld timer to time the inter-trial intervals.
20. A clicker or long-wired computer mouse that has a long wire, to be able to start the video-tracking at the same time the rat is placed in the MWM (Optional) (*see* **Note 5**).
21. Temperature-control (optional)—The main concern regarding water temperature is to ensure that the rats do not experience

hypothermia (usually a concern with mice more than rats); which may affect their swimming ability and visual motor coordination. The water temperature should be between 22 and 25 °C. If the investigator feels that the water temperature is a concern, then water temperature can be increased to 25 ± 1 °C. Some commercially available water tanks are sold with a temperature control function.

2.3 Rotarod

1. Rotarod apparatus. The apparatus is commercially available and consists of a motorized cylindrical rod that rotates with controlled speed.
   (a) The rod is about 6–10 cm in diameter and is elevated about 1 ft above the surface.
   (b) Most rotarod devices have dividers that provide multiple lanes (about 8–10 cm wide). These lanes allow the user to test more than one rat at the same time. In some devices, the lane width is adjustable and caution must be taken to keep the lane width narrow enough to avoid the rats from turning around on the rotarod and start walking backwards instead of moving forward.
   (c) Even with an apparatus that has multiple lanes, handling more than two rats at a time can be very challenging and may cause a distraction to other rats on the rotarod. This could cause rats being tested to fall off the rotarod for reasons that are obviously not related to vestibulomotor function.
   (d) The time it takes for the rat to fall off the rotarod is usually noted by a timer that is built into the device. In some devices, separate timers are connected to the floor panel below that lane, such that the user starts the timer when placing the rat on the device and when the rat falls on to the floor panel, the timer stops. In other devices, infrared beams detect the presence of the rat on the rotarod and when the rat falls to the floor panel, the timer stops as the beams detect that the rat is not there.
2. Handheld timer to time inter-trial periods.
3. 10% ethanol.
4. Sponge.
5. Video-camera (optional).

3 Methods

3.1 Animal Adaptation/Preparation

For all behavioral protocols detailed below, it is important to note the following:

1. After the arrival of the rats at the animal facility, the rats should acclimate to the new environment for 1 week. During this week, it is important for the experiment performer (denoted as user in this chapter) to become familiar with the rats and remove the stress factor associated with handling. Consistent animal handling is of great importance in behavioral experiments [30, 31]. Recommended handling for the animals is a daily interaction about 2–3 min/rat that will include picking up the rats from under their shoulders with one hand and gently keeping the other hand under their four paws. Petting them lightly and allowing them to walk on the user's arm before returning them to their cage. should be repeated a couple of times a day during the 1 week period.
2. Noise must be kept to a minimum during the entire time the behavioral experiment is taking place, from the start of the first animal until the last animal. Loud noises and changing cages can result in short term stress that might affect the behavioral outcomes being assessed [32, 33].
3. Rats must be handled according to the approved protocol procedures at the research institution as well as the funding agency, if applicable.
4. For additional information on rat handling, the reader is referred to the Guide for the care and use of laboratory animals [33] and the animal care and use courses at the American Association for Laboratory and Animal Science Library (www.aalaslearninglibrary.org).

3.2 Detecting Anxiety Using the Elevated Plus Maze (EPM)

The elevated plus maze (EPM) is a very convenient method to assess anxiety and is preferred by many investigators due to its dependence on natural behavioral preferences of the rat. It does not involve a trigger (acoustic or temperature change) nor a fearful stimulus (predator odor or foot shock) or any form of motivated or conditioned responses (levers and food rewards) for it to occur. The main natural behavior assessed is the rat's preference to stay in the dark areas (especially when anxious) and its curiosity to explore novel areas when not experiencing anxiety. As explained in materials, the EPM consists of four arms in the shape of a plus sign, the closed arms provide the dark areas where rats prefer to spend more time in the enclosed darker space when anxious. The cross section where the arms meet is the center zone and the video-tracking software uses the user-assigned area borders to calculate the number of entries into and out of each arm/zone, the time spent in each arm as well as their mobility while on the EPM among many other measures.

3.2.1 Procedure

1. All the rats to be tested should be moved from their housing area into the vicinity of the testing area. Allow the rats to adapt to the new environment for 20–30 min in their cages before starting the experiment.

2. The user should become familiar with the video-tracking software and set up the protocols ahead of time (2–3 days before the experiment starts), to avoid mistakes during the testing session.
3. All surfaces of the EPM must be cleaned with 10% ethanol solution before placing each rat. A sponge dipped in 10% ethanol, with excess solution squeezed out, i.e., not drenched, is very convenient, cleans well, easily dries up and does not have a strong distinguishable smell.
4. In addition to the software tracking and calculations of the behavior being measured, the user can also record live videos of the tests for each rat (*see* **Note 10**).
5. If being used, the clicker/mouse should be placed in a handy position; take the rat out of the cage, hold it in one hand and the clicker/mouse in the other hand. As the rat is placed on the center zone facing the open arm, the clicker should be pressed to start the program simultaneously. Alternatively, the newer versions of the software allow users to use the autostart feature once the experimenter leaves the view of the camera.
6. The experimenter needs to quickly leave the room (or stand behind the divider) and watch the rat through the computer screen.
7. Another set of rat behavioral measures that the experimenter can track through the software is rearing, grooming and head-dipping over the edge of the open arms. The user can set up the software in a manner that will allow the user to press a button and hold it down for the period of that behavior every time the rat does it. In other words, whenever the rat rears (stands on two hind paws), press R. Whenever the rat stops to groom (rubs paws on nose or hind paws on their body), press G. Whenever the rat dips his head over the edge of the arm to look at the floor, press H. The button is released whenever the rat ends that behavior. The software will then calculate the number of times and duration the rat displayed rearing, grooming and head-dipping behavior respectively.
8. The rat is allowed to explore the EPM for 5 min and the software automatically stops tracking the rat.
9. Once the session is done, the user will take the rat from the EPM, place it in the extra cage and then stop the video recording manually, if applicable (*see* **Note 11**).
10. Clean the EPM from urine and feces then wipe all surfaces with 10% ethanol and allow the alcohol to dry (*see* **Note 12**).
11. Repeat **steps 3–11** for the next rat, and then return both rats to their housing cage (*see* **Note 13**).
12. Once all rats are done, they are returned to their housing facility and the software is used to analyze the results further.

3.2.2 Precautions

1. Some EPM equipment have an additional rail at the edge of the open arms to prevent rats from falling; however that may also be a distraction and cause the rats to spend more time in the open arms than they usually would without the rail being there. If the rat does fall, it should be excluded from the analysis and cannot be rerun. The EPM allows us to measure anxiety due to it being a new environment to the rat, however, once the rat has been exposed to the EPM it is not a new environment any longer. The rat may display less anxious behavior and that will complicate the interpretation of the results.
2. All surfaces of the EPM should be smooth and clean; that includes the walking areas and the walls/sides of the EPM. Any residual smell of previous rats or minor protrusions from nails in the EPM could be a potential distraction for the rat and will cause it to spend more time in that arm for reasons not related to anxiety, thereby skewing the results.
3. There have been some reports of rats' freezing on the EPM and resulting in immobility measures that can be quantified. The user should continue with the full duration of the experiment without any interference.
4. Things that can distract the rat should be blocked, whether it is the tile on the floor, objects, or closets on the wall. White shower curtains are useful to conceal anything that might be a distraction for the rat or might make it spend more time in one of the arms versus the other.

3.2.3 Data Analysis

Overall, if the rat is displaying normal behavior, he will spend time in both the open and closed arms during the 5 min trial period. Time spent in the closed arms is usually more than the time spent in the open arms. A subset of rats will show anxiety-like behavior post TBI and will spend much less time in the open arms compared to non-anxious rats and this often represents with a significant increase in the time spent in the closed arms compared to sham rats. Figure 3 demonstrates the anxiety-like behavior using track-plots derived from the video-tracking software (ANY-maze, Stoelting Inc.) for each rat, such that TBI rats spent most of their time in the closed arms and the center zone, with very little exploration of the open arms compared to the sham rats that explored both open and closed arms.

The most common measures for comparison between the sham rats and TBI rats are the following:

1. Time spent in closed arms.
2. Time spent in open arms.
3. Number of entries into the closed arms.
4. Number of entries into the open arms.

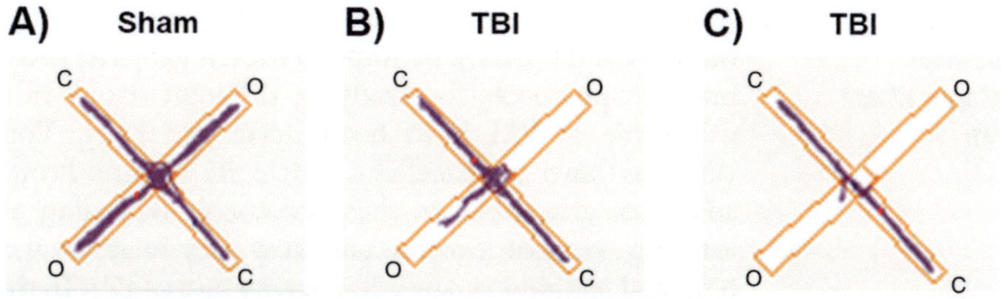

Fig. 3 Track-plots representing anxiety-like behavior post TBI on the EPM. These representative track plots were determined using the ANY-MAZE software for both sham and blast-induced TBI male Sprague-Dawley rats tested 9 days post injury. Panel (**a**) Representative track-plot of a sham rat displaying normal behavior exploring both open (O) and closed (C) arms. Panels (**b** and **c**) Representative track-plot of blast induced TBI rats displaying anxiety-like behavior by exploring the closed (C) arms more than the open (O) arms. *Orange lines* represent the perimeter of the apparatus. The *white area* represents the walking surface of the EPM. The *violet lines* represent the tracking path of the center of the animal's body over the 5 min trial period

5. Mean duration of time spent in closed arms.
6. Mean duration of time spent in open arms.
7. Distance traveled in closed arms.
8. Distance traveled in open arms.
9. Number of times and duration of rearing, grooming, and head dipping (each behavior presented separately).

Different rats may have different baseline activity, and therefore it is important to compare rats of the same group/treatment to ensure consistency of the behavior measured within that group. If baseline activity is different, the user has the option of presenting their data as a fraction or percentage instead of the raw values. The time spent in the central zone will differ for each rat, and in many cases the data is presented as a fraction value = [Total time in closed arms/(Time spent in open arm + Time spent in closed arms)]. Data can be presented by multiple measures and it is at the user's discretion to show the best representation of the rats' behavioral data. Data has also been presented as an anxiety index randing from 0 to 1 using this equation = 1 − [{(Time spent in Open arms/ Total time on maze) +, (Number of entries into Open arms/Total number of entries into any arm)}/2]. Higher numbers indicated increased anxiety-like behavior [39]. Data is often presented as mean ± standard error value for each group of animals for the measurements explained above. In some cases, minute by minute analysis can also be useful. Statistics used for EPM data analysis vary with the design of the experiment. Our data analysis was done using a one-way ANOVA with post-hoc analysis. Please refer to Carobrez and Bertoglio's review [13] and Walf and Frye's protocols [12] for more details. It is noteworthy to mention that newer methods of analysis have been developed to study exploratory behavior of rats on the EPM [34].

3.3 Detecting Cognitive Deficits Using the Morris Water Maze (MWM)

1. The Morris Water Maze was initially established for assessing memory and learning by Richard Morris [15, 16] and since then multiple protocols for studying different forms of cognition using the MWM have been developed [22]. Vorhees and Williams have an excellent article in Nature Protocols that addresses a number of these protocols regarding acquisition learning, reversal learning and how they relate to spatial mapping and learning versus reference memory [17]. In this section, protocols for the most relevant TBI-induced cognitive deficits are explained: learning memory acquisition and working memory matching-to-sample protocols. The concept behind the water maze experiment is to determine the learning memory for rats to locate a hidden platform in a water tank using distal cues that are placed around the water maze after repeated trials over a certain period of time. The integration of spatial memory is dependent on spatial cues and it is of utmost importance to keep in mind that rats can use other methods to locate the platform, such as the dependency on proximal cues or simply a strategy of taking consecutive turns to find the platform. Therefore, it is very important to study the paths of the randomized entry points to the island and to make sure that a specific learning strategy or pattern that the rats may develop is avoided.

3.3.1 Procedures

General Swimming Trial Procedure

1. All the rats to be tested should be moved from their housing area into the vicinity of the testing area. Allow the rats to adapt to the new environment for 20–30 min while they are in their cages before starting the experiment.
2. Place the distal cues in the room and keep them posted on the wall or curtain until the last day of the experiment period. Cued Navigation paradigm is an exception where the distal cues are removed.
3. Remix the water to ensure that the Tempera paint is mixed well and is opaque. Water must be clean and at temperatures between 22 and 25 °C (*see* **Note 14**).
4. Set up the video-tracking software protocols (*see* **Note 10**).
5. The clicker should be placed in a handy position, take the rat out of the cage, hold it in one hand and the clicker in the other hand. Once the rat is placed gently into the water maze and released facing the wall of the water tank at the designated entry point for that trial, the program should be started simultaneously. Alternatively, the newer versions of the software allow users to use the autostart feature once the experimenter leaves the view of the camera.
6. The user should quickly leave the room (or stand behind a divider) and watch the rat on the computer screen. Whenever the rat finds the platform/island, the software will automatically stop tracking the data for that rat, however the video camera will keep recording until the user stops it manually.

7. Each trial is 120 s long, with some exceptions noted, and the trial ends once the rat reaches the platform or the trial duration ends.
8. The handheld timer is started for 15 s with the rat sitting on the platform. This step allows the rat to remember the spatial settings for the island location (*see* **Note 15**).
9. If the trial time ends and the rat does not find the platform, then the user should guide the rat to the platform and allow it to sit on the island for the inter-trial interval of 15 s.
10. The rat is now ready for the next step/trial. This step varies depending on the MWM paradigm being used as explained below.

Acquisition Learning MWM Paradigm

Acquisition learning is a measure of how quickly the rats remember the location of the hidden platform placed in a *fixed* location over a period of 5–6 days until their performance reaches asymptote. In the example below, the location of the island is in the SE location.

1. Distal cues are present in this protocol.
2. The island location is stationary through the first 6 days of the experiment and is hidden 2 cm under the water level.
3. Each rat will have four different "general swimming trials" each day as explained in the previous section "General Swimming Trial Procedure." Each trial starts from a different entry point relative to the platform as indicated in Table 1.
4. Inter-trial intervals are 15 s on the island.

Table 1
Entry points and island location details for performing the acquisition learning paradigm in the MWM

		Entry points			
Days post-TBI	**Island location**	**Trial 1**	**Trial 2**	**Trial 3**	**Trial 4**
Day 1	SE	W	N	NE	SW
Day 2	SE	NE	W	SW	N
Day 3	SE	SW	NE	N	W
Day 4	SE	N	SW	W	NE
Day 5	SE	W	NE	N	SW
Day 6	SE	N	SW	W	NE
Day 7	No island	NW	Not applicable		

Every rat undergoes a total of 24 trials over a period of 6 days, 4 trials per day. Each trial is 120 s as explained above and inter-trial intervals are 15 s. The probe test is performed on day 7 in the absence of the island. N, E, W, S, SE, NE, SW, and NW are the arbitrary entry points assigned to the water tank (*see* Fig. 2), and the island location in this example is in the SE quadrant of the MWM

5. After the inter-trial interval is over, pick up the rat from the island with one hand and keep the clicker in the other hand. Place the rat gently into the water tank facing the tank's wall at the correct entry point and press the clicker/mouse to start the tracking program simultaneously.
6. After the fourth trial is over, it is recommended to use a towel under the rat when picking it up to avoid getting the user's clothes wet (*see* **Note 16**).
7. Place the rat in a cage dedicated for drying the rats after they finish their trials for the day. This rat cage will have a heat-warming lamp on top and towels inside it. This will allow the rat to dry up faster (10 min) and avoid the possibility of developing hypothermia.
8. Before starting the trial for the next rat, any feces that are floating should be removed since they may act as a proximal cue for the rats. This can easily be done using a skimmer pool net.
9. On day 7, the probe test is a single 60 s trial in which the island is removed from the pool. Alternatively, the island height can be reduced to almost floor level if using a retractable platform. This probe test will determine whether the rats remembered the location of the island.

Cued Navigation MWM Paradigm

Cued navigation is a paradigm that is used as a control measure to ensure that each rat tested has the necessary visual and motor skills to perform the MWM task without distal spatial cues. In this paradigm, the platform is made visible such that it extends 1 cm above the surface of the water. In many cases, the platform is made more visible using white hard foam board to cover the island and a flag or colored ball could be attached to the platform such that it is 7 cm above the island using a copper wire. Distal cues are absent in this paradigm. Each rat receives four trials to reach the visible platform in the MWM from four different entry points. Trials end after 120 s or upon finding the platform. Cued navigation is monitored and recorded as latency to reach the island using the video-tracking software ANY-maze (Stoelting Inc.). Each animal is allowed to remain on the island for 15–30 s after reaching the island then tested for the next trial after a 15–30 s interval. For more details, please refer to Vorhees and Williams reference [17].

Precautions:

1. Rats may develop motor deficits following a brain injury, however, it is important to note that motor function required for movement on land is different from that required for swimming. Nonetheless, the cued navigation is one method to establish inclusion criteria for learning and memory paradigms in which rats have demonstrated intact visual and motor skills to perform the task.

2. When using a white foam board to cover the island, the video-tracking program will find it difficult to detect the entry of the rat to the target zone because both the target zone and the rat are white. In this case, the user can manually stop the program when the rat locates the island.

Spatial Working Memory: Matching-to-Sample MWM Paradigm

The spatial working memory matching-to-sample MWM test consists of two trials. The rats undergo the first trial (trial 1) to find the location of the hidden platform followed by a second trial (trial 2) to repeat the exact same task, with a 30 s inter-trial interval where the rat sits on the island. This method determines how well the rats remember the previous task and can use their spatial working memory to recall that information from the first task. This experiment consists of 3 phases; acquisition phase (21 days), random phase (6 days) and test phase (12 days). In the random phase, the island location and entry point are different between trial 1 and trial 2. The purpose of this phase is to erase any non-spatial strategy or swimming pattern that the rats may have developed. The test phase, also called the reinstatement phase, follows the random phase and confirms that differences seen in working memory of TBI rats are solely due to spatial working memory impairment.

1. The island is hidden 2 cm below the surface of the water (this can be confirmed by placing a stick with a 2 cm mark on the island).
2. The platform location and entry point are fixed within the day's trials, but are changed from 1 day to the other.
3. Each rat will have two identical "general swimming trials" each day with an intertrial interval of 30 s such that the second trial is the matching-to-sample working memory trial as indicated in Table 2.
4. The trial ends after 120 s or upon the rat finding the platform. If the rat does not locate the platform, they should be manually guided to the platform.
5. The distal visual cues are present.

3.3.2 Data Analysis

In general, the rats are initially learning the task of finding the island during their first trial and may take a few laps around the water maze before locating the hidden platform. Within a few trials, the rats that have intact spatial memory will locate the platform more rapidly (<10 s) as shown in the track-plot diagram in Fig. 4a. On the other hand, rats with impaired cognitive function post-TBI will often take longer to find the platform (Fig. 4b), and in some cases the trial duration (120 s) ends without them finding the location (Fig. 4c). The latency to reach the island is a common measure that is used to establish whether the rat has learned to locate the hidden island. In addition to latency, the distance traveled from

Table 2
Entry points and island location details for performing the matching-to-sample spatial working memory paradigm in the MWM

	Trial 1		Trial 2	
Day	**Island location**	**Entry point**	**Island location**	**Entry point**
Acquisition phase				
1	NE	W	NE	W
2	NW	N	NW	N
3	SE	E	SE	E
4	NE	S	NE	S
5	NW	E	NW	E
6	SW	W	SW	W
7	NW	S	NW	S
8	NE	N	NE	N
9	SW	S	SW	S
10	NE	E	NE	E
11	SE	N	SE	N
12	SE	W	SE	W
13	SW	N	SW	N
14	NW	S	NW	S
15	NE	W	NE	W
16	SE	E	SE	E
17	NW	W	NW	W
18	SW	E	SW	E
19	SW	N	SW	N
20	SE	S	SE	S
21	NE	W	NE	W
Random phase				
1	SW	E	NE	E
2	SW	N	NE	N
3	NE	S	SE	S
4	NW	W	SE	W
5	SE	N	NW	N
6	SE	E	SW	E

(continued)

Table 2 (continued)

	Trial 1		Trial 2	
Day	Island location	Entry point	Island location	Entry point
Test phase				
1	NE	W	NE	W
2	NW	N	NW	N
3	SE	E	SE	E
4	NE	S	NE	S
5	NW	S	NW	S
6	NE	N	NE	N
7	SW	S	SW	S
8	NE	E	NE	E
9	SE	N	SE	N
10	SE	W	SE	W
11	SW	N	SW	N
12	NW	S	NW	S

Each trial is 120 s as explained above and inter-trial intervals are 15 s. Trial 2 is identical to trial 1. The island is hidden and its location changes every day. N, E, W, S, SE, NE, SW, and NW are the arbitrary entry points assigned to the water tank (*see* Fig. 2)

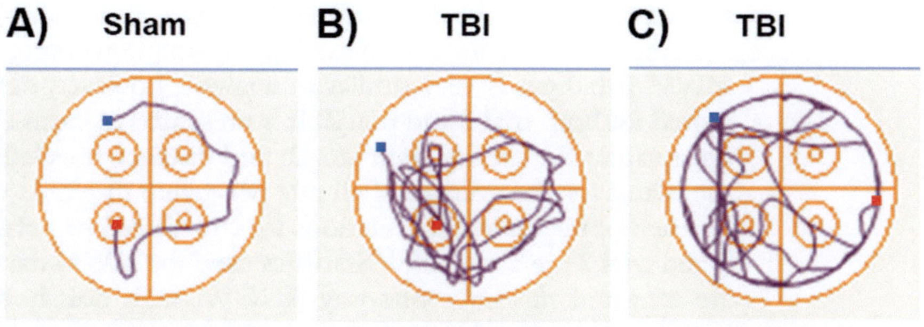

Fig. 4 Track-plots representing impaired cognitive function post TBI using the MWM. These representative track plots were determined using the ANY-MAZE software for both sham and blast-induced TBI male Sprague-Dawley rats tested on their fourth trial 1 day post injury. The hidden island was located in the SW quadrant. Panel (**a**) Representative track-plot of a sham rat displaying intact spatial memory acquisition by locating the island rapidly. Panels (**b** and **c**) Representative track-plots of blast-induced TBI rats displaying impaired spatial memory acquisition by taking longer to locate the island (**b**) or failing to locate the island (**c**). The rat in panel **c** displays a thigmotaxic swimming pattern nearby the walls of the MWM as demonstrated by the *circular tracks*. The *blue square* represents the entry point for that trial and the *red square* represents the end point upon locating the island or at the end of the 120 s trial duration

the point of entry to the location of the platform is also indicative of how well the rat remembers the location of the hidden island. The video-tracking software also calculates the average speed, the number of entries into quadrants, island target zone or the circle surrounding the island target zone as well as the average amount of time spent in each quadrant among many other measures.

The latency and the distance traveled to reach the island are dependable measures for cognitive function post TBI. TBI rats with impaired cognitive function will show increased latency and distance traveled to reach the island compared to the sham rats. Mild TBI rats rarely have a complete loss of spatial memory and therefore the data often shows a delay in memory and reaching performance asymptote. For the acquisition learning MWM paradigm, data should be presented as mean ± standard error value of the average of all four trials in one day for each group of animals. The data is plotted for each day tested. Furthermore, data from the probe test on day 7 of the MWM acquisition paradigm is presented as the number of entries into the target zone and the amount of time spent in the target quadrant where the island was located. The 60 s probe trial is often divided into two 30 s intervals for analysis. Memory retention from the probe test shows that the rats with intact cognitive function will spend more time in the correct quadrant looking for the hidden island and will cross over the island location more times than the rats with impaired cognitive function post-TBI. Interestingly, blast-injured rats using our blast TBI model showed memory retention by spending more time in the target quadrant in the second 30 s; with the first 30 s being similar to the sham rats. However, the sham rats started looking elsewhere in the second 30 s of the probe test while the blast-injured rats showed perseverance and kept searching for the platform in the target quadrant [35].

Data collected for the Matching-to-Sample working memory MWM paradigm is very similar in analysis, however, data is presented for both trial 1 and trial 2. It is presented as mean ± standard error values for latency, path length and distance traveled to reach the island for the average of all rats per group in blocks of 3 days for each of its phases. Calculations for the difference between trial 1 and trial 2 are also useful. Statistics used for MWM data analysis are repeated measures one-way ANOVA with post-hoc analysis between groups and that may vary with the design of the experiment. More details on the presentation of this working memory paradigm data and its analysis in our blast injured rat model have been published [36]. Additionally, readers could refer to Vorhees and Williams protocol for more details [17].

3.4 Detecting Vestibulomotor Deficits Using the Rotarod Apparatus

Vestibular pathology is a behavioral deficit that is detected in 30–40 % of TBI patients and is more common in blast TBI than blunt TBI [37]. Vestibular pathology affects the daily life routine of TBI patients and can prevent them from operating machinery, driving vehicles and working at elevated areas to mention a few [18, 19].

The rotarod apparatus (initially introduced by Dunham and Miya [38]) has been shown by Robert Hamm to be the most sensitive method for determining vestibulomotor function in rats following TBI. His group has shown that the rotarod could detect deficits in vestibulomotor function even when other tests such as the beam balance and beam walking procedures, appear to show normal vestibulomotor function [20, 21]. The rotarod differs from other motor functional tests in that it also tests grip strength, coordination and balance of the animal. All of which contributes to its sensitivity, such that it can detect 'low-level' injuries with mild TBI [21].

There are many versions of the rotarod protocol in the literature that differ from the initial experiments described by Hamm. These variations can be in the speed setting of the rotarod, number of trials per day and total time allowed on the rotarod during the test period. Below is a description of a modified rotarod protocol that we have found to have a reliability score of 0.97 ± 0.07, when measuring test-pre-test values on the last 2 days of performance; days 7 and 8 post-TBI.

3.4.1 Procedure

Rats are pretrained on the rotarod for 4 days prior to TBI, and the test results on the fourth day of training are chosen as the pre-TBI rotarod performance value (*see* **Note 17**). The rats are tested on the rotarod for a total of 6 days in an 8 day period post-TBI.

Pre-TBI Training

1. *Day 1* is mainly to familiarize the rats with the apparatus and the task of balancing on the rotarod. Each rat is placed on the rotarod for 30 s without any rotational movement, i.e., it is switched off (*see* **Note 18**).
2. The rotarod is then switched on at 3 rpm (very slow rotation) while the rat is on the rotarod. The goal from this session is to allow the rat to maintain balance on the rotarod for 3 min continuously. If the rat spends 3 min on the rotarod without falling, the rat is taken back to its cage. If the rat falls, he is picked up and placed on the rotarod again for another 3 min trial. A total of 10 min combined is given for these failed attempts.
3. The time spent on the rotarod is documented in seconds.
4. All the rotarod surfaces must be cleaned from urine or feces, wiped with 10% alcohol solution using a sponge and allowed to dry between each rat trial (*see* **Note 12**).
5. *Day 2*: The rats receive an initial trial at 3 rpm, three attempt trials at 10 rpm with 5 min inter-trial intervals and one final trial at 20 rpm. The main goal is to allow the rats to balance on the rotarod without falling for 3 continuous minutes. Therefore, if the rat spends 3 min on the rotarod during any of the three attempts at 10 rpm, the rat is given a 5 min rest and then given a final attempt at 20 rpm. Time spent on the rotarod for each trial is documented.

Table 3
Test trial duration and speed on the accelerating rotarod apparatus

Length of time and rotarod speed	Time on timer
Start with rat standing still on rod:	
5 s at 3 rpm	0–5 s
Increase to 10 rpm within 5 s	5–10 s
Keep at 10 rpm for 15 s	10–25 s
Increase to 15 rpm within 5 s	25–30 s
15 rpm for 15 s	30–45 s
Increase to 20 rpm within 5 s	45–50 s
20 rpm for 15 s	50 s to 1 m 5 s
Increase to 25 rpm within 5 s	1 m 5 s to 1 m 10 s
25 rpm for 15 s	1 m 10 s to 1 m 25 s
Increase to 30 rpm within 5 s	1 m 25 s to 1 m 30 s
30 rpm until 3 min total time	1 m 30 s to 3 m

6. *Days 3 and 4*: At this point, the rats are familiar with the apparatus and the task of increasing rotation on the rotarod. Each rat will undergo 3 test trials with increasing speeds (from 3 to 30 rpm) and 15 min inter-trial intervals as explained in the Table 3 (*see* **Note 19**).

Selection Criteria

1. Rotarod performance values from the pretraining day 4 are used to determine grouping selection and exclusion criteria for the next step of post-TBI testing. Rats that fall off the rotarod within the first 45 s on any two trials on day 4 are considered non-performers and are excluded from the study.
2. Next, the mean value from the three trials is taken as the representative value for time spent on the rotarod for each rat.
3. Rats are then group-matched based on their weight and pre-TBI rotarod performance.
4. Post-TBI, the rats are tested on days 1, 2, 3, 4, 7, and 8. The tests are the exact same 3 test trials with increasing speed having a 15 min inter-trial interval as explained above.

3.4.2 Data Analysis

It is very important to make sure that pre-TBI rotarod performance is comparable between the groups being studied. A good number of rats will not perform well with the accelerated speeds on the rotarod [21]. The median value ± S.E.M. of the average time each rat spent on the rotarod for all three trials on the final test day pre-TBI and on days 1–4, 7, and 8 post-TBI is calculated per group. Variability is very high between individual rats and therefore depends on the data distribution. Nonparametric statistics are recommended for non-Gaussian distribution and repeated measures one-way ANOVA for more normal distribution data when comparing rotarod performance on individual days. Rotarod performance post-TBI has been presented as the mean ± S.E.M. for all trials over the entire 5–6 days.

4 Notes

1. The lighting in the room is very important and the recommended amount of light for the open arms is 300 lx and for the closed arms is 20 lx [12].
2. For all the behavioral apparatuses: the dimensions described in this chapter for rats are different from those used for mice.
3. The video camera's autofocus or auto-tracking function must be deactivated or turned off.
4. Some protocols assign the entry of all four paws of the rat into an arm as their criteria for entry or exit into a zone. This may be applicable for manual data collection but is very difficult for the video-tracking software to accurately detect the paws with the camera placed directly above the rat.
5. Alternatively, one person could place the animal onto the EPM and another person starts the program on the computer simultaneously.
6. The tank should be uniform in appearance and color all around and free of any dents/angles/decorations or seams or marks that could become proximal visual cues for the rats. This will allow them to develop a non-spatial learning strategy to locate the island without depending on spatial visual cues.
7. For non-albino species, the water tank will have to be painted white and the water will have to be mixed with nontoxic Tempera white paint.
8. The hidden island will not be visible via the camera, and therefore, the user should find a way to confirm the island's location at the beginning of every experiment. A round heavy object can be used for this purpose (e.g., a flask weight: round lead ring with vinyl coating, 10 mm in diameter).
9. A piece of white foam board attached to a string could be used as a dummy rat to test the contrast, lighting and efficiency of the tracking software in detecting the center of the white body as it moves around.

10. Video-recording each rat is recommended; this will allow the experimenter to observe a specific rat's performance if needed or if a computer glitch occurs. If video-recording will happen, press the record button right before taking the rat out of its cage.
11. The rat should not be returned to its housing cage so that it will not startle the cage mate and cause increased anxiety right before testing the cage mate. Some protocols recommend that rats be placed in individual cages during the entire testing period. Keep in mind that rat isolation might have an acute short-lived stress effect on the rats that could add to their anxiety-like behavior.
12. Rats will urinate and defecate, especially the anxious rats, as a natural reflex to the new environment. Therefore, it is important to keep the apparatus clean. Some protocols recommend cleaning the surfaces with quatricide. Quatricide has a strong smell that can irritate the experimenter and might influence the rat's behavior. Therefore, it is recommended to use 10% ethanol to clean surfaces between each rat test and use quatricide as a final cleaning up step after the experiment is over to properly disinfect the apparatus.
13. It is very important to maintain the cage mates together and not mix up the housing conditions of the rats. Mixing the rats will very likely cause the rats to experience intruder social stress.
14. The water temperature must be checked and noted daily at the beginning of the experiment, half way through the experiment and at the end of the experiment. If the temperature is not within the acceptable range, then the user must add hot or cold water until the desired temperature is achieved and the water level readjusted to ensure the island height is correct relative to the water level.
15. The inter-trial interval varies between protocols, some use 15 s and others use 30 s. Fifteen seconds is usually a sufficient time for normal rats to remember the location and reach asymptote performance within the last 2 days of the paradigm. However, if the normal rat is not reaching asymptote performance, the inter-trial time may be increased to 30 s in an effort to help the rats learn better.
16. Water will stain clothes with a black color from the tempera paint added to the water. Users should consider dedicating specific clothes for this experiment.
17. Some protocols recommend 3 days, however, we have found better consistency and reliability in rotarod performance with 4 days of pretraining.
18. If the rat jumps off, carry it and place it back on the rotarod.

19. If a rat falls due to obvious reasons not related to motor balance and coordination (e.g., a sneeze or sudden loud noise in the environment, or the rat holds on to the side of the divider/separator and falls, or if the rat turns around and starts walking backwards), then redo the trial and consider the new value as the correct one.

Acknowledgements

The author would like to thank Kelly M. Standifer, for her support and mentorship on this project that was funded by the Department of the Army W81XWH-09-1-0443 (KMS); Larry P. Gonzalez for his expertise in behavioral sciences and training mentorship in these experimental methods; Paul Tompkins for operating the blast wave generator to induce TBI; and Charles Vorhees for his helpful correspondence regarding MWM paradigms.

References

1. Blennow K, Hardy J, Zetterberg H (2012) The neuropathology and neurobiology of traumatic brain injury. Neuron 76:886–899
2. Masel BE, DeWitt DS (2010) Traumatic brain injury: a disease process, not an event. J Neurotrauma 27:1529–1540
3. Ling GS, Ecklund JM (2011) Traumatic brain injury in modern war. Curr Opin Anaesthesiol 24:124–130
4. Bazarian JJ, Cernak I, Noble-Haeusslein L, Potolicchio S, Temkin N (2009) Long-term neurologic outcomes after traumatic brain injury. J Head Trauma Rehabil 24:439–451
5. Cernak I, Noble-Haeusslein LJ (2010) Traumatic brain injury: an overview of pathobiology with emphasis on military populations. J Cereb Blood Flow Metab 30:255–266
6. Soble JR, Spanierman LB, Fitzgerald Smith J (2013) Neuropsychological functioning of combat veterans with posttraumatic stress disorder and mild traumatic brain injury. J Clin Exp Neuropsychol 35:551–561
7. Carlson KF, Kehle SM, Meis LA, Greer N, Macdonald R, Rutks I, Sayer NA, Dobscha SK, Wilt TJ (2011) Prevalence, assessment, and treatment of mild traumatic brain injury and posttraumatic stress disorder: a systematic review of the evidence. J Head Trauma Rehabil 26:103–115
8. Nelson NW, Hoelzle JB, McGuire KA, Ferrier-Auerbach AG, Charlesworth MJ, Sponheim SR (2011) Neuropsychological evaluation of blast-related concussion: illustrating the challenges and complexities through OEF/OIF case studies. Brain Inj 25:511–525
9. Warden DL, French L (2005) Traumatic brain injury in the war zone. N Engl J Med 353:633–634
10. Silver JM, McAllister TW, Arciniegas DB (2009) Depression and cognitive complaints following mild traumatic brain injury. Am J Psychiatry 166:653–661
11. Kalueff AV (2007) Neurobiology of memory and anxiety: from genes to behavior. Neural Plast 2007:78171
12. Walf AA, Frye CA (2007) The use of the elevated plus maze as an assay of anxiety-related behavior in rodents. Nat Protoc 2:322–328
13. Carobrez AP, Bertoglio LJ (2005) Ethological and temporal analyses of anxiety-like behavior: the elevated plus-maze model 20 years on. Neurosci Biobehav Rev 29:1193–1205
14. Ponsford J, Downing M, Olver J, Ponsford M, Acher R, Carty M, Spitz G (2014) Longitudinal follow-up of patients with traumatic brain injury: outcome at 2, 5, and 10-years post-injury. J Neurotrauma 31:64–77
15. Morris R (1984) Developments of a water-maze procedure for studying spatial learning in the rat. J Neurosci Methods 11:47–60
16. Morris RG, Hagan JJ, Rawlins JN (1986) Allocentric spatial learning by hippocampectomised rats: a further test of the "spatial mapping" and "working memory" theories of hippocampal function. Q J Exp Psychol B 38:365–395
17. Vorhees CV, Williams MT (2006) Morris water maze: procedures for assessing spatial and

related forms of learning and memory. Nat Protoc 1:848–858

18. Scherer MR, Burrows H, Pinto R, Littlefield P, French LM, Tarbett AK, Schubert MC (2011) Evidence of central and peripheral vestibular pathology in blast-related traumatic brain injury. Otol Neurotol 32:571–580
19. Scherer MR, Schubert MC (2009) Traumatic brain injury and vestibular pathology as a comorbidity after blast exposure. Phys Ther 89:980–992
20. Hamm RJ (2001) Neurobehavioral assessment of outcome following traumatic brain injury in rats: an evaluation of selected measures. J Neurotrauma 18:1207–1216
21. Hamm RJ, Pike BR, O'Dell DM, Lyeth BG, Jenkins LW (1994) The rotarod test: an evaluation of its effectiveness in assessing motor deficits following traumatic brain injury. J Neurotrauma 11:187–196
22. Fujimoto ST, Longhi L, Saatman KE, Conte V, Stocchetti N, McIntosh TK (2004) Motor and cognitive function evaluation following experimental traumatic brain injury. Neurosci Biobehav Rev 28:365–378
23. Hamm RJ, White-Gbadebo DM, Lyeth BG, Jenkins LW, Hayes RL (1992) The effect of age on motor and cognitive deficits after traumatic brain injury in rats. Neurosurgery 31:1072–1077, discussion 1078
24. Reid WM, Rolfe A, Register D, Levasseur JE, Churn SB, Sun D (2010) Strain-related differences after experimental traumatic brain injury in rats. J Neurotrauma 27:1243–1253
25. Finnie J (2001) Animal models of traumatic brain injury: a review. Aust Vet J 79:628–633
26. Morganti-Kossmann MC, Yan E, Bye N (2010) Animal models of traumatic brain injury: is there an optimal model to reproduce human brain injury in the laboratory? Injury 41(Suppl 1):S10–S13
27. O'Connor WT, Smyth A, Gilchrist MD (2011) Animal models of traumatic brain injury: a critical evaluation. Pharmacol Ther 130:106–113
28. Xiong Y, Mahmood A, Chopp M (2013) Animal models of traumatic brain injury. Nat Rev Neurosci 14:128–142
29. Cernak I (2005) Animal models of head trauma. NeuroRx 2:410–422
30. Brown GM, Martin JB (1974) Corticosterone, prolactin, and growth hormone responses to handling and new environment in the rat. Psychosom Med 36:241–247
31. Kvetnansky R, Sun CL, Lake CR, Thoa N, Torda T, Kopin IJ (1978) Effect of handling and forced immobilization on rat plasma levels of epinephrine, norepinephrine, and dopamine-beta-hydroxylase. Endocrinology 103:1868–1874
32. Claassen V (1994) Neglected factors in pharmacology and neuroscience research: biopharmaceutics, animal characteristics, maintenance, testing conditions. Techniques in the behavioral and neural sciences. Elsevier, Amsterdam, New York. xiv, 486 p
33. National Research Council (U.S.) Committee for the Update of the Guide for the Care and Use of Laboratory Animals., Institute for Laboratory Animal Research (U.S.), and National Academies Press (U.S.) (2011) Guide for the care and use of laboratory animals, 8th ed. National Academies Press, Washington, DC. xxv, 220 p
34. Tejada J, Bosco GG, Morato S, Roque AC (2009) Characterization of rat behavior in the elevated plus-maze using a directed graph. J Neurosci Methods 184:251–255
35. Tompkins P, Tesiram Y, Lerner M, Gonzalez LP, Lightfoot S, Rabb CH, Brackett DJ (2013) Brain injury: neuro-inflammation, cognitive deficit, and magnetic resonance imaging in a model of blast induced traumatic brain injury. J Neurotrauma 30:1888–1897
36. Awwad HO, Gonzalez LP, Tompkins P, Lerner M, Brackett DJ, Awasthi V, Standifer KM (2015) Blast overpressure waves induce transient anxiety and regional changes in cerebral glucose metabolism and delayed hyperarousal in rats. Front Neurol 6: 132, doi:10.3389/fneuro.2015.00132. eCollection 2015
37. Gottshall K (2011) Vestibular rehabilitation after mild traumatic brain injury with vestibular pathology. NeuroRehabilitation 29:167–171
38. Dunham NW, Miya TS (1957) A note on a simple apparatus for detecting neurological deficit in rats and mice. J Am Pharm Assoc Am Pharm Assoc (Baltimore) 46:208–209
39. Cohen H, Liu T, Kozlovsky N, Kaplan Z, Zohar J, Mathe AA (2012) The neuropeptide Y (NPY)-ergic system is associated with behavioral resilience to stress exposure in an animal model of post-traumatic stress disorder. Neuropsychopharmacology 37:350–363

Chapter 32

Advanced and High-Throughput Method for Mitochondrial Bioenergetics Evaluation in Neurotrauma

Jignesh D. Pandya, Patrick G. Sullivan, Lai Yee Leung, Frank C. Tortella, Deborah A. Shear, and Ying Deng-Bryant

Abstract

Mitochondrial dysfunction is one of the key posttraumatic neuropathological events observed in various experimental models of traumatic brain injury (TBI). The extent of mitochondrial dysfunction has been associated with the severity and time course of secondary injury following brain trauma. Critically, several mitochondrial targeting preclinical drugs used in experimental TBI models have shown improved mitochondrial bioenergetics, together with cortical tissue sparing and cognitive behavioral outcome. Mitochondria, being a central regulator of cellular metabolic pathways and energy producer of cells, are of a great interest for researchers aiming to adopt cutting-edge methodology for mitochondrial bioenergetics assessment. The traditional way of mitochondrial bioenergetics analysis utilizing a Clark-type oxygen electrode (aka. oxytherm) is time-consuming and labor-intensive. In the present chapter, we describe an advanced and high-throughput method for mitochondrial bioenergetics assessments utilizing the Seahorse Biosciences XF^{e}24 Flux Analyzer. This allows for simultaneous measurement of multiple samples with higher efficiency than the oxytherm procedure. This chapter provides helpful guidelines for conducting mitochondrial isolation and studying mitochondrial bioenergetics in brain tissue homogenates following experimental TBI.

Key words Neurotrauma, Traumatic brain injury, Mitochondria, Mitochondrial bioenergetics, Brain metabolism, Oxidative phosphorylation, Therapeutic drugs preclinical screening

1 Introduction

Mitochondria are present in all eukaryotic cells and play a central role in the production of cellular energy in the form of adenosine triphosphate (ATP) through the metabolism of carbohydrates, fats, and/or proteins. In the adult brain, glucose remains the preferential energy substrate, which is converted to pyruvate via glycolysis in the cytosol. Subsequently, pyruvate is transported across mitochondrial membranes and enters into Krebs cycle for further metabolism. After the completion of the Krebs cycle, the resulting high-energy pools of reducing equivalents (i.e., NADH and $FADH_2$) transfer their electrons to the mitochondrial electron transport chain (ETC). As the mitochondrial ETC (i.e., complex I

Firas Kobeissy et al. (eds.), *Injury Models of the Central Nervous System: Methods and Protocols*, Methods in Molecular Biology, vol. 1462, DOI 10.1007/978-1-4939-3816-2_32,

through IV) transfers electrons to the final acceptor oxygen, the protons are transferred from mitochondrial matrix to the intermembrane space (IMS), generating a mitochondrial membrane potential (ΔΨm). This process of the breakdown of substrates in the presence of oxygen is known as mitochondrial respiration. When the respiration process is linked to the synthesis of ATP by complex V of the ETC, it is known as oxidative phosphorylation (OXPHOS). Additionally, mitochondria maintain redox homeostasis with antioxidant activities to scavenge reactive oxygen/nitrogen species (ROS/RNS) that are generated as byproducts of OXPHOS [1, 2].

Under physiological conditions during ATP production, mitochondria also maintain calcium (Ca^{2+}) homeostasis, and regulate mitochondrial permeability transition pore (mPTP) formation. At the expense of ΔΨm, mitochondria maintain cellular Ca^{2+} levels by the uptake of cytosolic Ca^{2+} up to a certain threshold, while limiting mPTP opening. However, excessive cytosolic Ca^{2+} influx due to excitotoxicity after neurotrauma can lead to mitochondrial Ca^{2+} overload and mPTP opening. The activation of mPTP releases many pro-inflammatory, apoptotic, and necrotic signaling factors that are responsible for cell death. Therefore, mitochondria are also considered as the "biological on/off switch" that determines the fate of cells in response to noxious stimuli [3–6].

Given its critical role in maintaining normal cellular function in the mammalian brain, mitochondrial dysfunction following TBI has been shown to be devastating for neuronal cell survival [7–10]. Following the primary mechanical insult of a TBI, the secondary injury cascades exacerbate over time, where mitochondrial bioenergetics failure occurs. Time course studies of mitochondrial bioenergetics in preclinical TBI models suggest that mitochondrial energy failure (i.e., ATP synthesis) is the key pathological event that is initiated immediately (e.g., within 30 min), and remains evident for a prolonged period of time (e.g., ≥48 h) after injury [9, 11, 12]. During this period, mitochondrial bioenergetic impairments can also lead to mitochondrial Ca^{2+} overload, loss of ΔΨm, excessive productions of free radicals, decreased levels of antioxidants, and increased oxidative stress [8, 13–19]. Notably, several drugs that target mitochondrial bioenergetics and related mitochondrial functions have shown therapeutic potential when they were administered during the post-injury excitotoxicity period. The mitochondrial function targeted therapeutics, such as mitochondrial uncouplers (e.g., FCCP, DNP) [8, 12, 20, 21], alternative food supplements (e.g., creatine, ketone bodies) [22, 23], mPTP inhibitors (e.g., cyclosporine A and its structural analog NIM811) [24–30], redox scavengers (e.g., Tempol, Phenelzine) [31–33], antioxidant (e.g., *N*-acetyl cysteine amide) [34], have shown significant efficacy in the recovery of mitochondrial functions, together with improvements in cortical tissue sparing and behavioral outcomes following experimental TBI. Therefore, mitochondrial bioenergetic parameters can serve as useful tools to evaluate early impairments in brain metabolism, and provide therapeutic targets for testing and screening preclinical novel drug therapies following brain trauma.

The traditional method of mitochondrial bioenergetic analysis employing a Clark-type oxygen electrode (aka. oxytherm) has been reported previously [35–39]. Oxytherm procedures require up to ~50–100 μg of mitochondrial protein and at least 15 min analysis time per sample. Additionally, samples can only be analyzed in a sequential manner, limiting the total sample size examined per day. Therefore, the oxytherm method offers limited throughput for the routine assessments of mitochondrial bioenergetics. In contrast, the currently available Seahorse Bioscience XF^e24 Flux Analyzer has overcome the oxytherm-related analytical limitations and provides increased efficiency for mitochondrial bioenergetic analysis. In this chapter, we describe an advanced and high-throughput method for mitochondrial bioenergetics assessment using the Seahorse analyzer [34, 40–42]. This method allows simultaneous measurement of multiple mitochondrial samples (i.e., up to 10 in duplicates) with higher sensitivity (~5 μg each) and higher efficiency (i.e., within 30 min). Therefore, the focus of this chapter is to provide detailed methods for the isolation and assessment of mitochondrial bioenergetics utilizing the Seahorse analyzer, which will be useful for investigating brain metabolism and assessing targeted drug therapies following neurotrauma.

2 Materials

2.1 Animals and Surgical Procedures

This research has been conducted at the Walter Reed Army Institute of Research (WRAIR) and at the University of Kentucky (UK). All surgical procedures for experimental TBI and protocols involving animals have been approved by individual Institutional Animal Use and Care Committees (IACUC). Male adult Sprague Dawley rats (250–300 g; Charles River Labs, Raleigh, VA, USA) were used for the experiment. Animals were housed individually under a 12 h of light–dark cycle and have access of ad libitum food and water.

2.2 Instruments Required

1. Seahorse Bioscience Flux XF^e24 Analyzer (Seahorse Bioscience, North Billerica, MA).
2. Ultracentrifuge (Beckman Coulter Optima L-100 XP Ultracentrifuge) with SW-55Ti rotor and ultracentrifuge tubes (Beckman Coulter, IN, USA).
3. Nitrogen gas based cell disruptor Instrument (model 4639; Parr Instrument Co, Moline, IL, USA).
4. BioTek Synergy HT Mx 96-well microplate reader (BioTek, Winooski, VT, USA).

2.3 Laboratory Supplies

1. Seahorse Extracellular XF^{e}24 Flux Cartridges (Seahorse Bioscience, North Billerica, MA).
2. Eppendorf tubes and microplates centrifuges (Eppendorf, Hauppauge, NY).
3. Potter-Elvehjem tissue homogenizer (Wheaton, Millville, NJ).
4. 37 °C water bath (Fisher Scientific, Hanover Park, IL).
5. Shaker (Fisher Scientific, Hanover Park, IL).
6. Tissue dissection tools (Fine Surgical Instruments Inc, Hempstead, NY).
7. Other tools and laboratory supplies, i.e., Eppendorf tubes, tips, conical tubes, reagents, chemicals (Fisher Scientific, Hanover Park, IL).

2.4 Reagents/Buffers Stocks

Chemicals for mitochondrial isolation and bioenergetics assessments were ordered from Sigma-Aldrich (St. Louis, MO). Prepare following stock solutions in advance and keep them at 4 °C for up to 6 months for routine use.

1. 3 M KCl.
2. 0.5 M EGTA, pH 7.2.
3. 1 M HEPES.
4. 1 M KH_2PO_4.
5. 1 M $MgCl_2$.
6. 20% Ficoll solution.
7. 1 M sucrose.
8. 1 M tris–HCl buffer at pH 7.2 (*see* **Note 1**).

2.5 Reagents/Buffers

1. Mitochondrial Isolation Buffer (MIB^+): MIB^+ is composed of 215 mM mannitol, 75 mM sucrose, 0.1% BSA, 20 mM HEPES, 1 mM EGTA at pH 7.2.
2. Prepare MIB without EGTA (MIB^-): MIB^- is composed of 215 mM mannitol, 75 mM sucrose, 0.1% BSA, 20 mM HEPES at pH 7.2. This buffer is used to wash away EGTA from mitochondrial samples after MIB^+ buffer was used during mitochondrial isolation procedure.
3. Respiration Buffer (RB): RB is composed of 125 mM KCl, 2 mM $MgCl_2$, 2.5 mM KH_2PO_4, 0.1% BSA, 20 mM HEPES at pH 7.2.
4. Ficoll solution: Ficoll working solutions should be prepared fresh daily. On the day of an experiment, prepare 13% Ficoll solution by mixing proportional amount of previously made stocks (22.5 ml of 20% Ficoll, 12 ml of 1 M sucrose, 37.5 μl of 0.1 M tris–HCl and 15 μl of 0.5 M EGTA).

5. Ficoll gradient: Use 13% Ficoll solution and MIB⁺ buffer to prepare 10% and 7.5% Ficoll solutions. Layer 2 ml of 7.5% Ficoll over 2 ml of 10% Ficoll into ultracentrifuge tubes (total capacity = 6 ml) to generate Ficoll density gradient.

3 Methods

3.1 Mitochondrial Isolation

1. Mitochondrial isolation procedure should be performed at 4 °C on ice and all required reagents should be kept at 4 °C during isolation and bioenergetics assessment period.
2. Quickly remove brain tissues and submerge them into homogenizer containing 10× volume of MIB^+ buffer (to prepare ~10% of homogenate).
3. Carefully avoid generating bubbles while homogenizing tissues (about 10 gentle strokes up and down per sample).
4. Centrifuge each homogenized tissue sample twice at 1300 × *g* for 3 min to obtain nuclei-free supernatant.
5. Discard pellets (which consist of cell membrane and nuclear debris) and collect supernatants in new 2 ml Eppendorf tubes.
6. Centrifuge supernatants at 13,000 × *g* for 10 min to obtain crude mitochondrial pellets.
7. Discard supernatants which consist of cytosolic fraction and myelin.
8. The crude mitochondrial pellets (consist of mitochondria plus synaptosomes and myelin) can be purified further by Ficoll density gradient based ultracentrifugation procedure.
9. Resuspend crude mitochondrial pellets in 500 μl of MIB^+ and place them into a nitrogen gas based cell disruption chamber at 1200 psi for 10 min.
10. Quickly release nitrogen gas pressure of the chamber to release trapped mitochondria in the samples from the synaptosomes.
11. Prepare Ficoll gradient in the ultracentrifuge tubes (place 2 ml of 7.5% Ficoll solution on top of 2 ml of 10% Ficoll solution).
12. Place the collected 0.5 ml mitochondrial fraction on top of preloaded discontinuous Ficoll gradient in the ultracentrifuge tube and ultracentrifuge at 100,000 × *g* for 30 min.
13. Carefully remove ultrapure mitochondrial pellets at the bottom of the tube, devoid of synaptosomal membrane and myelin debris that are separated at the interface of 10% and 7.5% Ficoll gradient.
14. Resuspend and wash the mitochondrial pellets in MIB^- buffer and centrifuge at 10,000 × *g* for 10 min to remove Ficoll and EGTA from the samples.

15. Carefully collect the washed mitochondrial protein pellets and resuspend them in MIB^- to get a final concentration no less than 10 mg/ml for each sample.
16. Determine absolute protein concentration using the bicinchoninic acid (BCA) protein assay kit (Fisher Scientific, Hanover Park, IL).

3.2 Respiratory Substrates/Inhibitors Stocks

1. Mitochondrial bioenergetics assessment chemicals were ordered from Sigma-Aldrich (St. Louis, MO). Prepare all mitochondrial substrates and inhibitors stocks concentrations before the actual day of an experiment and store them at −20 °C.
2. Pyruvate plus malate: Prepare 500 mM pyruvate plus 250 mM malate mixture stock in de-ionized water (diH_2O) containing 20 mM HEPES at pH 7.2.
3. Adenosine 5′-diphosphate (ADP): 30 mM ADP in diH_2O containing 20 mM HEPES, pH 7.2.
4. Oligomycin A: 1 mg/ml oligomycin in methanol.
5. FCCP: 1 mM FCCP in 100% ethanol.
6. Rotenone: 1 mM rotenone in 100% ethanol.
7. Succinate: 1 M succinate in diH_2O containing 20 mM HEPES at pH 7.2.

3.3 Calibration Cartridges Preparation

1. The day before the planned experiment, add 1 ml of Seahorse XF Calibrant solution (provided by Seahorse Bioscience) in each well of a 24-well dual-analyte sensor cartridge (XF^e24 Extracellular flux assay kit).
2. Place the sensor cartridge plate (with lid) overnight in a 37 °C non-CO_2 incubator.
3. Allow complete hydration and saturation of the oxygen and pH sensors that are present on the sensor cartridge overnight at 37 °C under non-CO_2 incubation.
4. On the day of the experiment, prepare working stocks of substrates/inhibitors in RB.
5. Prepare and load working stocks of pyruvate + malate + ADP (e.g., load 8× concentration in port A), oligomycin (e.g., load 9× concentration in port B), FCCP (e.g., load 10× concentration in port C), and succinate + rotenone (e.g., load 11× concentration in port D).
6. All working stocks concentrations in each port should be precalculated based on the final concentrations of substrates/inhibitors required for the experiment.
7. Fill all four injection ports (A–D) of the sensor cartridge with 75 μl of freshly prepared working stocks.

8. Once the sensor cartridge is loaded with experimental reagents, remove any top and sandwich plastic covers of the calibration cartridge before placing it in the Seahorse analyzer for an experiment. The final cartridge should contain the green top tray with all ports filled, situating on the lower tray filled with the calibrant solution.
9. Place calibration cartridge into the Seahorse XF^{e}24 Flux Analyzer for an automated calibration.
10. After the automated calibration is over, insert the sample plate immediately for mitochondrial respiration measurements.
11. After the automated calibration is over, insert the sample plate immediately for mitochondrial respiration measurements.

3.4 Sample Cartridges Preparation

1. Use XF^{e}24 cell culture microplate for mitochondrial bioenergetics assessments.
2. Out of 24 wells, assign 4 wells for blanks (50 μl of RB without mitochondria) to serve as internal control wells for temperature and oxygen level calibration.
3. Use the remaining 20 wells for mitochondrial bioenergetics assessments.
4. Dilute mitochondrial proteins in RB prior to the experiment to achieve a final concentration of 5 μg/50 μl.
5. Centrifuge the XF^{e}24 cell culture plate at 2,100 × *g* for 4 min at room temperature.
6. Pre-warm RB at 37 °C for 10 min and very gently add 475 μl of RB in all wells. The total volume of assay system will remain 525 μl before adding any reagents through ports A–D.
7. Immediately after calibration, place the sample plate into Seahorse XF ^{e}24 Flux Analyzer.

3.5 Respiration Parameters

1. Create experimental protocol for each experiment (Table 1). After calibration command in the protocol, the automated Seahorse machine follows sequential command steps as listed in Table 1, and measure mitochondrial oxygen consumption rates (OCR) in multiple samples while in the presence of various mitochondrial substrates/inhibitors (*see* **Notes 2–4**).
2. In the protocol, the OCR rates before any substrate addition are considered as basal OCR.
3. The basal OCR values should remain near 0–50. They will be higher if the isolated mitochondrial samples are not pure enough and show non-mitochondrial respiration.
4. Port A: pyruvate + malate + ADP addition measures State III OCR. The State III OCR measures complex I (NADH dehydrogenase) driven ATP synthesis rates due to coupled electron transport through ETC with OXPHOS.

Table 1
Seahorse instrument protocol for standard mitochondrial analysis

• Start command on automated Seahorse machine
• Place calibrant plate for probe calibration (~17 min)
• Place assay plate for assay (~ 20 min) and follow sequential commands given below
• Mix for 1 min
• Wait time 30 s
• Measure for 2 min (**Base line OCR**)
• Mix for 30 s
• Inject port A (pyruvate + malate + ADP)
• Mix for 30 s
• Measure for 2 min (**State III, OCR**)
• Mix for 1 min
• Inject port B (oligomycin)
• Mix for 30 s
• Measure for 2 min (**State IV, OCR**)
• Mix for 1 min
• Inject port C (FCCP)
• Mix for 30 s
• Measure for 2 min (**State V1, OCR**)
• Mix for 1 min
• Inject port D (rotenone/succinate)
• Mix for 30 s
• Measure for 2 min (**State V2, OCR**)
• End command to remove both sample plate and calibration plate and save data

5. Port B: oligomycin addition measures State IV OCR. The State IV OCR measures complex I (NADH dehydrogenase) driven mitochondrial oxygen consumption rates (in presence of mitochondrial complex V inhibitor oligomycin). The higher State IV OCR values indicate proton leak from ETC.
6. The OCR value of State III/State IV is considered as Respiratory Control Ratio (RCR). The control brain mitochondrial sample

shows RCR greater than 5 due to well coupled and healthy status of mitochondria during preparation and an assessment period.

7. Port C: FCCP addition measures State V_{FCCP} OCR. The State V_{FCCP} OCR measures complex I (NADH dehydrogenase) driven maximum respiratory capacity of mitochondria (in the presence of ETC uncoupler FCCP and Complex I substrates pyruvate + malate in assay).
8. Port D: rotenone + succinate addition measures State V_{Succ} OCR. The State V_{Succ} OCR measures complex II (Succinate dehydrogenase) driven maximum respiratory capacity of mitochondria (in presence of ETC uncoupler FCCP, Complex II substrates succinate, and complex I inhibitor rotenone).
9. The current protocol steps listed here in Table 1 measures complex I driven ATP synthesis rates (State III OCR) together with other mitochondrial respiratory parameters. Similarly, ATP synthesis rates in the presence of other mitochondrial substrates can also be measured.

3.6 Data Analysis

1. Use the XF^{e}24 (Seahorse Bioscience) wave software for data analysis.
2. Data can be viewed as oxygen consumption rates (OCR, pmol/min) for baseline, State III, State IV and State V_{FCCP} and State V_{Succ} rates.
3. Display OCR rates as group means ± standard error (SEM)/ standard deviation (SD).
4. Together, Seahorse Analyzer records oxygen tension (O_2 mmHg) data which determines oxygen level during the time of protocol operation.
5. The Seahorse XF^{e}24 analyzer creates a transient, 7 μl microchamber in specialized assay plates that allows for the determination of oxygen and proton concentrations (dual probe sensor) together in real time.
6. The results represented here in Fig. 1 shows mitochondrial respiration measured under optimized assay conditions. We have tested 2.5 μg up to 10 μg of isolated mitochondrial samples/well. The 5 μg mitochondria/well works optimally for brain mitochondrial bioenergetics assessments.
7. The experimental results in Fig. 2 shows that mitochondrial bioenergetics parameters were significantly impaired following mild traumatic brain injury (TBI). In this study, adult male Sprague Dawley rats were subjected to either a sham-operation or a mild (1.5 mm TBI) unilateral controlled cortical impact (CCI) injury. After 24 h post-TBI, mitochondria were isolated using Ficoll purification procedure. Both TBI and sham group

mitochondria (5 μg/well) were loaded on the Seahorse $XF^{e}24$ assay plate and the mitochondrial bioenergetics assay was performed in the presence of mitochondrial substrates. Data from the respiration parameter measurements indicated that the complex I (pyruvate + malate) driven State III OCR rates (ATP synthesis) and State V_{FCCP} (Maximum complex I driven respiratory capacity) respiration rates were significantly impaired ($p^* < 0.05$) following TBI as compared to the sham group, whereas no significant changes were observed in State IV OCR and State V_{Succ} OCR rates.

4 Notes

1. Adjust mitochondrial stocks and buffers to pH 7.2 with 1 M HCl or 1 M KOH. Do not use NaOH. Use paraformaldehyde free glassware during buffer preparation, mitochondrial isolation and assessment. Both NaOH and paraformaldehyde interfere with mitochondrial bioenergetics.
2. It is recommended to optimize mitochondrial protein concentration and substrate concentrations before performing an experiment. In general, the final concentration of mitochondrial substrates such as pyruvate, malate, succinate, and ADP, should be kept in such a way that they remain in excess quantity and should not be rate limiting during the experimental assay period.
3. For assessing optimal mitochondrial concentration, run mitochondrial concentration curve and pick the optimal concentration that gives linear OCR rates. It should remain below the dynamic detection limit of the instrument (OCR rates, ~2000 pmol/min). Typically, mitochondrial concentration range (0–10 μg) should be evaluated for an assay optimization. As shown in Fig. 1a, the identified optimal concentration was 5 μg for mitochondria isolated from rat brain tissues using Ficoll purification procedure.
4. Additionally, check oxygen tension (mmHg O_2) graph simultaneously during mitochondrial concentration curve optimization. This level should be maintained and remain near 150 mmHg O_2 during the entire assay condition as shown in Fig. 1b. During measurement cycle in the protocol, the oxygen tension falls transiently within the normal range, and returns to normal levels before the addition of substrates. The oxygen tension should not reach values near zero during the experimental period since it can potentially lead to mitochondrial damage due to anoxia. Reduce mitochondrial protein load per well when this occurs during assay optimization.

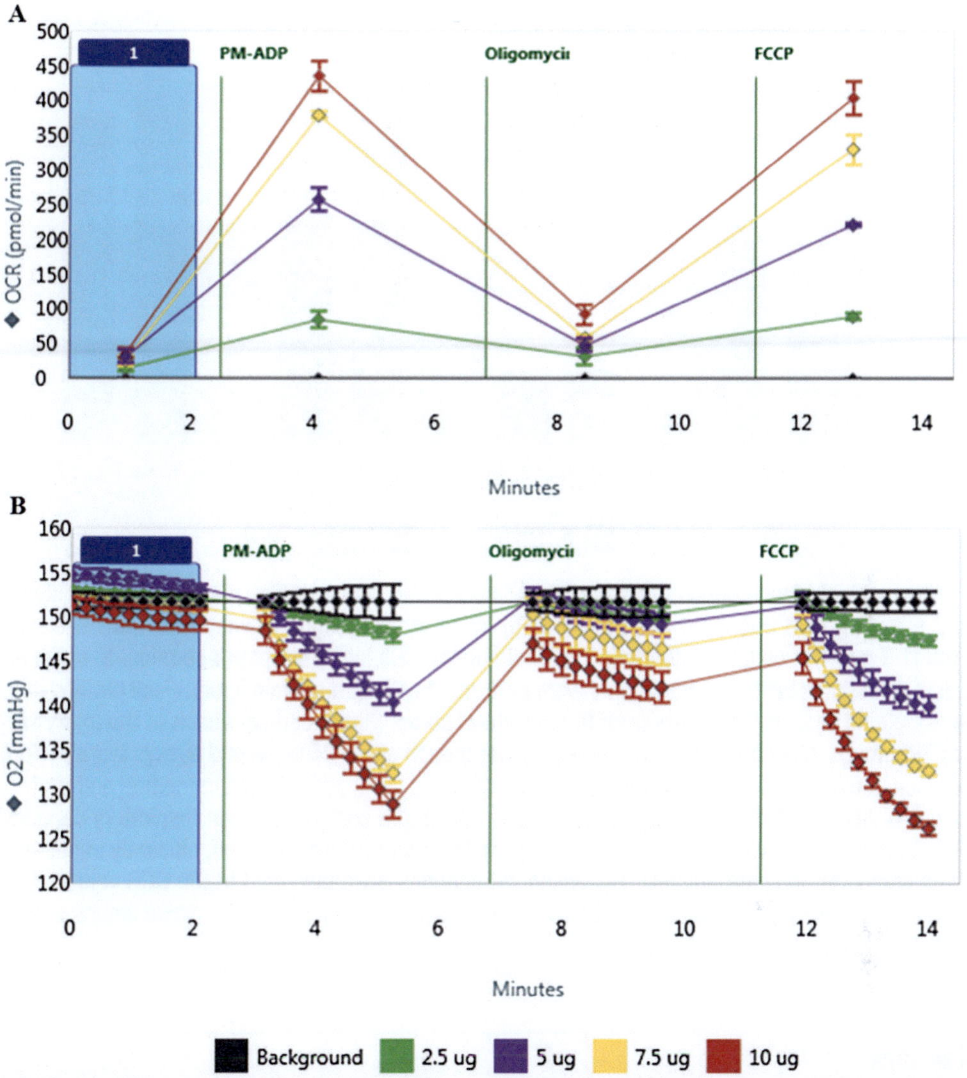

Fig. 1 Mitochondrial concentration curve optimization. Ficoll purified mitochondria were isolated from the cortex of adult male Sprague Dawley rats. Different concentration of mitochondria (2.5–10 μg) per well loaded on XF^{e}24 well and mitochondrial substrates were added through ports (**A–C**) as indicated in method and protocol. The optimal concentration range of cortical mitochondrial respiration was determined after real time measurements of Oxygen Consumption Rates (OCR) and absolute O_2 tension parameters in mitochondria incubated in micro-chambers (**A**, **B**). The OCR rates of State III (OCR rates in presence of PM-ADP substrates) and State V_{FCCP} (OCR rates in presence of FCCP) follow linear increase with increasing mitochondrial concentration (2.5–7.5 μg) and reached plateau at 10 μg. The Respiratory Control Ratio (RCR) remained higher (RCR > 5) for 5–7.5 μg of mitochondria in wells. The background wells (no mitochondria in wells) show OCR values near zero (**A**). The oxygen tension (O_2 mmHg) in the micro-chamber drops down immediately after substrates addition through ports and it reaches normal level in mitochondria containing wells. No changes were observed in background wells. The O_2 tension drops transiently after substrates addition through ports and reached back to up to background O_2 level in mitochondria (2.5–7.5 μg); however, it remains lower at 10 μg concentration (**B**). From these observations, the optimal concentration (5 μg mitochondria/well) was determined for bioenergetics assessments. Data are represented as *horizontal colored lines* with *dotted point* (group means) and error bars (SEM) in triplicates (N = 2 animals). The *vertical lines* indicate substrates addition time through ports

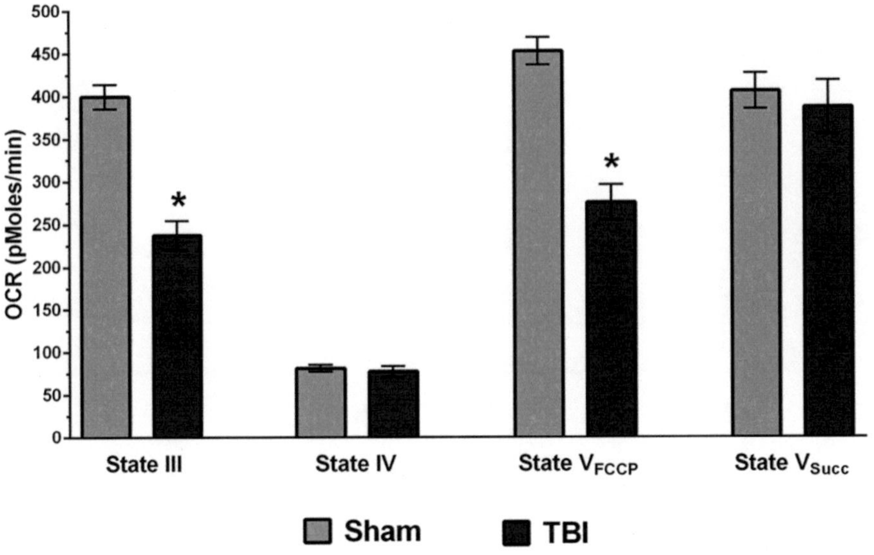

Fig. 2 Mitochondrial bioenergetics following TBI. Mitochondrial bioenergetics parameters were significantly impaired following mild traumatic brain injury (TBI). Adult male Sprague Dawley rats were subjected to a moderate (1.5 mm TBI) unilateral controlled cortical impact (CCI) injury or sham operation. In sham animals, all surgical procedures including craniotomy were performed without any brain injury. Ficoll purified mitochondria were isolated at 24 h post-injury from the ipsilateral cortex (5 mm cortical punch at the injury site) from sham or TBI groups. Mitochondria (5 μg per well) were loaded on XF[e]24 wells and assays were performed in the presence of mitochondrial substrates added through ports (**a**–**d**) as indicated in method and protocol. The OCR values of State III (ATP synthesis) and State V_{FCCP} (Maximum complex I driven respiratory capacity) indicated that complex I (pyruvate plus malate) driven mitochondrial respiration was significantly impaired following TBI as compared to sham animals. No significant changes were observed in the OCR values of State IV (proton leak) and State V_{Succ} (complex II driven maximum respiration) rates. Bars are group means ± SEM ($N = 3$ animals/group, unpaired t-test, $^*p < 0.05$)

Disclaimers

The views of the authors do not purport or reflect the position of the Department of the Army or the Department of Defense (para 4-3, AR 360-5). The authors declare that there are no conflicts of interest regarding this protocol. This research is funded by Combat Casualty Care Research Program (H_026_2014_WRAIR), and NIH/NINDS grant NS 48191 (PGS).

References

1. Berg JM, Tymoczko J, Stryer L (2002) Biochemistry, 5th edn. W. H. Freeman and Company, New York, NY
2. Chance B, Williams GR (1956) The respiratory chain and oxidative phosphorylation. Adv Enzymol Relat Subj Biochem 17:65–134
3. Crompton M (1999) The mitochondrial permeability transition pore and its role in cell death. Biochem J 341(Pt 2):233–249
4. Halestrap AP (2009) Mitochondrial calcium in health and disease. Biochim Biophys Acta 1787:1289–1290

5. Halestrap AP (2009) What is the mitochondrial permeability transition pore? J Mol Cell Cardiol 46:821–831
6. Sullivan PG, Rabchevsky AG, Waldmeier PC, Springer JE (2005) Mitochondrial permeability transition in CNS trauma: cause or effect of neuronal cell death? J Neurosci Res 79:231–239
7. Lifshitz J, Sullivan PG, Hovda DA, Wieloch T, McIntosh TK (2004) Mitochondrial damage and dysfunction in traumatic brain injury. Mitochondrion 4:705–713
8. Pandya JD, Pauly JR, Nukala VN, Sebastian AH, Day KM, Korde AS, Maragos WF, Hall ED, Sullivan PG (2007) Post-Injury Administration of Mitochondrial Uncouplers Increases Tissue Sparing and Improves Behavioral Outcome following Traumatic Brain Injury in Rodents. J Neurotrauma 24:798–811
9. Singh IN, Sullivan PG, Deng Y, Mbye LH, Hall ED (2006) Time course of post-traumatic mitochondrial oxidative damage and dysfunction in a mouse model of focal traumatic brain injury: implications for neuroprotective therapy. J Cereb Blood Flow Metabol 26:1407–1418
10. Xiong Y, Gu Q, Peterson PL, Muizelaar JP, Lee CP (1997) Mitochondrial dysfunction and calcium perturbation induced by traumatic brain injury. J Neurotrauma 14:23–34
11. Gilmer LK, Roberts KN, Joy K, Sullivan PG, Scheff SW (2009) Early mitochondrial dysfunction after cortical contusion injury. J Neurotrauma 26:1271–1280
12. Pandya JD, Pauly JR, Sullivan PG (2009) The optimal dosage and window of opportunity to maintain mitochondrial homeostasis following traumatic brain injury using the uncoupler FCCP. Exp Neurol 218:381–389
13. Ansari MA, Roberts KN, Scheff SW (2008) A time course of contusion-induced oxidative stress and synaptic proteins in cortex in a rat model of TBI. J Neurotrauma 25:513–526
14. Ansari MA, Roberts KN, Scheff SW (2014) A time course of NADPH-oxidase up-regulation and endothelial nitric oxide synthase activation in the hippocampus following neurotrauma. Free Radic Biol Med 77:21–29
15. Deng Y, Thompson BM, Gao X, Hall ED (2007) Temporal relationship of peroxynitrite-induced oxidative damage, calpain-mediated cytoskeletal degradation and neurodegeneration after traumatic brain injury. Exp Neurol 205:154–165
16. Hall ED, Bryant YD, Cho W, Sullivan PG (2008) Evolution of post-traumatic neurodegeneration after controlled cortical impact traumatic brain injury in mice and rats as assessed by the de Olmos silver and fluorojade staining methods. J Neurotrauma 25:235–247
17. Opii WO, Nukala VN, Sultana R, Pandya JD, Day KM, Merchant ML, Klein JB, Sullivan PG, Butterfield DA (2007) Proteomic identification of oxidized mitochondrial proteins following experimental traumatic brain injury. J Neurotrauma 24:772–789
18. Thompson SN, Gibson TR, Thompson BM, Deng Y, Hall ED (2006) Relationship of calpain-mediated proteolysis to the expression of axonal and synaptic plasticity markers following traumatic brain injury in mice. Exp Neurol 201:253–265
19. Reed TT, Owen J, Pierce WM, Sebastian A, Sullivan PG, Butterfield DA (2009) Proteomic identification of nitrated brain proteins in traumatic brain-injured rats treated postinjury with gamma-glutamylcysteine ethyl ester: insights into the role of elevation of glutathione as a potential therapeutic strategy for traumatic brain injury. J Neurosci Res 87:408–417
20. Sullivan PG, Dube C, Dorenbos K, Steward O, Baram TZ (2003) Mitochondrial uncoupling protein-2 protects the immature brain from excitotoxic neuronal death. Ann Neurol 53:711–717
21. Sullivan PG, Springer JE, Hall ED, Scheff SW (2004) Mitochondrial uncoupling as a therapeutic target following neuronal injury. J Bioenerg Biomembr 36:353–356
22. Sullivan PG, Geiger JD, Mattson MP, Scheff SW (2000) Dietary supplement creatine protects against traumatic brain injury. Ann Neurol 48:723–729
23. Sullivan PG, Rippy NA, Dorenbos K, Concepcion RC, Agarwal AK, Rho JM (2004) The ketogenic diet increases mitochondrial uncoupling protein levels and activity. Ann Neurol 55:576–580
24. Mbye LH, Singh IN, Sullivan PG, Springer JE, Hall ED (2008) Attenuation of acute mitochondrial dysfunction after traumatic brain injury in mice by NIM811, a non-immunosuppressive cyclosporin A analog. Exp Neurol 209:243–253
25. Readnower RD, Pandya JD, McEwen ML, Pauly JR, Springer JE, Sullivan PG (2011) Post-injury administration of the mitochondrial permeability transition pore inhibitor, NIM811, is neuroprotective and improves cognition after traumatic brain injury in rats. J Neurotrauma 28:1845–1853
26. Scheff SW, Sullivan PG (1999) Cyclosporin A significantly ameliorates cortical damage following experimental traumatic brain injury in rodents. J Neurotrauma 16:783–792
27. Sullivan PG, Rabchevsky AG, Hicks RR, Gibson TR, Fletcher-Turner A, Scheff SW (2000) Dose-response curve and optimal dosing regimen of cyclosporin A after traumatic brain injury in rats. Neuroscience 101:289–295

28. Sullivan PG, Sebastian AH, Hall ED (2011) Therapeutic window analysis of the neuroprotective effects of cyclosporine A after traumatic brain injury. J Neurotrauma 28:311–318
29. Sullivan PG, Thompson M, Scheff SW (2000) Continuous infusion of cyclosporin A postinjury significantly ameliorates cortical damage following traumatic brain injury. Exp Neurol 161:631–637
30. Sullivan PG, Thompson MB, Scheff SW (1999) Cyclosporin A attenuates acute mitochondrial dysfunction following traumatic brain injury. Exp Neurol 160:226–234
31. Deng-Bryant Y, Singh IN, Carrico KM, Hall ED (2008) Neuroprotective effects of tempol, a catalytic scavenger of peroxynitrite-derived free radicals, in a mouse traumatic brain injury model. J Cereb Blood Flow Metabol 28: 1114–1126
32. Singh IN, Gilmer LK, Miller DM, Cebak JE, Wang JA, Hall ED (2013) Phenelzine mitochondrial functional preservation and neuroprotection after traumatic brain injury related to scavenging of the lipid peroxidation-derived aldehyde 4-hydroxy-2-nonenal. J Cereb Blood Flow Metabol 33:593–599
33. Singh IN, Sullivan PG, Hall ED (2007) Peroxynitrite-mediated oxidative damage to brain mitochondria: protective effects of peroxynitrite scavengers. J Neurosci Res 85: 2216–2223
34. Pandya JD, Readnower RD, Patel SP, Yonutas HM, Pauly JR, Goldstein GA, Rabchevsky AG, Sullivan PG (2014) N-acetylcysteine amide confers neuroprotection, improves bioenergetics and behavioral outcome following TBI. Exp Neurol 257:106–113
35. Pandya JD, Grondin R, Yonutas HM, Haghnazar H, Gash DM, Zhang Z, Sullivan PG (2015) Decreased mitochondrial bioenergetics and calcium buffering capacity in the basal ganglia correlates with motor deficits in a nonhuman primate model of aging. Neurobiol Aging 36:1903–1913
36. Pandya JD, Nukala VN, Sullivan PG (2013) Concentration dependent effect of calcium on brain mitochondrial bioenergetics and oxidative stress parameters. Front Neuroenerg 5:10
37. Sullivan PG, Krishnamurthy S, Patel SP, Pandya JD, Rabchevsky AG (2007) Temporal characterization of mitochondrial bioenergetics after spinal cord injury. J Neurotrauma 24:991–999
38. Clark LC Jr, Helmsworth JA, Kaplan S, Sherman RT, Taylor Z (1953) Polarographic measurement of oxygen tension in whole blood and tissues during total by-pass of the heart. Surg Forum 4:93–96
39. Clark LC Jr, Wolf R, Granger D, Taylor Z (1953) Continuous recording of blood oxygen tensions by polarography. J Appl Physiol 6:189–193
40. Patel SP, Sullivan PG, Pandya JD, Goldstein GA, VanRooyen JL, Yonutas HM, Eldahan KC, Morehouse J, Magnuson DS, Rabchevsky AG (2014) N-acetylcysteine amide preserves mitochondrial bioenergetics and improves functional recovery following spinal trauma. Exp Neurol 257:95–105
41. Rogers GW, Brand MD, Petrosyan S, Ashok D, Elorza AA, Ferrick DA, Murphy AN (2011) High throughput microplate respiratory measurements using minimal quantities of isolated mitochondria. PLoS One 6:e21746
42. Sauerbeck A, Pandya J, Singh I, Bittman K, Readnower R, Bing G, Sullivan P (2011) Analysis of regional brain mitochondrial bioenergetics and susceptibility to mitochondrial inhibition utilizing a microplate based system. J Neurosci Methods 198:36–43

Chapter 33

Determination of Vascular Reactivity of Middle Cerebral Arteries from Stroke and Spinal Cord Injury Animal Models Using Pressure Myography

Mohammad A. Anwar and Ali H. Eid

Abstract

Stroke and other neurovascular derangements are main causes of global death. They, along with spinal cord injuries, are responsible for being the principal cause of disability due to neurological and cognitive problems. These problems then lead to a burden on scarce financial resources and societal care facilities as well as have a profound effect on patients' families. The mechanism of action in these debilitating diseases is complex and unclear. An important component of these problems arises from derangement of blood vessels, such as blockage due to clotting/embolism, endothelial dysfunction, and overreactivity to contractile agents, as well as alteration in endothelial permeability. Moreover, the cerebro-vasculature (large vessels and arterioles) is involved in regulating blood flow by facilitating auto-regulatory processes. Moreover, the anterior (middle cerebral artery and the surrounding region) and posterior (basilar artery and its immediate locality) regions of the brain play a significant role in triggering the pathological progression of ischemic stroke particularly due to inflammatory activity and oxidative stress. Interestingly, modifiable and non-modifiable cardiovascular risk factors are responsible for driving ischemic and hemorrhagic stroke and spinal cord injury. There are different stroke animal models to examine the pathophysiology of middle cerebral and basilar arteries. In this context, arterial myography offers an opportunity to determine the etiology of vascular dysfunction in these diseases. Herein, we describe the technique of pressure myography to examine the reactivity of cerebral vessels to contractile and vasodilator agents and a prelude to stroke and spinal cord injury.

Key words Brain injury, Spinal cord injury, Stroke, Pressure myography, Middle cerebral artery, Basilar artery, Spinal cord injury, Middle cerebral artery, Peripheral arterial disease

1 Introduction

Brain injuries and related neuropsychological and cognitive deficits remain the focus of interest for academic and pharma research departments, health care professionals, as well as national and international policy makers. Stroke is ranked second as the most prominent cause of global mortality, following ischemic heart disease [1–3]. Stroke is subdivided into two broad categories, ischemic and hemorrhagic, the latter being inclusive of ischemic and

Firas Kobeissy et al. (eds.), *Injury Models of the Central Nervous System: Methods and Protocols*, Methods in Molecular Biology, vol. 1462, DOI 10.1007/978-1-4939-3816-2_33, © Springer Science+Business Media New York 2016

subarachnoid hemorrhages. Ischemic stroke is more common and accounts for 80 % of the total cases of stroke patients. Clinically, the middle cerebral artery (MCA) is found to be most often occluded in stroke patients. This is largely due to the fact that it arises from the internal carotid artery as well as due to its anatomical branching further downstream. Inherently, the bifurcation geometry of the cerebral arteries predicts a risk for disturbed flow and hence a future potential for pathological milieu [4].

The incidence of spinal cord injuries due to complications from surgery for aortic aneurysms gives rise to spinal cord ischemia caused by a prolonged aortic cross-clamping time, or ligation of intercostal or lumber arteries [5]. This has an effect on thoracic, lumbar, or sacral regions and may lead to paraplegia (hind limb paralysis). Victims of vehicle and sporting accidents, in addition to victims of gunshot wounds, may also suffer from paraplegia as a consequence of spinal cord injuries (SCIs). Altogether, this has a significant impact on the peripheral circulation and the cerebrovascular system. Indeed, subjects with SCI are at an increased risk prediction of cardiovascular disturbances which appear to be the cause of 40 % of deaths in these SCI individuals [6]. Contextually, the normalized intima-media thickness for the structural inward remodeling of lower extremity arteries (superficial femoral and popliteal arteries) is greater in SCI individuals than the control group. This is indicative of subclinical prevalence of lower limb peripheral arterial disease (PAD) [7]. Interestingly, a greater incidence of lower limb PAD associated with higher burden of stroke has also been reported [8]. The aforementioned literature offers an explanation for the observation that patients with SCI are at an increased risk of future ischemic stroke [9]. Among survivors, stroke-driven brain damage is the foremost cause of disability and almost half of them remain permanently disabled [10]. Brain (stroke and vascular dementia) and spinal cord injuries are a cause of significant neurological or psychological deficits, and are significant contributors to the burden of illness, including health care costs, loss of productive years, and reduced life-span (Germany—2006–2025: Euros109 billion, U.S.A.—2005–2050: $2.2 trillion) [11, 12].

1.1 Cerebral Circulation

Blood is transported to the cerebral circulation via the internal carotid artery and vertebral arteries. This latter network coalesces into the basilar artery, which anastomoses with other communicating connections of the internal carotid arteries to constitute the circle of Willis [13, 14]. As part of the anterior cerebral circulation, the middle cerebral artery (MCA) is integral to the regulation of blood flow. Any form of MCA dysfunction may lead to loss of auto-regulatory control [4, 15].

The cerebral circulation has distinctive features, particularly as related to the distribution of resistance within its vessels. Indeed, large cerebral arteries and pial arterioles (short in length) contribute to approximately 55 % of total vascular resistance. Another 35 % of vascular resistance is distributed over the neurovascular unit, which comprises arterioles (short in length), capillaries, and post-capillary venules located in the brain parenchyma. The remaining portion of the resistance resides in the venular network [15]. However, an exquisitely coordinated cellular cross talk between the components of the vasculature (endothelial and vascular smooth muscle cells), circulating blood cells (erythrocytes and leukocytes), neurons, astrocytes, and glial cells regulate pressure and cerebral blood flow (autoregulation), which is maintained relatively constant over a moderate range of arterial blood pressures [4, 15]. This is critical for maintenance of brain homeostatic processes.

Hindbrain or posterior circulation occlusion is the most devastating form of ischemic strokes. Abnormalities associated with cerebral and spinal cord circulation have been implicated in stroke: ischemic [4, 15], hemorrhagic [16], SCI [17, 18], and other cerebral-vascular diseases such as vascular dementia [19]. Our knowledge on this vertebrobasilar vasculature region certainly lags behind our understanding of the pathophysiological mechanisms of MCA stroke [20, 21]. Yet, 25 % of all strokes are located in this domain. The neurovascular compartment of this region is unique compared to the forebrain, in terms of histology, anatomy, and neuronal makeup [21]. Therefore, further studies are warranted to address this void and tailor any personalized therapeutic approaches targeted at specified regions within the brain in prevention and recovery from stroke.

1.2 Risk Factors

Over the years, an accumulating list of risk factors that impair cerebrovascular function have been reported to cause stroke (Fig. 1), including inflammatory markers, biomarkers [raised levels of C-reactive protein (CRP), tumor necrosis factor-α (TNF-α), interleukin 6 (IL-6)], elevated total plasma homocysteine, leukocytes, and abnormal blood coagulation [1, 22]. Approximately 70% of stroke-related afflictions are apportioned to these modifiable risk factors [23]. Directly targeting the major traditional risk factors (Fig. 1) for stroke such as smoking, poor diet, hypertension, and hyperlipidemia has led to the prevention and management of this debilitating disease [24, 25]. However, human stroke is an etiologically heterogeneous disorder, and its multi-factor mechanistic nature calls for advanced approaches such as monotherapy for multiple targets, the use of polytherapies, or combinatorial therapy [10]. Further, comorbid studies, particularly in relation to regional differences in brain (anterior versus posterior), are required for complete comprehension of the heterogeneity that exists within stroke patients.

1.3 Cerebrovascular Signalopathy

The underlying mechanism for stroke and spinal cord injury is far from clear and not properly characterized [4, 15, 26]. In stroke-related patients, the only therapeutic modality with a moderately

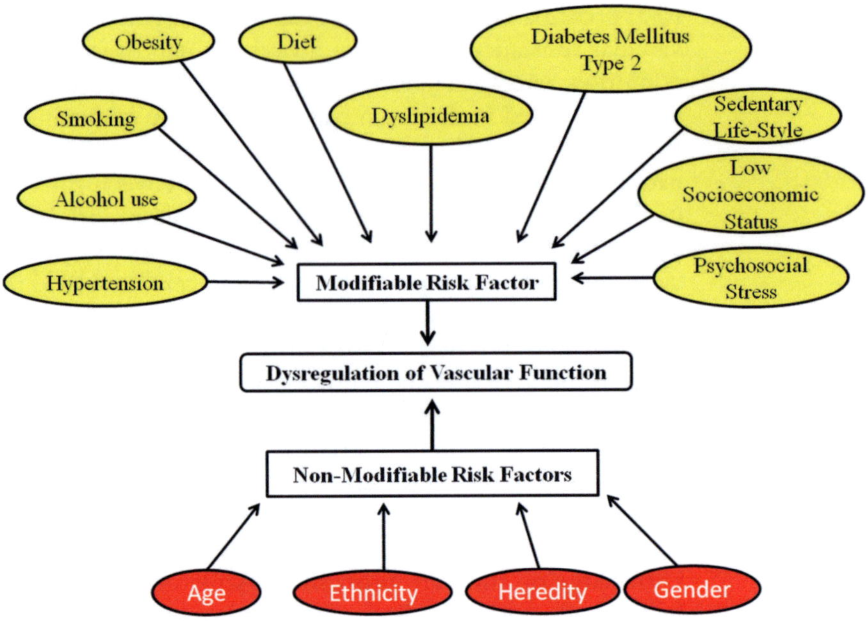

Fig. 1 Modifiable and non-modifiable risk factors for stroke and spinal cord injury, leading to vascular dysfunction

successful record has been recombinant tissue-type plasminogen activator (rt-PA, Alteplase) and even this has a success rate only when applied within a limited time frame of 4.5 h following the onset of stroke [27, 28]. Evidently, there is an urgent need to translate any novel, successful potential therapy from the bench to the bedside for stroke patients [4, 15]. Again, in animal models, application of these new therapeutic agents can be assessed for normalization/restoration of vascular function by use of pressure myography. This tool would definitely help our understanding of the effective mechanism of action, and hence therapeutic approach.

Cardiovascular risk factors studied in combination synergistically interact to accelerate vascular dysfunction as the cerebrovasculature is markedly predisposed to any homeostatic malfunctions. It must be stressed that our knowledge in this area of brain and SCI research is limited by the paucity of data [15].

The endothelial cell layer is at the interface between the circulating blood and the surrounding tissues (smooth muscle cells, adventitial layer, astrocytes, and neurons) (*see* Fig. 2). Therefore, it should not come as a surprise that endothelial impairment promotes stroke [15, 29] and spinal cord injuries [17, 30]. Indeed, activation of endothelial nitric oxide synthase (eNOS) releases nitric oxide, which regulates vascular tone [31]. However, cardiovascular derangement ensues if the activity/expression of eNOS pathway is impaired. This appears to be the most common transduction route for cardiovascular dysregulation throughout the vascular network, including stroke [15].

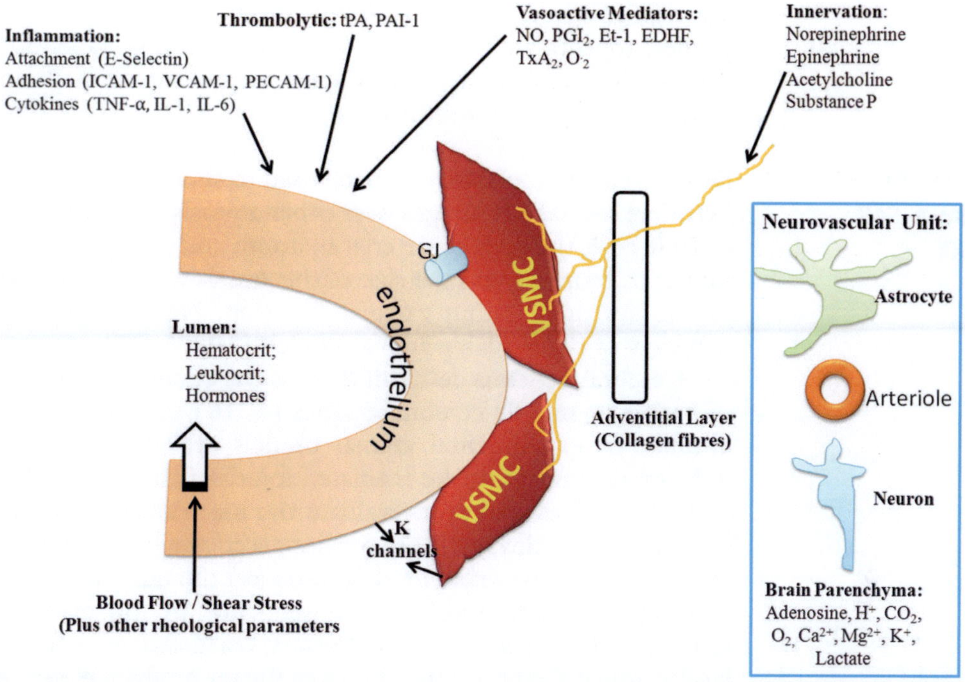

Fig. 2 A cartoon displaying components of an artery: endothelial cell, vascular smooth muscle cell (VSMC), and part of the adventitial layer, including innervation. Also, showing constituents of the neurovascular unit (*right-hand side*): arteriole, neuron, and astrocyte. *GJ*: gap junction, *tPA* tissue plasminogen activator, *PAI-1* plasminogen activator inhibitor, *AII* angiotensin II, *Vasp* vasopressin, K^+ potassium ions, *TNF-α* tumor necrosis factor – α, *IL-6* interleukin 6, *IL-1* interleukin 1, *ICAM-1* intercellular adhesion molecule 1, VCAM-1:; PECAM-1:; E-selectin

Other mechanistic pathways that may be affected during progress of stroke include renin-angiotensin system, oxidative stress (NADPH oxidase, mitochondria, and cyclooxygenases), inflammatory pathways (nuclear factor-κB, cytokines, toll-like receptors, and matrix metalloproteinases), and cell death pathways (necrosis, apoptosis, and autophagy), rho-kinase/ROCK and ion channels (e.g. calcium and potassium channels) [4, 15, 26]. Evidently, similar mechanistic pathways are implicated in SCI [30, 32, 33]. On the other hand, antioxidants, interleukin-10 (IL-10), dimethylarginine dimethylaminohydrolase (DDAH), and peroxisome proliferator-activated receptor-γ (PPARγ) appear to be neurovascular protective [15]. In addition, stroke-mediated vascular insult develops into cerebral inflammation as a result of recruitment and stimulation of white blood cells [22], which secrete proinflammatory molecules (TNF-α, IL-6). Hypertension is a predominant risk factor for stroke and it has been reported that macrophage depletion of stroke-prone spontaneously hypertensive rats (SPSHR) normalizes endothelial function and mitigates MCA remodeling [34].

A clear understanding of all the known triggers, modulators, mediators, and others, as yet to be discovered, may help ameliorate the risk of stroke and SCI. Arising from these pharmacological studies, appropriate targets will be identified and therapeutic strategies can be applied at an early stage to prevent the pathological state and restore the healthy one.

1.4 Rat Models for Stroke, SAH, and SCI

Different animal species have been used to simulate stroke, but the rat offers several advantages over other animals, including the reasonable cost, the cerebro-arterial anatomy, and physiology which is similar to humans, and the size of the vessels which is manageable for studies quantitating vascular function. Figure 3 provides an overview of the available animal models [10, 35–37].

Cerebral ischemia ensues if there is any circulatory impairment of basilar or middle cerebral arteries [4, 15]. Luckily, there are a number of experimental animal models available to assess the induced severity of stroke-mediated injuries [10, 35, 36, 38] and SCI [37, 39] as well as to evaluate the mechanism of the pathophysiological pathways. From previous sections, it would be advantageous to use comorbid models to dissect the unique mechanistic pathway that may arise as a consequence of the experimental insults, such as middle cerebral artery occlusion or embolism of basilar artery. Certainly, the choice of the rat model will depend on the paradigm to be addressed, particularly with regard to the mechanism or signaling pathway of interest and ongoing research

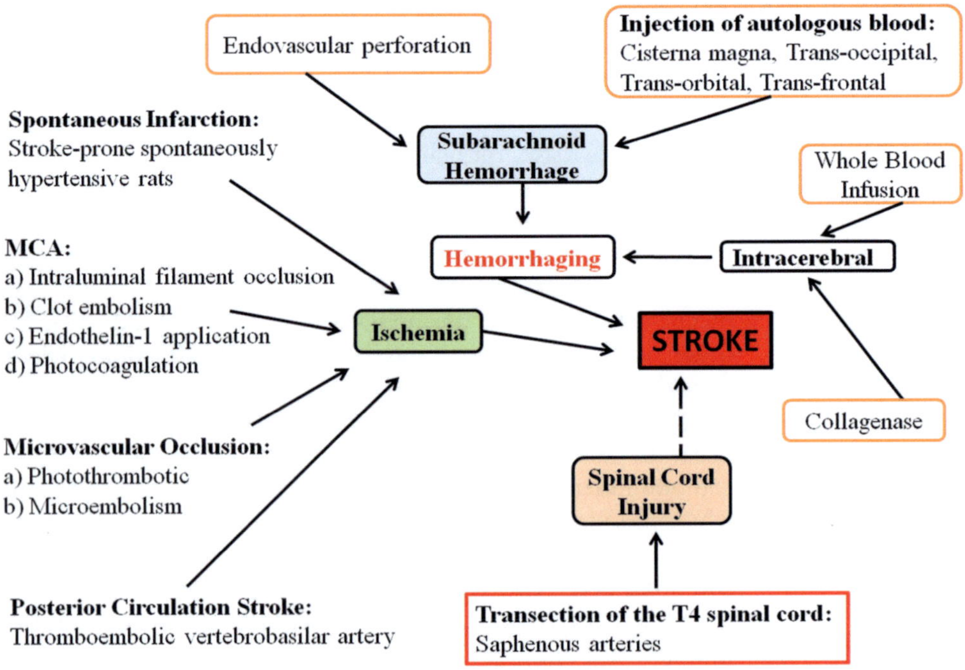

Fig. 3 An overview of the rat models used for induction of stroke (ischemic and hemorrhagic) and spinal cord injury

projects at the time of experimentation. Also, the extent of ischemic injury can be determined by noninvasive techniques such as magnetic resonance imaging (MRI) and positron-emission tomography (PET) [36, 40].

1.5 Different Systems Are Available for Assaying Vascular Function of Isolated Vessels

There are several commercially available setups for evaluating physio-pharmacological functions of isolated vessels, including organ bath systems for large vessels (rings and strips of arteries and veins), wire myographs (small- to medium-size vessels, ≥ 60 μm internal diameter), and pressure myograph arrangements (micro- to small-size vessels ≤ 200 μm) [41, 42]. Changes in intravascular pressure are more physiological since it reflects the transmural pressure, which in turn is intimately related to the myogenic response and longitudinal extension. Therefore, the pressure myograph system is more suitable for characterizing the properties of arteries and veins, and closely simulates the in vivo behavior of the specified vessels. Various modalities, such as the lifestyle factors illustrated in Fig. 1, can be examined as the basic protocol, followed by the use of different stroke models as given in Fig. 2. Below, we describe the use of a pressure myograph system (Danish Myo Technology (DMT)'s Model 110P; Aarhus, Denmark) to characterize pharmacological responses (vascular reactivity) of rat isolated middle cerebral artery from healthy control animals.

2 Materials

2.1 Preparation of Solution

1. Physiological saline solution (PSS) (in mM): 119 NaCl, 4.7 KCl, 2.5 $CaCl_2 \cdot 2H_2O$, 1.17 $MgSO_4 \cdot 7H_2O$, 25 $NaHCO_3$, 1.18 KH_2PO_4, 0.027 Na_2EDTA, and 5.5 glucose, maintained at 37 °C; bubbled with 5 % CO_2 in 95 % O_2 to achieve a pH of 7.4.
2. Potassium PSS (KPSS): Exchange the sodium chloride in PSS with an equimolar concentration of potassium chloride.

2.2 Instruments for Vessel Isolation

1. Dissecting stereomicroscope (Zeiss Stemi 2000).
2. Cold Light Source/Fiber Optic Illumination (NCL 150, Volpi).
3. Micro-surgical scissors (Sigma-Aldrich).
4. Micro-forceps with sharp ends (Sigma-Aldrich).
5. Sylgard-coated glass petri dishes (Danish Myo Technology; Denmark).
6. Dissection pins (0.1, 0.2, and 0.5 mm).
7. Spring scissors (12.5 cm long, 9 mm, straight).
8. Multi-purpose Iris scissors (11 cm long, straight).

2.3 Instruments for Studying Vessel Reactivities

2.3.1 Pressure Myograph System: DMT 110P Model

This pressure myograph system is an excellent apparatus that can be used to determine structural and functional features/changes of small blood vessels under conditions that are near-physiological. Among other parameters, diameter of these isolated vessel sections can be determined in response to various physiological or pharmacological stimuli. Importantly, this system encompasses a built-in heating unit that maintains the chamber temperature. For an easier cleaning of the setup (*see* **Note 14**), DMT makes the chamber of acid-resistant stainless steel material. Continuous recording of measurements is performed by a computer-assisted dimension analysis software called MyoView™.

2.3.2 MyoView Software

MyoView™ is a software-based interface that can be used for analysis of studies involving vascular reactivities of small blood vessels mounted on the DMT 110P model above. Using MyoView™, the researcher can perform analysis as well as utilize data logging entities for recording dynamic events. Importantly, one can also design, implement, and save experimental protocols, which can be accessed and used at any time in the future.

2.4 Animals

1. Male Wistar or Sprague-Dawley rats (250–350 g body weight) maintained in standard cages in a thermo-regulated (22 ± 1 °C) colony room with fixed photoperiod of 12-h light and 12-h darkness, humidity of 60%, and access to water and rodent chow ad libitum (*see* **Notes 1** and **2**).

3 Methods

3.1 Determination of Vascular Reactivity in Rat Isolated Middle Cerebral Artery

1. Rats are anesthetized with halothane and killed by cervical dislocation with an iron bar (*see* **Note 3**).
2. Brains are rapidly removed and placed in ice-cold, oxygenated PSS.
3. Segments of right and left MCAs (~200 μm internal diameter) are gently dissected from the surface of the brain (*see* **Note 4**); both segments are placed in ice-cold, oxygenated PSS contained in a 10 cm glass Petri dish layered at the bottom with SYLGARD 184 elastomer. Arterial segments of each branch are to be held at proximal and distal ends with dissecting pins (*see* **Note 5**). The segments are cleaned of any adherent connective tissue and cut into rings of 2–3 mm in length (*see* **Note 6**).
4. The three-way valves, metal pipes, silicone tubing, and glass cannulas are filled with cold PSS (5% CO_2/95% O_2) using a 10 ml syringe at a stable, low flow rate so as to avoid damaging the sensitive pressure transducers and ensure that all air bubbles are removed (*see* **Note 7**). Subsequently, three-way inlet and outlet valves are closed.

5. Loosely tie single strands of nylon suture on each of the cannulas. Place MCA ring in myograph unit, mount the proximal end of the artery (*see* **Note 5**), with the help of fine forceps, on the right-hand cannula, and securely ligate with a single strand of the nylon suture already on the cannula. Again using the inlet valve, carefully and gently wash/flush the inside of the artery with ice-cold PSS to ensure the removal of any blood or residual material, and then close the inlet valve.
6. With the help of micro-adjustment positioners (horizontal, vertical, and longitudinal), carefully align and bring the right cannula close to the left cannula. Again, with the assistance of fine micro-forceps gently glide the vessel over the left cannula, taking into consideration as not to stretch the artery, and ligate with the loosely secured suture already on the cannula. At either end of the artery, cut off any extraneous thread beyond the suture knots (Fig. 4).
7. Fill the myograph chamber with 10 ml of PSS. To ensure that the vessel is intact and does not leak, carefully inject cold PSS through the artery using a 10 ml syringe attached to the three-way inlet valve. If any leak is found (*see* **Note 8**), tie off any side branches or repeat the above steps (from step 3 onwards).
8. Install the myograph unit on the inverted microscope stage (Zeiss AX10, Vert.A1). Turn on the gas cylinder (95 % O_2/5 % CO_2) to equilibrate the PSS (400 ml) in the myo-interface and the chamber in the myograph unit. Ensure that the inlet on the three-way valve is closed; connect the valve with tube P1 from the PSS reservoir, allow the solution and any bubbles to pass

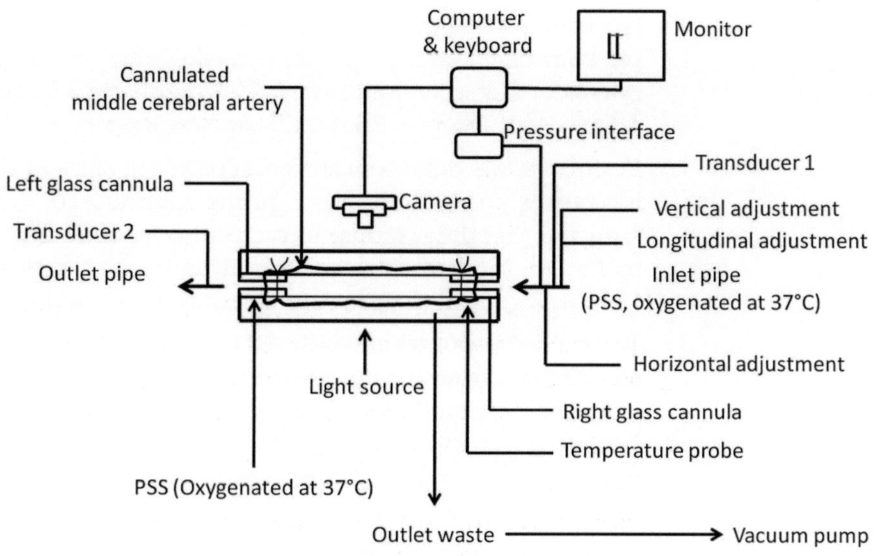

Fig. 4 Pictorial representation of perfusion myograph setup

through the tube, and ensure that no bubbles remain or solution is free of any bubbles (*see* **Note 7**). Open the valve.

9. At the distal end of the vessel, insert tube P2 from the pressure regulator onto the left valve of the myograph unit, ensuring that there are no bubbles or leakage anywhere along the system (*see* **Notes 7** and **8**).
10. Turn on the pump (ensuring that the flow is at off) either at the myo-interface or via MyoView II (software), (*see* **Note 9**).
11. Using MyoView, press collect, and artery appears on the monitor. Adjust the focus for optimum resolution to record, monitor, and assay the vessel geometry (lumen, diameter, and wall thickness).
12. Increase the intra-luminal pressure by incremental steps of 10 mmHg till 100 mmHg by selecting P1 pressure (inlet valve) either at the myo-interface or Myoview, ensuring at each stage that the artery remains straight and parallel along the horizontal axis of cannulas (avoiding any contortions/buckling) by means of micropositioners (vertical or longitudinal—for any strain or tension in the vessel) (*see* **Note 10**).
13. Equilibrate the artery at 70 mmHg for 1 h (37 °C) and wash vessel every 15 min with pre-warmed PSS during equilibration.
14. To "awaken" the artery in the myograph chamber, empty the vessel bath by pressing the vacuum pump on and add 10 ml of pre-warmed 3 μM of phenylephrine to the chamber, followed by thorough washes till baseline pressure is reached. This is followed by two exposures to 125 mM KPSS, again with comprehensive washing with PSS in between and subsequent to the final application of KPSS till baseline pressure is attained (*see* **Note 11**).
15. A relaxation response to acetylcholine (10 μM) stimulation of pre-constricted arteries (3 μM phenylephrine) is recognized as evidence of intact endothelium (~90%) (*see* **Note 12**). Wash every 10 min with PSS (3×, 30 ml) till baseline pressure is stable.
16. Perform cumulative concentration response curves (CCRC) to increasing concentrations of phenylephrine (10^{-10}–10^{-4} M), ensuring that the response to each concentration is stable prior to the addition of subsequent concentration. Wash with PSS (10 ml, 3×) until the pressure is restored to baseline.
17. Vasomotor responses are analyzed from the arterial lumen diameter changes from the pre-recorded tracings and as a percentage from the baseline pressure. These are used to determine EC_{50} (logarithm of the concentration producing 50% of the maximum response and E_{Max}) (the maximum constriction).
18. To examine the functional response of the endothelium (*see* **Note 13**), pre-constrict the MCA with phenylephrine (10 μM) and carry out a CCRC to increasing concentrations of acetylcholine

(10^{-10}–10^{-4} M), again taking into consideration that a plateau is reached before the application of higher concentration.

19. Clean the myobath system via the myograph unit (*see* **Note 14**).
20. From pre-recorded tracings, calculate the lumen diameter for each concentration of acetylcholine added, which is expressed as percentage change from the phenylephrine derived constriction.
21. Draw cumulative concentration response curves using Prism (GraphPad Software, La Jolla, California, USA), using iterative nonlinear regression analysis fitted to a sigmoid equation. Using the data analysis automatically generated by the software, the EC_{50} and E_{Max} values are calculated.

4 Notes

1. Rodents (rats and mice) are increasingly being used for experiments on brain- and spinal cord-related injuries. The mouse offers unique opportunities for using a broad array of transgenic and gene-deficient models to address various paradigms [43–45].
2. The gender, age, and weight of the animals, including the species and the strain, in addition to the vendor and their location must be clearly stated in the methods section of all manuscripts.
3. The various models used to induce stroke, subarachnoid hemorrhage, and spinal cord injury use different anesthetics (halothane, isoflurane, ketamine, propofol, and xenon). These are considered to be not only neurovascular protective [46–49], but may also have a profound influence on the blood pressure, blood gasses, body temperature, and ultimately the outcome and severity of the induced stroke [10, 47]. Therefore, the duration of the applied anesthesia is of considerable importance and must be time limited and closely monitored.
4. Internal arterioles of very small diameter are difficult to handle and may cause problems for unexperienced vascular practitioners.
5. The endothelium is highly sensitive to the direction of flow. Therefore, the proximal end (in situ, in vivo) of the artery must face the inflow (inlet end) of PSS; otherwise the endothelium may get damaged.
6. This in vitro setup of the isolated artery avoids any confounding factors such as circulating blood cells, neurons, and astrocytes.
7. Air bubbles are considered the bane of this technique and therefore it is imperative to ensure that all air bubbles have been removed from the perfusion circuit (pathway of PSS), particularly at tubing connections and prior to actual start of experiment. This cannot be overemphasized; it is to make sure that the endothelium is not damaged and hence endothelium-dependent functional responses can be tested.

8. It is essential to express that there should be no leakage in the perfusion circuit.
9. Vessel dimensions can be examined at a wide range of intravascular pressures (or if using the flow circuit, at broad range of shear stresses) to comprehend the pathophysiological mechanisms.
10. For the duration of the experiment, the intrinsic topology of the vessel is maintained (three-dimensional shape and length).
11. Agonists and antagonists applied on the abluminal (adventitial) surface.
12. Endothelium-independent (denuded vessels) or endothelium-mediated (intact vessels) actions/activities can be determined.
13. Arterial segments can be denuded of endothelium through a number of ways to assess endothelium-independent responses, namely by the passage of air bubbles through the vessel lumen, use of human or horse hair, a hypotonic solution (water), and a stainless steel wire covered in cotton wool (less than the internal diameter) which is repeatedly passed (using to-and-fro motion) through the vessel lumen [50, 51]. In all cases, the removal of endothelium is confirmed by the loss of relaxation response (70%) to acetylcholine (10 μM).
14. Clean the myograph instrument following the vendor's (Danish Myo Technology, DMT) instructions.

Acknowledgements

This publication was made possible by grant # NPRP 4-571-3-171 and NPRP 5-409-3-112 from the Qatar National Research Fund (a member of Qatar Foundation). The statements made herein are solely the responsibility of the authors.

Conflict of Interest
There are no potential conflicting interests to declare in the preparation of this chapter.

References

1. Mackay J, Mensah G (2004) The atlas of heart disease and stroke, 1st edn. World Health Organization, Geneva
2. Krishnamurthi RV, Feigin VL, Forouzanfar MH, Mensah GA, Connor M, Bennett DA, Moran AE, Sacco RL, Anderson LM, Truelsen T, O'Donnell M, Venketasubramanian N, Barker-Collo S, Lawes CM, Wang W, Shinohara Y, Witt E, Ezzati M, Naghavi M, Murray C (2013) Global and regional burden of first-ever ischaemic and haemorrhagic stroke during 1990-2010: findings from the Global Burden of Disease Study 2010. Lancet Glob Health 1:e259–e281
3. Feigin VL, Forouzanfar MH, Krishnamurthi R, Mensah GA, Connor M, Bennett DA, Moran AE, Sacco RL, Anderson L, Truelsen T, O'Donnell M, Venketasubramanian N, Barker-Collo S, Lawes CM, Wang W, Shinohara Y, Witt E, Ezzati M, Naghavi M, Murray C (2014) Global and regional burden of stroke during 1990-2010: findings from the

Global Burden of Disease Study 2010. Lancet 383:245–254
4. Moskowitz MA, Lo EH, Iadecola C (2010) The science of stroke: mechanisms in search of treatments. Neuron 67:181–198
5. Hollier LH, Money SR, Naslund TC, Proctor CD Sr, Buhrman WC, Marino RJ, Harmon DE, Kazmier FJ (1992) Risk of spinal cord dysfunction in patients undergoing thoracoabdominal aortic replacement. Am J Surg 164:210–213, discussion 213-214
6. Garshick E, Kelley A, Cohen SA, Garrison A, Tun CG, Gagnon D, Brown R (2005) A prospective assessment of mortality in chronic spinal cord injury. Spinal Cord 43:408–416
7. Bell JW, Chen D, Bahls M, Newcomer SC (2011) Evidence for greater burden of peripheral arterial disease in lower extremity arteries of spinal cord-injured individuals. Am J Physiol Heart Circ Physiol 301: H766–H772
8. Criqui MH, Langer RD, Fronek A, Feigelson HS, Klauber MR, McCann TJ, Browner D (1992) Mortality over a period of 10 years in patients with peripheral arterial disease. N Engl J Med 326:381–386
9. Wu JC, Chen YC, Liu L, Chen TJ, Huang WC, Cheng H, Tung-Ping S (2012) Increased risk of stroke after spinal cord injury: a nationwide 4-year follow-up cohort study. Neurology 78:1051–1057
10. Macrae IM (2011) Preclinical stroke research--advantages and disadvantages of the most common rodent models of focal ischaemia. Br J Pharmacol 164:1062–1078
11. Flynn RW, MacWalter RS, Doney AS (2008) The cost of cerebral ischaemia. Neuropharmacology 55:250–256
12. Olesen J, Gustavsson A, Svensson M, Wittchen HU, Jonsson B (2012) The economic cost of brain disorders in Europe. Eur J Neurol 19:155–162
13. Levick RJ (2009) An introduction to cardiovascular physiology, 5th edn. CRC Press Taylor and Francis Group, Boca Raton, FL
14. Cipolla M (2009) The cerebral circulation. In: Granger DN, Granger JP (eds) Colloquium series on integrated systems physiology: from molecular to function to disease. Morgan & Claypool Life Sciences, San Rafael, CA, pp 1–59
15. Faraci FM (2011) Protecting against vascular disease in brain. Am J Physiol Heart Circ Physiol 300:H1566–H1582
16. Sehba FA, Hou J, Pluta RM, Zhang JH (2012) The importance of early brain injury after subarachnoid hemorrhage. Prog Neurobiol 97:14–37
17. Popa C, Popa F, Grigorean VT, Onose G, Sandu AM, Popescu M, Burnei G, Strambu V, Sinescu C (2010) Vascular dysfunctions following spinal cord injury. J Med Life 3:275–285
18. Sharma HS (2011) Early microvascular reactions and blood-spinal cord barrier disruption are instrumental in pathophysiology of spinal cord injury and repair: novel therapeutic strategies including nanowired drug delivery to enhance neuroprotection. J Neural Transm 118:155–176
19. Gorelick PB, Scuteri A, Black SE, Decarli C, Greenberg SM, Iadecola C, Launer LJ, Laurent S, Lopez OL, Nyenhuis D, Petersen RC, Schneider JA, Tzourio C, Arnett DK, Bennett DA, Chui HC, Higashida RT, Lindquist R, Nilsson PM, Roman GC, Sellke FW, Seshadri S (2011) Vascular contributions to cognitive impairment and dementia: a statement for healthcare professionals from the american heart association/american stroke association. Stroke 42:2672–2713
20. Henninger N, Eberius KH, Sicard KM, Kollmar R, Sommer C, Schwab S, Schabitz WR (2006) A new model of thromboembolic stroke in the posterior circulation of the rat. J Neurosci Methods 156:1–9
21. Lekic T, Ani C (2012) Posterior circulation stroke: animal models and mechanism of disease. J Biomed Biotechnol 2012:587590
22. Rodrigues SF, Granger DN (2014) Leukocyte-mediated tissue injury in ischemic stroke. Curr Med Chem 21:2130–2137
23. Hankey GJ (2006) Potential new risk factors for ischemic stroke: what is their potential? Stroke 37:2181–2188
24. McArthur K, Lees KR (2010) Advances in emerging therapies 2009. Stroke 41:e67–e70
25. Bos MJ, Koudstaal PJ, Hofman A, Ikram MA (2014) Modifiable etiological factors and the burden of stroke from the Rotterdam study: a population-based cohort study. PLoS Med 11:e1001634
26. Filosa JA, Iddings JA (2013) Astrocyte regulation of cerebral vascular tone. Am J Physiol Heart Circ Physiol 305:H609–H619
27. Lees KR, Bluhmki E, von Kummer R, Brott TG, Toni D, Grotta JC, Albers GW, Kaste M, Marler JR, Hamilton SA, Tilley BC, Davis SM, Donnan GA, Hacke W, Allen K, Mau J, Meier D, del Zoppo G, De Silva DA, Butcher KS, Parsons MW, Barber PA, Levi C, Bladin C, Byrnes G (2010) Time to treatment with intravenous alteplase and outcome in stroke: an updated pooled analysis of ECASS, ATLANTIS, NINDS, and EPITHET trials. Lancet 375:1695–1703

28. Dhillon S (2012) Alteplase: a review of its use in the management of acute ischaemic stroke. CNS Drugs 26:899–926
29. Volpe M, Iaccarino G, Vecchione C, Rizzoni D, Russo R, Rubattu S, Condorelli G, Ganten U, Ganten D, Trimarco B, Lindpaintner K (1996) Association and cosegregation of stroke with impaired endothelium-dependent vasorelaxation in stroke prone, spontaneously hypertensive rats. J Clin Invest 98:256–261
30. Fassbender JM, Whittemore SR, Hagg T (2011) Targeting microvasculature for neuroprotection after SCI. Neurotherapeutics 8:240–251
31. Toda N, Ayajiki K, Okamura T (2009) Cerebral blood flow regulation by nitric oxide: recent advances. Pharmacol Rev 61:62–97
32. Mautes AE, Weinzierl MR, Donovan F, Noble LJ (2000) Vascular events after spinal cord injury: contribution to secondary pathogenesis. Phys Ther 80:673–687
33. Sinescu C, Popa F, Grigorean VT, Onose G, Sandu AM, Popescu M, Burnei G, Strambu V, Popa C (2010) Molecular basis of vascular events following spinal cord injury. J Med Life 3:254–261
34. Pires PW, Girgla SS, McClain JL, Kaminski NE, van Rooijen N, Dorrance AM (2013) Improvement in middle cerebral artery structure and endothelial function in stroke-prone spontaneously hypertensive rats after macrophage depletion. Microcirculation 20:650–661
35. Durukan A, Tatlisumak T (2007) Acute ischemic stroke: overview of major experimental rodent models, pathophysiology, and therapy of focal cerebral ischemia. Pharmacol Biochem Behav 87:179–197
36. Hossmann KA (2008) Cerebral ischemia: models, methods and outcomes. Neuropharmacology 55:257–270
37. Cheriyan T, Ryan DJ, Weinreb JH, Cheriyan J, Paul JC, Lafage V, Kirsch T, Errico TJ (2014) Spinal cord injury models: a review. Spinal Cord 52:588–595
38. Strbian D, Durukan A, Tatlisumak T (2008) Rodent models of hemorrhagic stroke. Curr Pharm Des 14:352–358
39. Rummery NM, Tripovic D, McLachlan EM, Brock JA (2010) Sympathetic vasoconstriction is potentiated in arteries caudal but not rostral to a spinal cord transection in rats. J Neurotrauma 27:2077–2089
40. Durukan A, Strbian D, Tatlisumak T (2008) Rodent models of ischemic stroke: a useful tool for stroke drug development. Curr Pharm Des 14:359–370
41. Anwar MA, Schwab M, Poston L, Nathanielsz PW (1999) Betamethasone-mediated vascular dysfunction and changes in hematological profile in the ovine fetus. Am J Physiol 276:H1137–H1143
42. Eid AH, Maiti K, Mitra S, Chotani MA, Flavahan S, Bailey SR, Thompson-Torgerson CS, Flavahan NA (2007) Estrogen increases smooth muscle expression of alpha2C-adrenoceptors and cold-induced constriction of cutaneous arteries. Am J Physiol Heart Circ Physiol 293:H1955–H1961
43. Kuraoka M, Furuta T, Matsuwaki T, Omatsu T, Ishii Y, Kyuwa S, Yoshikawa Y (2009) Direct experimental occlusion of the distal middle cerebral artery induces high reproducibility of brain ischemia in mice. Exp Anim 58:19–29
44. Myers SA, DeVries WH, Andres KR, Gruenthal MJ, Benton RL, Hoying JB, Hagg T, Whittemore SR (2011) CD47 knockout mice exhibit improved recovery from spinal cord injury. Neurobiol Dis 42:21–34
45. Cheng MH, Lin LL, Liu JY, Liu AJ (2012) The outcomes of stroke induced by middle cerebral artery occlusion in different strains of mice. CNS Neurosci Ther 18:794–795
46. Kawaguchi M, Furuya H, Patel PM (2005) Neuroprotective effects of anesthetic agents. J Anesth 19:150–156
47. Braeuninger S, Kleinschnitz C (2009) Rodent models of focal cerebral ischemia: procedural pitfalls and translational problems. Exp Transl Stroke Med 1:8
48. Traystman RJ (2010) Effect of anesthesia in stroke models. In: Dirnagl U (ed) Rodent models of stroke. Humana Press, Totowa, NJ, p 285
49. Altay O, Hasegawa Y, Sherchan P, Suzuki H, Khatibi NH, Tang J, Zhang JH (2012) Isoflurane delays the development of early brain injury after subarachnoid hemorrhage through sphingosine-related pathway activation in mice. Crit Care Med 40:1908–1913
50. Osol G, Cipolla M, Knutson S (1989) A new method for mechanically denuding the endothelium of small (50-150 microns) arteries with a human hair. Blood Vessels 26:320–324
51. Anwar MA, Ford WR, Herbert AA, Broadley KJ (2013) Signal transduction and modulating pathways in tryptamine-evoked vasopressor responses of the rat isolated perfused mesenteric bed. Vascul Pharmacol 58:140–149

Chapter 34

Assessment of Basilar Artery Reactivity in Stroke and Subarachnoid Hemorrhage Using Wire Myograph

Crystal M. Ghantous, Zeina Azrak, Farah Abdel Rahman, Hana A. Itani, and Asad Zeidan

Abstract

Blood flow regulation of normal cerebral arteries is a critical and important factor to supply the brain tissue with nutrients and oxygen. Stroke insult results in a disruption or reduction in cerebral arteries' blood flow with subsequent brain tissue damage. Hemorrhagic stroke is one type of stroke and accounts for about 13% of all of stroke insults. In this type of stroke, the cerebral artery breaks open and causes bleeding in or surrounding the brain. Subsequently, this bleeding causes blood vessels to constrict in a process called vasospasm, in which the vessels narrow and impede the blood flow to brain tissue. Hemorrhagic stroke is the major cause of prolonged constriction of cerebral arteries. This leads to partial brain damage and sometimes death in patients with aneurysmal subarachnoid hemorrhage. Among the key delicate techniques to assess small blood vessel functionality is the wire myograph, which can be utilized in several cerebral injury models including stroke. The wire myograph is a device that provides information about the reactivity, stiffness, and elasticity of small blood vessels under isometric conditions. In this book chapter, we describe the techniques involved in wire myography assessment and the different measures and parameters recorded; we describe the utility of this technique in evaluating the effects of subarachnoid hemorrhage on basilar artery sensitivity to different agonists.

Key words Basilar artery, Wire myograph, Vessel viability, Cumulative concentration-response curves

1 Introduction

1.1 Structure of Blood Vessels

The wall of most blood vessels consists of three layers: the inner layer (intima) made of endothelial cells; the muscle layer (media), which is the thickest part and composed of vascular smooth muscle cells (VSMC); and the outer layer (adventia) containing fibroblasts and connective tissue. The main function of VSMC is to control blood flow, primarily by regulating the diameter of small resistance vessels. The VSMC can sense and react to changes in serum composition, extracellular fluid composition (due to bleeding), and intraluminal blood pressure [1].

Firas Kobeissy et al. (eds.), *Injury Models of the Central Nervous System: Methods and Protocols*, Methods in Molecular Biology, vol. 1462, DOI 10.1007/978-1-4939-3816-2_34,

1.2 Microvasculature Structure and Function

It is well known that the main drop in hydrostatic pressure occurs at the level of resistance arteries (small blood vessels); these vessels have a crucial role in the regulation of blood pressure and blood flow. Resistance arteries comprise the small resistance arteries (<300 μm of lumen diameter), arterioles (<100 μm of lumen diameter), and capillaries (about 7 μm of lumen diameter) [2, 3]. Thus, any structural alteration that takes place at the level of these microvessels will have an impact on the regulation of blood pressure and blood flow. An increase in blood contraction or the thickness of the arterial wall with a reduced lumen causes an increase in vascular resistance [4, 5]; this alteration may define the remodeling process involved in blood remodeling. Poiseuille's law states that resistance is inversely proportional to the radius to the fourth power, which makes vascular resistance highly sensitive to small alterations in the lumen of arteries [6].

1.3 Tone of Cerebral Arteries

Being a highly metabolically active organ, the brain relies heavily on a constant delivery of nutrients and oxygen, whether in a state of wakefulness or sleep. Blood flow to the brain is maintained by the tone of cerebral arteries [7]. Tone is regulated by vasoconstriction or vasodilation; when the VSMCs of cerebral arteries contract, the arteries constrict and reduce blood flow to the region of the brain they supply. This could result in an immediate neural damage [8] which may be preceding the pathophysiology of certain neurological diseases, such as Alzheimer's disease or stroke [8].

1.4 Cerebral Artery Function

Cerebral arteries supply blood to the brain (Fig. 1). The main arteries that originally irrigate the brain are the anterior, middle, and posterior cerebral arteries. The anterior cerebral artery (ACA) brings blood to the medial surface of the frontal and parietal lobes, the anterior section of the basal ganglia and internal capsule, the olfactory bulb and tract, and the anterior part of the corpus callosum [9]. The middle cerebral artery (MCA), the largest of the cerebral arteries, irrigates almost all of the basal ganglia, the posterior and anterior internal capsules, and the lateral sides of the temporal, frontal, and parietal lobes [10]. The posterior cerebral artery (PCA) supplies most of the hypothalamus, thalamus, midbrain, splenium of the corpus callosum, medial part of the temporal lobe (including the hippocampal formation), posterior medial parietal lobe, and the medial and inferior parts of the occipital lobe [11].

The large cerebral arteries contribute significantly to the total cerebral vascular resistance, and their tone considerably determines the overall blood flow and the microvascular pressure [12]. Thus, abnormal changes in the tone of these arteries could contribute to the development of pathological conditions observed in several neurological conditions [12].

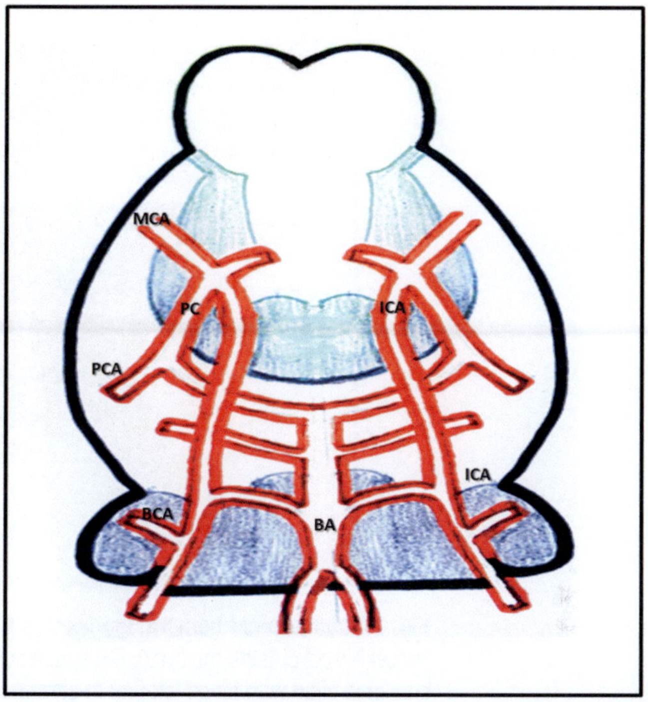

Fig. 1 Schematic diagram of the ventral rat brain surface showing the location of the basilar artery. *MCA* middle cerebral artery, *PC* posterior communicating artery, *ICA* internal carotid artery, *PCA* posterior cerebral artery, *BA* basilar artery

1.5 Cerebrovascular Diseases Caused by Changes in Cerebral Artery Tone

Maintaining a normal cerebrovascular tone is crucial for sustaining a sufficient perfusion to the brain. When blood flow to the brain is reduced, the lack of oxygen and nutrients may cause neurons to die. Thus, disruptions in tone contribute to the pathophysiology of several neurological conditions, like Alzheimer's disease and hemorrhagic and ischemic stroke. In this section, these conditions are discussed.

1.5.1 Hemorrhagic Stroke

Intracerebral hemorrhage occurs when a weak blood vessel ruptures inside the brain. When blood starts to leak, the pressure inside the brain increases causing damage to the surrounding milieu. An adaptive mechanism, although an unsafe complication, is constriction of the involved cerebral artery and vasospasm [13]. These changes in the vascular tone reduce the blood flow to the part of the brain supplied by the injured artery, increasing the risk of ischemic stroke occurrence [14].

In the case of subarachnoid hemorrhage (Fig. 2), a blood vessel bursts outside the brain, causing a buildup of blood between the pia mater and the arachnoid membrane [13]. Vasospasm also occurs in this case and may also cause neuronal damage and ischemic stroke [14, 15].

Shortly after a hemorrhagic stroke, cerebral autoregulation is impaired, making cerebral vessels incapable of responding adequately

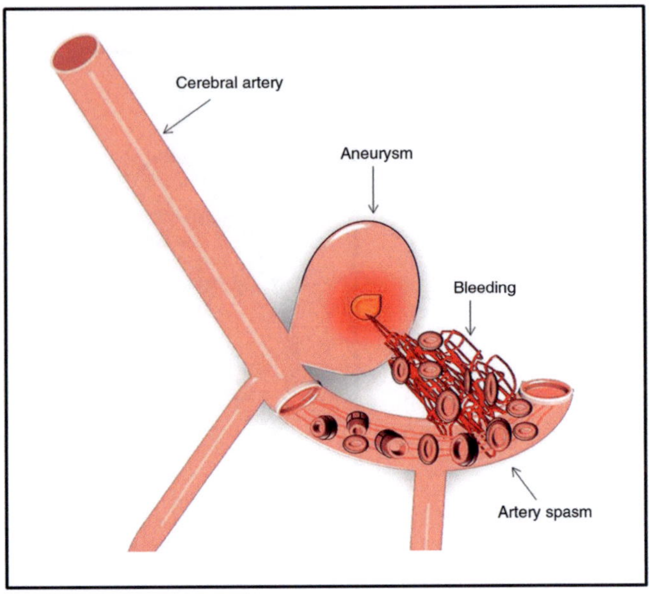

Fig. 2 Subarachnoid hemorrhage leading to cerebral artery spasm. When blood vessel bursts outside the brain, the space surrounding the brain (the subarachnoid space) is filled with blood leading to cerebral blood vessel vasospasm (narrowing)

to changes in transmural pressure [16]. Moreover, vasoconstrictor receptors, like those for serotonin and endothelin, are upregulated [17]. These events, in turn, could contribute to subsequent cerebrovascular conditions like ischemia or hemorrhage.

1.5.2 Ischemic Stroke

Ischemic stroke takes place when a cerebral artery is obstructed, either by atherosclerosis or blood clots forming thrombosis or embolism [13]. When the blood flow to the particular part of the brain supplied by the blocked artery becomes drastically reduced, the lack of adequate oxygen supply causes neuronal damage and death.

In the acute phase after ischemic stroke, the VSMC and endothelial cells become less responsive [18, 19]. In the chronic phase, atherosclerosis may occur due to an upregulation in angiogenic and apoptotic factors in the endothelium [20]. Both of these events decrease blood flow to the brain, inducing even further damage and possibly increasing the risk of a secondary stroke.

1.5.3 Alzheimer's Disease

Although Alzheimer's disease develops as a result of neuronal injury induced by the accumulation of β-amyloid peptide [21], vasculopathy and changes in cerebrovascular tone also contribute to an increased susceptibility to this disease [22–25]. For example, changes in vascular tone caused by hypertension or atherosclerosis increase the risk for Alzheimer's disease [26]. Moreover, the β-amyloid peptide has been shown to be associated with vasculopathy by upregulating endothelial responses [27] and impairing cerebrovascular autoregulation [28].

1.6 Regulation of Cerebral Artery Tone

Regulating the tone of cerebral arteries is essential in maintaining sufficient cerebral blood flow. When the perfusion pressure increases in the brain, the cerebral vessels constrict in order to maintain a relatively stable blood flow [29]. This change in resistance is largely caused by the myogenic tone, which significantly affects autoregulation of the blood flow in the brain. Several mechanisms that govern regulation of the cerebrovascular tone include calcium-induced vasoconstriction, calcium sensitization, excitation-contraction coupling, endothelial responses, and potassium channels. Thus, understanding these mechanisms will pave the way to developing therapeutic strategies in preventing the pathological events that arise from a disruption in cerebrovascular tone.

1.6.1 Calcium-Mediated VSMC Contraction

VSMC contraction is dependent on calcium. When agonists that promote contraction, such as angiotensin II, endothelin I, or norepinephrine bind to their G-protein-coupled receptors, phospholipase C (PLC) is activated. This enzyme catalyzes the conversion of phosphatidylinositol 4,5-bisphosphate (PIP_2) into the secondary messengers inositol trisphosphate (IP_3) and diacylglycerol (DAG).

IP_3 binds to ligand-gated calcium channels on the sarcoplasmic reticulum, resulting in calcium release into the cytosol. Calcium then binds to calmodulin to form a complex that activates myosin light chain (MLC) kinase, which in turn phosphorylates myosin light chain. Phosphorylated light chain of myosin then binds to the actin filament in a crossbridge and causes contraction. Hence, the degree of VSMC contraction is dependent on the level of myosin phosphorylation.

To stop the contraction of VSMC, the enzyme MLC phosphatase is activated. In turn, the MLC phosphatase removes the phosphate group of the light chain of myosin and thus terminates the VSMC contraction [30]. Of note, calcium also enters VSMC from extracellular stores through ligand-gated calcium channels in the plasma membrane.

DAG also promotes VSMC contraction by activating protein kinase C (PKC), which in turn phosphorylates specific proteins like L-type calcium channels that further regulate the crossbridge cycle. The L-type calcium channels are voltage-gated channels that open in response to membrane depolarization, which is brought on by stretch of the VSMC.

1.6.2 Calcium Sensitization via the RhoA/ROCK Pathway

When the blood pressure against VSMC increases, mechanotransduction results in the activation of the RhoA/ROCK pathway, which in turn promotes remodeling of the actin cytoskeleton and calcium sensitization. RhoA is a guanine nucleotide (GTP)-binding protein that is activated by guanine nucleotide exchange factors (GEF) that convert the inactive GDP-bound RhoA into the active GTP-bound form. RhoA activates Rho kinase (ROCK), which in turn activates LIM kinase-2 (LIMK2). This leads to the phosphorylation and subsequent inactivation of cofilin, and thus a depletion of G-actin and accumulation of F-actin [31–34]. These changes in the actin cytoskeleton cause VSMC remodeling.

VSMC contraction is regulated by either the cytosolic calcium concentration or the calcium sensitivity. Calcium sensitivity is regulated by inhibition of MLC phosphatase, thus promoting the phosphorylated form of the light chain of myosin. After getting activated by RhoA, ROCK phosphorylates MLC phosphatase and inactivates it, thus increasing calcium sensitization and promoting vasocontraction. Specific inhibitors of ROCK, such as Y-27632 or fasudil, induce VSMC relaxation and reduce blood pressure and cerebral resistance [35, 36].

1.6.3 Excitation-Contraction Coupling in VSMC

Unlike skeletal muscle cells, VSMC do not require neuronal input or an action potential to contract. VSMC contract to other kinds of stimuli like hormones (angiotensin II, endothelin I, epinephrine), metabolic substances (adenosine, hydrogen ions, potassium ions), or physical interactions (mechanical stretch, shear stress). In these cases, there is no excitation-contraction coupling.

Neuronal stimuli that stimulate VSMC contraction are autonomic. VSMC are innervated mainly by the sympathetic nervous system. The main neurotransmitter used by the sympathetic nervous system is norepinephrine, which binds to its metabotropic adrenergic receptors to activate VSMC to contract by raising intracellular calcium concentrations. This pathway then involves binding to calmodulin and activating MLC kinase as described in the previous section. The parasympathetic input employs the neurotransmitter acetylcholine which binds to muscarinic receptors on the VSMC to promote contraction. However, both sympathetic and parasympathetic inputs could promote vasodilation, depending on the receptors activated on the VSMC.

1.6.4 Endothelium-Mediated Regulation of Cerebrovascular Tone

The endothelial cells of the cerebral arteries release vasoactive factors that regulate the tone of the vessels. The most important ones are endothelium-derived hyperpolarizing factor (EDHF), prostacyclin (PGI_2), and nitric oxide (NO).

EDHF is secreted by endothelial cells after endothelial intracellular calcium increases [37] or through myoendothelial gap junctions [38, 39]. It causes VSMC to relax and thus causes vasodilation. On the other hand, PGI_2 is a prostanoid produced by the enzyme cyclooxygenase (COX) from arachidonic acid. PGI_2 binds to its cell surface receptor on VSMC to induce an increase in 3′–5′-cyclic adenosine monophosphate (cAMP), which activates protein kinase A (PKA) and causes vasodilation [40, 41].

NO is a small molecule produced by endothelial (eNOS), inducible (iNOS), and neuronal (nNOS) nitric oxide synthase. This gaseous molecule activates guanylate cyclase to form cyclic guanosine monophosphate (cGMP), which in turn activates protein kinase (PKG) [42]. PKG activates MLC phosphatase [43, 44] and induces VSMC relaxation and vasodilation [30, 45, 46].

1.6.5 Potassium Channels Involved in Regulation of Cerebrovascular Tone

When potassium channels open and potassium ions start to efflux out of the cell, the plasma membrane undergoes hyperpolarization, inducing the closure of voltage-gated calcium channels, like the L-type calcium channels, and vasorelaxation. Potassium channel activity majorly influences vascular tone and vasorelaxation [47, 48].

Adenosine triphosphate (ATP)-dependent potassium channels (K_{ATP}) are linked to metabolic activity. When the intracellular ratio of ATP to adenosine diphosphate (ADP) decreases, the K_{ATP} channels open and induce vasorelaxation [29]. This adaptive mechanism is needed in cases of insufficient oxygenation, and inhibition of K_{ATP} prevents hypoxic vasorelaxation in cerebral and peripheral blood vessels [49, 50].

The inward-rectifier potassium ion channels (K_{ir}) selectively allow potassium to pass more easily into the VSMC cell than out of the cell [51, 52] and promote vasorelaxation. In fact, K_{ir} has been shown to couple cerebral blood flow to neuronal activity in rats since active neurons release potassium ions into the extracellular space [53].

Voltage-gated potassium channels (K_V) also help regulate cerebral artery tone. K_V channels are opened by depolarization of the VSMC and promote vasorelaxation. This negative feedback mechanism occurs in response to arterial constriction [48]. Inhibition of K_V induces vasoconstriction of cerebral arteries [54, 55].

1.7 Assessment of Blood Vessel Reactivity

Small blood vessels' (less than 100 μm) isometric responsiveness to hormones and other agonists can be measured by using the wire myograph (Mulvany myograph). This device was developed by Mulvany and Halpern [56] to investigate the active and passive tension of blood vessels from a variety of arteries, including cerebral arteries [57, 58]. The wire myograph system is a highly sensitive tool used to investigate the responsiveness of blood vessels to different stimuli and hormones. In this chapter, we explain the steps of myograph measurement which consist of blood vessel isolation, mounting, normalization, assessment of tissue viability, and construction of a cumulative concentration response curve [59].

2 Materials

2.1 Buffers and Solutions (See Note 1)

1. Physiological salt solution (PSS): mmol/L: 115 NaCl, 4.7 KCl, 1.4 $MgSO_4$, 7.0 H_2O, 5 $NaHCO_3$, 1.2 K_2HPO_4, 1.1 Na_2HPO_4, 1.0 $CaCl_2$, 20 HEPES, and 5 glucose; pH 7.35.
2. High-potassium solution (123 mmol/L K^+): KCl 123.70 mmol/L, $MgSO_4$. H_2O 1.17 nmol/L, KH_2PO_4 1.18 mmol/L, $CaCl_2$ $2H_2O$ 2.5 mmol/L, $NaHCO_3$ 25 mmol/L, EDTA 0.03 mmol/L, glucose 5.5 mmol/L.

3. Acetylcholine (C3389, Sigma) 10 mmol/L stock solution.
4. Norepinephrine (A7257, Sigma) 1 mmol/L stock solution.

2.2 Tools and Instruments

2.2.1 Dissection Tools (Please Refer to Fig. 3)

1. Large dissection dish (93 mm diameter × 22 mm H; DD-90-S-BLK, Living Systems Instrumentation) (*see* **Note 2**).
2. Deep dissection dish (100 mm diameter × 50 mm H; DD-100-D-BLK, Living Systems Instrumentation).
3. Stereo microscope (C-FMBN, Nikon) (*see* **Note 3**).
4. Cold Light Source/Fiber Optic Illumination (NCL 150, Volpi).
5. Multi-purpose Iris scissors (11 cm long, straight).
6. Vannas scissors (8 cm long, straight, 3 mm blades).
7. Spring scissors (12.5 cm long, 9 mm, straight).
8. Dressing tissue forceps.
9. Dumont Tweezers #5 (11 cm long; 0.025 × 0.005 mm).
10. Dissection pins (0.1, 0.2, and 0.5 mm).

2.2.2 Multi-Wire Myograph System and Data Recording

1. Multi-wire myograph system (620 M, DMT, Aarhus, Denmark) (Fig. 4).
2. Stainless steel wires (diameter = 40 μm) (*see* **Note 4**).
3. Mounting jaws (one jaw is connected to the false transducer, the other one is mobile and is connected to a micrometric screw). The head is suspended in the myograph chamber (Fig. 5).
4. PowerLab 4/35 (AD Instruments).

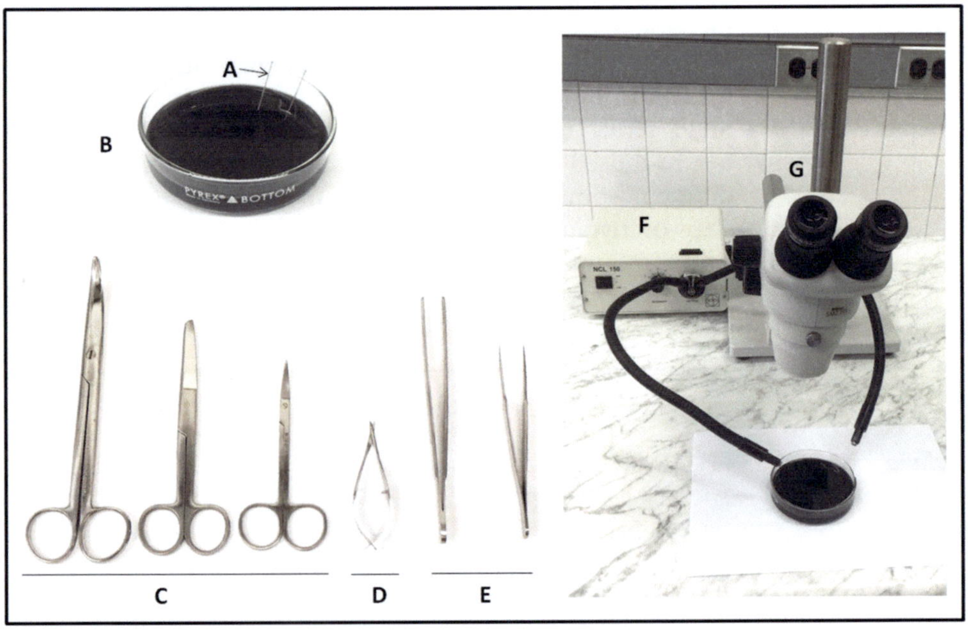

Fig. 3 Dissection tools and microscope. (**a**) Dissection pin; (**b**) coated plate; (**c**) scissors; (**d**) micro-scissors; (**e**) forceps; (**f**): fiber-optic illuminator; (**g**) stereo zoom microscope

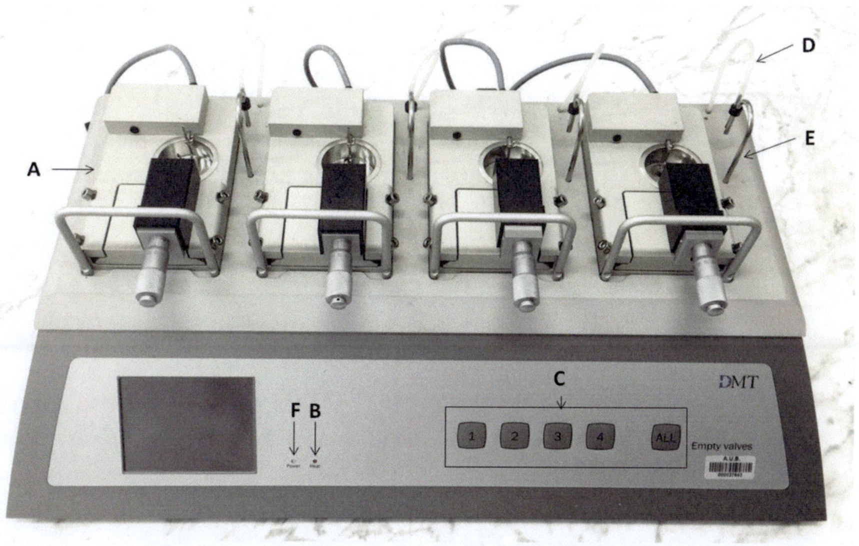

Fig. 4 Multi-wire myograph system (620 M, DMT, Aarhus, Denmark). The system is composed of (**a**) multi-wire myograph unit, (**b**) heat indication, (**c**) valve buttons myo-interface display, (**d**) gas tube, (**e**) suction tube

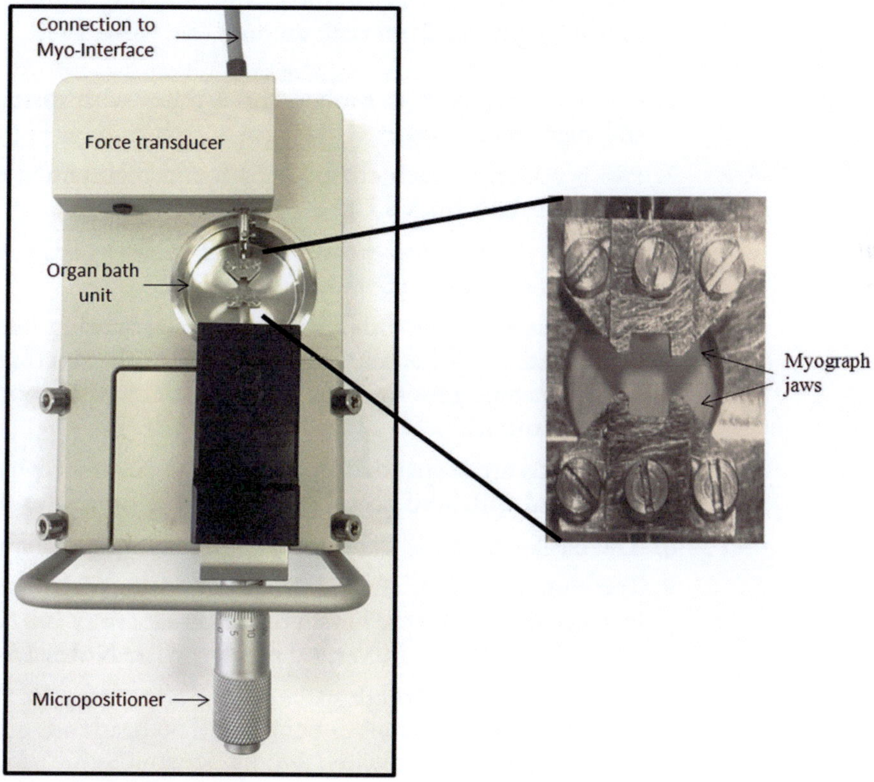

Fig. 5 Multi-wire myograph unit

3 Methods

3.1 Dissection Procedure for Rat Basilar Artery

1. Male Sprague-Dawley rats (200–250 g) are used for the study (*see* **Note 5**).
2. Cerebral injury rat models are used or subarachnoid hemorrhage is induced in rats by injection of 250 μL autologous blood into the basal cisterns [60].
3. After 3 days, the rats are euthanized by CO_2.
4. The brains are quickly removed and placed in deep dissection dish containing cold PSS (*see* **Note 6**).
5. The basilar artery (Fig. 1) is isolated under the microscope and then transferred into 5 mL large dissection dish containing ice-cold PSS (*see* **Notes 7–10**).
6. The artery is pinned up on a coated bottom Sylgard material (*see* **Note 11**).
7. The dissection and cleaning are done with the help of a pair of trabecular scissors and a pair of fine forceps (*see* **Notes 12** and **13**).
8. The basilar arteries are cut into 1.5–2 mm ring segments (three to four segments from each animal) (*see* **Note 14**).

3.2 Mounting of the Vessel on the Wire Myograph

1. The chamber bath is washed three times with distilled H_2O and twice with PSS.
2. PSS is added to each chamber bath and maintained at 37 °C and perfused with 5 % CO_2.
3. Two 40 μm stainless steel segments (4 cm) of the wires are isolated (*see* **Notes 15** and **16**).
4. The wires are fixed with screws on two mounting hits. One is connected to a force transducer while the other is connected to a micrometer screw, so it can move (to stretch the blood vessels) (Figs. 4 and 5).
5. The heads are suspended in a myograph chamber which is initially filled with cold PSS. Keep the wire as straight as possible.
6. The basilar artery is transferred to the myograph chamber.
7. The mounting jaws are widely spread apart. With fine forceps, insert one wire into the lumen of the basilar artery (*see* **Note 17**). Maintain the filled wire vertical in the gap (*see* **Notes 18** and **19**).
8. In order to have enough space for the artery to pass, the heads are moved apart and wires bent out. The heads are now gently pulled further up the wire (Fig. 5).
9. The vessel is adjusted to bring the distal part into the gap between the mounting heads. With the screwdriver pressed against the wire, the vessel is held back and positioned in the

gap. The artery should be able to move freely in the longitudinal direction and therefore should not be clamped between the lower parts of the support.

10. The wire is tightened with a pole bent to the lower left screw and fixed in place. It is important that the wire is tight and follows the support all the way (*see* **Note 20**).
11. In order to get a second wire into the lumen from the top end, remove the part of the vessel that is not in the gap between the jaws.
12. In order to close the jaws, use a pair of forceps to rub the tissue away.
13. The wires are adjusted in order to have parallel wires and simultaneously horizontal levels (*see* **Note 21**).
14. Turn on the heating and gas pump. The system is now ready for use.
15. Rotate the micrometer counterclockwise until the pre-tension is achieved.

3.3 Measuring Isometric Response

Contractile responses of basilar artery are examined in a wire myograph (Danish Myo Technology, 620) by recording isometric tension (*see* **Notes 22** and **23**) (Fig. 6).

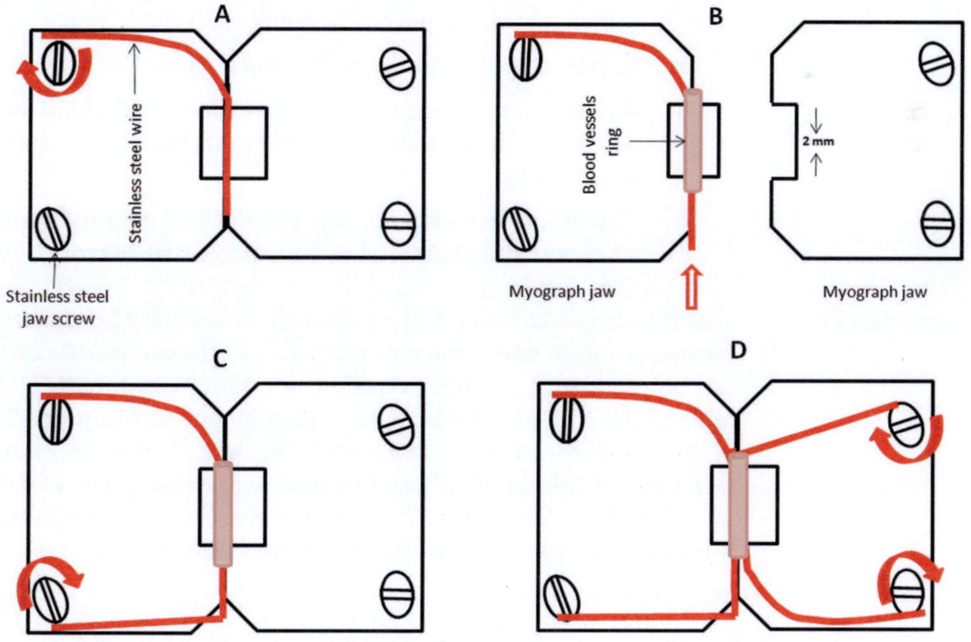

Fig. 6 Mounting of basilar artery: (**a**) Stainless steel wire is attached to the stainless steel jaw screw. (**b**) Stainless steel wire is inserted through basilar artery. (**c**) The wire is attached to the second screw. (**d**) Repeat same procedure with the second wire

3.3.1 Resting Tension Internal Circumference Relationship

1. The sensitivity of blood vessels to different agonists is dependent on the degree of muscle stretch, so it is important to normalize the blood vessel dimensions.
2. It is useful to have optimal internal circumference (optimal stretch) that gives the maximum contraction. This can be calculated using active tension/internal circumference relationship.
3. Blood vessel lumen circumference is considered L_{100} when it is fully relaxed under a transmural pressure of 100 mmHg. L_{100} is calculated using resting tension-internal circumference curve.
4. Another advantage of resting tension-internal circumference curve is that the blood vessel's wall thickness at L_{100} (μm) can be calculated using microscopy. The wall thickness of the mounted basilar artery is measured using maximum magnification of the dissection microscope.
5. Resting tension-internal circumference curve is used also to normalize the vessel's dimensions.
6. The maximum active tension can be achieved at $0.9 \times L_{100}$. Any segment giving more than 10 mN is used in the experiments.

Procedure

1. Fill the chamber with 1 mL PSS maintained at 37 °C and perfused with 5 % CO_2 (*see* **Notes 24** and **25**).
2. Basilar artery is subjected to stretch using the micropositioner (about 10 μm) after being held for 1 min.
3. Repeat step 2 until the wall tension gives 1 mN/mm.
4. Basilar artery is subjected to relaxation (as done in **steps 2** and **3** but the muscle is relaxed) until the tension becomes 0 mN/mm.
5. The points between tension and internal circumference are recorded and expressed as an exponential curve.

3.3.2 Vessel Viability

This is an important step and is necessary to test whether the vascular muscle is intact and contracts normally. Different vasoconstrictors can be used, such as high-potassium solution (123 mmol/L; NaCl replaced by KCl on a molar basis), norepinephrine (10 μmol/L), or high-potassium solution containing 10 μmol/L noradrenaline. In order to test the ability of vessel to contract normally, five blood vessel stimulations are applied (lasting 2 min each), each followed by a 5-min washout period with PSS (*see* **Notes 26–29**).

Procedure

1. Fill the chamber with 1 mL PSS maintained at 37 °C and perfused with 5 % CO_2.
2. A pre-tension of 2 mN/mm is applied to each basilar artery (*see* **Note 30**).

3. Basilar artery is subjected to a 45-min period of normalization.
4. High-potassium solution (123 mm/L K^+) is added to the baths. Wait for 2 min (*see* **Note 29**).
5. Wash the blood vessel with PSS two times and incubate it in PSS to allow the blood vessel to relax and reach the baseline.
6. Repeat **steps 4** and **5**.
7. Add norepinephrine (10 μmol/L) to the chamber bath for 2 min.
8. Repeat **step 5**.
9. Add high-potassium solution containing 10 μmol/L noradrenaline to the chamber and wait for 2 min.
10. All contractions are recorded and measured using a PowerLab unit and LabChart software (ADInstruments).

3.3.3 Checking Endothelial Function (Vasodilator Response Curve)

It is very important to test the integrity of endothelium in the basilar artery after blood vessel isolation, cleaning, or mounting (*see* **Note 31**). This procedure is used to test the success of endothelium removal (*see* **Note 32**). In order to have vasodilator response curve, blood vessels should be pre-contracted using a suitable agonist such as noradrenaline (0.1–3 μM) or (*see* **Notes 33** and **34**) to produce 60–80% of the response.

Procedure

1. Fill the chamber with 1 mL PSS maintained at 37 °C and perfused with 5% CO_2.
2. Slowly stretch the vessel to its normalized micrometer setting (pretension of 2 mN) and let the vessel equilibrate for 45 min.
3. Pre-constriction with a sub-maximal concentration (3 μmol/L) of noradrenaline is induced (*see* **Note 35**).
4. 2 μL of acetylcholine (5 μmol/L) is added in order to have 10^{-8} mol/L acetylcholine.
5. Once the vessel starts to relax (after 2–3 min), 0.18 μL of acetylcholine (50 μmol/L) is added to give a final concentration of 0.1 μmol/L acetylcholine in total volume within the chamber.
6. Add 0.18 μL of acetylcholine (0.5 mmol/L) to give 1 μmol/L acetylcholine and wait for 3 min.
7. Add 0.18 μL of acetylcholine (5 μmol/L) stock solution to give 10 μmol/L.
8. After 3 min, add 0.9 μL of acetylcholine (10 mmol/L; stock solution in freezer) to give 0.1 mmol/L acetylcholine. Wait for 4 min.
9. Wash the blood vessels four times with PSS to reach initial tension.

10. Relaxation responses of basilar artery are recorded using a PowerLab 4/35 unit and LabChart software (ADInstruments) and expressed as a percentage of the induced noradrenaline-induced contraction.

3.3.4 Cumulative Concentration-Response Curves

A standard cumulative concentration-response curve is defined by four parameters: the baseline tension, the maximum tension (*see* **Note 36**), the slope of the curve, and the agonist concentration that provokes a response halfway between baseline and maximum tension (EC_{50}). The EC_{50} is the molar concentration of an agonist that produces 50% of the maximum contraction for that agonist. Different agonists with different concentrations are used in order to create cumulative concentration-response curves (*see* **Notes 37** and **38**). The well-known vasoconstrictors are 5-hydroxytryptamine type 1B ($5\text{-}HT_{1B}$ receptor agonist), angiotensin II (AT_1 and AT_2 receptor agonist), and endothelin-1 type A and B (ET_A and ET_B receptor agonist, 10^{-14}–0.3 μmol/L).

Procedure

1. Fill the chamber with 1 mL PSS maintained at 37 °C and perfused with 5% CO_2.
2. Allow the basilar artery to equilibrate in PSS for 45 min.
3. Blood vessels are normalized. L_{100} is calculated and the internal circumference of 0.9 L_{100} is used in order to have maximum active tension.
4. Cumulative concentration-response curves are obtained to the following agonists: $5\text{-}HT_{1B}$ (10^{-11}–30 μmol/L), AT_1 and AT_2 (10^{-12}–3 μmol/L), and ET_A and ET_B (10^{-14}–0.3 μmol/L).
5. Following each concentration-response curve, the basilar artery is washed with PSS and allowed to relax (around 30 min) before the next vasoconstrictor is applied.
6. Contractile responses of basilar artery are recorded using a PowerLab unit and LabChart software (ADInstruments). This software will automatically generate the EC_{50}.
7. The contractile responses to different agonists are expressed as increase in isometric tension divided by the normalized lumen diameter of blood vessels (active wall tension (mN/mm)).
8. The EC_{50} for each cumulative concentration-response curve is calculated.
9. The sensitivity of basilar artery to each vasoconstrictor is expressed as the - log EC_{50}.
10. It is very important to check the integrity of the endothelium at the end of the experiment with the silver nitrate stain (or repeat Subheading 3.3.2).

3.3.5 Denuding the Endothelium

In order to assess the endothelial dependence of the actions of different vasoconstrictors on the basilar artery, the endothelium is removed mechanically by passing a 50–75 μm human hair through the basilar artery lumen and rubbing the intimal surface. Endothelial denudation is considered successful if the precontracted basilar artery has no response to 10 μmol/L acetylcholine.

4 Notes

1. The solutions should be prepared freshly for each experiment. If left for several weeks (even at 4 °C), the buffers might produce fungi, which will affect the experiment and results.
2. Dissection dishes are lined with black Sylgard material for tissue fixation during dissection procedures.
3. It is preferable to use an inverted microscope instead of an upright microscope, since accessories like pH meter or oxygen electrode might be used in the bath. An inverted microscope has more space for these accessories.
4. Stainless steel wires are better than tungsten wires due to oxidation.
5. The rats are maintained at stable room temperature, humidity, and sleep cycle, with regular access to uncontaminated drinking water and food in order to maintain stable physiologic cycles of metabolism.
6. After euthanizing the animal, remove the brain and isolate the basilar arteries as quickly as possible. If cerebral arteries are left on the brain for too long, toxins released from the dying neurons will affect the blood vessels and reduce blood vessel reactivity.
7. It is easy to differentiate between arteries and veins. The wall (tunica media) of arteries is thicker than that of veins. The lumen of an artery is narrower than that of a vein.
8. This technique requires a careful grip of the scissors. Move the blades directly up and make the cuts parallel to the vessel wall, in order to avoid cutting the vessel itself.
9. Work fast and carefully to avoid damaging endothelial cells in the vessels.
10. Branched blood vessels are challenging to work with, so the middle cerebral arteries are not often used. The superior cerebellar arteries or posterior cerebral arteries (specifically the third branch) are easier to isolate because they have longer segments that are not branched. However, if a branched vessel is the one you need to isolate, it is possible, but tedious. Ligatures may be required to isolate a non-branchy segment and more effort will be made in the mounting process.

11. It is important and necessary to stretch the blood vessels during dissection to aid the microdissection.
12. Most of the fat is removed firstly from one side of the vessel and then from the other side, and finally from the smaller branches.
13. The vessel should be as clean as possible because excessive fat on the vessel affects the muscle contractility.
14. Interestingly, the blood vessels can be left in the fridge for a couple of days and then actually be used in the wire myograph experiment. However, it is crucial to leave the vessels in the fridge at 4 °C, and not at −20 °C or −80 °C.
15. In general, small arteries with internal diameters between 100 and 500 μm can be mounted on 40 μm stainless steel wires. In case the arterioles have an internal diameter of less than 60 μm, the 25 μm tungsten wires are used.
16. Two wires of approximately 2 cm length, with blunt ends and as straight as possible, are required.
17. The lumen of the vessel often closes when it is cut. To open the blood vessels, grab it firmly with the forceps as close to the proximal end as possible. The wire tip can also be used to open the lumen.
18. Once the wire is inside the lumen, the vessel is slowly and carefully drawn up the wire to avoid damaging the endothelial layer and vascular wall. Take special care when the wire tip is passing the branching points.
19. Remember that the transducer is very sensitive to pressure at the transducer head level, so do not apply too much force when you tighten the screws. This also applies any time you have the heads together.
20. Screws are turned in clockwise so as to tighten the wires. When tightening the screws, always place the wire clockwise around the screw head.
21. If the wires are not parallel, move the wires with the use of forceps. Moving the heads back and forth makes it easier to see how the wires are positioned.
22. For beginners that are new to this technique, take your time to practice the entire procedure before beginning the actual research experiments.
23. In the process of practicing, it is advisable to use mesenteric arteries and to keep repeating the same experiment, such as KCl-induced constriction or phenylephrine concentration-response curve, until consistent results are obtained.

24. If you note any contraction in the vessel, do not force it; let it relax, and then change the PSS in the chamber bath.
25. In case problems arise, first check the pH, oxygenation, and temperature in the bath.
26. Since force production and sensitivity of vessels to different agonists are dependent on the extent of stretch (according to active tension-length relationship), it is essential to conduct contraction studies under isometric conditions to prevent compliance of the mounting wires.
27. You can use only segments that developed more than 2 mN force to high potassium with intact endothelium.
28. In the case of endothelium-denuded segment, you can use segments that developed 1.5 mN force to high-potassium solution.
29. It is very important to have a fixed time interval between each dose.
30. Basilar arteries may vary with length and size, and it is important to have maximum contraction for normalization. High-potassium solution (123 mmol/L) or norepinephrine (10 μmol/L) can be used to normalize experimental data as a percent of high potassium or norepinephrine-induced contraction.
31. If the endothelium is intact, acetylcholine will induce relaxation on the blood vessels (vasodilator curve).
32. Before starting vasodilator response curve experiments, make sure that the blood vessel has cholinergic receptors (placental vessels lack cholinergic receptors).
33. Phenylephrine or norepinephrine produces a sustained constriction (tonic) which is stabilized after 3 min.
34. This tonic contraction is uniform along the length of the basilar artery.
35. This will produce a contraction of 80% of the maximum response to high-potassium solution.
36. Concentration-response curve can be obtained by using large ranges of concentrations (until receptor saturation) of an agonist.
37. Concentration-response curve is usually logarithmic. The *X*-axis plots the concentration of an agonist while the *Y*-axis plots response (tension mN/mm).
38. Some agonists bind irreversibly to their receptor and cannot be washed out with physiological salt solution. Irreversible agonists should be used at the end of experiments.

5 Acknowledgments

The present work was supported by Medical Practice Plan (MPP), Faculty of Medicine at AUB, and the National Council for Scientific Research (CNRS) in Lebanon to Asad Zeidan.

Conflict of interests
The authors have NO affiliations with or involvement in any organization or entity with any financial interest or nonfinancial interest in the materials discussed in this book chapter.

References

1. Lee MY, Griendling KK (2008) Antioxid Redox Signal 10:1045–1059
2. Intengan HD, Schiffrin EL (2000) Hypertension 36:312–318
3. Egginton S, Hudlicka O (1999) J Physiol 515(Pt 1):265–275
4. Rizzoni D, Castellano M, Porteri E, Giacche M, Ferrari P, Cusi D, De Ciuceis C, Boari GE, Rosei EA (2009) Hypertens Res 32:581–585
5. Rizzoni D, De Ciuceis C, Porteri E, Paiardi S, Boari GE, Mortini P, Cornali C, Cenzato M, Rodella LF, Borsani E, Rizzardi N, Platto C, Rezzani R, Rosei EA (2009) J Hypertens 27:838–845
6. Rizzoni D, Agabiti-Rosei E (2012) Intern Emerg Med 7:205–212
7. Peterson EC, Wang Z, Britz G (2011) Int J Vasc Med 2011:823525
8. Pagnussat AS, Faccioni-Heuser MC, Netto CA, Achaval M (2007) J Anat 211:589–599
9. Gomes FB, Dujovny M, Umansky F, Berman SK, Diaz FG, Ausman JI, Mirchandani HG, Ray WJ (1986) Surg Neurol 26:129–141
10. Pai SB, Varma RG, Kulkarni RN (2005) Neurol India 53:186–190
11. Parraga RG, Ribas GC, Andrade SE, de Oliveira E (2011) World Neurosurg 75:233–257
12. Faraci FM, Heistad DD (1990) Circ Res 66:8–17
13. Rink C, Khanna S (2011) Antioxid Redox Signal 14:1889–1903
14. Qi M, Hang C, Zhu L, Shi J (2011) Neurol Sci 32:551–557
15. Kasprowicz M, Czosnyka M, Soehle M, Smielewski P, Kirkpatrick PJ, Pickard JD, Budohoski KP (2012) Neurocrit Care 16:213–218
16. Diedler J, Sykora M, Rupp A, Poli S, Karpel-Massler G, Sakowitz O, Steiner T (2009) Stroke 40:815–819
17. Edvinsson LI, Povlsen GK (2011) J Cereb Blood Flow Metab 31:1554–1571
18. Cipolla MJ, McCall AL, Lessov N, Porter JM (1997) Stroke 28:176–180
19. Kontos HA (2001) Stroke 32:2712–2716
20. Fagan SC, Hess DC, Hohnadel EJ, Pollock DM, Ergul A (2004) Stroke 35:2220–2225
21. Kelley BJ, Petersen RC (2007) Neurol Clin 25:577–609, v
22. Iadecola C (2004) Nat Rev Neurosci 5:347–360
23. Chew SH, Meighan Smith Tomic M, Cheung AT (2010) Clin Hemorheol Microcirc 46:69–73
24. Roher AE, Esh C, Kokjohn TA, Kalback W, Luehrs DC, Seward JD, Sue LI, Beach TG (2003) Arterioscler Thromb Vasc Biol 23:2055–2062
25. Skoog I, Gustafson D (2006) Neurol Res 28:605–611
26. Iadecola C, Davisson RL (2008) Cell Metab 7:476–484
27. Iadecola C, Zhang F, Niwa K, Eckman C, Turner SK, Fischer E, Younkin S, Borchelt DR, Hsiao KK, Carlson GA (1999) Nat Neurosci 2:157–161
28. Niwa K, Kazama K, Younkin L, Younkin SG, Carlson GA, Iadecola C (2002) Am J Physiol Heart Circ Physiol 283:H315–H323
29. Faraci FM, Heistad DD (1998) Physiol Rev 78:53–97
30. Toda N, Ayajiki K, Okamura T (2009) Pharmacol Rev 61:62–97
31. Kaibuchi K, Kuroda S, Amano M (1999) Annu Rev Biochem 68:459–486
32. Lawler S (1999) Curr Biol 9:R800–R802
33. Zeidan A, Nordstrom I, Albinsson S, Malmqvist U, Sward K, Hellstrand P (2003) Am J Physiol Cell Physiol 284:C1387–C1396
34. Zeidan A, Paylor B, Steinhoff KJ, Javadov S, Rajapurohitam V, Chakrabarti S, Karmazyn M (2007) J Pharmacol Exp Ther 322:1110–1116
35. Feletou M, Vanhoutte PM (2000) Acta Pharmacol Sin 21:1–18

36. Uehata M, Ishizaki T, Satoh H, Ono T, Kawahara T, Morishita T, Tamakawa H, Yamagami K, Inui J, Maekawa M, Narumiya S (1997) Nature 389:990–994
37. Fukao M, Hattori Y, Kanno M, Sakuma I, Kitabatake A (1995) Br J Pharmacol 115:987–992
38. Ujiie H, Chaytor AT, Bakker LM, Griffith TM (2003) Stroke 34:544–550
39. Xu HL, Santizo RA, Baughman VL, Pelligrino DA (2002) Am J Physiol Heart Circ Physiol 283:H1082–H1091
40. Stitham J, Arehart EJ, Gleim SR, Douville KL, Hwa J (2007) Prostaglandins Other Lipid Mediat 82:95–108
41. Myren M, Olesen J, Gupta S (2011) Vascul Pharmacol 55:50–58
42. Schmidt HH, Walter U (1994) Cell 78:919–925
43. Yuen S, Ogut O, Brozovich FV (2011) J Biol Chem 286:37274–37279
44. Surks HK (2007) Circ Res 101:1078–1080
45. Salom JB, Barbera MD, Centeno JM, Orti M, Torregrosa G, Alborch E (1999) Nitric Oxide 3:85–93
46. Salom JB, Barbera MD, Centeno JM, Orti M, Torregrosa G, Alborch E (1998) Pharmacology 57:79–87
47. Brayden JE (2002) Clin Exp Pharmacol Physiol 29:312–316
48. Nelson MT, Quayle JM (1995) Am J Physiol 268:C799–C822
49. Liu J, Lai ZF, Wang XD, Tokutomi N, Nishi K (1998) J Cardiovasc Pharmacol 31:558–567
50. Reid JM, Paterson DJ, Ashcroft FM, Bergel DH (1993) Pflugers Arch 425:362–364
51. Quayle JM, McCarron JG, Brayden JE, Nelson MT (1993) Am J Physiol 265:C1363–C1370
52. Johnson TD, Marrelli SP, Steenberg ML, Childres WF, Bryan RM Jr (1998) Am J Physiol 274:R541–R547
53. Filosa JA, Bonev AD, Straub SV, Meredith AL, Wilkerson MK, Aldrich RW, Nelson MT (2006) Nat Neurosci 9:1397–1403
54. Knot HJ, Nelson MT (1995) Am J Physiol 269:H348–H355
55. Zhong XZ, Harhun MI, Olesen SP, Ohya S, Moffatt JD, Cole WC, Greenwood IA (2010) J Physiol 588:3277–3293
56. Mulvany MJ, Halpern W (1977) Circ Res 41:19–26
57. Mulvany MJ, Nyborg N (1980) Br J Pharmacol 71:585–596
58. Mulvany MJ, Warshaw DM (1979) J Gen Physiol 74:85–104
59. Zeidan A, Nordstrom I, Dreja K, Malmqvist U, Hellstrand P (2000) Circ Res 87:228–234
60. Edvinsson L, Povlsen GK, Ahnstedt H, Waldsee R (2014) J Neuroinflammation 11:207

Chapter 35

Magnetic Resonance Imaging in Experimental Traumatic Brain Injury

Qiang Shen, Lora Tally Watts, Wei Li, and Timothy Q. Duong

Abstract

Traumatic brain injury (TBI) is a leading cause of death and disability in the USA. Common causes of TBI include falls, violence, injuries from wars, and vehicular and sporting accidents. The initial direct mechanical damage in TBI is followed by progressive secondary injuries such as brain swelling, perturbed cerebral blood flow (CBF), abnormal cerebrovascular reactivity (CR), metabolic dysfunction, blood–brain-barrier disruption, inflammation, oxidative stress, and excitotoxicity, among others. Magnetic resonance imaging (MRI) offers the means to noninvasively probe many of these secondary injuries. MRI has been used to image anatomical, physiological, and functional changes associated with TBI in a longitudinal manner. This chapter describes controlled cortical impact (CCI) TBI surgical procedures, a few common MRI protocols used in TBI imaging, and, finally, image analysis pertaining to experimental TBI imaging in rats.

Key words MRI, Traumatic brain injury, ADC, CBF, DWI, PWI, BBB, Experimental TBI model, Rodents, fMRI

1 Introduction

Traumatic brain injury (TBI) is a leading cause of death and disability, affecting 3.5 million American civilians, with an annual cost exceeding $60 billion [1]. In addition, more than 270,000 US service members have been diagnosed with TBI since the beginning of the global war on terrorism (http://dvbic.dcoe.mil). Common causes of TBI include falls, violence, injuries from wars, and vehicular and sporting accidents. The initial direct mechanical damage in TBI is followed by progressive secondary damage such as brain swelling, perturbed cerebral blood flow (CBF), abnormal cerebrovascular reactivity (CR), metabolic dysfunction, blood–brain-barrier disruption, inflammation, oxidative stress, and excitotoxicity, among others [2, 3]. Ischemic-like events (such as membrane depolarization, ion dysregulation, oxidative stress, excitotoxicity, and inflammation, among others) subsequently lead to apoptotic and necrotic cell death. The multidimensional cascades

Firas Kobeissy et al. (eds.), *Injury Models of the Central Nervous System: Methods and Protocols*, Methods in Molecular Biology, vol. 1462, DOI 10.1007/978-1-4939-3816-2_35, © Springer Science+Business Media New York 2016

of secondary brain injury in TBI offer many potential targets for therapeutic intervention.

Magnetic resonance imaging (MRI) offers the means to noninvasively probe many of these secondary injuries. MRI has been used to image anatomical, physiological, and functional changes associated with TBI in a longitudinal manner [4, 5]. T_1 and T_2 anatomical MRI allows for visualization of gross structural damage, possible hemorrhage, and vasogenic edema in TBI [6, 7]. Diffusion-weighted imaging, in which contrast is based on the apparent diffusion coefficient (ADC) of water, can detect ischemic brain injury within minutes. ADC reduction is often attributed to cytotoxic edema, a common feature of TBI [8–10]. Diffusion tensor imaging (DTI) provides both magnitude and directional information about water diffusion and thus has been used to detect changes in fiber tracts following TBI [11]. Fractional anisotropy (FA), an index of anisotropy of water motion from DTI measurements, is sensitive to white-matter injury, microstructural changes [12, 13], and diffuse axonal injury [13–17] in TBI.

Disruptions in CBF and CR could be the result of either direct mechanical injury to vessels or indirect reduction of local perfusion pressure by elevated intracranial pressure. CBF can be measured using an exogenous intravascular contrast agent or by magnetically labeling the endogenous water in blood [18, 19]. The former is efficient, but it is incompatible with dynamic CBF fMRI as the long half-life of the contrast agent allows only one CBF measurement per bolus injection. Arterial spin labeling (ASL) techniques, on the other hand, are totally noninvasive, and the labeled water has a short half-life (~blood T_1), making it possible to perform multiple repeated measurements that can be used to augment spatial resolution and/or signal-to-noise ratio.

Blood–brain-barrier (BBB) disruption is common following TBI [20, 21]. Traditional histological Evans Blue extravasation has been widely used to measure BBB leakage following TBI [20, 22, 23]. However, this method requires the sacrifice of the animals and does not allow for longitudinal assessments. Dynamic contrast-enhanced (DCE) MRI can longitudinally measure the transport coefficient K^{trans} which reflects BBB permeability [24].

Rodent models have been widely utilized to study TBI; they include the controlled cortical impact (CCI), fluid percussion, acceleration-impact or weight drop, Marmarou, Feeney, and blast injury models [25, 26]. The common areas of impact included somatosensory/motor, auditory, parietal, and visual cortices. Outcomes and lesion sizes are highly variable due to the use of different experimental models and/or different injury parameters [25, 26].

In this chapter, we describe the controlled cortical impact (CCI) TBI surgery procedures, a few common MRI protocols used in TBI imaging, and, finally, image analysis pertaining to experimental TBI imaging in rats.

2 Materials

2.1 TBI Modeling

1. Rats (200–250 g) (Taconic Farms, Hudson, NY; Charles River, Wilmington, MA).
2. Anesthetics (isoflurane or pentobarbital, etc.) (VetOne, MWI, Boise, ID).
3. Common surgical tools and supplies (Fine Science Tools, Foster City, CA; Integra Miltex, Plainsboro, NJ; World Precision Instruments, Sarasota, FL; Ethicon, Somerville, NJ).
4. Pneumatic controlled cortical impactor (Precision Systems and Instrumentation, Fairfax Station, VA).
5. Stereotaxic frame equipped with tooth and ear bars (myNeurolab.com, St. Louis, MO).
6. Bone wax (Ethicon, Somerville, NJ).
7. Antibiotic ointment (McKesson, Richmond, VA).
8. Analgesic (Buprenex) (Henry Schein, Melville, NY).
9. Warm pad, temperature feedback monitoring, and other monitoring equipment to ensure normal animal physiology (Fisher Scientific, Pittsburgh, PA; Cole-Palmer, Vernon Hills, IL).
10. Cresyl violet acetate for Nissl staining (Sigma, St. Louis, MO).

2.2 MRI

1. Bruker 7 T scanner (Billerica, MA).
2. 40-G/cm BGA12 gradient insert (ID = 12 cm, 120 μs rise time).
3. Animal holder (custom-made).
4. Custom-made RF transmitter and receiver coils for brain imaging.
5. Custom-made RF transmitter coil for arterial spin labeling.
6. Actively decoupled switch box to detune RF coils.
7. Other magnet, gradient, and RF coil configurations should also work.

2.3 Peripheral MRI-Compatible Monitor Equipment and Animal Supports

1. Oximetry (heart rate, arterial oxygen saturation) (Mouse Ox, STARR Life Sciences, Oakmont, PA).
2. Blood pressure (invasive with artery catheterization) (Biopac/Acknowledge, Goleta, CA).
3. Respiration rate via force transducer (Biopac/Acknowledge, Goleta, CA).
4. Forepaw stimulation device—home-made device or Grass stimulators.
5. Circulating warm water bath (Haake water bath, Rheology Solutions, Bacchus Marsh, Victoria, Australia; Cole Palmer, Vernon Hills, IL).

6. Temperature feedback regulator (Digi-Sense, Cole-Palmer, Vernon Hills, IL).
7. Anesthetic delivery, such as vaporizer (Universal Vaporizer Support, Foster City, CA).

3 Methods

3.1 TBI Surgery [24, 27–32]

1. Male rats (250–350 g) are anesthetized with isoflurane (~2 %). Other anesthetics can also be used. Male rats are often used to avoid the effects of female hormones on outcome. Female rats are also widely studied; some female hormones have been found to have neuroprotective effects.
2. Aseptic preparations (Betadine and ethanol washes) should be performed to prevent infection and immunological responses that could affect outcome.
3. The animal is secured in a stereotaxic frame with ear and tooth bars; an incision is made at the level of the cerebellum as posterior from the impact site as possible to prevent artifacts during MRI acquisition. The periosteum is removed over the impact site. A Ø5mm craniotomy is created over the left S1FL (+0.25 mm anterior and 3.5 mm lateral to bregma), exposing the dura matter. The intact dura matter is impacted using a pneumatic-controlled cortical impactor (Precision Systems and Instrumentation, LLC, Fairfax Station, VA) fitted with a Ø3mm tip (5.0 m/s, 250 μs dwell time, 1 mm depth) to produce a mild focal TBI. Following the impact, the cranial opening is sealed with bone wax, the scalp sutured closed, and antibiotic ointment applied. Saline is injected under the skin to facilitate the removal of air pockets between the scalp and the skull, thus minimizing artifacts during MRI acquisition (*see* **Note 1**). Buprenex (0.05 mg/kg) is given subcutaneously every 12 h for 3 days as needed for pain relief.
4. The right femoral artery is catheterized for blood-gas sampling, continuous blood pressure, and heart-rate monitoring. These physiological parameters are important, because deviations could affect TBI outcome causing an increase in statistical scatters.
5. Rats are secured in a supine position on an MR-compatible rat stereotaxic headset, and anesthesia is reduced to ~1.1 % isoflurane. Rats breathe spontaneously. Mechanical ventilation can also be used. Rectal temperature should be maintained at 37.0 ± 0.5 °C. It is strongly suggested that heart rate, respiration rate, mean arterial blood pressure, and oxygen saturation (from oximetry) are monitored. Blood gas should be sampled once during a break between imaging scans. All recorded physiological parameters are within normal physiological ranges. MRI is acquired on the day of the TBI procedure

(1–3 h post-TBI), and again on days 1, 2, 7, and 14 after TBI onset.

6. Rectal temperature is maintained at 36.5–37.5 °C and respirations are recorded throughout the study. Body core temperature is critical, because it could affect the outcome.
7. Nissl staining is used to measure morphological changes within and around the impact following TBI [27, 33]. Briefly, anesthetized rats are perfused on day 14 post-TBI with ice-cold heparinized phosphate-buffered saline, followed by ice-cold 4% buffered paraformaldehyde (PFA). Brains are removed and post-fixed with 4% PFA and subsequently cryopreserved (30% sucrose) for 48 h at 4 °C. Coronal section measuring 25-μm are affixed to gelatin-coated slides and dried overnight at 37 °C. Slides are hydrated through a series of graded alcohols to distilled water, followed by 0.1% cresyl violet acetate for 7 min. Brain sections are then dehydrated, cleared in Histo-clear, xylene and cover-slipped with Permount mounting medium. Images are acquired on an Olympus BX60F microscope equipped with an Olympus DP70 camera. A 5× objective is utilized to acquire images for mosaic full brain images assimilated MRI using Microsoft ICE software. Intact neurons are defined as nonbasophilic neurons with pale nuclei and discrete nucleoli and that had an intact neuronal body. Dark-stained neurons are defined as neurons with abnormal morphology, such as hyperbasophilic neurons and neurons with shrunken morphology.

3.2 MRI

1. *Position of RF coil*: Position the RF coil as central to the region of interest as possible. For a surface coil, avoid pressing the coil too hard on the animal's head as it would increase "loading," which decreases SNR.
2. *Tune and match RF coil*: Tune and match RF coil by adjusting the capacitors to 1 H resonance frequency and 50 Ω.
3. *Position scan*: Position the scan on *x*, *y*, and *z* to ensure that the subject is centered. Open up the FOV if needed.
4. *Shimming*: Run autoshim or manual shim as needed.
5. *Calibrate RF pulses*: Calibrate RF pulses for given pulse shapes and durations. This can be set up to be done automatically.
6. *Pilot scan*: Perform a pilot scan, using a 2D gradient echo FLASH or RARE sequence (10–30 s). Based on the pilot scan, plan 5–8 1.0 mm coronal slices to cover the region of interest.
7. *T2 map (9.5 min)*: T_2-weighted images are acquired using a fast spin-echo sequence with four effective echo times (18, 54, 90, and 126 ms), and four echo train lengths, where the center of the echo train is taken as the effective TE for T_2 calculation. Typical parameters are spectral width 30–50 kHz, TR = 2–3 s (90° flip angle), pulse shape Gaussian or Sinc3, and pulse duration 1–2 ms.

8. *Diffusion tensor-weighted MRI* (*3.5 min*): DTIs are obtained with five $b = 0$ and 30 directions with a 1200 s/mm² b_{max}-value. Echo planar imaging (EPI) scans with partial Fourier (3/4) are also acquired using the following settings: seven 1.0 mm coronal images, field of view (FOV) = 2.56 × 2.56 cm, matrix 96 × 96 and reconstructed to 128 × 128, single shot, repetition time (TR) = 3 s, echo time (TE) = 32 msec, separate between diffusion gradient $\Delta = 14$ ms, diffusion gradient duration $\delta = 5$ ms, and 2 transients for signal averaging.
9. *CBF*: There are two methods to measure CBF, namely continuous arterial spin labeling (cASL) or dynamic susceptibility contrast (DSC) MRI with Magnevist (Gd-DTPA) or Omiscan (another contrast agent). With the latter, measurement can only be made once every hour or so because intravascular half-life of MRI is on the order of 6 min. In TBI, the contrast agent is often trapped; a longer wait time may be necessary.

   For the cASL technique, single-shot, gradient-echo, and echo-planar-imaging (EPI) acquisition is used. Paired images are acquired alternately—one with arterial spin labeling and the other without (control). MR parameters are as follows: data matrix = seven 1.0 mm coronal images, field of view (FOV) = 2.56 × 2.56 cm, matrix 96 × 96 and reconstructed to 128 × 128, single shot, repetition time (TR) = 3 s, and echo time (TE) = 10.2 msec. Continuous arterial spin labeling employed a 2.7-s square radiofrequency pulse to the labeling coil in the presence of 1.0 G/cm gradient along the flow direction, such that the condition of adiabatic inversion is satisfied. The sign of the frequency offset is switched for control (non-labeled) images. Number of averages is typically 60–100 depending on the SNR needed.

   Parameters for the DSC technique are single-shot, gradient-echo, echo-planar-imaging (EPI) acquisition with matrix = 64 × 64, FOV = 2.56 cm × 2.56 cm, 3–5 slices of 1.0 mm, TE = 20 ms, and TR = 0.333 s (22° flip angle). Preload the IV line with 0.15–0.2 ml of Magnevist or Omiscan (typically 3 ft long of PE-50 tubing will hold such volume). About 20 s into the acquisition, deliver the contrast agent in a single bolus flush of saline. Continue DSC acquisition for another 40 s. Note that if DSC is used, Gd-DTPA has a non-negligible intravascular half-life. cASL and fMRI studies cannot be done immediately after Gd-DTPA injection (*see* **Note 2**).
10. *fMRI*: Combined CBF and BOLD measurements are made using the continuous arterial spin-labeling technique with single-shot, gradient-echo, and echo-planar-imaging (EPI) acquisition. Paired images are acquired alternately—one with arterial spin labeling and the other without (control). MR parameters are as follows: seven 1.0 mm coronal images, field

of view (FOV) = 2.56 × 2.56 cm, matrix 96 × 96 and reconstructed to 128 × 128, single shot, repetition time (TR) = 3 s, and echo time (TE) = 10.2 ms. Continuous arterial spin labeling employed a 2.7-s square radiofrequency pulse to the labeling coil in the presence of 1.0 G/cm gradient along the flow direction, such that the condition of adiabatic inversion is satisfied. The sign of the frequency offset is switched for control (non-labeled) images. For each set of fMRI measurements, 4 min of baseline data acquisition, 3 min of data acquisition during hypercapnic challenge, and 3 min of data acquisition during the post-stimulation surveillance period. Hypercapnic challenges use a premixed gas of 5 % CO_2 with 21 % O_2 and balance N_2.

11. *Ktrans-MRI* (*19.4 min*): K^{trans} MRI data are acquired using a 2D multi-slice FLASH sequence. A prescan module is used to determine the flip angle and M_0 distribution, which includes 3 FLASH scans with different TRs: 64 ms (scan 1), 200 ms (scan 2), and 3000 ms (scan 3). The rest of imaging parameters are as follows: five 1.0 mm coronal slices, TE = 2 ms, FOV = 2.56 × 2.56 cm^2, 128 × 128 data matrix, and 30° nominal flip angle. The prescan takes 7.4 min. Dynamic scans use a TR of 64 ms, and otherwise identical sequence parameters. After baseline data are acquired for 2 min, a bolus (0.2 ml/kg) of gadodiamide (GE Healthcare, USA) is injected intravenously through the tail vein, during which the dynamic scan is continued. A total of 90 dynamic images are acquired with a temporal resolution of 8 s, lasting 12 min total. Because of using gadodiamide, cASL scan cannot be performed after K^{trans} scan (*see* **Note 2**).
12. *Quantitative susceptibility mapping* (*QSM*): QSM data are acquired using a 3D multi-gradient echo (MGE) sequence with the following parameters: FOV = 2.56 × 2.56 × 2.56 cm^3, data matrix = 256 × 192 × 128, TE1 = 4.4 ms, echo spacing = 6.1 ms, 4 echoes, and TR = 28 ms.

3.3 Image Analysis

Image calculation and co-registration can be done using codes written in Matlab (MathWorks Inc, Natick, MA) [34, 35]. In addition to Matlab programs, we also use the STIMULATE (University of Minnesota) software for display and plotting. There are also many other free software programs available to calculate and display MRI images.

3.3.1 Map Calculations

1. ADC

ADC maps with intensity in unit of mm^2/s are calculated pixel by pixel by using [36, 37],

$$\text{ADC} = -\ln\left(S_1 / S_0\right) / \left(b_1 - b_0\right)$$

where $b_i = \gamma^2 G_i^2 \delta^2(\Delta - \delta/3)$, ln is the natural logarithm, and S_0 and S_1 are the signal intensities obtained with b_0 and b_1, respectively. The *b*-value is proportional to the gradient strength (G), magnetogyric ratio (γ), duration of each gradient pulse (δ), and the time (Δ) between applications of the two gradient pulses. ADC maps are calculated at each time point.

2. Fractional anisotropy (FA):

   FA can be calculated using either Matlab code or other MRI analysis toolboxes (e.g., DTIStudio, BrainVoyager). Three eigenvalues (λ_1, λ_2, and λ_3) are derived from DTI data and FA is calculated using

$$\text{FA} = \frac{\sqrt{3\left((\lambda_1 - \mathbb{E}[\lambda])^2 + (\lambda_2 - \mathbb{E}[\lambda])^2 + (\lambda_3 - \mathbb{E}[\lambda])^2\right)}}{\sqrt{2(\lambda_1^2 + \lambda_2^2 + \lambda_3^2}}$$

3. CBF

   For ASL images, CBF images (S_{CBF}) with intensity in units of ml/gram/min are calculated [38, 39] pixel by pixel using

$$S_{\text{CBF}} = l / T_1 (S_c - S_L) / (S_L + (2a - 1) S_c),$$

   where S_C and S_L are signal intensities of the control and labeled images, respectively. λ—0.9 ml/g—is the partition coefficient [40]. α is the labeling efficiency which is measured to 0.75–0.9 in animal models.

   For DSC-CBF calculation, the transverse relaxation rate (ΔR_2^*) is calculated using the equation $\Delta R_2^*(t) = -\ln(S(t)/S_0)/\text{TE}$, where $S(t)$ is the signal intensity at time t, S_0 is the precontrast baseline signal intensity, and TE is the sequence echo time. A CBF map is then generated by deconvolving the change in tissue concentration over the first pass of contrast agent with an arterial input function using singular value decomposition [41, 42]. Mean transit time and cerebral blood volume can also be obtained with this analysis and they may be useful for stroke analysis. This software can be obtained from many sources.

4. T_2

   T_2 maps can be calculated from T_2-weighted MRI with two or more echo times (TE). When using two echoes, T_2 can be calculated directly using

$$T_2 = -\ln(S_{\text{TE2}} / S_{\text{TE1}}) / (\text{TE}_2 - \text{TE}_1),$$

   where S_{TE2} and S_{TE1} are the signal intensities obtained with TE_2 and TE_1, respectively. If using more than two echoes, a linear fit needs to be used to fit

$$\ln(S_{\text{TE}}) = \ln(S_0) - \text{TE} / T_2,$$

   where S_0 is the maximum signal from the sample.

5. K^{trans} mapping [24]

The steady-state spoiled gradient echo (GRE, acquired using the 2D FLASH sequence) signal amplitude for a given TR (M_{TR}) can be related to M_0, R_1, and α as follows:

$$M_{TR} = M_0\left(r, s_{coil}\right) \cdot \sin\alpha \frac{1-\exp\left(-TR \cdot R_1\right)}{1-\exp\left(-TR \cdot R_1\right) \cdot \cos\alpha} \cdot \exp\left(-TE / T_2^*\right) \quad (1)$$

where M_0 is a function of both spatial location r and coil sensitivity s_{coil}. To minimize the R_2^* signal decay, a single-echo GRE scan is used with minimum TE and a low dose of gadolinium. As an approximation, $\exp\left(-TE / T_2^*\right) \approx 1$ is used in the subsequent calculation. A prescan module composed of three GRE scans is used to determine flip angle (α) and tissue magnetization (M_0).

For flip-angle mapping, simulations using Eq. 1 show that the ratio of GRE magnitudes at TRs of 200 and 64 ms depends strongly on α, but very weakly on T_1. Given the narrow T_1 distribution of brain tissue, a raw α (or B_1^+) map can be obtained from the ratio map with a fixed T_1 value. Assuming that α is a smooth function of location, the final α map can be obtained after smoothing.

For M_0 determination, according to Eq. 1, the GRE signal becomes less dependent on T_1 with increasing TR. With a long TR of 3000 ms, GRE magnitude depends primarily on $\sin(\alpha)$ for small to medium flip angles ($\alpha < 45°$), but very weakly on T_1. Given the normal flip angle of 30°, the actual flip angle of the brain tissue is within the range of 15–45°; therefore, M_0 can be determined from the long TR scan and the flip angle distribution as follows:

$$M_0 = M_{3000ms} / \sin\alpha \quad (2)$$

Given the smooth α distribution, high-SNR M_0 map, and dynamic scans using sensitive surface coil at high field, dynamic R_1 map can be obtained using Eq. 1 with sufficient SNR.

The AIF is determined using the data and the scaling approach by Ewing and colleagues [43, 44]. Briefly, the mean AIF (AIF_{mean}) is measured in a group of male rats (approximately 300 g) using the custom-synthesized radiolabeled Gd-DTPA [43]. Assuming that the plasma volume is 1 % and there is no BBB leakage in the contralesional side of caudate-putamen (CPU), the AIF is determined using the following scaling:

$$AIF(t) = AIF_{mean} \cdot \left[100 \cdot \int_{t=3min}^{9min} R_{1,CPU}(t)\,dt\right] / \left[\int_{t=3min}^{9min} AIF_{mean}\,dt\right] \quad (3)$$

The R_1 and the AIF are then used to fit the extended Kety model:

$$C_t(t) = K^{\text{trans}} \int_{\tau=0}^{t} C_p(\tau) e^{-k_{ep}(t-\tau)} d\tau + v_p C_p(t) \quad (4)$$

where $C_t(t)$ and $C_p(t)$ are tissue and plasma gadolinium concentrations, and k_{ep} is the reversible mass transfer coefficient. In this study, the R_1 and scaled AIF values are directly used for data fitting without further conversion to concentrations using the relativity of contrast agent. Since it is difficult to obtain acceptable K^{trans} maps by simultaneously fitting all three parameters, we use the model selection approach by Ewing and colleagues [44] to select a simpler model that could sufficiently describe the dynamic contrast change. Data is fit to the extended Kety model with the two following assumptions as described by Ewing and colleagues [44]:

Irreversible leakage (two-parameter model, or Model 2):

$$K^{\text{trans}} > 0, k_{ep} = 0 \quad (5)$$

Reversible leakage (three-parameter model, or Model 3):

$$K^{\text{trans}} > 0, \; k_{ep} > 0$$

For model selection, the irreversible model is selected, unless the reversible model yields a statistically significant better fit. The cutoff criteria use the F-statistic, which can be calculated using the summed squared residues (SSE) and the number of samples (N) as follows:

$$F = \left[(\text{SSE}_3 - \text{SSE}_2)/1 \right] / \left[\text{SSE}_3 / (N-3) \right] \quad (6)$$

where the subscript represents the corresponding model. The final selection mask is determined as

$$M_{\text{select}} = F > F_0 \quad (7)$$

The threshold for F-statistic (F_0) is set as 10 ($P < 0.05$). The final K^{trans} map is then determined as

$$K_{\text{final}}^{\text{trans}} = K_{\text{Model1}}^{\text{trans}} \cdot (1 - M_{\text{select}}) + K_{\text{Model2}}^{\text{trans}} \cdot M_{\text{select}} \quad (8)$$

All the calculations are performed using Matlab R2011b (Mathworks, Natick, MA).

6. Quantitative susceptibility mapping:

QSM is performed using the STI Suite [45]. Briefly, the brain image is reconstructed with 3D fast Fourier transform. The complex data is separated into magnitude and phase. The magnitude image is used for the extraction of the brain tissue. The phase images are unwrapped using Laplacian-based unwrapping [46] and summed. The background phase is removed using the V-SHARP method [47]. The tissue susceptibility is determined using the LSQR method [48]. The resulting magnetic susceptibility values are directly used for comparison [49].

3.3.2 Abnormal Tissue Volume Analysis

Abnormal tissue volume can be determined using either manually ROI drawing or threshold methods. Manually region of interest (ROI) drawing method usually needs more than two data analyzers to manually delineate the border of the abnormal tissue and then count the number of pixels inside the ROI. Commonly used threshold method is using the mean value ± 2 standard deviations of the homologous contralesional region (*see* **Note 3**). Depending on the data, it could be two or three standard deviations (*see* **Note 4**). In TBI abnormal tissue volume analysis, ADC, CBF, FA, T_2, and K^{trans} can all be analyzed using these two methods.

3.3.3 Hypercapnic Responses

Images obtained during the transition period between baseline and stimulus onset (30 s for CO_2 challenge) are discarded. BOLD images are obtained from the control (non-labeled) images of the CBF measurements. BOLD and CBF magnitude and percent changes relative to baselines are calculated on a pixel-by-pixel basis.

3.4 Additional Advanced Analysis

3.4.1 Data Co-registration

Image maps of individual subjects are co-registered across time points via the application of a transformation matrix generated by QuickVol and MRIAnalysisPak software [35, 50].

3.4.2 Spatial-temparal Changes of MRI Measures (ADC, FA, CBF, T2, K^{trans}, and Hypercapnic Responses)

Three regions of interest (ROIs) are defined along the ipsilesional cortex, along with three ROIs in homologous regions of the contralesional hemisphere. The same ROIs are used to tabulate the MRI measures across all time points using Stimulate software (University of Minnesota). Values of the ipsilesional ROI are normalized with respect to those of the contralesional ROI.

3.4.3 Quantification of White Matter Damage

To evaluate MRI contrast along the white matter tracts and to eliminate the subjectivity of manual region of interest selection, we project the maximum values of magnetic susceptibility, and the minimum values of DTI FA along the corpus callosum. Such an approach is used to minimize the potential partial volume effect due to the small fraction of white matter in rodent brains. The projected values are compared across different time points.

4 Notes

1. Following the TBI impact, the cranial opening needs to be sealed with bone wax and then the scalp sutured closed. The blood, if any, needs to be cleared. Saline should be injected under the skin to facilitate removal of air pockets between the scalp and the skull to minimize artifacts during MRI acquisition. These steps (using bone wax and saline) are very important to avoid MRI artifacts.
2. Gd-DTPA or gadodiamide will change the relaxation time (T_1 and T_2) of blood and tissue. Therefore CBF will not be accurately obtained using cASL after using Gd-DTPA or gadodiamide.
3. In TBI, some parameters can change in the positive or negative direction at the same time due to heterogeneity within the tissue. For example, both decreased and increased ADC can be observed in acute TBI. When calculating abnormal ADC tissue volume, both types of tissues need to be counted.
4. The variances of different parameters are different. For small variance data, the threshold needs to use mean value plus higher times of standard deviations (SD). For example, for T_2 maps, two SDs would be good. For ADC, which has small variance, it may be needed to use mean plus three SDs as the threshold.

Acknowledgements

This work is supported in part by NIH/NINDS R01 NS45879, a TL1 and KL2 TR001118, and Mike Hogg Fund via the Clinical Translational Science Awards (CTSA, parent grants UL1TR000149, TL1TR001119, and KL2TR001118). We thank our former and current colleagues who have participated on these TBI projects.

References

1. Nortje J, Menon DK (2004) Traumatic brain injury: physiology, mechanisms, and outcome. Curr Opin Neurol 17:711–718
2. Werner C, Engelhard K (2007) Pathophysiology of traumatic brain injury. Br J Anaesth 99:4–9
3. Blennow K, Hardy J, Zetterberg H (2012) The neuropathology and neurobiology of traumatic brain injury. Neuron 76:886–899
4. Gardner A, Kay-Lambkin F, Stanwell P, Donnelly J, Williams WH, Hiles A, Schofield P, Levi C, Jones DK (2012) A systematic review of diffusion tensor imaging findings in sports-related concussion. J Neurotrauma 29:2521–2538
5. Albensi BC, Knoblach SM, Chew BGM, O'Reilly MP, Faden AI, Pekar JJ (2000) Diffusion and high resolution MRI of traumatic brain injury in rats: time course and correlation with histology. Exp Neurol 162:61–72
6. Iwamoto Y, Yamaki T, Murakami N, Umeda M, Tanaka C, Higuchi T, Aoki I, Naruse S, Ueda S (1997) Investigation of morphological change of lateral and midline fluid percussion injury in rats, using magnetic resonance imaging. Neurosurgery 40:163–167

7. Van Putten HP, Bouwhuis MG, Muizelaar JP, Lyeth BG, Berman RF (2005) Diffusion-weighted imaging of edema following traumatic brain injury in rats: effects of secondary hypoxia. J Neurotrauma 22:857–872
8. Ito J, Marmarou A, Barzo P, Fatouros P, Corwin F (1996) Characterization of edema by diffusion-weighted imaging in experimental traumatic brain injury. J Neurosurg 84:97–103
9. Barzo P, Marmarou A, Fatouros P, Hayasaki K, Corwin F (1997) Contribution of vasogenic and cellular edema to traumatic brain swelling measured by diffusion-weighted imaging. J Neurosurg 87:900–907
10. Liu AY, Maldjian JA, Bagley LJ, Sinson GP, Grossman RI (1999) Traumatic brain injury: diffusion-weighted MR imaging findings. Am J Neuroradiol 20:1636–1641
11. Kou ZF, Wu Z, Tong KA, Holshouser B, Benson RR, Hu JN, Haacke EM (2010) The role of advanced MR imaging findings as biomarkers of traumatic brain injury. J Head Trauma Rehabil 25:267–282
12. Rutgers DR, Toulgoat F, Cazejust J, Fillard P, Lasjaunias P, Ducreux D (2008) White matter abnormalities in mild traumatic brain injury: a diffusion tensor imaging study. Am J Neuroradiol 29:514–519
13. Inglese M, Makani S, Johnson G, Cohen BA, Silver JA, Gonen O, Grossman RI (2005) Diffuse axonal injury in mild traumatic brain injury: a diffusion tensor imaging study. J Neurosurg 103:298–303
14. Nakayama N, Okumura A, Shinoda J, Yasokawa YT, Miwa K, Yoshimura SI, Iwama T (2006) Evidence for white matter disruption in traumatic brain injury without macroscopic lesions. J Neurol Neurosur Psychiat 77:850
15. Mac Donald CL, Dikranian K, Bayly P, Holtzman D, Brody D (2007) Diffusion tensor imaging reliably detects experimental traumatic axonal injury and indicates approximate time of injury. J Neurosci 27:11869–11876
16. van de Looij Y, Mauconduit F, Beaumont M, Valable S, Farion R, Francony G, Payen JF, Lahrech H (2012) Diffusion tensor imaging of diffuse axonal injury in a rat brain trauma model. NMR Biomed 25:93–103
17. Arfanakis K, Haughton VM, Carew JD, Rogers BP, Dempsey RJ, Meyerand ME (2002) Diffusion tensor MR imaging in diffuse axonal injury. Am J Neuroradiol 23:794–802
18. Barbier EL, Lamalle L, Decorps M (2001) Methodology of brain perfusion imaging. J Magn Reson Imaging 13:496–520
19. Calamante F, Thomas DL, Pell GS, Wiersma J, Turner R (1999) Measuring cerebral blood flow using magnetic resonance imaging techniques. J Cereb Blood Flow Metab 19:701–735
20. Baskaya MK, Rao AM, Dogan A, Donaldson D, Dempsey RJ (1997) The biphasic opening of the blood-brain barrier in the cortex and hippocampus after traumatic brain injury in rats. Neurosci Lett 226:33–36
21. Chodobski A, Zink BJ, Szmydynger-Chodobska J (2011) Blood-brain barrier pathophysiology in traumatic brain injury. Transl Stroke Res 2:492–516
22. Adelson PD, Whalen MJ, Kochanek PM, Robichaud P, Carlos TM (1998) Blood brain barrier permeability and acute inflammation in two models of traumatic brain injury in the immature rat: a preliminary report. Acta Neurochir Suppl 71:104–106
23. Dempsey RJ, Baskaya MK, Dogan A (2000) Attenuation of brain edema, blood-brain barrier breakdown, and injury volume by ifenprodil, a polyamine-site N-methyl-D-aspartate receptor antagonist, after experimental traumatic brain injury in rats. Neurosurgery 47:399–404, discussion 404-396
24. Li W, Long JA, Watts LT, Jiang Z, Shen Q, Li YX, Duong TQ (2014) A quantitative MRI method for imaging blood-brain barrier leakage in experimental traumatic brain injury. PLoS One 9:e114173
25. Cernak I (2005) Animal models of head trauma. NeuroRx 2:410–422
26. Xiong Y, Mahmood A, Chopp M (2013) Animal models of traumatic brain injury. Nat Rev Neurosci 14:128–142
27. Watts LT, Long JA, Chemello J, Van Koughnet S, Fernandez A, Huang SL, Shen Q, Duong TQ (2014) Methylene blue is neuroprotective against mild traumatic brain injury. J Neurotrauma 31:1063–1071
28. Long JA, Watts LT, Chemello J, Huang SL, Shen Q, Duong TQ (2015) Multiparametric and longitudinal MRI characterization of mild traumatic brain injury in rats. J Neurotrauma 32:598–607
29. Long JA, Watts LT, Li W, Shen Q, Muir ER, Huang S, Boggs RC, Suri A, Duong TQ (2015) The effects of perturbed cerebral blood flow and cerebrovascular reactivity on structural MRI and behavioral readouts in mild traumatic brain injury. J Cereb Blood Flow Metab 35:1852
30. Talley Watts L, Long JA, Manga VH, Huang S, Shen Q, Duong TQ (2015) Normobaric oxygen worsens outcome after a moderate traumatic brain injury. J Cereb Blood Flow Metab 35:1137–1144

31. Talley Watts L, Shen Q, Deng S, Chemello J, Duong TQ (2015) Manganese-enhanced magnetic resonance imaging of traumatic brain injury. J Neurotrauma 32:1001–1010
32. Watts LT, Long JA, Boggs RC, Manga H, Huang S, Shen Q, Duong TQ (2016) Methylene blue improves lesion volume, multi-parametric quantitative MRI measurements, and behavioral outcome following TBI. J Neurotrauma 33:194
33. Deitch AD, Moses MJ (1957) The Nissl substance of living and fixed spinal ganglion cells. II. An ultraviolet absorption study. J Biophys Biochem Cytol 3:449–456
34. Shen Q, Ren H, Cheng H, Fisher M, Duong TQ (2005) Functional, perfusion and diffusion MRI of acute focal ischemic brain injury. J Cereb Blood Flow Metab 25:1265–1279
35. Liu ZM, Schmidt KF, Sicard KM, Duong TQ (2004) Imaging oxygen consumption in forepaw somatosensory stimulation in rats under isoflurane anesthesia. Magn Reson Med 52:277–285
36. Stejskal EO, Tanner JE (1965) Spin diffusion measurements: spin echoes in the presence of a time-dependent field gradient. J Chem Phys 42:288–292
37. Shen Q, Meng X, Fisher M, Sotak CH, Duong TQ (2003) Pixel-by-pixel spatiotemporal progression of focal ischemia derived using quantitative perfusion and diffusion imaging. J Cereb Blood Flow Metab 23:1479–1488
38. Silva AC, Lee S-P, Yang C, Iadecola C, Kim S-G (1999) Simultaneous blood oxygenation level-dependent and cerebral blood flow functional magnetic resonance imaging during forepaw stimulation in the rat. J Cereb Blood Flow Metab 19:871–879
39. Duong TQ, Silva AC, Lee S-P, Kim S-G (2000) Functional MRI of calcium-dependent synaptic activity: cross correlation with CBF and BOLD measurements. Magn Reson Med 43:383–392
40. Herscovitch P, Raichle ME (1985) What is the correct value for the brain-blood partition coefficient for water? J Cereb Blood Flow Metab 5:65–69
41. Ostergaard L, Sorensen AG, Kwong KK, Weisskoff RM, Gyldensted C, Rosen BR (1996) High resolution measurement of cerebral blood flow using intravascular tracer bolus passages. Part II: Experimental comparison and preliminary results. Magn Reson Med 36:726–736
42. Ostergaard L, Weisskoff RM, Chesler DA, Gyldensted C, Rosen BR (1996) High resolution measurement of cerebral blood flow using intravascular tracer bolus passages. Part I: Mathematical approach and statistical analysis. Magn Reson Med 36:715–725
43. Nagaraja TN, Karki K, Ewing JR, Divine GW, Fenstermacher JD, Patlak CS, Knight RA (2010) The MRI-measured arterial input function resulting from a bolus injection of Gd-DTPA in a rat model of stroke slightly underestimates that of Gd-[14C]DTPA and marginally overestimates the blood-to-brain influx rate constant determined by Patlak plots. Magn Reson Med 63:1502–1509
44. Ewing JR, Bagher-Ebadian H (2013) Model selection in measures of vascular parameters using dynamic contrast-enhanced MRI: experimental and clinical applications. NMR Biomed 26:1028–1041
45. Li W, Avram AV, Wu B, Xiao X, Liu C (2014) Integrated Laplacian-based phase unwrapping and background phase removal for quantitative susceptibility mapping. NMR Biomed 27:219–227
46. Langkammer C, Schweser F, Krebs N, Deistung A, Goessler W, Scheurer E, Sommer K, Reishofer G, Yen K, Fazekas F, Ropele S, Reichenbach JR (2012) Quantitative susceptibility mapping (QSM) as a means to measure brain iron? A post mortem validation study. Neuroimage 62:1593–1599
47. Wu B, Li W, Guidon A, Liu CL (2012) Whole brain susceptibility mapping using compressed sensing. Magn Reson Med 67:137–147
48. Li W, Wu B, Liu CL (2011) Quantitative susceptibility mapping of human brain reflects spatial variation in tissue composition. Neuroimage 55:1645–1656
49. Li W, Wu B, Batrachenko A, Bancroft-Wu V, Morey RA, Shashi V, Langkammer C, De Bellis MD, Ropele S, Song AW, Liu C (2014) Differential developmental trajectories of magnetic susceptibility in human brain gray and white matter over the lifespan. Hum Brain Mapp 35:2698–2713
50. Schmidt KF, Ziu M, Schmidt NO, Vaghasia P, Cargioli TG, Doshi S, Albert MS, Black PM, Carroll RS, Sun Y (2004) Volume reconstruction techniques improve the correlation between histological and in vivo tumor volume measurements in mouse models of human gliomas. J Neurooncol 68:207–215

Chapter 36

A Simplified Workflow for Protein Quantitation of Rat Brain Tissues Using Label-Free Proteomics and Spectral Counting

Angela M. Boutté, Shonnette F. Grant, and Jitendra R. Dave

Abstract

Mass spectrometry-based proteomics is an increasingly valuable tool for determining relative or quantitative protein abundance in brain tissues. A plethora of technical and analytical methods are available, but straightforward and practical approaches are often needed to facilitate reproducibility. This aspect is particularly important as an increasing number of studies focus on models of traumatic brain injury or brain trauma, for which brain tissue proteomes have not yet been fully described. This text provides suggested techniques for robust identification and quantitation of brain proteins by using molecular weight fractionation prior to mass spectrometry-based proteomics. Detailed sample preparation and generalized protocols for chromatography, mass spectrometry, spectral counting, and normalization are described. The rat cerebral cortex isolated from a model of blast-overpressure was used as an exemplary source of brain tissue. However, these techniques may be adapted for lysates generated from several types of cells or tissues and adapted by the end user.

Key words Brain tissue, Electrophoresis, Mass spectrometry, Spectral counting, Traumatic brain injury, Biomarkers

1 Introduction

Traumatic brain injury (TBI) remains a serious medical issue worldwide. In the USA alone, more than 2.5 million patients suffered a TBI in 2010. Military personnel and athletes are among high-risk groups. More than 27,000 military personnel were diagnosed with a TBI in 2013 alone (http://www.cdc.gov/traumaticbraininjury) and sports-related TBI (e.g., concussions) have become a major health concern between both juvenile and adult athletes [1]. Although TBI has become increasingly important to the medical research community at-large and has been gaining more public attention, little is known about how the brain is affected at the molecular level. Application of a discovery-based approach, such as mass spectrometry-based proteomics analysis of animal models (e.g., rats or mice), may fill this current knowledge gap.

Firas Kobeissy et al. (eds.), *Injury Models of the Central Nervous System: Methods and Protocols,* Methods in Molecular Biology, vol. 1462, DOI 10.1007/978-1-4939-3816-2_36,

Mass spectrometry-based proteomics allows discreet analysis of brain tissues in an unbiased manner. Defining global protein abundance changes that occur in the injured brain is critical to understanding mechanisms of injury and may lead to identifying molecular based targets of therapy. Prior to determining the effect of a TBI, the uninjured brain proteome must first be defined and used as a reference. The source of brain tissue used within this exemplary analysis was derived from the control, e.g., sham treated, group of a rodent blast overpressure model [2, 3]. The method described herein offers a flexible strategy used to determine protein abundance in the rat cerebral cortex defined by label-free mass spectrometry and spectral counting. Intended for use as a control within a larger dataset, TBI-related proteins which were detectable in these sham samples are shown. The strategy defined herein may be easily adapted to study alterations in the rodent brain proteome as a consequence of TBI or neurological diseases using animal models as well as several types of mass spectrometry platforms.

2 Materials

2.1 General Reagents, Equipment, and Supplies

1. 20–200 μL Pipettes.
2. 5–200 μL Disposable, low-retention pipette tips (*Fisher Scientific*).
3. Optima LC-MS-grade H_2O (*Fisher Scientific*).
4. DL-dithiothreitol 99% (DTT) (*Sigma-Aldrich*).
5. 1.5 mL Low-retention Eppendorf tubes (*Fisher Scientific*).
6. 1–200 μL Low-retention gel loading tips (*Fisher Scientific*).
7. Refrigerated, tabletop centrifuge.
8. Microplate incubator shaker, 0–65 °C.
9. Tabletop vortex.
10. Variable-temperature heating block for Eppendorf tubes.
11. Speedvac.
12. Personal protective equipment: Nitrile gloves, lab coat, safety glasses.
13. High-performance personal computer with high-resolution monitor.

2.2 Isolation and Quantitation of Brain Tissue Proteins

1. Probe sonicator.
2. RIPA lysis buffer (*EMD Millipore*).
3. Protease and phosphatase inhibitors (*Sigma-Aldrich*).
4. Standard or micro BCA protein assay kit with bovine γ-globulin or serum albumin standards (*Thermo/Pierce*).

2.3 In-Gel Digest and Isolation of Tryptic Peptides

1. Power supply (e.g., *VWR or Fisher*).
2. Gel electrophoresis chamber, e.g., XCell SureLock® Mini-Cell (Thermo/Thermo/Life Technologies).
3. Accurate, broad-range molecular weight marker, typically Novex® Sharp Prestained Protein Standards, (3.5–260 kDa) (Thermo/Life Technologies).
4. 4–12% NuPage® Bis-Tris pre-cast gradient gels (Thermo/Life Technologies).
5. Novex MES running buffer (Thermo/Life Technologies).
6. NuPage® Lithium dodecyl sulfate (LDS) buffer (Thermo/Life Technologies).
7. Glass or plastic basins for mini-gels.
8. Optima-grade methanol (MeOH).
9. Acetic acid (CH_3COOH).
10. Optima-grade acetonitrile (ACN).
11. Novex® Colloidal Coomassie Kit (Thermo/Life Technologies).
12. Razor blades.
13. Backlight platform or light box.
14. α-Iodoacetamide (IAA) (*Sigma-Aldrich*).
15. Ammonium bicarbonate (Ambic, NH_4HCO_3) (*Sigma-Aldrich*).
16. Demethylated porcine trypsin (*Sigma-Aldrich*).
17. Formic acid (HCO_2H, FA) (*Sigma-Aldrich*).

2.4 Reversed-Phase Liquid Chromatography and Tandem Mass Spectrometry (RP-/LC-MS/MS)

1. Mass spectrometer capable of nano-electrospray ionization, fitted with an auto-sampler with variable temperature settings and an ion trap such as:
   (a) Easy nLC-II system (*Thermo Fisher Scientific*) with the
   (b) LTQ Velos dual-pressure linear ion *trap system (Thermo Fisher Scientific*).
2. Fused silica nanospray emitter tips (*New Objective*).
3. C18 resin (*Sigma-Aldrich*).
4. Mobile-phase buffers.
   (a) Solvent A (0.1% FA *in Optima* H_2O).
   (b) Solvent B (99.9% Optima ACN with 0.1% FA).

2.5 Database Searching, Peptide Assembly, and Generation of Spectral Count Data with Myrimatch and IDPicker

1. The most recent rat protein database in FASTA format (http://www.uniprot.org/).
2. Software packages
   (a) Bumberdash and IDPicker (http://fenchurch.mc.vanderbilt.edu/software.php).
   (b) Database Maker, e.g., from the COMPASS Suite (http://www.chem.wisc.edu/~coon/software.php, [4]).
   (c) ReAdW (Thermo) for converting .RAW MS/MS files into to .mzML or .mzXML, if needed for database searching. Most mass spectrometer vendors provide file converting tools …
   (d) Prism (Graphpad) or other statistical software package capable of viewing, normalizing, and analyzing count data (e.g., R http://cran.r-project.org/, R Studio http://www.rstudio.com/, or Matlab http://www.mathworks.com/products/matlab/, The Mathworks, Inc.).
   (e) Excel (Microsoft Corporation) or Open Office (https://www.openoffice.org/, The Apache Software Foundation) for normalization and viewing count data and for viewing direct exports of IDPicker generated data.

3 Methods

The proteome of any tissue or cell population is inherently complex and contains thousands of proteins and their variants (*see* **Notes 1** and **2**). Therefore, reliable semiquantitative data that contains appropriate proteome coverage must be established by using reliable methodology. Further, it is important that these methods provide error rates (e.g., false-positive rates (FDR) or *p*-values) and are accessible to other end users, such that studies may provide reproducible data with reasonable confidence in peptide and protein identifications.

Protein and/or peptide fractionation prior to reverse-phase liquid chromatography and mass spectrometry (LC-MS/MS) improve proteome coverage [5, 6]. Molecular weight fractionation via one-dimensional poly-acrylamide gel electrophoresis (1D-PAGE) followed by peptide generation and extraction using in-gel tryptic digests of biological replicates provides a straightforward approach. After LC-MS/MS, data may then be analyzed by spectral counting. Spectral counting is the frequency with which an MS/MS output containing a peptide's amino acid sequence has been detected. This approach has been shown to be directly proportional to protein abundance [7–9]. Using an exemplary data set and proposed schema (Fig. 1), the Myrimatch and IDPicker platforms allow (1) parsimonious peptide-protein assembly and (2) determination of FDR calculation with the use of a decoy (e.g., false positives) appended database [10].

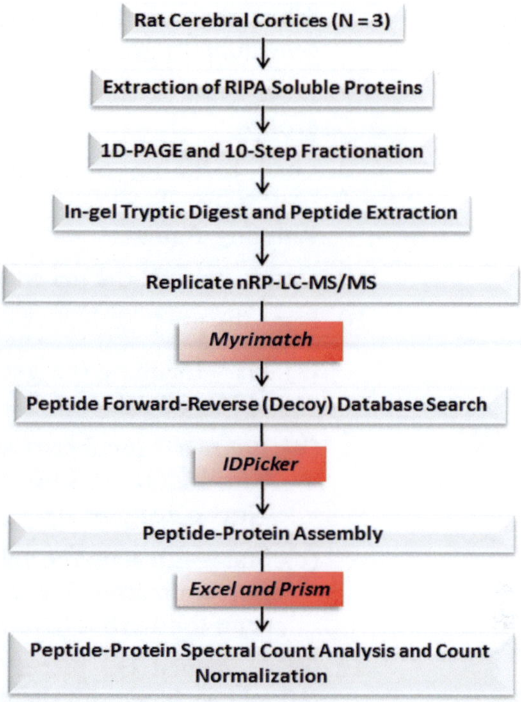

Fig. 1 Label-free proteomics and spectral counting schema for analysis of rat cerebral cortex brain tissues. An overview of sample preparation, 1D-PAGE protein fractionation, peptide generation, and extraction with subsequent proteomics analysis is indicated. After mass spectrometry and database searching with Myrimatch, resulting peptide-protein assembly is conducted with IDPicker. Protein lists containing spectral count data is then exported into spreadsheet and/or statistical analysis packages, e.g., Excel and Prism, respectfully

For TBI and other neurobiological studies that utilize animal models, establishment of the proteome based upon a control dataset, rather than relying upon the entire rat genome, is crucial to establishing a baseline of protein abundance. This control proteome may then be used as a reference to determine which peptides and proteins exhibit differential abundance as a consequence of TBI.

3.1 Extraction and Quantitation of Brain Tissue Proteins (See Notes *3–6)*

1. Gently lyse tissue by sonication in cold 1× RIPA buffer supplemented with protease/phosphatase inhibitors (1/5000, each).
2. Centrifuge 10–15k × *g* at 4 °C for 20 min.
3. Collect supernatant and determine protein concentration in triplicate by BCA method using 1/20–1/50 sample dilutions prepared in Optima H_2O.
4. Prepare stock solution of the supernatants at a final concentration of 1 μg/μL in 1× LDS buffer supplemented with 125 mM DTT.
5. Denature the samples by placing into a heating block at 95 °C; denature for 5 min.

6. Allow the samples to cool to room temperature and briefly centrifuge to prevent evaporation.
7. Load 2 μL of the molecular weight marker between sample groups.
8. Load 5–20 μg of sample per lane onto 4–12% gradient NuPAGE® Bis-Tris gels using 1× NuPAGE® MES-SDS Running Buffer.
9. Gently remove the gel and briefly wash with Optima H_2O.
10. Stain the gel with Novex Colloidal Blue Staining kit.
    (a) Fix—Incubate the gel for 10 min fixing solution (40 mL H2O, 50 mL MeOH, and 10 mL acetic acid).
    (b) Pretreat—Incubate the gel for 10 min in Stainer A missed with MeOH (55 mL H2O, 20 mL methanol, and 20 mL Novex Reagent A).
    (c) Stain—To stain the gel, add Reagent B directly to the solution containing Stainer A (5 mL B per every 100 mL A) to stain for 4–16 h (overnight).
11. Wash stained gel pieces with Optima H_2O, two times for 20 min each or until the background is clear.
12. Scan the gel and store images for reference.

3.2 In-Gel Digest and Isolation of Tryptic Peptides (*See* Note 7)

3.2.1 Protein Band Excision

1. Use a clean razor blade to excise the band of interest from the gel. Cut as close to the band as possible.
    (a) Molecular weight ranges (kDa) used to establish 10 fractions for each sample were <10, 10–20, 20–30, 30–40, 50–60, 60–70, 70–80, 80–100, 100–200, >250 (top).
    (b) This step serves as the first gradient used in this protocol based on observed molecular weight.
2. Cut the gel band into four cubes that are approximately 1 mm^3. Place the gel cubes in 1.5 mL low-retention Eppendorf tubes containing 100 μL Optima LC-MS H_2O.
3. Remove the liquid using a pipette with a gel loading tip and discard.
4. Add 100 μL Optima LC-MS, vortex for 5 min, and discard H_2O.
5. Wash gel pieces with 100 μL 50% 100 mM NH_4HCO_3/50% ACN for 30 min.
6. At 5–10-min intervals, mix the samples by vortexing. Remove the wash and discard.
7. Repeat **steps 3–4** until gel pieces are colorless.
8. Dehydrate gel pieces with 20 μL Optima ACN.
9. Speedvac the samples to dryness, without heat, for 10 min.
10. Store at −80 °C or continue to the following steps.

3.2.2 Reduction and Alkylation

1. Rehydrate gel pieces with three washes of 150 μL 100 mM Ambic and discard washes.
2. Add 150 μL 100 mM Ambic supplemented with 5 μL of 100 mM DTT.
3. Incubate for 15 min at 50 °C.
4. Allow tubes to cool to room temperature and discard liquid.
5. Add 10 μL 100 mM IAA in 100 mM NH_4HCO_3.
6. Incubate for 15 min, in the dark, at room temperature.
7. Discard the liquid and add 100 μL of 10 mM NH_4HCO_3.
8. Vortex for 15 min.
9. Repeat **steps 6–7** three times.
10. Dehydrate gel pieces with 20 μL ACN for 5 min.
11. Remove liquid and speedvac, without heat, for 10 min.
12. Store at −80 °C or continue to the following steps.

3.2.3 Generation of Peptides with Tryptic Digests

1. Rehydrate gel pieces with 15 μL of 12.5 ng/μL trypsin solution in 25 mM NH_4HCO_3.
2. Incubate at 4 °C for 30 min.
3. Add additional 20 μL of 25 mM NH_4HCO_3.
4. Incubate overnight at 37 °C in a heating block.

3.2.4 Peptide Extraction and Reconstitution

1. Centrifuge the samples at 200–300 × *g* for 15 min.
2. Appropriately label a new set of 1.5 mL low-retention Eppendorf microcentrifuge tubes.
3. Transfer the supernatant into the newly labeled tubes.
4. Add 30 μL 50% ACN/50% Optima H_2O with 0.1% formic acid to the gel cubes.
5. Centrifuge for 20 min at 200–300 × *g*.
6. Transfer the supernatant into the newly labeled tubes.
7. Repeat **step 4–6**, pooling the extracted sample.
8. Speedvac without heat to complete dryness (or to <10 μL).
9. Resuspend the dried sample in 20 μL of Optima LC-MS-grade H_2O with 0.1% formic acid and mix by pipetting.
10. Centrifuge tubes for 15 min at 1500 rpm (269 × *g*; radius: 10.70 cm).
11. Carefully transfer supernatant into properly labeled HPLC vials. Avoid pipetting any residue or aggregates, if they occur, from the bottom of the Eppendorf tube.

3.3 Reversed-Phase Nanospray Liquid Chromatography Tandem Mass Spectrometry (nRP-LC/MS-MS) (See Note 6)

1. Set up mobile-phase buffers consisting of solvent A (99.9% H_2O supplemented with 0.1% formic acid) and solvent B (99.9% ACN supplemented with 0.1% formic acid).
2. Set up the gradient to separation which was achieved using a run time of 70 min.
   (a) First linear gradient—from 2 to 40% B over 60 min.
   (b) Second linear gradient—from 40 to 90% B over 5 min and held for 5 min before returning to the initial mobile-phase composition (2% B).
3. Inject 2 μL of each sample (e.g., derived from each gel piece) via the auto-sampler onto a trap column (EASY-Column 2 cm, ID 100 μm, 5 μm, C18-A).
4. Direct the injected sample onto an analytical column (EASY-Column, 10 cm, ID 75 μm, 3 μm, C18-A2) at a flow rate of 250 nL/min.
5. Acquire the MS/MS spectra on the top ten most abundant ions at a given chromatographic time point by data-dependent scanning.
6. Repeat the injection for each sample, such that each fraction is analyzed at least 2–3 times.

3.4 Database Searching, Peptide Assembly, and Generation of Spectral Count Data with Myrimatch and IDPicker (See Note 8)

1. With the Database Maker program, prepare a concatenated forward-reverse rat protein database using the most recent assembly rat.fasta file from Uniprot.org.
2. Save the resulting file in FASTA format.
3. If desired, convert the .RAW files to a Set up the database search using your FASTA database and specified input files containing MS/MS data (.RAW or mzXML).
4. Build search parameters and search the MS/MS files against the concatenated database inclusive of notation for decoy proteins.
   (a) Typically, search parameters are determined by the methods used during mass spectrometry data acquisition and specify amino acids modifications, e.g., addition of iodoacetyl cysteine (+57, carboxamidomethyl, from addition of IAA) and oxidation of methionine (+16, from typical sample preparation).
   (b) Detailed usage of Myrimatch and IDPicker is described [11].
   (c) Comprehensive lists of amino acid modifications are available (http://www.abrf.org/index.cfm/dm.home).
5. After searching, import the resulting .pepXML (e.g. IDPicker input) file directly into IDPicker.
   (a) Files may be imported as groups such that data is grouped by injury type, replicate injection, subject, control vs. treated, etc. Sorting the data in this manner facilitates viewing and exporting of spectral count data within IDPicker.

(b) Filter the data upon import such that the FDR limit is 5 % for proteins and 2 % for peptides.

(c) Due to the complexity of the samples, the exemplary data shown herein was sorted by animal (biological replicate, Rat01, 02, 03) and replicate injection (technical replicate, rep01, 02, or 03) (Fig. 2).

6. Choose the option in IDPicker to view the number of spectra for each protein, experimental group, biological replicate (subject or animal), and technical replicate (e.g., duplicate or triplicate LC-MS/MS injections).
7. Export the data as a .csv or spread sheet-compatible file. IDPicker also allows .html exports.
8. Import the data into a spreadsheet program (e.g., Excel or Open Office) and into a statistical analysis program (e.g., Prism). An example of data containing a subset of proteins identified in rat brain tissues that was exported from IDPicker into a spreadsheet program is shown (Table 1) For larger protein datasets, programs such as R or Matlab may be considered. The total number of proteins identified via this strategy and forward-reverse (decoy) database searching is indicated and annotated with the calculated FDR (Fig. 2).

3.5 Normalization and Visualization of Spectral Count Data (*See* Note 9)

1. First determine the mean, standard error, or coefficient of variance and range of each technical replicate based on the spectral count derived from IDPicker results.
2. Manual normalization in Excel

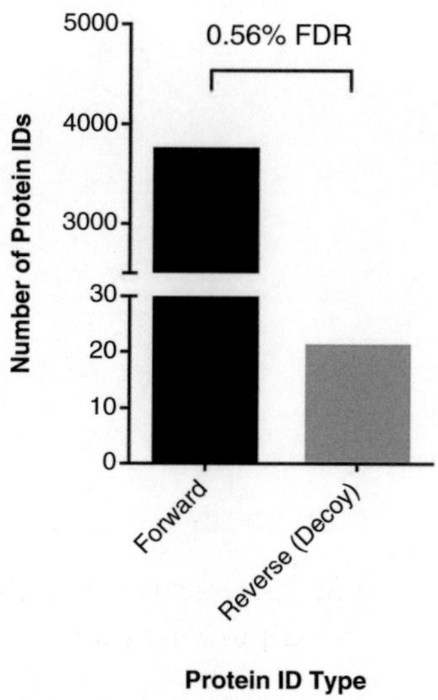

Fig. 2 The number of proteins and false discovery rat identified. The total number of proteins identified from the forward and reverse (decoy) databases. The calculated FDR based on the dataset is indicated

Table 1
Modified spectral count view of proteins detected in the rat cerebral cortex

Accession	Cluster	Coverage	Protein group	Distinct peptides	Distinct matches	Filtered spectra	Description	/ SS_ Groups/	Rat01_ Rep01	Rat01_ Rep02	Rat02_ Rep01	Rat02_ Rep02	Rat03_ Rep01	Rat03_ Rep02
sp|P16086|SPTN1_RAT	69	64.6	1989	180	214	1180	Spectrin alpha chain, non-erythrocytic 1 OS = Rattus norvegicus GN = Sptan1 PE = 1 SV = 2	1180	256	249	171	28	240	236
sp|Q6P9V9|TBA1B_RAT	169	44.8	1138	18	32	1084	Tubulin alpha-1B chain OS = Rattus norvegicus GN = Tuba1b PE = 1 SV = 1	1084	208	188	203	179	163	143
sp|P69897|TBB5_RAT	169	50.5	451	17	37	746	Tubulin beta-5 chain OS = Rattus norvegicus GN = Tubb5 PE = 1 SV = 1	746	127	110	133	111	135	130
sp|Q5XIF6|TBA4A_RAT	169	43.8	1239	16	29	715	Tubulin alpha-4A chain OS = Rattus norvegicus GN = Tuba4a PE = 2 SV = 1	715	144	127	137	118	105	84
sp|B0LPN4|RYR2_RAT	69	9.3	2098	36	36	59	Ryanodine receptor 2 OS = Rattus norvegicus GN = Ryr2 PE = 1 SV = 2	59	17	18	14	0	3	7

<table>
<tr><td>sp|P85515|ACTZ_RAT</td><td>1083</td><td>29.0</td><td>50</td><td>9</td><td>11</td><td>58</td><td>Alpha-centractin OS = Rattus norvegicus GN = Actr1a PE = 1 SV = 1</td><td>58</td><td>8</td><td>9</td><td>6</td><td>6</td><td>15</td><td>14</td></tr>
<tr><td>sp|Q794F9|4 F2_RAT</td><td>721</td><td>33.8</td><td>371</td><td>13</td><td>15</td><td>58</td><td>4 F2 cell-surface antigen heavy-chain OS = Rattus norvegicus GN = Slc3a2 PE = 1 SV = 1</td><td>58</td><td>13</td><td>10</td><td>7</td><td>10</td><td>9</td><td>9</td></tr>
<tr><td>sp|Q03626|MUG1_RAT</td><td>265</td><td>14.4</td><td>342</td><td>14</td><td>15</td><td>36</td><td>Murinoglobulin-1 OS = Rattus norvegicus GN = Mug1 PE = 2 SV = 1</td><td>36</td><td>5</td><td>9</td><td>6</td><td>8</td><td>6</td><td>2</td></tr>
<tr><td>sp|Q03626-2|MUG1_RAT</td><td>265</td><td>14.4</td><td>342</td><td>14</td><td>15</td><td>36</td><td>Isoform 2 of Murinoglobulin-1 OS = Rattus norvegicus GN = Mug1</td><td>36</td><td>5</td><td>9</td><td>6</td><td>8</td><td>6</td><td>2</td></tr>
<tr><td>sp|P07943|ALDR_RAT</td><td>1325</td><td>25.6</td><td>1551</td><td>8</td><td>10</td><td>35</td><td>Aldose reductase OS = Rattus nor vegicus GN = Akr1b1 PE = 1 SV = 3</td><td>35</td><td>6</td><td>4</td><td>7</td><td>7</td><td>2</td><td>9</td></tr>
<tr><td>sp|P20788|UCRI_RAT</td><td>1416</td><td>21.5</td><td>1757</td><td>5</td><td>6</td><td>17</td><td>Cytochrome b-c1 complex subunit Rieske, mitochondrial OS = Rattus norvegicus GN = Uqcrfs1 PE = 1 SV = 2</td><td>17</td><td>3</td><td>0</td><td>3</td><td>5</td><td>5</td><td>1</td></tr>
</table>

(continued)

Table 1
(continued)

Accession	Cluster	Coverage	Protein group	Distinct peptides	Distinct matches	Filtered spectra	Description	/ SS_ Groups/	Rat01_ Rep01	Rat01_ Rep02	Rat02_ Rep01	Rat02_ Rep02	Rat03_ Rep01	Rat03_ Rep02
sp|Q5U316|RAB35_RAT	1606	26.4	1849	5	5	15	Ras-related protein Rab-35 OS = Rattus norvegicus GN = Rab35 PE = 2 SV = 1	15	8	0	2	4	1	0
sp|P61589|RHOA_RAT	997	29.0	100	6	7	15	Transforming protein RhoA OS = Rattus norvegicus GN = Rhoa PE = 1 SV = 1	15	1	8	4	2	0	0
sp|P62959|HINT1_RAT	1478	18.3	1692	2	2	12	Histidine triad nucleotide-binding protein 1 OS = Rattus norvegicus GN = Hint1 PE = 1 SV = 5	12	3	2	2	4	1	0
sp|Q5PQL2|RCD1_RAT	1238	7.7	147	2	2	4	Cell differentiation protein RCD1 homolog OS = Rattus norvegicus GN = Rqcd1 PE = 1 SV = 1	4	1	0	0	1	1	1

<table>
<tr><td></td><td></td><td></td><td></td><td></td><td></td><td></td><td></td><td></td><td></td><td></td><td></td><td></td><td></td></tr>
<tr><td>sp|Q4V8B7|HSDL1_RAT</td><td>1236</td><td>7.9</td><td>1629</td><td>2</td><td>2</td><td>4</td><td>Inactive hydroxysteroid dehydrogenase-like protein 1 OS = Rattus norvegicus GN = Hsdl1 PE = 2 SV = 1</td><td>4</td><td>0</td><td>1</td><td>2</td><td>1</td><td>0</td><td>0</td></tr>
<tr><td>sp|Q496Z0|ELP2_RAT</td><td>1234</td><td>2.4</td><td>907</td><td>2</td><td>2</td><td>4</td><td>Elongator complex protein 2 OS = Rattus norvegicus GN = Elp2 PE = 2 SV = 1</td><td>4</td><td>1</td><td>1</td><td>2</td><td>0</td><td>0</td><td>0</td></tr>
<tr><td>sp|Q07009|CAN2_RAT</td><td>1232</td><td>5.6</td><td>777</td><td>3</td><td>4</td><td>4</td><td>Calpain-2 catalytic subunit OS = Rattus norvegicus GN = Capn2 PE = 1 SV = 3</td><td>4</td><td>0</td><td>0</td><td>1</td><td>2</td><td>0</td><td>1</td></tr>
<tr><td>DECOY_sp|Q7TT49|MRCKB_RAT</td><td>895</td><td></td><td>1965</td><td>2</td><td>2</td><td>2</td><td>Decoy</td><td>2</td><td>0</td><td>0</td><td>1</td><td>0</td><td>1</td><td>0</td></tr>
<tr><td>DECOY_sp|Q6AYA1|GAR1_RAT</td><td>8</td><td></td><td>2174</td><td>2</td><td>2</td><td>2</td><td>Decoy</td><td>2</td><td>0</td><td>2</td><td>0</td><td>0</td><td>0</td><td>0</td></tr>
</table>

Data generated by IDPicker was exported to an Excel spreadsheet

A subset of proteins with high, medium, and low abundance is shown

The accession number (a), cluster (b), percent of sequence coverage (c), protein group (d), distinct peptides (e), distinct matches (f), filtered spectra for an entire dataset (g), and description (h) are indicated. The cumulative number of filtered spectra for a group within an entire dataset is shown as "/SS_Groups/" (i). Decoy proteins are included and have a null entry within the concatenated database and are indicated by "Decoy_" notation and an empty disruption filed

The number of spectra is shown for each rat and replicate (Rat01_Rep01, etc.)

Empty fields are indicative of no detectable spectra and were manually annotated with "0" values

(a) Determine the average spectral count for each subject and replicate (subject 1, replicate 1 = S1R1; subject n and replicate m = SnRm).

(b) Determine the average spectral count across all replicates (e.g., the entire dataset) = X.

(c) Divide the number of spectral counts in each replicate by the average spectral count value derived from the entire dataset, such that (SnRm/X) = Y.

(d) Now, divide the original spectral count by the value generated "Y."

(e) The normalized spectral counts in each replicate should now be nearly equivalent. Spectral count data shown before and after manual normalization (Fig. 3b).

(f) Additional normalization strategies may be used. Often, some are included in mass spectrometry software packages. An overview of options is provided [12–15].

3. The normalized spectral count data of the proteome may now be imported into programs for further statistical analysis or protein-protein interactions. The spectral count of each biological and technical replicate is shown before and after normalization (Fig. 3a). In addition, normalized spectral counts of select TBI-related proteins (amyloid beta A4 protein, myelin basic protein, α-III tubulin, α-actinin, Tau, light chain of neurofilament, glial fibrillary acidic protein, strathmin, and β-III spectrin) identified by this schema are shown (Fig. 3b).

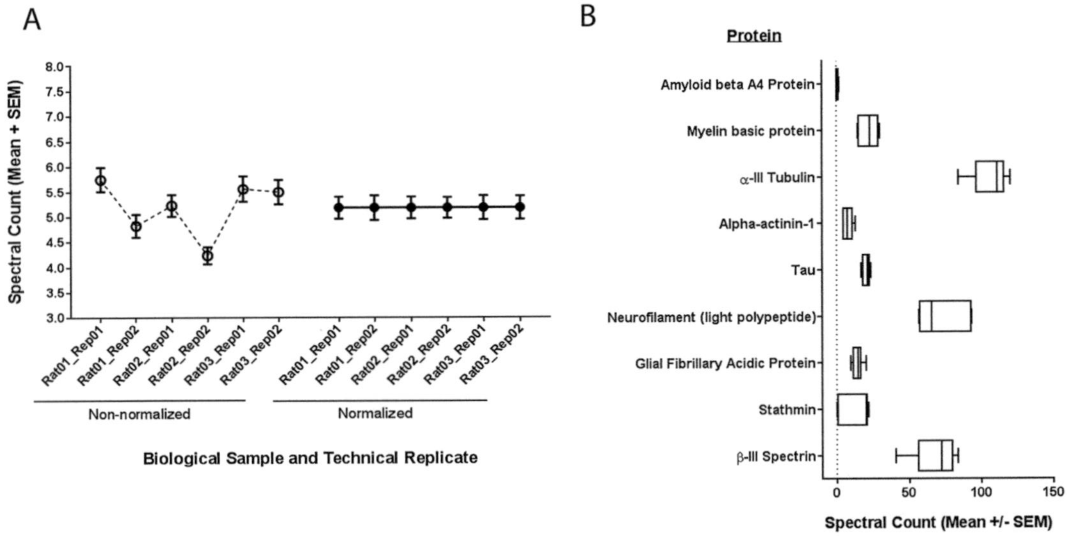

Fig. 3 Spectral count normalization and protein quantitation. (**a**) The spectral count (mean ± SEM) of all proteins detected by IDPicker within the rat cerebral cortex proteome is indicated before and after manual normalization. Each biological and technical replicate is detailed. (**b**) Box-plot showing the spectral count (Tukey) of several individual TBI-related proteins after normalization

4 Notes

1. Rodentia offer flexibility in several studies that require models for traumatic brain injury or neurological diseases. In the above example, rat strains offer the benefit of larger cerebrum size and a greater amount of starting material. Mice and guinea pigs are easily studied using this protocol. However, greater care in brain dissection techniques must be followed for smaller species and/or smaller brain regions.
2. This protocol is based on using male Sprague-Dawley rats weighing 300–350 g [3, 16]. For any manuscript, the age, weight, gender, and time elapsed after injury or induction of pathological models must be clearly stated in the manuscript.
3. For models that feature protein aggregation, further steps to isolate proteins that may not be readily soluble must be taken. Enrichment of insoluble proteins may be accomplished by incorporation of urea, *n*-laurel sarkosyl, greater concentrations of reducing agents, or isolation of proteins in formic acid. Due to changes in pH caused by formic acid, these samples must be neutralized or subjected to buffer exchange [17–20]. High salt and detergent concentrations can be removed with PAGE (as described), buffer exchange, or detergent removal resins. Samples must then be resuspended in appropriate buffer for PAGE prior to in-gel digests. Consideration must be given to the occurrence of sample loss each step; thus controls must be designed accordingly.
4. Increased sample handling and manipulation may increase protein oxidation and presence of contaminating keratins. Care must be taken to account for contaminants and detection of oxidation as a posttranslational modification and keratins (or similar peptides therein) during search and peptide-protein assembly, respectfully. The spectral count of contaminants should be very low or non-detectable after adjusting for false positives and appropriate statistical filters.
5. Several algorithms for searching, peptide-protein matching, and spectral counting exist and it is important to strongly consider allowing peptides to match to several proteins. Thus, isoforms and proteins with similar peptides will be included within a full dataset. In addition, semiquantitative analysis may be augmented by the addition of peptide peak intensity within spectral counting protocols [21].
6. This protocol relies upon in-gel digests to detect proteins initially isolated by molecular weight. Although detailed mass spectrometry is not the focus of this chapter, mass range for peptide detection should be designed by the end user. The mass window for each molecular weight range can be adjusted for

expected molecular weight ranges during the MS1 scans in order to avoid longer scan times and generation of (potentially) unnecessarily large files. However, this strategy must be used with caution. Protein migration may be altered due to incomplete denaturation and/or reduction, particularly in analysis of biological samples isolated after a pathological state or variability in total protein vs. moles of detergent or reducing agent.

7. All instruments and supplies should be cleaned according to standards required in an analytical chemistry lab and proper precautions should be made to protect the sample from contamination by the user and, more importantly, the user from chemical exposure.
8. Exporting data with tools such as IDPicker generally results in very large files when studying many replicates/samples and larger proteomes. A >4 gHz processor and >8 GB memory is suggested.
9. A common approach to manual normalization is presented. Other options as stated are perfectly suitable. However, it is diligent to manually check a few proteins and/or samples manually to ensure that data is reliable.

Acknowledgements

The authors would like to thank David Tabb, Matt Chambers, and Jay Holman, for advice as well as development and public licensing of the Bumberdash Suite and IDPicker; Joy Cagmat-Guingab for her expertise in analytical chemistry and mass spectrometry; Mr. Eric Maudlin-Jeronimo, SGT Myint, SPC Vincent Donkor, for exemplary technical skills; and Raymond Genovese, Stephen Ahlers, David W. Johnson, and Kara E. Schmid for advice and support. This research is funded by Combat Casualty Care Research Program and Congressionally Directed Medical Research Program (Contract #: W81XWH-12-2-0134).

Conflict of Interest

There are no potential conflicting interests to declare in the preparation of this chapter.

Disclaimer

This material has been reviewed by the Walter Reed Army Institute of Research. There is no objection to its presentation and/or publication. The views of the authors do not purport or reflect the position of the Department of the Army or the Department of Defense (para 4-3, AR 360-5).

References

1. McKee AC, Daneshvar DH, Alvarez VE, Stein TD (2014) The neuropathology of sport. Acta Neuropathol 127:29–51
2. Yarnell AM, Shaughness MC, Barry ES, Ahlers ST, McCarron RM, Grunberg NE (2013) Blast traumatic brain injury in the rat using a blast overpressure model. Jacqueline N. Crawley et al. (eds.). Current protocols in neuroscience Chapter 9, Unit 9 41
3. Genovese RF, Simmons LP, Ahlers ST, Maudlin-Jeronimo E, Dave JR, Boutte AM (2013) Effects of mild TBI from repeated blast overpressure on the expression and extinction of conditioned fear in rats. Neuroscience 254:120–129
4. Wenger CD, Phanstiel DH, Lee MV, Bailey DJ, Coon JJ (2011) COMPASS: a suite of pre- and post-search proteomics software tools for OMSSA. Proteomics 11:1064–1074
5. Antberg L, Cifani P, Sandin M, Levander F, James P (2012) Critical comparison of multidimensional separation methods for increasing protein expression coverage. J Proteome Res 11:2644–2652
6. Jafari M, Primo V, Smejkal GB, Moskovets EV, Kuo WP, Ivanov AR (2012) Comparison of in-gel protein separation techniques commonly used for fractionation in mass spectrometry-based proteomic profiling. Electrophoresis 33:2516–2526
7. Liu H, Sadygov RG, Yates JR 3rd (2004) A model for random sampling and estimation of relative protein abundance in shotgun proteomics. Anal Chem 76:4193–4201
8. Hoehenwarter W, Wienkoop S (2010) Spectral counting robust on high mass accuracy mass spectrometers. Rapid Commun Mass Spectrom 24:3609–3614
9. Lundgren DH, Hwang SI, Wu L, Han DK (2010) Role of spectral counting in quantitative proteomics. Expert Rev Proteomics 7:39–53
10. Eisenacher M, Kohl M, Turewicz M, Koch MH, Uszkoreit J, Stephan C (2012) Search and decoy: the automatic identification of mass spectra. Methods Mol Biol 893:445–488
11. Tabb DL, Fernando CG, Chambers MC (2007) MyriMatch: highly accurate tandem mass spectral peptide identification by multivariate hypergeometric analysis. J Proteome Res 6:654–661
12. Dillies MA, Rau A, Aubert J, Hennequet-Antier C, Jeanmougin M, Servant N, Keime C, Marot G, Castel D, Estelle J, Guernec G, Jagla B, Jouneau L, Laloe D, Le Gall C, Schaeffer B, Le Crom S, Guedj M, Jaffrezic F, French StatOmique C (2013) A comprehensive evaluation of normalization methods for Illumina high-throughput RNA sequencing data analysis. Brief Bioinform 14:671–683
13. Fermin D, Basrur V, Yocum AK, Nesvizhskii AI (2011) Abacus: a computational tool for extracting and pre-processing spectral count data for label-free quantitative proteomic analysis. Proteomics 11:1340–1345
14. Gokce E, Shuford CM, Franck WL, Dean RA, Muddiman DC (2011) Evaluation of normalization methods on GeLC-MS/MS label-free spectral counting data to correct for variation during proteomic workflows. J Am Soc Mass Spectrom 22:2199–2208
15. Carvalho PC, Fischer JS, Chen EI, Yates JR 3rd, Barbosa VC (2008) PatternLab for proteomics: a tool for differential shotgun proteomics. BMC Bioinformatics 9:316
16. Boutte AM, Yao C, Kobeissy F, May Lu XC, Zhang Z, Wang KK, Schmid K, Tortella FC, Dave JR (2012) Proteomic analysis and brain-specific systems biology in a rodent model of penetrating ballistic-like brain injury. Electrophoresis 33:3693–3704
17. Crowell AM, MacLellan DL, Doucette AA (2015) A two-stage spin cartridge for integrated protein precipitation, digestion and SDS removal in a comparative bottom-up proteomics workflow. J Proteomics 118:140–150
18. Liebler DC, Ham AJ (2009) Spin filter-based sample preparation for shotgun proteomics. Nat Methods 6:785, author reply 785-786
19. Manza LL, Stamer SL, Ham AJ, Codreanu SG, Liebler DC (2005) Sample preparation and digestion for proteomic analyses using spin filters. Proteomics 5:1742–1745
20. Sharma R, Dill BD, Chourey K, Shah M, VerBerkmoes NC, Hettich RL (2012) Coupling a detergent lysis/cleanup methodology with intact protein fractionation for enhanced proteome characterization. J Proteome Res 11:6008–6018
21. Arike L, Peil L (2014) Spectral counting label-free proteomics. Methods Mol Biol 1156:213–222

Chapter 37

Phenotypic Screening of Small-Molecule Inhibitors: Implications for Therapeutic Discovery and Drug Target Development in Traumatic Brain Injury

Hassan Al-Ali, Vance P. Lemmon, and John L. Bixby

Abstract

The inability of central nervous system (CNS) neurons to regenerate damaged axons and dendrites following traumatic brain injury (TBI) creates a substantial obstacle for functional recovery. Apoptotic cell death, deposition of scar tissue, and growth-repressive molecules produced by glia further complicate the problem and make it challenging for re-growing axons to extend across injury sites. To date, there are no approved drugs for the treatment of TBI, accentuating the need for relevant leads. Cell-based and organotypic bioassays can better mimic outcomes within the native CNS microenvironment than target-based screening methods and thus should speed the discovery of therapeutic agents that induce axon or dendrite regeneration. Additionally, when used to screen focused chemical libraries such as small-molecule protein kinase inhibitors, these assays can help elucidate molecular mechanisms involved in neurite outgrowth and regeneration as well as identify novel drug targets. Here, we describe a phenotypic cellular (high content) screening assay that utilizes brain-derived primary neurons for screening small-molecule chemical libraries.

Key words High-content screening, Primary neurons, Cell-based assay, Axon regeneration, CNS injury, Kinase inhibitor, Drug discovery

1 Introduction

According to the Centers for Disease Control and Prevention (CDC), at least 1.7 million cases of traumatic brain injury (TBI) occur every year and contribute to one third of all injury-related deaths in the USA [1]. The mechanism of TBI progression often involves diffuse axonal injury, associated with axon shearing from mechanical trauma, neurodegeneration, and formation of amyloid protein aggregate in afflicted areas [2]. Therefore, regeneration of axons to reestablish interrupted connections is a critical step for functional recovery from TBI. The adult CNS, however, is characterized by a markedly low capacity for axonal regeneration [3]. This involves an intrinsic loss of regenerative ability in mature neurons, which is correlated with changes in expression

Firas Kobeissy et al. (eds.), *Injury Models of the Central Nervous System: Methods and Protocols*, Methods in Molecular Biology, vol. 1462, DOI 10.1007/978-1-4939-3816-2_37, © Springer Science+Business Media New York 2016

of regeneration-associated genes [4, 5]. Furthermore, extension of axons from either mature or newly formed neurons is repressed by the presence of glial scars around damaged areas in the brain [6]. These scars contain growth-repressive molecules secreted by reactive astrocytes, of which chondroitin sulfate proteoglycans (CSPGs) are the best characterized [7–11]. Additionally, several myelin-associated proteins have been shown to negatively affect axon growth [7, 12–15]. Thus, identifying agents that can induce axon growth or overcome growth-repressive signals within the injury site could have significant clinical impact on the treatment of, and recovery from, TBI.

Over the past decade, phenotypic drug screening has been recognized as an effective method for discovering first-in-class drugs with new molecular modes of action [16]. In contrast to target-based screening, where the most selective ligands for a target involved in disease etiology are identified, phenotypic screening relies on complex cellular readouts [17, 18]. Phenotypic screening is therefore agnostic with regard to targets and molecular mechanisms. While this allows for the discovery of less selective, but potentially more effective drugs [19], the absence of prior knowledge regarding targets makes it difficult to optimize lead compounds or develop backups in case initial leads fail in clinical trials. However, phenotypic screening of focused and highly annotated chemical libraries, such as libraries of small molecules with verified activities against intracellular kinases, can identify effective pharmacological targets [20]. Kinases are proven drug targets in a multitude of diseases [21–26] and appear to be desirable targets in neurological disorders [20, 27–30]. Once a few candidate targets are identified, follow-up validation experiments, such as knockdown or overexpression, can be carried out to verify the relevance of the suggested targets and gain insight into their biological involvement [20, 31–33].

In this chapter, we detail a high-content bioassay suitable for low- to medium-throughput screening in the current reported format. It utilizes brain-derived primary neurons from E18 rat pups cultured on poly-d-lysine (PDL) and identifies protein kinase inhibitors that can promote neurite outgrowth (Fig. 1). The assay has an excellent Z′-factor of 0.73 and a relatively low false discovery rate of 7% [20, 34].

2 Materials

2.1 Cell Preparation and Culture

1. Hanks' Balanced Salt Solution (HBSS; Life Technologies cat. no. 14170-112).
2. 1 M HEPES (Life Technologies cat. no. 15630-080).
3. Glass Pasteur pipettes, flame polished.

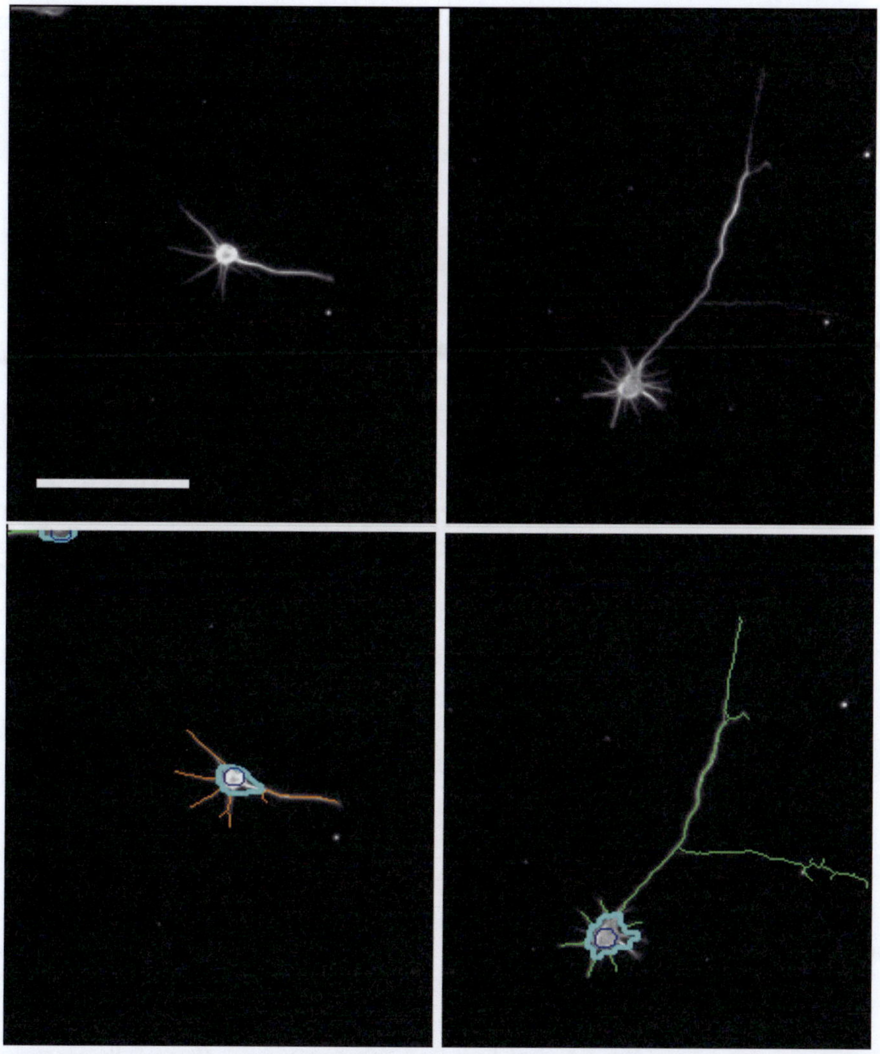

Fig. 1 High-content analysis of neurons in culture. Neurons cultured on PDL and immunostained for βIII-tubulin (cell bodies and neurites) and nuclei (Hoechst). *Top left*: Image of a neuron in a low-density culture treated with DMSO (control treatment). *Top right*: Image of a neuron in a low-density culture treated with a kinase inhibitor that promotes neurite outgrowth. *Bottom right* and *left*: Automated tracing of top images by the Cellomics Neuronal Profiling Bioaspplication software yields dozens of phenotypic measurements for each neuron. Scale bar 100 μm

4. Hibernate®-E (Life Technologies cat. no. A12476-01).
5. NbActiv4 Cell-culture media (BrainBits cat. no. Nb4-500).
6. NeuroCult™ SM1 Supplement (Stemcell cat. no. 05711): Prepare 1 ml aliquots and store at −20 °C.
7. 2.5 % (wt/vol) trypsin: Aliquot for single use (0.5 ml) and store at −20 °C.
8. DNase 30 mg/ml in ddH_2O (Sigma-Aldrich cat. no. D-5025): Make 0.1 ml aliquots and store at −20 °C.

9. Poly-d-lysine (PDL) solution, 0.5 mg/ml in HBSS prepared from PDL (molecular weight 30,000–70,000 kDa; Sigma-Aldrich, cat. no. P2636): Prepare stock at 1 mg/ml and store at –20 °C.

2.2 Immunostaining Reagents

1. Phosphate-buffered saline (PBS) [add 0.02% NaN_3 for long-term storage of plates and solutions].
2. 4% Paraformaldehyde (PFA) in PBS: Make stock solution at 16% and store at –20 °C. Dilute immediately before use and do not refreeze.

   PFA is a fixative and carcinogen: Exercise care when handling PFA and perform all steps inside a fume hood with sufficient protection. Be sure to properly dispose of waste PFA solution. To make 16% PFA stock, dissolve 16 g paraformaldehyde (wear mask during weighing) in 90 ml ddH_2O. Heat to 60 °C and stir until solution turns milky white. Add 0.1 M NaOH dropwise and continue stirring for at least 10 min after each drop. Repeat until solution is clear (keep at 60 °C and be patient). Check pH with pH strip and make sure that it is around 7.0.
3. Blocking and permeabilization buffer (0.2% fish gelatin, 0.03% Triton X-100, in PBS).
4. Anti-tubulin antibody: An antibody raised against βIII-tubulin [35] is recommended to avoid staining non-neuronal cells.
5. Hoechst solution (10 mg/ml).
6. Alexa Fluor® 488 Goat Anti-Mouse IgG (H + L) (Life Technologies cat. no. A-11029).

2.3 Equipment

1. Biological safety cabinet, big enough to accommodate a dissecting microscope.
2. Tissue culture incubator at 37 °C with humidified 5% CO_2 atm. Ensure that building and equipment (especially air pump) vibrations do not affect the incubator. These vibrations are often variable and can add noise to neurite outgrowth experiments. It may be necessary to set the incubator on vibration isolation blocks (such as VibraSystems ZA-124) to mitigate these effects.
3. Water bath at 37 °C.
4. Dissecting microscope.
5. Sterilized dissecting tools: Fine-tipped forceps, micro-dissecting scissors.
6. Hemocytometer or automated cell counter.
7. Sterile plasticware: 5, 10 and 25 ml serological pipettes, Falcon™ 96-Well Flat- and Round-Bottom Plates (VWR cat. no. 80086-578), bacteriological dishes, 15 and 50 ml conical centrifuge tubes.
8. Sterile cotton-plugged glass Pasteur pipettes.
9. Sterile filter-plugged pipette tips (1200, 200, and 20 μl).

3 Methods

3.1 Pre-coating Culture Plates

1. On the day before the experiment, pre-coat the 96-well plates with PDL (*see* **Note 1**) by plating 50 μl of PDL solution in each well. Since only the central 48 wells will be used for cell culture, it is recommended to save PDL by leaving the perimeter wells empty (*see* Subheading 3.3 and Fig. 2).
2. Wash plates the next morning four or five times with HBSS or PBS (150 μl/rinse), and then leave in buffer until cell plating.

3.2 Preparing the Cells

1. Euthanize pregnant rats timed to carry E18 embryos using an IACUC-approved method.
2. In a laminar flow hood, remove the embryos and place in a petri dish containing HBSS with 20 mM HEPES, pH 7.3.
3. Dissect pup brains [36] and collect hippocampi in 15 ml conical tube containing *x* ml of Hibernate E with SM1 (2% v/v) (*see* **Note 2**).
4. Prepare dissociation media by combining 4.5 ml of Hibernate E (without SM1) with 0.5 ml of trypsin and 100 μl of DNase solution.
5. Carefully remove the medium over the hippocampi and then add dissociation solution.
6. Incubate at 37 °C for 15–20 min, occasionally swirling the tube.
7. Using flame-polished cotton-plugged Pasteur pipettes, remove the dissociation media and then add 5 ml of Hibernate E containing SM1. Swirl the tube to thoroughly wash the tissue. Allow the tissue to settle to the bottom of the tube and then carefully remove the rinse solution. Repeat this step five times to dilute out the trypsin and DNase and remove any debris from lysed cells.
8. Remove final rinse media from the tube and then add 1 pipette-full (1–2 ml) of Hibernate E with SM1.
9. Using the flame-polished Pasteur pipette (pre-wetted with rinse media), triturate until all cells are dissociated and no visible chunks of tissue remain. It is recommended to perform fewer than ten triturations as these adversely affect cell viability (usually six is enough) (*see* **Note 3**).
10. Bring volume up to 8–12 ml using Hibernate E containing SM1, then mix well (tap tube; do not vortex), and determine cell concentration.

3.3 Plating

1. Dilute cells in culture media to a final concentration of 10,000 cells/ml.

A01	A02	A03	A04	A05	A06	A07	A08	A09	A10	A11	A12
B01	B02	150 μl	150 μl	150 μl	150 μl	150 μl	150 μl	150 μl	150 μl	B11	B12
C01	C02	150 μl	150 μl	150 μl	150 μl	150 μl	150 μl	150 μl	150 μl	C11	C12
D01	D02	150 μl	150 μl	150 μl	150 μl	150 μl	150 μl	150 μl	150 μl	D11	D12
E01	E02	150 μl	150 μl	150 μl	150 μl	150 μl	150 μl	150 μl	150 μl	E11	E12
F01	F02	150 μl	150 μl	150 μl	150 μl	150 μl	150 μl	150 μl	150 μl	F11	F12
G01	G02	150 μl	150 μl	150 μl	150 μl	150 μl	150 μl	150 μl	150 μl	G11	G12
H01	H02	H03	H04	H05	H06	H07	H08	H09	H10	H11	H12

Fig. 2 Cell plate layout: Load cells in indicated wells at 1500 cells per well in 150 μl of culture media

2. Aspirate HBSS from PDL-coated plates.
3. Load 150 μl of cell solution in the middle 48 wells (Fig. 2) at 1500 cells per well (*see* **Notes 4** and **5**). It is advisable to load culture media in the outer wells to decrease variations in cellular behavior from edge-wells effects.
4. Allow cells to adhere for 2 h in tissue culture incubator prior to treatment.

3.4 Preparing Compound Dilution Plates

1. Place culture medium in a 96-well plate (new—uncoated) as shown in Fig. 3 below (solution A = culture medium, solution B = culture medium + 0.8 % DMSO).
2. Add 1 μl of compound stock solution (10 mM in DMSO) to the wells in row B containing 124 μl (Fig. 3). This will produce a final DMSO concentration of 0.8 % and a final compound concentration of 80 μM. Control wells receive 1 μl of DMSO.
3. Using a multichannel manual P200 pipette, mix the solutions in the top row (3–5 times) and then move 25 μl to the row below. This results in a 1:5 dilution of compound while DMSO is held constant at 0.8 %. Repeat serial dilutions until the last row. This format allows for screening three compounds per plate at six concentrations with duplicate wells per condition.
4. Equilibrate dilution plates in tissue culture incubator for 1–2 h.

3.5 Quality Control (QC)

It is important to calculate the Z′-factor for any bioassay, both for quantifying the dynamic range of the assay [37] and for performing regular QC (*see* **Note 6**). Include an additional plate (or more) with the corresponding treatment format below (Fig. 4). (C_H is a compound that produces the high bioassay readout—in this case neurite outgrowth promotion, and C_L is a compound that produces the opposite (low) effect—or neurite outgrowth repression. Both compounds must be prepared at 4× the corresponding concentration where their maximal effects are observed. See below.) In the current assay, ML-7 and Torin-2 can be used as promoter and repressor controls, respectively [34].

A01	A02	Compound 1		Compound 2		Compound 3		DMSO		A11	A12
B01	B02	124 (A)	124 (A)	124 (A)	124 (A)	124 (A)	124 (A)	124 (A)	124 (A)	B11	B12
C01	C02	100 (B)	100 (B)	100 (B)	100 (B)	100 (B)	100 (B)	100 (B)	100 (B)	C11	C12
D01	D02	100 (B)	100 (B)	100 (B)	100 (B)	100 (B)	100 (B)	100 (B)	100 (B)	D11	D12
E01	E02	100 (B)	100 (B)	100 (B)	100 (B)	100 (B)	100 (B)	100 (B)	100 (B)	E11	E12
F01	F02	100 (B)	100 (B)	100 (B)	100 (B)	100 (B)	100 (B)	100 (B)	100 (B)	F11	F12
G01	G02	100 (B)	100 (B)	100 (B)	100 (B)	100 (B)	100 (B)	100 (B)	100 (B)	G11	G12
H01	H02	H03	H04	H05	H06	H07	H08	H09	H10	H11	H12

Fig. 3 Compound masterplate layout: Load media in indicated wells at indicated volumes. A is the medium without DMSO and B is the medium with DMSO added

A01	A02	A03	A04	A05	A06	A07	A08	A09	A10	A11	A12
B01	B02	4X C_H	4X C_H	4X C_H	4X C_H	4X C_L	4X C	4X C	4X C_L	B11	B12
C01	C02	4X C_H	4X C_H	4X C_H	4X C_H	4X C_L	4X C_L	4X C_L	4X C_L	C11	C12
D01	D02	4X C_H	4X C_H	4X C_H	4X C_H	4X C_L	4X C_L	4X C_L	4X C_L	D11	D12
E01	E02	4X C_H	4X C_H	4X C_H	4X C_H	4X C_L	4X C_L	4X C_L	4X C_L	E11	E12
F01	F02	4X C_H	4X C_H	4X C_H	4X C_H	4X C_L	4X C_L	4X C_L	4X C_L	F11	F12
G01	G02	4X C_H	4X C_H	4X C_H	4X C_H	4X C_L	4X C_L	4X C_L	4X C_L	G11	G12
H01	H02	H03	H04	H05	H06	H07	H08	H09	H10	H11	H12

Fig. 4 High/low control masterplate layout: Masterplates for high and low controls are prepared as indicated at 4× of final concentration where maximal effect is observed

3.6 Treatment

1. Using a multichannel manual P200 pipette, aspirate 50 μl from the compound dilution plate and add to the corresponding row in the cell plate (*see* **Notes 7** and **8**). Final compound concentrations in assay: 0.0064, 0.032, 0.16, 0.8, 4, and 20 μM (in 0.2% DMSO).
2. Culture for 48 h.

3.7 Fixing the Cultures

1. Remove medium from plates and immediately replace with 100–200 μl of warm (37 °C) PFA solution (*see* Subheading 4 for details about handling the plates).
2. Fix for 15–20 min at room temperature.
3. Rinse with PBS (200 μl/well × 3) (*see* **Note 9**).

3.8 Staining and Imaging

1. Remove PBS and add 100 μl of blocking/permeabilization buffer (PBS, 0.2% fish gelatin, 0.03% Triton X-100, 0.02% NaN_3) and incubate overnight at 4 °C (can be stored over the weekend) (*see* **Note 10**).
2. Add 100 μl of primary antibody solution (mouse anti-Beta III tubulin in blocking buffer) and incubate overnight at 4 °C (*see* **Note 11**).
3. Rinse wells with PBS (200 μl × 3).

4. Remove PBS and add 100 μl of secondary antibody solution (goat anti-mouse Alexa 488, 10 μg/ml Hoechst 33342, 0.2% fish gelatin, 0.02% azide, in PBS). Shake gently on a rotating shaker for 2 h.
5. Rinse wells with PBS (200 μl × 5).
6. Image plates using a Cellomics ArrayScan VTI in two different channels for nuclear staining (Hoechst) and cell body/neurite staining (βIII-tubulin). Typically, nine fields per well are imaged with a 5× objective and automatically traced by the Neuronal Profiling Bioapplication. To get reproducible results, at least 200–300 valid neurons (see below) should be measured per condition.

3.9 Data Analysis

1. Export plate data in Excel sheet format.
2. Filter out artifacts, cells that died upon plating, debris, etc. This can usually be achieved by setting inclusion cutoffs on phenotypic parameters. A set that works well for this assay includes the following cutoffs (NeuriteTotalLength > 10 μm, NeuriteMaxLengthWithoutBranching < 500 μm, MinCellBodyArea > 100 μm^2, MaxCellBodyArea < 3000 μm^2, MaxNeuriteBranching < 50). Cutoffs for additional parameters (such as for nuclear or cell body average intensity) can be derived empirically and added as required. This leaves only the *valid neurons* in each well to be analyzed (*see* **Note 12**)
3. For each drug condition, normalize the data to the controls in the corresponding row. For example, to calculate the effect of Compound 1 on neurite total length (NTL) at 20 μM, use the following formula:

$$\text{Compound1 \%NTL}(20\mu\text{M}) = \frac{\text{AVG}(\text{B3}_{\text{NTL}} + \text{B4}_{\text{NTL}})}{\text{AVG}(\text{B9}_{\text{NTL}} + \text{B10}_{\text{NTL}})} \times 100$$

where B3NTL and B4NTL are the averages of neurite total length in wells B3 and B4, and B9NTL and B10NTL are the averages of neurite total length in wells B9 and B10.

4. Calculate the Z-score for each attribute using the following formula:

$$\text{Compound1 Z-score NTL}(20\mu\text{M}) = \frac{\text{AVG}(\text{B3}_{\text{NTL}} + \text{B4}_{\text{NTL}}) - \text{AVG}(\text{B9}_{\text{NTL}} + \text{B10}_{\text{NTL}})}{\sigma_{\text{controls}}}$$

where σ_{controls} is the standard deviation for NTL in all DMSO wells within the plate (columns 9[B → G] and 10[B → G]).

5. Calculate the Z′-factor for the assay using the high and low control (C_H/C_L) plate(s) according to the following formula:

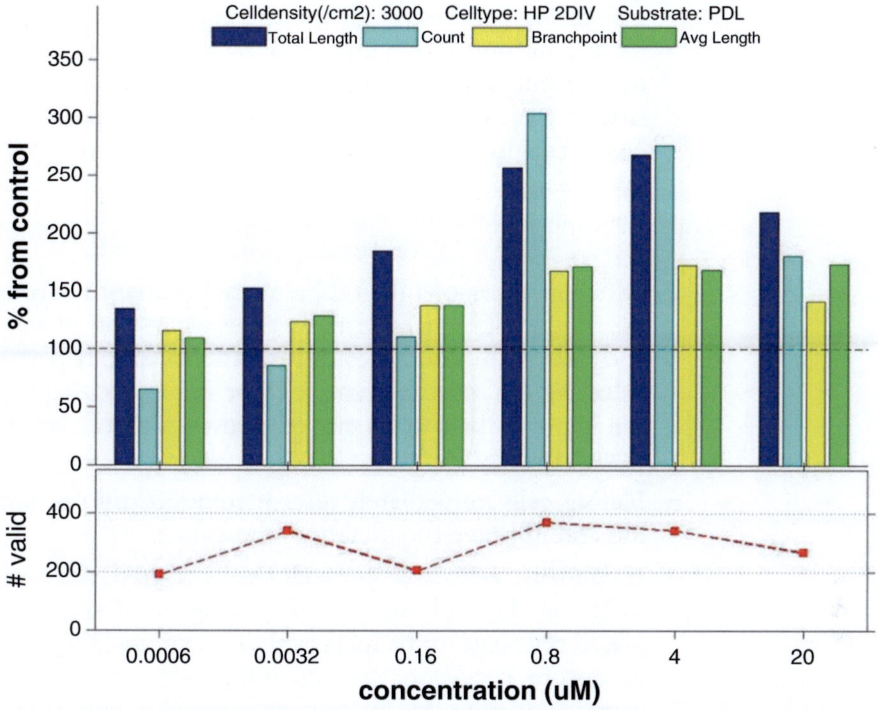

Fig. 5 Phenotypic profile of ML-7, a kinase inhibitor that strongly promotes neurite outgrowth in cultured neurons: Bar chart: HCS data for ML-7 collected for four different parameters: Neurite total length, neurite count, neurite branching, and neurite average length and processed as described above to yield a phenotypic profile for ML-7 at six treatment doses. Lineplot: Total number of valid neurons at each data point

$$Z' \text{ - factor} = \frac{(3\sigma C_H + 3\sigma C_L)}{\mu C_H - \mu C_H}$$

where μC_H and μC_L are the means for the high and low (neurite outgrowth promoter and neurite outgrowth repressor) control wells, respectively, and σC_H and σC_L are the respective standard deviations.

6. Hits are identified as compounds that fulfill the following criteria: %NTL > 130 %, NTL Z-score > 1.5, % valid neurons > 60 %, and valid neurons Z-score > –4 (Fig. 5). Hits must be confirmed in two independent screens. These criteria usually yield an average false discovery rate of ~10 % .

4 Notes

1. The coating substrate can significantly influence the behavior of cells and growing neurites and thus the dynamic range and suitability of the assay for screening. PDL is a good substrate for detecting neurite outgrowth promoters, as it is only

marginally permissive for neurite outgrowth, leaving a large margin for outgrowth induction by screened agents. If, on the other hand, a neurite outgrowth-promoting substrate (such as laminin) is used, most cells will extend long neurites, leaving less room for outgrowth induction by screened agents and decreasing the dynamic range of the experiment. However, that may make laminin a more suitable substrate when screening for agents that decrease neurite outgrowth.

2. Different types of fully differentiated primary neurons exhibit significant differences in their gene expression profiles, which correlate with differences in response to various perturbagens. Selecting the most relevant cell type for a screening experiment can improve the success rate of follow-up hit validation prior to in vivo testing.
3. Plating cells immediately after trituration will prevent clumping and improve the quality of the culture.
4. If culturing cells in larger well size is desired, the assay can be scaled up. The volume of PDL coating solution and number of plated cells have to be increased proportionally to the well area to ensure consistent neurite outgrowth profiles. Increasing neuronal plating density can cause growing neurites to overlap before the experiment is terminated, thus making automated tracing difficult. Decreasing the plating density too much can result in decreased viability.
5. When plating the cells, use a manual multichannel pipette and load gently. Alternatively, use the slowest speed setting on an electronic pipette. This helps avoid patterning in the wells, promote even spreading of the cells, and decrease variation from local changes in cell density.
6. Variations in media components across different lots can induce enough variability to interfere with hit selection and increase the yield of false negatives. It is critical to run positive controls alongside each screen to ensure that the behavior of the cells is consistent across multiple experiments.
7. When adding compound solutions to the cell plates, start with the final dilution (row G) and slowly add to the corresponding row of recipient cell plate (4× dilution, 50 μl of compound solution into 150 μl of cell culture). Repeat going up the rows (increasing concentration). For the Z′-factor plates, the row order for loading the treatment does not matter, since all rows have the same concentration. However, it is helpful to follow a consistent pattern of treatment for all plates.
8. When taking cell plates out of the incubator for treating with compounds, do not take out more than five plates at a time. It is important to minimize the time the plates spend out of the incubator as exposure to room air can lead to changes in both temperature and pH of culture media, reducing viability.

9. To avoid mechanical damage to delicate neurites and cellular structures, it is advisable to remove the rinse media by gently inverting the plates rather than using suction. If an automated plate washer is being used, make sure to program the washer to leave 50 ul in the wells at all times to avoid shearing and detachment of neural processes or cell bodies.
10. If faster permeabilization and blocking are desirable, cells can be permeabilized in a higher Triton X-100 concentration (0.3% in PBS) for 1 h and then blocked for 1 h.
11. The primary antibody used for tubulin staining was produced in-house and is specific for the neuronal Beta-III tubulin isoform. Many acceptable anti-tubulin antibodies are commercially available, though most recognize all isoforms of beta tubulin and will thus stain all cells, not just neurons.
12. Considering the large size of datasets generated from this kind of experiment, it is advisable to automate the data analysis process, for example using MatLab routines, provided that sufficient attention is given to QC and validation of all steps is performed.

References

1. Centers for Disease Control and Prevention (2013) Injury prevention and control: traumatic brain injury. Centers for Disease Control and Prevention, Washington, DC
2. DeKosky ST, Blennow K, Ikonomovic MD, Gandy S (2013) Acute and chronic traumatic encephalopathies: pathogenesis and biomarkers. Nat Rev Neurol 9:192–200
3. Cafferty WBJ, Mcgee AW, Strittmatter SM (2008) Axonal growth therapeutics: regeneration or sprouting or plasticity? Trends Neurosci 31:215–220
4. Blackmore MG, Moore DL, Smith RP, Goldberg JL, Bixby JL, Lemmon VP (2010) High content screening of cortical neurons identifies novel regulators of axon growth. Mol Cell Neurosci 44:43–54
5. Blackmore MG, Moore DL, Smith RP, Goldberg JL, Bixby L, Lemmon VP (2011) Regulators of axon growth. Mol Cell 44:43–54
6. Itoh T, Satou T, Nishida S, Hashimoto S, Ito H (2007) Immature and mature neurons coexist among glial scars after rat traumatic brain injury. Neurol Res 29:734–742
7. Sivasankaran R, Pei J, Wang KC, Zhang YP, Shields CB, Xu X-M, He Z (2004) PKC mediates inhibitory effects of myelin and chondroitin sulfate proteoglycans on axonal regeneration. Nat Neurosci 7:261–268
8. Shen Y, Tenney AP, Busch SA, Horn KP, Cuascut FX, Liu K, He Z, Silver J, Flanagan JG (2009) PTPsigma is a receptor for chondroitin sulfate proteoglycan, an inhibitor of neural regeneration. Science 326:592–596
9. Shields LBE, Zhang YP, Burke DA, Gray R, Shields CB (2008) Benefit of chondroitinase ABC on sensory axon regeneration in a laceration model of spinal cord injury in the rat. Surg Neurol 69:568–577, discussion 577
10. Monnier PP, Sierra A, Schwab JM, Henke-Fahle S, Mueller BK (2003) The Rho/ROCK pathway mediates neurite growth-inhibitory activity associated with the chondroitin sulfate proteoglycans of the CNS glial scar. Mol Cell Neurosci 22:319–330
11. Lingor P, Teusch N, Schwarz K, Mueller R, Mack H, Bähr M, Mueller BK (2007) Inhibition of Rho kinase (ROCK) increases neurite outgrowth on chondroitin sulphate proteoglycan in vitro and axonal regeneration in the adult optic nerve in vivo. J Neurochem 103:181–189
12. Lee JK, Zheng B (2012) Role of myelin-associated inhibitors in axonal repair after spinal cord injury. Exp Neurol 235:33. doi:10.1016/j.expneurol.2011.05.001
13. Lee JK, Chow R, Xie F, Chow SY, Tolentino KE, Zheng B (2010) Combined genetic atten-

uation of myelin and semaphorin-mediated growth inhibition is insufficient to promote serotonergic axon regeneration. J Neurosci 30:10899–10904
14. Filbin MT (2003) Myelin-associated inhibitors of axonal regeneration in the adult mammalian CNS. Nat Rev Neurosci 4:703–713
15. Wang H, Shen J, Xiong N, Zhao H, Chen Y (2011) Protein kinase B is involved in Nogo-66 inhibiting neurite outgrowth in PC12 cells. Neuroreport 22:733–738
16. Swinney DC, Anthony J (2011) How were new medicines discovered? Nat Rev Drug Discov 10:507–519
17. Giuliano KA (1997) High-content screening: a new approach to easing key bottlenecks in the drug discovery process. J Biomol Screen 2:249–259
18. Al-Ali H, Blackmore M, Bixby JL, Lemmon VP (2013) High content screening with primary neurons. Assay Guid Man
19. Frantz S (2005) Drug discovery: playing dirty. Nature 437:942–943
20. Al-Ali H, Lee D-H, Danzi M, Nassif H, Gautam P, Wennerberg K, Zuercher B, Drewry D, Lee JK, Lemmon V, Bixby J (2015) Rational polypharmacology: systematically identifying and engaging multiple drug targets to promote axon growth. ACS Chem Biol 10:1939
21. Klebl BM, Müller G (2005) Second-generation kinase inhibitors. Expert Opin Ther Targets 9:975–993
22. Pearce LR, Komander D, Alessi DR (2010) The nuts and bolts of AGC protein kinases. Nat Rev Mol Cell Biol 11:9–22
23. Pettus LH, Wurz RP (2008) Small molecule p38 MAP kinase inhibitors for the treatment of inflammatory diseases: novel structures and developments during 2006-2008. Curr Top Med Chem 8:1452–1467
24. Duffy P, Schmandke A, Schmandke A, Sigworth J, Narumiya S, Cafferty WBJ, Strittmatter SM (2009) Rho-associated kinase II (ROCKII) limits axonal growth after trauma within the adult mouse spinal cord. J Neurosci 29:15266–15276
25. Zhang J, Yang PL, Gray NS (2009) Targeting cancer with small molecule kinase inhibitors. Nat Rev Cancer 9:28–39
26. Don ASA, Tsang CK, Kazdoba TM, D'Arcangelo G, Young W, Zheng XFS (2012) Targeting mTOR as a novel therapeutic strategy for traumatic CNS injuries. Drug Discov Today 17:861–868
27. Bermel C, Tönges L, Planchamp V, Gillardon F, Weishaupt JH, Dietz GPH, Bähr M, Lingor P (2009) Combined inhibition of Cdk5 and ROCK additively increase cell survival, but not the regenerative response in regenerating retinal ganglion cells. Mol Cell Neurosci 42:427–437
28. Watanabe K, Ueno M, Kamiya D, Nishiyama A, Matsumura M, Wataya T, Takahashi JB, Nishikawa S, Nishikawa S, Muguruma K, Sasai Y (2007) A ROCK inhibitor permits survival of dissociated human embryonic stem cells. Nat Biotechnol 25:681–686
29. Mueller BK, Mack H, Teusch N (2005) Rho kinase, a promising drug target for neurological disorders. Nat Rev Drug Discov 4:387–398
30. Lu Q, Longo FM, Zhou H, Massa SM, Chen Y-H (2009) Signaling through Rho GTPase pathway as viable drug target. Curr Med Chem 16:1355–1365
31. Buchser WJ, Slepak TI, Gutierrez-Arenas O, Bixby JL, Lemmon VP (2010) Kinase/phosphatase overexpression reveals pathways regulating hippocampal neuron morphology. Mol Syst Biol 6:391
32. Loh SHY, Francescut L, Lingor P, Bähr M, Nicotera P (2008) Identification of new kinase clusters required for neurite outgrowth and retraction by a loss-of-function RNA interference screen. Cell Death Differ 15:283–298
33. Welsbie DS, Yang Z, Ge Y, Mitchell KL, Zhou X, Martin SE, Berlinicke CA, Hackler L, Fuller J, Fu J, Cao L-H, Han B, Auld D, Xue T, Hirai S-I, Germain L, Simard-Bisson C, Blouin R, Nguyen JV, Davis C-HO, Enke RA, Boye SL, Merbs SL, Marsh-Armstrong N, Hauswirth WW, Diantonio A, Nickells RW, Inglese J, Hanes J, Yau K-W, Quigley HA, Zack DJ (2013) Functional genomic screening identifies dual leucine zipper kinase as a key mediator of retinal ganglion cell death. Proc Natl Acad Sci U S A 110:4045–4050
34. Al-Ali H, Schürer SC, Lemmon VP, Bixby JL (2013) Chemical interrogation of the neuronal kinome using a primary cell-based screening assay. ACS Chem Biol 8:1027–1036
35. Lerch JK, Kuo F, Motti D, Morris R, Bixby JL, Lemmon VP (2012) Isoform diversity and regulation in peripheral and central neurons revealed through RNA-Seq. PLoS One 7:e30417
36. Meberg PJ, Miller MW (2003) Culturing hippocampal and cortical neurons. Methods Cell Biol 71:111–127
37. Zhang JJ-H, Chung T, Oldenburg K (1999) A Simple Statistical Parameter for Use in Evaluation and Validation of High Throughput Screening Assays. J Biomol Screen 4:67–73

Chapter 38

Postnatal Neural Stem Cells in Treating Traumatic Brain Injury

Hussein Gazalah*, Sarah Mantash*, Naify Ramadan, Sawsan Al Lafi, Sally El Sitt, Hala Darwish, Hassan Azari, Lama Fawaz, Noël Ghanem, Kazem Zibara, Rose-Mary Boustany, Firas Kobeissy, and Jihane Soueid

Abstract

Traumatic brain injury (TBI) is one of the leading causes of death and disabilities worldwide. It affects approximately 1.5 million people each year and is associated with severe post-TBI symptoms such as sensory and motor deficits. Several neuro-therapeutic approaches ranging from cell therapy interventions such as the use of neural stem cells (NSCs) to drug-based therapies have been proposed for TBI management. Successful cell-based therapies are tightly dependent on reproducible preclinical animal models to ensure safety and optimal therapeutic benefits. In this chapter, we describe the isolation of NSCs from neonatal mouse brain using the neurosphere assay in culture. Subsequently, dissociated neurosphere-derived cells are used for transplantation into the ipsilateral cortex of a controlled cortical impact (CCI) TBI model in C57BL/6 mice. Following intra-cardiac perfusion and brain removal, the success of NSC transplantation is then evaluated using immunofluorescence in order to assess neurogenesis along with gliosis in the ipsilateral coronal brain sections. Behavioral tests including rotarod and pole climbing are conducted to evaluate the motor activity post-treatment intervention.

Key words Neural stem cells, Neonatal mouse brain, Neurosphere, Controlled cortical impact, Traumatic brain injury, Perfusion, Immunofluorescence, Rotarod, Pole climbing

1 Introduction

Traumatic brain injury (TBI) is the damage of the brain tissues and structures, which is caused by various outside mechanical forces that strike the head such as in car accidents, falls, or being struck by or against a moving or stationary object [1]. The injury is characterized by both morphological and physiological changes, where TBI can cause disturbances at the level of cellular integrity and/or

*Hussein Gazalah and Sarah Mantash contributed equally with all other contributors.

Firas Kobeissy et al. (eds.), *Injury Models of the Central Nervous System: Methods and Protocols*, Methods in Molecular Biology, vol. 1462, DOI 10.1007/978-1-4939-3816-2_38, © Springer Science+Business Media New York 2016

affect the homeostasis of brain tissues and cells, thus causing reversible or irreversible cellular dysfunction or death [2]. Due to the fact that the brain is composed of different populations of cells, including neurons, and a closely interacting population of supporting cells, TBI is considered to be complex. Recently, scientists have demonstrated the existence of an interaction between cells of the central nervous system (CNS) and those of the immune system, thus raising the degree of TBI complexity [3]. For these reasons, TBI is considered one of the leading causes of death and morbidity [1]. Experimental models of TBI have been developed to mimic human TBI. Such models allow evaluation and understanding of the morphological, physiological, and behavioral changes associated with TBI. The most widely used model is the controlled cortical impact (CCI) experimental brain injury model, proposed by Smith et al. [4]. It has the advantages of being able to control the severity and the magnitude as well as level of injury. The CCI model mimics focal brain injury and cortical damage as well as axonal injury along with neuronal cell loss [4]. These characteristics will be discussed fully in Chapters 4, 5, and 11.

TBI is associated with a wide range of post-injury affliction, and scientists are currently trying to find fully effective neurotherapies aiming to ameliorate TBI symptoms. One promising strategy is the use of neural stem cells (NSCs) originating from embryonic, neonatal or adult brains [5] whereby NSCs are harvested from the sub-ventricular zone (SVZ) of neonatal mouse brain. These cells are a promising tool for the study of regenerative therapy in TBI, since they are able to become mature and functional cells to replace degenerated ones [6].

A significant amount of work has focused on NSCs. Transplanted cells into the site of injured mouse brain show significant cell differentiation, migration, and long-term survival post-transplantation. Also, the motor and spatial learning functions of the injured animal improve [7].

When neuronal replacement is the primary goal of a cell therapy regimen, the efficacy of NSC-based therapy could be increased via promoting neuronal differentiation. Application of poly-unsaturated fatty acids (PUFAs) [8–10] shows promising results in this regard. Docosahexaenoic acid (DHA) is one of the main PUFAs that promote hippocampal neuronal development and synaptic function in the developing hippocampus [8]. As NSC transplantation in a hostile environment can lead to severe gliosis [11], it may be beneficial to treat NSCs with PUFAs before transplantation and/or co-transplanting NSCs with the neurogenic promoting PUFAs.

In this chapter, we describe the methodology of culturing and harvesting NSCs, and, inducing TBI using the CCI mouse model. In addition, we describe the process of intra-cardial perfusion in mice prior to brain removal. We also explain the detailed methodology for the detection of neurogenic markers using immunofluorescence assays (IF) on brain sections, and, hematoxylin and eosin (H&E) staining to evaluate the injury. Finally, we illustrate rotarod and pole climbing tests used for assessing motor activity post-NSC transplantation.

2 Materials

2.1 Harvesting Neural Stem Cells from Neonatal Mouse Brain

2.1.1 NSC Isolation

1. 15 ml Tubes.
2. 50 ml Tubes.
3. Eppendorf tubes.
4. Petri dishes.
5. Pads (towel-like mesh, placed on the bench used to absorb blood and liquid).
6. DMEM-F12 HAM, 7 ml (Sigma, D8437 Ca, USA)
7. Ethanol 75 %.
8. Distilled water.
9. Blade (Paragon, 0086, USA).
10. Straight Iris scissors (RS-5650—Roboz Surgical Instrument Co.).
11. Curved narrow pattern forceps (RS-5671—Roboz Surgical Instrument Co.).
12. Narrow pattern forceps (Roboz, Switzerland).

2.1.2 NSC Culture and Neurosphere Formation

1. Tissue Dissociation Solution (ATV; Aqueous Trypsin and Versene EDTA chelating agent):
   (a) NaCl 8 g/l (Sigma, 71383, Switzerland)
   (b) KCL 0.4 g/l (Sigma, 746436, Switzerland)
   (c) Glucose 1 g/l (Merck, K22651037, Germany)
   (d) $NaHCO_3$ 0.85 g/l (Fisher Scientific, S233-500, USA)
   (e) Trypsin 2 g/l (Lonza, BE02-007E, Belgium)
2. Blocking solution:
   (a) DMEM/F12 HAM (Sigma, D8437 Ca, USA)
   (b) Fetal bovine serum 10 % (FBS) (Sigma, F9665, USA)
   (c) DNase 20 μg/ml (Roche, 10104159001, Germany)
3. Complete medium:
   (a) DMEM/F12 (Sigma, D8437 Ca, USA)
   (b) Insulin 50 mg/ml (Humulin R (U100), USA)
   (c) B27 (Gibco, 17504-044, USA)
   (d) N2 (Gibco, 17502-048, USA)
   (e) Glucose 16.25 mM (Merck, K22651037, Germany)
   (f) HEPES 1 M (Amresco, PH = 7, J848)
   (g) Penicillin/streptomycin (Lonza, DE17-602, Belgium)

4. Trypan blue (Sigma, RNBC2340, Germany)
5. EGF 20 ng/ml (Sigma, E4127, USA)
6. FGF 20 ng/ml (Sigma, F5392, USA)

2.2 Controlled Cortical Impact (CCI) Injury Model

1. Ketamine (Panpharma, 30692)/Xylazine (Interchemie) (90 mg/kg and 10 mg/kg, respectively).
2. Betadine solution (Basle, Mundipharma Ag, 12H145K3, Switzerland).
3. (0.9 %) Saline.
4. Blade.

2.3 Neural Stem Cell Transplantation

1. Ketamine/xylazine (90 mg/kg and 10 mg/kg, respectively).
2. Betadine solution.
3. 0.9 % Saline
4. Hamilton syringe (10 μl) (Hamilton Company, Reno, Nevada, USA).
5. DMEM/F12 medium (Gibco, 31331-028, USA).

2.4 Perfusion, Brain Removal, and Brain Slicing

2.4.1 Intra-cardial Perfusion

1. 10 ml Syringe and 27 G needle for anesthesia.
2. Anaesthetics: Ketamin and xylazine.
3. Ethanol 75 %.
4. PBS 1× (Lonza, BE17-517Q, Belgium).
5. Paraformaldehyde (PFA, 4 %) (Sigma, 16005, Germany).
6. Butterfly catheter (23 G) with blunted needle.
7. Straight Iris scissors.
8. Curved Iris scissors (Germany, RS-5671).
9. Chemical fume hood.

2.4.2 Brain Removal

1. Freshly prepared PFA (4 %).
2. Sucrose (30 %) (Sigma, 16104, Germany).
3. Curved narrow pattern forceps (RS-4980).

2.4.3 Brain Slicing

1. Ethanol 75 %.
2. PBS 1×.
3. Cryopath.
4. 6-Well plate.
5. Filter paper.

2.5 Immunofluorescence (IF) Assay

1. PBST (PBS-1 %; Triton), Triton (100×) (Sigma, T8787-250ML, USA).
2. PBS.
3. Blocking solution: 1 ml FBS (Sigma, F9665, USA) + 9 ml PBST.

4. Antibody dilution: 1 ml Blocking solution + 9 ml PBST.
5. Primary antibodies:
   (a) GFAP: rabbit polyclonal antibody, 1/1000 (Abcam, 7260, USA)
   (b) NeuN: rabbit Polyclonal antibody, 1/500 (Abcam, 104225, USA)
   (c) DCX: goat Polyclonal antibody, 1/500 (Santa Cruz, sc-8066, USA)
   (d) Iba1: rabbit Polyclonal antibody, 1/200 (Wako, 091-19741, USA)
6. Secondary antibodies: Anti-rabbit, anti-mouse and anti-goat (Molecular probes).
7. Mounting solution (Sigma, F4680, USA).
8. Anti-fade (Life Technologies, P36930, USA).
9. Coated slides (OMEGALAB, 217105, USA).

2.6 Hematoxylin and Eosin Stain (H&E)

1. Coated slides.
2. Distilled water.
3. Hematoxylin (Merck, Hx247326, Germany).
4. Eosin (0.5 %) (Kaltek, 1123, USA).
5. Ethanol (95 %; 100 %).
6. Xylol (# 3905, Belgium).
7. Mounting solution.

3 Methods

3.1 Animals

All animal experimentation was performed in compliance with Institutional Animal Care and Use Committee (IACUC) guidelines at the American University of Beirut. C57BL6 mice were obtained from Charles Rivers Laboratories maintained in the animal care facility at the American University of Beirut (AUB). All animals were handled under pathogen-free conditions and fed chow diet. Female mice were coupled with males for one night before removal of the male. Females were then tested for vaginal plug and positive mice were considered pregnant (0.5 day of gestation). Females gave birth to ~5–10 pups at days 20–21 of pregnancy. A total of ten pups were used for harvesting NSCs for each experimental group.

3.2 Harvesting Neural Stem Cells from Neonatal Mouse Brain

A thin layer of tissue is isolated from the sub-ventricular zone (SVZ) and the septal region surrounding the lateral ventricles of mouse pup brain (Fig. 1). Tissues from 7 to 10 pups are pooled to generate NSCs for each experimental group. For further details on how to harvest SVZ tissue, please refer to the video protocol by Azari et al. [12].

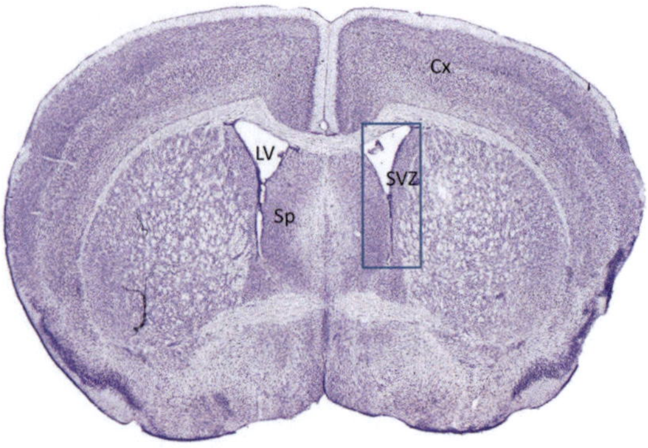

Fig. 1 Coronal section of a mouse brain showing the SVZ for harvesting NSCs. The SVZ and the septal region are marked by a *blue rectangle*. The brain is sliced into coronal sections using a blade. Image is modified from "The Mouse Brain in Stereotaxic Coordinates," George Paxinos and Keith B. J. Franklin, Second Edition, 2001. Cx; cortex, LV; lateral ventricle, Sp; septum, SVZ; subventricular zone

3.2.1 NSC Isolation

1. Sterilize the work area in order to prevent any contamination (*see* **Note 1**).
2. Prepare a couple of Petri dishes containing DMEM/F12 (Sigma-Aldrich) in which the removed brains and sections would be placed (*see* **Note 2**).
3. Anesthetize pups on ice before cutting their heads using scissors. Each brain is removed from the skull and then a coronal section is performed at the level of optic chiasm using a razor blade. The front part of the brain that contains the SVZ can be either micro-dissected as a whole to harvest the SVZ from the lateral walls of the lateral ventricles or it can be further cut into coronal sections of about 350 μm in thickness. The resulting sections are transferred under the microscope to micro-dissect the SVZ tissue containing NSCs.
4. Dissect under the microscope a rectangle that frames the SVZ including the septal region surrounding the lateral ventricles, then drop the tissue in a 15 ml tube containing DMEM/F12 (*see* **Note 3**).
5. Repeat the same procedure until all the brain tissues are harvested.

3.2.2 NSC Culture and Neurosphere Formation

1. Remove excess DMEM using a pipette cone.
2. Put the harvested tissues in a sterile petri dish and gently mince using a sterile scalpel blade. Then, add 2.5 ml ATV solution, collect all tissues in a 15 ml conical tube, and incubate for 10–15 min at 37 °C with gentle shaking (*see* **Note 4**).
3. Inhibit the activity of ATV solution by adding 5 ml of DMEM, 10% FBS (500 μL/5 ml DMEM), and DNase (200 μL/5 ml DMEM) for a total of 5 min (*see* **Note 5**).

4. Centrifuge the mixture for 5 min at 110×*g* and remove the supernatant.
5. Add 1 ml complete media to the pellet; gently triturate cells by pipetting up and down to make sure that the cells are resuspended in the media.
6. Count the cells using a hematocytometer by taking 50 μL of cell suspension and diluting it in 30 μL DMEM and 20 μL trypan blue in an Eppendorf tube. Mix well and place 20 μL on both sides of the counting chamber (*see* **Note 6**).
7. Seed cells in T25 flasks to a total of 100,000 cells per flask (5 ml complete media/flask). For example, if 2×10^6 (2 million cells/1 ml of complete media) are obtained, prepare three flasks needed, each containing 1×10^5 cells as described in the next step.
8. Prepare a total of 3×10^5 cells in 15 ml complete media (5 ml/flask × 3 flasks); take 150 μL of the cell suspension into 15 ml complete media, and then distribute to the three flasks (10^5/5 ml/flask).
9. Incubate the flasks at 37 °C and 5 % CO_2.

   At this stage the cells will give rise to primary neurospheres (**P0/D0**) (Fig. 2a); it takes 5–6 days to observe neurospheres in the flask.
10. Transfer the supernatant into new labeled flasks the next day (**D1**).
11. Repeat the above transfer for two consecutive days (**D3**) and then add 2.5 ml of complete media containing 0.5 μL EGF and 0.25 μL FGF (2.5 ml/flask) (*see* **Note** 7).
12. Repeat the addition of 2.5 ml complete media after 3 days (**D6**), and supplement with growth factors in the same way. Starting on **D7** up till **D10**, 1–4 days after the last addition of complete media and growth factors, neurospheres should be ready for the first passage. Ideally, neurospheres should have an average size of 110 μm, thus ranging in size between 70 and 140 μm before the passage. Neurospheres should not be grown for longer periods as this may lead to increased cell density at the center of the spheres (dark region) and, eventually, cell death and/or differentiation.
13. Collect the cell suspensions from all flasks in one 15 ml or 50 ml tube and centrifuge the neurospheres at 110×*g* for 5 min (*see* **Note 8**).
14. Transfer the supernatant into a new 15 ml tube after filtering it using a syringe and a blue strainer (0.20 μM). The collected suspension is called conditioned media.
15. Add 1 ml ATV to the neurosphere pellet for about 5–10 min at 37 °C with gentle shaking (*see* **Note 9**).

16. Prepare 5 ml DMEM containing 500 μL of 10% FBS and 200 μL DNase. Make sure to filter as indicated above. Stop ATV activity by adding the mix directly onto the neurosphere pellet containing dissociated cells and ATV.
17. Wait for 5 min to ensure complete inactivation of ATV activity, and then centrifuge the suspension at 110 × *g* for 5 min.
18. Remove supernatant and tap the base of the tube.
19. Add 1 ml of conditioned media, then take 50 μL of the cell suspension, and add it to 30 μl DMEM and 20 μl trypan blue.
20. Take 20 μL to perform cell counts using a hematocytometer.
21. Depending on the number of cells obtained after dissociation, distribute 1×10^5 cells in each T25 flask (*see* **Note 10**).
22. Incubate the flasks at 37 °C and 5% CO_2.

At this stage, cells grow into secondary neurospheres (**P1/D10**). It takes 4–5 days to observe neurospheres in the flask.

23. Three days after passage 1 (**D13**), add 1.5 ml complete media to each flask supplemented with growth factors (0.1 μL EGF, 0.15 μL FGF).
24. Five days after passage 1 (**D15**), add 2.5 ml complete media for each flask and supplement with factors (0.5 μL EGF, 0.25 μL FGF).
25. Seven days after passage 1 (**D17**), neurospheres are ready for the second passage (repeat **step 13** to **24**). **TBI** is performed during that same day as described below.
26. At this stage, tertiary neurospheres are obtained (**P2/D20**) (Fig. 2b) and are ready to be dissociated and injected 2 days (**D22**) after the last addition of media and growth factors [13].

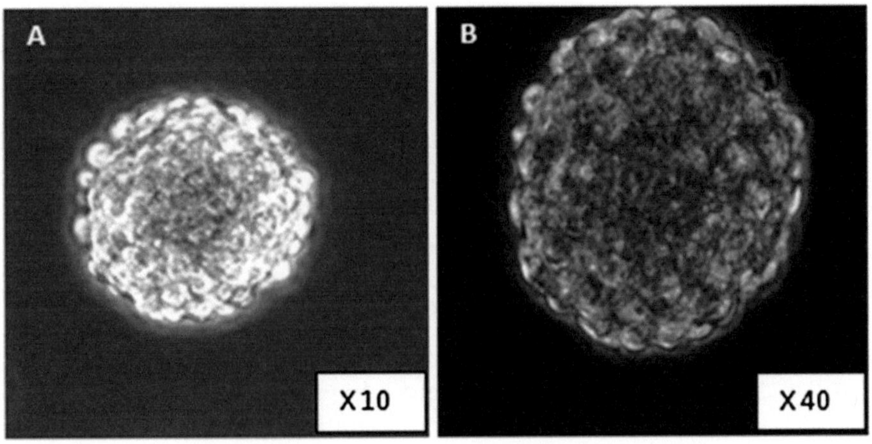

Fig. 2 Different stages of neurosphere culture with little morphological differences. Primary (**a**) and tertiary neurosphere (**b**) at different magnifications

3.3 Controlled Cortical Impact Injury Model (CCI)

An open head injury (penetrating), known as CCI, is performed to induce TBI (Fig. 3), using a Leica Angle Two system (Leica Microsystem, UK). It is a computer-assisted model that allows delivery of an easy and accurate impact to the mouse cortex.

1. Using the mouse coronal brain atlas software of the CCI machine, set the target region parasagitally between Bregma and Lambda (somatosensory area of the parietal cortex) (Fig. 4).
2. Adjust the velocity of the probe to 2 m/s and the dwell time to 150 ms.
3. Anesthetize the mouse by intramuscular injection of ketamin/xylazine (3 μL/g and 0.5 μL/g, respectively) (*see* **Note 11**) and position it in a stereotaxic frame, using the nose clamp and the ear bars.
4. Trim the scalp using scissors and apply Betadine to the underlying skin. A midline incision is made to expose the skull.
5. Locate Bregma and Lambda points manually, using the 1 mm impactor tip (*see* **Note 12**).
6. Move the impactor to "zero" in the instrument medio-lateral (ML) and antero-posterior (AP) coordinates. At this point, the impactor is right above the target site. Using the dorso-ventral (DV) drive, the impactor is lowered until it reaches the skull, confirmed by the contact sensor.
7. The skull is then marked at this particular point and the impactor is retracted in order to perform manual craniotomy, using a dental drill.
8. Lower the probe again until it reaches the brain, then retract it, and lower it again for 1 mm depth (DV axis) in order to per-

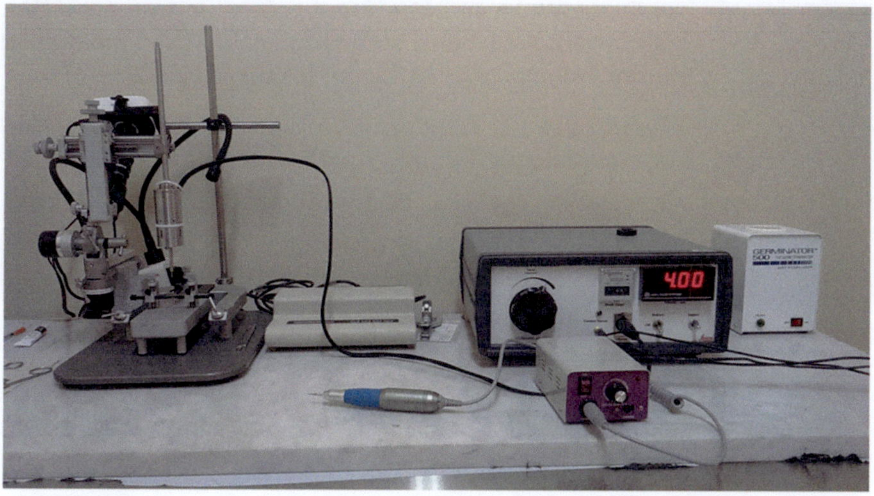

Fig. 3 Controlled cortical impact (CCI) apparatus. A computer-assisted model that induces TBI in the cortex

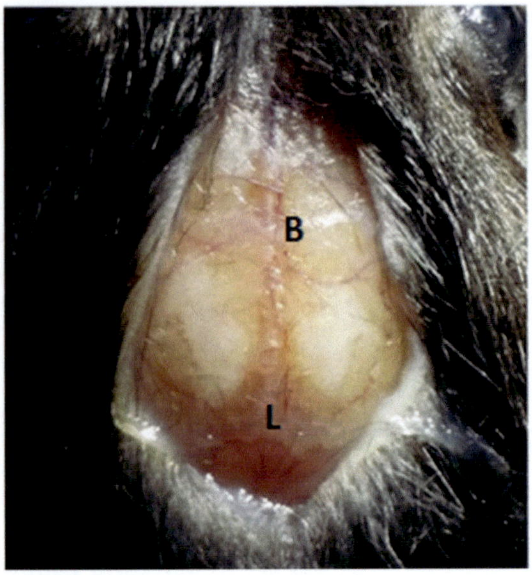

Fig. 4 Bregma and Lambda location. The mouse is positioned in a stereotaxic frame to induce injury

form the impact injury. An injury to the depth of 1 mm is induced to simulate a mild traumatic brain injury.

9. Remove the mouse rapidly from the stereotaxic frame, suture the skin, and keep in a holding cage until recovery from anesthesia (*see* **Note 13**).

TBI is performed on day 17 of cell culture so that the cells would be ready for injection 1 week after inducing TBI.

3.4 Neural Stem Cell Transplantation

The derived neurospheres, supplemented with 1.5 ml and 2.5 ml complete media and associated growth factors on D19 and D21, respectively, are labeled with Hoechst [14] and ready to be injected into the mouse brain at the injury site on D22.

1. Add 2.4 μL Hoechst (stock) to the flask containing 9 ml of complete media and growth factors at this stage.
2. Incubate the flask for 15–30 min.
3. Wash by transferring the suspension into 15 ml tube, add 5 ml PBS (1×) to the neurospheres pellet, and centrifuge for 5 min at 110×*g*
4. Repeat **step 3** three times.
5. Remove supernatant (PBS), then add 1 ml ATV solution, and incubate for 10 min at 37 °C until neurospheres are dissociated.
6. Add 5 ml of DMEM containing 500 μL FBS and 200 μL DNase using a syringe and a blue filter directly above the pellet to stop ATV activity.

7. Keep the tube in the hood for 5 min until the ATV activity is completely inhibited. Then centrifuge for 5 min at 110 × *g*.
8. Remove supernatant, tap well the base of the tube, and add 1 ml of DMEM.
9. For cell counting, take 50 μL of the cell suspension and add 30 μL trypan blue and 20 μL DMEM (refer to Subheading 3.2.2) (*see* **Note 14**).
10. After cell counting, prepare a total of 3 μL of DMEM containing 1×10^5 cells to be injected in each mouse (refer to **Note 14**).
11. In order to avoid cell death, cells in Eppendorf tubes should be placed on ice and injected immediately.
12. Anesthetize the mouse, wait until the reflex (???elaborate which one or describe??? dissappears, and place the mouse in the stereotaxic frame using the same technique used in CCI.
13. Replace the impactor on the CCI machine with the Hamilton syringe, and fill it with the 3 μL suspension containing 1×10^5 cells.
14. A midline incision is made to expose the skull, and then Bregma and Lambda points are located manually using the 10 μL Hamilton syringe. The syringe is then moved to "zero" in the instruments medio-lateral (ML) and antero-posterior (AP) coordinates; at this point, the syringe is right above the target site. Using the dorso-ventral (DV) drive, the syringe is lowered until it reaches the brain (site of TBI). This is followed by lowering by 1 mm (DV axis) in order to perform the injection (Fig. 5) (*see* **Note 15**).

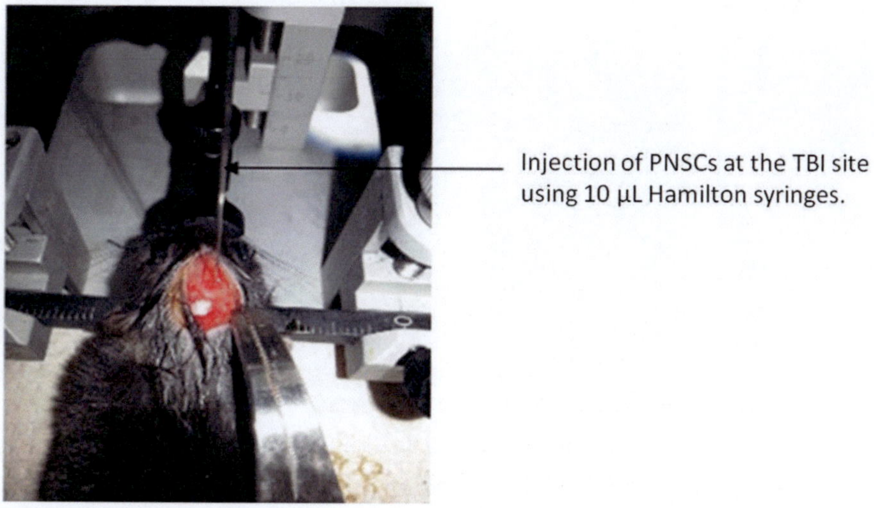

Fig. 5 Transplantation of NSCs using the CCI apparatus. The mouse is placed in a stereotaxic position. The Hamilton needle is placed on the surface of the brain, directly at the TBI site. The DV is lowered 1 mm and the cells are transplanted in the vicinity of the site of injury

3.5 Perfusion, Brain Removal, and Brain Slicing

One week after injecting NSCs at the site of TBI, brains are perfused, removed and sectioned into brain slices.

3.5.1 Intra-cardial Perfusion of Mice

1. Inject the mouse with ketamine/xylazine mixture. The suitable dose to avoid lethality is 90 mg/kg and 10 mg/kg, respectively [15].
2. Place the mouse in the cage until it is completely anesthetized and no reflex is observed at the pelvic limbs.
3. Place the mouse on its back and rinse the abdomen with 70% ethanol (Fig. 6a).

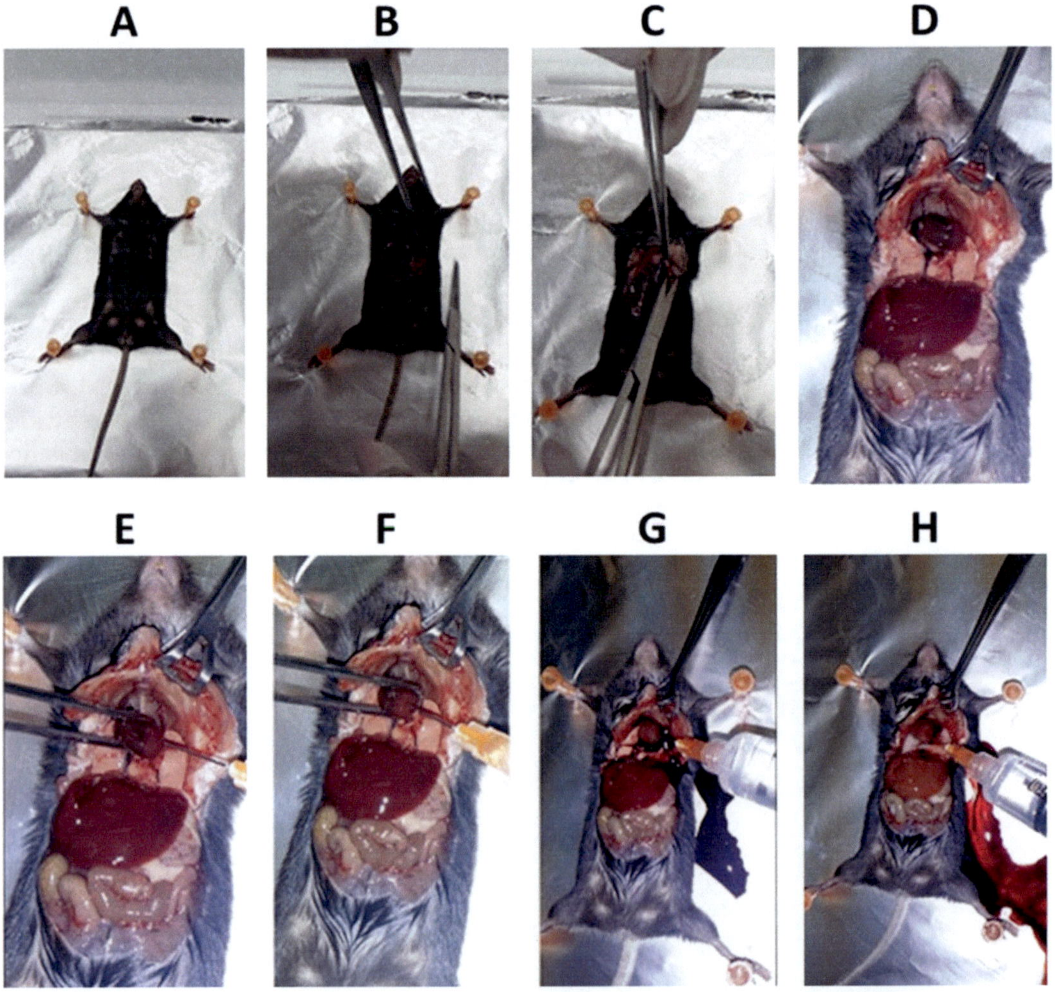

Fig. 6 Perfusion procedure. (**a**) Place the mouse on its back and rinse the abdomen with 70% ethanol. (**b**) A midline incision is made from the thoracic inlet to the pelvis, where the abdomen is opened using scissors. (**c**) Use forceps to grasp the xiphoid (white tip of the sternum), where an incision is made through the diaphragm, and then down the costal cartilage. (**d**) Flip the sternum to visualize the heart well. (**e**, **f**) Introduce a butterfly needle into the apex of the left ventricle. (**g**) Press on the syringe allowing PBS buffer to enter circulation, and then cut the right auricle immediately allowing perfusate to exit from the circulation. (**h**) When the mouse is clear of blood, perfuse using 4% PFA

4. Perform a midline incision from the thoracic inlet to the pelvis.
5. Open the abdomen using scissors (Fig. 6b).
6. Grasp the xiphoid (white tip of the sternum) using forceps, where an incision is made through the diaphragm, and then down the costal cartilage (Fig. 6c).
7. Flip the sternum to visualize the heart well (Fig. 6d).
8. Introduce a butterfly needle into the apex of the left ventricle, which is thicker and of lighter pink than the right ventricle (Fig. 6e, f).
9. Press on the syringe allowing PBS (1×) buffer to enter circulation, and then cut immediately the right auricle allowing the perfusate to exit the circulation (Fig. 6g) (*see* **Note 16**).
10. Once the mouse is clear of blood, perfuse using 4% paraformaldehyde (PFA) (Fig. 6h) (*see* **Note 17**).

3.5.2 Brain Removal

1. Make a midline incision using the scissors and flip the skin of the head to expose the skull (Fig. 7a).
2. Cut the optic nerves and remove the eyes using the scissors (Fig. 7b).
3. Cut along the sagittal suture after inserting Iris scissors caudally to the interparietal bone (Fig. 7c, d) (*see* **Note 18**).
4. Tilt one side of the parietal bone; again tilt the other side of the bone to reveal the brain using curved narrow pattern forceps (Fig. 7e–g).
5. Cut through the most anterior part of the skull between the eyes in order to expose the frontal lobe. This will ease the removal of the brain.
6. Cut the meninges beneath the skull, insert the curved forceps under the anterior part of the brain (olfactory bulbs), and then tilt the brain gently upward (Fig. 7h).
7. Separate the brain from the underlying tissue, by cutting the optic and cranial nerves (Fig. 7i).
8. Gently, lift the brain out of the skull (Fig. 7j).
9. Place the removed brain in 50 mL conical tube containing 10 mL PFA and place at 4 °C.

3.5.3 Brain Slicing

The next day, remove the brain from PFA and place it in 10 mL sucrose (30%) for 2–3 days. Change sucrose solution every day until the brain is ready for slicing (*see* **Note 19**). The aim of brain slicing is to perform histological staining such as IF and H&E.

1. Cut the cerebellum and the most anterior part of the brain using a blade.
2. Adjust the microtome temperature to −40 °C, speed to obtain 40 μm brain slices.

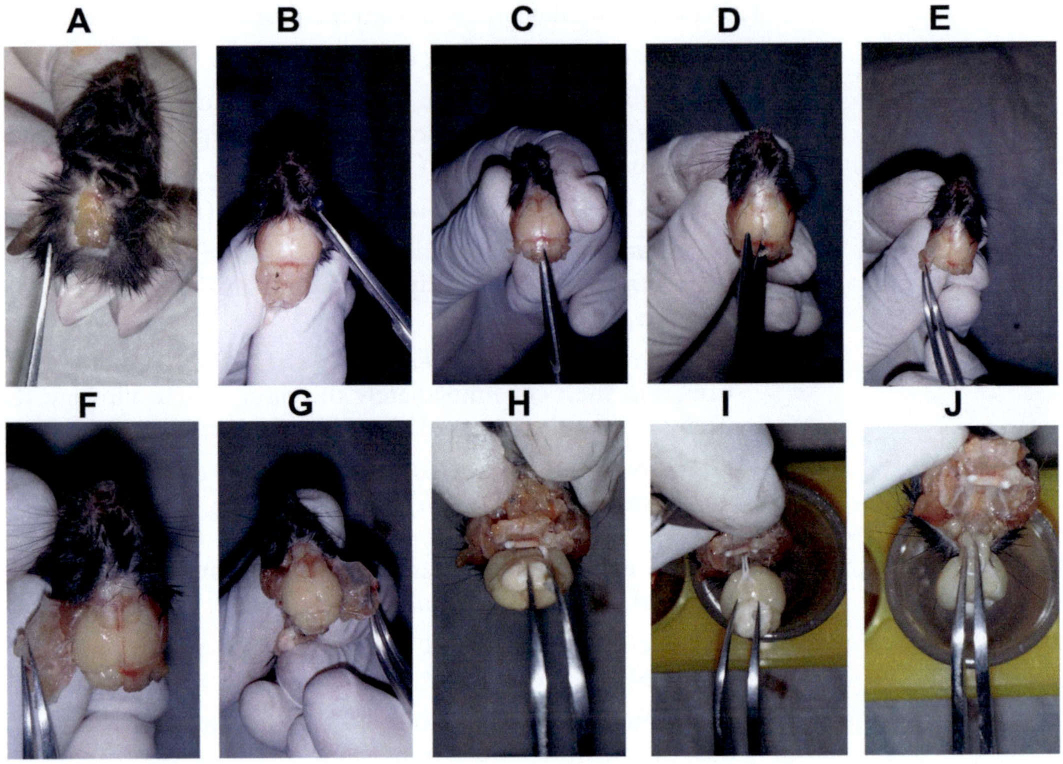

Fig. 7 Brain removal. (**a**) Use the scissors to make a midline incision and flip the skin to expose the skull. (**b**) Cut and remove the eyes. (**c**, **d**) Cut along the sagittal suture after inserting Iris scissors caudally to the inter-parietal bone. (**e**, **f**) Tilt one side of the parietal bone using curved narrow pattern forceps. (**g**) Tilt the other side of the bone to reveal the brain. Again, cut through the most anterior part of the skull between the eye lobes to expose the frontal lobe. This will ease the removal of brain. (**h**) Lift the brain out of the skull gently, and cut the cranial nerves. (**i**) Separate the brain from underlying tissue by cutting the optic nerve. (**j**) Cut the meninges and the most anterior part of the brain (olfactory bulbs). The removed brain is placed in 50 mL conical tube containing 10 mL PFA and stored at 4 °C overnight

3. Soak filter paper in PBS (1×), fold it into half, and place it on the metal of the microtome to avoid contact of brain tissue with the metal.
4. Place the brain on its posterior side (cerebellum side) on the filter paper in a straight position and spray the tissue with cryopath for instant freezing.
5. Start slicing the brain and distribute the slices into 6-well plates containing 1× PBS (*see* **Note 20**).

3.6 Immuno-fluorescence (IF) Assay

This technique uses specific antibodies raised against protein antigens in order to target fluorescent dyes to specific biomolecular targets within a cell, and therefore allows visualization of protein expression and distribution throughout the sample.

1. Wash the 40 μm brain slices previously placed in 6-well plates two times with PBST for 5 min at room temperature (RT) with gentle shaking (*see* **Note 21**).
2. Remove PBST after the second wash and add blocking solution (1 mL/well) for 1 h with continuous shaking.
3. Remove the blocking solution and add primary antibody (500 μL/well is enough to cover brain slices) overnight at 4 °C (*see* **Note 22**).
4. Remove primary antibody and perform three washes with PBST (15 min each, RT, on shaker).
5. Add 500 μL of the secondary antibody (specific for primary antibodies) for 1 h at RT with gentle shaking in the dark (*see* **Note 23**).
6. Remove secondary antibody and perform three washes with PBST (15 min each, RT, on shaker), followed by two washes with PBS (1×) (5 min each, RT, on shaker, in the dark) (*see* **Note 24**).
7. Mount the slices on microscope slides (star frost) using a mounting solution (2–3 drops).
8. Add a cover slip and make sure not to introduce any air bubbles (*see* **Note 25**).
9. Visualize using an upright fluorescent microscope.
10. Make sure to include a negative control (lacking primary antibody) for each assay in the immunofluorescence procedure (*see* **Note 26**).

3.7 Hematoxylin and Eosin (H&E) Stain

Most cells are colorless and transparent, and therefore histological sections are stained in order to make them visible. The techniques used can either be nonspecific, thus staining most of the cells in much the same way e.g. using H&E, or specific, selectively staining particular chemical groupings or molecules within cells or tissues e.g. by IF. Staining usually works by using a dye that stains a bright color some of the cellular components together with a counter stain that stains the rest of the cell by a different color. There are six wells per plate: five wells are used for IF assays and one well for H&E.

1. Translocate brain slices from Petri dishes into coated slides (star frost); tissues should be organized from anterior to posterior. This step is performed as slowly as possible to avoid tissue rupture (*see* **Note 27**).
2. Soak filter paper in PBS (1×) to stick slices on the slide and the slide is left on bench for 1 day until it dries.
3. Place the prepared slides in distilled water (dH_2O) for 3 min for hydration.

4. Transfer to hematoxylin for up to 1 min. Hematoxylin is a dark blue or violet stain that is basic (positively charged) and binds to basophilic substances (DNA/RNA) that are acidic and negatively charged.
5. Place the slides under running tap water. The use of tap water (as opposed to dH_2O) is to provide alkalinity necessary for "bluing" process.
6. Immerse the slides in eosin for up to 1 min. Eosin is a pink stain that is acidic (negatively charged) and binds to acidophilic substances (proteins) that are basic and positively charged.
7. Place slides in 95 % ethanol two times, 3 min each.
8. Immerse in 100 % ethanol for 5 min to prevent cell lysis.
9. Finally place slides in xylol for 1 min. Xylol acts as a clearing solvent to remove alcohol from tissues.
10. Add 2–3 drops of mounting solution followed by cover slip; avoid formation of bubbles.
11. Visualize the slides under a bright field microscope.

3.8 Behavioral Tests

In order to compare changes in behavior with or without stem cell transplantation, two behavioral tests are used for the study of mouse motor activity and coordination: the rotarod performance test and the pole climbing test, respectively.

3.8.1 Rotarod Performance Test

This test is based on a rotating rod with forced motor activity being applied, usually by a rodent (Fig. 8). The test measures parameters such as riding time (in seconds). Some of the other functions associated with the test include evaluating balance, grip strength and motor coordination of the subjects, especially after TBI.

1. Adjust the time and speed of rotarod apparatus to 5 min and 4 m/s, respectively. The speed can be increased to 40 m/s as time progresses.
2. Perform rotarod test four times for each group (R0, R1, R2, and R3) where R0 is performed before the injury, while the rest is performed after TBI.
3. Perform each test for 4 consecutive days. The first 3 days consist of three trials each with an inter-trial time of 15 min. The fourth day is called the challenging session which consists of one trial. The differences between the first 3 days and the fourth day are the speed of rotating rod and the time set.
4. During the first 3 days, adjust the speed to 4 m/s and gradually increase it with time to reach 40 m/s for 5 min.
5. In the challenging session, adjust the speed at 4 m/s and then accelerate to 60 m/s for 8 min.

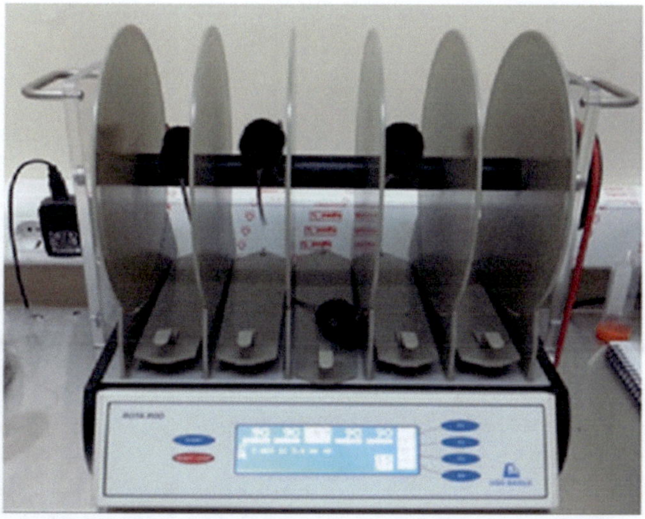

Fig. 8 Rotarod test. The mice are placed on a rotating rod; the speed and time at which the mice fall off the rod are recorded by the apparatus

6. R0 is performed 4 days before inducing TBI, R1 after TBI by 3 days, R2 after R1 by 4 days and R3 is performed after R2 by 4 days. When the R3 test is finished, the pole climbing test is performed immediately (*see* **Note 28**).

3.8.2 Pole Climbing Test

After R3 (rotarod test), the pole climbing is performed to study motor coordination in rodents.

1. Each mouse is trained for three consecutive trials during this test before starting with the experimental trials.
2. Place the mouse on the tip of a vertical rod where the mouse has to walk down the rod to reach the surface (Fig. 9).
3. Record four time points in this test, the t-turn (time at which the mouse turns on the rod), the t-half (time at which the mouse reaches half of the rod), the t-stop (time at which the mouse stops on the rod), and the total-t (time at which the mouse reaches the surface).
4. Repeat the test for five consecutive trials without break (*see* **Note 29**).

4 Conclusion

In conclusion, we have described in this chapter the detailed techniques and methods used in order to treat TBI in a CCI mouse model. We started with harvesting and isolating NSCs from the neonatal SVZ, then growing primary, secondary and tertiary

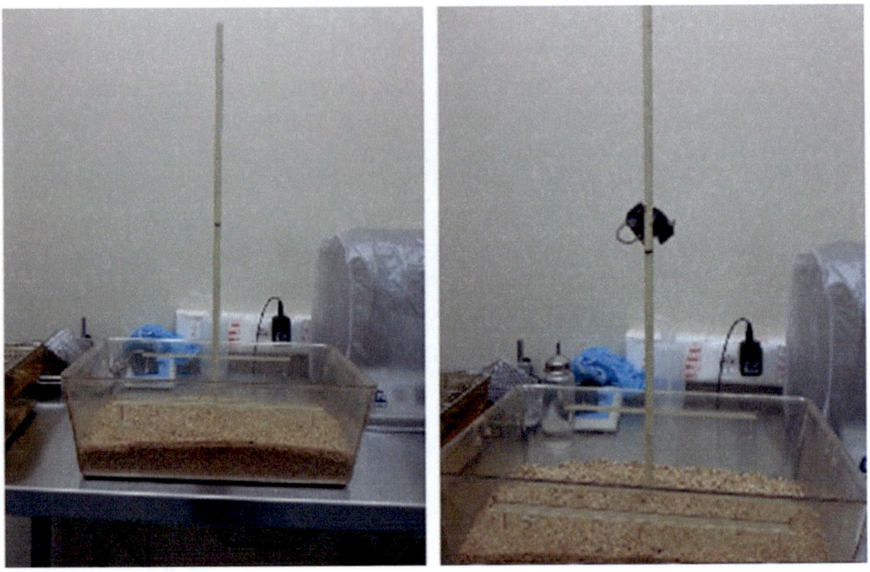

Fig. 9 Pole climbing test. The mouse is placed on the tip of a vertical rod. The t-turn thet-half, the t-stop and the t-total time are recorded as the mouse descends on the rod (see text for detail)

neurospheres. These tertiary neurospheres are then immunostained in order to confirm their undifferentiated state before injection. Mice underwent TBI using the CCI model, which is a highly reproducible brain injury model. One week following injury, NSCs are transplanted at the vicinity of the site of injury rather than in deep brain areas. Following transplantation, motor coordination and balance are evaluated using behavioral tests. The rotarod test is used to assess balance, grip strength and motor coordination of mice after TBI. It consists of placing the mouse on a rotating rod under continuous acceleration and recording the time it took to fall off. The pole climbing test was performed to study motor coordination of the mouse after training for three consecutive trials, followed by five consecutive experimental trials. The mouse was placed on the tip of a vertical rod (height = 60 cm, diameter = 1 cm) and the time at which the mouse reached the bottom (t-total) was recorded. This chapter also described how to sacrifice, perfuse and remove mice brains. Cardiac Perfusion is the method used to introduce a fixative in order to fix and preserve brain tissues. This is followed by removing the brains and splicing them into coronal sections using a microtome. This method is necessary for optimal results through the use of immunofluorescence assays in order to detect proteins in brain sections, as it demonstrates the presence of antigens (Ag) within tissue sections by means of specific antibodies (Abs). Immune reaction is then visualized by fluorescence after Ab-Ag coupling.

5 Notes

1. Isolation of NSCs should be done inside a laminar flow hood, where stringent sterile conditions need to be maintained to keep the harvested cells free of any contamination. Disinfection is performed by rinsing the area and the hood with sterilizing agent (Clorox), followed by distilled water and finally 70% ethanol. Moreover, all required equipment (microscissors, scissors, blade, curved forceps, and forceps) should be sterilized prior to use.
2. The first Petri dish is used to place removed brains and the second one is used for brain sections containing SVZ. Removal of brains should occur one by one.
3. Once the tissue is dropped into the tube, it should be placed on ice until all tissues are harvested from all pups. This is important to preserve tissues; otherwise, stem cells will be lost.
4. The base of the tube should be tapped gently every 2 min for 10–15 min, thus allowing dissociation of the tissue. Alternatively, the tube can be placed inside a shaker at 37 °C with gentle shaking.
5. To stop the activity of ATV, prepare 5 ml DMEM containing FBS and DNase in a 15 ml tube. Add the mixture to 5 ml syringe having a blue filter on its tip, and then filter the mixture directly above the tissue and ATV. Adding serum-containing medium might interfere with neurosphere formation or cause attachment of NSCs to the plate. It is highly recommended to add soybean trypsin inhibitor to the trypsin solution to inactivate it. Stopping the activity of ATV can be done by gently mixing the two mixtures via pipetting to ensure trypsin inactivation and also dissociation of tissue into single-cell suspension.
6. Count eight squares on both sides of the hematocytometer. In order to obtain the number of cells in the suspension, divide the obtained number of cells by eight and multiply by the dilution factor (two) and again multiply by ten to the power four ($X/8 \times 2 \times 10^4$). Another method to count the cells is to take 10 μl of the resuspended cells, mix with 90 μl of trypan blue, then transfer 10 μl from this mixture to a hematocytometer, and perform cell counting.
7. Since we have three flasks, and by referring to the example in **step 8**, a total of 7.5 ml of complete media should be prepared (2.5 ml/flask) in a 15 ml tube to which 1.5 μL EGF and 0.75 μL FGF are added.
8. Before collecting neurospheres in suspension, use 5–10 ml pipette (depending on suspension volume) with pipette aid in order to wash the base of the flask and hence collect the highest number of neurospheres.

9. After adding ATV solution, place the tube in the incubator at 37 °C for 7–10 min for complete dissociation. Tap the base of the tube regularly every 2 min to ensure dissociation and then observe dissociation under the microscope.
10. For example, suppose that 2×10^6 cells were obtained in 1 ml conditioned media and only three flasks of cells are needed with a total of 15 ml (5 ml/flask). A mix of conditioned media and complete media are prepared in one tube (10.5 ml complete media + 4.5 conditioned media + 3 μL EGF, 1.5 μL FGF) (3.5 ml complete media/5 ml; 1.5 ml conditioned media/5 ml; 1 μL EGF/5 ml and 0.5 μL FGF/5 ml). The volume taken from the cell suspension is dependent on the number of cells; so, if two million cells are collected and only 10^5 are needed for 5 ml, a total of 0.15 ml (0.05 ml × 3 the number of flasks = 0.15 ml) of suspension is added to 15 ml complete media + conditioned media. The 15 ml are then distributed equally to the three flasks (5 ml/flask).
11. Different ratios of ketamine and xylazine were tried on mice and caused immediate death after anesthesia. Therefore, changing the ratio will be a limiting step.
12. Bregma and Lambda points are used by the software to calculate the distance that the manipulator must move along each axis to reach the target site where TBI should be performed.
13. When the injury is sutured, antibiotics should be rubbed over the skin in order to prevent any inflammatory response. Moreover, the temperature of the cage should be kept around ~25–30 °C to prevent death of the mouse.
14. Example: Suppose we get 268 cells, so $(268/8) \times 2 \times 10^4 = 6.7 \times 10^5$ in 1 ml DMEM:

    (a) x $10^5 \rightarrow 1$ ml ⟹

    (b) $105 \rightarrow$ x X = 0.15 μL of suspension placed in 3 μL DMEM/mouse
15. The volume and the rate of injection are adjusted (3 μL; 0.30 μL/min, 0.45 μL/min, or 0.50 μL/min, respectively). Depending on the adjusted rate, injection takes between 6 and 12 min. Therefore, after injecting 3 μL, keep the Hamilton syringe in the brain for 5 min to avoid loss of suspension. Then, the Hamilton syringe is withdrawn slowly from the brain and the lesion is sutured.
16. The flow of PBS into the circulation should be slow to avoid high pressure on the blood vessels; otherwise, the brain will not be well perfused. The volume of PBS used is usually 30–40 mL until no blood is observed through the auricle.

17. Since PFA is a hazardous chemical, perfusion must be done in a chemical fume hood for the best personal protection. The volume of PFA is 30–40 mL, where muscle contraction and white liver are indicators of good perfusion. PFA and other fixatives must be collected after the perfusion and disposed of appropriately as hazardous chemical waste.
18. Incline the scissors 45° to avoid cutting through the brain.
19. Sucrose is used to dehydrate the tissue and cryoprotect it, thus and preventing freeze artifact and loss of tissue architecture.
20. Coronal sections are made allowing matching levels along the rostr-caudal axis of the brain to be examined so that comparisons can be made between littermate controls and experimental animals. Coronal sections of the entire brain should be examined in order to detect small abnormalities. The distribution of brain slices should be in a well-organized manner (serial collection) so that one well can represent the whole brain. For instance, the first slice should be placed in the first well, the second slice in the second well, and so on until the brain is completely sliced. For long-term storage, brain slices are placed in PBS supplemented with sodium azide.
21. If the brain slices are kept in PBS with sodium azide, wash slices twice with PBS (1×) for 10 min each to remove the chemical and then transfer to PBST.
22. The primary antibody can be kept for 1–2 h on shaker at 4 °C. The primary antibodies used are GFAP (rabbit), NeuN (rabbit), Map2 (mouse), DCX (goat) and Iba1 (rabbit). The volume of primary antibody may vary depending on the dilution factor. The antibody stock should be diluted in antibody dilution buffer or block solution.
23. The secondary antibody should be added in the dark since it is light sensitive and the plates should be covered with aluminum paper during the whole experiment. The secondary antibodies are anti-rabbit, anti-mouse and anti-goat. The volume of secondary antibody can vary depending on the dilution factor (1:400 usually). Finally the antibody stock should be diluted using antibody dilution buffer or block solution.
24. To label the cells with Hoechst before injection, we usually add 1 mL Hoechst diluted in PBS (1/10,000) for 5 min, followed by washing with PBS (1×) two times (5 min, RT, on shaker, in the dark).
25. The slides should be placed in a slide-specific book at 4 °C.
26. The brain slices should be distributed into wells in a way that each well represents the brain from anterior to posterior (as described above). In order to avoid losing tissue, it is recom-

mended to use a 100–1000 pipette while washing brain tissues. A negative control is performed for each assay in addition to the experimental plates. In this control, no fluorescence should be observed; tissues are incubated with antibody dilution buffer and secondary antibody. On the other hand, experimental tissues are labeled with both antibodies.

27. The tissues should be placed in a certain order; for example all TBI samples should be on the right.
28. The behavioral tests (rotarod and pole climbing) should be done by the same manipulator and in a silent room to avoid stressful stimuli that may affect animal behavior.
29. After performing the pole climbing test, mice should be perfused and their brains removed. Data for both behavioral tests should be transferred to a computer program (Excel, GraphPad, etc.) in order to draw graphs and compare the results.

References

1. Xiong Y, Mahmood A, Chopp M (2013) Animal models of traumatic brain injury. Nat Rev Neurosci 14:128–142
2. Maxwell WL (2013) Damage to myelin and oligodendrocytes: a role in chronic outcomes following traumatic brain injury? Brain Sci 3:1374–1394
3. Bigler ED (2013) Neuroinflammation and the dynamic lesion in traumatic brain injury. Brain 136:9–11
4. Smith DH, Soares HD, Pierce JS, Perlman KG, Saatman KE, Meaney DF, Dixon CE, McIntosh TK (1995) A model of parasagittal controlled cortical impact in the mouse: cognitive and histopathologic effects. J Neurotrauma 12:169–178
5. Decimo I, Bifari F, Krampera M, Fumagalli G (2012) Neural stem cell niches in health and diseases. Curr Pharm Des 18:1755–1783
6. Kindler V (2005) Postnatal stem cell survival: does the niche, a rare harbor where to resist the ebb tide of differentiation, also provide lineage-specific instructions? J Leukoc Biol 78:836–844
7. Lam PK, Lo AW, Wang KK, Lau HC, Leung KK, Li KT, Lai PB, Poon WS (2013) Transplantation of mesenchymal stem cells to the brain by topical application in an experimental traumatic brain injury model. J Clin Neurosci 20:306–309
8. Cao D, Kevala K, Kim J, Moon HS, Jun SB, Lovinger D, Kim HY (2009) Docosahexaenoic acid promotes hippocampal neuronal development and synaptic function. J Neurochem 111:510–521
9. Desai A, Kevala K, Kim HY (2014) Depletion of brain docosahexaenoic acid impairs recovery from traumatic brain injury. PLoS One 9:e86472
10. Hamedi A, Ghanbari A, Razavipour R, Saeidi V, Zarshenas MM, Sohrabpour M, Azari H (2015) Alyssum homolocarpum seeds: phytochemical analysis and effects of the seed oil on neural stem cell proliferation and differentiation. J Nat Med 69:387–396
11. Sofroniew MV (2009) Molecular dissection of reactive astrogliosis and glial scar formation. Trends Neurosci 32:638–647
12. Azari H, Rahman M, Sharififar S, Reynolds BA (2010) Isolation and expansion of the adult mouse neural stem cells using the neurosphere assay. J Vis Exp
13. Han J, Calvo CF, Kang TH, Baker KL, Park JH, Parras C, Levittas M, Birba U, Pibouin-Fragner L, Fragner P, Bilguvar K, Duman RS, Nurmi H, Alitalo K, Eichmann AC, Thomas JL (2015) Vascular endothelial growth factor receptor 3 controls neural stem cell activation in mice and humans. Cell Rep 10:1158–1172
14. Ma J, Yu LJ, Ma RD, Zhang YP, Fang JZ, Zhang XY, Yu TX (2007) Repair of glutamate-induced excitotoxic neuronal damage mediated by intracerebroventricular transplantation of neural stem cells in adult mice. Neurosci Bull 23:209–214
15. Flecknell PA (1993) Anaesthesia of animals for biomedical research. Br J Anaesth 71:885–894

Chapter 39

A Novel Biopsy Method for Isolating Neural Stem Cells from the Subventricular Zone of the Adult Rat Brain for Autologous Transplantation in CNS Injuries

Hadi Aligholi, Gholamreza Hassanzadeh, Ali Gorji, and Hassan Azari

Abstract

Despite all attempts the problem of regeneration in damaged central nervous system (CNS) has remained challenging due to its cellular complexity and highly organized and sophisticated connections. In this regard, stem cell therapy might serve as a viable therapeutic approach aiming either to support the damaged tissue and hence to reduce the subsequent neurological dysfunctions and impairments or to replace the lost cells and re-establish damaged circuitries. Adult neural stem/progenitor cells (NS/PCs) are one of the outstanding cell sources that can be isolated from the subventricular zone (SVZ) of the lateral ventricles. These cells can differentiate into neurons, astrocytes, and oligodendrocytes. Implanting autologous NS/PCs will greatly benefit the patients by avoiding immune rejection after implantation, better survival, and integration with the host tissue. Developing safe and efficient methods in small animal models will provide us with the opportunity to optimize procedures required to achieve successful human autologous NS/PC transplantation in near future. In this chapter, a highly controlled and safe biopsy method for harvesting stem cell containing tissue from the SVZ of adult rat brain is introduced. Then, isolation and expansion of NS/PCs from harvested specimen as well as the techniques to verify proliferation and differentiation capacity of the resulting NS/PCs are discussed. Finally, a method for assessing the biopsy lesion volume in the brain is described. This safe biopsy method in rat provides a unique tool to study autologous NS/PC transplantation in different CNS injury models.

Key words Neural stem cells, Autologous NS/PC transplantation, Subventricular zone, Biopsy, Rat

1 Introduction

Central nervous system (CNS) disorders are one of the leading causes of death and disabilities in societies [1]. Despite the huge advance in science and technologies, no definitive cures are yet available for CNS diseases. In recent decades, numerous studies focused on cell therapy as a promising approach to improve regeneration and decrease functional impairments related to CNS diseases [2–4]. In this regard, autograft, allograft, and xenograft cellular transplants can be used among which autologous transplantation appears more advantageous as it causes lesser deleterious immune response [5]

Firas Kobeissy et al. (eds.), *Injury Models of the Central Nervous System: Methods and Protocols*, Methods in Molecular Biology, vol. 1462, DOI 10.1007/978-1-4939-3816-2_39, © Springer Science+Business Media New York 2016

and thus leads to a better graft survival and integration with the host tissue. Adult neural stem/progenitor cells (NS/PCs) are an outstanding cell source for CNS cell therapy strategies as they are capable of generating all major cell types of the CNS including neurons, astrocytes, and oligodendrocytes [6]. One of the main niches of NS/PCs is located in the subventricular zone (SVZ) of lateral wall of lateral ventricles [7]. Previous studies have indicated that the SVZ-derived stem cells can be isolated and cultured in vitro and used for cell-based treatment programs [8, 9].

Before proceeding to any clinical application of autologous NS/PCs, it is essential to optimize the related technologies to ensure their safety and efficiency in harvesting and propagating high-quality NS/PCs in large scales to meet the requirements and standards of clinical settings. To this end, using animal models is of great importance in which all procedures from the surgical removal of stem cell containing tissue to cell implantation, evaluating graft survival and integration, and also behavioral and functional improvement in any particular CNS injury models could be successfully tested. Rats represent as one of the best animals for biological researches and specifically to study stem cell therapy in CNS injury models.

To date, very few studies have been conducted to harvesting autologous SVZ tissue from adult rat brain. In some cases, an aspiration needle was used for tissue sampling of SVZ in rats [10, 11]. Unfortunately, this method is uncontrollable and its safety has not been evaluated; thus developing a reproducible, safe, and controlled surgery method to harvest a defined size of SVZ tissue from a known coordinate of rat brain without any noticeable damage that could lead to any behavioral and cognitive changes is highly required.

In this chapter, we explain in detail a stereotactic surgery method to harvest a small piece of SVZ that contains NS/PCs from adult rat brain. Then, we describe the isolation and cultivation of NS/PCs as neurospheres. Afterwards, we discuss methods to assess the quality of the harvested NS/PCs in terms of proliferation and differentiation. Finally, a histologic method is introduced to measure biopsy size to ensure consistency in safe surgical removal of SVZ specimens when the same animals are to be used for stem cell transplantation studies.

2 Materials

2.1 Stereotactic Surgery

1. Male Wistar rats (150–180 g) should be purchased from the approved vendors and kept separately under a 12-h light/dark cycle and allowed to have free access to food and water.
2. Ketamine (alfasan, The Netherlands).
3. Xylazine (alfasan, The Netherlands).
4. 1 ml Syringe and 27 G needle.
5. Stereotactic instrument (World Precision Instruments, USA).

6. Betadine.
7. Surgical clamps.
8. Narrow pattern forceps.
9. 0.9 % Saline.
10. Surgical knife.
11. Dental drill.
12. 16-gauge semi-automatic biopsy needle (TSK Laboratory, Japan).
13. Bone wax.
14. 2-0 silk.
15. Surgical gloves.

2.2 SVZ Tissue Collection

1. Phosphate buffer saline (PBS, Invitrogen, USA) or complete neurosphere medium.
2. Penicillin/streptomycin (pen/strep, Invitrogen, USA).
3. 1.5 ml Micro tube.

2.3 Tissue Dissociation

1. Glass petri dish.
2. PBS or complete neurosphere medium containing 1 % pen/strep.
3. Surgical knife.
4. 1.5 ml Eppendorf tube.
5. 0.05 % Trypsin/EDTA (Invitrogen, USA).
6. Soybean trypsin inhibitor (Sigma, USA).
7. 0.22 μm Filter (Millipore, Germany).
8. Dulbecco's Phosphate-Buffered Saline (D-PBS, Sigma, USA).

2.4 Neurosphere Culture

1. Neurosphere medium including Dulbecco's modified Eagle's medium/F12 (Invitro-gen, USA), 1 % N2 supplement (Invitrogen, USA), 3 % B27 supplement (Invitrogen, USA), 2 μg/ml heparin (Sigma, USA), 1 % pen/strep, 1 % glutamax (Invitrogen, USA), 10 ng/ml basic fibroblast growth factor (bFGF; Millipore, Germany), and 20 ng/ml epidermal growth factor (EGF; Milteny Biotech, Germany).
2. 24-Well plates.

2.5 Passaging and Propagating the Neurospheres

1. PBS.
2. 15 ml Eppendorf tubes.
3. Accutase (Sigma, Germany).
4. Neurosphere medium.
5. 24-Well plates.
6. Trypan blue (Sigma, Germany).

2.6 Proliferation Assay

2.6.1 Number of Neurospheres

1. Inverted light microscope.

2.6.2 Diameter of Neurospheres

1. Inverted light microscope.
2. Digital camera.
3. Infinity software for image analysis.

2.6.3 Cell Counting

1. Microtube.
2. Trypan blue (Sigma, Germany).
3. Inverted light microscope.
4. Hemocytometer chamber.

2.6.4 MTS Assay

1. MTS cell proliferation colorimetric assay kit (Biovision, USA).
2. 96-Well plate.
3. ELISA plate reader.

2.7 Differentiation Assay

2.7.1 Monolayer Cell Culture

1. 6-, 12-, or 24-well plate.
2. Cover slips.
3. Poly-L-lysine (PLL).
4. Sterile H_2O.
5. PBS.
6. Differentiation medium:
   (a) DMEM f12
   (b) 1 % N2 supplement
   (c) 1 % B27
   (d) 5 % Fetal bovine serum
   (e) 1 % Pen/strep

2.7.2 Immunofluorescence

1. 4 % Paraformaldehyde in PBS (pH 7.4).
2. Triton X-100.
3. Blocking solution:
   (a) 1 % Bovine serum albumin (BSA)
   (b) 0.3 mol Glycine
   (c) 5 % Normal gout serum
   (d) PBST (PBS+ 0.1 % Tween 20)
4. Antibody diluent solution:
   (a) 1 % BSA
   (b) PBS

5. Primary antibodies:
   (a) Rabbit anti-GFAP (Astrocyte marker, Abcam, USA)
   (b) Rabbit anti-βIII-tubulin (Neuron marker, Abcam, USA)
   (c) Rabbit anti-O4 (Oligodendrocyte marker, Abcam, USA)
6. Goat anti-rabbit FITC secondary antibody (Abcam, USA).
7. 4′,6-Diamidino-2-phenylindole (DAPI).
8. Mounting medium.
9. Fluorescent microscope.
10. Infinity software.

2.8 Assessment of the Biopsy Lesion Volume

2.8.1 Cardiac Perfusion and Brain Removal

1. Ketamine (Alfasan, The Netherlands)
2. Xylazine (Alfasan, The Netherlands)
3. Surgical knife.
4. 0.9% Saline.
5. 4% Paraformaldehyde in PBS (pH 7.4).
6. 20-gauge angiocatheter.
7. Scissors.
8. Micro scissors.
9. Forceps.
10. Curved forceps.

2.8.2 Tissue Preparation

1. 50 ml Conical tube.
2. 4% Paraformaldehyde.
3. Sucrose (Sigma, USA).
4. PBS.

2.8.3 Cryosectioning

1. Cryostat instrument (Leica, Germany).
2. Surgical knife.
3. Optimal cutting temperature (OCT) compound (Biooptica, Italy).
4. Microscope slides.
5. Chromium potassium sulfate (Sigma, USA).
6. Gelatin (Sigma, USA).

2.8.4 Nissl Staining

1. PBS.
2. 100% Ethanol.
3. Cresyl violet (Sigma, USA).
4. Xylene (Merk, Germany).
5. Cover slip.

6. Mounting solution (Sigma, USA).
7. Stereo microscope (Leica, Germany).
8. Digital camera.

2.8.5 Image Analysis

1. Infinity software.

3 Methods

3.1 Stereotactic Surgery

All procedures should be performed in accordance with institutional guidelines for animal care and use.

1. Anesthetize the animal with ketamine (80 mg/kg, i.p.) and xylazine (10 mg/kg, i.p.).
2. Place the animal in a stereotactic instrument (*see* **Note 1**).
3. Remove the scalp hairs by the shaving device.
4. Clean the scalp by Betadine and sterile normal saline (*see* **Note 2**).
5. Make a skin incision in mid-sagittal line of the scalp using a surgical knife. To see the bregma and lambda points, the length of the incision should be from eyes to ears.
6. Remove the connective tissue above the skull by a blade.
7. Determine the anterior–posterior (A–P) coordinate for bregma.
8. Determine the A–P coordinate for lambda.
9. Calculate the distance between bregma and lambda (as an index of brain size) by subtracting the A–P of bregma from the A–P of lambda. The distance between bregma and lambda for the rats used in Paxinos and Watson atlas (2006) is 9 mm. In other words, the coordinates in the atlas were calculated for an animal with a defined skull size. Consequently, you must correct the coordinates in the brain atlas when using them for another animal with a different distance between bregma and lambda. The spatial characteristics of SVZ according to Paxinos and Watson are A–P = +1.56 mm; medial–lateral, M–L = +1.2 mm; and dorso-ventral, D–V = −4.2 mm (Fig. 1).

   For example, if the distance between bregma and lambda is 7 mm based on your measurements, then the corrected A–P is calculated by the following formula:

   Corrected A-P = $(1.56 \times 7)/9 = 1.21$ mm.

   Accordingly, the corrected M-L and D-V coordinates would be 0.93 mm and 3.27 mm, respectively.
10. Using the corrected A–P and M-L, determine the right point on the skull.

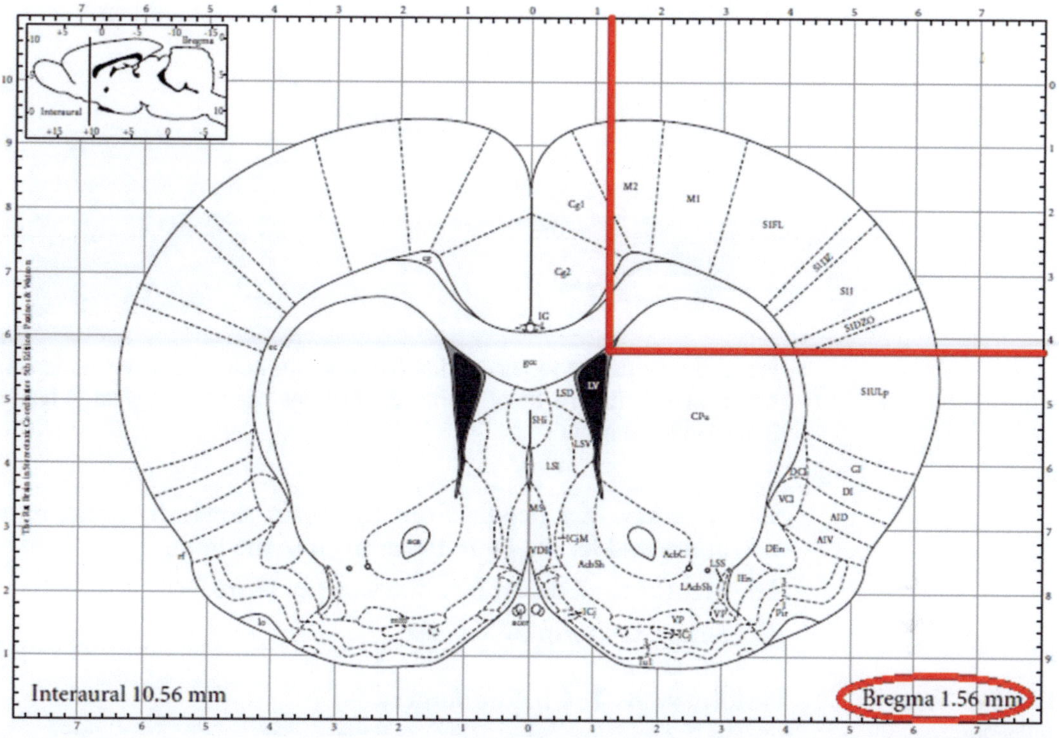

Fig. 1 Determining the location of SVZ using brain atlas. As demonstrated, the SVZ is seen in the coronal section 1.56 mm in front of the bregma (A–P). In addition, by drawing perpendicular lines from SVZ to horizontal and vertical axes, the values of M-L and D-V are obtained as 1.2 mm and 4.2 mm, respectively

11. Make a burr hole at the above point using a dental drill (0.5 mm tip, *see* **Note 3**).
12. Insert a 16-gauge semi-automatic biopsy needle into the hole (dorso-ventral, D-V = −4.2 mm according to Paxinos and Watson rat brain atlas). The 16-gauge semi-automatic biopsy needle consists of a stylet with a side-cutting window as well as a cannula. When the toggle button is pressed, the cannula is rapidly released and a piece of tissue is trapped in the side-cutting window of the stylet. Because it has not been commercially made for biopsy in rat brain tissue and its side-cutting window is very large, one should redesign its tip. Our studies indicated that for harvesting SVZ tissue from the rat brain, the optimum size for the cutting window was 3 mm. Thus, one should cut both stylet and the cannula at above of primary cutting window followed by making a new cutting window in the tip of stylet (Fig. 2). In addition, you should adjust the toggle button to decrease the range of motion (*see* **Note 4**).
13. Push the needle button to cut the SVZ specimen.
14. Wait for 15 s after pushing the bottom.

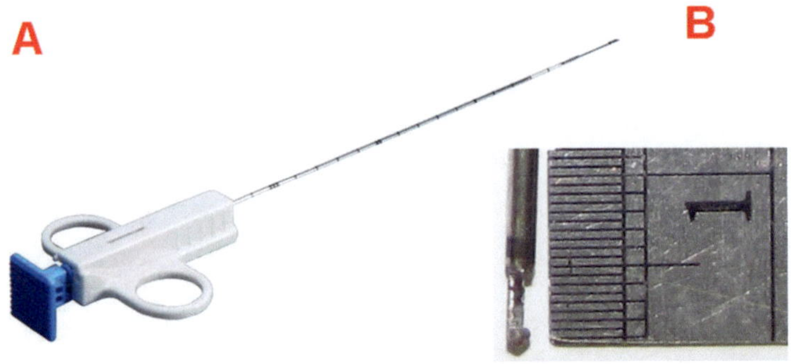

Fig. 2 Biopsy needle to harvest the SVZ from rat brain. (**a**) A semi-automatic biopsy needle. (**b**) The tip of the needle that has been re-designed to have a 3 mm cutting window

15. Withdraw the needle from the brain slowly to avoid more injury and let the brain tissue to close the hole.
16. Seal the skull defect by bone wax (*see* **Note 5**).
17. Suture the scalp by 2-0 silk.

3.2 SVZ Tissue Collection

Perform **steps 1–4** before surgery.

1. Prepare sterile PBS.
2. Add 10% pen/strep to PBS.
3. Pour 1 ml of the PBS containing 10% pen/strep into a sterile microtube.
4. Keep the microtube in 4 °C until using.
5. Put the specimen into the microtube (*see* **Note 6**).
6. Transfer the microtube to culture room quickly (*see* **Note 7**).

3.3 Tissue Dissociation

1. Transfer the microtube under a laminar hood in the culture room.
2. Transfer the specimen to a glass petri dish.
3. Wash the specimen with cold PBS containing 10% pen/strep (*see* **Note 8**).
4. Transfer the specimen to a new glass petri dish (*see* **Note 9**).
5. Mince the specimen using a surgical blade for 1 min as follows. Using the dominant hand, move the surgical blade up and down. Simultaneously, rotate slowly the glass petri dish by the other hand.
6. Transfer the specimen into a tube containing 200 μl of pre-warmed (37 °C) 0.05% trypsin/EDTA.
7. Incubate the tube at 37 °C for 3 min (*see* **Note 10**).
8. Add 200 μl of trypsin inhibitor (*see* **Note 11**).

9. Mix the contents of the tube by pipetting 2–3 times gently (*see* **Note 12**).
10. Centrifuge the tube at 110 × *g* for 5 min (*see* **Note 13**).
11. Remove the supernatant gently (*see* **Note 14**).

3.4 Neurosphere Culture

1. Add 500 μl of pre-warmed (37 °C) neurosphere medium to the tube.
2. Mix the contents of the tube by pipetting 5–7 times gently. Try to provide a homogenous mixture.
3. Transfer the 500 μl neurosphere medium containing cells from the tube to a well of a 24-well plate containing 500 μl neurosphere medium (*see* **Note 15**).
4. Incubate the cells in an incubator at 37 °C with a 5 % CO_2.
5. Add growth factors and heparin to the medium 4 days later (*see* **Note 16**).
6. Transfer the growing neurospheres to a new well at day 7 as follows (*see* **Note 17**):
   (a) Tap lightly to the bottom of the well two times (see Note 18).
   (b) Transfer the contents of the well to a tube.
   (c) Add 1 ml of PBS to the well, wash it two times by pipetting up and down, and then transfer it to the tube (see Note 19).
   (d) Centrifuge the tube at 110 × g for 5 min.
   (e) Remove the supernatant.
   (f) Add 500 μl of fresh neurosphere medium.
   (g) Pipet three times gently.
   (h) Transfer the contents of the tube to a new well of a 24-well plate (see Note 20).
   (i) Add the growth factors and heparin every 2 days.

3.5 Passaging and Propagating the Neurospheres

The neurosphere size increases due to cell proliferation. When the diameter of neurospheres becomes bigger than 150 μm, the center of neurosphere becomes dark (Fig. 3). In other words, access of cells in the center of the sphere to medium is decreased. In this condition, the spheres should be dissociated as single cells by passaging procedure. Approximately, the neurospheres are ready for passaging at day 15.

1. Tap lightly to the bottom of the well two times.
2. Transfer the contents of the well to a tube.
3. Add 1 ml of PBS to the well, wash it two times by pipetting up and down, and then transfer it to the tube.
4. Centrifuge the tube at 110 × *g* for 5 min.
5. Remove the supernatant.

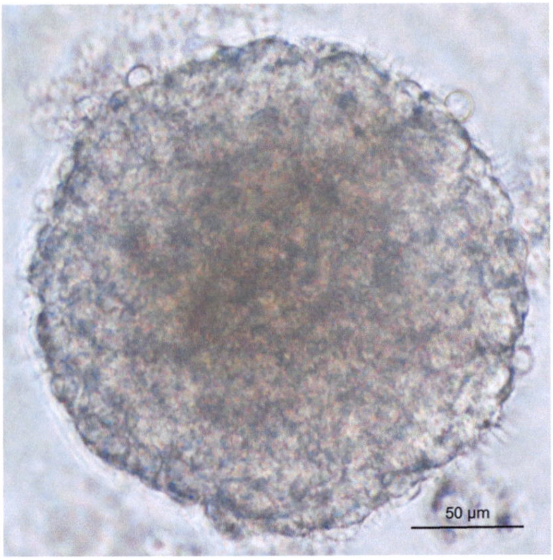

Fig. 3 A representative neurosphere picture. Proliferation of NS/PCs leads to formation of spheroidal structures named as the neurosphere. The microproceses can be seen around the neurosphere. Because the sphere is large (bigger than 150 μm in diameter), its center becomes dark. The spheres should be passaged before reaching this stage

6. Incubate the neurospheres with 200 μl of accutase at 25 °C for 10 min (*see* **Note 21**).
7. Add 500 μl of fresh neurosphere medium with 37 °C temperature to the tube (*see* **Note 22**).
8. Pipette two times gently.
9. Centrifuge the tube at 110 × *g* for 5 min.
10. Remove the supernatant.
11. Add 500 μl of fresh neurosphere medium to the tube.
12. Pipette three times gently.
13. Perform a cell count (*see* Subheading 3.6.3).
14. Plate the cells at the density of 5×10^5 cell per ml of medium.
15. Add the growth factors and heparin every 2 days.

3.6 Proliferation Assay

Proliferation is one of the main characteristics of NS/PCs. To evaluate this property, one can analyze the number and diameter of neurospheres as well as the number of cells. Moreover, MTS assay can be utilized.

3.6.1 Number of Neurospheres

For analyzing the number of neurospheres as an index of cell proliferation, the following steps can be performed before passaging (*see* **Note 23**).

1. Use 20× objective lens of an inverted microscope.
2. Start counting from the top right corner and end at the bottom left corner. All neurospheres with the size of about 50 μm or bigger in diameter should be counted. Each well should be thoroughly examined field by field.
3. The numbers could be presented as the neurosphere forming frequency per brain.

3.6.2 Diameter of Neurospheres

For calculating the diameter of neurospheres as another index of cell proliferation, you can perform the following steps:

1. Use 40× objective lens of an inverted microscope.
2. Photograph neurospheres throughout the entire plate using a systematic random sampling method.
3. Use infinity software to calculate the diameter (*see* **Note 24**).
4. Open the image via select File; Open.
5. First, calibrate the software as follows (*see* **Note 25**):

   Using Adjust; Micrometer; Calibrate draw a line in size of scale bar. Then in the micrometer calibration box: write the length of the scale bar in the section related to length. Write 1 in the magnification section. Write μm in the linear unit section. Click ok.
6. Select Measure; Caliper. Draw a diameter for the sphere, and then double right click to see the length of the diameter.
7. Repeat **step 6** for a diameter perpendicular to the first one (Fig. 4).
8. Calculate the average of these two diameters as diameter of the neurosphere (*see* **Note 26**).
9. Report the mean neurosphere diameter per brain.

3.6.3 Cell Counting

You can perform cell counting when the spheres are dissociated into single cells after passaging.

1. Dilute 50 μL of cell suspension with 30 μl medium, and 20 μl trypan blue into a micro tube.
2. Mix the micro tube contents by gentle pipetting.
3. Place 20 μl of the mixture on both sides of the hemocytometer.
4. Count eight squares of both sides of the hemocytometer. Conventionally, the cells located exactly on two edges of the square are counted and those locating on other two edges of the square are not counted.
5. Divide the obtained number of cells by 8.
6. Multiply the obtained number by 2 as dilution factor (*see* **Note 27**).
7. Multiply the obtained number by 10^4.
8. The obtained number is the number of cells in 1 ml cell suspension (*see* **Note 28**).

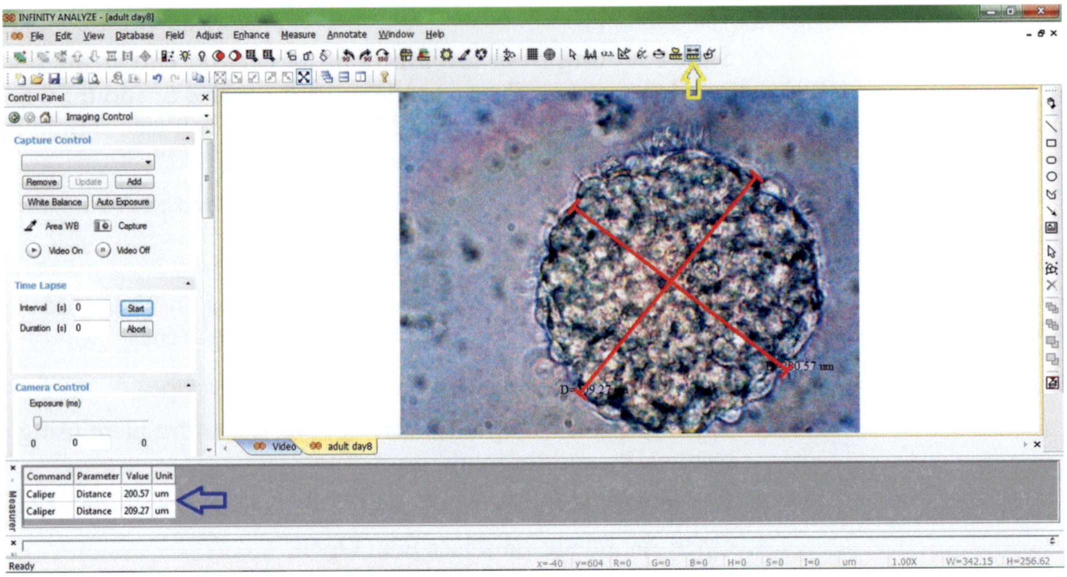

Fig. 4 Calculating neurosphere diameter. In the infinity software, use caliper option (*yellow arrow*) and draw two perpendicular diameter of the neurosphere. You can observe the values in the right corner (*blue arrow*) of the picture

3.6.4 MTS Assay

MTS cell proliferation assay is a colorimetric method for sensitive quantification of viable cells in a proliferation assay. The method is based on the reduction of MTS tetrazolium compound by viable cells to generate a colored formazan product that is soluble in cell culture media.

1. Culture cells (50×10^3/well) in a 96-well plate in a final volume of 200 μl/well of culture medium supplemented with growth factors.
2. Incubate cells for a defined period of time.
3. Add 20 μl/well MTS reagent into each well and incubate for 4 h at 37 °C in incubator.
4. Shake the plate briefly on a shaker.
5. Measure the absorbance of the cells by a plate reader at 490 nm (*see* **Note 29**).

3.7 Differentiation Assay

To assess another main property of NS/PCs, differentiation, one should culture the cells in differentiation condition for a desired time and then evaluate the cells by specific markers of neural cells. NS/PCs should be able to differentiate into neurons, astrocytes, and oligodendrocytes. At first, you should culture the cells as monolayer and then perform immunofluorescence.

3.7.1 Monolayer Cell Culture

1. Insert cover slips into 6-well plate and then coat them with poly-L-lysine (PLL) for 2 h at room temperature (*see* **Note 30**).
2. Wash the cover slips with sterile H_2O three times for 5 min.
3. Dry and sterilize the cover slips under UV light.
4. Culture the single cells obtained by passaging neurospheres on the cover slips with differentiation medium for 7 days (*see* **Note 31**).

3.7.2 Immuno-fluorescence

1. Wash the cells gently with PBS.
2. Remove the PBS and fix the cells with 4 % paraformaldehyde in PBS (pH 7.4) for 15 min at room temperature.
3. Rinse the cells with PBS three times and leave the PBS on samples for 5 min each time.
4. Incubate the cells with 0.25 % Triton X-100 for 5 min (*see* **Note 32**).
5. Wash the cells by PBS three times and leave the PBS on samples for 5 min each time.
6. Incubate the cell in blocking solution for 30 min.
7. Incubate the cells in the primary antibody solution for 1 h at room temperature or overnight at 4 °C.
8. Wash the cells by PBS three times and leave the PBS on samples for 5 min each time.
9. Incubate the cells in secondary antibody for 1 h at room temperature (*see* **Note 33**).
10. Wash the cells by PBS three times and leave the PBS on samples for 5 min each time.
11. Incubate the cells in DAPI for 15 s.
12. Remove DAPI and rinse the sample with PBS three times.
13. Mount the cover slip with mounting medium (*see* **Note 34**).
14. Use a fluorescent microscope to evaluate the slides.
15. Capture images by a camera.
16. Analyze the pictures by Infinity software as follows (Fig. 5):
    (a) File; Open
    (b) Measure; Counting
    (c) Select immunopositive cells
    (d) Double right click to show the number of immunopositive cells
17. To report the percentage of immunopositive cells/field, use the following equation:

Percentage of cells expressing a defined marker = (The number of immunopositive cells × 100)/the number of DAPI-positive nuclei (Fig. 5).

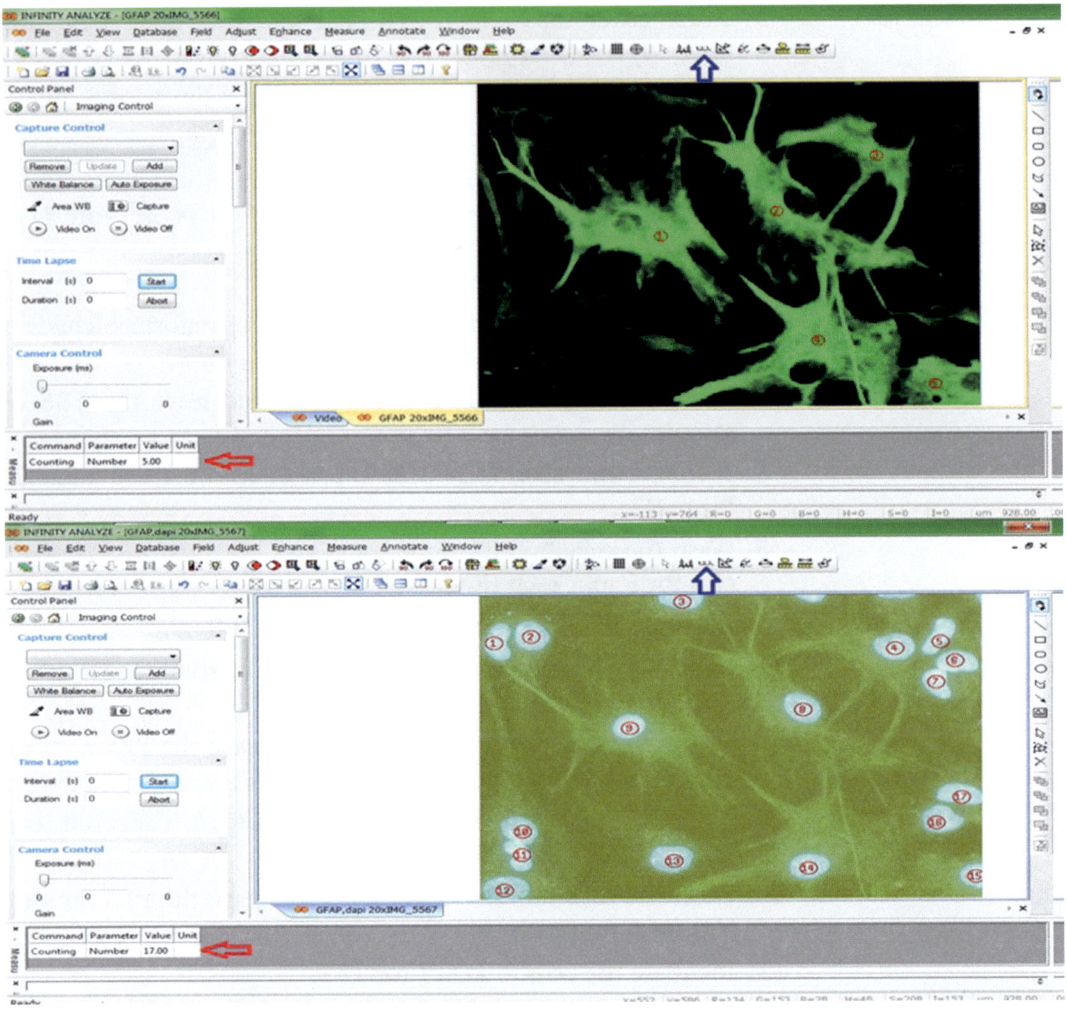

Fig. 5 Calculating the number of immunopositive cells. In the infinity software, use counting option (*blue arrow*) to calculate the number of GFAP-positive cells (*green*, above) or nuclei stained with DAPI (*blue*, below). The results are indicated in the right corner (*red arrow*)

3.8 Assessment of the Lesion Volume

To evaluate the size of biopsy lesion, you should prepare brain sections, stain them, and analyze the injury site as follows:

3.8.1 Cardiac Perfusion and Brain Removal

1. Anesthetize the animal with ketamine (80 mg/kg, i.p.) and xylazine (10 mg/kg, i.p.).
2. Place the animal in supine position. Make a midline incision in the abdomen skin.
3. Grasp the xiphoid by a forceps and open the diaphragm. Be careful not to damage the lungs when opening the diaphragm.
4. Open the chest by cutting the ribs bilaterally.
5. Insert the angiocatheter into the apex of the heart, remove the metal cannula from it, and connect it to a bottle of 0.9 % saline (*see* **Note 35**).

6. Cut the right auricle by a micro scissors (*see* **Note 36**).
7. Perfuse the body by 300 ml 0.9% saline during 30 min (*see* **Note 37**).
8. Perfuse the body by 300 ml 4% paraformaldehyde during 30 min (*see* **Note 38**).
9. Expose the skull by removing the skin.
10. Using scissors, make a midline incision in the skull rostrally.
11. Remove the parietal and then frontal bones to expose the brain using curved narrow pattern forceps.
12. Cut the meninges by micro scissors.
13. Remove the brain from the skull gently by a curved forceps. Start from anterior part of the brain (olfactory bulbs). To disconnect the brain completely, you should cut the optic nerves under the brain.

3.8.2 Tissue Preparation

1. Place the brain in 50 ml conical tube containing 10 ml 4% paraformaldehyde for 24 h at 4 °C.
2. Place the brain in 50 ml conical tube containing 10 ml 10% sucrose in PBS for 24 h at 4 °C (*see* **Note 39**).
3. Place the brain in 50 ml conical tube containing 10 ml 20% sucrose in PBS for 24 h at 4 °C.
4. Place the brain in 50 ml conical tube containing 10 ml 30% sucrose in PBS at 4 °C till the brain sinks.

3.8.3 Cryosectioning

1. Adjust the temperature of the cryostat instrument at −17 °C (*see* **Note 40**).
2. Remove the cerebellum by a surgical knife (*see* **Note 41**).
3. Cast the brain using a proper mold and OCT compound (*see* **Note 42**).
4. Make serial coronal sections of the brain with 10 μm thickness from the beginning of the lesion to end.
5. Mount the obtained brain sections on microscope slides (*see* **Note 43**).
6. Keep the slides at −20 °C until using.

3.8.4 Nissl Staining

To evaluate the morphology of the brain tissue and lesion size, you can stain CNS cells by a special staining such as Cresyl violet, which stains nissl substance in the neurons and the cell nuclei. Choose slides with 10-section interval.

1. Incubate the slides in PBS for 5 min.
2. Incubate the slides in 100% ethanol for 2 min.
3. Incubate the slides in 90% ethanol for 1 min.

4. Incubate the slides in 70% ethanol for 1 min.
5. Incubate the slides in water for 1 min.
6. Incubate the slides in cresyl violet solution for 1–2 min.
7. Incubate the slides in water for 1 min.
8. Incubate the slides in 70% ethanol for 30 s.
9. Incubate the slides in 90% ethanol for 30 s.
10. Incubate the slides in 95% ethanol for 30 s.
11. Incubate the slides in 100% ethanol for 1 min.
12. Incubate the slides in xylene for 1–2 min
13. Incubate the slides in PBS for 5 min
14. Mount the slides with mounting solution and cover the slides by cover slip.
15. Under a loop, take a picture including the whole of the section.

3.8.5 Image Analysis

You can use infinity software to calculate the size of tissue damage (Fig. 6).

1. File; Open.
2. Adjust; Micrometer; Calibrate. Draw a line in size of scale bar. In the micrometer calibration box: write the length of the scale

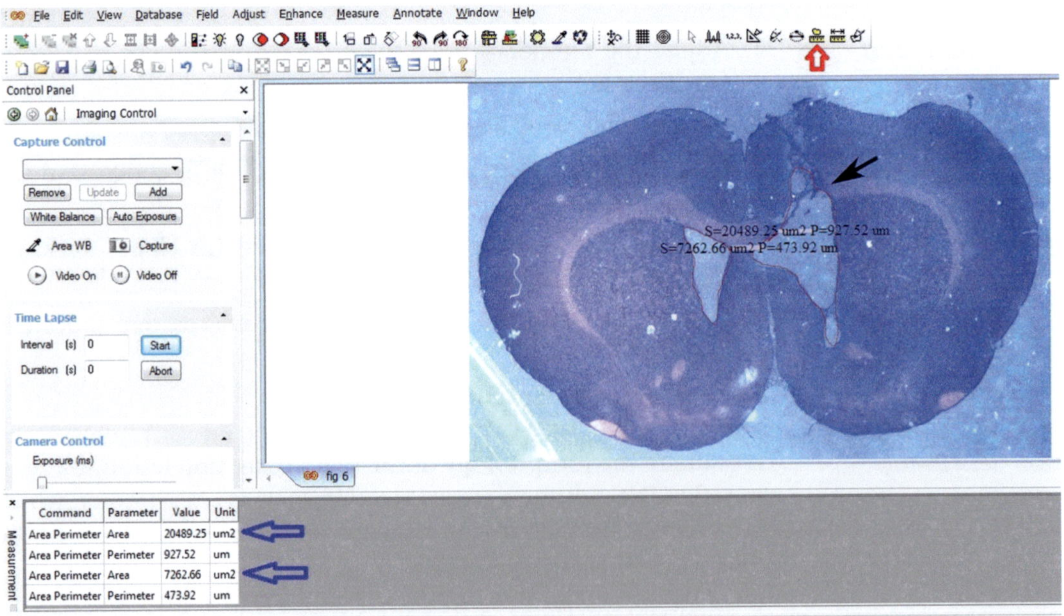

Fig. 6 Calculating the area of the lesion resulted from biopsy procedure. In infinity software, after calibration, use area perimeter option (*red arrow*) and draw a line (*red line* in the picture) around the ventricles separately. The results are represented in the right corner (*blue arrows*)

bar in the section related to length. Write 1 in the magnification section. Write μm in the linear unit section. Click ok.

3. Select Measure; Area perimeter. Draw a line around the right ventricle. Double right click to see area of the right ventricle (*see* **Note 42**).
4. Perform **step 3** for the left ventricle.
5. The area of the cavity produced by biopsy procedure is calculated by subtraction of the area of the left ventricle (damaged hemisphere) from the area of the right ventricle (intact hemisphere).
6. The total lesion volume is calculated using the following formula: $0.5D\ (\mathrm{A}1+\mathrm{A}n)+\mathrm{D}(\mathrm{A}2+\mathrm{A}3+\ \ldots\ +\mathrm{A}n-1)$, where A is the area of the cavity and D is the distance between sections (100 μm).

4 Conclusion

In this chapter, a safe, efficient, and controlled biopsy method for harvesting SVZ from rat brain was introduced. By using the semi-automatic biopsy needle, a defined size of brain tissue is cut without more damage to other neighboring tissues. In addition, we described neurosphere assay to culture one of the most delicate stem cells in a serum-free condition. This method is used previously in some literatures and here we adjusted it for culturing very small pieces of SVZ biopsy. By applying the explained methods including calculation of number and diameter of neurospheres, MTS assay, and immunocytochemistry, you can evaluate two major characteristics of NS/PCs including proliferation and differentiation. If you want to assess the size of tissue damage following the biopsy method, you can perform the histologic analysis described in the last part of this chapter (*see* **Note 44**).

Our previous study showed that this biopsy procedure was safe and did not have any adverse effect on neurological status, spatial learning and memory, passive avoidance task, and exploratory behaviors [12]. In addition, this method provides enough stem cells for autologous transplantations. Using this method opens a new window in front of researchers for future autologous transplantation studies in rats, which could be translated to clinical approaches.

5 Notes

1. Insert the animal head in the stereotaxic instrument correctly. The head should be in the horizontal line without any deviation to right, left, up, or down directions.

2. To avoid transferring any skin contamination to culture medium, you should decontaminate the skin carefully.
3. Drill the skull gently to avoid any damage to the brain. Note that the sagittal sinus is near the hole in the midline of the skull and massive bleeding will be occurred if this sinus is cut.
4. For sterilization of the tip of biopsy needle, you can insert it in boiling water for 30 min or use a metal tool sterilizer before reusing.
5. The bone wax helps to avoid brain contamination and CSF leakage.
6. Note that the specimen is inserted into the liquid phase but not above it.
7. It is better that the dissection room is located near the culture room to put the samples in culture as quickly as possible, which enhances neurosphere-forming frequency.
8. You should clean the specimen from any blood to avoid culture blood cells.
9. It is better not to use plastic petri dishes instead of glass ones because the surgical knife may damage the surface of the petri dish and produce small plastic particles.
10. Do not pipet the content of the tube in this step. You just can shake the tube by a single flip.
11. For preparing 1 L of trypsin inhibitor first add 0.14 g of trypsin inhibitor to 800 ml of HEM. Then add 10 ml of DNase solution to the solution and mix it. Once all ingredients have dissolved, make up to 1 L volume with HEM. Finally, filter the solution and store in freezer.
12. Do not pipet the content of the tube more than three times to prevent any damage to the cells.
13. The "g" force can be converted to revolutions per minute (RPM) by the following formula: $g = 1.12 \times R \times (\text{RPM}/1000)^2$ where R is the radius of rotation measured in millimeters.
14. Be careful not to eliminate the cell plate when you remove the supernatant.
15. Filling the plate with medium prior to adding the cell mixture helps homogenous distribution of cells in the plate and prevents from aggregation of the cells in one site of the plate.
16. Do not move the plate during the first 4 days after primary culture. Let the cells to find their new environment and start to contact with their neighbors by sending as well as receiving molecular signals.
17. Sometimes, neurospheres are hidden within the cell aggregates or tissue debris and their detection is difficult. By changing their well, they become visible due to elimination of debris.

18. Some spheres may attach to the plate slightly. Tapping the plate would release them.
19. Add PBS to the plate quickly to prevent the plate from drying.
20. Do not use the previous well, because some debris have been attached to the plate and have not been eliminated.
21. Using accutase is a much more gentle method compared to using trypsin/EDTA. Do not pre-warm accutase solution. Use it at room temperature.
22. Accutase auto-inhibits at 37 °C without the need for a neutralizing solution as used with trypsin.
23. When a comparative study is running among different groups, counting should be performed by a person blinded to the culture condition.
24. You can use other standard image analyzing software instead of infinity software.
25. Calibration should be done for all images separately.
26. Because the neurosphere may not be completely spherical, we use the mean of two perpendicular neurosphere diameters.
27. Because in this way the volume of the cell mixture became doubled (from 50 to 100 μl), the dilution factor is 2.
28. Note that in this way the number of cells in 1 ml cell suspension is obtained. It means if the volume of cell suspension is not 1 ml, an additional calculation is needed. For example, if the cell suspension volume is 500 μl, and the obtained cell number is 100×10^4 cell/ml, the exact number of cells is 50×10^4 cells.
29. You should perform **step 4** immediately after **step 3**. Alternatively, to measure the absorbance later, add 10 μl of 10% SDS to each well to stop the reaction. Store SDS-treated plates in a dark humidified chamber at room temperature for up to 18 h.
30. Any other coating materials might be used depending on your experimental design. Carefully follow the procedure described in reagent specification sheet.
31. When NS/PCs are plated in differentiation medium, they start to change their morphology and you can see their processes elongated far from the cell body. In addition, the morphology of the cell body changes from fusiform to polyhedral.
32. Triton X-100 acts as a detergent and helps the cell membrane to become permeable. Do not use triton X-100 when the antigen is located in the cell membrane.
33. From now on, you should perform the procedure at darkroom when the secondary antibody is fluorescent.

34. As negative control, perform the protocol in which antibody dilution is used instead of primary antibody. You should not observe any positive reaction.
35. Using angiocatheter is safer than utilizing a butterfly needle because it has plastic cannula.
36. The cardiovascular system should be cleared from blood. The amount of saline in milliliter is proportional to the animal weight in grams.
37. If you just need the brain, you can close the inferior aorta by a clamp to save fluids.
38. When paraformaldehyde circulates in the body, you can see muscle tremors especially in the extremities.
39. Because we want to perform cryosectioning, we use sucrose as a cryoprotectant agent.
40. Use a new and special blade for the cryostat instrument to prevent any damage to the tissue. Let the instrument to pre-cool enough.
41. Do not cut the brain lopsidedly.
42. Do not mold the brain crookedly.
43. To gelatinize the slides first add 3.75 g gelatin to 250 ml pre-warmed (45 °C) water on a shaker to dissolve. Then, add 0.375 g chromium potassium sulfate and insert the slides in the above mixture for 10 min. Finally, let the slides dry.
44. When the SVZ biopsy is performed correctly, a piece of the lateral wall of lateral ventricle is removed and consequently the ventricle volume increases.

References

1. Feigin VL, Barker-Collo S, Krishnamurthi R, Theadom A, Starkey N (2010) Epidemiology of ischaemic stroke and traumatic brain injury. Best Pract Res Clin Anaesthesiol 24:485–494
2. Lindvall O, Kokaia Z, Martinez-Serrano A (2004) Stem cell therapy for human neurodegenerative disorders–how to make it work. Nat Med 10(Suppl):S42
3. Kim SU, De Vellis J (2009) Stem cell-based cell therapy in neurological diseases: a review. J Neurosci Res 87:2183–2200
4. Huang H, Chen L, Sanberg P (2010) Cell therapy from bench to bedside translation in CNS neurorestoratology era. Cell Med 1:15
5. Chen Z, Palmer TD (2008) Cellular repair of CNS disorders: an immunological perspective. Hum Mol Genet 17:R84–R92
6. Delcroix GJ-R, Schiller PC, Benoit J-P, Montero-Menei CN (2010) Adult cell therapy for brain neuronal damages and the role of tissue engineering. Biomaterials 31:2105–2120
7. Reynolds BA, Weiss S (1992) Generation of neurons and astrocytes from isolated cells of the adult mammalian central nervous system. Science 255:1707–1710
8. Azari H, Osborne GW, Yasuda T, Golmohammadi MG, Rahman M, Deleyrolle LP, Esfandiari E, Adams DJ, Scheffler B, Steindler DA (2011) Purification of immature neuronal cells from neural stem cell progeny. PLoS One 6:e20941
9. Azari H, Rahman M, Sharififar S, Reynolds BA (2010) Isolation and expansion of the adult mouse neural stem cells using the neurosphere assay. J Vis Exp (45)

10. Pfeifer K, Vroemen M, Caioni M, Aigner L, Bogdahn U, Weidner N (2006) Autologous adult rodent neural progenitor cell transplantation represents a feasible strategy to promote structural repair in the chronically injured spinal cord. Regen Med 1:255

11. Muraoka K-I, Shingo T, Yasuhara T, Kameda M, Yuan W, Hayase H, Matsui T, Miyoshi Y, Date I (2006) The high integration and differentiation potential of autologous neural stem cell transplantation compared with allogeneic transplantation in adult rat hippocampus. Exp Neurol 199:311–327

12. Aligholi H, Hassanzadeh G, Azari H, Rezayat SM, Mehr SE, Akbari M, Attari F, Khaksarian M, Gorji A (2014) A new and safe method for stereotactically harvesting neural stem/progenitor cells from the adult rat subventricular zone. J Neurosci Methods 225:81–89

Part VI

Future Directions in Neurotrauma Research

Chapter 40

Challenging the Paradigms of Experimental TBI Models: From Preclinical to Clinical Practice

Frank C. Tortella

Abstract

Despite prodigious advances in TBI neurobiology research and a broad arsenal of animal models mimicking different aspects of human brain injury, this field has repeatedly experienced collective failures to translate from animals to humans, particularly in the area of therapeutics. This lack of success stems from variability and inconsistent standardization across models and laboratories, as well as insufficient objective and quantifiable diagnostic measures (biomarkers, high-resolution imaging), understanding of the vast clinical heterogeneity, and clinically centered conception of the TBI animal models. Significant progress has been made by establishing well-defined standards for reporting animal studies with "preclinical common data elements" (CDE), and for the reliability and reproducibility in preclinical TBI therapeutic research with the Operation Brain Trauma Therapy (OBTT) consortium. However, to break the chain of failures and achieve a therapeutic breakthrough in TBI will probably require the use of higher species models, specific mechanism-based injury models by which to theranostically targeted treatment portfolios are tested, more creative concepts of therapy intervention including combination therapy and regeneration neurobiology strategies, and the adoption of dosing regimens based upon pharmacokinetic—pharmacodynamic (PK-PD) studies and guided by the injury severity and TBI recovery process.

Key words Traumatic brain injury, Animal models, Neuroprotection, Clinical trials, Experimental TBI, Animal modeling

1 Introduction

Throughout the course of this book, the fundamental principles of preclinical model development have been captured across the spectrum of TBI research. While the majority of TBI models have a clearly defined purpose, the key issue is "do they truly capture the clinically relevant features embodied in the complex spectrum of this injury"? Be they common data elements or simply fundamental domains of modeling a disease state, the goal of this collection of chapters has been to engage leaders in the field in an exercise that ultimately defines the state-of-the-art research within, arguably, one of the most complex and overwhelming injury paradigms challenging clinical medicine today. Understanding and exercising with true

Firas Kobeissy et al. (eds.), *Injury Models of the Central Nervous System: Methods and Protocols*, Methods in Molecular Biology, vol. 1462, DOI 10.1007/978-1-4939-3816-2_40,

fidelity preclinical TBI research with a goal towards improving diagnostics, management and care across the spectrum of concussion through severe TBI in many ways represents the Holy Grail for TBI researchers, and brain injury researchers at large.

Considering the collective failures our field has repeatedly experienced translating from animals to humans, particularly in the area of TBI therapeutics, is it really so perplexing once we consider the amazing fidelity of our animal models as compared to the uneven human dimensions of this complex disease state? We experimentally create TBIs in animals (mostly rodents) in the most sterile environments and using controlled protocols, models that are highly reproducible and predictable allowing for the most manageable group sizes possible. Beyond the primary injury/insult our models rarely, if ever, mimic the real-life circumstances of a TBI. Consider our rodent models of TBI (which comprise the majority of our neurotrauma research platforms) where we routinely witness even the most severely injured of surviving subjects appearing outwardly normal across the spectrum of behavioral, physiological, and cognitive domains within days, if not hours post-injury—something we rarely, if ever, encounter in the human dimension of this severe TBI population. Arguably (and a debate deserving consideration but in a different context than here), a completely different clinical dynamic might occur in higher species models, particularly nonhuman primates, but then again between (1) the extraordinary expense (likely approaching a total expense of $20–25K per subject!) required for conducting optimally designed and appropriately powered primate studies for neurotherapeutic outcomes and (2) the unfortunate but always looming threat of animal rights groups to the scientist, our neurotrauma society will forever be reliant on the rodent species. This problem of preclinical to clinical translation carries an even greater burden when one considers the amazing complexity of the TBI animal models currently used in the laboratory, and yet the great distance (unintended, or otherwise) that laboratories have placed between their model designs and interlaboratory implementation. Ultimately we are working in a research field where it has become almost impossible to compare data between laboratories employing two seemingly identical animal models and/or protocol designs, hence the critical requirement to establish a set of acceptable "preclinical common data elements (CDEs)" [1] complementary to, and similar in content and structure, the clinical CDEs established previously for our human study counterparts [2]. In this context, and with regard to the success of the now 5-year-old Operation Brain Trauma Therapy (OBTT) research consortium [3], among its numerous contributions OBTT has indeed brought significant attention and new insights into this problem of reliability and reproducibility in preclinical TBI therapeutic research.

Numerous manuscripts and review articles published herein and elsewhere over the past 20 years have eloquently and accurately described TBI as a complex disease process resulting from a plethora of pathobiological injury mechanisms. Without question, when it comes to the issue of "TBI clinical drug trials and why have they all failed" it is easy to point the proverbial finger at our animal models as the main reason for the lack of transition. However, while the fallibility of our animal models certainly represents a potential weak link, they alone are not the problem. Clinical trial failures are virtually assured as long as we continue to fail to account for (1) the vast clinical heterogeneity presented by this injured population that has had a stranglehold on our clinical trial designs, (2) the absence of sensitive and validated outcome metrics (accepted by the FDA) in our clinical trial designs, and (3) our insistence on a one-size-fits-all study design. (I will comment more on each of these three points later in this chapter.) Having said this, as reviewed in great detail in the chapters of this book, and as captured with significant clarity and purpose in several very recent review articles [4–11] our field recognizes the critical need to explore ways by which to embellish these heretofore "sterile" gold-standard animal models of TBI with more clinically and pathophysiologically relevant injury domains in an effort to establish more specific mechanism-based injuries by which to theranostically target treatment portfolios. For example, let us take the case of moderate-severe TBI: How few of these clinically diagnosed injuries are indeed isolated TBIs with no other bodily injury? And even where non-sports-related mTBIs are concerned, such as those encountered in military combat and training environments, or civilian motor vehicle, pedestrian, and/or workplace accidents, the incidence and clinical influence of a *polytrauma* is significant [4, 6]. Where isolated mTBIs/concussions are of concern, acutely the diagnostic focus must be more comprehensive and objective and ensure that attention is being given to quantifiable measures, be they serum proteins, high-resolution imaging, and vestibular as well as ocular dysfunction [12, 13]. Chronically, the major concern is, of course, the potential problems associated with repeat concussions and the more long-term sequelae of clinical and pathological syndromes consistent with chronic neurological diseases that are likely to ensue such as Alzheimer's, Parkinson's, and, most recently, chronic traumatic encephalopathy [14, 15].

A comprehensive review of the excellent contributions outlined by this book assures that all hope is not lost, and in fact there is reason for great optimism by our neurotrauma community as we now recognize these challenges with far greater clarity than ever before. As alluded to earlier, the most recent publication on the subject of common data elements for preclinical TBI research co-authored by numerous leaders in the field of TBI research [1] will hopefully go a long way toward unifying our animal models and therapy protocol designs across laboratories, while high-profile clinical research initiatives like the “TBI Endpoints Development” project bring us closer to the holy grail of a set of FDA-approved outcome metrics acceptable across the spectrum of TBI severities. (https://tbiendpoints.ucsf.edu).

On this basis, and with the collective knowledge published in the chapters herein, we all should agree that taking a more challenging and clinically centric path with our animal models would yield better direction for clinical trial design and therapeutic strategies. Also, we must recognize that we have created models of primary injury where the injury, in reality, can only be prevented, not treated. Therapy research therefore can only be truly relevant to the secondary injury, regardless of its basis (i.e., be it metabolic, cellular, molecular, or neurofunctional). Our experimental path moving forward therefore must be focused not only on embellishing our current animal models with higher fidelity injury protocols complicated by secondary injury cascades and mechanisms, but also on more creative concepts of therapy intervention to include not only drug studies but also combination therapy strategies, regeneration neurobiology strategies, and nontraditional therapies such as hypothermia, hypobaria, and enriched environments/rehabilitation.

One final thought on the subject of therapeutic strategies and treatment protocols which I offer for your consideration for both preclinical TBI studies and Phase III clinical trial designs. To date, all of the therapies tested in the heretofore failed TBI clinical trials (including hypothermia) that had shown great promise in preclinical models of acute TBI have targeted a wide range of putative mechanisms of action; and, in fact, many of the tested therapies were indeed designed to target multiple mechanisms of neuronal injury or dysfunction. However, virtually all of these preclinical studies relied on a classical neuroprotection study paradigm: that is acute, single-dose, or short-term multiple dosing protocols started very early (within 30 min to 1 h) after experimental injury and where the outcomes were also measured as acute responses to neuronal injury. Rarely, if ever, did these protocols include exploration of pharmacokinetic-pharmacodynamic (PK-PD) characteristics of drug action as it relates to the evolution of TBI pathology and symptoms. In contrast, Phase III clinical trials have invariably used “delayed” outcome metrics such as the Glasgow Outcome Scale or Glasgow Outcome Scale-Extended (GOS-e) as primary outcome measures; yet these same trials adopted dosing regimens based upon the acute preclinical neuroprotection studies such that therapy never continued to be given throughout the course of the TBI recovery process.

Against the backdrop of a 100 % trial failure rate, I would argue (or propose) for a departure from the orthodoxy of the classical neuroprotection study design where both preclinical TBI models and subsequent clinical trials could potentially benefit from study protocols that utilize dosing regimens based on administration of the therapy until a pre-specified outcome is achieved (e.g., death, neurofunctional recovery, neurocognitive recovery, GOS-e) where the maximum dose and duration of treatment are defined by toxicokinetics, tolerability, the natural time course of TBI recovery within each patient (personalized therapy), and, perhaps, practical considerations (e.g., drug cost, availability of systemic and oral formulations of a given therapy, compliance). If these principles were applied to preclinical studies of candidate therapeutics, it would provide a richer data set to inform the design of subsequent clinical trials.

Given our ever-expanding knowledge base in neurotrauma and what we have learned about the neurobiology of TBI from our preclinical models and clinical studies the "one-size-fits-all" approach to neuroprotection study design in animal model and human studies is no longer justifiable. If, indeed, we are failing in our translation studies because we are losing sight of the basic principles of pharmacokinetics/pharmacodynamics of drug action, creating better prediction models and algorithms that account for injury heterogeneity, biochemical, genetic, and/or proteomic diversity associated with ongoing processes of damage and recovery may indeed lead to more personalized and tailored treatment approaches. Continued debate and discussion of all of these concepts will (and must) continue too such that our current view and implementation of experimental and clinical TBI expand along with this knowledge and, hopefully, better position the field to achieve successful clinical trial outcomes and deliver to the TBI patient and TBI community at large hopes for improved patient outcome.

References

1. Smith DH, Hicks RR, Johnson VE, Bergstrom DA, Cummings DM, Noble LJ, Hovda D, Whalen M, Ahlers ST, LaPlaca M, Tortella FC, Duhaime AC, Dixon CE (2015) Preclinical traumatic brain injury common data elements: toward a common language across laboratories. J Neurotrauma 32:1725
2. Manley GT, Diaz-Arrastia R, Brophy M, Engel D, Goodman C, Gwinn K, Veenstra TD, Ling G, Ottens AK, Tortella F, Hayes RL (2010) Common data elements for traumatic brain injury: recommendations from the biospecimens and biomarkers working group. Arch Phys Med Rehabil 91: 1667–1672
3. Kochanek PM, Bramlett HM, Dixon CE, Shear DA, Dietrich WD, Schmid KE, Mondello S, Wang KKW, Hayes RL, Povlishock JT, Tortella FC (2016) Operation brain trauma therapy: approach to modeling, therapy evaluation, drug selection, and biomarker assessments, for a multi-center pre-clinical drug screening consortium for acute therapies in severe traumatic brain injury. J Neurotrauma 33:513
4. Tortella FC, Leung LY (2015) Traumatic brain injury and polytrauma in theaters of combat: the case for neurotrauma resuscitation? Shock 44(Suppl 1):17–26
5. Briones TL (2015) Chapter 3 animal models of traumatic brain injury: is there an optimal model that parallels human brain injury? Annu Rev Nurs Res 33:31–73
6. Wang HC, Sun CF, Chen H, Chen MS, Shen G, Ma YB, Wang BD (2014) Where are we in the modelling of traumatic brain injury? Models complicated by secondary brain insults. Brain Inj 28:1491–1503
7. Bondi CO, Semple BD, Noble-Haeusslein LJ, Osier ND, Carlson SW, Dixon CE, Giza CC, Kline AE (2015) Found in translation: understanding the biology and behavior of experimental traumatic brain injury. Neurosci Biobehav Rev 58:123

8. Petraglia AL, Dashnaw ML, Turner RC, Bailes JE (2014) Models of mild traumatic brain injury: translation of physiological and anatomic injury. Neurosurgery 75(Suppl 4):S34–S49
9. Panzer MB, Wood GW, Bass CR (2014) Scaling in neurotrauma: how do we apply animal experiments to people? Exp Neurol 261:120–126
10. Xiong Y, Mahmood A, Chopp M (2013) Animal models of traumatic brain injury. Nat Rev Neurosci 14:128–142
11. Marklund N, Hillered L (2011) Animal modelling of traumatic brain injury in preclinical drug development: where do we go from here? Br J Pharmacol 164:1207–1229
12. Papa L, Ramia MM, Edwards D, Johnson BD, Slobounov SM (2015) Systematic review of clinical studies examining biomarkers of brain injury in athletes after sports-related concussion. J Neurotrauma 32:661–673
13. Mucha A, Collins MW, Elbin RJ, Furman JM, Troutman-Enseki C, DeWolf RM, Marchetti G, Kontos AP (2014) A brief vestibular/ocular motor screening (VOMS) assessment to evaluate concussions: preliminary findings. Am J Sports Med 42:2479–2486
14. Goldstein LE, McKee AC, Stanton PK (2014) Considerations for animal models of blast-related traumatic brain injury and chronic traumatic encephalopathy. Alzheimers Res Ther 6:64
15. Ojo JO, Mouzon BC, Crawford F (2016) Repetitive head trauma, chronic traumatic encephalopathy and tau: challenges in translating from mice to men. Exp Neurol 275:389

Index

Firas Kobeissy et al. (eds.), *Injury Models of the Central Nervous System: Methods and Protocols*, Methods in Molecular Biology, vol. 1462, DOI 10.1007/978-1-4939-3816-2, © Springer Science+Business Media New York 2016

O

P

R

S

T

U

V

W

MIX
Papier aus verantwortungsvollen Quellen
Paper from responsible sources
FSC® C105338

If you have any concerns about our products,
you can contact us on
ProductSafety@springernature.com

In case Publisher is established outside the EU,
the EU authorized representative is:
Springer Nature Customer Service Center GmbH
Europaplatz 3, 69115 Heidelberg, Germany

Printed by Libri Plureos GmbH
in Hamburg, Germany